ARCHITECTURAL
DRAFTING AND DESIGN

ARCHITECTURAL
DRAFTING AND DESIGN

Third Edition

Alan Jefferis
David Madsen

The authors are architectural designers and instructors in
drafting technology at Authorized Autodesk Training Center,
Clackamas Community College, Oregon City, Oregon

Delmar Publishers

I(T)P™ An International Thomson Publishing Company

Albany · Bonn · Boston · Cincinnati · Detroit · London · Madrid
Melbourne · Mexico City · New York · Pacific Grove · Paris
San Francisco · Singapore · Tokyo · Toronto · Washington

NOTICE TO THE READER

Publisher does not warrant or guarantee any of the products described herein or perform any independent analysis in connection with any of the product information contained herein. Publisher does not assume, and expressly disclaims, any obligation to obtain and include information other than that provided to it by the manufacturer.

The reader is expressly warned to consider and adopt all safety precautions that might be indicated by the activities herein and to avoid all potential hazards. By following the instructions contained herein, the reader willingly assumes all risks in connection with such instructions.

The publisher makes no representation or warranties of any kind, including but not limited to, the warranties of fitness for particular purpose or merchantability, nor are any such representations implied with respect to the material set forth herein, and the publisher takes no responsibility with respect to such material. The publisher shall not be liable for any special, consequential, or exemplary damages resulting, in whole or part, from the readers' use of, or reliance upon, this material.

Cover images courtesy of Lindal Cedar Homes, Inc., and Digital Stock™   Cover design by Spiral Design Studio

Delmar Staff
Publisher: *Robert Lynch*
Administrative Editor: *John Anderson*
Developmental Editor: *Kathleen Tatterson*
Senior Project Editor: *Christopher Chien*
Production Manager: *Larry Main*
Art and Design Coordinator: *Lisa Bower*
Editorial Assistant: *John Fisher*

COPYRIGHT © 1996
By Delmar Publishers
A division of International Thomson Publishing Inc.

The ITP logo is a trademark under license.

Printed in the United States of America

For more information, contact:

Delmar Publishers' Online Services
To access Delmar on the World Wide Web, point your browser to: **http://www.delmar.com/delmar.html**
To access through Gopher: **gopher://gopher.delmar.com**
(Delmar Online is part of "thomson.com©", an Internet site with information on more than 30 publishers of the International Thomson Publishing Organization.)
For information on our products and services:
email: **info@delmar.com** or call **800-347-7707**

Delmar Publishers
3 Columbia Circle
Box 15015
Albany, New York 12212-5015

International Thomson Publishing Europe
Berkshire House 168–173, High Holborn
London, WC1V 7AA England

Thomas Nelson Australia
102 Dodds Street
South Melbourne, 3205
Victoria, Australia

Nelson Canada
1120 Birchmont Road
Scarborough, Ontario
Canada, M1K 5G4

International Thomson Editores
Campos Eliseos 385, Piso 7
Col Polanco
11560 Mexico D F, Mexico

International Thomson Publishing GmbH
Konigswinterer Strasse 418
53227 Bonn, Germany

International Thomson Publishing Asia
221 Henderson Road
#05–10 Henderson Building
Singapore 0315

International Thomson Publishing—Japan
Hirakawacho Kyowa Building, 3F
2-2-1 Hirakawacho
Chiyoda-ku, Tokyo 102 Japan

6 7 8 9 10 XXX 01 00 99 98

Library of Congress Cataloging-in-Publication Data

Jefferis, Alan.
    Architectural drafting and design / Alan Jefferis, David Madsen.—3rd ed.
      p. cm.
    Includes index.
    ISBN 0-8273-6749-X
    1. Architectural drawing. 2. Architectural design. I. Madsen, David A. II. Title.
NA2700.J44 1995   95-43569
720'.28'4—dc20   CIP

Contents

SECTION II   RESIDENTIAL DESIGN

Preface

Architectural Drafting and Design is a practical, comprehensive textbook that is easy to use and understand. The content may be used as presented, by following a logical sequence of learning activities for residential and light commercial architectural drafting, or the chapters may be rearranged to accommodate alternate formats for traditional or individualized instruction. This text is the only reference students of architectural drafting need.

▼ APPROACH
Practical

Architectural Drafting and Design provides a practical approach to architectural drafting as it relates to current common practices. The emphasis on standardization is an excellent and necessary foundation of drafting training as well as implementing a common approach to drafting nationwide. After students become professional drafters, this text will serve as a valuable desk reference.

Realistic

Chapters contain professional examples, illustrations, step-by-step layout techniques, drafting problems, and related tests. The examples demonstrate recommended drafting presentation with actual architectural drawings used for reinforcement. The correlated text explains drafting techniques and provides useful information for skill development. Step-by-step layout methods provide a logical approach to beginning and finishing complete sets of working drawings.

Practical Approach to Problem Solving

The professional architectural drafter's responsibility is to convert architects', engineers', and designers' sketches and ideas into formal drawings. The text explains how to prepare formal drawings from design sketches by providing the learner with the basic guidelines for drafting layout, and minimum design and code requirements in a knowledge-building format; one concept is learned before the next is introduced. The concepts and skills learned from one chapter to the next allow students to prepare complete sets of working drawings in residential and light commercial drafting. Prob-

lem assignments are presented in order of difficulty and in a manner that provides students with a wide variety of architectural drafting experiences.

The problems are presented as preliminary designs or design sketches in a manner that is consistent with actual architectural office practices. It is not enough for students to duplicate drawings from given assignments; they must be able to think through the process of drawing development with a foundation of how drawing and construction components are implemented. The goals and objectives of each problem assignment are consistent with recommended evaluation criteria based on the progression of learning activities. The drafting problems and tests recommend that work be done using drafting skills on actual drafting materials with either professional manual or computer drafting equipment. A problem solution or test answer should be accurate and demonstrate proper drafting technique.

▼ FEATURES OF THE TEXT
Applications

Special emphasis has been placed on providing realistic drafting problems. Problems are presented as design sketches or preliminary drawings in a manner that is consistent with industry practices. The problems have been supplied by architectural designers and architects. Each problem solution is based on the step-by-step layout procedures provided in the chapter discussions.

Problems are given in order of complexity so that students may be exposed to a variety of drafting experiences. Problems require students to go through the same thought and decision-making processes that a professional drafter faces daily, including scale and paper size selection, view layout, dimension placement, section placement, and many other activities. Problems may be solved using manual or computer drafting, as determined by individual course guidelines. Chapter tests provide a complete coverage of each chapter and may be used for student evaluation or as study questions.

Illustrations

Drawings and photos are used liberally throughout this text to amplify the concepts presented. Two-color treatment enhances the learning value. Abundant step-by-step illustrations take students through the drafting process and clarify the detailed stages. This new edition also incorporates many CADD-drawn illustrations.

Computer-Aided Design Drafting (CADD)

CADD is presented as a valuable tool that is treated no differently from manual drafting tools. The complete discussion of computer graphics introduces the workstation, environment, terminology, drafting techniques, and sample drawings. While individual course guidelines may elect to solve architectural drafting problems using either computer or manual drafting equipment, the concepts remain the same; the only difference is the method of presentation.

Systems Drafting

Systems drafting techniques that are commonly used in the architectural drafting industry, such as overlay, repetitive elements, and drawing restoration, are covered early in the text content so that individual procedures may be implemented as appropriate to course guidelines.

Codes and Construction Techniques

The 1994 model building codes UBC, SBC, and BOCA are introduced and compared throughout the text as they relate to specific instructions and applications. Construction techniques differ throughout the country. This text clearly acknowledges the difference in construction methods and introduces the student to the format used to make complete sets of working drawings for each method of construction. Students may learn to prepare drawings from each construction method or, more commonly, for the specific construction techniques that are used in their locality. The problem assignments are designed to provide drawings that involve a variety of construction alternatives.

▼ FEATURES OF THE NEW EDITION

○ Revised order of topics to reflect the development of architectural drawings.
○ Updated code material based on 1994 model codes.
○ Expanded coverage of interior design features.
○ Separation of material to be presented on a floor plan and the electrical and framing plans.
○ Additional framing methods, including advanced framing techniques, rigid foam panels, timber construction, concrete masonry construction, steel framing, and modular construction.
○ Additional framing material, including laminated beams, open web joist, laminated veneer lumber, oriented strand board, waferboard, vapor barriers, expanded plywood coverage, and expanded truss information.

○ Expanded coverage of loads resisted by a structure, including the stresses from hurricanes, floods, tornadoes, and earthquakes.
○ Revised standards for joist and rafter selection based on 1994 wood values.
○ New material for the drawing of a framing plan separate from the floor plan to show structural materials based on regional demands.
○ Expanded coverage of concrete slab foundations, including prestressed and posttensioned slabs.
○ Expanded coverage of wall sections and stock details.
○ Additional photos showing current commercial building methods.

Coverage of commercial construction will be completely covered in the new Delmar publication, *Commercial Drafting and Detailing*, to be available in the summer of 1996. The book will feature development of structures using wood, timber, masonry, steel, and concrete. Emphasis will be placed on drafting and detailing of the drawings required for the architectural and structural drawings.

Revised Step-by-Step Illustrations

These are an important learning tool in this text. Many of these steps have been completely redrawn for refined clarity. Students will find these new drawings easy to interpret. Previous steps are printed in gray, with the new step added in black; all new steps are called out in blue.

▼ ORGANIZING YOUR COURSE

Architectural drafting is the primary emphasis of many technical drafting curricula, whereas other programs offer only an exploratory course in this field. This text is appropriate for either application, as its content reflects the common elements in an architectural drafting curriculum.

Prerequisites

An interest in architectural drafting, plus basic arithmetic, written communication, and reading skills are the only prerequisites required. Basic drafting skills and layout techniques are presented as appropriate. Students with an interest in architectural drafting who begin using this text will end with the knowledge and skills required to prepare complete sets of working drawings for residential and light commercial architectural drafting.

Fundamental Through Advanced Coverage

This text may be used in an architectural drafting curriculum that covers the basics of residential architecture in a one- two- or three-semester sequence. In this application, students use the chapters directly associated with the preparation of a complete set of working drawings for a residence, where the emphasis is on the use of fundamental skills and techniques. The balance of the text may remain as a reference for future study or as a valuable desk reference.

The text may also be used in the comprehensive architectural drafting program where a four- to six-semester sequence of residential and light commercial architectural drafting and design is required. In this application, students may expand on the primary objective of preparing a complete set of working drawings for the design of residential and light commercial projects with the coverage of any one or all of the following areas: systems drafting, energy-efficient construction techniques, solar and site orientation design applications, heating and cooling thermal performance calculations, structural load calculations, and presentation drawings.

Section Length

Chapters are presented in individual learning segments that begin with elementary concepts and build until each chapter provides complete coverage of every topic. Instructors may choose to present lectures in short, 15-minute discussions or divide each chapter into 40- to 50-minute lectures.

Drafting Equipment and Materials

Identification and use of a manual and computer-aided drafting equipment is given. Students will require an inventory of equipment available for use as listed in the chapters. Professional drafting materials are explained, and it is recommended that students prepare problem solutions using actual drafting materials.

▼ SUPPLEMENTS

Instructor's Resource Guide

The instructor's resource guide for the main text contains learning objectives, test and problem solutions, and supplementary text materials.

Workbooks and CADD Disks

We have developed a workbook and CADD workdisks to correlate with *Architectural Drafting and Design.* This workbook is correlated with the main text and contains "survival information" related to the topics covered. The plentiful problem assignments take the beginning architectural drafting student from the basics of line and lettering techniques to drawing a complete set of residential plans. Problems may be done manually or on CADD.

The workbook also provides the student with a series of complete residential plans presented as architectural design problems. The plans range from simple to complex and give the student an opportunity to vary the design elements of the home. Survival information is again correlated with the main text to provide additional instruction. Problems may be done on CADD or manually.

This workbook is for the advanced student and provides a variety of complete light commercial drafting and design projects. The types of projects include multifamily residential, tilt-up concrete, steel, and heavy timber construction. Survival information is correlated with the commercial chapters of the main text. This information leads the student into the comprehensive drafting and design projects.

▼ ACKNOWLEDGEMENTS

We would like to thank and acknowledge the many professionals who reviewed the manuscript to help us publish our architectural drafting text. A special acknowledgment is due the following instructors who reviewed the chapters in detail:

Richard Anderson
ITT Technical Institute

Danny Barrett
ITT Technical Institute

Francesco Escobar
El Paso Community College

Glen Geissinger
Northwest Technical Institute

F. Ron Kirkey
J. Sargeant Community College

Irving Krupke
Bangor High School

Ralph Liebling, R.A.
ITT Technical Institute

Bart Lynn
Utah Valley Community College

Brice Palmer
SOWELA

Ron Pergl
Clayton State College

Arnold Radman
Uniersity of North Texas

Carmine Ruocco
Jefferson State Community College

Eli Tippetts
ITT Technical Institute

Errol Ward
Four C College

Larry Weese
Modesto Junior College

Bruce Williams
Northern Alberta Institute of Technology

Patrick Wynorny
ITT Technical Institute

The quality of this text is also enhanced by the support and contributions from architects, designers, engineers, and vendors. The list of contributors from architects, designers, engineers, and vendors. The list of contributors is extensive and acknowledgment is given at each illustration; however, the following individuals and companies gave an extraordinary amount of support:

Dan Kovac
Piercy & Barclay Designers, Inc.
Alan Mascord
Alan Mascord Design Associates, Inc.
Charles Talcott
Home Planners, Inc.
Ken Smith
Ken Smith & Associates, A.I.A.
Bob Ringland
Southland Corporation
Bob D. Heinrich, P. E.
International Conference of
   Building Officials
Michael J. Baker
Trus Joist MacMillan
Richard Wallace
Southern Forest Product Association
Vincent A. Trippy
Kathleen McCarthy
Vincent Donofrio, Jr.
Joan Fleming
KOH-I-NOOR Rapidograph, Inc.
Cynthia Ann Murphy
Danier Partner
Computervision Corporation
Walter Stein
Home Building Plan Service, Inc.
Steve Webb
Lennox Industries, Inc.
Renee Randall
CALCOMP
Peggy Siglan
Nontoxic Environments
Howard Kaufman
John Doleva
Chartpak
Jim Mackay
Berol Corporation
Pat Bowlin
American Plywood Association
Richard Branham
National Concrete Masonry Association
E. Henry Fitzgibbon, AIA
Soderstrom Architects

Approximately 180 illustrations are reproduced from *Engineering Drawing and Design*, by Madsen, Shumaker, and Stewart, from Delmar Publishers. Special thanks to the students who provided step-by-step drawings: Cheryl Day, Andy Baethke, Ken Prouty, Steve Bloedel, Mark Hartman, Doug Major, April Muilenburg, and Mike Wilson. Photos were provided by John Black, Michelle Cartwright, Janice Jefferis, Sherrie Long, Saundra Powell, and Angie Renner.

▼ TO THE STUDENT

Architectural Drafting and Design is designed for you, the student. The development and format of the presentation have been tested in both conventional and individualized classroom instruction. The information presented is based on architectural standards, drafting room practice, and trends in the drafting industry. This text is the only architectural drafting reference that you will need. Use the text as a learning tool while in school, and take it along as a desk reference when you enter the profession. The amount of written text is complete but kept to a minimum. Examples and illustrations are used extensively. Drafting is a graphic language, and most drafting students learn best by observation of examples. Here are a few helpful hints.

1. *Read the text.* The text content is intentionally designed for easy reading. Sure, it doesn't read the same as an exciting short story, but it does give the facts in as few, easy-to-understand words as possible. Don't pass up the reading, because the content will help you to understand the drawings clearly.

2. *Look carefully at the examples.* The figure examples are presented in a manner that is consistent with drafting standards. Look at the examples carefully in an attempt to understand the intent of specific applications. If you are able to understand why something is done a certain way, it will be easier for you to apply those concepts to the drawing problems and, later, to the job. Drafting is a precise technology based on rules and guidelines. The goal of a drafter is to prepare drawings that are easy to interpret. There will always be situations when rules must be altered to handle a unique situation. Then you will have to rely on judgment based on your knowledge of accepted standards. Drafting is often like a puzzle; there may be more than one way to solve a problem.

3. *Use the text as a reference.* Few drafters know everything about drafting standards, techniques, and concepts, so always be ready to use the reference if you need to verify how a specific application is handled. Become familiar with the definitions and use of technical terms. It would be difficult to memorize everything noted in this text, but after considerable use

of the concepts, architectural drafting applications should become second nature.

4. *Learn each concept and skill before you continue to the next.* The text is presented in a logical learning sequence. Each chapter is designed for learning development, and chapters are sequenced so that drafting knowledge grows from one chapter to the next. Problem assignments are presented in the same learning sequence as the chapter content and also reflect progressive levels of difficulty.

5. *Practice.* Development of good manual and computer drafting skills depends to a large extent on practice. Some individuals have an inherent talent for manual drafting, and some people are readily compatible with computers. If you fit into either group, great! If you don't, then practice is all you may need. Practice manual drafting skills to help improve the quality of your drafting presentation, and practice communicating and working with a computer. A good knowledge of drafting practice is not enough if the manual skills are not satisfactory. When the computer is used, however, most manual skills are not needed.

6. *Use sketches or preliminary drawings.* When you are drawing manually or with a computer, the proper use of a sketch or preliminary drawing can save a lot of time in the long run. Prepare a layout sketch or preliminary layout for each problem. This will give you a chance to organize thoughts about drawing scale, view selection, dimension and note placement, and paper size. After you become a drafting veteran, you may be able to design a sheet layout in your head, but until then, you will be sorry if you don't use sketches.

7. *Use professional equipment and materials.* For the best possible learning results and skill development, use the professional drafting equipment, supplies, and materials that are recommended.

Alan Jefferis
David A. Madsen

ARCHITECTURAL
DRAFTING AND DESIGN

SECTION I

INTRODUCTION TO ARCHITECTURAL DRAFTING

FRONT ELEVATION
SCALE: 1/4" = 1' - 0"

CHAPTER 1

Professional Architectural Careers, Office Practice, and Opportunities

As you begin working with this text, you are opening the door to many exciting careers. Each career in turn has many different opportunities within it. Whether your interest lies in theoretical problem solving, artistic creations, or working with your hands creating something practical, a course in residential architecture will help prepare you to satisfy that interest. An architectural drafting class can lead to a career as a drafter, designer, architect, or engineer. Once you have mastered the information and skills presented in this text, you will be prepared for each of these fields as well as many others.

▼ DRAFTER

A drafter is the person who draws the drawings and details for another person's creations. It is the drafter's responsibility to use the proper line and lettering quality and to lay out properly the required drawings necessary to complete a project. Such a task requires a great attention to detail as the drafter redraws the supervisor's sketches. Because the drafter could possibly be working for several architects or engineers within an office, the drafter must be able to get along well with others.

The Beginning Drafter

As a beginning or junior drafter, your job will generally consist of making corrections to drawings drawn by others. There may not be a lot of mental stimulation to making changes, but it is a very necessary job. It is also a good introduction to the procedures and quality standards within an office.

As your line and lettering quality improve, your responsibilities will be expanded. Typically your supervisor will give you a sketch and expect you to draw the required drawing. Figure 1–1 shows an engineer's sketch. Figure 1–2 shows the drawing created by a drafter. As you gain an understanding of the drawings that you are making and gain confidence in your ability, the sketches that you are given generally will become more simplified.

Eventually your supervisor may just refer you to a similar drawing and expect you to be able to make the necessary adjustments to fit it to the new application.

The decisions involved in making drawings without sketches require the drafter to have a good understanding of what is being drawn. This understanding does not come just from a textbook. To advance as a drafter, you will need to spend time at construction sites observing buildings being constructed. An even better way to gain an understanding of what you are drafting is to spend time working at a construction site. Understanding what a craftsman must do as a result of what you have drawn is necessary if you are to advance as a drafter.

Depending on the size of the office where you work, you may also spend a lot of your time as a beginning drafter running prints, making deliveries, obtaining permits, and other such office chores. Don't get the idea that a drafter does only the menial chores around an office. You do need to be prepared, as you go to your first drafting job, to do things other than drafting. Many entry-level drafters expect that they will go to their first job and start designing great works for humanity. As an entry-level drafter it may not be so.

The Experienced Drafter

Although your supervisor may prepare the basic design for a project, experienced drafters are expected to make construction design decisions. These might include determining structural sizes, connection methods for intersecting beams, drawing renderings, making job visitations, and supervising beginning drafters. The types of drawings that you will be working on are also affected as you gain experience. Instead of drawing plot plans, cabinet elevations, or roof plans or revising existing details, an experienced drafter may be working on the floor and foundation plans, elevations, and sections. Probably your supervisor will still make the initial design drawings but will pass these drawings on to you as soon as a client approves the preliminary drawings.

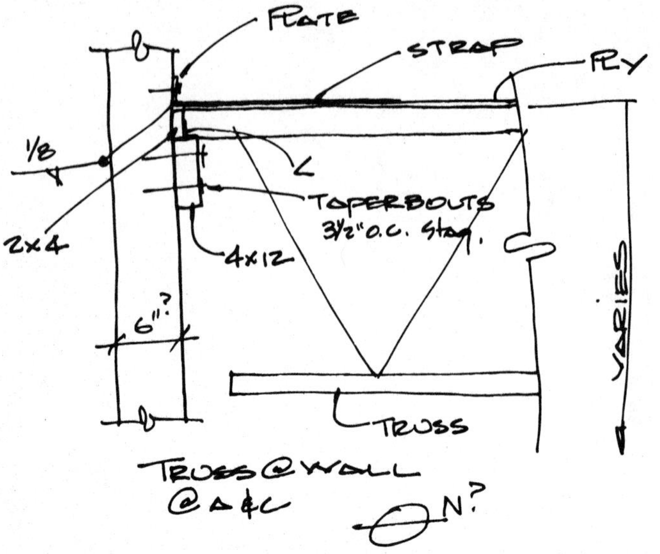

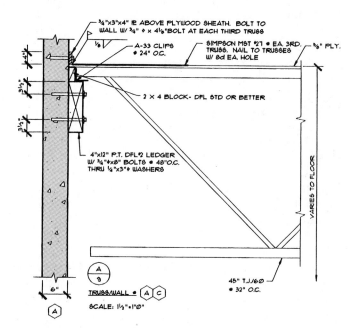

Figure 1–1 A sketch is usually given to a junior drafter to follow for a first drawing. The drafter can find information for completing the drawing by examining similar jobs in the office.

In addition to drafting, you may work with the many city and state building departments that govern your work. This will require you to do research in the many codes that govern the building industry. Not only will you be required to do research with building codes, but you will also need to become familiar with vendor catalogs. The most common of these catalogs is the Sweet's Catalog. Sweet's, as it is known, is a series of books that contain product information on a wide variety of building products. Information within these catalogs is listed by manufacturer, trade name, and type of product.

Employment Opportunities

As a drafter, you may find employment in firms of all sizes. Designers, architects, and engineers all require entry-level and advanced drafters to help produce their drawings. Many drafters are also employed by suppliers of architectural equipment. This work might include drawing construction details for a steel fabricator, making layout drawings for a cabinet shop, or designing ductwork for a heating and air-conditioning installer. Many manufacturing companies hire drafters with an architectural background to help draw and sometimes sell a product. Drafters might be called upon to draw installation diagrams for instruction booklets or sales catalogs. Drafters are also employed by many government agencies. These jobs include working in planning, utility, or building departments, survey crews, or other related municipal jobs.

Educational Requirements

In order to get your first drafting job you will need a solid education, good line and lettering quality, and the ability to *sell* yourself to an employer. Good linework and lettering will come as a result of practice. The education required for a drafter can range from one or more years in a high school drafting program, a diploma from a one-year accredited technical school, a degree from a two-year college program, all the way to a master's degree in architecture.

Helpful areas of study for an entry-level drafter would include math, writing, and drawing. As a drafter, the math required ranges from simple addition to calculus. Although the drafter may spend most of the day adding dimensions expressed in feet and inches, a knowledge of geometry will be helpful for solving many building problems. Writing skills will also be very helpful. As a drafter, you are often required to complete the paperwork that accompanies any set of plans, such as permits, requests for variances, written specifications, or environmental impact reports. In addition to standard drafting classes, classes in photography, art, surveying, and construction could be helpful to the drafter.

▼ DESIGNER

The meaning of the term *designer* varies from state to state. Many states restrict the use of the term *designer* by requiring anyone using the term to have had formal training and to have passed a competency test. A designer's responsibilities are very similar to those of an experienced drafter and are usually based on both education and experience. A designer is usually the coordinator of a team of drafters. The designer may work under the direct supervision of an architect or engineer or both and supervise the work schedule of the drafting team.

Figure 1–2 A detail drawn by a drafter using the sketch shown in Figure 1–1.

In addition to working in a traditional architectural office setting, designers often have their own office practice in which they design residential, multifamily, and some types of light commercial buildings. State laws vary on the types and sizes of buildings a designer may work on without requiring the stamp of an architect or engineer. Students wishing more information about this career can obtain information from:

The American Institute of Building Design (AIBD)
P.O. Box 1148
Pacific Palisades, CA 64057

Students can also contact the American Design and Drafting Association, the U.S. Department of Labor, or the U.S. Office of Education.

▼ ARCHITECT

An architect performs the tasks of many professionals, including designer, artist, project manager, and construction supervisor. Few architects work full time in residential design. Although many architects design some homes, most devote their time to commercial construction projects such as schools, offices, and hospitals. An architect is responsible for the design of a structure and for the way the building relates to the environment. The architect often serves as a coordinator on a project to ensure that all aspects of the structure blend together to form a pleasing relationship. This coordinating includes working with the client, the contractors, and a multitude of engineering firms that may be working on the project.

Use of the term *architect* is legally restricted to use by individuals who have been licensed by the state where they practice. Obtaining a license requires passing several written tests. There are two routes a person can take to prepare for these tests. A drafter or designer can prepare to take the licensing test through practical work experience. Although standards vary for each state, five to seven years of experience under the direct supervision of a licensed architect or engineer usually is required. An alternative to this method is to obtain a bachelor's degree from an accredited college and three years of professional experience with a licensed architect or engineer.

Education

High school and two-year college students can prepare for a degree program by taking classes in fine arts, math, science, and social science. Many two-year drafting programs offer drafting classes that can be used for credit in four- or five-year architectural programs. A student planning to transfer to a four-year program should verify with the new college which classes can be transferred.

Fine arts classes such as drawing, sketching, design, and art along with architectural history will help the future architect develop an understanding of the cultural significance of structures and help transform ideas into reality. Math and science, including algebra, geometry, trigonometry, and physics, will provide a stable base for the advanced structural classes that will be required. Sociology, psychology, cultural anthropology, and classes dealing with human environments will help develop an understanding of the people who will use the structure. Because architectural students will need to read, write, and think clearly about abstract concepts, preparation should also include literature and philosophy courses.

In addition to formal study, students should discuss with local architects the opportunities and possible disadvantages that may await them in pursuing the study and practice of architecture.

Areas of Study

The study of architecture is not limited to the design of buildings. Although the architectural curriculum typically is highly structured for the first two years of study, students begin to specialize in an area of interest during the third year of the program.

Students in a bachelor's program may choose courses leading to a degree in several different areas of architecture such as urban planning, landscape architecture, and interior architecture. Urban design is the study of the relationship among the components within a city. Interior architects work specifically with the interior of a structure to ensure that all aspects of the building will be functional. Landscape architects specialize in relating the exterior of a structure to the environment. Students wishing further information on training or other related topics can contact:

American Institute of Architects
1735 New York Avenue, NW
Washington, DC 20006
American Society of Landscape Architects
1750 Old Meadow Road
McLean, VA 22101
American Institute of Landscape Architects
501 E. San Juan Avenue
Phoenix, AZ 85012
Association of Collegiate Schools of Architecture
1735 New York Avenue, NW
Washington, DC 20006

▼ ENGINEER

The term *engineer* covers a wide variety of professions. In the construction fields, structural engineers are the most common, although many jobs exist for electrical, mechanical, and civil engineers. Structural engineers typically specialize in the design of structures built of steel or concrete. Many directly supervise drafters and designers in the design of multifamily and light commercial structures.

Figure 1–3 Drawings are often prepared by an architectural illustrator to help convey design concepts. *Drawing by Huxley, Courtesy Structureform Masters, Inc.*

Electrical engineers work with architects and structural engineers and are responsible for the design of lighting and communication systems. They supervise the design and installation of specific lighting fixtures, telephone services, and requirements for computer networking.

Mechanical engineers are also an instrumental part of the design team. They are responsible for the sizing and layout of the heating, ventilation, and air-conditioning systems (HVAC) and plan how treated air will be routed throughout the project. They work with the project architect to determine the number of occupants of the completed building and the heating and cooling load that will be generated.

Civil engineers are responsible for the design and supervision of a wide variety of construction projects, such as highways, bridges, sanitation facilities, and water treatment plants. They are often directly employed by construction companies to oversee the construction of large projects and to verify that the specifications of the design architects and engineers have been carried out.

Similar to the requirements for becoming an architect, a license is required to function as an engineer. The license can be applied for after several years of practical experience, or after obtaining a bachelor's degree and three years of practical experience.

Success in any of the engineering fields requires a high proficiency in math and science. Similar to the requirements for becoming an architect, to become an engineer typically requires five years of education at an accredited college or university, followed by successful completion of a state-administered examination. Certification can also be accomplished by training under a licensed engineer and then successfully completing the examination. Additional information can be obtained about the field of engineering by writing to:

American Society of Civil Engineers
345 E. 47th Street
New York, NY 10017

▼ RELATED FIELDS

So far only the opportunities that are similar because of their drawing, design, and creative nature have been covered. In addition to these careers, there are many related careers that require an understanding of drafting principles to be successful. Some of these include illustrator, model maker, specification writer, inspector, and construction-related trades.

Illustrator

Many drafters, designers, and architects have the basic skills to draw architectural renderings. Very few, though, have the expertise to make this type of drawing rapidly. Most illustrators have a background in art. By combining their artistic talent with a basic understanding of architectural principles, the illustrator is able to produce drawings that realistically show a proposed structure. Figure 1–3 shows a drawing that was prepared by an architectural illustrator. Section XI provides an introduction to presentation drawings.

Model Maker

In addition to presentation drawings, many architectural offices use models of a building or project to help convey design concepts. Models such as the one shown in Figure 1–4 are often used as a public display to help gain support for large projects. Model makers need basic drafting skills to help interpret the plans required to build the actual project. Model makers may be employed within a large architectural firm or for a company that only makes models for architects.

Specification Writer

Specifications are written instructions for methods, materials, and quality of construction. Written specifications are introduced in Section XII.

A specification writer must have a thorough understanding of the construction process and have a good ability to read plans. Generally, a writer will have had classes in technical writing at the two-year-college level.

Inspector

Building departments require that plans and the construction process be inspected to ensure that the required codes for public safety have been met. A plan examiner must be licensed by the state to certify minimum understanding of the construction process. In most states, there are different levels of examiners. An experienced drafter or designer may be able to qualify as a low-level or residential-plan inspector. Generally, a degree in engineering or architecture is required to advance to an upper-level position.

The construction that results from the plans must also be inspected. Depending on the size of the building department, the plan examiner may also serve as the building inspector. In large building departments, one group inspects plans and another inspects construction. To be a construction inspector requires an exceptionally good

Figure 1–4 Models such as this one of the KOIN Center of Portland, Oregon, are often used in public displays to convey design ideas. *Courtesy KOIN Center, Olympia & York Properties (Oregon), Inc.*

understanding of codes limitations, print reading, and construction methods. Each of these skills has its roots in a beginning drafting class.

Jobs in Construction

Many drafters are employed directly by construction companies. The benefits of this type of position have already been discussed. These drafters typically not only do drafting, but also work part time in the field. Many drafters give up their jobs for one of the high-paying positions in the construction industry. The ease of interpreting plans as a result of a background in drafting is of great benefit to any construction worker.

▼ DESIGN BASICS

Designing a home for a client can be an exciting but difficult process. Very rarely can an architect sit down and dash off a design that meets the needs of the client perfectly. The time required to design a home can range from a few days to several months. It is important for you to understand the design process and the role the drafter plays in it. This requires an understanding of basic design, financial considerations, and common procedures of design. In the rest of this chapter, the terms *architect* and *designer* can be thought of as synonymous.

Typically the designing and much of the drafting will already be done before the project is given to the junior drafter. As experience and confidence are gained, the drafter enters the design process at earlier stages.

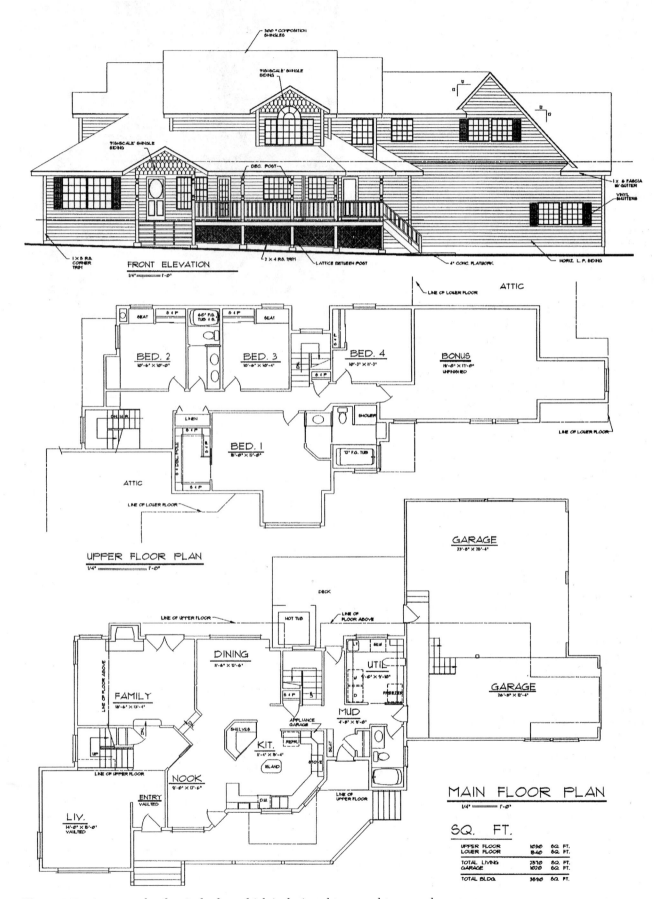

Figure 1–5 An example of a stock plan which is designed to appeal to many buyers.

Financial Considerations

Both designers and drafters need to be concerned with costs. Finances influence decisions made by the drafter about framing methods and other structural considerations. Often the advanced drafter must decide between methods that require more materials with less labor or those that require fewer materials but more labor. These are decisions of which the owner may never be aware, but they can make the difference in the house's being affordable or not.

The designer makes the major financial decisions that affect the cost and size of the project. The designer needs to determine the client's budget at the beginning of the design process and work to keep the project within these limits.

Through past experience and contact with builders, the designer should be able to make an accurate estimate of the finished cost of a house. This estimate is often made on a square footage basis in the initial stages of design. For instance, in some areas a modest residence can be built for approximately $50 per sq ft. In other areas this same house may cost approximately $90 per sq ft. A client wishing to build a 2500 sq ft house could expect to pay between $125,000 and $225,000, depending on where the home will be built. Keep in mind that these are estimates. A square footage price tells very little about the home. In the design stage, square footage estimates help set parameters for the design. The final cost of the project is determined once the house is completely drawn and a list of materials is prepared. With a list of materials, contractors are then able to make accurate decisions about cost.

The source of the finances can also affect the design process of the project. Certain lending institutions may require some drawings that the local building department may not. Federal Housing Administration (FHA) and Veterans Administration (VA) loans often require extra forms, drawings, and specifications that need to be taken into account in the initial design stages.

The Client

Most houses are not designed for one specific family. In order to help keep costs down, houses are often built in subdivisions and designed to appeal to a wide variety of people. Often one basic plan can be built with several different options, thus saving the contractor the cost of paying for several different complete plans. Some families may make minor changes to an existing plan. These modified *stock plans* allow the prospective home buyer a chance to have a personalized design at a cost far below that of custom-drawn plans. See Figure 1–5. If finances allow, or if a stock plan cannot be found to meet their needs, a family can have a plan custom designed. It does not matter if the residence is designed for mass appeal or for a specific family; the design and drafting procedure is similar.

The Design Process

The design of a residence can be divided into several stages. These generally include initial contact, preliminary design studies, initial working drawings, final design considerations, completion of working drawings, permit procedures, and job supervision.

Initial Contact.   A client often approaches the designer to obtain background information on him or her. Design fees, schedules, and the compatibility of personalities are but a few of the basic questions to be answered. This initial contact may take place by telephone conversation or a personal visit.

The questions asked are important to both the architect and the client. The client needs to pick a designer who can work within budget and time limitations. The designer needs to screen clients to determine if the client's needs fit within the office schedule.

Once an agreement has been reached, the preliminary design work can begin. This selection process usually begins with the signing of a contract to set guidelines that identify which services are to be provided and when payment is expected. This is also the time when the initial criteria for the project will be determined. Generally, clients have a basic size and a list of specifics in mind, a sketch of proposed floor plans, and a file full of pictures of items that they would like in their house. During this initial phase of design it is important to become familiar with the lifestyle of the client as well as the site where the house will be built.

Preliminary Design Studies

Once a thorough understanding of the client's lifestyle, design criteria, and financial limits have been developed,

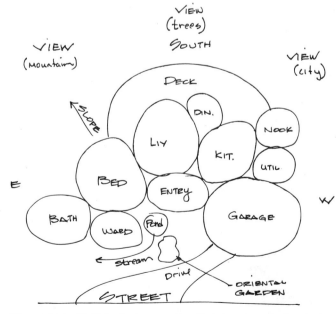

Figure 1–6 Bubble designs are the first drawings in the design process. These drawings are used to explore room relationships.

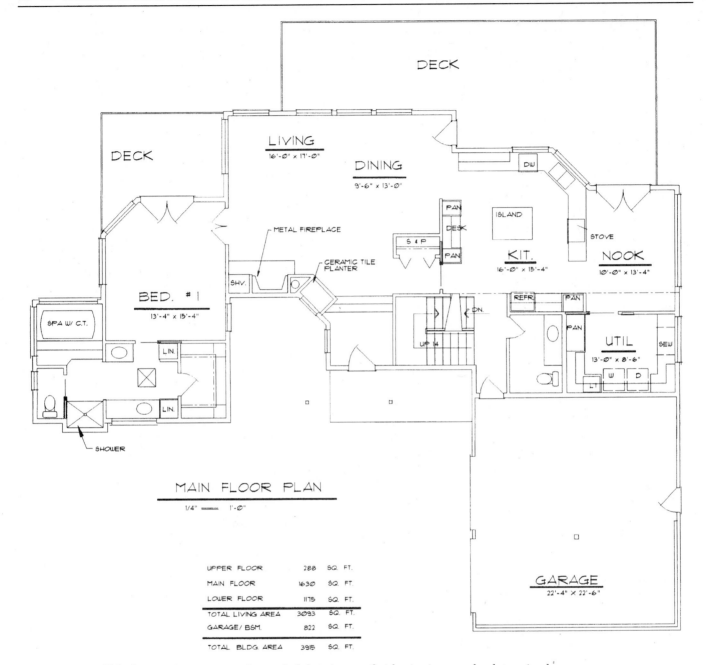

Figure 1–7 Bubble drawings are converted to scaled drawings so that basic sizes can be determined.

the preliminary studies can begin. These include research with the building and zoning departments that govern the site, investigation of the site, and discussions with any board of review that may be required. Once this initial research has been done, preliminary design studies can be started. The preliminary drawings usually take two stages: bubble drawings and scaled sketches.

Bubble drawings are freehand sketches used to help determine room locations and relationships. See Figure 1–6. It is during this stage of the preliminary design that consideration is given to the site and energy efficiency of the home.

Once a satisfactory layout has been sketched, these shapes are transformed into scaled sketches. Figure 1–7 shows a preliminary floor plan. Usually several sets of sketches are developed to explore different design possibilities. Consideration is given to building code regulations and room relationships and sizes. After the options to the design are explored, the designer selects a plan to prepare for the client. This could be the point where a drafter first becomes involved in the design process. Depending on the schedule of the designer, a senior drafter might prepare the refined preliminary drawings, which can be seen in Figure 1–8. These would include

Figure 1–8 The front elevation is drawn to explore design options based on the preliminary floor plans.

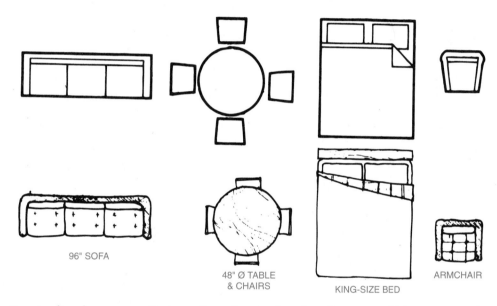

Figure 1–9 Furniture is often drawn on preliminary floor plans to show how the space within a room can be utilized.

floor plans and an elevation. These drawings are then presented to the client. Changes and revisions are made at this time. Once the plans are approved by the client, the preliminary drawings are ready to be converted to design drawings.

Room Planning. Room usage must be considered throughout the design process. Chapter 8 will introduce major design concepts to be considered. Typically the designer will have talked with the owners about how each room will be used and what types of furniture will be included. Occasionally, placement of a family heirloom will dictate the entire layout. When this is the case, the owner should specify the size requirements for a particular piece of furniture.

When placing furniture outlines to determine the amount of space in a room, the drafter can use transfer shapes known as rub-ons, templates, freehand sketching methods, or blocks if working with a CAD (computer-aided design) program. Rub-ons provide a fast method of representing furniture, but they can also prove to be expensive. Furniture can be drawn with a template or sketched. Figure 1–9 shows examples of the outlines of common pieces of furniture. Figure 1–10 shows typical sizes for common pieces of furniture. Figure 1–11 shows a preliminary floor plan with the furniture added.

COMMON FURNITURE SIZES					
Living Room and Family Room					
Sofas	34 × 76	End Tables	18 × 18	Arm Chairs	18 × 21
	34 × 90		18 × 24		18 × 24
	34 × 96		18 × 30		22 × 24
	34 × 102		18 × 36		28 × 32
			24 × 24		32 × 34
Sectional Pieces	26 × 30		26 × 26	Recliners	30 × 29
	26 × 53		24 hex		to 66
	26 × 62		26 hex		
				Piano	24 × 56
Love Seats	32 × 50	Coffee Tables	18 × 36	Grand	
	30 × 66		20 × 52	Piano	58 × 75
			20 × 60		
			20 × 75		
Dining Room					
Oval Tables	54 × 76		Hutch	18 × 36 × 72	
Round Tables	30		Chairs	18 × 18	
	32			18 × 20	
	36				
	42		Tea Cart	16 × 24	
	48			18 × 33	
Rectangular Tables	30 × 34				
	36 × 44 to 108				
	42 × 48 to 108				
Bedroom					
Beds:			Dressers	18 × 30 to 82	
Cribs	20 × 50				
	30 × 54		Desks	24 × 30	
				24 × 36	
Twin	44 × 80			24 × 42	
				32 × 42	
Double	54 × 75			32 × 48	
				32 × 60	
Queen	65 × 80				
			Night	12 × 15	
King	72 × 84		Stands	15 × 15 to 21	

Figure 1–10 Common furniture sizes.

Initial Working Drawings

With the preliminary drawings approved, a drafter can begin to lay out the working drawings. The procedure will vary with each office, but generally each of the drawings required for the project will be started. These would include the foundation, plot, roof, electrical, cabinet, and framing plans. At this stage the drafter must rely on past experience for drawing the size of beams and other structural members. Beams and other structural material will be located, but exact sizes are not usually determined until the entire project has been laid out.

Final Design Considerations

Once the drawings have been laid out, the designer will generally meet with the client again to get information on flooring, electrical needs, cabinets, and other finish materials. This conference will result in a set of marked drawings that the drafter will use to complete the working drawings.

Completion of Working Drawings

The complexity of the residence will determine which drawings are required. Most building departments require a plot plan, floor plan, foundation plan, elevations, and one cross section as the minimum drawing to get a building permit. On a complicated plan, a wall framing plan, roof framing plan, grading plan, and construction details may be required. Depending on the lending institution, interior elevations, cabinet drawings, and finish specifications may be required.

The skills of the drafter will determine his or her participation in preparing the working drawings. As an entry-level drafter, you will often be given the job of making corrections on existing drawings or drawing plot plans, cabinets, or other drawings. As you gain skill you will be given more drawing responsibility. With increased ability, you will start to share in the design responsibilities.

Working Drawings. The drawings that will be provided vary for each residence and within each office. Figures 1–12 through 1–18 show what are typically considered the *architectural drawings*. Figures 1–19 through 1–20 show the *electrical drawings*. These are drawings that show finishing materials. Figures 1–21 through 1–28 show what are typically called the *structural drawings*. Many architectural firms number each page based on the type of drawing on that page. Pages would be represented by:

A—Architectural
S—Structural
M—Mechanical
E—Electrical
P—Plumbing

Figure 1–12 shows the plot plan that was required for the residence in Figure 1–7. Typically the plot plan is completed by a junior drafter from a sketch provided by the senior drafter or designer. A grading plan is not required for every residence, but because of the amount of dirt to be excavated for the residence shown in Figure 1–7, a grading plan was provided. See Figure 1–13. The junior drafter may draw the base drawing showing the structure and plot plan, and the designer or senior drafter will usually complete the grading information. Typically the owner is responsible for hiring a surveyor, who will provide the topography and site map, although the designer may coordinate the work between the two offices. Section III will provide insight into how these two drawings are developed.

Using the preliminary drawing that was presented in Figure 1–8, the balance of the working elevations can be completed. Depending on the complexity of the structure, they may be completed by either the junior or senior

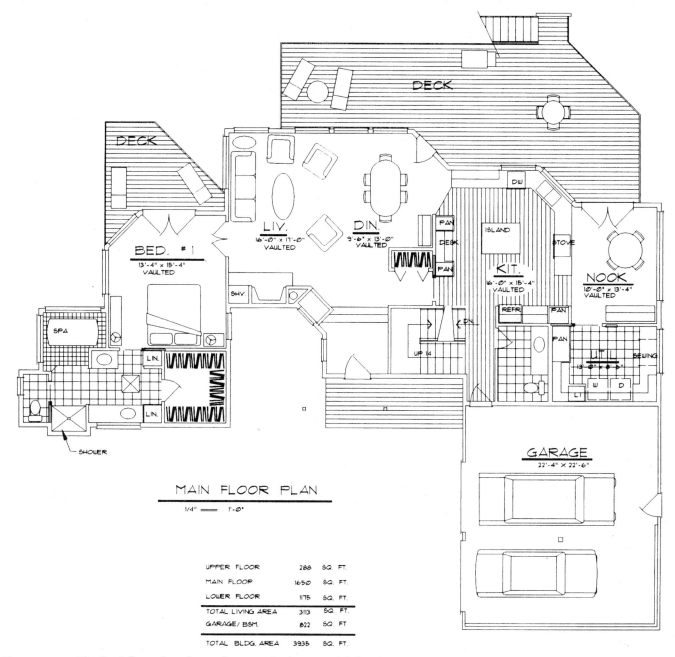

DECK

DECK

LIV.
16'-0" x 17'-0"
VAULTED

DIN.
9'-6" x 13'-0"
VAULTED

BED. #1
13'-4" x 15'-4"
VAULTED

ISLAND

DW

PAN

DESK

PAN

KIT.
16'-0" x 15'-4"
VAULTED

STOVE

NOOK
10'-0" x 13'-4"
VAULTED

SPA

SHV.

LIN.

LIN.

REFR

PAN

PAN

UP 14

DN

LT

SEWING

W   D

SHOWER

MAIN FLOOR PLAN
1/4" = 1'-0"

GARAGE
22'-4" x 22'-6"

UPPER FLOOR	288	SQ. FT.
MAIN FLOOR	1650	SQ. FT.
LOWER FLOOR	1175	SQ. FT.
TOTAL LIVING AREA	3113	SQ. FT.
GARAGE/ BSM.	822	SQ. FT
TOTAL BLDG. AREA	3935	SQ. FT.

Figure 1–11 The final floor plan shows the owner's changes with furniture.

drafter. Figure 1–14 shows the working elevations for this structure. Section VII will provide information needed to complete working elevations, and Section XI will introduce presentation drawings.

Figures 1–15 through 1–18 show the completed plan views of this structure. Figure 1–15 shows the view from flying over the structure. The roof plan is often completed by the junior drafter using sketches provided by the senior drafter. Section VI will provide information on roof plans. The plan views represented in Figures 1–16 through 1–18 usually are completed by the senior drafter. These drawings provide information related to the room arrangements and interior finishes. Junior drafters also

work on the plan views by making corrections or adding notes, which are typically placed on a marked-up set of plans provided by the designer or senior drafter. Section V will provide information for completing floor plans.

Figures 1–19 and 1–20 show the electrical plan. Depending on the complexity of the residence, electrical information may be placed directly on the floor plan. A junior drafter typically completes the electrical drawings by working from marked-up prints provided by the designer. Chapter 16 will provide information for completing the electrical plans.

Figures 1–21 through 1–23 represent the framing plans and notes associated with these drawings. The notes

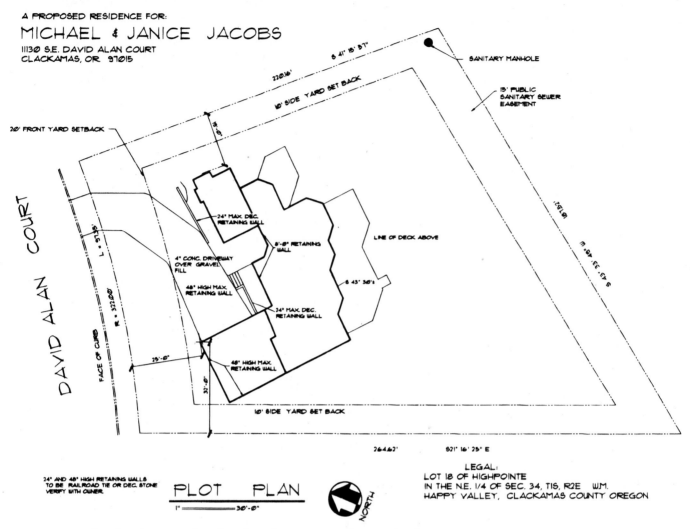

A PROPOSED RESIDENCE FOR:

MICHAEL & JANICE JACOBS
11130 S.E. DAVID ALAN COURT
CLACKAMAS, OR. 97015

SANITARY MANHOLE

15' PUBLIC SANITARY SEWER EASEMENT

S 41° 15' 37"

10' SIDE YARD SET BACK

20' FRONT YARD SETBACK

24" MAX. DEC. RETAINING WALL

8'-0" RETAINING WALL

LINE OF DECK ABOVE

4" CONC. DRIVEWAY OVER GRAVEL FILL

48" HIGH MAX. RETAINING WALL

S 43° 30's

24" MAX. DEC. RETAINING WALL

DAVID ALAN COURT

FACE OF CURB

48" HIGH MAX. RETAINING WALL

10' SIDE YARD SET BACK

24" AND 48" HIGH RETAINING WALLS TO BE RAILROAD TIE OR DEC. STONE VERIFY WITH OWNER.

PLOT PLAN
1" = 30'-0"

NORTH

S21° 16' 25" E

LEGAL:
LOT 18 OF HIGHPOINTE
IN THE N.E. 1/4 OF SEC. 34, T1S, R2E W.M.
HAPPY VALLEY, CLACKAMAS COUNTY OREGON

Figure 1–12   The plot plan is used to show how the structure relates to the site.

shown with Figure 1–21 are standard notes from the UBC/CABO that specify the nailing for all framing connections. Depending on the complexity of the structure, framing information may be placed on the floor plan. Because of the information to be shown regarding seismic and wind problems, separate framing plans have been provided. Typically these drawings would be completed by the designer and the senior drafters. Notice on Figure 1–22 that notes have been provided to specify the minimum code standards used to design the residence. An index is also provided to help find other structural information. Sections VI, VIII, and IX will provide information needed to complete the framing plans.

Figures 1–24 through 1–28 contain the section views required to show the vertical relationships of materials specified on the framing plans. Office practice varies greatly on what drawings will be provided and the scale and detail that will be shown in the sections. Section X will provide information for drawing sections. Sections

are typically drawn by the designer and the senior drafter and completed by the junior drafters.

Figure 1–28 shows what is typically the final drawing in a set of residential plans, although the foundation plan is one of the first drawings used at the construction site. Because of the importance of this drawing, it is typically completed by the designer or the senior drafter, although junior drafters may work on corrections or add notes to the drawing.

As you progress through this text, you will be exposed to each type of working drawing. It is important to understand that a drafter would rarely draw one complete drawing and then go on to another drawing. Because the drawings in a set of plans are so interrelated, often one drawing is started, and then another drawing is started so that relationships between the two can be studied before the first drawing is completed. For instance, floor plans may be laid out, and then a section may be drawn to work out the relationships between floors or any headroom problems.

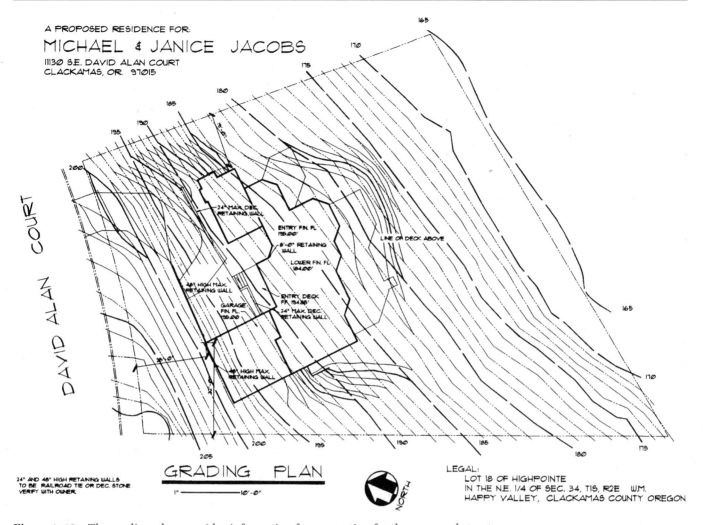

Figure 1–13 The grading plan provides information for excavation for the proposed structure.

When the plans are completely drawn, they must be checked. Dimensions must be checked carefully and cross referenced from one floor plan to another and to the foundation plan. Bearing points from beams must be followed from the roof down through the foundation system. Perhaps one of the hardest jobs for a new drafter is coordinating the drawings. As changes are made throughout the design process, the drafter must be sure that those changes are reflected on all affected drawings. When the drafter has completed checking the plans, the drafting supervisor will again review the plans before they leave the office.

Permit Procedures

When the plans are complete, the owner will ask several contractors to estimate the cost of construction. Once a contractor is selected to build the house, a construction permit is obtained. Although the drafter is sometimes responsible for obtaining the permits, this is usually done by the owner or contractor. The process for obtaining a permit varies depending on the local building department and the complexity of the drawings. Permit reviews generally take several weeks. Once the review is complete either a permit is issued or required changes are made to the plans.

Job Supervision

In residential construction, job supervision is rarely done when working for a designer, but it is quite often provided when an architect has drawn the plans. Occasionally a problem at the job site will require the designer or drafter to go to the site to help find a solution.

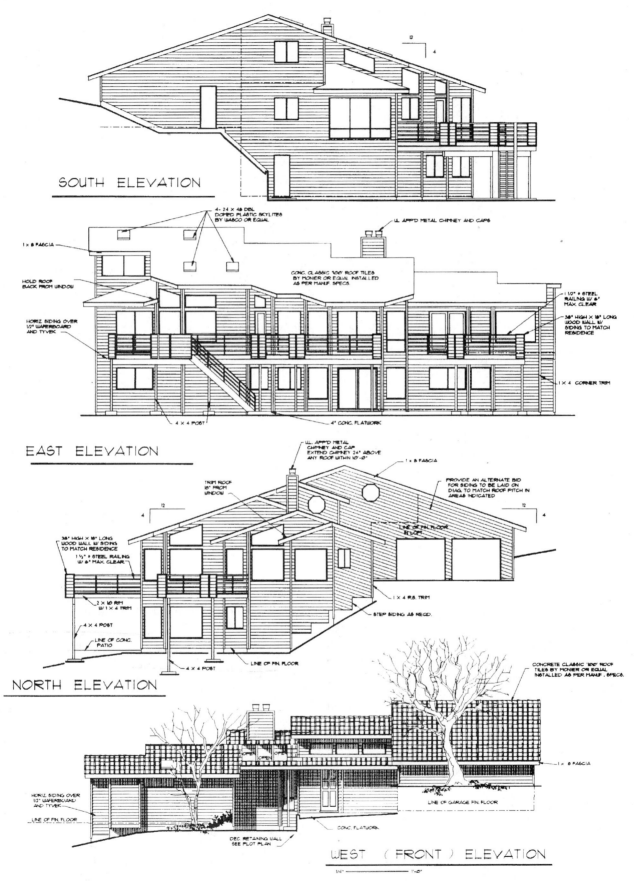

Figure 1–14 The working elevations provide a view of each side of the structure and show the shape as well as the exterior materials.

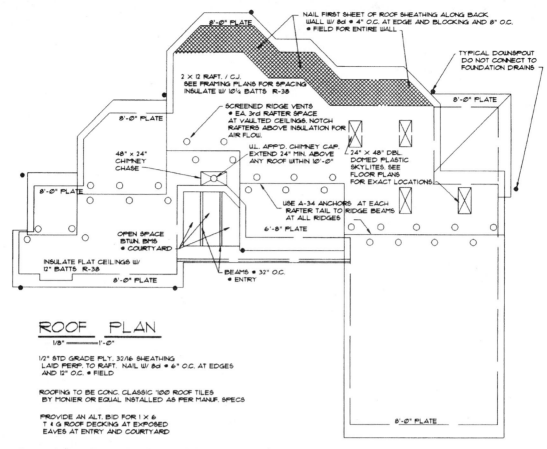

Figure 1–15 The roof plan shows the shape of the roof structure as well as vents and drains.

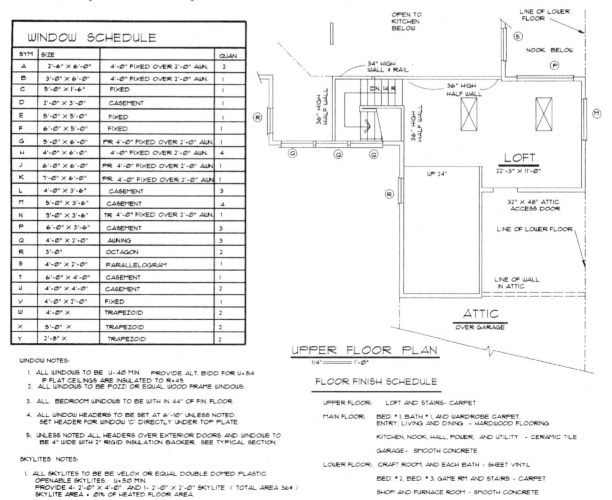

Figure 1–16 The upper floor plan shows the layout of the upper level and the outline of the lower levels.

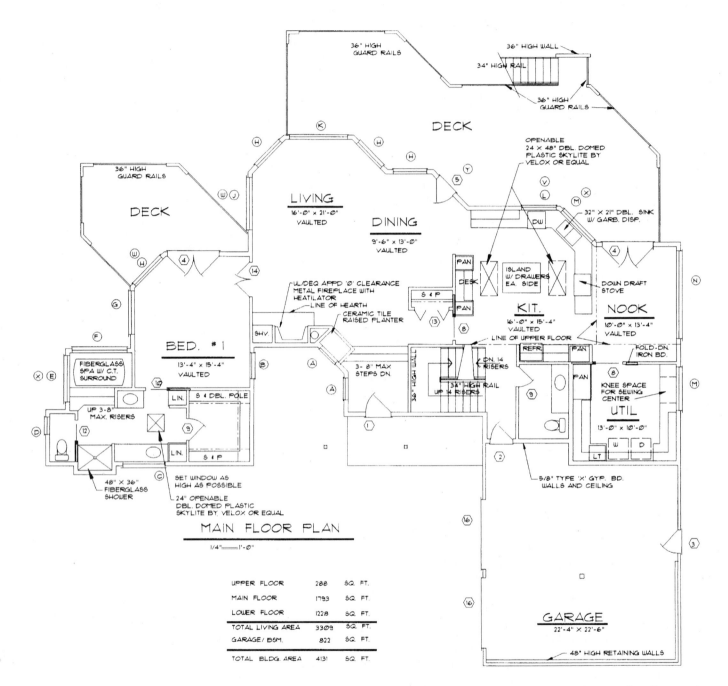

Figure 1–17 The main floor plan shows the layout of the rooms on the main level as well as how they relate to other floors.

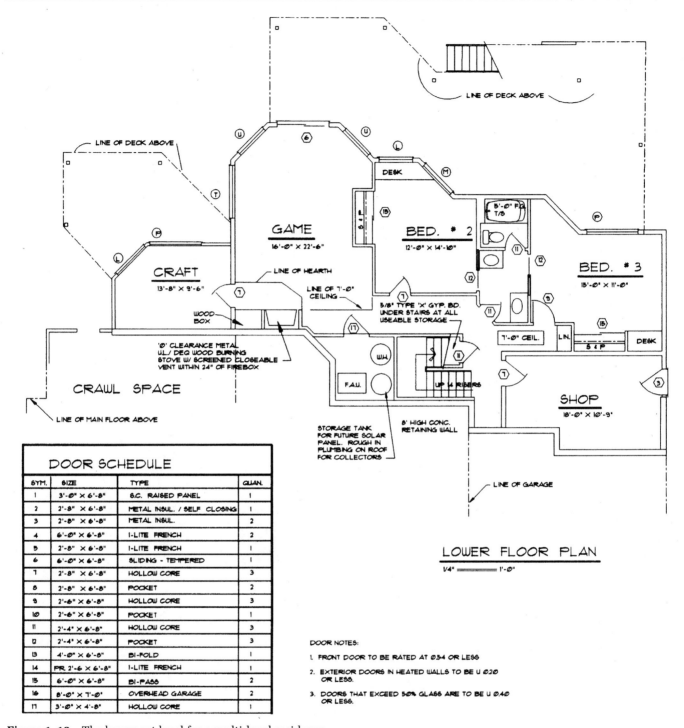

GAME
16'-0" X 22'-6"

CRAFT
13'-8" X 9'-6"

BED. # 2
12'-0" X 14'-10"

BED. # 3
15'-0" X 11'-0"

SHOP
18'-0" X 10'-9"

LINE OF DECK ABOVE

LINE OF DECK ABOVE

DESK

DESK

5'-0" F.
T/S

7'-0" CEIL.

LIN.

LINE OF HEARTH

LINE OF 7'-0"
CEILING

5/8" TYPE 'X' GYP. BD.
UNDER STAIRS AT ALL
USEABLE STORAGE

WOOD
BOX

'0' CLEARANCE METAL
U.L./ DEQ WOOD BURNING
STOVE W/ SCREENED CLOSEABLE
VENT WITHIN 24" OF FIREBOX

CRAWL SPACE

LINE OF MAIN FLOOR ABOVE

WH

F.A.U.

UP 14 RISERS

STORAGE TANK
FOR FUTURE SOLAR
PANEL. ROUGH IN
PLUMBING ON ROOF
FOR COLLECTORS

8' HIGH CONC.
RETAINING WALL

LINE OF GARAGE

LOWER FLOOR PLAN
1/4" ===== 1'-0"

DOOR SCHEDULE

SYM.	SIZE	TYPE	QUAN.
1	3'-0" X 6'-8"	S.C. RAISED PANEL	1
2	2'-8" X 6'-8"	METAL INSUL. / SELF CLOSING	1
3	2'-8" X 6'-8"	METAL INSUL.	2
4	6'-0" X 6'-8"	1-LITE FRENCH	2
5	2'-8" X 6'-8"	1-LITE FRENCH	1
6	6'-0" X 6'-8"	SLIDING - TEMPERED	1
7	2'-8" X 6'-8"	HOLLOW CORE	3
8	2'-8" X 6'-8"	POCKET	2
9	2'-6" X 6'-8"	HOLLOW CORE	3
10	2'-6" X 6'-8"	POCKET	1
11	2'-4" X 6'-8"	HOLLOW CORE	3
12	2'-4" X 6'-8"	POCKET	3
13	4'-0" X 6'-8"	BI-FOLD	1
14	PR 2'-6 X 6'-8"	1-LITE FRENCH	1
15	6'-0" X 6'-8"	BI-PASS	2
16	8'-0" X 7'-0"	OVERHEAD GARAGE	2
17	3'-0" X 4'-8"	HOLLOW CORE	1

DOOR NOTES:

1. FRONT DOOR TO BE RATED AT Ø.54 OR LESS

2. EXTERIOR DOORS IN HEATED WALLS TO BE U Ø.20
   OR LESS.

3. DOORS THAT EXCEED 50% GLASS ARE TO BE U Ø.40
   OR LESS.

Figure 1–18 The basement level for a multi-level residence.

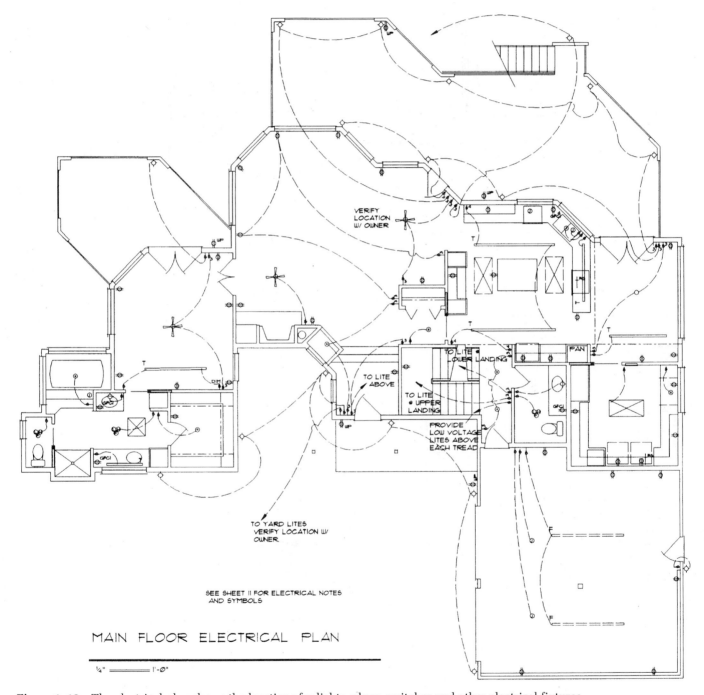

VERIFY
LOCATION
W/ OWNER

TO LITE
ABOVE

TO LITE
LOWER LANDING

TO LITE
● UPPER
LANDING

FAN

TO LITE

PROVIDE
LOW VOLTAGE
LITES ABOVE
EACH TREAD

TO YARD LITES
VERIFY LOCATION W/
OWNER

SEE SHEET 11 FOR ELECTRICAL NOTES
AND SYMBOLS

MAIN FLOOR ELECTRICAL PLAN

¼" ════ 1'-0"

Figure 1–19 The electrical plan shows the locations for lights, plugs, switches and other electrical fixtures.

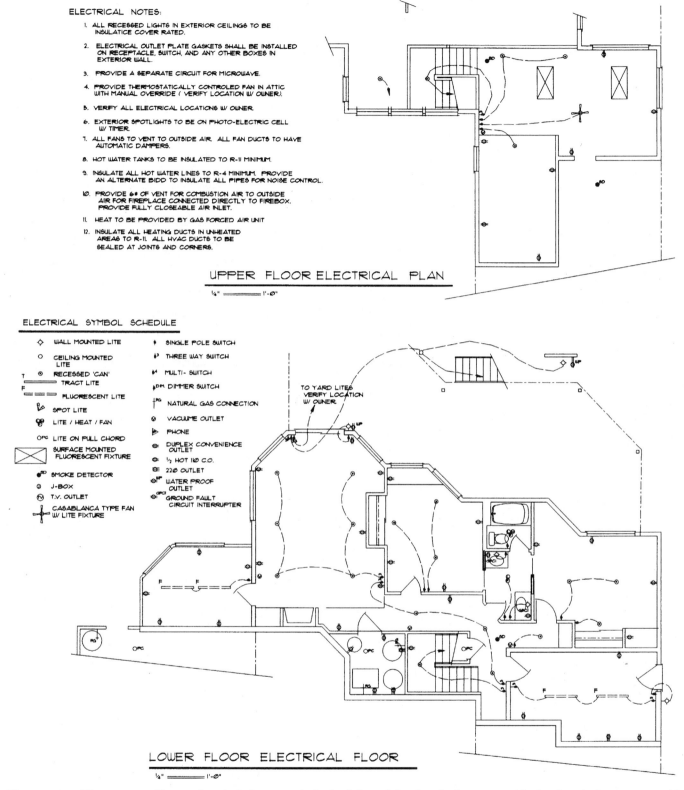

ELECTRICAL NOTES:

1. ALL RECESSED LIGHTS IN EXTERIOR CEILINGS TO BE INSULATICE COVER RATED.

2. ELECTRICAL OUTLET PLATE GASKETS SHALL BE INSTALLED ON RECEPTACLE, SWITCH, AND ANY OTHER BOXES IN EXTERIOR WALL.

3. PROVIDE A SEPARATE CIRCUIT FOR MICROWAVE.

4. PROVIDE THERMOSTATICALLY CONTROLED FAN IN ATTIC WITH MANUAL OVERRIDE (VERIFY LOCATION W/ OWNER).

5. VERIFY ALL ELECTRICAL LOCATIONS W/ OWNER

6. EXTERIOR SPOTLIGHTS TO BE ON PHOTO-ELECTRIC CELL W/ TIMER.

7. ALL FANS TO VENT TO OUTSIDE AIR ALL FAN DUCTS TO HAVE AUTOMATIC DAMPERS.

8. HOT WATER TANKS TO BE INSULATED TO R-11 MINIMUM.

9. INSULATE ALL HOT WATER LINES TO R-4 MINIMUM. PROVIDE AN ALTERNATE BIDD TO INSULATE ALL PIPES FOR NOISE CONTROL.

10. PROVIDE 6# OF VENT FOR COMBUSTION AIR TO OUTSIDE AIR FOR FIREPLACE CONNECTED DIRECTLY TO FIREBOX. PROVIDE FULLY CLOSEABLE AIR INLET.

11. HEAT TO BE PROVIDED BY GAS FORCED AIR UNIT

12. INSULATE ALL HEATING DUCTS IN UNHEATED AREAS TO R-11. ALL HVAC DUCTS TO BE SEALED AT JOINTS AND CORNERS.

UPPER FLOOR ELECTRICAL PLAN

¼" ══════ 1'-0"

ELECTRICAL SYMBOL SCHEDULE

◇ WALL MOUNTED LITE

○ CEILING MOUNTED LITE

⊙ RECESSED 'CAN'

═══ TRACT LITE

F ╌╌╌ FLUORESCENT LITE

SPOT LITE

LITE / HEAT / FAN

OPC LITE ON PULL CHORD

⊠ SURFACE MOUNTED FLUORESCENT FIXTURE

SMOKE DETECTOR

J-BOX

T.V. OUTLET

CASABLANCA TYPE FAN W/ LITE FIXTURE

SINGLE POLE SWITCH

THREE WAY SWITCH

MULTI- SWITCH

DM DIMMER SWITCH

JPG NATURAL GAS CONNECTION

VACUUME OUTLET

PHONE

DUPLEX CONVENIENCE OUTLET

½ HOT 110 C.O.

220 OUTLET

WATER PROOF OUTLET

GROUND FAULT CIRCUIT INTERRUPTER

TO YARD LITES VERIFY LOCATION W/ OWNER

LOWER FLOOR ELECTRICAL FLOOR

¼" ══════ 1'-0"

Figure 1–20 The upper and lower electrical plans, symbol schedule and the electrical notes provide the electrical contractor with a plan for installing all electrical fixtures.

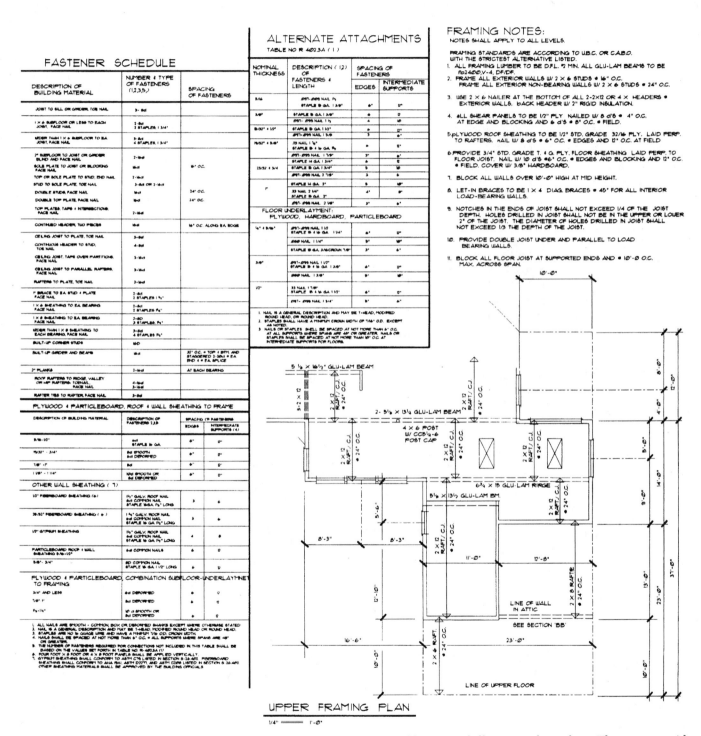

Figure 1–21 The framing plans provide the framing crew with the size and location of all structural members. The notes provide information for connecting structural members.

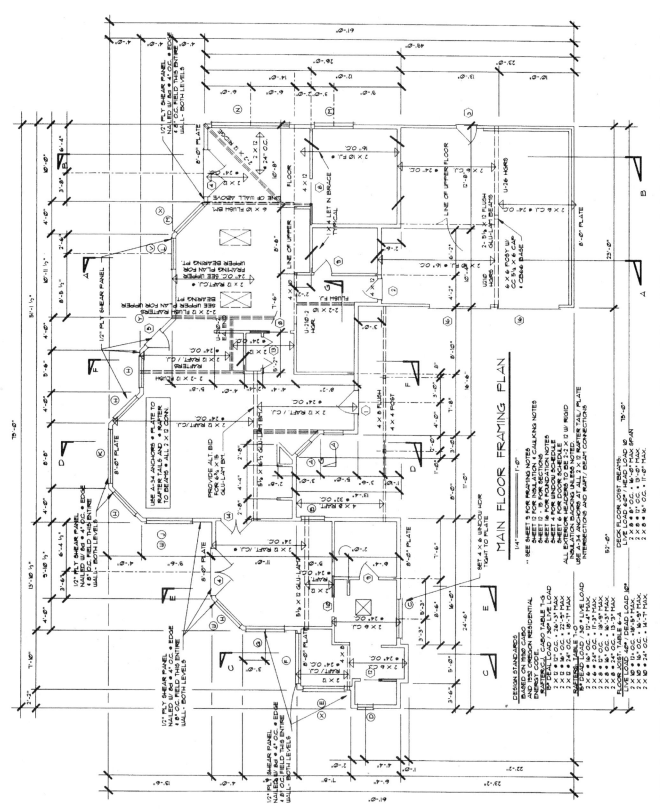

Figure 1-22 In addition to showing materials needed to frame the structure, materials needed to resist the forces of wind, seismic and other forces of nature are shown.

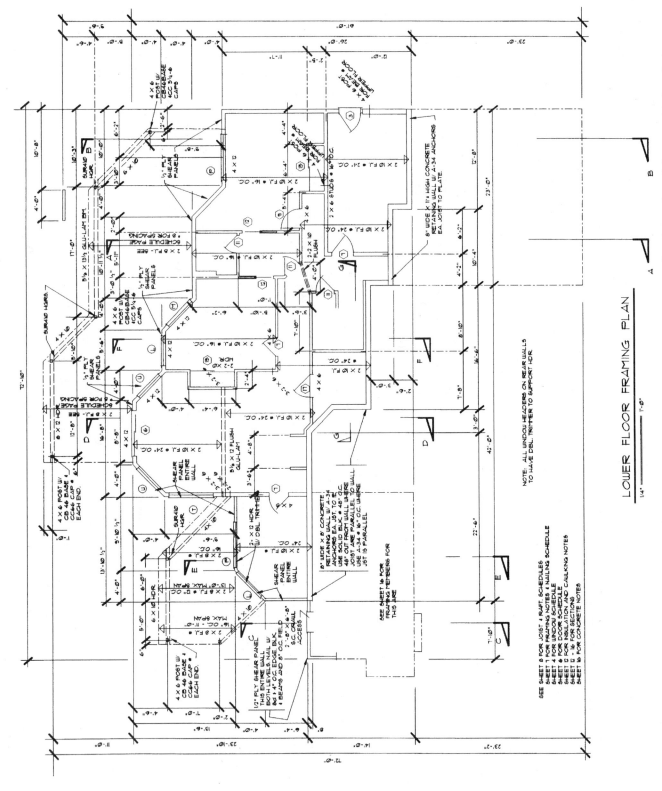

Figure 1–23 Each level of the framing plan also shows how loads from upper levels are to be supported and transferred down to the foundation.

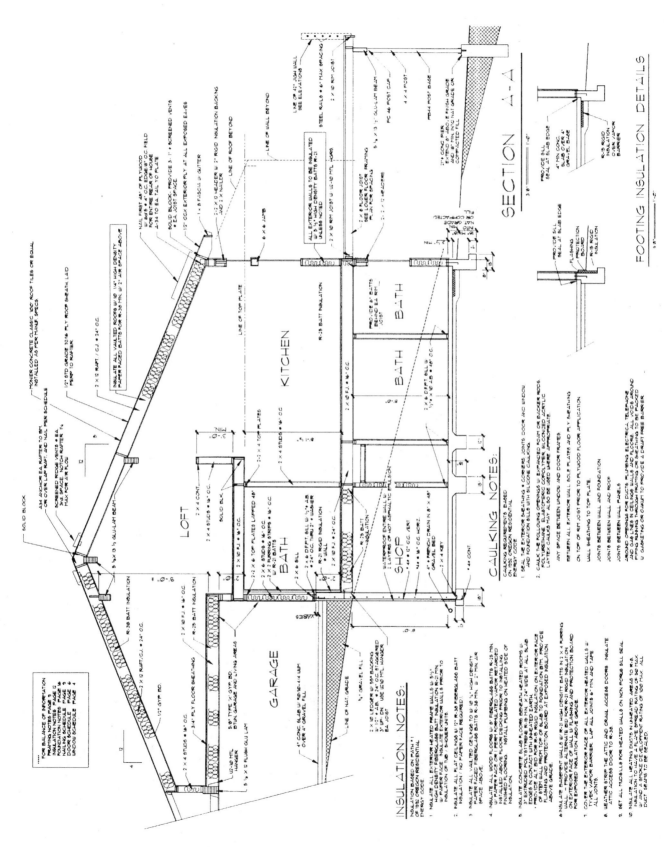

Figure 1-24 Sections are used to show the vertical relationships of the structural members.

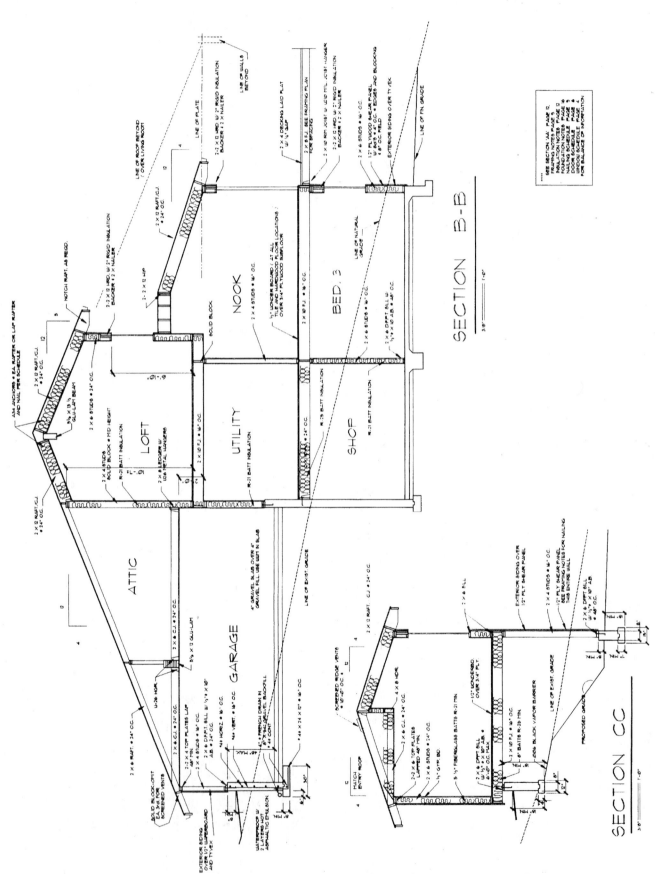

Figure 1–25 A section is typically drawn to show each major shape within the structure.

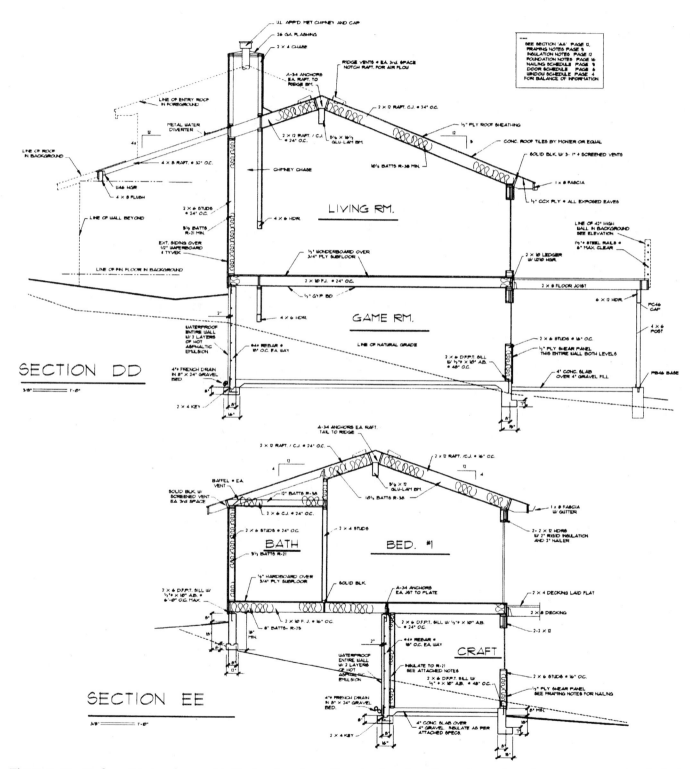

Figure 1–26   Each section is referenced to the framing plan with a viewing plane line to show where the structure is 'cut' and which direction the viewer is looking.

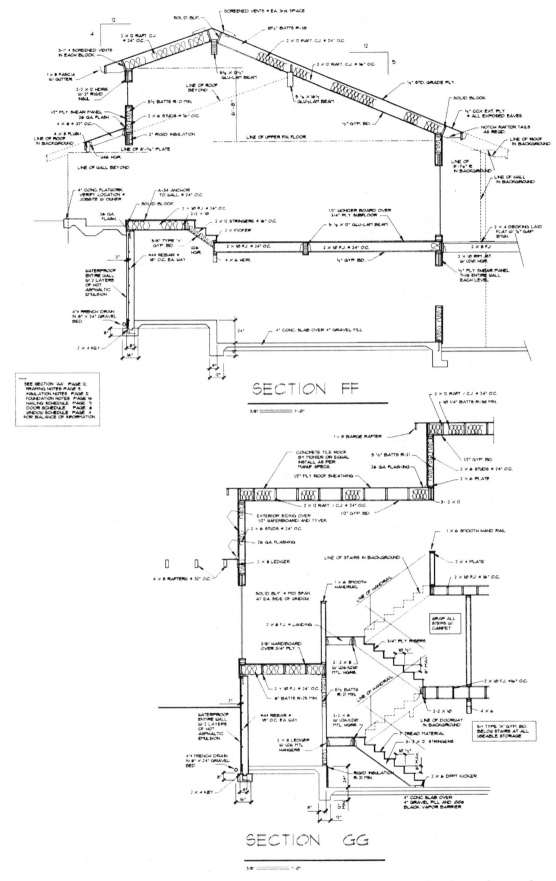

SECTION FF

SECTION GG

Figure 1–27 In addition to showing the shape of a structure, sections also show the materials to be used to insulate the structure.

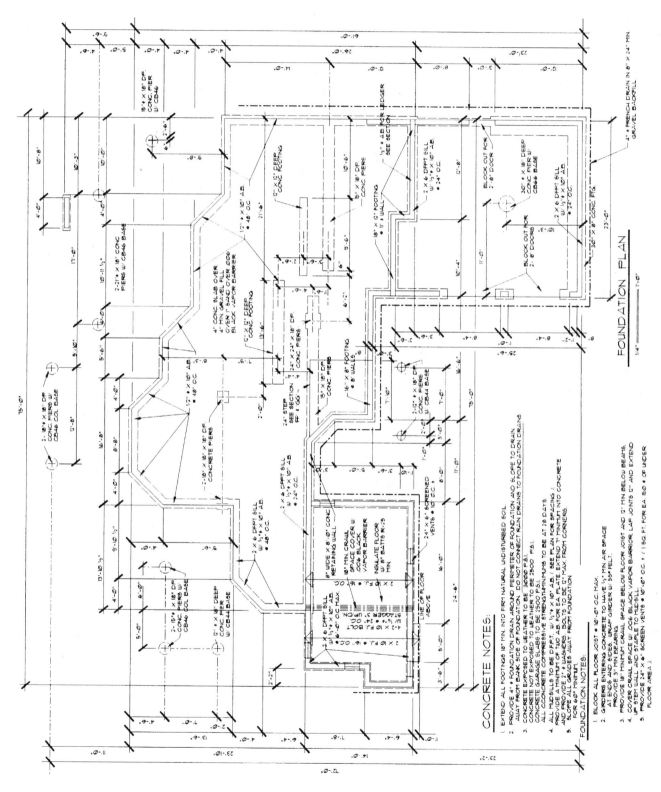

Figure 1-28 The foundation plan shows how each of the loads will be supported. The foundation not only resists the loads caused by gravity, but also resists the loads from seismic, wind and flooding.

CHAPTER

Professional Architectural Careers, Office Practice, and Opportunities Test

DIRECTIONS

Answer the problems with short complete statements or drawings as needed on an 8½ × 11 sheet of notebook paper as follows:

1. Letter your name, chapter number, and the date at the top of the sheet.
2. Letter the question number and provide the answer. You do not need to write out the question. Answers may be prepared on a word processor if appropriate with course guidelines.

PROBLEMS

Problem 1–1 List five types of work that a junior drafter might be expected to perform.

Problem 1–2 What three skills are usually required of a junior drafter for advancement?

Problem 1–3 What types of drawings should a junior drafter expect to be drawing?

Problem 1–4 Describe what the junior drafter might be given to assist in making drawings.

Problem 1–5 List four sources of written information that a drafter will need to be able to use.

Problem 1–6 List and briefly describe some different careers in which drafting would be helpful.

Problem 1–7 What is the purpose of a bubble drawing?

Problem 1–8 Why do you think furniture placement should be considered in the preliminary design process?

Problem 1–9 What would be the minimum drawings required to get a building permit?

Problem 1–10 List five additional drawings that may be required for a complete set of house plans in addition to the five basic drawings.

Problem 1–11 List and describe the steps of the design process.

Problem 1–12 What are the functions of the drafter in the design process?

Problem 1–13 Following the principles of this chapter, prepare a bubble sketch for a home with the following specifications:
a. 75′ × 120′ lot with a street on the north side of the lot
b. a gently sloping hill to the south
c. south property line is 75′ long
d. 40′ oak trees along the south property line
e. 3 bedrooms, 2½ baths, living, dining, with separate eating area off kitchen
f. exterior style as per your choice

Explain why you designed the house that you did.

CHAPTER 2

Architectural Drafting Equipment

Drafting tools and equipment are available from a number of vendors that sell professional drafting supplies. For accuracy and long life, always purchase high-quality equipment.

▼ DRAFTING SUPPLIES

Whether equipment is purchased in a kit or by the individual tool, the following items are normally needed:

○ One mechanical leadholder with 4H, 2H, 4, and F grade leads. This pencil and lead assortment allows for individual flexibility of line and lettering control, or use in sketching. (The lead holder and leads may be considered optional if automatic pencils are used.)
○ One 0.3-mm automatic drafting pencil with 4H, 2H, and H leads.
○ One 0.5-mm automatic drafting pencil with 4H, 2H, and F leads.
○ One 0.7-mm automatic drafting pencil with 2H, H, and F leads.
○ One 0.9-mm automatic drafting pencil with H, F, and HB leads. Automatic pencils and leads may be considered optional if lead holders and leads are used.
○ Lead sharpener (not needed if the lead holder is omitted).
○ 6″ bow compass.
○ Dividers.
○ Eraser.
○ Erasing shield.
○ 30°–60° triangle.
○ 45° triangle.
○ Irregular curve.
○ Adjustable triangle (optional).
○ Scales:
    1. Triangular architect's scale.
    2. Triangular civil engineer's scale.
○ Drafting tape.
○ Architectural floor plan template, ¼″ = 1′–0″ (for residential plans).
○ Circle template (small holes).
○ Lettering guide.
○ Sandpaper sharpening pad.
○ Dusting brush.

Not all of the items listed here are used in every school or workplace. Additional equipment and supplies may be needed depending on the specific application, for example:

○ Technical pen set.
○ Drafting ink.
○ Lettering template.
○ Assorted architectural drafting templates.

▼ DRAFTING FURNITURE
Tables

Drafting tables are generally sized by the dimensions of their tops. Standard table-top sizes range from 24 × 36″ (610 × 915 mm) to 42 × 84″ (1067 × 2134 mm).

The features to look for in a good-quality, professional table include:

○ One-hand tilt control.
○ One-hand or foot height control.
○ The ability to position the board vertically.
○ An electrical outlet.
○ A drawer for tools and/or drawings.

Most offices commonly cover drafting-table tops with smooth, specially designed surfaces. The material, usually vinyl, provides the proper density for effective use under normal drafting conditions. Drafting tape is commonly used to adhere drawings to the table top, although some drafting tables have magnetized tops and use magnetic strips to attach drawings.

Chairs

Good drafting chairs have these characteristics:

○ Padded or contoured seat design.
○ Height adjustment.
○ Foot rest.
○ Fabric that allows air to circulate.
○ Sturdy construction.

▼ DRAFTING PENCILS AND LEADS
Mechanical Pencils

The term *mechanical pencil* is applied to a pencil that requires a piece of lead to be manually inserted. Special sharpeners are available to help maintain a conical point.

Automatic Pencils

The term *automatic pencil* refers to a pencil with a lead chamber that, at the push of a button or tab, advances the lead from the chamber to the writing tip. Automatic pencils are designed to hold leads of one width that do not need to be sharpened. These pencils are available in 0.3, 0.5, 0.7, and 0.9 mm lead thickness sizes. Many drafters have several automatic pencils, each with a different grade of lead hardness, used for a specific application.

Lead Grades

The leads that you select for line work depend on the amount of pressure you apply and other technique factors. You should experiment until you identify the leads that give you the best line quality. Leads commonly used for thick lines range from 2H to F; leads for thin lines range from 4H to H, depending on individual preference. Construction lines for layout and guidelines are very lightly drawn with a 6H or 4H lead. Figure 2–1 shows the different lead grades. Generally, softer leads are used for sketching than those used for making formal lines. Good sketching leads are H, F, or HB.

Polyester and Special Leads

Polyester leads, also known as *plastic leads,* are for drawing on polyester drafting film, often called by its trade name, *Mylar.* Plastic leads come in grades equivalent to F, 2H, 4H, and 5H and are usually labeled with an "S," as in 2S. Some companies make a combination lead for use on both vellum and polyester film. See Chapter 3 for information about the techniques and use of polyester leads.

Colored leads have special uses. Red or yellow lead is commonly used to make corrections. Red or yellow lines appear black on a print. Blue lead can be used on the original drawing for information that is not to show on a print such as for notes, lines, and outlines of areas to be corrected. Blue lines do not appear on the print when the original is run through a Diazo machine but will reproduce in the photocopy process if not very lightly drawn. Some drafters use light blue lead for layout work and guidelines.

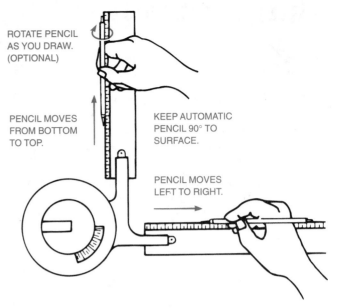

Figure 2–2 Basic pencil motions.

Basic Pencil Motions

Keep your automatic pencil straight from side to side and tilted about 40° with the direction of travel. Try not to tilt the lead under or away from the straightedge. Automatic pencils do not require rotation, although some drafters feel better lines are made by rotating the pencil. Provide enough pressure to make a line dark and crisp. Figure 2–2 shows some basic pencil motions.

▼ TECHNICAL PENS AND ACCESSORIES

Technical Pens

Technical pens have improved over the past few years in quality and ability to produce excellent inked lines. These pens function on a capillary action; a needle acts as a valve to allow ink to flow from a storage cylinder through a small tube, which is designed to meter the ink so a specific line width is created. Figure 2–3 shows a comparison of some of the different line widths available with technical pens.

Technical pens are available in different price ranges as determined by the kind of material used to make the point. Each kind of point has a recommended use. Stainless steel points are made for use on vellum; they wear out very rapidly when used on polyester film. Tungsten carbide points are recommended for polyester film and

9H 8H 7H 6H 5H 4H	3H 2H H F HB B	2B 3B 4B 5B 6B 7B
HARD	MEDIUM	SOFT
4H AND 6H ARE COMMONLY USED FOR CONSTRUC-TION LINES.	H AND 2H ARE COMMON LEAD GRADES USED FOR LINE WORK. H AND F WOULD BE FOR LETTER-ING AND SKETCH-ING.	THESE GRADES ARE FOR ART WORK. THEY ARE TOO SOFT TO KEEP A SHARP POINT AND THEY SMUDGE EASILY.

Figure 2–1 The range of lead grades.

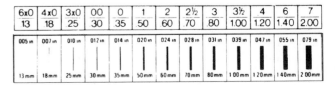

Figure 2–3 Technical pen line widths. *Courtesy Koh-I-Noor, Inc.*

can be used on vellum. Jewel points provide the longest life and can be used on vellum or polyester film with the best results. They also provide the smoothest ink flow on polyester film.

Technical pens may be used with templates to make circles, arcs, and symbols. Compass adapters to hold technical pens are also available. Technical pen tips are designed to fit into scribers for use with lettering guides. This concept is discussed further in Chapter 5.

Some symptoms to look for when pens need cleaning are discussed below.

Pen Cleaning

○ Ink constantly creates a drop at the pen tip.
○ Ink tends to flow out around the tip holder.
○ The point plunger does not activate properly. If you do not hear and feel the plunger when you shake the pen, it is probably clogged with thick or dried ink.
○ When drawing a line, ink does not flow freely.
○ Ink flow starts with difficulty.

The ink level in the reservoir should be kept between one-quarter to three-quarters full.

Read the cleaning instructions that come with the pen you purchase. Some pens require disassembly for cleaning while others should not be taken apart. The main parts of a technical pen are the cap, nib, pen body, ink cartridge (or reservoir), and pen holder as seen in Figure 2–4. The nib contains a cleansing wire with drop weight and a safety plug. Most manufacturers recommend that the nib remain assembled during cleaning because the cleansing wire is easily damaged. To fill the technical pen, unscrew the cap, holder, and clamp ring from the pen body. Remove the cartridge and fill it three-quarters full with ink. Slowly replace the cartridge and assemble the parts. Pens should be cleaned before each filling or before being stored for a long period of time. Clean the technical pen nib, cartridge, and body separately in warm water or special cleaning solution.

Ink

Drafting inks should be opaque or have a matte or semiflat black finish that will not reflect light. The ink should reproduce without hot spots or line variation. Drafting ink should have excellent adhesion properties for use on paper or film. Certain inks are recommended for use on film in order to avoid peeling, chipping, or cracking. Inks recommended for use in technical pens also have nonclogging characteristics. This property is especially important for use in high-speed computer-graphics plotters.

▼ ERASERS AND ACCESSORIES

The common shapes of erasers are rectangular and stick. The stick eraser works best in small areas. Select an eraser that is recommended for the particular material used.

Erasers Used for Drafting

White Pencil Eraser. The white pencil eraser is used for removing pencil marks from paper. It contains a small amount of pumice (an abrasive material) but can be used as a general-purpose pencil eraser.

Pink Pencil Eraser. A pink pencil eraser contains less pumice than the white pencil eraser. It can be used to erase pencil marks from most kinds of office paper without doing any damage to the surface.

Soft-Green or Soft-Pink Pencil Eraser. The soft-green and soft-pink erasers contain no pumice and are designed for use on even the most delicate types of paper without doing damage to the surface. The color difference is maintained only to satisfy long-established individual preferences.

Kneaded Eraser. Kneaded erasers are soft, stretchable, and nonabrasive. They can be formed into any size you need. Kneaded erasers are used for erasing pencil and charcoal; they are also used for shading. In addition, kneaded erasers are handy for removing smudges from a drawing and for cleaning a drawing surface.

Vinyl Eraser. The vinyl or plastic eraser contains no abrasive material. This eraser is used on drafting film to erase plastic lead or ink.

India Ink Erasing Machine Refill. The India ink refill eraser is for removing India ink from vellum or film. It contains a solvent that dissolves the binding agents in the ink and the vinyl, then removes the ink. This eraser does not work well with pencil lead.

Erasing Tips

When erasing, the idea is to remove an unwanted line or letter, not the surface of the paper or Mylar. Erase only hard enough to remove the unwanted line. However, you must bear down enough to eliminate the line completely.

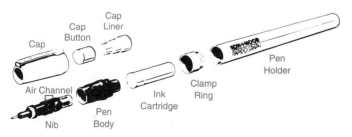

Figure 2–4 Parts of a technical pen. *Courtesy Koh-I-Noor, Inc.*

If the entire line does not disappear, ghosting results. A *ghost* is a line that should have been eliminated but still shows on a print. Lines that have been drawn so hard as to make a groove in the drawing sheet can cause a ghost too. To remove ink from vellum, use a pink or green eraser, or an electric eraser. Work the area slowly. Do not apply too much pressure or erase in one spot too long or you will go through the paper. On polyester film, use a vinyl eraser and/or a moist cotton swab. The inked line usually comes off easily, but use caution. If you destroy the mat surface of either a vellum or film sheet, you will not be able to redraw over the erased area.

Electric Erasers

Professional drafters use electric erasers, available in cord and cordless models. When working with an electric eraser, you do not need to use very much pressure because the eraser operates at high speed. The purpose of the electric eraser is to remove unwanted lines quickly. Use caution; these erasers can also remove paper quickly! Eraser refills for electric units are available in all types.

Erasing Shield

Erasing shields are thin metal or plastic sheets with a number of differently shaped holes in them. They are used to erase small, unwanted lines or areas. For example, if you have a corner overrun, place one of the slots of the erasing shield over the area to be removed while covering the good area. Erase the overrun through the slot, as shown in Figure 2–5.

Cleaning Agents

Special eraser particles are available in a shaker-top can or a pad. Both types are used to sprinkle the eraser particles on a drawing to help reduce smudging and to keep the drawing and your equipment clean. The particles also help float triangles, straightedges, and other drafting equipment to reduce line smudging. Use this

Figure 2–5 Erasing shield. *Courtesy Koh-I-Noor, Inc.*

material sparingly since too much of it can cause your lines to become fuzzy. Cleaning powders are not recommended for use on ink drawings or on polyester film.

Dusting Brush

Use a dusting brush to remove eraser particles from your drawing. Doing so helps reduce the possibility of smudges. Avoid using your hand to brush away eraser particles because the hand tends to cause smudges, which reduces drawing neatness. A clean, dry cloth works better than your hand for removing eraser particles, but a brush is preferred.

▼ DRAFTING INSTRUMENTS

Kinds of Compasses

Compasses are used to draw circles and arcs. However, using a compass can be time-consuming. Use a template, whenever possible, to make circles or arcs quickly. A compass is especially useful for large circles.

There are several basic types of drafting compasses:

1. A *drop-bow compass,* mostly used for drawing small circles. The center rod contains the needle point and remains stationary while the pencil, or pen, leg revolves around it.
2. The *center-wheel bow compass,* most commonly used by professional drafters. This compass operates on the screwjack principle by turning the large knurled center wheel, as shown in Figure 2–6.
3. A *beam compass,* a bar with an adjustable needle and a pencil or pen attachment for swinging large arcs or circles. Also available is a beam that is adaptable to the bow compass. Such an adapter works only on bow compasses that have a removable break point, not on the fixed point models.

Compass Use

Keep both the compass needle point and lead point sharp. The points are removable for easy replacement. The better compass needle points have a shoulder on them. The shoulder helps keep the point from penetrating the paper more than necessary. Compare the needle points in Figure 2–7.

The compass lead should, in most cases, be one grade softer than the lead you use for straight lines because less pressure is used on a compass than a pencil. Keep the compass lead sharp. An elliptical point is commonly used with the bevel side inserted away from the needle leg. Keep the lead and the point equal in length. Figure 2–8 shows properly aligned and sharpened points on a compass.

Use a sandpaper block to sharpen the elliptical point. Be careful to keep the graphite residue away from your drawing and off your hands. Remove excess graphite

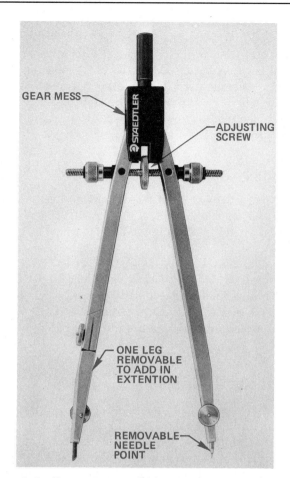

Figure 2–6   Bow compass and its parts. *Courtesy J. S. Staedtler, Inc.*

drafting tape at the center point for protection. There are small plastic circles available for just this purpose. Place one at the center point; then pierce the plastic with your compass.

Dividers

Dividers are used to transfer dimensions or to divide a distance into a number of equal parts.

Some drafters prefer to use a bow divider because the center wheel provides the ability to make fine adjustments easily. Also, the setting remains more stable than with standard friction dividers.

A good divider should not be too loose or too tight. It should be easily adjustable with one hand. In fact, you should control dividers with one hand as you lay out equal increments or transfer dimensions from one feature to another. Figure 2–9 shows how the dividers should be handled when used.

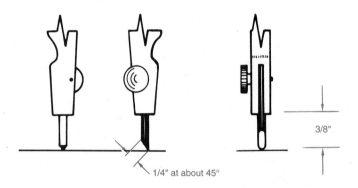

Figure 2–8   Properly sharpened and aligned elliptical compass point.

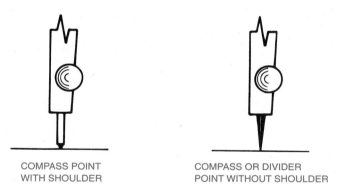

Figure 2–7   Common compass points.

from the point with a tissue or cloth after sharpening. Sharpen the lead often.

Some drafters prefer to use a conical point in their compass. This is the same point used in a mechanical pencil. If you want to try this point, sharpen a piece of lead in a mechanical pencil, then transfer it to your compass.

If you are drawing a number of circles from the same center, you will find that the compass point causes an ugly hole in your drawing sheet. Reduce the chance of making such a hole by placing a couple of pieces of

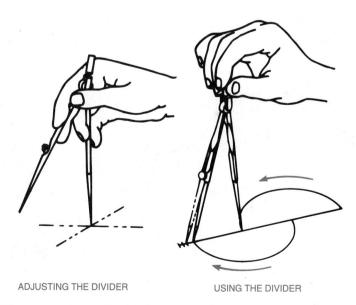

Figure 2–9   Using the divider.

Proportional Dividers

Proportional dividers are used to reduce or enlarge an object without the need of mathematical calculations or scale manipulations. The center point of the dividers is set at the correct point for the proportion you want. Then measure the original size line with one side of the proportional dividers, and the other side automatically determines the new reduced or enlarged size.

Parallel Bar

The parallel bar slides up and down the drafting board on cables that are attached to pulleys mounted at the top and bottom corners of the table (see Figure 2–10). The function of the parallel bar is to allow you to draw horizontal lines. Vertical lines and angles are made with triangles in conjunction with the parallel bar (Figure 2–11). Angled lines are made with standard or adjustable triangles.

The parallel bar is commonly found in architectural drafting offices because architectural drawings are frequently very large. Architects and architectural drafters often need to draw straight lines the full length of their boards, and the parallel bar is ideal for such lines. The parallel bar may be used with the board in an inclined position. Triangles and the parallel bar must be held securely when used in any position. Professional units have a brake that may be used to lock the parallel bar in place during use.

Triangles

There are two standard triangles. One has angles of 30°–60°–90° and is known as the 30°–60° triangle. The other has angles of 45°–45°–90° and is known as the 45° triangle.

Triangles may be used as a straightedge to connect points for drawing lines. Triangles are used individually or in combination to draw angled lines in 15° increments (Figure 2–12). Also available are adjustable triangles with built-in protractors that are used to make angles of any degree.

Architectural Templates

Templates are plastic sheets that have standard symbols cut through them for tracing. There are as many standard architectural templates as there are architectural symbols to draw. Templates may be scaled to match a particular application, such as floor plan symbols, or they may be unscaled for use on drawings that do not require a scale.

One of the most common architectural templates used is the residential floor plan template (Figure 2–13). This template is typically available at a scale of ¼" = 1'–0" (1:50 metric) which coincides with the scale usually used for residential floor plans. Features that are often found on a floor plan template include circles, door swings,

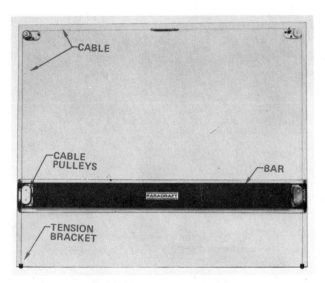

Figure 2–10 Parallel bar. *Courtesy Charvoz-Carsen Corporation.*

Figure 2–11 Drawing vertical lines with the parallel bar and triangle. *Courtesy Cascade Architectural and Engineering Supply.*

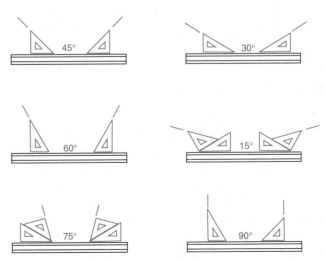

Figure 2–12 Angles that may be made with the 30°–60° and 45° triangles, individually or in combination.

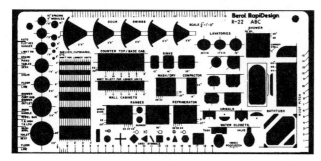

Figure 2–13 Architectural floor plan symbol template. *Courtesy Berol Corporation.*

sinks, bathroom fixtures, kitchen appliances, and electrical symbols. Other floor plan templates are available for both residential and commercial applications. Additional architectural templates are introduced in chapters where their specific applications are discussed.

Template Use. Always use a template when you can. Templates save time, are very accurate, and help standardize features on drawings. For most applications the template should be used only after the layout has been established. For example, floor plan symbols are placed on the drawing after walls, openings, and other main floor plan features have been established.

When using a template, carefully align the desired template feature with the layout lines of the drawing. Hold the template firmly in position while you draw the feature. Try to keep your pencil or pen perpendicular to the drawing surface for best results. When using a template and a technical pen, keep the pen perpendicular to the drawing sheet. Some templates have risers built in to keep the template above the drawing sheet. Without this feature, there is a risk of ink running under a template that is flat against the drawing. If your template does not have risers, purchase and add template lifters, use a few layers of tape placed on the underside of the template (although tape does not always work well), or place a second template with a large circle under the template you are using. See Figure 2–14.

Irregular Curves

Irregular curves are commonly called French curves. These curves have no constant radii. A selection of irregular curves is shown in Figure 2–15. Also available are ship's curves, which become progressively larger and, like French curves, have no constant radii. They are used for layout and development of ships' hulls. Flexible curves are made of plastic and have a flexible core. They may be used to draw almost any curve desired by bending and shaping.

Irregular curves are used for a variety of applications in architectural drafting. Some drafters use the curves to draw floor plan electrical runs or leader lines that connect notes to features on the drawing.

▼ DRAFTING MACHINES

Drafting machines may be used in place of triangles and parallel bars. The drafting machine maintains a horizontal and vertical relationship between scales, which also serve as straightedges, by way of a protractor component. The protractor allows the scales to be set quickly at any angle. There are two types of drafting machines: arm and track. Although both types are excellent tools, the track machine has generally replaced the arm machine in industry. A major advantage of the track machine is to allow the drafter to work with a board in the vertical position. A vertical drafting surface position is generally more comfortable to use than a horizontal table. For school or home use, the arm machine may be a good economical choice. When ordering a drafting machine, the size specified should relate to the size of the drafting board. For example, a $37\frac{1}{2} \times 60''$ machine would fit the same size table.

Arm Drafting Machine

The arm drafting machine is compact and less expensive than a track machine. The arm machine clamps to a table and through an elbow-like arrangement of supports allows the drafter to position the protractor head and scales anywhere on the board.

The arm drafting machine is shown in Figure 2–16.

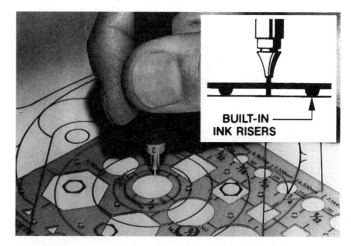

Figure 2–14 Using a template with built-in risers. *Courtesy Chartpak.*

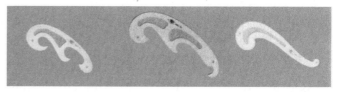

Figure 2–15 Irregular or French curves. *Courtesy Teledyne Post.*

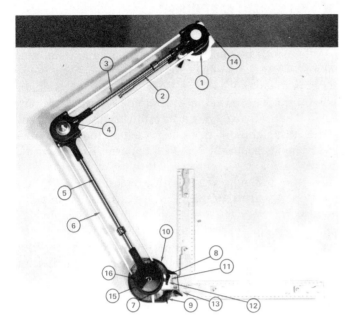

Figure 2–16 Arm-type drafting machine. *Courtesy Mutoh America, Inc.*

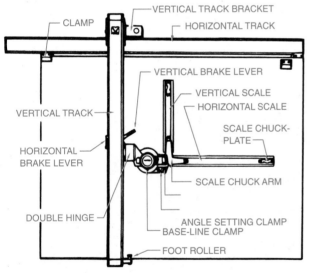

Figure 2–17 Track drafting machine and its parts. *Courtesy Mutoh America, Inc.*

Track Drafting Machines

A track drafting machine has a traversing arm that moves left and right across the table and a head unit that moves up and down the traversing arm. There is a locking device for both the head and the traversing arm. The shape and placement of the controls of a track machine vary with the manufacturer, although most brands have the same operating features and procedures. Figure 2–17 shows the component parts of a track drafting machine.

As with the arm machines, track drafting machines have a vernier head that allows the user to measure angles accurately to 5′ (minutes).

Controls and Machine Head Operation

The drafting machine head contains the controls for horizontal, vertical, and angular movement. Although each brand of machine contains similar features, controls may be found in different places on different brands. See Figure 2–18.

To operate the drafting machine protractor head, place your hand on the handle, and with your thumb, depress the index thumb piece. Doing so allows the head to rotate. Each increment marked on the protractor is 1 degree, with a label every 10° (Figure 2–19). As the vernier plate (the small scale numbered from 0 to 60) moves past the protractor, the zero on the vernier aligns with the angle that you wish to read. Figure 2–19 shows a reading of 10°.

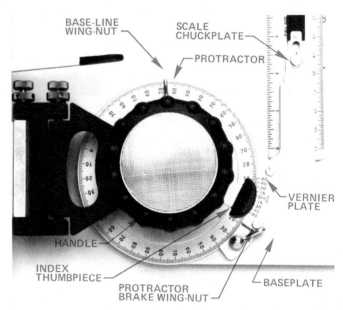

Figure 2–18 Drafting machine head, controls, and parts. *Courtesy Vemco.*

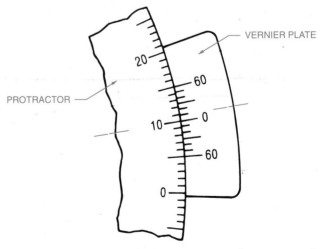

Figure 2–19 Vernier plate and protractor showing a reading of 10°.

As you rotate the handle, notice that the head automatically locks every 15°. To move the protractor past the 15° increment, you must again depress the index thumb piece. Figure 2–18 shows the index thumb piece.

Having rotated the protractor head 40° clockwise, the machine is in the position shown in Figure 2–20. The vernier plate at the protractor reads 40°, which means that both the horizontal and vertical scale moved 40° from their original position at 0° and 90°, respectively. The horizontal scale reads directly from the protractor starting from 0°. The vertical scale reading begins from the 90° position. The key to measuring angles is to determine if the angle is to be measured from the horizontal or vertical starting point. See the examples in Figure 2–21.

Measuring full degree increments is easy since you simply match the zero mark on the vernier plate with a full degree mark on the protractor. See the reading of 12° in Figure 2–22. The vernier scale allows you to measure angles as accurately as 5′ (minutes). Remember, 1 degree equals 60 minutes (1° = 60′) and 1 minute equals 60 seconds (1′ = 60″).

Reading and Setting Angles with the Vernier. To read an angle other than a full degree, we will assume the vernier scale is set at a *positive angle,* as shown in Figure 2–23. Positive angles read upward, and negative angles read downward from zero. Each mark on the vernier scale represents 5′. First see that the angle to be read is between

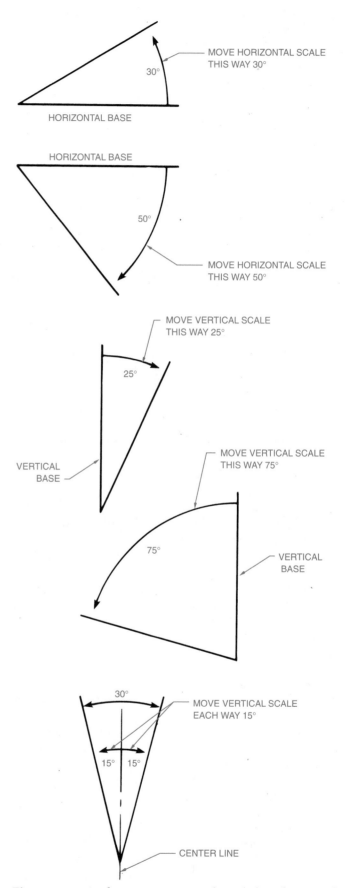

Figure 2–21 Angle measurements reference from horizontal or vertical.

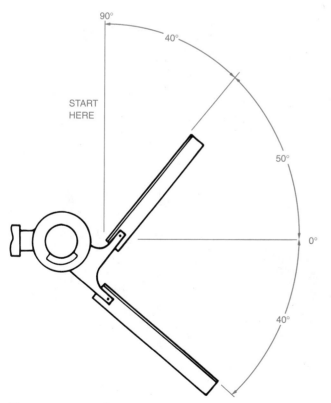

Figure 2–20 Angle measurement.

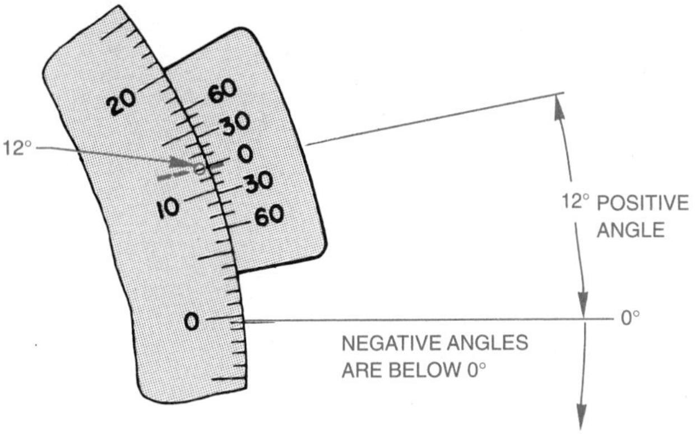

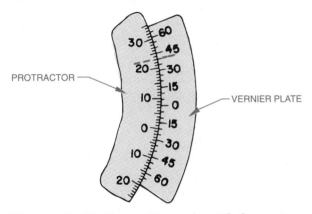

Figure 2–22   Measuring full degrees.

Figure 2–23   Reading positive angles with the vernier.

7° and 8°. Then find the 5′ mark on the *upper* half of the vernier (the direction in which the scale has been turned) that is most closely aligned with a full degree on the protractor. In this example, it is the 40′ mark. Add the minutes to the degree just past. The correct reading, then, is 7°40′.

The procedure for reading *negative* angles is the same, except to read the minute marks on the *lower* half of the vernier.

Machine Setup

To insert a scale in the baseplate chuck, place the scale flat on the board and align the scale chuckplate with the baseplate chuck on the protractor head. Firmly press but do not drive the scale chuckplate into the baseplate chuck. To remove a scale, use the scale wrench, as shown in Figure 2–24. With the pin side of the wrench pointing down, slip the wrench over screw C and turn clockwise, thus pressing curved section B strongly against section A of the baseplate chuck. Removing a scale by hand without the aid of a key could result in damage to the scale and/or machine.

Scale Alignment.   Before drawing with any drafting machine, check the scales for alignment and, if needed, adjust them at right angles to each other. For best results

with a track drafting machine, the scales should also be aligned with respect to the horizontal track. Both operations can be accomplished through the following procedure:

Step 1.   Tighten the flat-head screw nearest the end of the scale on each scale chuckplate. Insert the scales in the baseplate chuck and press them firmly into place. Release the inner scale-chuckplate lock-screw on the horizontal scale. Set the scale near the center of its angular range of adjustment, and tighten the lock-screw. Be careful not to overtighten any screw.

Step 2.   Draw a reference line parallel to the horizontal track by:
  a. Locking the vertical brake and releasing the horizontal brake.
  b. Placing a pencil point at zero on the horizontal scale, as shown in Figure 2–25, and moving both the pencil and protractor head together laterally along the board (Figure 2–26). *Caution:* Merely drawing the pencil along the scale does not ensure the line being parallel to the horizontal track.

Step 3.   Release the locknut on the micrometer baseline screw, and turn this screw until the scale is brought parallel to the reference line (Figure 2–27).

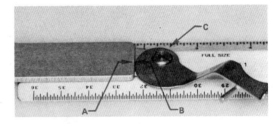

Figure 2–24   Scale removal. *Courtesy Vemco.*

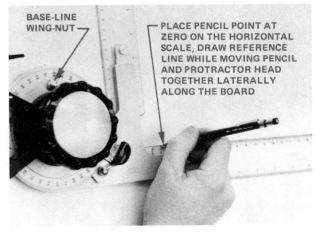

Figure 2–25   Step 2: Place the pencil point at zero on the horizontal scale, and move both the pencil and the scale together.

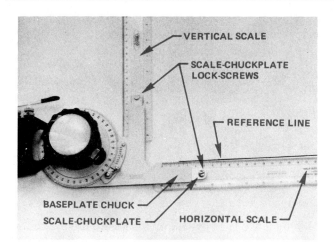

Figure 2–26 Step 2 (continued): Draw the horizontal reference line for scale alignment.

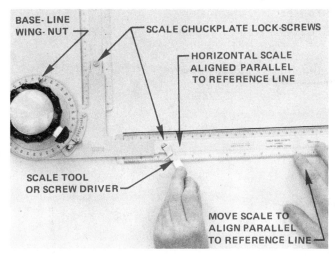

Figure 2–27 Steps 3 and 4: Align the horizontal scale with the reference line.

Tighten the locknut firmly. Some machines have a baseline wing nut. If yours does, release the baseline wing nut, and bring the scale parallel to the reference line. Tighten the baseline wing nut. For those machines that have a baseline zero on the protractor (usually found to the left of the handle), first loosen the baseline wing nut and align the arrow to 0°. Then lock the baseline wing nut. Finally, loosen the horizontal scale-chuckplate lock screw and adjust the horizontal scale to the reference line; then retighten the screw.

Step 4. Remove the horizontal scale, turn it 180°, and replace it. Loosen the scale-chuckplate lock screw, adjust the scale parallel to the reference line, and tighten the lock screw. Now the horizontal scale is properly aligned when inserted from either end.

Step 5. Move the head 90° clockwise, and adjust both ends of the vertical scale in the same manner as

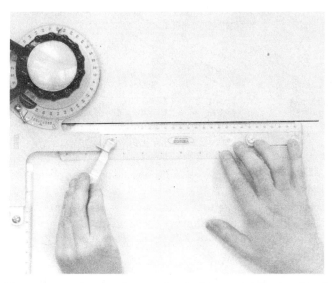

Figure 2–28 Step 5: Rotate the drafting machine head 90°, and align the vertical scale with the reference line.

the horizontal scale and along the same reference line drawn in Step 2b. See Figure 2–28.

After the horizontal scale has been properly adjusted, the vertical scale may be aligned 90° to the horizontal scale by placing the 90° angle of a 30°–60° or 45° triangle between the two scales. Be sure the drafting machine scales fit tightly against the triangle before tightening the screws.

By following this procedure, you have established a reference line setting that is parallel to the horizontal track and have adjusted the scales so they are parallel to the track and perpendicular to each other. For satisfactory results when drawing, the screws on the scales must be tight and the scale chuckplates pressed firmly into the chucks. Use good judgment when tightening any mechanism; too much force can cause components to break or wear out rapidly.

The alignment procedure of Steps 1 and 2 should be checked periodically, even daily. Doing so may seem like a lot of work and trouble, but once you have gone through the procedure several times, it becomes routine. By checking and adjusting the scale alignment often, you ensure that your drawings are accurate. One of a drafter's great frustrations is to prepare a layout and then discover that the machine was not properly aligned.

▼ SCALES
Scale Shapes

There are four basic scale shapes (Figure 2–29). The two-bevel scales are also available with chuckplates for use with standard arm or track drafting machines. These machine scales have typical calibrations, and some have no scale reading for use as a straightedge alone. Drafting machine scales are purchased by designating the length

TWO BEVEL   FOUR BEVEL   OPPOSITE BEVEL   TRIANGULAR

Figure 2–29   Scale shapes.

needed—12, 18, or 24″—and the scale calibration such as metric, engineer's full scale in tenths and half scale in twentieths or architect's scale ¼″ = 1′–0″ and ½″ = 1′–0″. Many other scales are also available.

Scale Notation

The scale of a drawing is usually noted in the title block or below the view that differs in scale from that given in the title block. Drawings are scaled so the object represented can be illustrated clearly on standard sizes of paper. It would be difficult, for example, to draw a house full size; thus, a scale that reduces the size of such large objects must be used.

The scale selected depends on:

○ The actual size of the structure.
○ The amount of detail to be shown.
○ The paper size selected.
○ The amount of dimensions and notes required.
○ Common practice that regulates certain scales.

The following scales and their notation are frequently used on architectural drawings.

⅛″ = 1′–0″	⅜″ = 1′–0″	1½″ = 1′–0″
³⁄₃₂″ = 1′–0″	½″ = 1′–0″	3″ = 1′–0″
³⁄₁₆″ = 1′–0″	¾″ = 1′–0″	12″ = 1′–0″
¼″ = 1′–0″	1″ = 1′–0″	

Some scales used in civil drafting are noted as follows:

1″ = 10′	1″ = 50′	1″ = 300′
1″ = 20′	1″ = 60′	1″ = 400′
1″ = 30′	1″ = 100′	1″ = 500′
1″ = 40′	1″ = 200′	1″ = 600′

Metric Scale

According to the American National Standards Institute (ANSI), the commonly used SI (International System of Units) linear unit used on drawings is the millimeter. On drawings where all dimensions are in either inches or millimeters, individual identification of units is not required. However, the drawing shall contain a note stating: Unless Otherwise Specified, All Dimensions Are in Inches [or Millimeters as applicable]. Where some millimeters are shown on an inch-dimensioned drawing, the millimeter value should be followed by the symbol, mm.

Where some inches are shown on a millimeter-dimensioned drawing, the inch value should be followed by the abbreviation, in.

Metric symbols are as follows:

millimeter = mm
centimeter = cm
decimeter = dm
meter = m
dekameter = dam
hectometer = hm
kilometer = km

Some metric-to-metric equivalents are the following:

10 millimeters = 1 centimeter
10 centimeters = 1 decimeter
10 decimeters = 1 meter
10 meters = 1 dekameter
10 dekameters = 1 kilometer

Some metric-to-U.S. customary equivalents are the following:

1 millimeter = 0.03937 inch
1 centimeter = 0.3937 inch
1 meter = 39.37 inches
1 kilometer = 0.6214 mile

Some U.S. customary-to-metric equivalents include the following:

1 mile = 1.6093 kilometers = 1609.3 meters
1 yard = 914.4 millimeters = 0.9144 meter
1 foot = 304.8 millimeters = 0.3048 meter
1 inch = 25.4 millimeters = 0.0254 meter

To convert inches to millimeters, multiply inches by 25.4 mm.

One advantage of metric scales is that any scale is a multiple of 10; therefore, reductions and enlargements are easily performed. In most cases, no mathematical calculations should be required when using a metric scale. Whenever possible, select a direct reading scale. If no direct reading scale is available, use multiples of other scales. To avoid the possibility of error, avoid multiplying or dividing metric scales by anything but multiples of ten. Some metric scale calibrations are shown in Figure 2–30. See "Using Metric" in Chapter 14.

Architect's Scale

The triangular architect's scale contains 11 different scales. On ten of them, each inch represents a foot and is subdivided into multiples of 12 parts to represent inches and fractions of an inch. The eleventh scale is the full scale with a 16 in the margin. The 16 means that each inch is divided into 16 parts, and each part is equal to ¹⁄₁₆

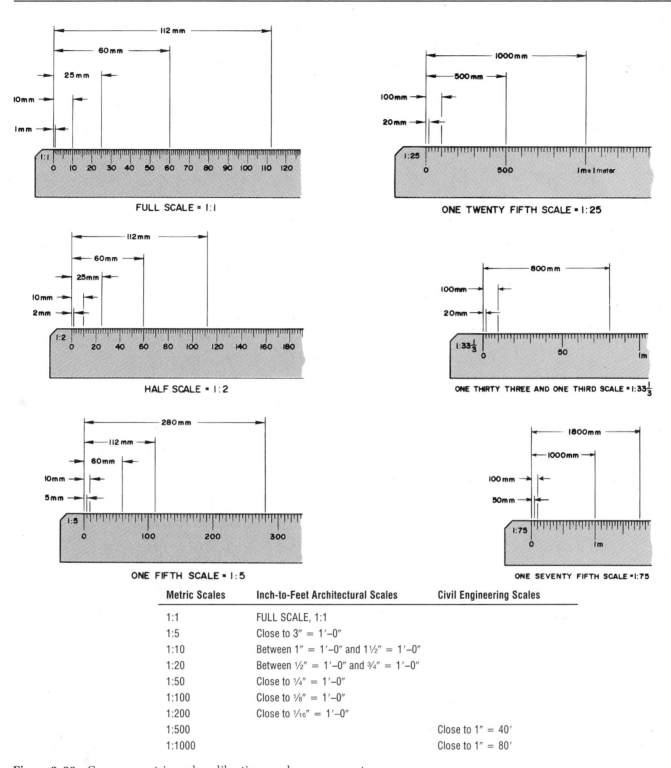

Metric Scales	Inch-to-Feet Architectural Scales	Civil Engineering Scales
1:1	FULL SCALE, 1:1	
1:5	Close to 3″ = 1′–0″	
1:10	Between 1″ = 1′–0″ and 1½″ = 1′–0″	
1:20	Between ½″ = 1′–0″ and ¾″ = 1′–0″	
1:50	Close to ¼″ = 1′–0″	
1:100	Close to ⅛″ = 1′–0″	
1:200	Close to ¹⁄₁₆″ = 1′–0″	
1:500		Close to 1″ = 40′
1:1000		Close to 1″ = 80′

Figure 2–30 Common metric scale calibrations and measurements.

of an inch. Figure 2–31 shows an example of the full architect's scale; Figure 2–32 shows the fraction calibrations.

Look at the architect's scale examples in Figure 2–33. Note the form in which scales are expressed on a drawing. The scale is expressed as an equation of the drawing size in inches or fractions of an inch to one foot. For example: 3″ = 1′–0″, ½″ = 1′–0″, or ¼″ = 1′–0″. The architect's scale commonly has scales running in both

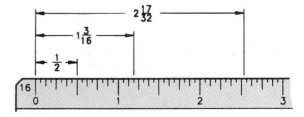

Figure 2–31 Full (1:1), or 12″ = 1′–0″, architect's scale.

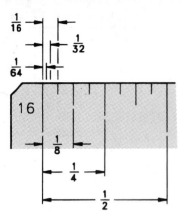

Figure 2–32   Enlarged view of architect's 16 scale.

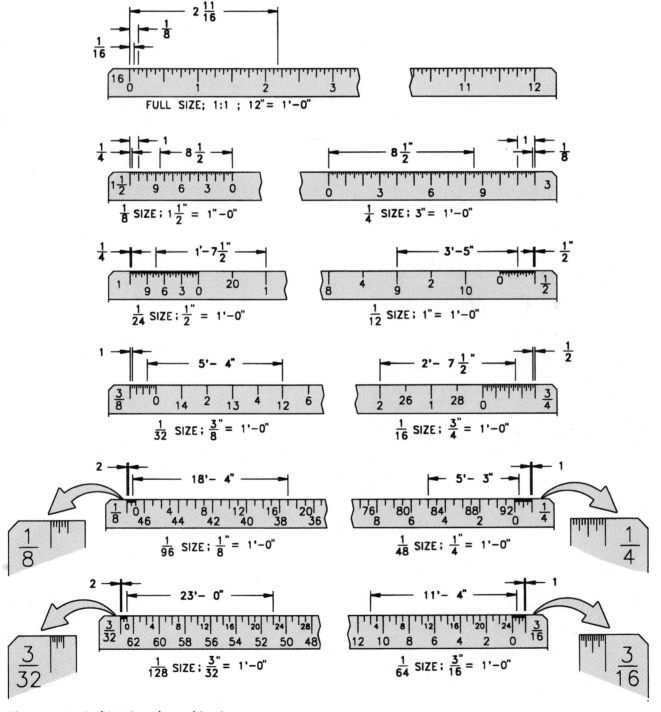

Figure 2–33   Architect's scale combinations.

directions along an edge. Be careful when reading a scale from left to right so as not to confuse its calibrations with the scale that reads from right to left.

When selecting scales for preparing a set of architectural drawings, there are several factors that must be considered. Some parts of a set of drawings are traditionally drawn at a certain scale. For example, the floor plans for most residential structures are drawn at ¼″ = 1′–0″. Although some applications may require a larger or smaller scale, commercial buildings are often drawn at ⅛″ = 1′–0″ when they are too large to draw at the same scale as a residence. Exterior elevations are commonly drawn at ¼″ = 1′–0″, although some architects prefer to draw the front or most important elevation at ¼″ = 1′–0″ and the balance of the elevations at ⅛″ = 1′–0″. Construction details and cross sections may be drawn at larger scales to help clarify specific features. Some cross sections can be drawn at ¼″ = 1′–0″ with clarity, but complex cross sections require a scale of ⅜″ = 1′–0″ or ½″ = 1′–0″ to be clear. Construction details may be drawn at any scale between ½″ = 1′–0″ and 3″ = 1′–0″, depending on the amount of information presented.

Civil Engineer's Scale

The architectural drafter may need to use the civil engineer's scale to draw plot plans, maps, or subdivision plats or to read existing land documents. Most land-related plans such as construction site plot plans are drawn using the civil engineer's scale, although some are drawn with an architect's scale. Common architect's scales used are ⅛″ = 1′–0″, ³⁄₃₂″ = 1′–0″, or ¹⁄₁₆″ = 1′–0″. The triangular civil engineer's scale contains six scales, one on each of its sides. The civil engineer's scales are calibrated in multiples of ten. The scale margin specifies the scale represented on a particular edge. Table 2–1 shows some of the many scale options available when using the civil engineer's scale. Keep in mind that any multiple of ten is available with this scale.

The 10 scale is used in civil drafting for scales of 1″ = 10′ or 1″ = 100′ and so on. See Figure 2–34. The 20 scale is used for scales such as 1″ = 2′, 1″ = 20′, or 1″ = 200.′

The remaining scales on the civil engineer's scale may be used in a similar fashion. The 50 scale is popular in civil drafting for drawing plats of subdivisions. Figure 2–35 shows scales that are commonly used on the civil engineer's scale, with sample measurements given.

▼ PROTRACTORS

When a drafting machine is not used or when combining triangles does not provide enough flexibility for drawing angles, a protractor is the instrument to use. Protractors can be used to measure angles to within ½° or even closer if you have a good eye.

Table 2–1
Civil Engineer's Scale

Divisions	Ratio	Scales Used with This Division			
10	1:1	1″ = 1″	1″ = 1′	1″ = 10′	1″ = 100′
20	1:2	1″ = 2″	1″ = 2′	1″ = 20′	1″ = 200′
30	1:3	1″ = 3″	1″ = 3′	1″ = 30′	1″ = 300′
40	1:4	1″ = 4″	1″ = 4′	1″ = 40′	1″ = 400′
50	1:5	1″ = 5″	1″ = 5′	1″ = 50′	1″ = 500′
60	1:6	1″ = 6″	1″ = 6′	1″ = 60′	1″ = 600′

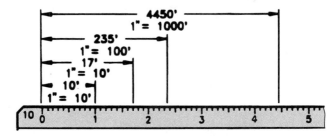

Figure 2–34 Civil engineer's scale, units of 10.

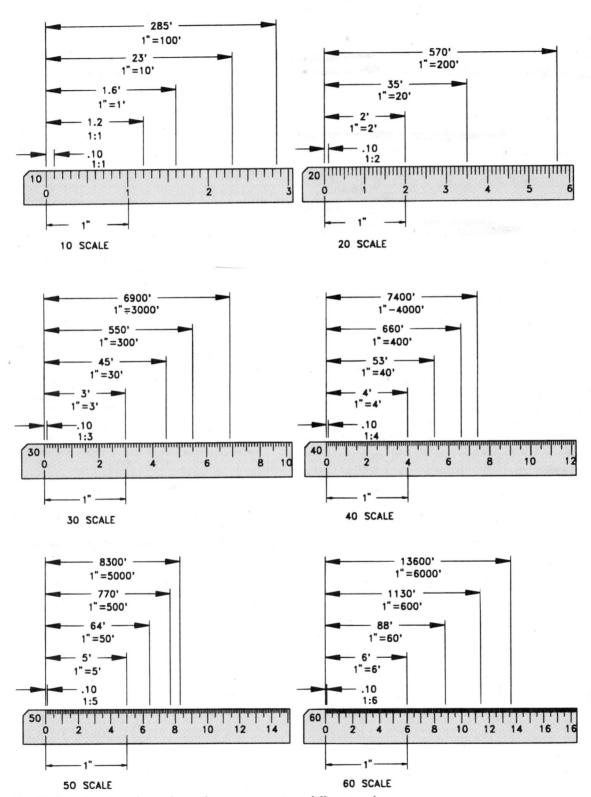

Figure 2–35 Civil engineer's scales and sample measurements at different scales.

CHAPTER 2

Architectural Drafting Equipment Test

DIRECTIONS

Answer the questions with short, complete statements or drawings as needed on 8½ × 11 lined paper, as follows:

1. Letter your name, Chapter 2 Test, and the date at the top of the sheet.
2. Letter the question number and provide the answer. You do not need to write out the question.

Answers may be prepared on a word processor if appropriate with course guidelines,

QUESTIONS

Question 2–1   Describe a mechanical drafting pencil.

Question 2–2   Describe an automatic drafting pencil.

Question 2–3   List three lead sizes available for automatic pencils.

Question 2–4   Identify how graphite leads are labeled.

Question 2–5   Name the type of technical pen tip that normally provides longer life.

Question 2–6   What precautions should be taken when disassembling a technical pen?

Question 2–7   List three characteristics recommended for technical pen inks.

Question 2–8   Describe one advantage and one disadvantage of using an electric eraser.

Question 2–9   Name the type of compass most commonly used by professional drafters.

Question 2–10   Why should templates be used whenever possible?

Question 2–11   Why should a compass point with a shoulder be used when possible?

Question 2–12   Identify two lead points that are commonly used in compasses.

Question 2–13   What is the advantage of placing drafting tape or plastic at the center of a circle before using a compass?

Question 2–14   A good divider should not be too loose or too tight. Why?

Question 2–15   List two uses for dividers.

Question 2–16   At what number of degree increments do most drafting machine heads automatically lock?

Question 2–17   How many minutes are there in 1°? seconds in 1′?

Question 2–18   Why should drafting machine scales be checked for alignment daily?

Question 2–19   In combination, the 30°–60° and 45° triangles may be used to make what range of angles?

Question 2–20   Show how the following scales are noted on an architectural drawing: ⅛ inch, ¼ inch, ½ inch, ¾ inch, 1½ inches, and 3 inches.

Question 2–21   Identify a civil engineer's scale that is commonly used for subdivision plats.

Question 2–22   What scale is usually used for drawing residential floor plans?

Question 2–23   List three factors that influence the selection of a scale for a drawing.

Determine the length of the following lines using both an architect's and civil engineer's scale. The scales to use when measuring are given above each line. The architect's scale is first; the civil engineer's scale is second; and the metric scale is third.

Question 2–24   a. Scale: ⅛″ = 1′–0″   b. 1″ = 10′–0″   c. 1:1 (metric)

Question 2–25   a. Scale: ¼″ = 1′–0″   b. 1″ = 20′–0″   c. 1:2 (metric)

Question 2–26   a. Scale: ⅜″ = 1–0″   b. 1″ = 30′–0″   c. 1:5 (metric)

Question 2–27   a. Scale: ½″ = 1′–0″   b. 1″ = 40′–0″   c. 1:10 (metric)

Question 2–28   a. Scale: ¾″ = 1′–0″   b. 1″ = 50′–0″   c. 1:20 (metric)

Question 2–29   a. Scale: 1″ = 1′–0″   b. 1″ = 60′–0″   c. 1:50 (metric)

Question 2–30   a. Scale: 1½″ = 1′–0″   b. 1″ = 100′   c. 1:75 (metric)

Question 2–31   a. Scale: 3″ = 1′–0″   b. 1″ = 200′   c. 1:75 (metric)

CHAPTER 3

Drafting Media and Reproduction Methods

This chapter covers the types of media that are used for the drafting and the reproduction of architectural drawings. Drawings are generally prepared on precut drafting sheets with printed graphic designs for company borders and title blocks. Drawing reproductions or copies are made for use in the construction process so the original drawings are not damaged. Alternate material options and reproduction methods are discussed as they relate to common practices.

▼ PAPERS AND FILMS
Selection of Drafting Media

Several factors influence the purchase and use of drafting media: durability, smoothness, erasability, dimensional stability, transparency, and cost.

Durability should be considered if the original drawing will have a great deal of use. Originals may tear or wrinkle, and the images become difficult to see if the drawings are used often.

Smoothness relates to how the medium accepts line work and lettering. The material should be easy to draw on so the image is dark and sharp without an excessive effort on the part of the drafter.

Erasability is important because errors need to be corrected and changes frequently made. When images are erased, ghosting should be kept to a minimum. *Ghosting* is the residue that remains when lines are difficult to remove. These unsightly ghost images reproduce in a print. Materials that have good erasability are easy to clean up.

Dimensional stability is the quality of the medium to maintain size due to the effects of atmospheric conditions such as heat and cold. Some materials are more dimensionally stable than others.

Transparency is one of the most important characteristics of drawing media. The diazo reproduction method requires light to pass through the material. The final goal of a drawing is good reproduction, so the more transparent the material is, the better is the reproduction, assuming that the image drawn is professional quality. Transparency is not a factor with photo copy reproduction.

Cost may influence the selection of drafting media. When the best reproduction, durability, dimensional stability, smoothness, and erasability are important, there may be few cost alternatives. If drawings are to have normal use and reproduction quality, then the cost of the drafting media may be kept to a minimum. The following discussion on available materials will help you evaluate cost differences.

Vellum

Vellum is drafting paper that is specially designed to accept pencil or ink. Lead on vellum is the most common combination used for manual drafting.

Vellum is the least expensive material, having good smoothness and transparency. Use vellum originals with care. Drawings made on vellum that require a great deal of use could deteriorate; vellum is not as durable a material as others. Also, some brands have better erasable qualities than others. Affected by humidity and other atmospheric conditions, vellum generally is not as dimensionally stable as other materials.

Polyester Film

Polyester film, also known by its trade name, Mylar®, is a plastic material that is more expensive than vellum but offers excellent dimensional stability, erasability, transparency, and durability. Drawing on polyester film is best accomplished using ink or special polyester leads. Do not use regular graphite leads; they smear easily. Drawing techniques that drafters use with polyester leads are similar to graphite leads except that polyester leads are softer and feel like a crayon when used.

Film is available with a single or double mat surface. *Mat* is surface texture. The double mat film has texture on both sides so drawing can be done on either side if necessary. Single mat film is the most common in use and has a slick surface on one side. When using polyester film, be very careful not to damage the mat by erasing. Erase at right angles to the direction of your lines, and do not use too much pressure. Doing so helps minimize damage to the mat surface. Once the mat is destroyed and removed, the surface will not accept ink or pencil. Also, be cautious about getting moisture on the polyester film surface. Moisture from your hands can cause ink to skip across the material.

Normal handling of drawing film is bound to soil it. Inked lines applied over soiled areas do not adhere well and in time will chip off or flake. It is always good practice to keep the film clean. Soiled areas can be cleaned effectively with special film cleaner.

Reproductions

The one thing most designers, engineers, architects, and drafters have in common is that their finished drawings are made with the intent of being reproduced. The goal of every professional is to produce drawings of the highest quality that give the best possible prints when they are reproduced.

Reproduction is the most important factor that influences the selection of media for drafting. The primary combination that achieves the best reproduction is the blackest, most opaque lines or images on the most transparent base or material. Each of the materials mentioned makes good prints if the drawing is well done. Some products have better characteristics than others. It is up to the individual or company to determine the combination that works best for its needs and budget. The question of reproduction is especially important when sepias must be made. (Sepias are second- or third-generation originals. See Sepias in this chapter.)

Figure 3–1 shows a magnified view of graphite on vellum, plastic lead on polyester film, and ink on polyester film. Judge for yourself which material and application provides the best reproduction. As you can see from the figure, the best reproduction is achieved with a crisp, opaque image on transparent material. If your original drawing is not good quality, it will not get better on the print.

▼ SHEET SIZES, TITLE BLOCKS, AND BORDERS

All professional drawings have title blocks. Standards have been developed for the information put into the title block and on the sheet next to the border so the drawing is easier to read and file than drawings that do not follow a standard format.

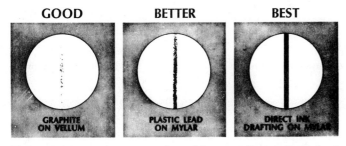

| GOOD | BETTER | BEST |

Figure 3–1 A comparison of graphite on vellum, plastic lead on Mylar, and ink on Mylar. *Courtesy Koh-I-Noor, Inc.*

Table 3–1
Standard Metric Drawing Sheet Sizes

Sheet Size	Dimensions (mm)	Dimensions (inches)
A0	1189 × 841	46.8 × 33.1
A1	841 × 594	33.1 × 23.4
A2	594 × 420	23.4 × 16.5
A3	420 × 297	16.5 × 11.7
A4	297 × 210	11.7 × 8.3

Sheet Sizes

Drafting materials are available in standard sizes, which are determined by manufacturers' specifications. Paper and polyester film may be purchased in cut sheets or in rolls. Architectural drafting offices generally use cut sheet sizes of 18 × 24″, 24 × 36″, 28 × 42″, 30 × 42″, 30 × 48″, or 36 × 48″. Metric sheet sizes vary from 420 × 594 mm to 841 × 1189 mm. Roll sizes vary in width from 18 to 48″. Metric roll sizes range from 297 to 420 mm wide.

The International Organization for Standardization (ISO) has established standard metric drawing sheet sizes (Table 3–1).

Zoning

Some companies use a system of numbers along the top and bottom margins, and letters along the left and right margins called *zoning*. Zoning allows the drawing to be read similar to a road map. For example, the reader can refer to the location of a specific item as D-4, which means the item can be found on or near the intersection of D across and 4 up or down. Zoning may be found on some architectural drawings, although it is more commonly used in mechanical drafting.

Architectural Drafting Title Blocks

Title blocks and borders are normally preprinted on drawing paper or polyester film to help reduce drafting time and cost.

Drawing borders are thick lines that go around the entire sheet. Top, bottom, and right-side border lines are usually between ⅜ and ½″ away from the paper edge, while the left border may be between ¾ and 1½″ away. This extra-wide left margin allows for binding drawing sheets.

Preprinted architectural drawing title blocks are generally placed along the right side of the sheet, although some companies place them across the bottom of the sheet. Each company uses a slightly different title block design, but the same general information is found in almost all blocks:

1. *Drawing number.* This may be a specific job or file number for the drawing.
2. *Company name, address, and phone number.*

3. *Project or client.* This is an identification of the project by company or client name, project title, or location.

4. *Drawing name.* This is where the title of the drawing may be placed—for example, MAIN FLOOR PLAN or ELEVATIONS. Most companies leave this information off the title block and place it on the face of the sheet below the drawing.

5. *Scale.* Some company title blocks provide a location for the drafter to fill in the general scale of the drawing. Any view or detail on the sheet that differs from the general scale must have the scale identified below the view title and both placed directly below the view.

6. *Drawing or sheet identification.* Each sheet is numbered in relation to the entire set of drawings. For example, if the complete set of drawings has eight sheets, then each consecutive sheet is numbered: 1 of 8, 2 of 8, 3 of 8, and so on, to 8 of 8.

7. *Date.* The date noted is the one on which the drawing or project is completed.

8. *Drawn by.* This is where the drafter, designer, or architect that prepared this drawing places his or her initials or name.

9. *Checked by.* This is the identification of the individual who approves the drawing for release.

10. *Architect or designer.* Most title blocks provide for the identification of the individual who designed the structure.

11. *Revisions.* Many companies provide a revision column where drawing changes are identified and recorded. After a set of drawings is released for construction, there may be some need for drawing changes. When this is necessary there is usually a written request for change that comes from the con-tractor, owner, or architect. The change is then implemented on the drawings that are affected. Where changes are made on the face of the drawing, a circle with a revision number accompanies the change. This revision number is keyed to a place in the drawing title block where the revision number, revision date, initials of the individual making the change, and an optional brief description of the revision are located. For further record of the change, the person making the change may also fill out a form, called a *change notice.* This change notice contains complete information about the revision for future reference.

Figure 3–2 shows several different architectural title blocks. Notice the similarities and differences among title blocks.

▼ DIAZO REPRODUCTION

Diazo Printer

Diazo prints, also known as *blue-line prints,* are made with a process that uses an ultraviolet light that passes through a translucent original drawing to expose a chemically coated paper or print material underneath. The light does not go through the dense, black lines on the original drawing. Thus, the chemical coating on the paper beneath the lines is not exposed. The print material is then exposed to ammonia vapor, which activates the remaining chemical coating to produce blue, black, or brown lines on a white or clear background. The print that results is a diazo, or blue-line print, not a blueprint.

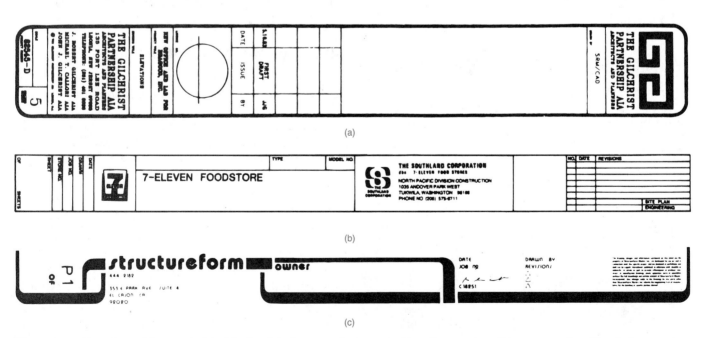

(a)

(b)

(c)

Figure 3–2 Sample architectural title blocks. (a) *Courtesy Summagraphics Corporation;* (b) *courtesy Southland Corporation;* (c) *courtesy Structureform.*

Figure 3–3 Feeding the original drawing and diazo material into the machine. *Courtesy Ozalid Corporation.*

The term *blueprint* is now a generic term that is used to refer to diazo prints even though they are not true blueprints.

Diazo is a direct print process, and the print remains relatively dry. There are other processes available that require a moist development. Blueprinting, for example, an old method of making prints uses a light to expose sensitized paper placed under an original drawing. The prints are developed in a water wash, which turns the background dark blue. The lines from the original are not fixed by the light and wash out, leaving the white paper to show. A true blueprint has a dark blue background with white lines.

The diazo process is less expensive and less time-consuming than most other reproductive methods. The diazo printer is designed to make quality prints from all types of translucent paper, film, or cloth originals. Diazo prints may be made on coated roll stock or cut sheets. The operation of all diazo printers is similar, but be sure to read the instruction manual or check with someone familiar with a particular machine if you are using a machine that is new to you.

Diazo Printer Speed. The operating speed of the machine is determined by the speed of the diazo material and the transparency of the original. For example, polyester film may be run faster than vellum. The speed of the diazo material also influences the quality of the print. Diazo materials that are classified as slow produce a higher-quality print than fast material. Some materials, such as fast-speed blue-line print paper, require a high- or fast-speed setting. Other materials, such as some sepias, require a low- or slow-speed setting.

Testing the Printer. Before you run a print using a complete sheet, run a test strip to determine the quality of the print you will get. Cut a sheet of diazo material into strips; then use a strip to get the proper speed setting. You know

you have the proper speed setting when the lines and letters are dark blue and the background is white. Once you are satisfied with the quality of the test strip, you are ready to run a full sheet to make a print.

Making a Diazo Print

To make a diazo print, place the diazo material on the feedboard, coated (yellow) side up. Position the original drawing, image side up, on top of the diazo material, making sure to align the leading edges. See Figure 3–3.

Using fingertip pressure from both hands, push the original and diazo material into the machine. The light will pass through the original and expose the sensitized diazo material, except where images (lines and lettering) exist on the original. When the exposed diazo material and the original drawing emerge from the printer section, remove the original and carefully feed the diazo material into the developer section, as shown in Figure 3–4.

In the developer section the sensitized material that remains on the diazo paper activates with ammonia vapor, and blue lines form. (Some diazo materials make black or brown lines.) Special materials are also available to make other colored lines, or to make transfer sheets and transparencies.

Storage of Diazo Materials

Diazo materials are light sensitive, so they should always be kept in a dark place until ready for use. Long periods of exposure to room light cause the diazo chemicals to deteriorate and reduce the quality of your print. If you notice brown or blue edges around the unexposed diazo material, it is getting old. Always keep it tightly

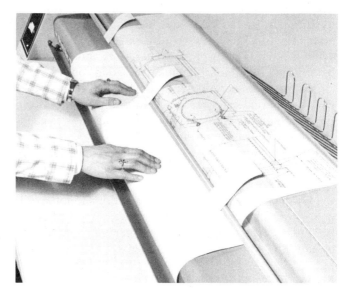

Figure 3–4 Feeding exposed, sensitized diazo material into the developer section. *Courtesy Ozalid Corporation.*

stored in the shipping package and in a dark place, such as a drawer. Some people prefer to keep all diazo material in a small, dark refrigerator. Doing so preserves the chemical for a long time.

Sepias

Sepias are diazo materials that are used to make secondary originals. A secondary, or second-generation, original is a print of an original drawing that can be used as an original. Changes can be made on the secondary original while the original drawing remains intact. The diazo process is performed with material called sepia. Sepia prints normally form dark brown lines on a translucent background; therefore, they are sometimes called brownlines. In addition to being used as an original to make alterations to a drawing without changing the original, sepias are also used when originals are required at more than one company location.

Sepias are made in reverse (reverse sepias) so corrections or changes can be made on the mat side (face side) and erasures can be made on the sensitized side. Sepia materials are available in paper or polyester. The resulting paper sepia is similar to vellum, and the polyester sepia is polyester film. Drawing changes can be made on sepia paper with pencil or ink, and on sepia polyester film with polyester lead or ink.

Computer-aided drafting is rapidly taking the place of sepia, because drawings are easily changed on the computer and a new original created.

▼ DIAZO SAFETY PRECAUTIONS

Ammonia

Ammonia has a strong, unmistakable odor. It may be detected, at times, while operating your diazo printer. Diazo printers are designed and built to provide safe operation from ammonia exposure. Some machines have ammonia filters; others require outside exhaust fans. When print quality begins to deteriorate and it has been determined that the print paper is in good condition, the ammonia may be old. Ammonia bottles should be changed periodically, as determined by the quality of prints. When the ammonia bottle is changed, it is generally time to change the filter also.

When handling bottles of ammonia or when replacing a bottle supplying a diazo printer, precautions are required.

Eye Protection.   Avoid direct contact of ammonia with your eyes. Always wear safety goggles, or equivalent eye protection, when handling ammonia containers directly or handling ammonia supply systems.

Ammonia and Filter Contact.   Avoid the contact of ammonia and the ammonia filter with your skin or clothing. Ammonia and ammonia residue on the filter can cause uncomfortable irritation and burns when in direct contact with your skin.

Ammonia Fumes Inhalation.   The disagreeable odor of ammonia is usually sufficient to prevent breathing harmful concentrations of ammonia vapors. Avoid prolonged periods of inhalation close to open containers of ammonia or where strong, pungent odors are present. The care, handling, and storage of ammonia containers should be in accordance with the suppliers' instructions and all applicable regulations.

First Aid.   If ammonia is spilled on your skin, promptly wash with plenty of water, and remove your clothing if necessary, to flush affected areas adequately. If your eyes are affected, irrigate with water as quickly as possible for at least 15 minutes, and, if necessary, consult a physician.

Anyone overcome by ammonia fumes should be removed to fresh air at once. Apply artificial respiration, preferably with the aid of oxygen if breathing is labored or has stopped. Obtain medical attention at once in the event of eye contact, burns to the nose or throat, or if the person is unconscious.

Ultraviolet Light Exposure

Under the prescribed operating instructions, there is no exposure to the ultraviolet rays emitted from illuminated printing lamps. However, to avoid possible eye damage, no one should attempt to look directly at the illuminated lamps under any circumstances.

▼ PHOTOCOPY REPRODUCTION

The photocopy engineering size printer makes prints up to 24″ wide, and up to 25′ long from originals up to 36″ wide by 25′ long. Prints can be made on bond paper, vellum, polyester film, colored paper, or other translucent materials. The reproduction capabilities also include reduction and enlargement on some models.

Almost any large original can be converted into a smaller-sized reproducible print, and the secondary original used to generate diazo or photocopy prints for distribution, inclusion in manuals, or more convenient handling. Also, a random collection of mixed-scale drawings can be enlarged or reduced and converted to one standard scale and format. Reproduction clarity is so good that halftone illustrations (photographs) and solid or fine line work has excellent resolution and density.

▼ MICROFILM

In many companies original drawings are filed in drawers by drawing number. When a drawing is needed, the drafter finds the original, removes it, and makes a copy. This process works well, although, depending on the size of the company or the number of drawings generated, drawing storage often becomes a problem. Sometimes an entire room is needed for drawing storage cabinets. Another problem occurs when originals are used over and over. They often become worn and damaged, and old vellum becomes yellowed and brittle. Also, in case of a fire or other kind of destruction, originals may be lost, and endless hours of drafting vanish. For these and other reasons many companies are using microfilm for the storage and reproduction of original drawings. Microfilm is used by industry for the photographic reproduction of original drawings into a film-negative format. Cameras used for microfilming are usually one of three sizes: 16, 35, or 70 mm. Microfilming processor cameras are now available that prepare film ready for use in about 20 seconds.

Aperture Card and Film Roll

The microfilm is generally prepared as one frame, or drawing, attached to an aperture card (Figure 3–5) or as many frames in succession on a roll of film (Figure 3–6). The aperture card becomes the engineering document and replaces the original drawing. A 35-mm film negative is mounted in an aperture card, which is a standard computer card. Some companies file and retrieve these cards manually; other companies punch the cards for computerized filing and retrieval. The aperture card, or data card as it is often called, is convenient to use. All drawings are converted to one size, regardless of the original dimensions. All engineering documents are filed together, instead of having originals ranging in size from 8½ × 11″ (297 × 420 mm) to 34 × 44″ (841 × 1189 mm). Then for easy retrieval each card is numbered with a sequential engineering drawing and/or project number.

When a drawing is prepared on microfilm, generally two negatives are made. One negative is placed in an active file, and the other is placed in safe storage in case the first is damaged or if an alternate is needed to make drawing changes and revisions. In many cases the original drawings are either stored out of the way or destroyed.

Microfilm Enlarger Reader-printer

Microfilm from aperture cards or rolls can be displayed on a screen for review, or enlarged and reproduced as a copy on plain bond paper, vellum, or offset paper-plates. Most microfilm printer readers can enlarge and reproduce prints. Prints are then available for disbursement to construction, sales, or other department personnel (Figure 3–7).

Computerized Microfilm Filing and Retrieval

The next step in the automation of drawing filing systems is the computerized filing and retrieval of microfilm

Figure 3–5 Aperture card with microfilm. *Courtesy 3M Company.*

Figure 3–6 Roll microfilm. *Courtesy 3M Company.*

Figure 3–7 Microfilm reader-printer. *Courtesy 3M Company.*

aperture cards. The implementation of this kind of system remains very expensive. Large companies and independent microfilming agencies are able to take advantage of this technology. Small companies can send their architectural documents to local microfilm companies known as job shops, where microfilm can be prepared and stored. Companies that are able to use computerized microfilm systems can have a completed drawing from the architect microfilmed, stored in a computer, and available in the construction office within a few minutes.

CHAPTER

3 Drafting Media and Reproduction Test

DIRECTIONS

Answer the questions with short, complete statements or drawings as needed on 8½ × 11 lined paper as follows:

1. Letter your name, Chapter 3 Test, and the date at the top of the sheet.
2. Letter the question number and provide the answer. You do not need to write out the question.

Answers may be prepared on a word processor if appropriate with course guidelines.

QUESTIONS

Question 3–1 List five factors that influence the purchase and use of drafting materials.

Question 3–2 Why is media transparency so important when reproducing copies using the diazo process?

Question 3–3 Describe vellum.

Question 3–4 Describe polyester film.

Question 3–5 What is another name for polyester film?

Question 3–6 Define mat, and describe the difference between single and double mat.

Question 3–7 Which of the following combinations would yield the best reproduction: graphite on vellum, plastic lead on polyester film, or ink on polyester film?

Question 3–8 What are the primary elements that give the best reproduction?

Question 3–9 Identify three standard sheet sizes that may commonly be used by architectural offices.

Question 3–10 Describe zoning.

Question 3–11 List six elements that normally may be found in a standard architectural title block.

Question 3–12 Identify two different locations where the scale may be located on an architectural drawing.

Question 3–13 What is another name for the diazo print?

Question 3–14 Is the diazo print the same as a blueprint? Explain.

Question 3–15 Describe how the diazo process functions.

Question 3–16 How does the speed of the diazo printer affect the resulting print?

Question 3–17 Describe how to run a test strip, and the importance of this test.

Question 3–18 What should a good quality diazo print look like?

Question 3–19 Describe the characteristics of slightly exposed or old diazo materials.

Question 3–20 Define sepias and their use.

Question 3–21 Why are sepias generally made in reverse (reverse sepia)?

Question 3–22 Discuss the safety precautions involving the handling of ammonia.

Question 3–23 Describe the recommended first aid for the following ammonia accidents: ammonia spilled on the skin, ammonia in the eyes, inhaling excess ammonia vapor.

Question 3–24 List four advantages of the photocopy reproduction method over the diazo process.

Question 3–25 Define the microfilm process.

Question 3–26 What is the big advantage of microfilm?

Question 3–27 What are the two common microfilm formats, and which microfilm format is the most popular, and why?

Question 3–28 Given the architectural title block shown in the illustration for this question, identify elements a–h.

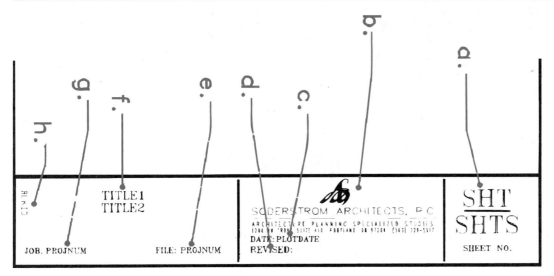

Question 3–28

CHAPTER 4

Sketching and Orthographic Projection

Sketching is freehand drawing, that is, drawing without the aid of drafting equipment. Sketching is convenient since all that is needed is paper, pencil, and an eraser. There are a number of advantages and uses for freehand sketching. Sketching is fast visual communication. The ability to make an accurate sketch quickly can often be an asset when communicating with people at work or at home. Especially when technical concepts are the topic of discussion, a sketch may be the best form of communication. Most drafters prepare a preliminary sketch to help organize thoughts and minimize errors on the final drawing. The computer-aided drafter usually prepares a sketch on graph paper to help establish the coordinates for drawing components. Some drafters use sketches to help record the stages of progress when designing until a final design is ready for implementation into formal drawings.

▼ TOOLS AND MATERIALS

The pencil should have a soft lead; a common number 2 pencil works fine. An automatic 0.7- or 0.9-mm pencil with F or HB lead is also good. The pencil lead should not be sharp. A dull, slightly rounded pencil point is best. Different thickness of line, if needed, can be drawn by changing the amount of pressure that you apply to the pencil. The quality of the paper is not critical either. A good sketching paper is newsprint, although most any kind of paper works. Paper with a surface that is not too smooth is best. Many architectural designs have been created on a napkin around a lunch table. Sketching paper should not be taped down to the table. The best sketches are made when you are able to move the paper to the most comfortable drawing position. Some people make horizontal lines better than vertical lines. If this is your situation, then move the paper so vertical lines become horizontal. Such movement of the paper may not always be possible, so it does not hurt to keep practicing all forms of lines for best results.

▼ STRAIGHT LINES

Lines should be sketched lightly in short connected segments as shown (Figure 4–1). If you sketch one long stroke in one continuous movement, your arm tends to make the line curved rather than straight. If you make a dark line, you may have to erase it if you make an error, whereas if you draw a light line, often there is no need to erase an error.

Use the following procedure to sketch a horizontal straight line with the dot-to-dot method:

Step 1. Mark the starting and ending positions, as in Figure 4–2. The letters A and B are only for instruction. All you need are the points.

Step 2. Without actually touching the paper with the pencil point, make a few trial motions between the marked points to adjust the eye and hand to the anticipated line.

Step 3. Sketch very light lines between the points by stroking in short, light strokes, 2 to 3″ long. With each stroke, attempt to correct the most obvious defects of the preceding stroke so that the finished light lines will be relatively straight. Look at Figure 4–3.

Step 4. Darken the finished line with a dark, distinct, uniform line directly on top of the light line. Usually the darkness can be obtained by pressing hard on the pencil. See Figure 4–4.

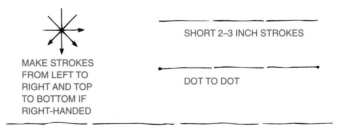

MAKE STROKES FROM LEFT TO RIGHT AND TOP TO BOTTOM IF RIGHT-HANDED

SHORT 2–3 INCH STROKES

DOT TO DOT

Figure 4–1 Sketching short line segments.

A •   • B

Figure 4–2   Step 1: Dot-to-dot.

A •———— ———— ———— ———— • B

Figure 4–3   Step 3: Short, light strokes.

A •————————————————————• B

Figure 4–4   Step 4: Darken to finish the line.

▼ CIRCULAR LINES

Figure 4–5 shows the parts of a circle. There are two sketching techniques to use when making a circle: the trammel and the hand compass method.

Trammel Method

Step 1.   Make a trammel. To sketch a 6-in. diameter circle, tear a strip of paper approximately 1″ wide and longer than the radius, 3″ On the strip of paper, mark an approximate 3″ radius with tick marks, such as A and B in Figure 4–6.

Step 2.   Sketch a straight line representing the circle diameter at the place where the circle is to be located. On the sketched line, locate with a dot the center of the circle to be sketched. Use the marks on the trammel to mark the other end of the radius line, as shown in Figure 4–7. With the trammel next to the sketched line, be sure point B on the trammel is aligned with the center of the circle you are about to sketch.

Step 3.   Pivot the trammel at point B, making tick marks at point A as you go, as shown in Figure 4–8, until you have a complete circle, as shown in Figure 4–9.

Step 4.   Lightly sketch the circumference over the tick marks to complete the circle; then darken it″ As shown in Figure 4–9.

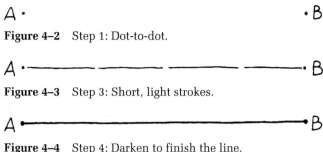

Figure 4–6   Step 1: Making a trammel.

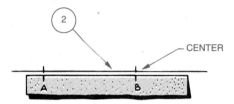

Figure 4–7   Step 2: Mark the radius with a trammel.

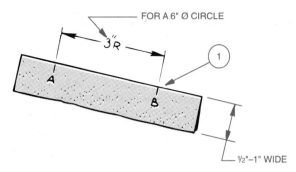

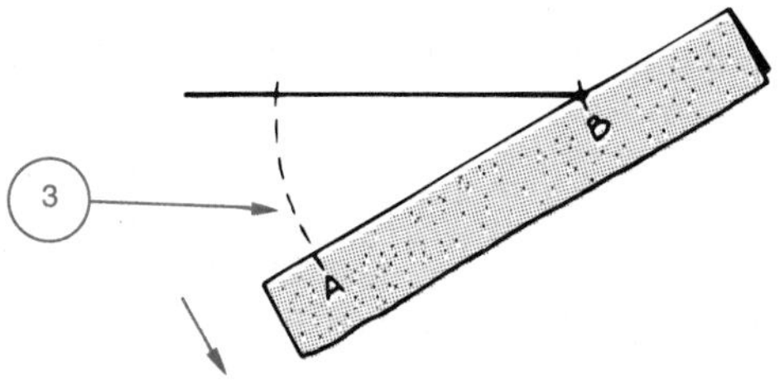

Figure 4–8   Step 3: Sketching the circle with the trammel.

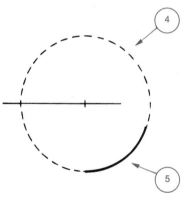

Figure 4–9   Step 4: Complete the circle.

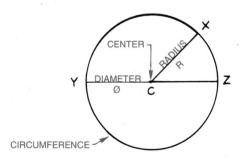

Figure 4–5   Parts of a circle.

Hand Compass Method

The hand compass method is a quick and fairly accurate method of sketching circles, although it takes some practice.

Step 1.   Be sure your paper is free to rotate completely around 360°. Remove anything from the table that might stop such a rotation.

Step 2.   To use your hand and a pencil as a compass, place the pencil in your hand between your thumb and the upper part of your index finger so your index finger becomes the *compass point* and the pencil becomes the *compass lead.* The other end of the pencil rests in your palm, as shown in Figure 4–10.

Step 3. Determine the circle radius by adjusting the distance between your index finger and the pencil point. Now, with the desired approximate radius established, place your index finger on the paper at the proposed center of the circle.

Step 4. With your desired radius established, keep your hand and pencil point in one place while you rotate the paper with your other hand. Try to keep the radius steady as you rotate the paper. Look at Figure 4–11.

Step 5. You can perform Step 4 very lightly and then go back and darken the circle or, if you have had a lot of practice, you may be able to draw a dark circle as you go. See Figure 4–11.

Another method generally used to sketch large circles is to tie a string between a pencil and a pin. The distance between the pencil and pin is the radius of the circle. Use this method when a large circle is to be sketched since the other methods may not work as well. This method is used with a nail driven at the center and a string connected to a pencil when drawing large circles at a construction site.

▼ MEASUREMENT LINES AND PROPORTIONS

When sketching objects, all the lines that make up the object are related to each other by size and direction. In order for a sketch to communicate accurately and completely, it must be drawn in the same proportion as the object. The actual size of the sketch depends on the paper size and how large you want the sketch to be. The sketch should be large enough to be clear, but the proportions of the features are more important than the size of the sketch.

Look at the lines in Figure 4–12. How long is line 1? How long is line 2? Answer these questions without mea-

Figure 4–10 Step 2: Holding the pencil for the hand compass method.

Figure 4–11 Steps 4 and 5: Sketching a circle with the hand compass method.

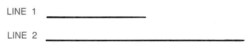

Figure 4–12 Measurement lines.

suring either line; instead relate each line to the other. For example, line 1 could be stated as being half as long as line 2, or line 2 called twice as long as line 1. Now you know how long each line is in relationship to the other (proportion), but you do not know how long either line is in relationship to a measured scale. No scale is used for sketching, so this is not a concern. So whatever line you decide to sketch first determines the scale of the drawing. This first line sketched is called the *measurement line.* You relate all the other lines in your sketch to the first line. This is one of the secrets in making a sketch look like the object being sketched.

The second thing you must know about the relationship of the two lines in the example is their direction and position to each other. Do they touch each other? Are they parallel, perpendicular, or at some angle to each other? When you look at a line ask yourself the following questions (for this example use the two lines given in Figure 4–13):

1. How long is the second line? *Answer:* Line 2 is about three times as long as line 1.
2. In what direction and position is the second line related to the first line? *Answer:* Line 2 touches the lower end of line 1 with about a 90° angle between them.

Carrying this concept a step further, a third line can relate to the first line or the second line and so forth. Again, the first line drawn (measurement line) sets the scale for the whole sketch.

Figure 4–13 Measurement line.

This idea of relationship can also apply to spaces. In Figure 4–14, the location of a table in a room can be determined by space proportions. A typical verbal location for the table in this floor plan might be as follows: The table is located about one-half the table width from the top of the floor plan or about two table widths from the bottom, and about one table width from the right side or about three table widths from the left side of the floor plan.

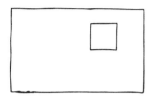

Figure 4–14   Space proportions.

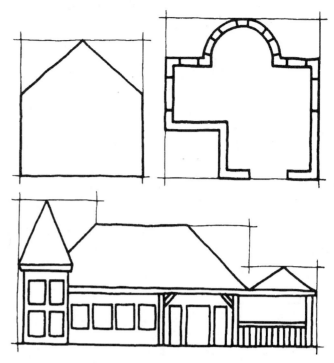

Figure 4–15   Block technique.

Using Your Pencil to Establish Measurements

Your pencil can be a handy tool for use when establishing measurements on a sketch. When you are sketching an object that you can hold, use your pencil as a ruler. To do this, place the pencil next to the feature to be sketched, and determine the length by aligning the pencil tip at one end of the feature and mark the other end on the pencil with your thumb. The other end may also be identified by a specific contour or mark on the pencil. Then transfer the measurement to your sketch. The two distances are the same.

A similar technique may be used when sketching a distant object. For example, to sketch a house across the street, hold your pencil at arm's length and align it with a feature on the house, such as the house width. With the pencil point at one end of the house, place your thumb on the pencil, marking the other end of the house. Transfer this measurement to your sketch in the same orientation as it was taken from the house. Repeat this technique for all of the house features until the sketch is complete. If you keep your pencil at arm's length and your arm straight, each measurement will have the same accuracy.

Block Technique

Any illustration of an object can be surrounded overall with a rectangle, as shown in Figure 4–15. Before starting a sketch, see the object to be sketched inside a rectangle in your mind's eye. Then use the measurement-line technique with the rectangle, or block, to help you determine the shape and proportions of your sketch.

Procedures in Sketching

Step 1.   When starting to sketch an object, try to visualize the object surrounded with a rectangle overall. Sketch this rectangle first with very light lines. Sketch it in the proper proportion with the measurement-line technique, as shown in Figure 4–16.

Step 2.   Cut sections out or away using proper proportions as measured by eye. Use light lines, as shown in Figure 4–17.

Step 3.   Finish the sketch by darkening the desired outlines for the finished sketch. See Figure 4–18.

Irregular Shapes

By using a frame of reference or an extension of the block method, irregular shapes can be sketched easily to their correct proportions. Follow these steps to sketch the freeform swimming pool shown in Figure 4–19.

Step 1.   Place the object in a lightly constructed box. See Figure 4–20.

Step 2.   Sketch several equally spaced horizontal and vertical lines, forming an evenly spaced grid, as shown in Figure 4–21. If you are sketching an object already drawn, draw your reference lines on top of the object lines to establish a frame of reference. Make a photocopy if the original may not be used. If you are sketching an object directly, you have to visualize these reference lines on the object you sketch.

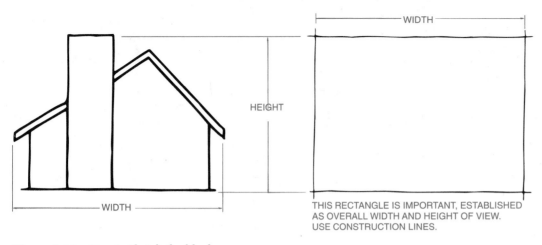

THIS RECTANGLE IS IMPORTANT, ESTABLISHED
AS OVERALL WIDTH AND HEIGHT OF VIEW.
USE CONSTRUCTION LINES.

Figure 4–16   Step 1: Sketch the block.

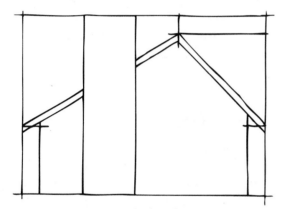

Figure 4–17   Step 2: Block technique.

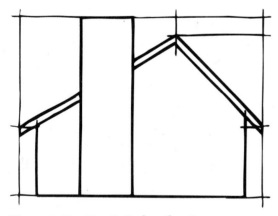

Figure 4–18   Step 3: Darken the view.

Figure 4–19   Freeform
swimming pool.

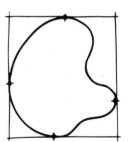

Figure 4–20   Imagi-
nary box.

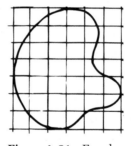

Figure 4–21   Evenly
spaced grid.

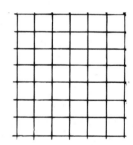

Figure 4–22   Propor-
tioned box with a regu-
lar grid.

Step 5.   Using the grid, sketch the small, irregular arcs
and lines that match the lines of the original, as
in Figure 4–23.

Step 6.   Darken the outline for a complete proportioned
sketch, as shown in Figure 4–24.

▼ **MULTIVIEW SKETCHES**

Multiviews, or multiview projection, is also known as
orthographic projection. In architectural drafting such
drawings are referred to as elevation views. Elevation
views are two-dimensional views of an object (a house,

Step 3.   On your sketch, correctly locate a proportioned
box similar to the one established on the original drawing or object, as shown in Figure 4–22.

Step 4.   Using the drawn box as a frame of reference,
include the grid lines in correct proportion, as
seen in Figure 4–22.

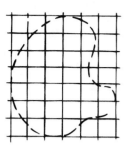

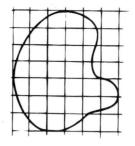

Figure 4–23 Sketching the shape using a regular grid.

Figure 4–24 Completely darken the object.

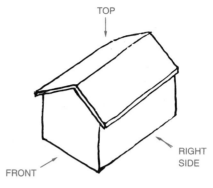

Figure 4–26 Pictorial view.

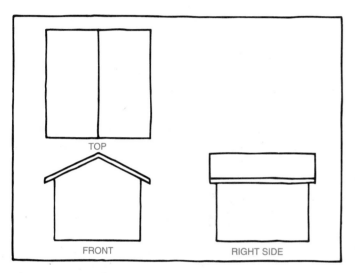

Figure 4–25 Multiviews.

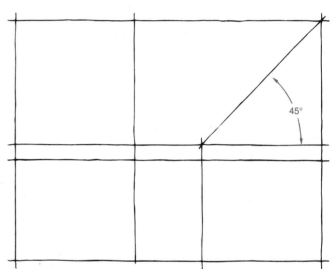

Figure 4–27 Step 1: Block out views and establish a 45° line.

for example) that are established by a line of sight that is perpendicular (90°) to the surface of the object. When you make multiview sketches, follow a systematic order. Learning to make these multiview sketches will help you later as you begin to prepare elevation drawings.

Multiview Alignment

To keep your drawing in a standard form, sketch the front view in the lower left portion of the paper, the top view directly above the front view, and the right-side view to the right side of the front view. See Figure 4–25. The views needed may differ depending on the object.

Multiview Sketching Technique

Step 1. Sketch and align the proportional rectangles for the front, top and right side of the object given in Figure 4–26. Sketch a 45° line to help transfer width dimensions. The 45° line is established by projecting the width from the top view across and the width from the right-side view up until the lines intersect, as shown in Figure 4–27.

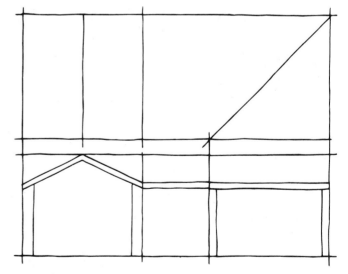

Figure 4–28 Step 2: Block out shapes.

Step 2. Complete the shapes within the blocks, as shown in Figure 4–28.

Step 3. Darken the lines of the object, as in Figure 4–29. Remember to keep the views aligned for ease of sketching and understanding.

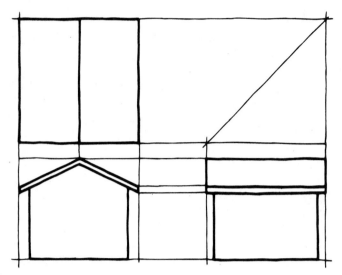

Figure 4–29 Step 3: Darken all lines.

▼ ISOMETRIC SKETCHES

Isometric sketches provide a three-dimensional pictorial representation of an object such as the shape of a building. Isometric sketches are easy to draw and make a fairly realistic exhibit of the object. Isometric sketches tend to represent the objects as they appear to the eye. Isometric sketches help in the visualization of an object because three sides of the object are sketches in a single three-dimensional view.

Establishing Isometric Axes

To establish isometric axes, you need four beginning lines: a horizontal reference line, two 30° angular lines, and one vertical line. Draw them as very light construction lines. Look at Figure 4–30.

Step 1. Sketch a horizontal reference line (consider this the ground-level line).

Step 2. Sketch a vertical line perpendicular to the ground line somewhere near its center. The vertical line is used to measure height.

Step 3. Sketch two 30° angular lines, each starting at the intersection of the first two lines, as shown in Figure 4–30.

Making an Isometric Sketch

The steps in making an isometric sketch from a multiview drawing or when viewing the real object are as follows:

Step 1. Select an appropriate view of the object to use as a front view, or study the front view of the multiview drawing.

Step 2. Determine the best position in which to show the object.

Step 3. Begin your sketch by setting up the isometric axes as just described.

Step 4. By using the measurement-line technique, draw a rectangular box to correct proportion, which could surround the object to be drawn. Use the object shown in Figure 4–31 for this explanation. Imagine the rectangular box in your mind. Begin to sketch the box by marking off the width at any convenient length (measurement line), as in Figure 4–32. Next estimate and mark the length and height as related to the measurement line. See Figure 4–33. Sketch the three-dimensional box by using lines parallel to the original axis lines. See Figure 4–33. The sketch of the box must be done correctly; otherwise your sketch will be out of proportion. All lines drawn in the same direction must be parallel.

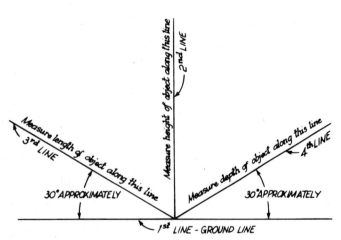

Figure 4–30 Isometric axes.

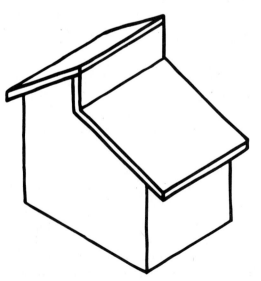

Figure 4–31 Given structure.

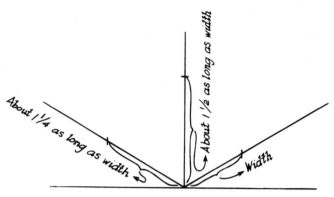

Figure 4–32 Step 4: Lay out the length, width, and height.

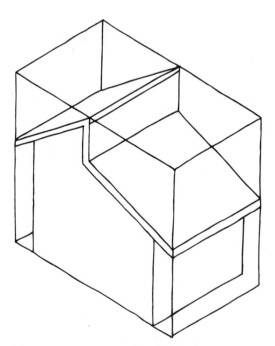

Figure 4–33 Step 4: Sketch the three-dimensional box.

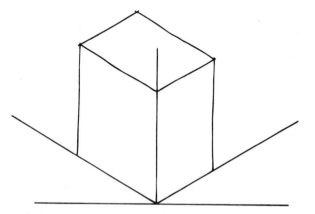

Figure 4–34 Step 5: Sketch the features of the house.

Step 5. Lightly sketch the features that define the details of the object. By estimating distances on the rectangular box, you will find that the features of the object are easier to sketch in correct

proportion than trying to draw them without the box. See Figure 4–34.

Step 6. To finish the sketch, darken all the outlines, as in Figure 4–35.

Nonisometric Lines

Isometric lines are lines that are on or parallel to the three original axes lines. All other lines are nonisometric lines. Isometric lines can be measured true length. Nonisometric lines appear either longer or shorter than they actually are. See Figure 4–35.

You can measure and draw nonisometric lines by connecting their end points. Find the end points of the nonisometric lines by measuring along isometric lines. To locate where nonisometric lines should be placed, you have to relate to an isometric line.

Sketching Isometric Circles

Circles and arcs appear as ellipses in isometric views. To sketch isometric circles and arcs correctly, you need to know the relationship between circles and the faces, or planes, of an isometric cube. Depending on which face the circle is to appear, isometric circles look like one of the ellipses shown in Figure 4–36. The angle the ellipse (isometric circle) slants is determined by the surface on which the circle is to be sketched.

Four-Center Method. The four-center method of sketching an isometric ellipse is simple to perform, but care must be taken to form the ellipse arcs properly so the ellipse does not look distorted.

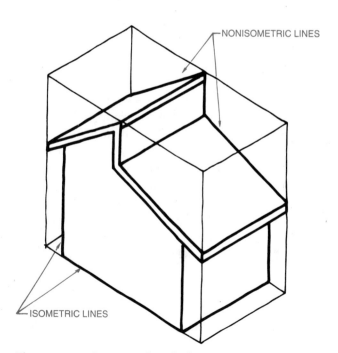

Figure 4–35 Step 6: Darken the house.

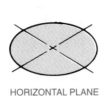
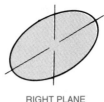

LEFT PLANE   HORIZONTAL PLANE   RIGHT PLANE

Figure 4–36   Isometric circles.

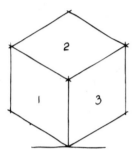

Figure 4–37   Step 1: Isometric cube.

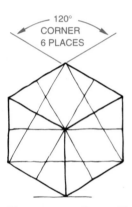

120°
CORNER
6 PLACES

Figure 4–38   Step 2: Four-center isometric ellipse construction.

Step 1.   Draw an isometric cube similar to Figure 4–37.

Step 2.   On each surface of the box draw line segments that connect the 120° corners to the centers of the opposite sides. See Figure 4–38.

Step 3.   With points 1 and 2 as the centers, sketch arcs that begin and end at the centers of the opposite sides on each isometric surface. See Figure 4–39.

Step 4.   On each isometric surface, with points 3 and 4 as the centers, complete the isometric ellipses by sketching arcs that meet the arcs sketched in step 3. See Figure 4–40.

Sketching Isometric Arcs

Sketching isometric arcs is similar to sketching isometric circles. First, block out the overall configuration of the object; then establish the centers of the arcs; finally, sketch the arc shapes. Remember that isometric arcs, just like isometric circles, must lie in the proper plane and have the correct shape.

▼ ORTHOGRAPHIC PROJECTION

Orthographic projection is any projection of features of an object onto an imaginary plane called a plane of projection. The projection of the features of the object is made by lines of sight that are perpendicular to the plane of projection. When a surface of the object is parallel to the plane of projection, the surface appears in its true size and shape on that plane. In Figure 4–41, the plane of

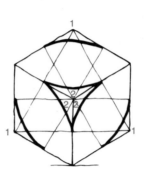

Figure 4–39   Step 3: Sketch arcs from points 1 and 2.

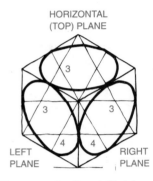

HORIZONTAL (TOP) PLANE

LEFT PLANE   RIGHT PLANE

Figure 4–40   Step 4: Sketch arcs from points 3 and 4.

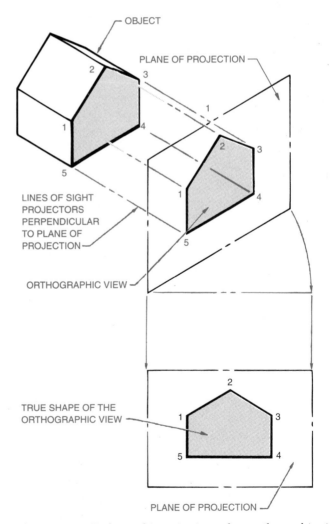

OBJECT

PLANE OF PROJECTION

LINES OF SIGHT PROJECTORS PERPENDICULAR TO PLANE OF PROJECTION

ORTHOGRAPHIC VIEW

TRUE SHAPE OF THE ORTHOGRAPHIC VIEW

PLANE OF PROJECTION

Figure 4–41   Orthographic projection to form orthographic view.

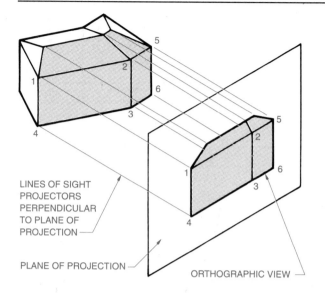

FORESHORTENED
ORTHOGRAPHIC VIEW
OF SURFACE 1, 2, 3, 4,

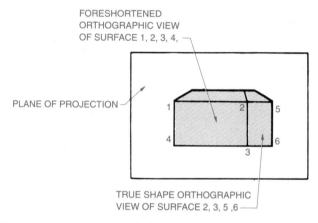

TRUE SHAPE ORTHOGRAPHIC
VIEW OF SURFACE 2, 3, 5 ,6

Figure 4–42 Projection of a foreshortened orthographic surface.

projection is parallel to the surface of the object. The line of sight (projection from the object) from the object is perpendicular to the plane of projection. Notice also that the object appears three dimensional (width, depth, and height), while the view on the plane of projection is two dimensional (width and height). In situations where the plane of projection is not parallel to the surface of the object, the resulting orthographic view is foreshortened, or shorter than the true length. See Figure 4–42.

▼ MULTIVIEW PROJECTION

Multiview projection establishes two or more views of an object as projected on two or more planes of projection by using orthographic projection techniques. The result of multiview projection is a multiview drawing. Multiview drawings represent the shape of an object using two or more views. Consideration should be given to the choice and number of views so, when possible, the surfaces of the object are shown in their true size and shape.

Elevations Are Multiviews

It is often easier to visualize a three-dimensional drawing of a structure than it is to visualize a two-dimensional drawing. In architectural drafting, however, it is common to prepare construction drawings that show two-dimensional exterior views of a structure that provide representations of exterior materials. These drawings are referred to as *elevations*. The method used to draw elevations is known as *multiview projection*. A more detailed discussion of elevation drawing is found in Chapters 21 and 22. Figure 4–43 shows an object represented by a three-dimensional drawing, called a pictorial drawing, and three two-dimensional views.

Glass Box

If you place the object in Figure 4–43 in a glass box so the sides of the box are parallel to the major surfaces of the object, you can project the surfaces of the object onto the sides of the glass box and create multiviews (Figure 4–44). Imagine the sides of the glass box are the planes of projection previously described. Look at Figure 4–45. If you look at all sides of the glass box, you see six views: front, top, right side, left side, bottom and rear. Now unfold the glass box as if the corners were hinged about the front view (except the rear view, which is attached to the left-side view), as demonstrated in Figure 4–46. These hinge lines are commonly called *fold lines*.

Completely unfold the glass box onto a flat surface, and you have the six views of an object represented in multiview. Figure 4–47 shows the glass box unfolded. Notice also that the views are labeled: front, top, right side, left side, rear, and bottom. It is in this arrangement that the views are always found when using multiviews. Analyze Figure 4–47 in detail so you see the features that are common among the views. Knowing how to identify the features of an object that are common among views will aid you later in the visualization of elevations.

Notice how the views are aligned. The top view is directly above and the bottom view is directly below the front view. The left side is directly to the left, while the right side is directly to the right of the front view. This alignment allows the drafter to project features from one view to the next to help establish each view.

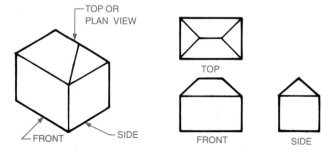

Figure 4–43 Comparison of a pictorial and multiview.

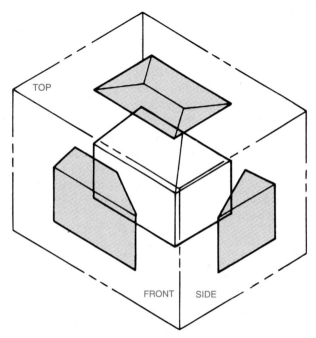

Figure 4–44   Glass box.

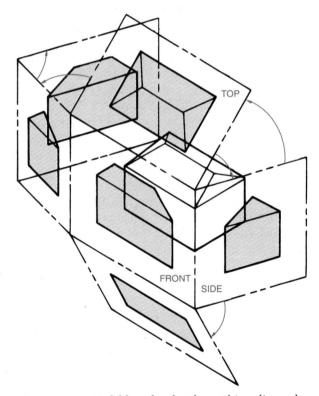

Figure 4–45   Unfolding the glass box at hinge lines, also called fold lines.

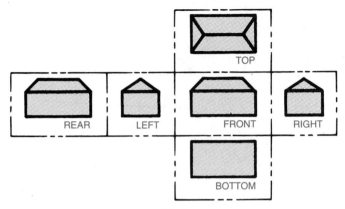

Figure 4–46   Glass box unfolded.

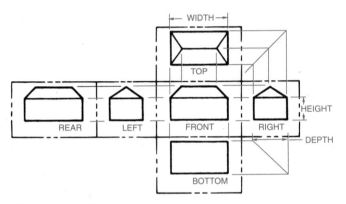

Figure 4–47   View alignment.

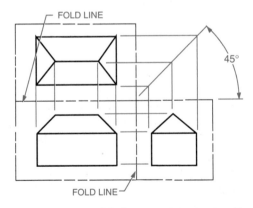

Figure 4–48   Establishing a 45° projection line.

Now look closely at the relationship of the front, top, and right-side views. A similar relationship exists using the left-side view. Figure 4–48 shows a 45° projection line established by projecting the fold, or reference line (hinge), between the front and side view up and the fold line between the front and top view over. All of the fea-

tures established on the top view can be projected to the 45° line and then down onto the side view. This is possible because the depth dimension is seen in both the top and side views. The reverse is also true. Features from the side view may be projected to the 45° line and then over to the top view.

The transfer of features achieved in Figure 4–48 using the 45° line can also be accomplished by using a compass with one leg at the intersection of the horizontal and vertical fold lines. The compass establishes the common relationship between the top and side views, as shown in Figure 4–49.

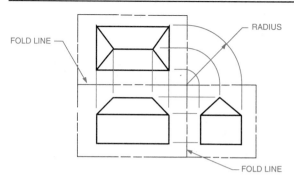

Figure 4–49 Projection with a compass.

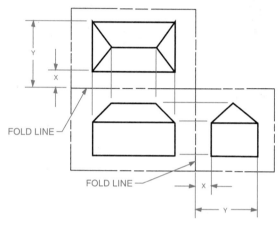

Figure 4–50 Using dividers to transfer view projections.

Another method that is commonly used to transfer the size of features from one view to the next is the use of dividers to transfer distances from the fold line at the top view to the fold line at the side view. The relationship between the fold lines and the two views is the same as shown in Figure 4–50.

The front view is usually the most important view since it is the one from which the other views are established. There is always one dimension common between adjacent views. For example, the width is common between the front and top views and the height between the front and side views. This knowledge allows you to relate information from one view to another. Take one more look at the relationship of the six views as shown in Figure 4–51.

View Selection

There are six primary views that you may select to describe a structure completely. In architectural drafting, the front, left side, right side, and rear views are used as elevations to describe the exterior appearance of a structure completely. Elevation drawings are discussed in more detail in Chapters 21 and 22. The top view is called the roof plan view. This view shows the roof of the struc-

ture and provides construction information and dimensions. Roof plans are discussed in detail in Chapters 19 and 20. The bottom view is often used in mechanical drafting but never in architectural drafting.

▼ PROJECTION OF FEATURES FROM AN INCLINED PLANE

Rectangular Features on an Inclined Plane

When a rectangular feature such as a skylight projects out of a sloped roof, the interesction of the skylight with the roof appears as a line when the roof also appears as a line. This intersection location may then be projected onto adjacent views, as shown in Figure 4–52.

Circles on an Inclined Plane

When the line of sight in a view is perpendicular to a circle, such as a round window, the window appears round, as shown in Figure 4–53. When a circle is projected onto an inclined surface, such as a round skylight projected onto a sloped roof, the view of the inclined circle is elliptical. See Figure 4–54.

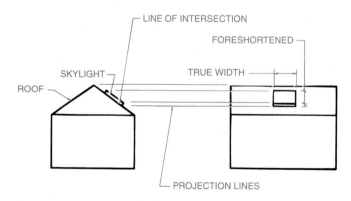

Figure 4–51 Multiview orientation.

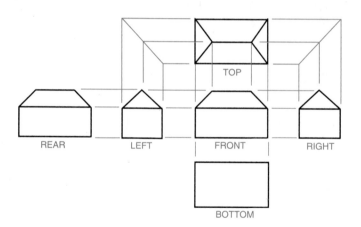

Figure 4–52 Rectangular features on an inclined plane.

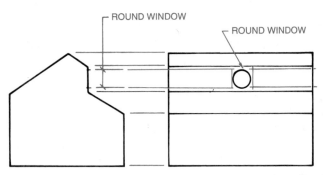

Figure 4–53 Round window is a circle when the line of sight is perpendicular.

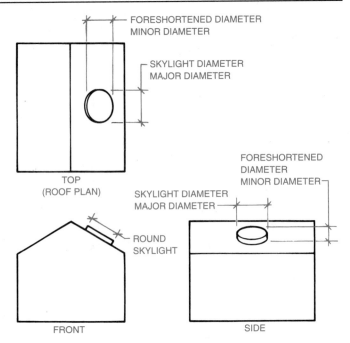

Figure 4–54 Circle projected onto an inclined surface appears as an ellipse.

CHAPTER 4

Sketching and Orthographic Projection Test

DIRECTIONS

Answer the questions with short, complete statements or drawings as needed on 8½ × 11 notebook paper as follows:

1. Letter your name, Chapter 4 Test, and the date at the top of the sheet.
2. Letter the question number and provide the answer. You do not need to write out the question.

Answers may be prepared on a word processor if appropriate with course guidelines.

QUESTIONS

Question 4–1 Define sketching.

Question 4–2 How are sketches useful as related to computer graphics?

Question 4–3 Describe the proper sketching tools.

Question 4–4 Should the paper be taped to the drafting board or table when sketching? Why?

Question 4–5 What kind of problem can occur if a long, straight line is drawn without moving the hand?

Question 4–6 What type of paper should be used for sketching?

Question 4–7 Name the two methods that can be used to sketch irregular shapes.

Question 4–8 Define an isometric sketch.

Question 4–9 What is the difference between an isometric line and a nonisometric line?

Question 4–10 What does the use of proportions have to do with sketching techniques?

Question 4–11 Define orthographic projection.

Question 4–12 What is the relationship between the orthographic plane of projection and the projection lines from the object or structure?

Question 4–13 When is a surface foreshortened in an orthographic view?

Question 4–14 How many principal multiviews of an object are possible?

Question 4–15 Give at least two reasons why the multiviews of an object are aligned in a specific format.

Question 4–16 In architectural drafting, what are the exterior front, right side, left side, and rear views also called?

Question 4–17 If a round window appears as a line in the front view and the line of sight is perpendicular to the window in the side view, what shape is the window in the side view?

Question 4–18 If a round skylight is positioned on a ⁵⁄₁₂ roof slope and appears as a line in the front view, what shape will the skylight be in the side view?

DIRECTIONS

Use proper sketching materials and techniques to solve the following sketching problems, on 8½ × 11 bond paper or newsprint. Use very lightly sketched construction lines for all layout work. Darken the lines of the object, but do not erase the layout lines.

PROBLEMS

Problem 4–1   Sketch the front view of your home or any other local single family residence using the block technique. Use the measurement line method to approximate proper proportions.

Problem 4–2   Use the box method to sketch a circle with an approximate 4″ diameter. Sketch the same circle using the trammel and hand compass methods.

Problem 4–3   Find an object with irregular shapes, such as a French (irregular) curve, and sketch a two-dimensional view using the grid method. Sketch the object to correct proportions without measuring.

Problem 4–4   Use the same structure you used for Problem 4–1, or a different structure, to prepare an isometric sketch.

Problem 4–5   Use the same structure you used for Problem 4–4 to sketch a front and right-side view.

Problem 4–6   Use a scale of ¼″ = 1′–0″ to draw a 38° acute angle with one side horizontal and both sides 8′–6″ long.

Problem 4–7   Given the sketch for this problem of a swimming pool, spa, and patio, resketch these elements large enough to fill most of an 8½ × 11 sheet of paper.

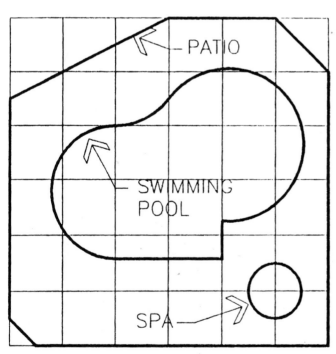

Problem 4–7

Problem 4–8   Given the top and side views shown in the sketch for this problem, redraw these views and draw the missing front view, filling most of an 8½ × 11 sheet of paper.

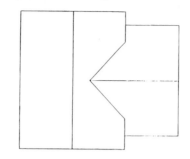

START VIEW HERE

Problem 4–8

Problem 4–9   Given the pictorial sketch for this problem, draw the front, top, and right-side views, filling most of an 8½ × 11 sheet of paper.

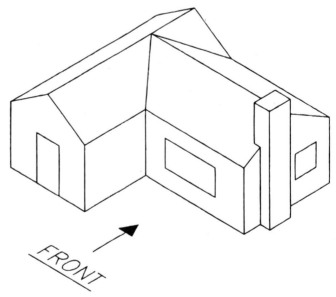

FRONT

Problem 4–9

Problem 4–10   Given the three views of the house shown in the sketches for this problem, sketch an isometric view, filling most of an 8½ × 11 sheet of paper.

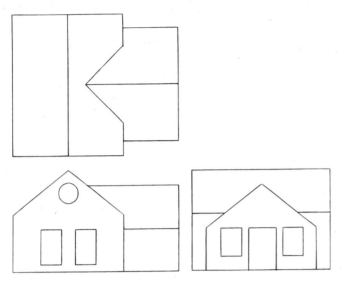

Problem 4–10

CHAPTER 5

Architectural Lines and Lettering

Drafting is a universal graphic language that uses lines, symbols, dimensions, and notes to describe a structure to be built. Lines and lettering on a drawing must be of a quality that reproduces clearly. Properly drawn lines are dark, crisp, sharp, and of a uniform thickness. There should be no variation in darkness, only a variation in thickness known as line contrast. Certain lines may be drawn thick so they stand out clearly from other information on the drawing. Other lines are drawn thin. Thin lines are not necessarily less important than thick lines, but they may be subordinate for identification purposes. In mechanical drafting recommended line thicknesses are much more defined than in architectural drafting. Architectural drafting line standards have traditionally been more flexible, and the creativity of the individual drafter or company may influence line and lettering applications. This is not to say that each drafter can do anything that he or she desires when preparing an architectural drawing. This chapter presents some general guidelines, styles, and techniques that are recommended for architectural drafters.

▼ TYPES OF LINES
Construction Lines and Guidelines

Construction lines are used for laying out a drawing. They are drawn very lightly so they will not reproduce and will not be mistaken for any other lines on the drawing. Construction lines are drawn with a very little amount of pressure using a 4H to 6H pencil, and if drawn properly will not need to be erased. Use construction lines for all preliminary work.

Guidelines are like construction lines and will not reproduce when properly drawn. Guidelines are drawn to guide your lettering. For example, if lettering on a drawing is ⅛″ (3 mm) high, then the lightly drawn guidelines are placed ⅛″ (3 mm) apart.

Some drafters prefer to use a light-blue lead rather than a graphite lead for all construction and guidelines. Light-blue lead will not reproduce in a diazo printer and is usually cleaner than graphite. Blue line will reproduce in a photocopy machine unless drawn very lightly.

Outlines

In mechanical drafting the outline lines, *or object lines* as they are commonly called, are a specific thickness so they stand out from other lines as they form the outline of views. In architectural drafting this concept is not quite so specific. Outline lines are used to define the outline and characteristic features of architectural plan components, but the method of presentation may differ slightly from one office to another. The following techniques may be alternatives for outline line presentation:

1. One popular technique is to enhance certain drawing features so they clearly stand out from outer items on the drawing. For example, the outline of floor plan walls and partitions, or beams in a cross section, may be drawn thicker than other lines so they are more apparent than the outer lines on the drawing. These thicker lines may be drawn with a 0.7- or 0.9-mm automatic pencil using a 2H, H, or F lead. See Figure 5–1. This technique may also use light shading to accentuate the walls of a floor plan. Dark shading would not be used because it would nullify the thicker outline lines.

2. Another technique is for all lines of the drawing to be the same thickness. This method does not differentiate one type of line from another, except that construction lines are always very lightly drawn. The idea of this technique is to make all lines medium thick to save drafting time. The drafter uses a lead that works best for him or her, although a 0.5-mm automatic pencil with 2H or H lead is popular. See Figure 5–2 (a). This technique may use dark shading to accentuate features such as walls in floor plans, as shown in Figure 5–2(b). The idea is to get all lines dark and crisp. If the lines are fuzzy, they will not reproduce well.

Dashed Lines

In mechanical drafting dashed lines are called *hidden lines*. In architectural drafting, dashed lines may also be considered as hidden lines since they are used to show drawing features that are not visible in relationship to the view or plan. These dashed features may also be subordinate to the main emphasis of the drawing.

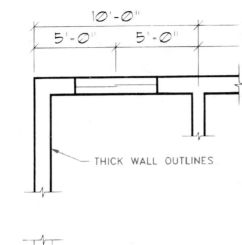

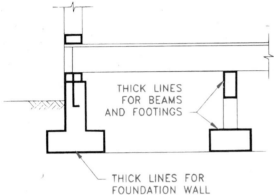

Figure 5–1 Thick outlines.

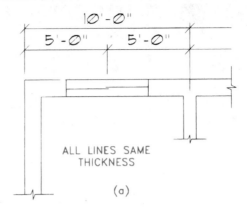

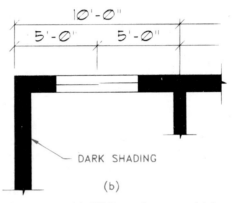

Figure 5–2 (a) All lines the same thickness; (b) accent with shading.

Dashed lines vary slightly from one office to the next. These lines are thin and generally drawn about ⅛ to ⅜″ (3–10 mm) in length with a space of ¹⁄₁₆ to ⅛″ (1.5–3 mm) between each dash. The dashes should be kept uniform in length on the drawing, for example, all ¼″ (6 mm) with approximately equal spaces. Dashed lines are thin, and the spacing should be by eye, never measured. Drawing dashed lines takes practice to do well. Recommended leads are a 0.5-mm automatic pencil of 2H or H hardness. Examples of dashed line representations include beams (Figure 5–3); headers (Figure 5–4); and upper kitchen cabinets, under-counter appliances (dishwasher), and electrical circuit runs (Figure 5–5). These concepts are discussed further in later chapters.

Extension and Dimensions Lines

Extension lines show the extent of a dimension, and dimension lines show the length of the dimension and terminate at the related extension lines with slashes, arrowheads, or dots. The dimension numeral in feet and inches is placed above and near the center of the solid dimension line. Figure 5–6 shows several dimension and extension line examples. Further discussion and examples are provided in later chapters. Extension lines are generally thin, dark, crisp lines that may be drawn with a sharp 0.5-mm automatic pencil using 2H lead.

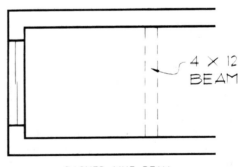

DASHED LINE BEAM

Figure 5–3 Dashed line beam.

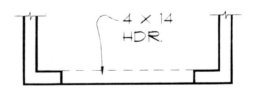

DASHED LINE HEADER

Figure 5–4 Dashed line header.

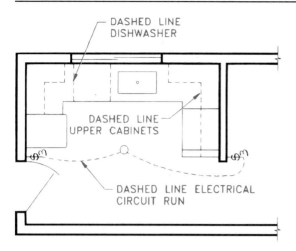

Figure 5–5 Dashed lines for upper kitchen cabinets, dishwasher, and electrical circuit run.

Leader Lines

Leader lines are also thin, sharp, crisp lines. These lines are used to connect notes to related features on a drawing. Leader lines may be drawn freehand or with an irregular curve. Do them freehand if you can do a good job but if they are not smooth, then use an irregular curve. The leader should start from the vertical center at the beginning or end of a note and be terminated with an arrowhead at the feature. Some companies prefer that leaders be straight lines that begin with a short shoulder and angle to the intended feature. Figure 5–7 shows several examples.

Break Lines

The two types of break lines are the long break line and the short break line. The type of break line that is nor-

mally associated with architectural drafting is the long break line. The break symbol is generally drawn freehand. Break lines are used to terminate features on a drawing when the extent of the feature has been clearly defined. Figure 5–8(a) shows several examples. The short break line may be found on some architectural drawings. This line, as shown in Figure 5–8(a) is an irregular line drawn freehand and may be used for a short area. Keep in mind that it is not as common as the long break line. Breaks in cylindrical objects such as steel bars and pipes are shown in Figure 5–8(b).

▼ LINE TECHNIQUES

When working on vellum or polyester film using pencil, polyester lead, or ink, use these basic techniques to help make the completion of the drawing easier:

1. Prepare a sketch to help organize your thoughts before beginning the formal drawing.
2. Do all layout work using construction lines. If you make an error, it is easy to correct.
3. Begin the formal drawing by making all horizontal lines from the top of the sheet to the bottom. Try to avoid going back over lines that have been drawn any more than is necessary.
4. If right handed, draw all vertical lines from right to left. If left handed, draw vertical lines from right to left.
5. Place all symbols on the drawing. Try to work from one side of the sheet to the other.
6. Do all lettering last. Place a clean piece of paper under your hand when lettering to avoid smudging lines or perspiring on the drawing.
7. Most important, keep your hands and equipment clean.

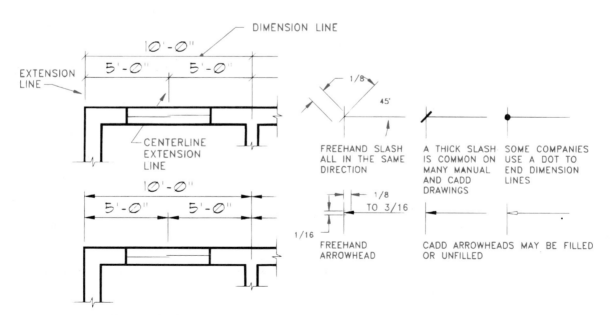

Figure 5–6 Dimension and extension lines.

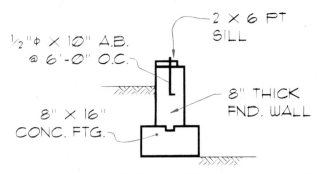

Figure 5–7 Sample leader lines. The leader style used should be the same throughout the drawing.

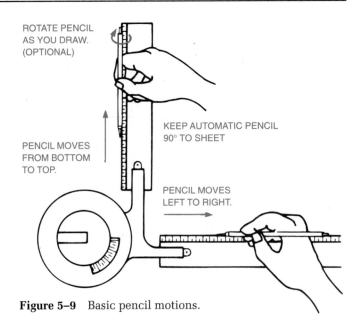

Figure 5–9 Basic pencil motions.

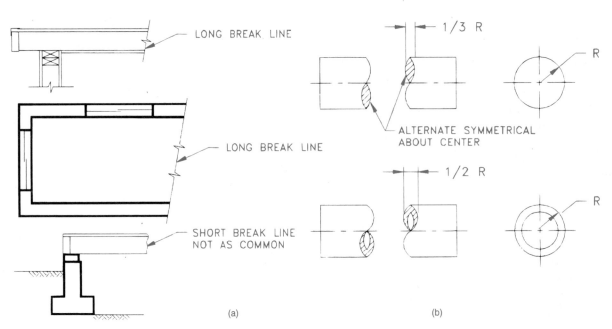

Figure 5–8 Solid break lines; cylindrical break lines.

Pencil Line Methods on Vellum

To see if your lines are dark and crisp enough, turn your drawing over and hold it up to the light or put it on a light table. The lines should have a dark, consistent density. Problems to look for are lines that you can see through or that have rough or fuzzy edges. The following hints may help you to make a proper line:

1. Use a lead of the proper hardness. A lead that is too hard will not make a dark line without a lot of extra work. A lead that is too soft causes fuzzy lines.
2. Draw with the automatic pencil perpendicular with the sheet.

3. When using an automatic pencil, you are not required to rotate it, although some drafters feel that rotation does help. See Figure 5–9.
4. Use enough pressure on the pencil. The amount of pressure depends on the individual. The suggested leads are only a recommendation. You may need to experiment with different lead hardnesses and pressures to find the right combination. Too much pressure can engrave the paper or break the point; not enough pressure may make fuzzy lines.
5. Some drafters may need to go over lines more than once to make them dark. Try to avoid this, as it slows you down and can cause an unwanted double line.

Polyester Lead Methods on Polyester Film

Some companies use polyester lead on film to help improve lines and lettering quality without using ink. Polyester lead is faster to use than ink and, in general, produces a better print than graphite on vellum. After working with graphite on vellum the use of polyester lead is similar to drawing with a crayon. Here are some basic techniques that may help:

1. Always draw on the mat side of the film.
2. Draw a single line in one direction. Retracing a line in both directions deposits a double line, which will smear and damage the mat.
3. Draft with a light touch. Drafting films require up to 40 percent less pressure than other media. Smearing and embossing can be reduced with less pressure.
4. Erase with a vinyl eraser. If an electric eraser is used, be very careful not to destroy the mat surface.

Inking Methods on Vellum or Polyester Film

When inking on either vellum or film, be sure to ink on the textured surface. On film the textured surface is the mat side. On vellum the inking surface has a water mark, printed label, or title block and border. Inking can be easy if you remember that ink is wet until it turns a dull color. If the ink is shiny, do not move your equipment over it. Follow the same recommended procedures as previously discussed. The following are some helpful hints to make inking easier.

1. When using technical pens, hold the pen perpendicular to the vellum or polyester film. Move the pen at a constant speed that is not too fast. Do not slow at the end of a line since this may cause widening of the line or a drop of ink to form at the end of a line. Do not apply any pressure to the pen. Allow the pen to flow easily. See Figure 5–10.
2. Care should be taken to provide a space between the instrument edge and the ink so it does not flow under the instrument and smear. Drafting machine scales

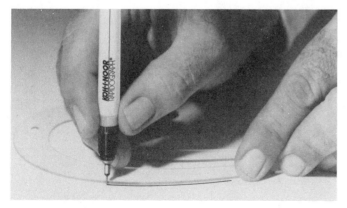

Figure 5–10   Inking with the technical pen. *Courtesy Koh-I-Noor, Inc.*

Figure 5–11   Templates with built-in risers. *Courtesy Chartpak.*

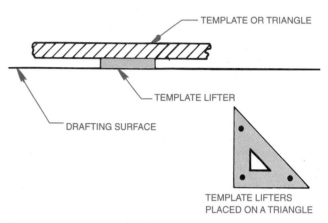

Figure 5–12   Templates lifters for inking.

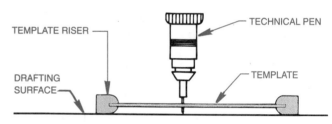

Figure 5–13   Template risers for inking.

have long been manufactured with edge relief for inking. Now templates, triangles, and other equipment are being made with ink risers built. See Figure 5–11. Other ways to keep instruments away from inked lines include the use of adhesive template lifters (Figure 5–12) or template risers, which are long plastic strips that fit on the template edges (Figure 5–13). If these devices are unavailable, it is possible to place a second template with a larger opening under the template being used.

3. Periodically check pens for leaks around the tip or a drop of ink at the end of the tip. Have a piece of tissue paper or cloth available to help keep the tip free of ink drops.
4. Keep technical pens clean and the reservoir between one-quarter and three-quarters full.
5. Shake technical fountain pens to get the ink started, but do not shake the pen over your drawing.

CADD Applications

DRAWING LINES WITH CADD

There are a variety of ways to draw lines using a CADD system. Drawing line segments is the simplest form of drafting on CADD. Each line is drawn between two points. This is referred to as *point entry*. The command name is LINE. Lines are drawn in CADD using one or a combination of the point entry methods.

The *Cartesian coordinate* point entry system, also referred to as *rectangular coordinates*, is based on selecting points that are a given distance from the horizontal and vertical axes, known, respectively, as the X, Y axes, the intersection of the axes is called the *origin*. This is where $X = 0$ and $Y = 0$, or 0, 0. Points are then identified by their distance from 0, 0. For example a point $X = 2$, $Y = 2$ is two units to the right and two units above the origin. This point is entered in CADD as 2, 2.

The rectangular coordinate system is divided into quadrants. Points in the upper right quadrant are X, Y. Points in the upper left quadrant are $-X$, Y. Points in the lower left quadrant are $-X$, $-Y$, and points in the lower right quadrant are X, $-Y$. Figure 5–14 shows the Cartesian coordinate system. Generally, the origin (0, 0) is in the lower left corner of the drawing limits (*limits* refers to the size of the CADD drawing area). You can relate this to sheet size. The point entry methods associated with the Cartesian coordinate system are absolute coordinate, relative coordinate, polar coordinate, or picking points with the screen cursor.

Points located using the *absolute coordinate* system are always measured from the origin (0, 0). To draw the lines shown in Figure 5–15 you enter these absolute coordinates:

○ First point: **2, 1**
○ Second point: **2, 4**
○ Third point: **4, 4**

Relative coordinates are always located from the previous point, or relative to the last point. To use this point entry method, you need to enter the @ symbol before entering the coordinates. The @ symbol is used in AutoCAD for this point entry system. Other CAD software may use other command entries. Check your users' guide for these applications. This tells the computer to place the next point a given distance from the previous point. To draw the lines shown in Figure 5–16, you enter these relative coordinates:

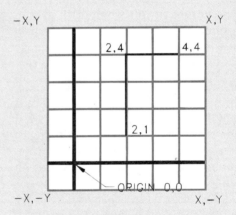

Figure 5–15 Points located using the absolute coordinate method.

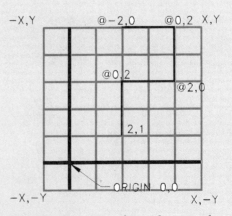

Figure 5–16 Points located using the relative coordinate method.

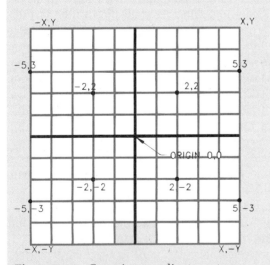

Figure 5–14 Cartesian coordinate system.

○ First point:
  2, 1
○ Second point:
  @0, 2
○ Third point:
  @2, 0
○ Fourth point:
  @0, 2
○ Fifth point:
  @ − 2, 0

Notice the fifth point. The point is @ − 2, 0 because it is negative two units along the *X* axis from the previous point.

Polar coordinates are located from the previous point using a distance and angle. The angular relationship of points from the origin in the Cartesian coordinate system is shown in Figure 5–17. Also shown is a point that is four units and 45° from the origin. When using the polar coordinate method, the symbol @ and < are used in AutoCAD. The @ symbol means from the previous point. The < symbol means from the previous point. The < symbol establishes an angular measurement to follow. Other command entries may be used with other CAD programs. Check your users' guide for these applications. To draw the lines shown in Figure 5–18, you enter these polar coordinates:

○ First point:
  2, 1
○ Second point:
  @2 < 90
○ Third point:
  @2 < 0
○ Fourth point:
  @2 < 90
○ Fifth point:
  @2 < 180

The screen cursor may be moved to any position on the screen and a point picked at that position. The CADD system has a screen grid system that lets you move the screen cursor accurately to any desired location. The grid is generally a pattern of equally spaced dots. In AutoCAD, the grid is set up using the GRID command. You determine the spacing based on the drawing.

For example, a floor plan may have the grid dots spaced 12″ apart. In AutoCAD the SNAP command forces the cursor to "snap" to these dots or any designated distance. The point coordinates are displayed in the upper right corner of the screen as you move the cursor to various positions on the screen. This is known as the *coordinate display* and constantly tells you where the cursor is located in relationship to the origin or to the previous point as desired. See Figure 5–19.

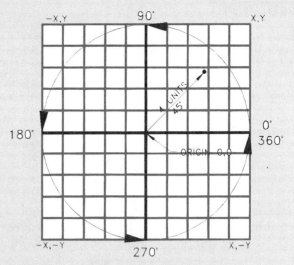

Figure 5–17 Angular relationship of points from the origin in the Cartesian coordinate system.

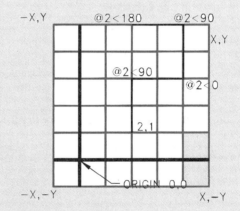

Figure 5–18 Points located using the polar method.

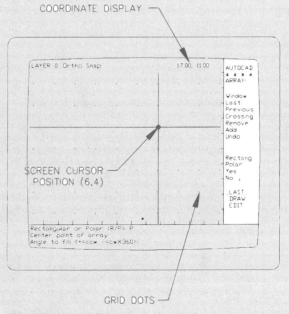

Figure 5–19 Picking points using the screen cursor.

▼ LETTERING

Information on drawings that cannot be represented graphically by lines may be presented by lettered dimensions, notes, and titles. It is extremely important that these lettered items be exact, reliable, and entirely legible in order for the user to have confidence in them and never have any hesitation as to their meaning. This is especially important when using reproduction techniques that require a drawing to be reduced in size such as with a photocopy or microfilm, or when secondary originals are made. *Secondary originals* are copies that may be used as originals to make changes and other copies. The quality of each generation of the secondary original is reduced, thus requiring the highest quality from the original drawing. Poor lettering will ruin an otherwise good drawing.

Rules for Metric Lettering

○ Unit names are given in lowercase—even those derived from proper names (e.g., millimeter, meter, kilogram, kelvin, newton, and pascal).
○ Use vertical text for unit symbols. Use lowercase text, such as mm (millimeter), m (meter), and kg (kilogram), unless the unit name is derived from a proper name, as in K (kelvin), N (newton), or Pa (pascal).
○ Use lowercase text for values less than 10. Use uppercase for values greater than 10.
○ Leave a space between a numeral and a symbol (e.g., 55 kg, 24 m, 38 °C).
○ Do not leave a space between a unit symbol and its prefix (e.g., kg, *not* k g).
○ Do not use plural unit symbols (e.g., 5 kg, *not* 5 kgs).
○ Use the plural of metric measurements (e.g., 125 meters).
○ Use *either* names or symbols for units; do not mix them. Symbols are preferred on drawings. Millimeters (mm) are assumed on architectural drawings unless otherwise specified. See Appendix A for other guidelines.

Single-stroke Gothic Lettering

The standard lettering that has been used for generations by mechanical drafters is vertical single stroke Gothic letters. The term *single stroke* comes from the fact that each letter is made up of single straight or curved line elements that make them easy to draw and clear to read. There are uppercase and lowercase Gothic letters, although the industry has traditionally used uppercase lettering. Figure 5–20 shows the recommended strokes that are used to form architectural letters and numbers.

Architectural designers and drafters do not use the same structured single-stroke Gothic lettering used by mechanical drafters. However, the architectural style is derived from the single-stroke concept. Architectural drafting is more individualized, and some people say it is more artistic than mechanical drafting. Architectural lettering styles range from standard Gothic forms to more avant-garde characters. Figure 5–21 shows a typical representation of architectural lettering.

As a beginning drafter, you would do well to be conservative. Too much emphasis on flair may cause unnatural lettering. One important point for lettering technique is to make letter forms consistent. Do not make a letter one way one time and another way the next time. Also, keep letters vertical and always use guidelines. Entry-level drafters should pay particular attention to the style used at the architectural firm where employed and match it. A professional drafter should be able to letter rapidly with clarity and neatness.

Lettering Size

Minimum lettering height should be ⅛″ (3 mm). Some companies use 5⁄32″ (4 mm). All dimension numerals, notes, and other lettered information should be the same height except for titles, drawing numbers, and other captions. Titles and subtitles, for example, may be 3⁄16 or ¼″ (4 or 6 mm) high. Verify the specific requirements with your instructor or employer.

Figure 5–20 Vertical straight element letters, curved element letters, and numerals and fractions.

A B C D E F G H I J K L M
N O P Q R S T U V W X Y Z
1 2 3 4 5 6 7 8 9 10

A B C D E F G H I J K L M
N O P Q R S T U V W X Y Z
1 2 3 4 5 6 7 8 9 0

FOUNDATION PLAN
SCALE: 1″ = 1′-0″

Figure 5–21 Architectural lettering examples.

Lettering Legibility

Lettering should be dark, crisp, and sharp for the best reproducibility. The composition of letters in words and the space between words in sentences should be such that the individual letters are uniformly spaced, with approximately equal background areas. To achieve such spacing usually requires that letters such as I, N, or S be spaced slightly farther apart than L, A, or W. A minimum recommended space between letters is approximately 1/16″ (1.5 mm). The space between words in a note or title should be about the same as the height of the letters. When a note is made up of more than one sentence, the space between separate notes should be twice the height of the letters.

When lettering notes, sentence, or dimensions that require more than one line, the space between the lines should be one-half to full height of the letters. Notes should be lettered horizontally on the sheet, or, in some cases where space dictates, notes may be lettered vertically so they read from the right side of the sheet. For some applications, some companies may prefer that lettering be done using a lettering template or guide.

Professional Lettering Hints

Vertical freehand lettering is the standard for architectural drafting. The ability to make good-quality lettering quickly is important. A common comment among employers hiring entry-level drafters is about their ability to do quality lettering and line work. The following suggestions may help establish good lettering skills:

1. Always use guidelines, which are lightly drawn horizontal or vertical lines spaced equal to the height of the letters. Guidelines should be so light that they do not reproduce.
2. Use a 0.5-mm automatic pencil with H, F, or HB lead. These pencils are usually easy to control for effective lettering and do not require sharpening.
3. Protect the drawing by resting your hand on a clean protective sheet placed over the drawing. This helps prevent smearing and smudging.
4. Lettering composition is spacing letters so the background areas look the same. This is done by eye, and there is no substitute for experience. In general, space vertical line letters farther apart than angled or curved line letters.
5. If your letters are wiggly or if you are nervous, try using a straightedge for vertical strokes as shown in Figure 5–22. Making each letter rapidly also helps some people. This tends to eliminate wiggly letters. Another option, if the problem continues, is to try using a softer lead and be sure that the lead does not extend too far out of the pencil tip.
6. Make sure your hand and arm are comfortable on the board.

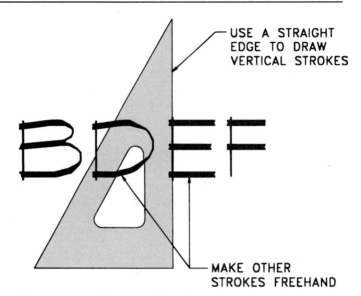

Figure 5–22 Using a straight edge to draw the vertical strokes of letters.

▼ MAKING GUIDELINES

A commonly used device for making guidelines is the Ames Lettering Guide. It is possible to draw guidelines and sloped lines for lettering from 1/16 to 2″ (1.5–51 mm) in height. Disk numbers from 10 to 2 give the height of letters in thirty-seconds of an inch. If 1/4″ high letters are required, simply rotate the disk so the 8 (8/32″ = 1/4″) is at the index mark on the bottom of the frame. See Figure 5–23. Instructions for use should be included when you purchase the guide. Metric guidelines may also be drawn.

Slanted or vertical guidelines can be drawn easily to help keep the letters vertical or slanted.

The numbers and set of six holes to the left of the disk relate to metric heights for guidelines. This column of six

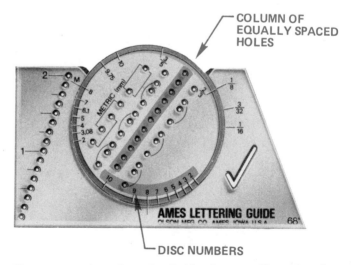

Figure 5–23 Ames Lettering Guide. *Courtesy Olson Manufacturing Co.*

holes offers the drafter the option of spacing guidelines equally (right brackets) or at half space (left brackets).

Other guideline lettering aids for equidistant spacing of lines have parallel slots ranging in width from ¹⁄₁₆ to ¼ in. (1.5 to 6 mm). These lettering guideline aids are not as complex as the Ames Lettering Guide, but they are not as versatile or as flexible for drawing guidelines.

▼ OTHER LETTERING STYLES

Some companies prefer *slanted lettering*. The general slant of these letters is 68°. Structural drafting is one field where slanted lettering may be commonly found. Figure 5–24 shows slanted uppercase letters.

ABCDEFGHIJKLM
NOPQRSTUVWXYZ
1234567890

Figure 5–24 Uppercase slanted letters and numbers.

▼ LETTERING GUIDE TEMPLATES

Although professional drafters are excellent at lettering, some companies prefer that drafters use lettering guides for uniformity. Standard lettering guide templates are available with vertical Gothic letters and numerals ranging in lettering height from ³⁄₃₂ to ³⁄₈″ (1–10 mm). Lettering guides are also available in many other lettering styles, including slanted Gothic, block, Futura, Old English, and Microfont letters in either uppercase or lowercase. Figure 5–25 shows a lettering guide in use.

▼ MECHANICAL LETTERING EQUIPMENT

Mechanical lettering equipment is available in kits with templates for letters and numerals in a wide range of sizes. A complete lettering equipment kit includes a

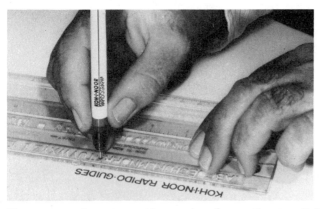

Figure 5–25 Lettering guide template in use. *Courtesy Koh-I-Noor, Inc.*

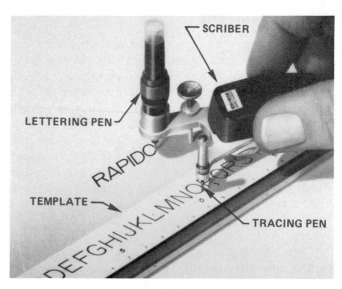

Figure 5–26 Components of a lettering equipment set. *Courtesy Koh-I-Noor, Inc.*

scriber plus templates, tracing pins and lettering pens. Figure 5–26 shows the component parts of a lettering equipment set. Mechanical lettering equipment sets generally contain instructions for use. You will need to practice to become good at using this equipment.

▼ MACHINE LETTERING

Lettering machines are available that produce a variety of fonts, styles, and letter sizes to prepare drawing titles, labels, or special headings. These types of lettering features are especially useful for making display letters, cover sheets, and drawing titles. Most machines use a typewriter keyboard for quick preparation of lettering. A personal computer may also interface with the lettering machine to increase speed and provide additional flexibility. Lettering machines prepare strips of lettering on clear adhesive-backed tape for placement on drawing originals. The tape is also available in a variety of colors for special displays and presentation drawings.

▼ TRANSFER LETTERING

A large variety of transfer lettering fonts, styles, and sizes are available on sheets. These transfer letters may be used in any combination to prepare drawing titles, labels, or special headings. They may be used to improve the quality of a presentation drawing or for titles on all drawings.

Transfer letters may be purchased as vinyl sheets; individual letters are removed from a sheet and placed on the drawing in the desired location. Transfer letters are also available on sheets; letters are placed on the drawing by rubbing with a burnishing tool (Figure 5–27). Rub-on letters, as they are often called, offer a high-quality lettering that is excellent for drawing titles and special displays.

Figure 5–27 Using rub-on letters. *Courtesy Chartpak.*

Special drawing transfers are also manufactured that can be used on presentation drawings. For example, a wide assortment of home furnishings or plant and office products may be purchased for use in preparing layout drawings. Other special items that are available are scales, call-outs, trees, landscaping items, representations of people, and transportation figures.

Drawing aids in the form of transfer tapes may be used to prepare borders, line drawings, or special symbols.

Some vendors prepare custom transfer templates for architectural customers. Such transfer templates may be used as standard drawing details.

CADD Applications

LETTERING WITH CADD

Lettering with a CADD system is one of the easiest tasks associated with computer drafting. It is just a matter of deciding on the style or font of lettering to use, and then locating the text where it is needed.

The CADD drafter often rejoices when the time comes to place text and notes on the drawing because no freehand lettering is involved. The computer places text of a consistent shape and size on a drawing in any number of styles, or fonts. The TEXT command is one of several that can be found in a section of a CADD menu labeled text, or text attributes. The drafter is also able to specify the height, width, and slant angle of characters (letters and numbers). Most systems maintain a certain size of text, called the *default size*, that is used if you do not specify one. The term *default* refers to any value that is maintained by the computer for a command or function that has variable parameters. The default text height may be ⅛" (3 mm), but you can change it.

You may need to scale your text height as related to the drawing scale. For example, text that is ⅛" high would be too small to see if you are working on a floor plan at ¼" = 1'–0" scale. This is because ⅛" at a ¼" = 1'–0" scale is very small. You need to size the text height in relation to the drawing scale. If you want the text on the final drawing to be ⅛" high, then you need to set the text height at 6" at a ¼" = 1'–0" scale. Six inches at this scale is actually ⅛" To do this properly, you need to determine the drawing scale factor. The *scale factor* establishes the relationship between the drawing scale and the desired text height. This should be determined when setting up the drawing as part of a prototype drawing. A *prototype drawing* is the basis for starting a drawing. It contains all of the standard elements that you need in the drawing format—for example, a border, title block, text style and scale factor, and drawing default values. The architectural drawing to be plotted at a ¼" = 1'–0" scale has a scale factor of 48, calculated as follows:

$$\frac{1}{4}'' = 1'-0'' \ (\tfrac{1}{4}'' = .25'')$$
$$.25'' = 12''$$
$$^{12}/_{.25} = 48$$

If your drawing is in millimeters where the scale is 1:1, the drawing scale factor may be converted to inches with the formula 1" = 25.4 mm. Therefore, the scale factor is 25.4. When the metric drawing is 1:2, the scale factor is 2 × 25.4 = 50.8. For floor-plan drawings scaled at 1:50 metric, the scale factor is 50 × 25.4 = 1270.

Many CADD systems possess a variety of lettering styles, or fonts. The drafter can select the style to use simply by pressing a menu-pad command or symbol, or by typing a command at the keyboard. Figure 5–28(a) shows some of the styles and sizes of characters that can be used in CADD. The size and styles of characters used is dictated by the nature of the drawing. Some CADD lettering font styles include symbols such as the mapping symbols shown in Figure 5–28(b).

The process of locating text has not changed. You still need to decide where to locate dimensions and notes, but with CADD the process of placing notes is a bit more technical. Most CADD systems provide for keyboard entry of location and size coordinates, thus allowing you to locate the text accurately. But using the keyboard to input text location coordinates is a tedious process. Often key strokes are combined with digitizer stylus commands to generate text.

How is text located? By pointing to one of several places on the lettering. Figure 5–29 shows an example of

ABCDEFGHIJKLMNOPQRSTUVWXYZ
1234567890

ABCDEFGHIJKLMNOPQRSTUVWXYZ
1234567890

𝔄𝔅ℭ𝔇𝔈𝔉𝔊𝔥𝔍𝔍𝔎𝔏𝔐𝔑𝔒𝔓𝔔𝔕𝔖𝔗𝔘𝔙𝔚
𝔛𝔜𝔷   1234567890

ABCDEFGHIJKLMNOPQRSTUVW
XYZ   1234567890

(a)

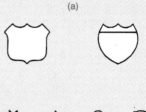

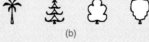

(b)

Figure 5–28 (a) Some text character styles and sizes that can be used with CADD; (b) special CADD mapping symbols.

+ TEXT INSERTION POINT

Figure 5–29 Some points on text that can be used for location purposes.

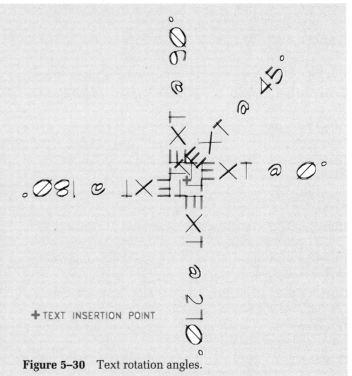

+ TEXT INSERTION POINT

Figure 5–30 Text rotation angles.

several points on the text that can be used for location purposes. Not all CADD systems use all the points shown, but most systems have a command known as TEXT that is used for placing written information on the drawing. Some systems may allow you to locate text between two points. The computer calculates the size of each letter to enable the text to fit in the desired space.

The first decision regarding text is to determine its height, width, and slant angle. Most CADD systems maintain a default text size that is used by the computer if the operator forgets or decides not to change it. The rotation angle direction, or text path is also determined by the drafter. The rotation angle is the angle from horizontal that the text is on. An example of rotation angle is shown in Figure 5–30. Text located on a horizontal line has a direction of 0°, and text that reads from the bottom up vertically has a rotation angle of 90°.

CHAPTER

5 Architectural Lines and Lettering Test

DIRECTIONS

Answer the questions with short complete statements or drawings as needed on 8½ × 11 notebook paper as follows:

1. Letter your name, Chapter 5 Test, and the date at the top of the sheet.
2. Letter the question number and provide the answer. You do not need to write out the question.

Answers may be prepared on a word processor if appropriate with course guidelines.

QUESTIONS

Question 5–1 Is there any recommended variation in line darkness, or are all properly drawn lines the same darkness?

Question 5–2 What are construction lines used for, and how should they be drawn?

Question 5–3 Discuss line uniformity and line contrast.

Question 5–4 Define guidelines.

Question 5–5 What is the recommended thickness of outlines?

Question 5–6 Identify two items that dashed lines represent on a drawing.

Question 5–7 Describe a situation where an extension line is the centerline of a feature.

Question 5–8 Extension lines are thin lines that are used for what purpose?

Question 5–9 Where should extension lines begin in relationship to the object and end in relationship to the last dimension line?

Question 5–10 Describe leaders.

Question 5–11 Describe and show an example of three methods used to terminate dimension lines at extension lines.

Question 5–12 How does architectural line work differ from mechanical drafting line work?

Question 5–13 What is the advantage of drawing certain outlines thicker than other lines on the drawing?

Question 5–14 If all lines on a drawing are the same thickness, how can walls and partitions, for example, be represented so that they stand out clearly on a floor plan?

Question 5–15 Describe the proper technique to use when drawing lines with a technical pen.

Question 5–16 How does architectural lettering compare to vertical single-stroke freehand Gothic lettering?

Question 5–17 What are the minimum recommended lettering heights?

Question 5–18 How should the letters within words be spaced?

Question 5–19 What is the recommended space between words in a note?

Question 5–20 What is the recommended horizontal space between lines?

Question 5–21 When should guidelines be use for lettering on a drawing?

Question 5–22 Why are guidelines necessary for freehand lettering?

Question 5–23 Describe the recommended leads and points used in automatic pencils for freehand lettering.

Question 5–24 Identify a method to use to help avoid smudging the drawing when lettering.

Question 5–25 When lettering fractions, what is the recommended relationship of the fraction division line?

Question 5–26 List two manual methods that can be used to make guidelines rapidly.

Question 5–27 Why should the mechanical lettering template be placed along a straightedge when lettering?

Question 5–28 Identify an advantage for using lettering guides.

Question 5–29 Describe a use for lettering machines.

Question 5–30 Describe four uses for transfer materials.

DIRECTIONS

1. Read all instructions carefully before you begin.
2. Use an 8½ × 11 vellum or bond paper drawing sheet for each drawing or lettering exercise.
3. Use the architect's scale of ¼″ = 1″–0″ for each drawing.
4. Draw the floor plan described in Problem 5–1 twice. On the first drawing, use thick lines for the walls and thin lines for extension, dimension, and symbol lines. On the second drawing, make all lines the same thickness. Each drawing will represent a line thickness technique discussed in this chapter.
5. Use guidelines for all lettering. Using architectural lettering, letter the title FLOOR PLAN, centered below the drawing in ¼″ high letters. In ⅛″ high letters and centered below the title, letter the following: SCALE: ¼″ = 1′–0″. Letter your name and all other

notes and dimensions using ⅛″ high architectural lettering.

6. Lightly lay out the drawing using construction lines. When you are satisfied with the layout, darken all lines.
7. For the lettering exercise, use ⅛″ guidelines with ⅛″ space between the lines.
8. Leave a ¾″ margin around the lettering exercise sheet.
9. Make a diazo print or photocopy of your original drawings or lettering exercise sheet as specified by your instructor.
10. Submit your copies and originals for grade evaluation.

PROBLEMS

Problem 5–1 Lines and lettering. Given the following information, draw the garage floor plan:
  1. Overall dimensions are 24′–0″ × 24′–0″. Dimensions are measured to the outside of walls.
  2. Make all walls 4″ thick.
  3. Center a 16′–0″ wide garage door in the front wall.
  4. Using dashed lines, draw a 4 × 12 header over the garage door and label it, 4 × 12 HEADER.
  5. Label the garage floor as 4″ CONC OVER 4″ GRAVEL FILL.

Problem 5–2 Lettering practice. Using uppercase architectural lettering with proper guidelines, letter the following statement as instructed:

YOUR NAME

MOST ARCHITECTURAL DRAWINGS THAT ARE NOT MADE WITH CADD EQUIPMENT ARE LETTERED USING VERTICAL FREEHAND LETTERING. THE QUALITY OF THE FREEHAND LETTERING GREATLY AFFECTS THE APPEARANCE OF THE ENTIRE DRAWING. MANY ARCHITECTURAL DRAFTERS LETTER WITH PENCIL ON VELLUM OR POLYESTER LEAD ON POLYESTER FILM. LETTERING IS COMMONLY DONE WITH A SOFT, SLIGHTLY ROUNDED LEAD IN A MECHANICAL PENCIL OR A .5-MILLIMETER LEAD IN AN AUTOMATIC PENCIL. LETTERS ARE MADE BETWEEN VERY LIGHTLY DRAWN GUIDELINES. GUIDELINES ARE PARALLEL AND SPACED AT A DISTANCE EQUAL TO THE HEIGHT OF THE LETTERS. GUIDELINES ARE REQUIRED TO HELP KEEP ALL LETTERS THE SAME UNIFORM HEIGHT. THE SPACE BETWEEN LINES OF LETTERING MAY BE BETWEEN ½ TO EQUAL THE HEIGHT OF THE LETTERS. ALWAYS USE GUIDELINES. WHEN LETTERING FREEHAND, LEARN TO RELAX SO THAT THE STROKES FOR EACH LETTER FLOW SMOOTHLY. WHEN YOUR HAND BECOMES TIRED, REST FOR A WHILE BEFORE BEGINNING AGAIN.

Problem 5–3 Given the partial floor plan shown for this problem, identify the line types labeled *a* through *k*.

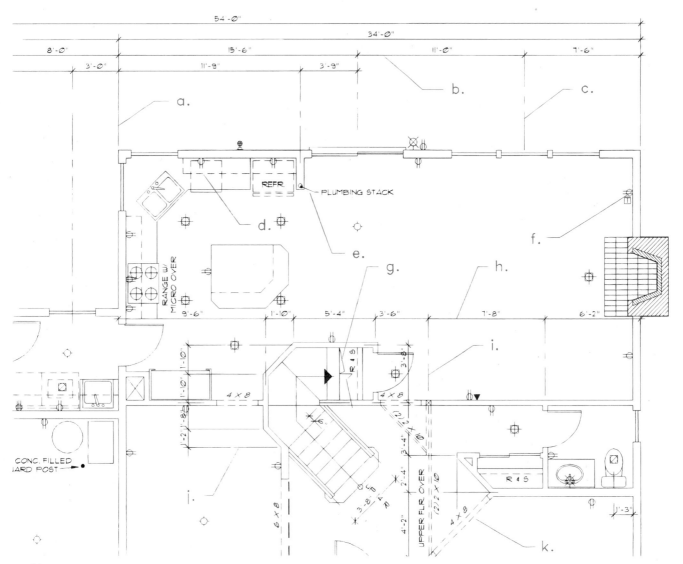

Problem 5–3

CHAPTER 6

Computer-aided Design and Drafting in Architecture

In Chapter 2 you learned about the traditional manual drafting workstation with modern conveniences such as an adjustable drafting table, and a drafting machine or parallel bar. Also available to enhance your speed and accuracy are automatic pencils, electric erasers, and templates of all kinds. As a professional drafter, designer, or an architect, you may continue to see the traditional equipment in many offices. However, drafting is changing. The concepts and theories are the same, but the tools are changing. Recently a panel of experts acknowledged that traditional manual drafting, in general, will convert to computer-aided design and drafting (CADD) within the next few years. This is not to say that the traditional manual drafting workstation that you are familiar with will not be around. However, CADD is now available in most drafting schools and is being used in many architectural firms.

▼ COMPUTER-AIDED DESIGN AND DRAFTING TERMINOLOGY

Attribute: Text or numeric information attached to a symbol or entity on a computer-aided drawing.

Background drawing: A base drawing in the CADD layers.

Bit: Smallest amount of information a computer can read—either 0 or 1.

Byte: Eight bits that make up letters or numbers.

CADD: Computer-aided design and drafting. CAD is often referred to as computer-aided design or computer-aided drafting.

Commands: Operator-supplied information that results in a task performed by the computer.

Compatible: Two or more computers or software can work together. The term *IBM compatible* means that non-IBM computers work just as IBM computers do.

Cursor: The cross, space, box, or other image on the screen that moves when the user moves the pointing device, such as a mouse, puck, or stylus.

Customize: To alter the software program to perform specific tasks. The standard AutoCAD program, for example, may be customized to have special screen, pull-down,

puck, or tablet menus that may be customized for drafting special applications or symbols.

Data: A collection of computer information.

Database: A collection of data stored in computer form.

Default value: One or more values established by the program. Alternates must be selected, or default values take precedence. For example, the default value for character height may be 0.125″ The computer will always output 0.125″ lettering unless the operator changes the size.

Digitizer tablet: An input device that converts graphic data (points) into X, Y coordinates for the computer to use. The tablet usually is equipped with an electronic cursor (sometimes called a puck) or a stylus.

Directory: Used to make hard or floppy disks work like a file cabinet, the main directory might be a design and drafting program such as AutoCAD; the subdirectories are used to store drawings such as ARCH (architectural), ELEC (electrical), or PLUMB (plumbing).

Disk drive: Used to write data to, and read from, a floppy disk.

Display: Information placed on the monitor or screen.

DOS: Acronym for disk operating system, generally meaning the operating system for IBM PCs and IBB compatibles.

Edit: To change an existing or new drawing.

Entity: A geometric element or single item of data on a computer-aided drawing, for example, a line, circle, or text.

File: A group of computer information. Individual CADD drawings are referred to as drawing files.

File server: The computer used for storing and transmitting information in a network system.

Floppy disk: A thin disk coated with magnetic material used to store data. Common floppy disk sizes are 3½″ and 5¼″ The 3½-in. disks are becoming more popular.

Hard copy: A drawing produced on paper or film.

Hard drive: The part of the computer where information is stored. Also known as a hard disk.

Hardware: The workstation components, such as the computer, monitor, digitizer, keyboard, and plotter.

Input: Information given from the operator to the computer.

Joystick: A control, similar to that used on video games, that, when shifted, moves a cross-hair target on the screen. Used for digitizing information. Another type on some units is two thumbwheels, which, when moved, control the *X* and *Y* axes of the cross hairs on the screen.

Laser: The acronym for light amplification from the stimulated emission of radiation. A laser light beam is focused very tightly and used for a variety of commercial purposes, such as laser printers.

Layers: In CADD, details of a design or different drafting information might be separated on layers, that is, one type of information over the other. Layers are generally different colors and have their own names for clarity. Layers may be kept together, or individual layers may be turned off or on as needed.

LISP: A programming language that means list programming. LISP is used to customize design and drafting software such as AutoCAD.

Load: To move a program from storage to the computer.

Memory: There are two kinds of memory. (1) Random access memory (RAM): In the computer, a program goes from the disk to the RAM, which is made up of computer chips. Information is held in the RAM until stored on the disk or power is removed. (2) Read-only memory (ROM): A computer chip with fixed data that are not lost when power is removed.

Menu: A group of commands that are related in some way and allow the operator to order items needed to make a drawing.

Modem: A device that lets computers and other electronic equipment communicate through telephone lines.

Mouse: A pointing device with a roller ball underneath that moves the screen cursor when the mouse is moved on a flat surface. The mouse also has a button or buttons for issuing commands or for picking items.

Network: A connection system that allows computers to communicate electronically with each other or with other peripherals.

Operating system: The software that controls the basic operation of the computer. Often referred to as OS.

Output: Information given from the computer to the operator.

PC: Acronym for personal computer; the trademark name of the first IBM personal computer.

Peripheral: Devices that are outside the computer, such as the keyboard, monitor, digitizer, and plotter.

Program: The system that runs the computer and a set of instructions.

Puck: A pointing device that works with a digitizer and normally has multiple buttons used to issue commands or pick items from the tablet menu. Movement of the puck on the digitizer moves the screen cursor correspondingly.

Reference file: A drawing file that can be displayed as a background but cannot be edited.

Relocate: A command that allows the operator to change the origin of a drawing or portion of a drawing.

Soft copy: The image on the screen or monitor.

Software: The programs or instructions that run the computer.

Stylus: A pointing device that works with a digitizer to pick points or items from the tablet menu. Movement of the puck on the digitizer moves the screen cursor correspondingly.

Symbol: A collection of CADD entities stored under a single name and available for use on other drawings. Referred to as a *block* or *wblock* in AutoCAD.

Text: Letters and numbers in a computer file.

Windows: A graphic system developed by Microsoft Corporation and similar to the Macintosh (by Apple), with symbols used on the screen to issue commands rather than typing everything at the keyboard. This is an easy system to use and allows the user to integrate several activities at the same time (called *multitasking*).

Other more specific terminology is defined throughout this textbook where related to the application.

▼ ARCHITECTURAL CADD SYMBOLS

Many companies have a CADD systems manager who is responsible for setting up the architectural CADD system and standards. This person and the staff often create custom architectural CADD menus and symbol libraries. The systems generally include custom screen, cursor, pulldown, icon, and tablet menus. Every time a new symbol or drawing detail is made, it can be saved for future use. In this way, it is never necessary to draw anything more than once with CADD. For example, if a designer or drafter draws a structural detail, as shown in Figure 6–1, this detail is saved for future use. A copy of the

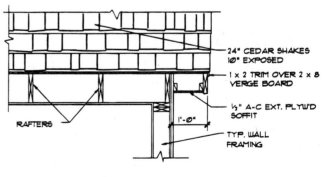

TYPICAL RAKE DETAIL

SCALE :   3/4" • 1' - 0"

Figure 6–1 Drawing a structural detail on CADD and saving it for future use. *Courtesy Piercy & Barclay Designers, Inc. CPBD.*

symbol or detail is kept in a symbols manual, and information about it is recorded.

Although every company saves symbols for future use, many companies purchase their core architectural program from a software developer. These are complete architectural CADD programs that are ready to use out of the box. Some of these packages provide commands for the design and layout of walls, symbols, stairs, roofs, structural grids, and plot plan symbols. Figure 6–2 shows a sample of pull-down, icon, and tablet menus available from an architectural software developer. Drawing can be created in two or three dimensions and switch back nd forth at any time. They may also provide the ability for you to render and animate three-dimensional models. These packages often have standard layer systems based on the American Institute of Architects (AIA) format.

Drawing Architectural Symbols

Symbols are drawn on plans by placing the puck or stylus pick on the table menu box, icon menu, or pull-down menu that contains the desired symbol. For example, if an exterior door is to be drawn, the door location is digitized, and the direction of rotation is specified. To do this, pick the symbol from the appropriate menu, and the symbol appears on the monitor, where you drag it into place and pick the insertion point. Each symbol has an established insertion point, as shown in Figure 6–3. Figure 6–4 shows an architectural floor plan with its many lines, symbols, and text drawn using CADD.

Creating Your Own CADD Symbols

You can create your own CADD symbols if a customized architectural menu is not available. The symbols that you make are called *blocks* or *wblocks* in AutoCAD. You can design an entire group of symbols for inserting into architectural drawings. As you create these symbols, prepare a symbols library or catalog, where each symbol is represented for reference. The symbol catalog should include the symbol name, insertion point, and other important information about the symbol.

▼ DRAWING ON CADD LAYERS

Layers have been used in architectural drafting for some time. This process is commonly referred to as *overlay* or *pin register drafting*. Drawings are made in layers, each perfectly aligned with the others. Each layer contains its own independent information. The layer may be reproduced individually, in combination, or together as one composite drawing. Figure 6–5 shows how layers may be used to share information among drawings.

Some architectural CADD systems automatically set drawing elements on separate layers; others require that you create your own layering system. These layers are part of the drawing and may be turned on or off as needed. In addition to the layer identification system, many companies set up other layer systems based on their own project needs. For example, layers may be identified with a layer identification system where the elements are divided into three groups: the major group, the minor group, and the basic group. The major group contains the type of drawing, such as Floor Plan. The minor group identifies the elements on the drawing, such as Electrical. The basic group labels the type of information, such as Lines or Text. An example of this system of layer identification might be: FL1ELECL, which means first floor (FL1), electrical (ELEC), and lines (L). This is a typical layout of the three groups:

Major Group	Minor Group	Basic Group
BSMT	HVAC	LINE
FL1	ELEC	TXT
FL2	PLUMB	DIM
GAR	WALLS	INSERT
SITE	DOORS	HATCH
SCHED	WINDOWS	
NOTES	APPL	
FDN		
ELEV		
ROOF		
REFLECT		

Layer names should be kept short, such as FDN for "foundation." Layer identification may also contain numbers specifying pen width used to produce the plot. For example, 1 = 0.25 mm, 2 = 0.35 mm, 3 = 0.5 mm, and 4 = 0.7 mm pen width.

The AIA recommends the use of one of two different layering systems: the *long format*, with layer names that contain six to sixteen characters, and the short format, with three to eight characters per name. Refer to the AIA standard, CAD Layer Guidelines, for complete information. The AIA recommended layering system has a major group and a minor group of information in the layer names. The major groups are identified by a letter and describe the project element, as follows:

A   Architecture, interiors, and facilities

S   Structural

M   Mechanical

P   Plumbing

F   Fire protection

E   Electrical

C   Civil engineering and site work

L   Landscape architecture

The minor groups reduce the layer names to type of information found in the major group, such as doors, windows, walls, ceilings, furniture, or equipment. Modifiers may also be added to the layer name for further definition. Layers might have modifiers such as NOTES, SYM (for "symbols"), DIMS (for "dimensions"), or TITLB

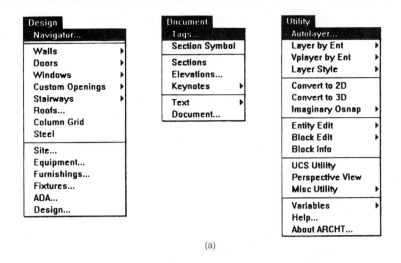

(a)

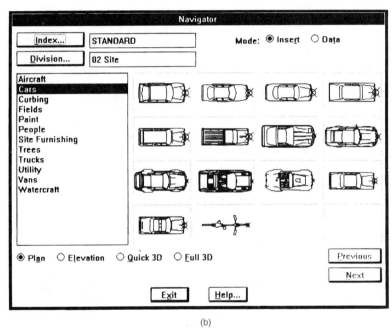

(b)

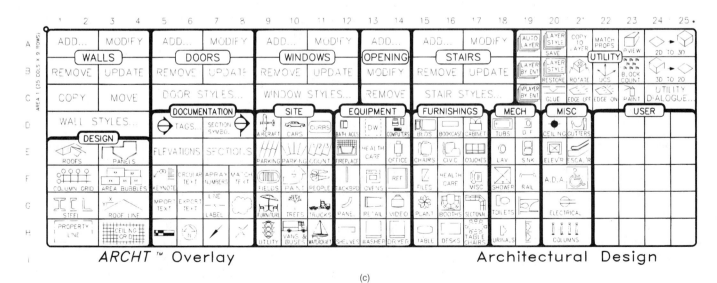

(c)

Figure 6–2 (a) Sample pull-down menu; (b) Sample icon menu; (c) sample tablet menu. *Courtesy k'ETIV Technologies, Inc.*

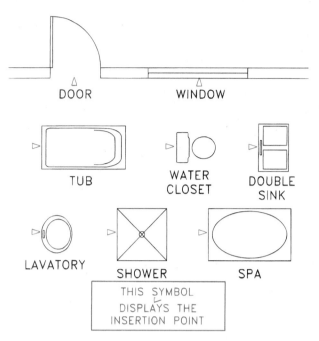

Figure 6–3 Insertion points for several architectural symbols.

(for "title block information"). The long format has the major group followed by a dash, the minor group followed by a dash, and the modifier is used. The long format also provides for user-defined characters that are entered after the modifier. The user-defined entries are for special project requirements. An example of the long format layer name is shown in Figure 6–6.

The *short format*, recommended by the AIA, combines the elements of the long format into abbreviated layer names, and the dashes are removed. The formula for this abbreviation is based on using the major group code followed by the first two letters of the minor group, and the first two letters of the modifier if used, as shown in Figure 6–7.

The complete layer naming system recommended by the AIA is displayed in Appendix B.

A big advantage of CADD is the ability to send drawing files on disk to different subcontractors involved in a project, an important capability in commercial applications, because the architect often creates the building design and has others design the electrical, plumbing, and mechanical systems. (The mechanical system encompasses heating, ventilating, and air-conditioning.) To make this system work effectively, the architect prepares a base drawing, which represents the floor plan and its bearing walls and partitions (Figure 6–8). A disk copy, modem, or network communication of this drawing is sent to the interior design department to create the layers containing the nonbearing facility layout, which includes the room layout with fixtures and appliances (Figure 6–9). The CADD drawing files then go to the plumbing, electrical, and HVAC contractors for design of their related elements. Figure 6–10 shows the plumbing layer

for the previous building. A composite drawing of all layers is displayed in Figure 6–11. Each layer is generally set up in its own color on the computer screen. However, the drawing in Figure 6–10 is the final plot of the layers using black ink on vellum.

CADD layers provide these advantages:

○ Reduced drafting time.
○ Improved project coordination by sharing information that is common to different elements of the project or common to several floors in a multistory building.
○ Ease of preparing design alternatives, which can share the same drawing by turning on or off as needed for display.

▼ MICROCOMPUTERS

The movement in CADD is toward microcomputers. These are desktop models that perform a multitude of tasks at a low installation cost. A complete workstation can be purchased for an amount between $5,000 and $15,000 depending on its capabilities. Some of the lower-cost units are much slower than the more expensive models. The availability of software, or programs, to operate microcomputers has increased greatly. Inexpensive software is available that performs all types of drafting tasks. The capabilities of microcomputers are improving as memory storage is expanded. Microcomputers are in virtually every business and industry. Placed on individual desktops, they can also be connected to interact between stations. Figure 6–12 shows an example of a typical microcomputer workstation.

▼ PRODUCTIVITY

There is agreement among CADD users that a drafter's productivity is increased with CADD over traditional manual drafting methods. While estimates range from a three-times productivity increase to a two-times increase, the reality is that any increase in productivity depends on the task, the system, and how quickly employees learn to use CADD. For many duties, CADD easily multiplies productivity several times. It is especially true that the computer performs redundant and time-consuming tasks much faster and more accurately than can be done by manual techniques. A great advantage of CADD is the time needed by the designer and drafter for creativity has increased while their time spent on the actual preparation of drawings has been reduced. Some of the business tasks related to architectural design and drafting, such as construction cost analysis, specifications writing, computations, materials inventory, time scheduling, and information storage, are also done better and faster on a computer.

With some units, a designer can look at several design alternatives at one time. The drafter draws in layers where one layer may be the plan perimeter, the next layer fixtures, the next electrical, and another layer dimensions

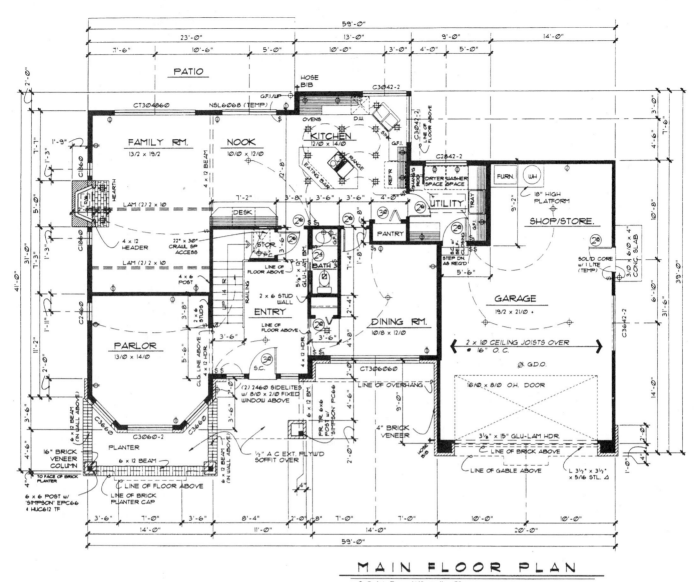

Figure 6–4   Architectural floor plan with its many lines, symbols, and text drawn using CADD. *Courtesy Piercy & Barclay Designers, Inc. CPBD.*

and notes. Each layer appears in a different color for easy comparison. Productivity is also related directly to the amount of time a company has had CADD in use, coupled with employee acceptance and experience. Most companies can expect more productivity after the users become comfortable with the capabilities of the equipment.

▼ DESIGNING

Architectural designers continue to make sketches even as the computer waits for input. The old methods of making circles to determine preliminary room arrangement continues to be with us. Designers still form small renderings in an effort to establish design ideas. The computer can help in the design process, but individual creativity may often take place before the computer is turned on.

There are a number of design functions that the CADD system can perform in a manner that often exceeds many expectations. Plan components, such as room arrangements, can be stored in a design file, which allows the designer to call on a series of these components to rearrange in a new design. Another factor that aids design is a given plan can be quickly reduced or enlarged in size while all components maintain their proper proportion. A particular room can be changed while the balance of the design remains the same. The design capabilities of drawing layers allows the architect to prepare a preliminary layout on one layer while changing and rearranging components on another layer.

Not only is creativity enhanced with CADD, but the repetitive aspects of design are handled more effectively. For example, in the design of an apartment or condominium

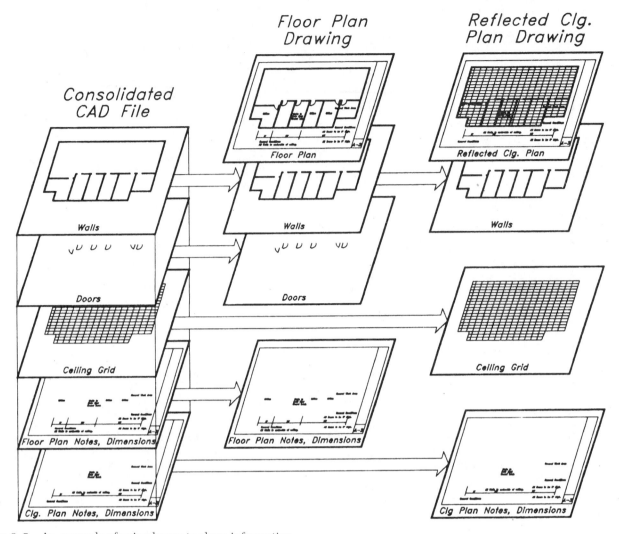

Figure 6–5   An example of using layers to share information.

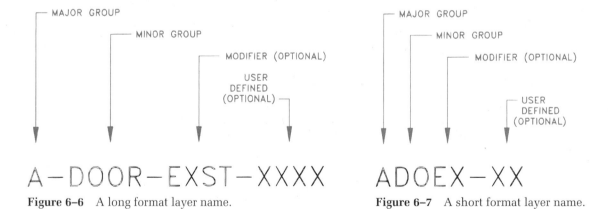

Figure 6–6   A long format layer name.

Figure 6–7   A short format layer name.

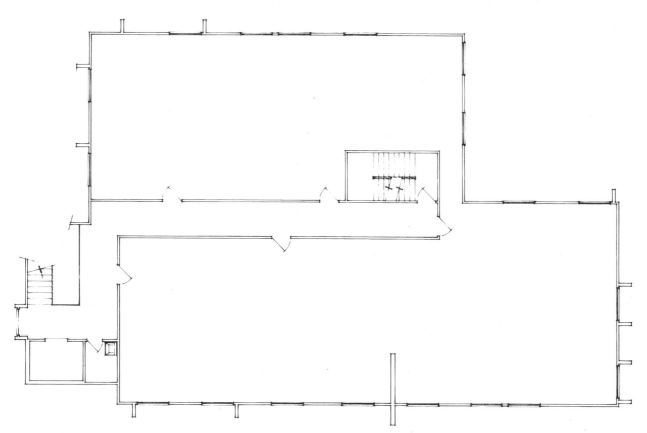

Figure 6–8   Exterior and support structure base layer sheet is unchanged through the project. *Courtesy Structureform.*

complex, an initial unit can be designed, and then any number of the same units can be attached to one another in any manner very quickly. Alternate designs can also be made. Some units can be expanded or reduced in size, or bedroom and bath alternatives can quickly be implemented.

▼ ERGONOMICS

Ergonomics is the study of a worker's relationship to physical and psychological environments. As drafters begin to make a transition from traditional drafting methods to CADD, there is concern about the effect of this new working environment on the individual. Some studies have shown that workers should not work continuously at a computer workstation for more than about four hours without a break. The various problems that are addressed by computer manufacturers should also be considered when purchasing a system.

Keyboard design in old mechanical typewriters was based on the importance of position. Keys were placed so they would not jam when used. Keyboard design can now be rearranged as electronics eliminates the old problem of jamming. Keys are arranged to help facilitate easier use. Also, a good keyboard has key tops that are dished to help reduce finger slippage. For some people, keyboards should either be adjustable or separate from the computer to provide more flexibility. The keyboard should be positioned, and arm or wrist supports may be considered to help reduce elbow and wrist tension. Also, when keys are depressed, a slight sound should be heard to give assurance that the key has made contact.

The monitor is one of the biggest concerns. The operator should be able to see images clearly without glare. One primary problem has been eyestrain and headache with extended use. If the position of the monitor is adjustable, the operator can tilt or turn the screen to help reduce glare from overhead or adjacent lighting. Some users have found that a small amount of background light is helpful. Monitor manufacturers have large, flat, nonglare screens available that help reduce eyestrain. Some CADD users have suggested changing screen background and text colors weekly to give variety and help reduce eyestrain.

The chair should be designed for easy adjustments to allow each operator to have optimum comfort. The chair should be comfortably padded. The user's back should be straight, feet should be flat on the floor, and the elbow-to-hand movement should be horizontal when working the keyboard or digitizer. The digitizer should be in close proximity to the monitor so movement is not strained and equipment use is flexible. The operator should not have to move a great deal to look directly over the cursor to activate commands.

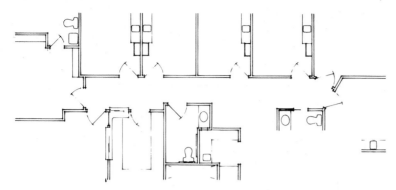

Figure 6–9 Interior design of nonbearing partitions may be changed as desired using overlays. *Courtesy Structureform.*

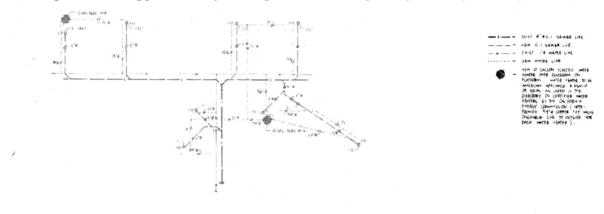

PLUMBING

Figure 6–10 Plumbing layer. *Courtesy Structureform.*

The plotter creates some noise and is best located in a separate room next to the workstation. Some companies put the potter in a central room, with small office workstations around the plotter. Others prefer to have plotters near the individual workstations that may be surrounded by acoustical partition walls or partial walls.

The working environment may be different from traditional drafting rooms in that air-conditioning and ventilation should be designed to accommodate the computers and equipment. Carpets should be antistatic. Noise should be kept to a minimum. If smoking is permitted, the room should have special ventilation. Smoke or other contaminants can affect computer equipment.

Drafters who are required to do tedious, repetitive tasks on the computer could develop symptoms of fatigue. Consideration should be given to allowing CADD employees to do a variety of jobs or take periodic breaks.

▼ THE CADD WORKSTATION

The CADD workstation is different from the traditional station because the tools have changed. For example, the new table is flat and contains a computer, monitor, keyboard, digitizer, and plotter on or available to the workstation.

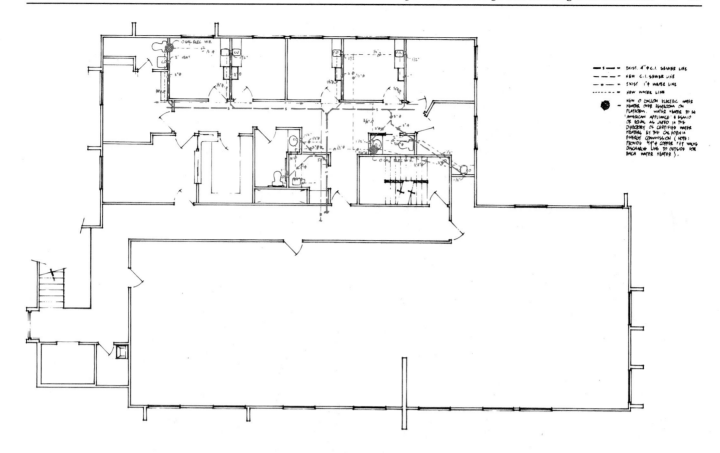

PLUMBING

Figure 6–11   Composite of base layer, interior design layer, and overlays. *Courtesy Structureform.*

Figure 6–12   Microcomputer graphics workstation. *Courtesy Applicon.*

The Computer

There are two basic computer-aided workstations. Some companies may have a large, centrally located computer that is generally in a special room or location. This *network computer*, as it is called, can serve several independent workstations. The network is a connection system that lets computers communicate electronically with each other, or to printers, plotters, and other devices. The network may be connected by way of a modem through telephone lines to computers at other companies or for a variety of services. The network may be set up to operate each workstation fully, or the network can support workstations. When computers are connected in an office environment, they are referred to as a *local area network* (LAN).

It is common for the individual workstations to be able to operate on a stand-alone basis. This means the workstation computer and its software may be used independent of the network computer. In this application, the network computer is often used, in part, to store drawing files and to access software programs. This is an advantage, because if there is a breakdown in the network computer, known as downtime, the workstation can continue to operate on its own without loss of valuable time. Downtime is also necessary periodically for repair and maintenance.

Whichever way the company selects to use CADD, the result is the same, and the computer is the core of the CADD workstation. Figure 6–13 shows a CADD workstation. The monitor is similar to a TV screen, and resting below or

Figure 6–13 Computer graphics workstation. *Courtesy Cal-Comp.*

attached to it is the keyboard. The monitor displays alphanumeric (characters and numbers) information from input given at the keyboard. *Input* is any data or information that the operator puts into the computer. The monitor also displays graphics (lines, symbols, or pictures) as input from the keyboard or digitizer.

The Digitizer

The *digitizer* is an electronically sensitized drafting board. The digitizer may be adjacent to the monitor or built into the unit. Digitizers may be equipped with a penlike instrument called a stylus, or a hand-held device called a cursor that contains a cross-hair lens and command buttons. Input is displayed on the monitor from the keyboard in the form of alphanumeric information or as graphics by providing the coordinates of the ends of a line or other geometric shape. The digitizer, however, allows the operator to perform the graphics tasks more rapidly and with greater ease. Provided with a list of symbols, lines, or words known as a *menu*, the operator can instantly input information on the monitor. The related group of commands is called a menu because each symbol or element can be ordered for immediate display with a touch of the stylus or cursor. Figure 6–14 shows a digitizer used to establish coordinate points for drawing lines.

Another method used to input graphics is to place a sketch over the digitizer surface or to work from a sketch or series of commands next to the digitizer. Here the stylus or cursor is used to establish coordinate points or geometric shapes by a touch on the active surface of the digitizer. The position of the command is transmitted

from the electronic sensing device inside the digitizer to the monitor. For example, press the stylus at the beginning and end points of a line, and the coordinates of the line ends are transmitted to the computer, and a line is shown as output. *Output* is any information the computer has to show or tell the operator. A CADD operator can digitize an entire drawing in this manner.

The Plotter and Printers

Once a drawing has been put into the computer and a desired monitor image has been established, then an actual drawing is needed. The *plotter* is the mechanism that makes the drawing. There are two basic types of pen plotters: flatbed (Figure 6–15) and drum (Figure 6–16). *Pen plotters* use liquid ink or felt or roller-tip pens to reproduce the computer image on plotter bond paper,

Figure 6–14 Digitizer used to establish coordinate points for drawing lines. *Courtesy Calcomp.*

Figure 6–15 Flatbed plotter. *Courtesy Calcomp.*

Figure 6–16 Drum plotter. *Courtesy Hewlett-Packard Company.*

vellum, or polyester film. The liquid ink pens, similar in construction to technical fountain pens, produce the highest-quality drawings. Liquid ink pens are available in disposable and refillable models.

Also available are *electrostatic plotters,* used where a large volume of plots must be made. Electrostatic plotters produce plots faster and more quietly than pen plotters. The electrostatic process uses a line of closely spaced electrically charged wire nibs to produce dots on coated paper. Pressure or heat is then used to attach ink permanently to the charged dot. These plotters are about ten times faster than pen plotters.

There are two basic types of printers that generate printed copy: impact and nonimpact printers. *Impact printers* operate by striking a ribbon to transfer ink from the ribbon to the paper. Two well-known members of this family are the daisy-wheel printer and the dot-matrix printer. The daisy-wheel printer produces characters similar to a typewriter and is considered to be a letter-quality printer. The dot-matrix printer creates images composed of tiny dots. Usually a dot-matrix printer that can strike 24 dots at one time (a 24-pin printer) is considered to be a letter-quality printer. Many companies use dot-matrix printers to run economical check prints. Non-impact printers are thermal, electrostatic, ink-jet, and laser printers. The thermal printer uses tiny heat elements to burn dot-matrix characters into treated paper. The ink-jet sprays droplets of ink onto the paper to produce dot-matrix images. The best-quality image comes from the laser printer. A laser draws lines on a revolving plate that is charged with a high voltage. The laser light causes the plate to discharge. An ink toner then adheres to the laser-drawn images. The ink is bonded to the paper by pressure or heat.

▼ DIGITIZING EXISTING DRAWINGS

Digitizing is a method of transferring drawing information from a digitizer into the computer. A digitizing tablet may be used to take existing drawings and digitize the information into the computer. Digitizer tablets range in size from 6″ × 6″ to 44″ × 60″. Most school and industry CAD departments use 11″ × 11″ digitizer tablets to input commands from standard and custom tablet menus. In most cases companies do not have the time to convert existing drawings to CAD, because the CAD systems are heavily used for new product drawings. Therefore, an increasing number of businesses digitize existing drawings for other companies.

▼ OPTICAL SCANNING OF EXISTING DRAWINGS

Scanning is a method of reproducing existing drawings in the form of computer drawings. Scanners work much the same way as taking a photograph of the drawing. One advantage of scanning over digitizing existing drawings is that the entire drawing, including dimensions, symbols, and text, is transferred to the computer. A disadvantage is that some scanned drawings require a lot of editing to make them presentable. The scanning process picks up visual graphic information from the drawing. What appears to be a dimension, for example, is only a graphic representation of the dimension picked up from the original drawing. What you get is an exact duplicate of the existing drawing. However, for the dimensional information to be technically accurate, the dimensions must be redrawn and edited on the computer. There are software packages available that make this editing process quick and easy.

After the drawing is scanned, the image is sent to a raster converter that translates information to digital or vector format. A *raster* is an electron beam that generates a matrix of pixels. *Pixels* make up the drawing image on the computer. A raster editor is then used to display the image for changes.

Companies involved in scanning can often reproduce existing drawings more efficiently than companies that digitize existing drawings.

One economic product, CAD/Camera, by Autodesk, Inc., is scanning software that runs on a microcomputer and works in conjunction with an electronic scanning camera. When an existing drawing has been transferred to the computer it becomes an AutoCAD drawing that can be edited. Houston Instrument has an optical scanning accessory that attaches to its DMP-60 series of drafting plotters. This system provides an easy-to-use and cost-effective way to enter up to 36″ × 48″ drawings into the computer system. The scanning accessory can automatically scan drawings from paper, vellum, film, or blue line and convert the hard-copy image into a raster data file.

Computer-Aided Design and Drafting in Architecture Test

DIRECTIONS

Answer the questions with short, complete statements or drawings as needed on 8½ × 11 notebook paper or as follows:

1. Place your name, Chapter 6 Test, and the date at the top of the sheet.
2. Letter the question number and provide the proper answer. You do not need to write out the question.

Use a CADD workstation or word processor to prepare the answers to the test questions if available.

QUESTIONS

Question 6–1 What does the abbreviation CADD mean?

Question 6–2 Define microcomputer in general terms.

Question 6–3 Identify three factors that influence an increase in productivity with CADD.

Question 6–4 Identify five nondrafting related tasks that may be performed by a computer.

Question 6–5 Describe how drawing layers can be prepared on a computer graphics system.

Question 6–6 Describe three ways architectural design functions are improved with a CADD system.

Question 6–7 Define ergonomics.

Question 6–8 Describe three factors that are potential disadvantages of CADD.

Question 6–9 Describe the following computer-aided drafting workstation components:
a. Computer
b. Monitor
c. Digitizer
d. Plotter

Question 6–10 Define the following CADD terminology:
a. Hardware
b. Software
c. Program
d. Commands
e. Soft copy
f. Default value
g. Disk drive
h. Storage
i. Floppy disk
j. Menu
k. DOS
l. File
m. Network
n. PC
o. Puck
p. Mouse
q. Stylus
r. Windows
s. Attribute
t. Font
u. LISP
v. Entity
w. Compatible
x. Text
y. Layers

CADD PROBLEMS

Write a report of 250 or fewer words on one or more of the following topics:

Problem 6–1 Productivity.

Problem 6–2 Ergonomics.

Problem 6–3 CADD workstation.

Problem 6–4 Customizing your own architectural symbols.

Problem 6–5 Architectural program by a software developer.

Problem 6–6 CADD layers.

Problem 6–7 Digitizing existing drawings.

Problem 6–8 Scanning existing drawings.

SECTION II

RESIDENTIAL DESIGN

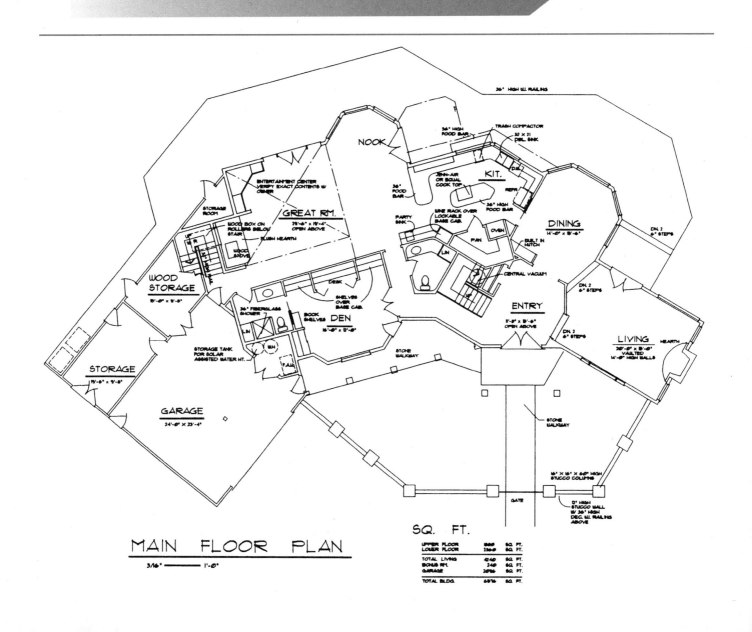

MAIN FLOOR PLAN

3/16" ———— 1'-0"

SQ. FT.

UPPER FLOOR	1995	SQ. FT.
LOWER FLOOR	1260	SQ. FT.
TOTAL LIVING	4240	SQ. FT.
BONUS RM.	340	SQ. FT.
GARAGE	2095	SQ. FT.
TOTAL BLDG.	6976	SQ. FT.

CHAPTER 7

Building Codes and Interior Design

Building codes are intended to protect the public by establishing minimum standards of building safety. Although landowners often assume they can build whatever they want because they own the property, building codes are intended to protect others from such short-sighted individuals. Consider the stories that you have seen in the national news media about structures that have been damaged by high winds, raging flood waters, hurricanes, tornadoes, mud slides, earthquakes, and fire. Although they do not make the national news, hundreds of other structures become uninhabitable because of inadequate foundation design, failed members from poor design, rot, or termite infestation. Building codes are designed to protect consumers by providing minimum guidelines for construction and inspection of a structure.

Each city, county, or state has the authority to write its own building codes to protect its population. Some areas have chosen to have no building codes. Some areas have codes but only on a voluntary compliance basis. Other areas have building codes but exempt residential construction. This wide variety of attitudes has produced over 2,000 different codes in the United States. Trying to design a structure can be quite challenging when faced with so many possibilities.

▼ NATIONAL CODES

Most states have adopted one of three national codes. The *Uniform Building Code* (UBC) has predominantly been adopted throughout the western and southwestern states. The *Basic National Building Code* (BNBC) has been adopted in many areas of the Midwest and Northeast. This code is often referred to as BOCA, an acronym for Building Officials and Code Administrators, which publishes this code. Many states have now adopted the *Standard Building Code Congess International* (SBCCI) as a model code.

To supplement the three national codes, the Council of American Building Officials has developed CABO, One- and Two-Family Dwelling Code. CABO combines parts of each major code to form one code. Areas that were once covered by a national code have adopted CABO for a single-family residence, while keeping another code for multifamily or commercial structures. In addition to these national codes, the Department of Housing and Urban Development (HUD), the Federal Housing Authority (FHA) and the Americans with Disabilities Act (ADA); each publishes its own guidelines for minimum property standards for residential construction.

The architect and engineer are responsible for determining which code will be used during the design process and ensuring that the structure complies with all required codes. Although the drafter is not expected to make decisions regarding the codes as they affect the design, drafters are typically expected to be knowledgeable of the content of the codes and how they affect the construction of a structure.

Each of the major codes is divided into similar sections that specify regulations covering

- ◯ Fire and Life Safety
- ◯ Structural
- ◯ Mechanical
- ◯ Electrical
- ◯ Plumbing

In addition to the major areas of the codes, the SBCCI has additional standards for construction in high-wind areas. The Fire and Life Safety and the Structural codes have the largest effect on the design and construction of a residence.

The UBC is sold as a three-volume set. Volume 1 contains the Administrative, Fire and Life Safety, and Field Inspections Provisions; Volume 2 the Structural Engineering Design Provisions; and Volume 3 the Material, Testing, and Installation Standards of materials that are addressed throughout each code. The effect of the codes on residential construction will be discussed throughout this text. Chapter 45 examines the effects of each code on commercial construction.

▼ FIRE AND LIFE SAFETY

Building codes have their major influence on construction methods rather than on design. (Code influences on construction methods will be discussed in Sections III, IV, V, VIII and X.) There are several areas that drafters and designers need to be familiar with in order to meet minimum design standards. Some of the major areas of the residential codes that you should be familiar with include exit facilities, room dimensions, and light, ventilation, and sanitation requirements.

Building codes divide structures into different categories or occupancies. Houses are a *Group R* occupancy.

This occupancy includes hotels, apartments, convents, single-family dwellings, and lodging houses. The *R* occupancy is further divided into three categories. A single-family dwelling and apartments with fewer than ten inhabitants are defined as R-3. This designation allows the least restrictive type of construction and fire rating group.

In addition to dividing structures into different occupancy ratings, the space within a home or dwelling unit is subdivided into habitable and nonhabitable space. A room is considered *habitable space* when it is used for sleeping, living, cooking, or dining purposes. *Nonhabitable spaces* include closets, pantries, bath or toilet rooms, hallways, utility rooms, storage spaces, garages, darkrooms, and other similar spaces.

Keep in mind that this is only an introduction to building codes. As a drafter and designer you will need to become familiar with and constantly use the building code which governs your area. Because of the similarities of the national codes, all references in this text will be to the UBC unless otherwise noted.

Exit Facilities

The major subjects to be considered in this area of the code are doors, emergency egress (exits), and stairs.

Doors. Each dwelling unit, as they are referred to in the codes, must have a minimum of one door that is at least 36 in. (914 mm) wide and 6′–8″ tall (203 mm). Any hallway adjacent to that door must also be a minimum of 36 in. (914 mm) clear. All other door sizes may be determined by the architect. Notice that the codes do not mandate that the front door be 36 in. (914 mm) wide; only that one exit door needs to have that dimension. Chapter 13 provides guidelines for determining door sizes throughout a residence.

A door cannot swing over a step that is more than 8″ (203 mm) UBC or 8½″ (216 mm) CABO below the level of the door. A minimum of a 3 ft. × 3 ft. (914 mm × 914 mm) landing is required on each side of the required egress door. For interior doors, a door may not swing over any step or change in elevation greater than 1½″ (38 mm). This often becomes a design problem when trying to place a door near a stairway. Figure 7–1 shows some common door and landing problems and their solutions.

Emergency Egress Openings. The term *egress* is used in most building codes to specify areas of access or exits. It is typically used in reference to doors, windows, and hallways. Windows are a major consideration when designing exits. All sleeping areas below the fourth floor must have at least one openable door or window that opens to the outside that has a minimum area of 5.7 sq. ft. (0.53 m²). This area must be 20″ (508 mm) minimum in width, 24″ (610 mm) in height, and be within 44″ (1118 mm) of the floor. This opening provides occupants in each sleeping area with a method of escape in case of a fire. See Figure 7–2.

When a sleeping room is in a basement, the room must be provided with a rescue window that has a window well. The window must be openable and provide a minimum accessible clear opening of 9 sq. ft. (0.84 m²) with a minimum dimension of 35″ (889 mm). Window wells that are deeper than 44″ (1118 mm) must be equipped with a permanently affixed ladder that can be used when the window is fully opened. The ladder cannot extend more than 6″ (152 mm) into the window well.

Smoke Detectors. Closely related to emergency egress is the use of smoke detectors to provide an opportunity for safe exit through early detection of fire and smoke. CABO requires one smoke detector on every floor near sleeping areas and on each additional floor over stairs. The UBC requires that a smoke detector be installed in each sleeping room, as well as at a point centrally located in a corridor that provides access to the bedrooms.

Smoke detectors must be located within 12″ (305 mm) of the ceiling or mounted on the ceiling. Most codes

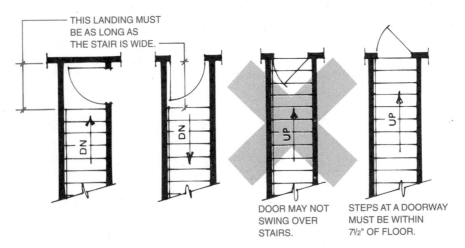

Figure 7–1 Placement of doors near stairs.

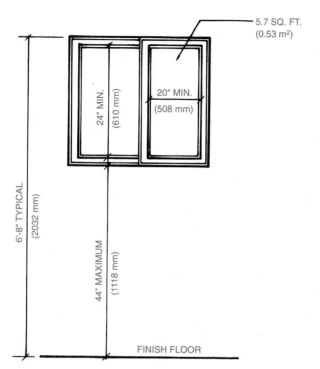

Figure 7–2 Minimum window size for emergency egress of bedrooms below the fourth floor.

require detectors to be connected to electrical wiring as well as having a battery-powered backup system. All smoke detectors must be interconnected so that if one alarm is activated, all will sound.

For a one-story residence, the smoke detector must be located at the start of every hall that serves a bedroom. For a multilevel residence, a detector is required on every level with a bedroom and must be located over the stairs leading to that level. Smoke detectors should not be placed in kitchens or near fireplaces because of the strong possibility of false alarms from small amounts of smoke.

Halls. Hallways must be a minimum of 36″ (914 mm) wide. This is very rarely a design consideration because hallways are often laid out to be 42″ (1067 mm) wide or wider to create an open feeling and to enhance accessibility from room to room.

Stairs. Stairs often can dictate the layout of an entire structure. Because of their importance in the design process, stairs must be considered early in the design stage. For a complete description of stair construction see Chapter 37. Code requirements for stair design vary

greatly (Figure 7–3). By following the minimum standards, stairs that are extremely steep, very narrow, and have very little room for foot placement would result. Good design practice would provide stairs with a width of between 36 and 42″ (914–1067 mm) for ease of movement. A common tread width is between 10 and 10½″ (254–267 mm) with a rise of about 7½″ (191 mm). Figure 7–4 shows the difference between the minimum stair layout and some common alternatives. The rise of a stairway can be no less than 4″ (102 mm). Within any flight of stairs, the largest step run cannot exceed the smallest step by ⅜″ (10 mm). The difference between the largest and smallest rise cannot exceed ⅜″ (10 mm).

When a spiral stair is the only stair serving an upper floor area, problems are created in getting furniture from one floor to another. Building codes place limitations on the size of floor area that spiral stairs may serve. Spiral stairs may not be used as an exit when the area exceeds 400 sq. ft. (37.16 m²). Spiral stair treads must provide a clear walking width of 26″ (660 mm) measured from the outer edge of the support column to the inner edge of the handrail. A run of 7½″ (191 mm) must be provided within 12″ (305 mm) of the narrowest part of the tread. The rise for a spiral stair must be sufficient to provide 6′–6″ (1981 mm) headroom, but no riser may exceed 9½″ (241 mm), and each riser must be equal.

Winding stairs may be used if the required run is provided at a point not more than 12″ (305 mm) from the narrowest side of the stairway. The narrowest portion of a winding stairway may be no less than 6″ (152 mm) minimum.

Headroom over stairs must also be considered as the residence is being designed. Stairs are required to have 6′–8″ (2032 mm) minimum headroom by each of the codes. Headroom of 6′–6″ (1980 mm) is allowed for spiral stairs. This headroom can have a great effect on wall placement on the upper floor over the stairwell. Figure 7–5 shows some alternatives in wall placement around stairs.

All stairways with three or more risers are required to have at least one smooth handrail that extends the entire length of the stair. The rail must be placed on the open side of the stairs and must be between 30 and 38″ (762–965 mm) above the front edge of the stair. UBC requires the rail to be within 34 to 38″ (864–965 mm) above the nose of the stair. Handrails are required to be 1½″ (38 mm) from the wall but may not extend into the required stair width by more than 3½″ (89 mm).

A guardrail must be provided at changes in floor elevation that exceed 30″ (762 mm). Guardrails are also

	UBC	CABO	BOCA	SBCCI
WIDTH (min.)	36″ (914 mm)	36″ (914mm)	36″ (914 mm)	36″ (914 mm)
RISE (max.)	8″ (203 mm)	8¼″ (210 mm)	8¼″ (210 mm)	7¾″ (197 mm)
TREAD (min.)	9″ (229 mm)	9″ (229mm)	9″ (229 mm)	10″ (254 mm)

*Treads and risers must be proportioned so that the sum of two risers and a tread (exclusive of tread nosing) must not be less than 24″ and not more than 25″.

Figure 7–3 A comparison of basic design values for stairs according to major building codes.

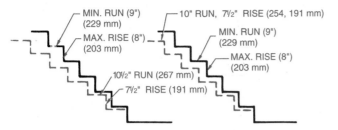

Figure 7–4 Stair layouts comparing minimum run and maximum rise with common design alternatives.

required to be 36" (914 mm) high, except around a stairwell, where the rail may be 34" (864 mm) high. Railings must be constructed so that a 4" (102 mm) diameter sphere cannot pass through any opening in the rail. The UBC allows the distance between railings to be 6" (152 mm).

Room Dimensions

Room dimension requirements affect the size and ceiling height of rooms. Every dwelling unit is required to have at least one room that has a minimum of 120 sq. ft. (11.2 m²) of floor area. Other habitable rooms except kitchens are required to have a minimum of 70 sq. ft. (6.5 m²) and shall not be less than 7' (2134 mm) in any direction. CABO requires kitchens to be a minimum of 50 sq. ft. (4.6 m²).

These code requirements rarely affect home design. One major code requirement affecting room size governs the space allowed for a toilet, which is typically referred to as a water closet. A space 30" (762 mm) wide must be provided for water closets. A distance of 21" (533 mm) is required in front of a toilet. This can often affect the layout of a bathroom.

Habitable rooms must have a ceiling height of 7'–6" (2286 mm) for 50% of the room, with no portion of the roof to be less than 7'–0" (2134 mm). The UBC requires a minimum ceiling height of 7'–6" (2286 mm) for all habitable rooms. Supporting beams are sometimes allowed to extend into this area. Bathrooms, kitchens, and hallways must have 7' (2134 mm)-high ceilings. This lowered ceiling can often be used for lighting or for heating ducts. In rooms that have a sloping ceiling, the minimum ceiling must be maintained in at least half of the room. The balance of the room height may slope to 5' (1524 mm) minimum. Any part of a room that has less than 5' (1524 mm)-high ceilings may not be included as habitable square footage. See Figure 7–6.

Light, Ventilation, Heating, and Sanitation Requirements

The light and ventilation requirements of building codes have a major effect on window size and placement. The heating and sanitation requirements are so minimal that they rarely affect the design process.

The codes covering light and ventilation have a broad impact on the design of the house. Many preliminary designs often show entire walls of glass to take advantage of beautiful surroundings. At the other extreme, some houses have very little glass and thus no view but very little heat loss. Building codes affect both types of designs.

All habitable rooms must have natural light provided by windows. Kitchens are habitable rooms but are not required to meet the light and ventilation requirements. Windows for other habitable rooms are required to open directly into a street, public alley, or yard on the same building site. Windows may open into an enclosed structure such as a porch as long as the area is at least 65

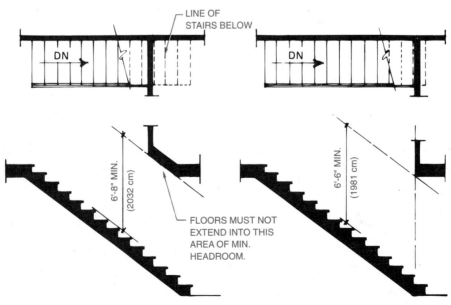

Figure 7–5 Wall placement over stairs.

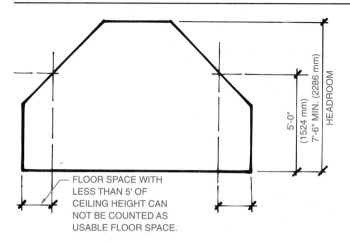

FLOOR SPACE WITH
LESS THAN 5' OF
CEILING HEIGHT CAN
NOT BE COUNTED AS
USABLE FLOOR SPACE.

Figure 7–6 Minimum ceiling heights for habitable rooms.

percent open. Required windows may also be provided by the use of skylights.

Figure 7–7 shows a comparison of the building code requirements for windows. A bedroom that is 9 × 10′ (2743 × 3050 mm) is required to have a window with a glass area of 7.2 sq. ft. (0.66 m²) to meet BOCA, CABO, and SBCCI standards, and 9 sq. ft. (0.83 m²) to meet the minimum standards of UBC. Maximum limits of glass vary between 17 and 22 percent of the exterior heated wall area of the entire house. Homes can have more glass, but typically it is required to be triple glazed to help cut down heat loss. Another limit to the amount of glass area would be in locations that are subject to strong winds or earthquakes. Because of the lateral movement created by winds and earthquakes, window and wall areas must be carefully proportioned to provide lateral stability.

Two alternatives to the light and ventilation requirements are typically allowed in all habitable rooms. Mechanical ventilation and lighting equipment can be used in place of openable windows in habitable rooms except bedrooms. The standards for emergency egress require bedrooms to have a window even if mechanical light and ventilation is provided. The second method allows floor areas of two adjoining rooms to be considered as one if half of the area of the common wall is open and unobstructed. This opening must also be equal to one-tenth of the floor area of the interior room or 25 sq. ft. (2.3 m²), whichever is greater.

Although considered a nonhabitable room, bathrooms and laundry rooms must be provided with an openable

window or a fan that can provide five air changes per hour. These fans must be vented to outside air.

Alternatives. Many municipalities are developing their own code variations to meet the needs of light and code restrictions. The 1992 Oregon Residential Energy Code, for example, allows nine different design alternatives to be used to develop an energy-efficient structure. Rather than strict percentages for windows, variable areas are allowed depending on the quality of insulation, the framing system used, and the quality of windows to be installed. Figure 7–8 contains a summary of this energy code.

Heating. The heating requirements for a residence are very minimal based on the residential codes. CABO requires a heating unit capable of producing and maintaining a room temperature of 68° F, or 20° C (70° F, 21°C UBC) at a point 3′ (914 mm) above the floor for all habitable rooms. Chapter 18 provides information on sizing heating units.

Sanitation. Code requirements for sanitation rarely affect the design of a structure. Each residence is required to have a toilet, a sink, and a tub or shower, but all are not required to have hot and cold water. All plumbing fixtures must be connected to a sanitary sewer or an approved private sewage disposal system. The room containing the toilet must be separated from the food preparation area by a tight-fitting door. Chapter 13 provides further information regarding the layout of each fixture, and Chapter 17 provides information regarding plumbing plans.

▼ CLIMATIC AND GEOGRAPHIC DESIGN CRITERIA

The major influence of each code on a structure covers the structural components of a building. Each building code contains chapters covering foundations, wall construction, wall coverings, floors, roof-ceiling construction, roof coverings, and chimney and fireplace construction. To use the information in the codes as it relates to each area of construction requires some basic climatic and geographic information about the area where the structure will be constructed. This includes knowledge of various factors of the building site:

Roof live load
Roof snow load
Wind pressure
Seismic condition by zone
Subject to damage by weathering
Subject to damage by termites
Subject to damage by decay
Frost line depth
Winter design temperature for heating facilities

	UBC	CABO	BOCA	SBC
LIGHT	10% 10 sq. ft. min.	8%	8%	8%
VENTILATION	5% 5 sq. ft. min.	4%	4%	4%

Figure 7–7 Light and ventilation requirements according to major building codes.

TABLE NO. 53-P [1, 2, 3]
PRESCRIPTIVE COMPLIANCE PATHS FOR RESIDENTIAL BUILDINGS

Building Components	PATH 1	PATH 2 Sun Tempered[4]	PATH 3	PATH 4 Sun Tempered[4]	PATH 5	PATH 6 Sun Tempered[4]	PATH 7 Sun Tempered[4]	PATH 8 House Size Limited[5]	PATH 9 Log Homes/ Solid Timber
Maximum Allowable Window Area[6]	No Limit	No Limit	No Limit	No Limit	No Limit	No Limit	No Limit	12%	No Limit
Window Class[7]	U=0.40	U=0.40	U=0.50	U=0.50	U=0.60	U=0.60	U=0.60	U=0.40	U=0.40
Doors, Other Than Main Entry	U=0.20	U=0.20	U=0.20	U=0.20	U=0.20	U=0.20	U=0.20	U=0.20	U=0.54
Main Entry Door, maximum 24 sq. ft.	U=0.54	U=0.54	U=0.20	U=0.20	U=0.20	U=0.20	U=0.54	U=0.20	U=0.54
Wall Insulation[8]	R-21[9]	R-15	R-21A	R-15A	R-24A	R-21A	R-21A	R-15	—[3]
Underfloor Insulation	R-25	R-21	R-25	R-21	R-30	R-21	R-25	R-21	R-30
Flat Ceilings[8]	R-38	R-49	R-49A	R-38	R-49A	R-49A	R-49A	R-49	R-49
Vaulted Ceilings[10, 11]	R-30	R-30	R-30	R-38	R-38	R-38	R-38	R-38	R-38
Skylight Class[7]	U=0.50	U=0.50	U=0.50	U=0.50	U=0.50	U=0.50	U=0.50	U=0.50	U=0.50
Skylight Area	< 2%	< 2%	< 2%	< 2%	< 2%	< 2%	< 2%	< 2%	< 2%
Basement Walls	R-21	R-21	R-21	R-21	R-21	R-21	R-21	R-21	R-21
Slab Floor Edge Insulation	R-15	R-15	R-15	R-15	R-15	R-15	R-15	R-15	R-15
Forced Air Duct Insulation	R-8	R-8	R-8	R-8	R-8	R-8	R-8	R-8	R-8

Notes:

[1] Path 1 is based on cost-effectiveness. Paths 2-7 are based on energy equivalence with Path 1. Cost-effectiveness of Paths 2-9 not evaluated.

[2] As allowed in current Chapter 5303(b), thermal performance of a component may be adjusted provided that overall heat loss does not exceed the total resulting from conformance to the required U-value standards. Calculations to document equivalent heat loss shall be performed using the procedure and approved U-values contained in Table No. 53-O.

[3] MINIMUM COMPONENT REQUIREMENTS: Walls R-15; Floors R-21; Flat Ceilings R-38; Vaults R-21; Basement Walls R-21; Slab Edge R-10; Duct Insulation R-8. R-values used in this table are nominal, for the insulation only and not for the entire assembly. Window and skylight U-values shall not exceed 0.65 (CL65). Door U-values shall not exceed 0.54 (Nominal R-2). The minimum wall component for Path 9 shall be an average solid log or timber wall thickness of 3.5 inches.

[4] The sun-tempered house shall have one lot line which borders on a street oriented within 30 degrees of true east-west and 50 percent or more of the total glazing area for the heated space on the south elevation.

[5] Path 8 applies only to homes with less than 1,500 sq. ft. heated floor space AND glazing area less than 12 percent of heated space floor area.

[6] Reduced window area may not be used as a trade-off criterion for thermal performance of any component, except as noted in Table No. 53-O.

[7] Window and skylight U-values shall not exceed the number listed. U-values may also be listed as "class" on some windows and skylights (i.e., CL40 is same as U=0.40).

[8] A= advanced frame construction as defined in Section 5303(c)(3) for walls and Section 5303(c)(4) for ceilings.

[9] R-19 Advanced Frame or 2 x 4 wall with exterior rigid insulation may be substituted if total nominal insulation R-value is 18.5 or greater.

[10] Partially vaulted ceilings and ceilings totaling not more than 150 square feet in area for dormers, bay windows or similar architectural features may be reduced to not less than R-21. When reduced, the cavity shall be filled (except for required ventilation spaces), and a 0.5 perm (dry cup) vapor retarder installed.

[11] Where ceiling/vault insulation upgrades are part of a prescriptive or trade-off path, vaulted area, unless insulated to R-38, may not exceed 50 percent of the total heated space floor area. This does not apply to Path 9, Log/Solid Timber construction.

Figure 7–8 Oregon Residential Energy Code summary.

Information for each of these categories can be obtained from a table in the applicable building code or from the governing building department. Building codes specify how a building can be constructed to resist the forces from wind pressure, seismic activity, freezing, decay, and insects. (Each of these areas will be discussed in detail in succeeding chapters.) In the initial design stage, an understanding of these factors will determine what types of materials will be required.

CHAPTER

Building Codes and Interior Design Test

DIRECTIONS

Answer the questions with short complete statements or drawings as needed on 8½ × 11 notebook paper, as follows:

1. Letter your name, and the date at the top of the sheet.
2. Letter the question number and provide the answer. You do not need to write out the question. Answers may be prepared on a word processor if appropriate with course guidelines.
3. Answers should be based on the code that governs your area unless noted.

QUESTIONS

Question 7–1 What is the minimum required size for an entry door?

Question 7–2 Give the maximum rise for stair treads for:
a. UBC
b. SBCCI
c. BOCA
d. CABO

Question 7–3 According to UBC, all habitable rooms must have what percent of floor area in glass for natural light?

Question 7–4 What is the required width for hallways?

Question 7–5 List the minimum width for residential stairs according to the CABO.

Question 7–6 What is the minimum ceiling height for a kitchen?

Question 7–7 What is the maximum height that bedroom windows can be above the finish floor?

Question 7–8 Toilets must have a space how many inches wide?

Question 7–9 What are the maximum areas of glass that are allowed without special provisions?

Question 7–10 List three sanitation requirements for a residence.

Question 7–11 What are the area limitations of spiral stairs?

Question 7–12 For a bedroom that is 10′ × 12′, what area is required to provide the minimum ventilation requirements?

Question 7–13 List the minimum square footage required for habitable rooms.

Question 7–14 For a sloping ceiling, what is the lowest height that is allowed for usable floor area according to SBCCI?

Question 7–15 What is the minimum size opening for an emergency egress?

CHAPTER 8

Room Relationships and Sizes

Architecture probably has more amateur experts than any other field. Wide exposure to houses causes many people to feel that they can design their own house. Because they know what they like, many people attempt to draw their own plans, only to have them rejected by building departments. Designing houses can be done by most anyone. Designing a home to meet a variety of family needs takes more knowledge.

▼ THE FLOOR PLAN

The floor plan is typically the first drawing that is started as a home is designed for a family. The floor plan must be designed to harmonize with the site and the inhabitants. Planning for the site is introduced in Chapters 10 through 12. The balance of this chapter will introduce both the design process and basic requirements for a home.

Clients typically provide the designer with many pictures collected from years of planning, which can be very helpful in planning for the family's needs and lifestyle. Many architectural firms have a written questionnaire to provide insight into how the structure will be used. Important design considerations include the following factors:

○ The number of inhabitants.
○ The ages and sex of children.
○ Future plans to add on to the dwelling.
○ A list of general family activities to be done in the home.
○ Entertainment habits.
○ Desired number of bedrooms and bathrooms.
○ Kitchen appliances desired.
○ Planned length of stay in the residence.
○ Live-in guest or requirements for people with disabilities.

It is helpful to have clients write out a specific list of minimum requirements: minimum number of rooms, minimum size requirements for each room, and a listing of how each room is to be furnished. The written list is an excellent way to have the clients agree on the design criteria before the design team spends hours pursuing unwanted features. Providing minimum room sizes is also much more helpful than allowing a client to request "three big bedrooms." An understanding of how the room is to be furnished will also help the designer be sure that the owner has realistic expectations for the size of the room. It is also helpful to have clients provide a "wish list" of items for their residence. These items can be eliminated if the budget is exceeded, but it may be possible to provide these items.

Once this information has been provided, the design process described in Chapter 1 can be started. The bubble drawings and preliminary sketches will need to integrate the living, sleeping, and service areas, as well as movement among these areas.

▼ LIVING AREA

The living area can be considered the living, dining, and family rooms, den, and nook. These are the rooms or areas of a house where family and friends will spend most of their leisure time. Typically each of the rooms of the living area will be clustered together near the entry to allow easy access for guests.

Living Room

If both a living room and a family room are provided, the living room is typically used to entertain guests or for quiet conversations. If the home has no family room, the living room will also need to provide room for recreation, hobbies, and relaxation.

The size of the room should be determined by the number of people who will typically use it, the furniture that is intended for it, and the budget for the entire construction project. Rectangular rooms are typically easier to plan for furniture arrangements. The closer a room is to being square, the harder furniture placement typically becomes. A room of 12–14' (3660–4270 mm) width is minimum to provide for the common activities. A room of 13 × 18' (3960–5500 mm) serves for most furniture arrangements and allows for easy furniture movement.

Within the room, an area of approximately 9' (2740 mm) diameter should be provided for the primary seating area. This area is typically arranged to take advantage of a view or centered around a fireplace. In smaller homes, the room can be made to seem larger if the living room is attached to the dining area. In larger residences, the living room is often set apart from other living areas to give it a more formal effect, as seen in Figure 8–1. It becomes a room where guests are entertained and a room for quiet conversation apart from the noisier activities of other rooms.

Figure 8–1 In larger homes, the living room is often set apart from the balance of the living areas to create a formal area. *Courtesy Marvin Windows and Doors.*

Figure 8–2 If the living room is open to other living areas a casual feeling is created. *Courtesy Marvin Windows and Doors.*

The living room is usually placed near the entry so that guests may enter it without having to pass through the rest of the house. When placed on the west side of the house, the room will have natural light in the evening, Figure 8–2 shows a casual living room that is open to both the rest of the home and to the yard.

Family Room

A family room is probably the most used area in the house. A multipurpose area, it is used for such activities as watching television, informal entertaining, sorting laundry, playing pool, or eating. With such a wide variety of possible uses, planning the family room can be quite difficult. The room needs to be separated from the living room but still be close enough to have easy access from the living and dining rooms for entertaining. Figure 8–3 shows a possible layout of living areas providing close

but separate layout of rooms. The family room needs to be near the kitchen since it is often used as an eating area. Placing the family room adjacent to the kitchen combines the two most used rooms into one area called a country kitchen. It is also common to have the family room near the service areas of the residence.

Sizes can vary greatly depending on the design criteria for the residence. When designed for a specific family, the size of the family room can be planned to meet their needs. When the room is being designed for a house where the owners are unknown, the room must be of sufficient size to meet a variety of needs. An area of about 13′ × 16′ (3960 × 4900 mm) should be the minimum size considered for a family room. This is a small room but would provide sufficient area for most activities. If a wood stove or fireplace is in the family room, the room should be large enough to allow for the heat from the stove. Often in an enclosed room where air does not circulate well, the area around a stove or fireplace is too warm to sit near comfortably.

Dining Room

Dining areas, depending on the size and atmosphere of the residence, can be treated in several ways. The dining area is often part of, or adjoining, the living area, as seen in Figure 8–4. For a more formal eating environment, the dining area will be near but separate from the living area. The two areas are usually adjoining so that guests may go easily from one area to the other without passing through other areas of the house. The dining room should also be near the kitchen for easy serving of meals but without

MAIN FLOOR 1420 sq. ft.

Figure 8–3 Living areas of this residence are close to each other for easy entertaining, but separated for noise control. *Courtesy Piercy & Barclay Designers, CPBD.*

Figure 8–4 The dining area can be combined with other living areas to create a spacious environment. *Courtesy Piercy & Barclay Designers, CPBD.*

Figure 8–5 Although room sizes are small, combining areas provides ample size for entertaining. *Courtesy Piercy & Barclay Designers, CPBD.*

providing a direct view into the work areas of the kitchen. Because of the need to be by the living room and kitchen, the dining room is typically placed between these two areas or in a corner between the two rooms, as seen in Figure 8–5.

Casual dining rooms can be as small as 9′ × 11′ if the area is open to another area. This area would allow for a table seating four to six people and a small storage hutch. A formal dining room should be about 11′ × 14′ (3350 × 4300 mm). This will allow room for a hutch or china cabinet as well as the table and chairs. It will also allow for chair placement and room to walk around the table. Allow for a minimum of 32″ (810 mm) from the table edge to any wall or furniture for placement of chairs and a minimal passage area. A space of approximately 42″

(1067 mm) will allow room for walking around a chair when it is occupied.

Nook

When space and finances will allow, a nook or breakfast area is often included in the design. This area is typically where the family will eat most of its meals or snacks (Figure 8–6). The dining room then becomes an area for formal eating only. Both the dining room and the nook need to be near the kitchen. If possible, the nook should also be near the family room.

Where space is at a premium, these areas are often placed together in one room called a grand or great room (Figure 8–7). In larger homes a patio or deck may be enclosed to provide a solarium (Figure 8–8).

Den/Study/Office

Although the title may vary, many families plan for a room to provide for quiet reading and study. This room is typically located off the entry and near the living room. The den often serves as a buffer between the living and sleeping areas of a residence. In many tract houses, the den is used as a spare bedroom or guest room.

Figure 8–6 The nook is a casual area for dining where a family will eat most of their meals. *Courtesy Marvin Windows and Doors.*

Figure 8–7 A grand room is a combination of a living, dining and family room. *Courtesy the National Oak Flooring Assoc.*

If the room is to be a true office within the home, the entrance should be directly off the entry. A separate entry from the residence is ideal. Entry to the office can be provided from a covered porch but should be distinctive enough so that clients will know where to enter. The size of the office depends on the equipment to be used and the number of clients to be accommodated. It should have room for a desk, computer work space, storage areas for books and files, and a separate area for phones, fax, and photocopying machines.

▼ SLEEPING AREA

The age, sex, and number of children will determine how many bedrooms will be required. Preteens can share a bedroom if space is limited. Teens of the same sex have even been known to survive sharing a room together, although this situation often results in distinctive markings of territory during tense times. The ideal situation is to provide a separate bedroom for each child. Each room should have space for sleeping, relaxation, study, storage, and dressing.

The sleeping area consists of the bedrooms. These rooms should be placed away from the noise of the living and service areas and out of the normal traffic patterns. Bedrooms are generally located with access from a hallway for privacy from living areas and to be near bathrooms. Care must be taken to keep the bathroom plumbing away from bedroom walls. One way to accomplish this is to place a closet between the bedroom and the bath. If plumbing must be placed in a bedroom wall, use insulation to help control noise.

Bedrooms

Bedrooms function best on the southeast side of the house. This location will place morning sunlight in the rooms. When a two-level layout is used, bedrooms are often placed on the upper level away from the living areas. Not only does this arrangement provide a quiet sleeping area, but bedrooms can often be heated by the natural convection of heat rising from the living area. Another option is placing the bedrooms in a daylight basement. Care must be taken in basement bedrooms to provide direct exits to the outside. An advantage of a bedroom in a daylight basement is the cool sleeping environment that the basement provides.

Size needs vary greatly with bedrooms. Spare or children's bedrooms are often as small as 9′ × 10′(2740 × 3050 mm). This might be adequate for a sleeping area but not if the room is also to be used as a study or play area. Each bedroom should have enough space for a single bed, a small bedside table, and a dresser. Clients often specify enough space for a twin or double bed. A twin bed may be within 6″ (150 mm) of a wall. Plan for a minimum of 24″ (600 mm) on each side of a bed, where space to walk is required. A space of approximately 36″ (900 mm) should be provided between dressers and any obstruction to allow for opening of drawers or doors. This requires a space of about 12′ × 14′ (3600 × 4300 mm) plus closet areas. The master bedroom should be at least 12′ × 14′ (3658 × 4267 mm) plus closet areas. An area of 13 × 16′

Figure 8–8 A sunroom or solarium can provide added square footage at a relatively low construction cost. *Courtesy Velux America, Inc.*

(4000 × 4900 mm) will make a spacious bedroom suite, although additional room is required for a sitting or reading area (Figure 8–9). A window seat can often be used to provide sitting room without greatly increasing the square footage of the room.

Try to arrange bedrooms so that at least two walls can be used for bed placement. This is especially important in the master bedroom to allow for periodic furniture movement.

Window planning is very important in a bedroom. Although morning sun is desirable, space should be planned for bed placement so that sunlight will not shine directly on it. Especially important in planning windows for a bedroom is to allow for emergency egress from the bedroom. (See Chapter 7 for a review of emergency window requirements.) For upper-level bedrooms, try to provide a roof, porch, or balcony for an escape route from a window for young children. A flexible escape ladder can be used for older children.

Closets

Each bedroom should have its own closet with a minimum length of 4′(1200 mm). The depth should be a minimum of 24 in. (610 mm). If the closet must be small, a double-pole system can be used to double the storage space or a premanufactured closet organizer can be used to provide storage. Closets can often be used as a noise buffer between rooms or located in what would be considered wasted space in areas with sloping ceilings. Master bedrooms often will have a walk-in closet. It should be a minimum of 6′ × 6′ (1800 × 1800 mm) to provide adequate space for clothes storage. A closet of 6′ × 8′ (1800 × 2400 mm) provides much better access to all clothes. Many clients request space for a freestanding

Figure 8–9 A master suite often includes a sitting area for quiet reading or conversation. *Courtesy Marvin Windows and Doors.*

Figure 8–10 A master suite often includes an enlarged tub or spa. *Courtesy Marvin Windows and Doors.*

armoire to provide additional storage. Space must also be provided near the bedrooms for linen storage. A space 2′ wide and 18″ deep would be minimal.

▼ SERVICE AREA

Bath, kitchen, and utility rooms and the garage are each considered a part of the service area. Notice that three of the four areas have plumbing in them. Because of the plumbing and the services that each provides, an attempt should be made to keep the service areas together. Another consideration in placing the service area is noise. Each of these four areas tends to have noises that will interrupt activities of the living and sleeping areas.

Bathrooms

Bathrooms are often placed so that access is gained from a short hallway in order to provide privacy from living areas. A house with only one bathroom must have it located for easy access from both the living and sleeping areas. Access to the bathroom should not require having to pass through the living or sleeping areas.

Options for bathrooms include a half-bath, three-quarter bath, full bath, and a bathroom suite. A *half-bath* has a lavatory and a water closet (a toilet). A *three-quarter* bath combines a shower with a toilet and a lavatory. A *full bath* has a lavatory, a toilet, and a tub or a combination tub-and-shower unit. A *bathroom suite* typically includes the features of a full bath with a separate tub and shower. The tub is often an enlarged tub or spa (Figure 8–10). If the tub is to be raised, skidproof steps should be provided. If windows are placed around a tub or spa, the glass is required to be tempered. The wardrobe, dressing area, and a sitting area are also usually part or adjoining the bathroom suite. Figure 8–11 shows a plan view for a master suite. Many custom homes feature a full bath with an adjoining exercise room, sauna, or steam room.

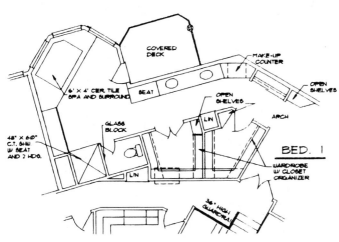

Figure 8–11 The master suite of many custom homes includes a bath, dressing area and wardrobe. *Courtesy Residential Designs, AIBD.*

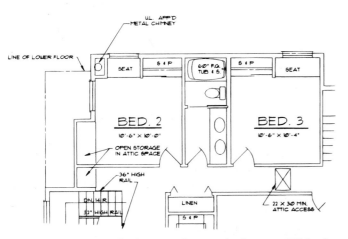

Figure 8–12 Many bedrooms share a bathroom. With the dressing area separated from the bathing area two people can use the room at once. *Courtesy Residential Designs, AIBD.*

The style of the house affects the number and locations of bathrooms. A single-story residence typically has one bathroom for the master bedroom and a second to serve the living room and the other bedrooms. A two-story house typically has two full bathrooms upstairs, with a minimum of a half-bath downstairs. Multilevel homes should have a bath on each level containing bedrooms or a half-bath on each level with any type of living area.

When a residence has two or more baths, they are often placed back-to-back or above each other to help reduce plumbing costs. A full bath is typically provided near the master bedroom, with a separate bathroom to serve the balance of the bedrooms. If space and budget allow, a bathroom for each bedroom is common. For families with young children, a tub is usually a must. Many bathrooms that are designed to be shared place the sinks in one room, with the toilet and tub in an adjoining room (Figure 8–12). For homes designed to appeal to a wide variety of families, a combination of a tub and shower is often used.

Homes designed for outside activities often have a three-quarter bath near the utility room or combine the laundry facilities with a bathroom to create what is often called a mud room.

Kitchen

A kitchen is used through most of a family's waking hours. It serves not only for meal preparation but often includes areas for eating, working, and laundry. It needs to be located close to the dining areas so that serving meals does not require extra steps. The kitchen should also be near the family room. This will allow those preparing meals to be part of family activities. When possible, avoid placing the kitchen in the southwest corner of the home. This location will receive the greatest amount of natural sunlight and could easily cause the kitchen to overheat. With the kitchen creating its own heat, try to place it in a cooler area of the house unless venting and shading precautions are taken. One advantage of a western placement is the natural sunlight available in the late evening.

When designing a house for families with young children, a kitchen with a view of indoor and outdoor play areas is a valuable asset. This will allow for the supervision of playtimes and control of traffic in and out of the house.

The kitchen is often closely related to the utility area. Because these are the two major workstations in the home, a kitchen close to the utility room can save valuable time and energy as the daily house chores are done. Also helpful in kitchen planning is to place the kitchen near the garage or carport. This location will allow groceries to be unloaded easily (Figure 8–13),

Not only must the kitchen location be given much thought, but the space within the kitchen also demands great consideration. Perhaps the greatest challenge facing the designer is to create a workable layout in a small

Figure 8–13 The kitchen must be near the dining, living, family utility and service entry for efficient living. *Courtesy Piercy & Barclay Designers, CPBD.*

kitchen. Layout within the kitchen includes the relationship of the appliances to the work areas. The main work areas of a kitchen are the storage area, the preparation and the cleaning area.

Storage Area. The storage area consists of the refrigerator, the freezer, and cabinet space for food and utensils. Most families prefer to have a refrigerator in the kitchen, with a separate freezer in the laundry or garage. Storage for cans and dried foods is typically in base cabinets 24″ (600 mm) wide or in a pantry unit. Upper units 12″ (300 mm) deep are also typically used for dishes and nonperishable foods. A minimum counter surface of 18″ (450 mm) wide should be provided next to the refrigerator to facilitate access to it. This area can also be a useful area for preparing snacks if the width is increased to approximately 36″ (900 mm) wide. As the size is increased, this counter can also be used to store appliances, such as mixers and blenders. This is a useful area for a breadboard and a drawer for silverware storage. Additional drawer space could be used for mixing bowls, baking pans, and other cooking utensils. Enclosed cabinet space should also be provided for storage of cookbooks, writing supplies, and telephone books.

Preparation Area. The main components of the preparation area are the sink and cooking units and a counter area of approximately 48″ (1200 mm) minimum. Large kitchens often have a small vegetable sink in the preparation area and a larger sink for cleaning utensils. A minimum sink of 16 × 16″ (400 × 400 mm) is typical for a vegetable sink.

The cooking appliances are usually a stove that includes both an oven and surface heating units or separate appliances for baking and cooking. Most kitchens also provide space for a microwave oven that can either be part of a built-in oven, be mounted below the upper cabinets, or be placed on the counter. A minimum counter space of 18″ (450 mm) should be placed on each side of a stove or oven unit to prevent burns and provide temporary storage while cooking.

Cleaning Center. The cleaning center typically includes the sink, garbage disposal, and dishwasher. Most clients prefer a double sink rather than a single sink. A typical double sink is 32 × 21″ (800 × 530 mm), but 36 (900 mm) and 42″ (1070 mm) wide units are also available. Double sinks are also available with the sinks at 90 degrees to each other for use in a corner. If the sink is connected to a public sewer, a garbage disposal is typically installed to one side of the sink to eliminate wet garbage. A garbage disposal should not be used on sinks connected to septic tanks. A minimum area of 36″ (900 mm) wide should be provided on one side of the sink for stacking dirty dishes, and about 24–30″ (600–700 mm) on the other side for clean dishes. The dishwasher can be placed on either side of the sink depending on the client's wishes. An upper cabinet should be provided near the dishwasher to store dishes.

The Work Triangle. The work triangle is formed by drawing a line from the center of the storage, preparation, and cleaning areas. This triangle outlines the main traffic area required to prepare a meal. Food will be taken from the refrigerator, cleaned at the sink, cooked at the microwave or stove, and then leftovers will be returned to the refrigerator. The work triangle should not be less than 15′ (4500 mm) or more than 22′ (6700 mm). Figure 8–14 shows common shapes for kitchen layout and the work triangle that results.

In addition to traffic within the kitchen, the relationship of the kitchen to other rooms also needs to be considered. If care is not taken in the design process the kitchen can become a hallway. The kitchen needs to be in a central location, but traffic must flow around the kitchen, not through it. Figure 8–15 shows the traffic pattern in three different kitchens.

Arrangement of Appliances. The kitchen sink is often placed under a window. This allows for supervision of outdoor activities and provides a source of light at this workstation. Because of the placement by a window, the sink is often located first as the kitchen is being planned. Provide a minimum of 48″ (1200 mm) between counters to accommodate someone working at the sink and another person walking behind him or her. Try to avoid placing the sink and dishwasher on different counters even if just around the counter from each other. Such a layout often leads to accidents in the kitchen from water dripping onto the floor. The open door of the dishwasher also hampers access to other areas of the kitchen.

Typically a space 36″ (900 mm) wide is provided for the refrigerator. If possible, it should be placed so that the door will not block traffic through the work triangle.

The refrigerator is usually placed at the end of a counter within five or six feet of the sink and stove. If it must be placed near a cabinet corner, allow a minimum counter width of 12″ (300 mm) in the corner to allow for access to the back of the counter.

The cooking units should be located near the daily eating area. The stove should be placed so that the person using it will not be standing in the path of traffic flowing through the kitchen. This will help eliminate the chance of hot utensils being knocked from the stove. Stoves should be placed so that there is approximately 18″ (450 mm) of counter space between the stove and the end of the counter to prevent burns to people passing by. Stoves should not be placed within 18″ (450 mm) of an interior cabinet corner. This precaution will allow the oven door to be open and still allow access to the interior cabinet. A walkway 48″ (1200 mm) wide should be provided to allow someone to walk behind a person working at the stove.

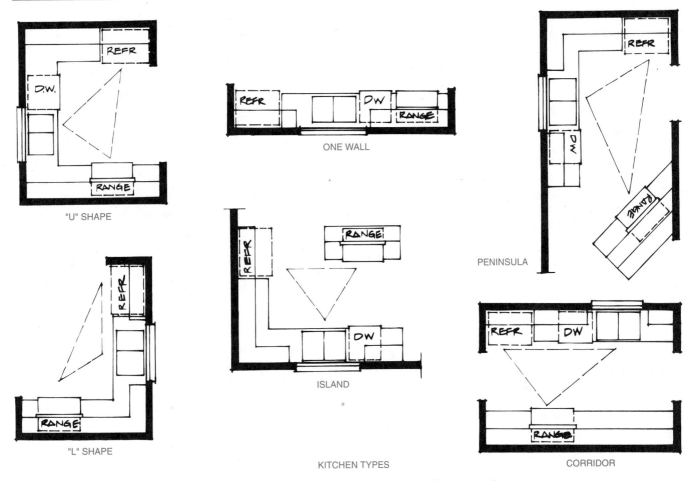

Figure 8–14   The work triangle must be between 15–22′ (4600–6700 mm) for efficient work patterns.

A cooktop and oven should be within three or four steps of each other if separate units are to be used. Cooktops are available in units 30, 36, 42, and 48″ (760, 900, 1070, and 1200 mm) wide. Oven units are typically 27, 30, or 33″ (680, 760, or 840 mm) wide. The bottom of the oven is typically placed at 30″ (760 mm) above the floor. Avoid placing a stove next to a refrigerator, a trash compactor, or a storage area for produce or breads.

A microwave unit may be placed above the oven unit, but the height of each unit above the floor should be considered in regard to the height of the clients. A microwave mounted at approximately 48″ (1200 mm) above the floor and near the daily eating area often proves to be very practical.

In addition to the major appliances of the kitchen, care must also be given to other details such as breadboards, counter workspaces, and specialized storage areas. The breadboard should be placed near the sink and the stove but not in a corner. Ideally a minimum of 5′ (1500 mm) of counter space can be placed between each appliance to allow for food preparation. Specialized storage often needs to be considered to meet clients' needs. These needs will be covered as cabinets are explained in Section VII.

A kitchen with an island is often desired (Figure 8–16). These are useful in gaining extra counter workspace as

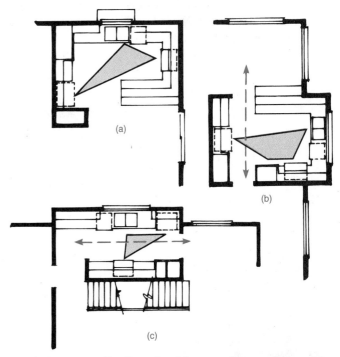

Figure 8–15   Traffic flow should never disrupt the working area of a kitchen as seen in plan A. The kitchen in plan B has traffic through the kitchen, but will not disrupt work. The traffic through plan C will be very disruptive.

well as an area for placing the stove near the sink and refrigerator. Islands are often used as an eating area as well as a workspace. Try to keep a minimum of 4' (1200 mm) between the island and other counters to allow for traffic around the island.

Utility Rooms

Utility rooms are often placed by either the bedrooms or kitchen area. There are advantages to both locations. Placing the utility room near the bath and sleeping areas places the washer and dryer near the primary source of the laundry. Care must be taken to insulate the sleeping area from the noise of the washer and dryer.

Placing the utilities near the kitchen allows for a much better traffic flow between the two major work areas of the home. With the utility room near the kitchen, space can often be provided in the utility room for additional kitchen storage. In smaller homes, the laundry facilities may be enclosed in a closet near the kitchen. In warm climates, laundry equipment may be placed in the garage or carport. Homes with a basement often place the laundry facilities near the water heater source. Many home owners find laundry rooms in a basement are too separated from the balance of the living areas. If bedrooms are on the upper floor, a laundry chute to the utility room can be a nice convenience. The utility room often has a door leading to the exterior. This allows the utility room to function as a mudroom. Entry can be made from the outside directly into the mudroom where dirty clothes can be removed. This allows for cleanup near the service entry and helps keep the rest of the house clean. Figure 8–17 shows this type of utility layout. Another common use for a utility room is to provide an area for sewing and ironing (Figure 8–18).

Figure 8–16 The island work counter and the number of windows create an open spacious kitchen. *Courtesy Marvin Windows and Doors.*

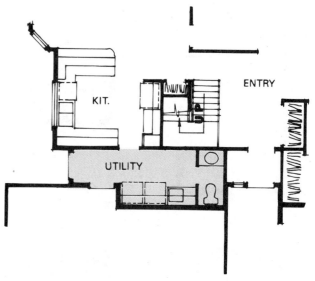

Figure 8–17 The utility room may serve as a service entry where dirty clothes can be removed as the home is entered.

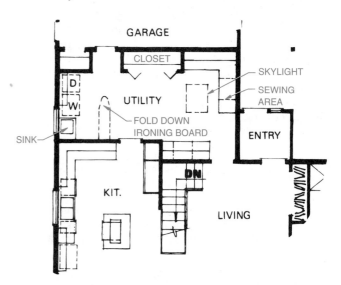

Figure 8–18 This utility room provides a spacious layout for home care chores.

Garage or Carport

Believe it or not, some people actually park their car in the garage. For those who do not, the garage often becomes a storage area, a second family room, or a place for the water heater and furnace. The minimum space for a single car is 11 × 20' (3400 × 6100 mm). A space for two cars should be 21 × 21' (6400 × 6400 mm). These sizes will allow minimal room to open doors or to walk around the car. Additional space should also be included for a workbench. A garage of 22 × 24' provides for parking, storage, and room to walk around. Many custom homes include enclosed parking for three cars. Consideration is often given for parking a boat, truck, camper, or recreational vehicle.

Space should be planned for the storage of a lawnmower and other yard maintenance equipment. Additional space must be provided if a post must be located near the center of the garage to support an upper floor. It will also be helpful if a required post can be located off center from front to back so that it will not interfere with the opening of car doors.

The garage location is often dictated by the site. Although access and size are important, the garage should be designed to blend with the residence. Many homeowners prefer that the garage doors not be seen from the street as the home is being approached. Clients usually have a preference for either one large single door or smaller doors for each parking space. A door 32″ (800 mm) wide should be installed to provide access to the yard.

In areas where cars do not need to be protected from the weather, a carport can provide an inexpensive alternative to a garage. Provide lockable storage space on one side of the carport if possible. This can usually be provided between the supports at the exterior end.

▼ TRAFFIC PATTERNS

A key portion of any design is the traffic flow between each area of the residence. *Traffic flow* is the route that people follow as they move from one area to another. Often this means hallways, but areas of a room are also used to aid circulation. Rooms should be of sufficient size to allow circulation through the room without disturbing the use of the room. By careful arrangements of doors, traffic can be arranged around furniture arrangements rather than through conversation or work areas.

Codes that cover the width of hallways were covered in Chapter 7. Good design practice will provide a width of 36 to 48″ (900–1200 mm) for circulation pathways. A space of 48″ (1200 mm) is appropriate for pathways that are used frequently, such as those that connect main living areas. Less frequently used pathways can be smaller. In addition to hallways, the entries of a residence are important in controlling traffic flow.

Guest Entry

The foyer is typically the first access to the residence for guests. In cold climates, many homes have an enclosure that leads to the foyer to prevent heat loss. The foyer should provide access to the living areas, the sleeping area, and the service areas of the house. The size of the foyer will vary with each home. Provide ample room to open a door completely and allow it to swing out of the way of people entering the home. A closet should be provided near the front door for storage of outdoor coats and sweaters. Provide access to the closet that will not be blocked by the opening of the front door.

Service Entry

The service entry is the entry used by the family as they return home. This is the typical pathway by which food and other parcels are carried from the garage to the kitchen. It may also be the entry used to provide access to yard areas. An area for storage of work clothes, coats, and boots should also be provided for the service entrance. A half- or three-quarter bath near the service entry is very practical if space and budget allow.

Stairs

In addition to considering traffic flow between rooms, it is important to determine how the levels of a home will relate to each other. The ideal stair location is off a central hallway. Many floor plans locate the stair in or near the main entry to provide an attractive focal point in the entry, as well as convenient access to upper areas. Stairs should be located for convenient flow from each floor level and should not require access through another room. The location will also be influenced by how often the stairs will be used. Stairs serving mechanical equipment in a basement will be used less frequently than stairs connecting the family room to other living areas.

▼ PROVIDING WHEELCHAIR ACCESSIBILITY

Well-planned, attractive, and accessible housing is essential for millions of people with disabilities—those with temporary injuries who walk with the aid of crutches or a walker or those who depend on a wheelchair for mobility. A house can be made accessible with just a few minor design changes based on the ADA, 1992.

Access

An accessible home has a level site for parking with paved walkways from parking areas to the main entry. A parking space of 9′–0″ (2700 mm) clear should be provided with a 42″ (1070 mm) minimum walkway between cars. A minimum of 9′–6″ (2896 mm) high ceilings are required for raised top vans. Any changes in elevation between the parking area and the front door can be no more than a 1:12, with a maximum rise of 30″ (760 mm) per run. Inclined ramps should be covered with a nonskid surface to provide sure footing for those walking and to help keep a wheelchair from going out of control. Ramps longer than 30′ (9100 mm) or higher than 30″ (760 mm) should have a landing 48″ (1200 mm) wide at the midpoint. Provide handrails that are 29 to 32″ (740–810 mm) high.

Interior Access

Doors with a 32-in. (813 mm) clear opening should be provided throughout the residence. Hinges are available that swing the door out of the doorway for maximum clearance. Hallways with a width of 42″ (1070 mm) will provide convenient traffic flow, but 48″ (1200 mm) will make turning from the hall to doorways much easier. Doors with a lever action handle or latches are typically easier to open than round door knobs. Windows should be within 24″ (600 mm) of the floor and should not require greater than 8 lb. of pressure to open the window.

An area 60″ (1500 mm) square will allow for a 360° turn in a wheelchair. An area of this size should be provided in each room. Floors covered with hardwood or tile provide excellent traction, although a low-pile carpet is also suitable. Avoid using small area rugs that move easily.

Work Areas

To provide a suitable kitchen work area, countertops should be 30 to 32″ (760–810 mm), high with pull-out cutting boards. Stoves should have all controls on a front panel rather than top mounted. A wall oven should be mounted between 30 and 42″ (762–1067 mm) above the floor. A side-by-side, frost-free refrigerator with door-mounted water and ice dispensers provides easy access.

Bathrooms

Doors that swing out of the bathroom will provide the most usable space in a small room such as a bathroom. Vanities should have a roll-under countertop with a lowered or tilted mirror. If possible, reinforce areas of each wall where a grab bar will be used with a ¾-in (20 mm) plywood backing material. Grab bars should be provided next to toilets and inside a tub or shower. Each tub and shower should have nonskid floor surfaces. Showers with a seat and a hand-held shower head will provide added safety for people with limited strength. A roll-in shower with no curb should be provided for clients using a wheelchair.

CHAPTER

Room Relationships and Sizes Test

DIRECTIONS

Answer the questions with short, complete statements or drawings as needed on an 8½ × 11 notebook sheet, as follows:

1. Letter your name, Chapter 8 Test, and the date at the top of the sheet.
2. Letter the question number and provide the answer. You do not need to write out the question.

Answers may be prepared on a word processor, if appropriate with course guidelines.

QUESTIONS

Question 8–1   What are the three main areas of a home?

Question 8–2   How much closet area should be provided for each bedroom?

Question 8–3   What rooms should the kitchen be near? Why?

Question 8–4   What are the functions of a utility room?

Question 8–5   Give the standard size for a two-car garage.

Question 8–6   List five functions of a family room.

Question 8–7   What are the advantages of bedrooms placed on an upper floor?

Question 8–8   What are the advantages of bedrooms placed on a lower floor?

Question 8–9   List four design criteria to consider when planning the dining room.

Question 8–10   What are the service areas of the home?

CHAPTER 9

Exterior Design Factors

The design of a house does not stop once the room arrangements have been determined. The exterior of the residence must also be considered. Often a client has a certain style in mind that will dictate the layout of the floor plan. In this chapter, ideas will be presented to help you better understand the design process. To design a structure properly, consideration must be given to the site, the floor plan style and shape, and exterior styles.

▼ SITE CONSIDERATIONS

Several site factors affect the design of a house. Among the most important to consider are the neighborhood and access to the lot. For a complete description of each item see Section 3.

Neighborhood

In the initial planning of a residence, the neighborhood in which it will be built must be considered. It is extremely poor judgment to design a $500,000 residence in a neighborhood of $150,000 houses. This is not to say the occupants will not be able to coexist with their neighbors, but the house will have poor resale value because of the lower value of the houses in the rest of the neighborhood. The style of the houses in the neighborhood should also be considered. Not all houses should look alike, but some unity of design can help keep the value of all the property in the neighborhood high.

Review Boards

In order to help keep the values of the neighborhood uniform, many areas have architectural control committees. These are review boards made up of citizens within an area who determine what may or may not be built. Although once found only in the most exclusive neighborhoods, review boards are now common in undeveloped subdivisions, recreational areas, and retirement areas. These boards often set standards for minimum square footage, height limitations, and the type and color of siding and roofing materials. The potential homeowner or designer is usually required to submit preliminary designs to the review board showing floor plans and exterior elevations.

Access

Site access can have a major effect on the design of the house. The narrower the lot is, the more the access will affect the location of the entry and the garage. Figure 9–1 shows typical access and garage locations for a narrow lot with access from one side. Usually only a straight driveway is used on interior lots because of space restrictions. When a plan is being developed for a corner lot, there is much more flexibility in the garage and house placement. When a residence is being planned for a rural site, weather and terrain can affect the access to the home site. Studying the weather patterns at the job site will help reveal areas of the lot that may be inaccessible during parts of the year due to poor water drainage or drifted snow. The shape of the land will determine where the access to the house can be placed.

▼ ELEMENTS OF DESIGN

The elements of design are the tools the architect or designer uses to create a structure that will be both functional and pleasing to the eye. These tools are line, form, color, and texture.

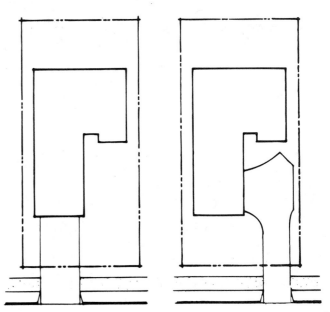

Figure 9–1 Access to an inner lot is limited by the street and garage location.

Line

The use of line provides a sense of direction or movement in the design of a structure and helps it to relate to the site and the natural surroundings. Lines may be curved, horizontal, vertical, or diagonal and are useful to accent or disguise features of a structure. Curved lines in a design tend to provide a soft, graceful feeling. Curves are often used in decorative arches, curved walls, and round windows and doorways. Figure 9–2 shows how curved surfaces can be used to accent a structure.

Horizontal lines, typically seen in long roof or floor lines, and balconies and siding patterns, can be used to minimize the height and maximize the width of a structure, as seen in Figure 9–3. Horizontal surfaces often create a sense of relaxation and peacefulness. Vertical lines create the illusion of height; they lead the eye upward and tend to provide a sense of strength and stability. Vertical lines are often used in columns, windows, trim, and siding patterns (Figure 9–4). Diagonal lines often can be used to create a sense of transition. Diagonal lines are typically used in roof lines and siding patterns. (Figure 9–5).

Figure 9–2 Curved lines are used throughout structures to provide a smooth transition between surfaces. *Courtesy California Redwood Association; design by Roland and Miller Associates, photo by Barbeau Engh.*

Figure 9–3 Horizontal lines can be used to accent the length, or hide the height of a structure. *Courtesy Western Red Cedar Lumber Association.*

Form

Lines are used to produce forms or shapes. Rectangles, squares, circles, ovals, and ellipses are the most common shapes found in structures. Often these forms are typically three-dimensional shapes, and the proportions between these shapes are important to design. The form of a structure should be dictated by the function for best design results. Forms are typically used to accent specific features of a structure. Figure 9–6 shows how form can be used to break up the length of the residence. Form also is used to create a sense of security. The use of large

Figure 9–4 Vertical lines can be used to accent the height of a structure. *Courtesy Louisiana Pacific Corporation.*

Figure 9–5 Diagonal lines are often used to create a sense of transition between surfaces of this modern Minneapolis home designed by the architect Meilus Wadsworth, New York City. *Courtesy Western Red Cedar Lumber Association.*

Figure 9–6 Forms such as rectangles, circles and ovals are used to provide interest. *Courtesy Louisiana Pacific Corporation.*

columns rather than thinner columns provides a greater sense of stability.

Color

The use of color is an integral part of interior design and decorating and serves to help distinguish exterior materials and accent shapes. A pleasing blend of colors creates a dramatic difference in the final appearance of any structure. Color is described by the terms *hue, value,* and *intensity.*

The *hue* represents what you typically think of as the color. Colors are divided into primary, secondary, and tertiary colors. Primary colors—red, yellow, and blue—cannot be created from any other color. All other colors are made by mixing, darkening, or lightening primary colors. Secondary colors—orange, green, and violet— are made from an equal combination of two primary colors. Mixing a primary color with a secondary color will produce a tertiary or intermediate color—for example, red-orange, yellow-orange, yellow-green, blue-green, blue-violet, and red-violet. Mixing each of the primary colors in equal amounts will produce black.

Mixing black with a color will darken the color, producing a shade of the original color. White is absent of color pigment, and gray is a mixture of black and white. Adding white to a color lightens the original color, producing a tint. The color *value* is the darkening or lightening of the hue.

Intensity is the brightness or strength of a specific color. A color is brightened as its purity is increased by removing neutralizing factors. Sports cars often have high-intensity colors. A color is softened by adding the color that is directly opposite the original color on the color wheel. Low-intensity colors such as mint green are used to create a calming effect.

Colors are also classified as warm or cool. Colors seen in warm objects, such as the reds and oranges of burning coals, are warm colors. Using warm colors tends to make objects appear larger or closer than they really are. Blues, greens, and violet are cool colors. Using these colors often makes objects appear farther away.

Texture

Texture, which refers to the roughness or smoothness of an object, is an important feature in selecting materials to complete a structure. Rough surfaces tend to create feelings of strength and security. Concrete masonry and rough-sawn wood have a rough texture. Rough surfaces give the illusion of reduced height. Resawn wood, plastic, glass, and most metals have a very smooth surface and create a sense of luxury. A smooth surface tends to give the illusion of increased height. Smooth surfaces also reflect more light and make colors seem brighter.

▼ PRINCIPLES OF DESIGN

Line, form, color, and texture are the tools of design. The principles of design affect how these tools are used to create an aesthetically pleasing structure. The basic principles to be considered are rhythm, balance, proportion, and unity.

Rhythm

Rhythm in music usually can be determined by most people. The beat of a drum gives a repetitive element that sets the foot tapping. In design, a repetitive element provides rhythm and leads the eye from one place to another in an orderly fashion through the design. Rhythm can also be created by a gradual change in materials, shape, and color. Gradation in materials could be from rough to small, with shape from large to small and with color from dark to light. Rhythm can also be created with a pattern that appears to radiate outward from a central point. A consistent pattern of shapes, sizes, or material can create a house that is pleasing to the eye as well as providing a sense of ease for the inhabitants and convey a sense of equilibrium. Figure 9–7 shows elements of rhythm in a structure.

Figure 9–7 Repetitive features lead the eye from form to form. *Courtesy Monier.*

Balance

Balance is the relationship between the various areas of the structure as they relate to an imaginary centerline. Balance may be formal or informal. Formal balance is symmetrical; one side of the structure matches the opposite side in size. The residence in Figure 9–8 is an example of formal balance. Both sides of the home are of similar mass. With informal balance, or a nonsymmetrical shape, balance can be achieved by placing shapes of different size in various positions around the imaginary centerline. This type of balance can be seen in Figure 9–9.

Proportion

Proportion is related to both size and balance. It can be thought of in terms of size, as in Figure 9–10, where one exterior area is compared to another. Many current design standards are based on the designs of the ancient Greeks. Rectangles using the proportions of 2:3, 3:5, and 5:8 have generally been accepted to be very pleasing.

Figure 9–8 Formal balance places features evenly along an imaginary center line.

Figure 9–9 Informal balance is achieved by moving the center line from the mathematical center of a structure and altering the sizes and shapes of objects on each side of the line. *Design by John J. Williams, Jr., of Knudson and Williams Architects. Photo by T. S. Gordon. Courtesy California Redwood Association.*

Proportion can also be thought of in terms of how the residence relates to its environment.

Proportions must also be considered inside the house. A house with a large living and family area needs to have the rest of the structure in proportion to these areas. In large rooms the height must also be considered. A 24′ × 34′ (7300 × 10,900 mm) family room is too large to have an 8′ high flat ceiling. This standard-height ceiling would be out of proportion to the room size. A 10′ (3000 mm) high or a vaulted ceiling would be much more in keeping with the size of the room.

Unity

Unity relates to rhythm, balance, and proportion. Unity ties a structure together with a common design or decorating pattern. Similar features that relate to each other can give a sense of well-being. Avoid adding features to a building that appear to be just there or tacked on. See Figure 9–11.

As an architectural drafter, you should be looking for these basic elements in residences that you see. Walk through houses at every opportunity to see how other designers have used these basics of design. Many magazines are available that feature interior and exterior home designs. Study the photos for pleasing relationships and develop a scrapbook of styles and layouts that are pleasing to you. This will provide valuable resource material as you advance in your drafting career and become a designer or architect.

▼ FLOOR PLAN STYLES

Many clients come to a designer with specific ideas about the kind and number of levels they desire in a house. Some clients want homes with only one level so that no stairs will be required. For other clients, the levels in the

Figure 9–10 Placing shapes that are related to others by size and shape creates a pleasing flow from form to form. *Photo courtesy Saundra Powell.*

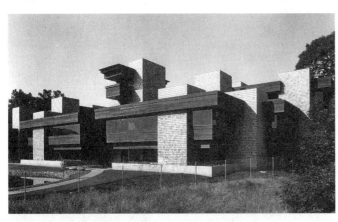

Figure 9–11 Unity is a blend of varied shapes and sizes into a pleasing appearance, as shown by this office structure in Santa Rosa, California, by Lawrence Simons and Associates, A.I.A. *Courtesy Shakertown Corp.*

Figure 9–12 A single-level residence is popular because of the ease of maintenance and its relatively low construction price. *Courtesy Michelle Cartwright.*

house are best determined by the topography of the lot. Common floor plan layouts are single, split level, daylight basement, two story, dormer, and multilevel.

Single Level

The single-level house (Figure 9–12) has become one of the most common styles built. It is a standard of many builders because it can be built with minimal expense. A one-level layout provides stair-free access to all rooms, which makes it attractive to people with limited mobility. It is also preferred by many homeowners because it is easy to maintain and can be used with a variety of exterior styles.

Split Level

The split level (Figure 9–13) is an attempt to combine the features of a one- and a two-story residence. This style is best suited to sloping sites, which allow one area of the house to be two stories and one area to be one story. Many clients like the reduced number of steps from one level to another that is found in the split-level design. The cost of construction for a split-level plan is usually greater than the cost of a single-level structure of the same size because of the increased foundation cost that the sloping lot will require.

Split-level plans may be split from side to side or front to back. The front entrance and the main living areas are typically located on one level, with the sleeping area

located over the garage and service areas on side-to-side split-level homes. In front-to-rear splits, the front of the residence is typically level with the street, with the rear portion of the structure stepping to match the contour of the building site.

Daylight Basement

Although called a daylight basement, this style of house could be either a one-story house over a basement or garage, or two complete living levels. This style of house is well suited for a sloping lot. From the high side of the lot, the house will appear to be a one-level structure. From the low side of the lot, both levels of the structure can be seen. See Figure 9–14.

Two Story

A two-story house (Figure 9–15) provides many options for families that don't mind stairs. Living and sleeping areas can be easily separated, and a minimum of land will be used for the building site. On a sloping lot, depending on the access, the living area can be on either the

Figure 9–13 Homes with a split in floor levels are ideal for gently sloping lots. *Courtesy Mike Jefferis.*

Figure 9–14 A home with a daylight basement allows living areas on the lower level to have emergency egress. *Courtesy Home Building Plan Service.*

Figure 9–15 Two-level homes provide a maximum amount of square footage at a lower cost per square footage than other styles of homes. *Courtesy Louisiana Pacific Corporation.*

upper or lower level. The most popular feature of a two-level layout is its ability to provide the maximum building area at a lower cost per square foot than other styles of houses. This savings results from less material being used on the foundation, exterior walls, and the roof of a two-story structure compared to other styles of houses.

Dormer

The dormer style provides for two levels, with the upper level usually about half of the square footage of the lower floor. This floor plan is best suited to an exterior style that incorporates a steep roof (Figure 9–16). The dormer level is formed in what would have been the attic area. The dormer has many of the same economic features of a two-story home.

Figure 9–16 Dormer-style home.

Figure 9–17 Multi-level home.

Multilevel

With this style of layout, the possibilities for floor levels are endless. Site topography will dictate this style as much as the owners' living habits. Figure 9–17 shows a multilevel layout. The cost for this type of home exceeds all other styles because of the problems of excavation, foundation construction, and roof intersections.

▼ EXTERIOR STYLES

Exterior style is often based on the styles of houses from the past. Some early colonists lived in lean-to shelters called wigwams, formed with poles and covered with twigs woven together and then covered with mud or clay. Log cabins were introduced by Swedish settlers in Delaware and soon were used throughout the colonies by the poorer settlers. Many companies now sell plans and pre-cut kits for assembling log houses. Many of the colonial houses were similar in style to the houses that had been left behind in Europe. Construction methods and materials were varied because of the weather and building materials available.

Colonial influence can still be seen in houses that are built to resemble the Georgian, saltbox, garrison, Cape Cod, federal, Greek revival, and southern colonial styles of houses. Other house styles from the colonial period include influences of the English, Dutch, French, and Spanish. Other popular home styles include farmhouse, ranch, Victorian, and contemporary.

Georgian

The Georgian style of design is a good example of a basic style that was modified throughout the colonies to meet the needs of available material and weather. This style receives its name for the kings of England who were in power as this style flourished. The Georgian style follows the classical principles of design used by the ancient Greeks. The principles of form and symmetry can be seen throughout the structure but are most evident in the front elevation. The front entry is centered on the wall, and equally spaced windows are placed on each side. The front entry is usually covered with a columned porch and the doorway trimmed with carved wood detailing. When constructed in the South, much of the facade is built of brick. In the northern states, wood siding is the major covering. Other common exterior materials are stucco or stone. An example of Georgian styling can be seen in Figure 9–18.

Figure 9–18 Georgian style homes have formal balance with the entry on the centerline. *Courtesy Janice Jefferis.*

Saltbox

One of the most common modifications of the Georgian style is the saltbox. The saltbox maintained the symmetry of the Georgian style but omitted much of the detailing. A saltbox is typically a two-story structure at the front but tapers to one story at the rear. Window areas are usually protected with shutters to provide protection from winter winds. Figure 9–19 shows features of saltbox styling in a residence.

Figure 9–19 A saltbox residence has a two-level front and a one-level rear. *Courtesy Home Planners, Inc.*

Figure 9–20 A garrison-style home has an upper level that overextends the lower level. *Courtesy Home Planners, Inc.*

Garrision

The garrison style combines the styling of the saltbox and Georgian houses with the construction methods of log buildings. The garrison was originally modeled after the lookout structures of early forts. The upper level extends past the lower level, and on the fort made the walls harder to scale. Originally, heavy timbers were used to support the overhang and were usually carved. Figure 9–20 shows the features of the garrison-style residence.

Cape Cod

Cape Cod styling is typically one level with a steep roof, which allows an upper-floor level to be formed throughout the center of the house. Dormers are typically placed on the front side of the roof to allow the upper floor area to be habitable space. Windows are symmetrically placed around the door and have shutters on the lower level. An example of Cape Cod styling can be seen in Figure 9–21. An offshoot of the Cape Cod style is the Cape Ann, which has many of the same features as a Cape Cod house but is covered by a gambrel roof.

Figure 9–21 A Cape Cod-style home features formal balance of other styles but adds dormers and shutters. *Courtesy Georgia Pacific.*

Federal

The Federal style of the late 1700s combines the Georgian with classical Roman and Greek styles to form a very dignified style. Homes were built of wood or brick. Major additions to the Georgian home include a high, covered

Figure 9–22 Common door and window pediments of American architecture.

Figure 9–23 Federal style combines features of Georgian and classical Roman and Greek architecture.

entry porch or portico supported on Greek-styled columns centered over the front door. The door is typically trimmed with arched trim, and windows are capped with a projected pediment. Figure 9–22 shows examples of common pediment styles found throughout American architecture. Figure 9–23 shows an example of a Federal-style residence.

Greek Revival

Homes of this styling are built using classic proportions and decorations of classical Greek architecture. Typically homes are large, rectangular, and very boxlike. Decoration is added by a two-story portico with a low, sloped gable roof supported on Greek columns centered on the residence. Figure 9–24 shows an example of Greek revival architecture.

Southern Colonial

Southern colonial homes are similar to the Georgian style with their symmetrical features. The southern colonial style usually has a flat, covered porch, which extends the length of the house to protect the windows from the summer sun. Figure 9–25 shows an example of southern colonial styling.

English

English-style houses are fashioned after houses that were built in England prior to the early 1800s. These houses feature an unsymmetrical layout and walls that are usually constructed of stone, brick, or heavy timber

and plaster. Window glass is typically diamond shaped rather than the more traditional rectangle. Figure 9–26 shows common features of English Tudor styling. The Elizabethan-style home has influence from the English and Dutch as well as Gothic styling. These homes shared the half-timber styling of Tudor homes but were typically irregularly shaped. Figure 9–27 shows features common to an Elizabethan manor.

Figure 9–24 Greek revival features classic Greek proportions and ornamentation.

Figure 9–25 Southern colonial or plantation-style homes reflect classic symmetry with a large, covered porch to provide protection from the sun.

Figure 9–26 English Tudor combines half-timber with brick, stone, and plaster. *Courtesy Piercy & Barclay Designers, Inc, CPBD.*

Figure 9–27 Elizabethan-style homes are similar to other English-style homes but are typically very irregularly shaped.

Figure 9–28 Dutch colonial homes often feature a gambrel roof. *Courtesy Mike Jefferis.*

Dutch

The Dutch colonial style has many of the same features of homes already described. The major difference is in the roof shape. A Dutch colonial style features a gambrel roof, also known as a barn roof. This roof is made of two levels. The lower level is usually very steep and serves as the walls for the second floor of the structure. The upper area of the roof is the more traditional gable roof. An example of Dutch colonial style can be seen in Figure 9–28. The gambrel roof will be described in Section VI.

French

French colonial styling is also a matter of roof design. Similar to the gambrel with its steep lower roof, French colonial styling uses a hipped or mansard roof to hide the upper floor area. A mansard roof is basically an angled wall and will be discussed in Section VI. Figure 9–29 shows an example of French colonial styling.

Later influences of the French are visible in the single-level French manor. Originally found in the northern states, these homes are usually a rectangle, with a smaller wing on each side. The roof is usually a mansard to hide the upper floor, but hip roofs were also used, as seen in

Figure 9–30. Another common style from France is the French city house, shown in Figure 9–31.

The French Normandy style first became popular in the 1600s. These homes were typically multilevel and framed with brick, stone or wood, and plaster with half-timber decoration. The roof is typically a gable, although some are now built with a hip roof. A circular turret is typically found near the center of the home. Figure 9–32 shows an example of a French Normandy residence.

In the southern states, the French Normandy style has been modified into the French plantation style. This home typically is two full floors with a wraparound porch covered with a hip roof. A third floor is often a common component, with dormers added to provide light and ventilation. Figure 9–33 shows a French plantation home. This style was modified around New Orleans into what is now known as the Louisiana French style. The

Figure 9–29 French colonial homes often hid the upper floor behind a mansard roof. *Courtesy Home Planners, Inc.*

Figure 9–30 A French manor typically features a wing on each side of a central rectangle. *Courtesy David Jefferis.*

Figure 9–31 A French city style residence typically has dormers and a hip roof covering a very symmetrical residence.

Figure 9–32 French Normandy homes typically feature a turret, which houses the stairs leading to an upper level. *Courtesy Ludowici Celadon Company.*

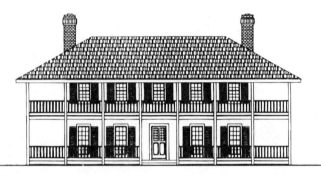

Figure 9–33 A French plantation home features a two-level wraparound porch covered by a hipped roof.

Figure 9–34 Spanish-style homes typically reflect the use of low-sloping tile roofs, arches, and window grills. *Courtesy Home Planners, Inc.*

size of the balconies is diminished, but the supporting columns and rails became fancier.

Spanish

Spanish colonial buildings were constructed of adobe or plaster and were usually one story. Arches and tiled roofs are two of the most common features of Spanish, or mission-style, architecture. Timbers are often used to frame a flat or very low-pitched roof. Windows with

grills or spindles and balconies with wrought iron railings are also common features. Figure 9–34 shows an example of Spanish style architecture.

Farmhouse

The farmhouse style of residence makes use of two-story construction and is usually surrounded by a covered porch. Trim and detail work, common on many other styles of architecture, are rarely found on this style of home. Figure 9–35 shows an example of farmhouse style.

Ranch

The ranch style of construction comes from the Southwest. This style is usually defined by a one-story, rambling layout, which is made possible because of the mild climate and plentiful land on which such houses are built. The roof shape is typically low pitched, with a large overhang to block the summer sun. The major exterior material is usually stucco or adobe. Figure 9–36 shows the ranch style.

Victorian

The Victorian and Queen Anne styles of house from the late 1800s are still being copied in many parts of the country. These styles feature irregularly shaped floor plans and very ornate detailing throughout the residence.

Figure 9–35 Farmhouse styles feature simple single-level structures with a covered porch. *Courtesy Home Building Plan Service.*

Figure 9–36 Ranch-style homes feature a one-level elongated floor plan covered by a low, sloping roof.

Figure 9–37 Victorian homes typically include very ornate, irregular shapes. *Courtesy Janice Ann Jefferis.*

Victorian houses often borrow elements from many other styles of architecture, including partial mansard roofs, arched windows, and towers. Exterior materials include a combination of wood and brick. Wrought iron is often used. Figure 9–37 shows an example of the Victorian style.

Contemporary

It is important to remember that a client may like the exterior look of one of the traditional styles, but very rarely would the traditional floor plan of one of those houses be desired. Quite often the floor plans of the older style of house produced small rooms with poor traffic flow. A designer must take the best characteristics of a particular style of house and work them into a plan that will best suit the needs of the owner.

Figure 9–38 Many contemporary homes feature clean lines with little or no trim to provide an attractive structure. *Design by Jack Smuckler, A.I.A., photo by Jerry Swanson. Courtesy California Redwood Association.*

Figure 9–39 Many contemporary homes feature a portico at the entry door to provide a warm, inviting entry. *Courtesy Georgia Pacific Corporation.*

Figure 9–40 Many contemporary homes feature an openness from room to room and from the interior to the exterior. *Design by John Rose, A.I.A., photo by Richard Mandelkorn. Courtesy Timberpeg.*

Figure 9–41 Many contemporary homes include passive solar heating features. *Courtesy Residential Designs, A.I.B.D.*

Contemporary, or modern, does not denote any special style of house. Some houses are now being designed to meet a wide variety of needs, and others reflect the particular lifestyle of the owner. Figures 9–38 through 9–43 show examples of the wide variety of contemporary houses.

Figure 9–42 Many homes are now being constructed that share a common property line. *Design by McCool, McDonald & Associates. Courtesy Shakertown Corp.*

Figure 9–43 Subterranean living provides an energy-efficient means of construction for severe climates. *Courtesy Weather Shield Mfg., Inc.*

CHAPTER 9

Exterior Design Factors Test

DIRECTIONS

Answer the questions with short complete statements or drawings as needed on an 8½ × 11 notebook sheet, as follows:

1. Letter your name, Chapter 9 Test, and the date at the top of the sheet.
2. Letter the question number and provide the answer. You do not need to write out the question.
3. Answers may be prepared on a word processor if appropriate with course guidelines.

QUESTIONS

Question 9–1   Explain how the neighborhood can influence the type of house that will be built.

Question 9–2   Describe how lines can be used in the design of a structure.

Question 9–3   Sketch a simple two-bedroom house in two of the following shapes: L, U, T, or V.

Question 9–4   List four functions of a review board.

Question 9–5   Describe four floor plan styles, and explain the benefits of each.

Question 9–6   What factors make a two-story house more economical to build than a single-story house of similar size?

Question 9–7   What house shape is the most economical to build?

Question 9–8   Photograph or sketch examples of the following historical styles found in your community:
a. Dutch colonial
b. Garrison
c. Saltbox
d. Victorian

Question 9–9   What forms are most typically seen in residential design?

Question 9–10   List and define the three terms used to describe color.

Question 9–11   Identify the primary and secondary colors.

Question 9–12   List the major features of the following styles of house.
a. Ranch
b. Tudor
c. Cape Cod
d. Spanish

Question 9–13   How is a tint created?

Question 9–14 Photograph or sketch three contemporary houses that have no apparent historic style.

Question 9–15 What period of architecture influenced the Georgian style?

Question 9–16 Sketch a Dutch colonial house.

Question 9–17 Sketch or photograph a house in your community built with a traditional influence, and explain which styles this house has copied.

Question 9–18 What are some of the drawbacks of having a traditional house style?

Question 9–19 What are the common proportions of classical Greek design?

Question 9–20 What style of residence has two full floors with a wraparound porch covered with a hip roof?

CHAPTER 10

Site Orientation

Site orientation is the placement of a structure on the property with certain environmental and physical factors taken into consideration. Site orientation is one of the preliminary factors that an architect or designer takes into consideration when beginning the design process. The specific needs of the occupants such as individual habits, perceptions, aesthetic values, and so on are important and need to be considered. The designer or architect has the responsibility of putting together the values of the owner and other factors that influence the location of the structure when designing the home or business. In some cases site orientation is predetermined. For example, in a residential subdivision where all of the lots are 50′ × 100′ (1524 × 3048 m), the street frontage clearly dictates the front of the house. The property line setback requirements do not allow much flexibility. In such a case, site planning has a minimal amount of influence on the design. This chapter presents factors that may influence site orientation: terrain, view, solar, wind, and sound.

▼ TERRAIN ORIENTATION

The terrain is the characteristic of the land on which the proposed structure will be placed. Terrain affects the type of structure to be built. A level construction site is a natural location for a single-level or two-story home. Some landscape techniques on a level site may allow for alternatives.

For example, the excavation material plus extra topsoil could be used to construct an earth *berm*, which is a mound or built-up area. The advantage of the berm is to help reduce the height appearance of a secondary story or to add earth insulation to part of the structure. See Figure 10–1.

Sloped sites are a natural location for multilevel or daylight basement homes. A single-level home is a poor choice on a sloped site due to the extra construction cost in excavation or building up the foundation. See Figure 10–2. Figure 10–3 shows how terrain can influence housing types. Subterranean construction has gained popularity due to some economical advantages in energy consumption. The terrain of the site is an important factor to consider in the implementation of these designs. See Figure 10–4.

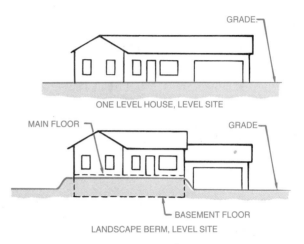

Figure 10–1   Level site.

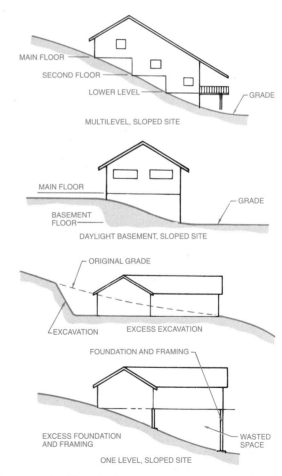

Figure 10–2   Sloped sites.

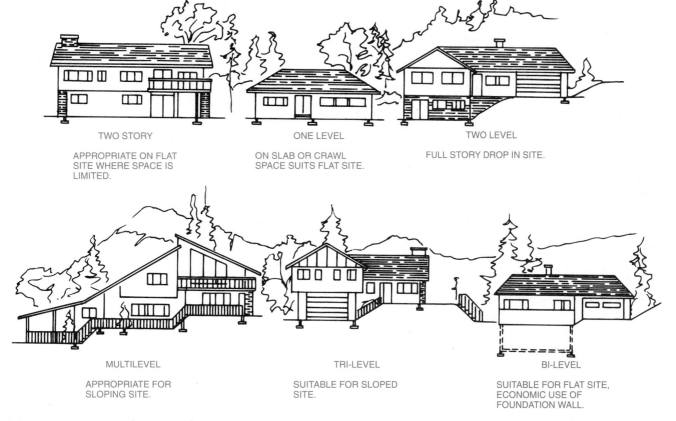

TWO STORY

APPROPRIATE ON FLAT SITE WHERE SPACE IS LIMITED.

ONE LEVEL

ON SLAB OR CRAWL SPACE SUITS FLAT SITE.

TWO LEVEL

FULL STORY DROP IN SITE.

MULTILEVEL

APPROPRIATE FOR SLOPING SITE.

TRI-LEVEL

SUITABLE FOR SLOPED SITE.

BI-LEVEL

SUITABLE FOR FLAT SITE, ECONOMIC USE OF FOUNDATION WALL.

Figure 10–3 Terrain determines housing types.

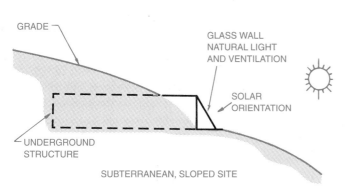

Figure 10–4 Subterranean construction site.

Figure 10–5 View orientation. *Courtesy Peachtree.*

▼ VIEW ORIENTATION

In many situations future homeowners purchase a building site even before they begin the home design. In a large number of these cases, the people are buying a view. The view may be of mountains, city lights, a lake, the ocean, or even a golf course. These view sites are usually more expensive than comparable sites without a view. The architect's obligation to the client in this situation is to provide a home design that optimizes the view. Actually the best situation is to provide an environment that al-

lows the occupants to feel as though they are part of the view. Figure 10–5 shows a dramatic example of a view as part of the total environment.

View orientation may conflict with the advantages of other orientation factors such as solar or wind. When a client pays a substantial amount to purchase a view site, such a trade-off may be necessary. When the view dictates that a large glass surface face a wind-exposed non-solar orientation, some energy saving alternatives should be considered. Provide as much solar-related glass as possible. Use small window surfaces in other areas to help minimize heat loss. Use triple-glazed windows in the exposed view surface as well as in the balance of the structure if economically possible. Insulate to stop air leakage.

▼ SOLAR ORIENTATION

The sun is an important factor in home orientation. Sites with a solar orientation allow for excellent exposure to the sun. There should not be obstacles such as tall buildings, evergreen trees, or hills that have the potential to block the sun. Generally a site located on the south slope has these characteristics.

When a site has southern exposure that allows solar orientation, a little basic astronomy can contribute to proper placement of the structure. Figure 10–6 shows how a southern orientation in relationship to the sun's path provides the maximum solar exposure.

Establishing South

A perfect solar site allows the structure to have unobstructed southern exposure. When a site has this potential, then true south should be determined. This determination should be established in the preliminary planning stages. Other factors that contribute to orientation, such as view, also may be taken into consideration at this time.

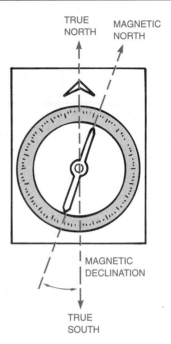

Figure 10–7 Magnetic declination.

If view orientation requires that a structure be turned slightly away from south, it is possible that the solar potential will not be significantly reduced.

True south is determined by a line from the North to the South Poles. When a compass is used to establish north-south, the compass points to *magnetic north*. Magnetic north is not the same as true north. The difference between true north and magnetic north is known as the *magnetic declination*. Figure 10–7 shows the compass relationship between true north-south and magnetic north. The magnetic declination differs throughout the country. Figure 10–8 shows a map of the United States with lines that represent the magnetic declination at different locations.

A magnetic declination of 18° east, which occurs in northern California, means that the compass needle points 18° to the east of true north, or 18° to the west of

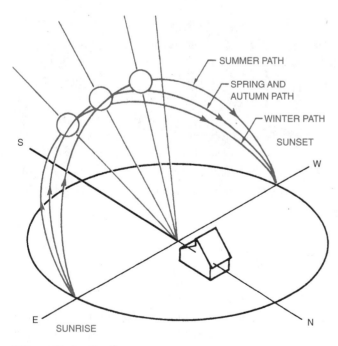

Figure 10–6 Southern exposure.

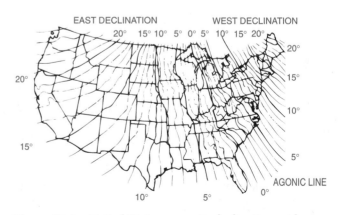

Figure 10–8 United States magnetic declination grid.

true south. In this location, as you face toward magnetic south, true south will be 18° to the left.

Solar Site Planning Tools

Instruments are available that calculate solar access and demonstrate shading patterns for any given site throughout the year. *Solar access* refers to the availability of direct sunlight to a structure or construction site.

Some instruments provide accurate readings for the entire year at any time of day, in clear or cloudy weather.

Solar Site Location

A solar site in a rural or suburban location where there may be plenty of space to take advantage of a southern solar exposure allows the designer a great deal of flexibility. Some other factors, however, may be considered when selecting an urban solar site. Select a site where zoning restrictions have maximum height requirements. This prevents future neighborhood development from blocking the sun from an otherwise good solar orientation. Avoid a site where large coniferous trees hinder the full potential of the sun exposure. Sites that have streets running east-west and 50′ × 100′ (1524 × 3048 m) lots provide fairly limited orientation potential. Adjacent homes can easily block the sun unless southern exposure is possible. Figure 10–9 shows how the east-west street orientation can provide the maximum solar potential. If the client loves the trees but also wants solar orientation, then consider a site where the home could be situated so southern exposure can be achieved with coniferous trees to the north and deciduous trees on the south side. The coniferous trees can effectively block the wind exposure without interfering with the solar orientation. The deciduous trees provide shade relief from the hot summer sun. In the winter, when these trees have lost their leaves, the winter sun exposure is not substantially reduced. Figure 10–10 shows a potential solar site with trees.

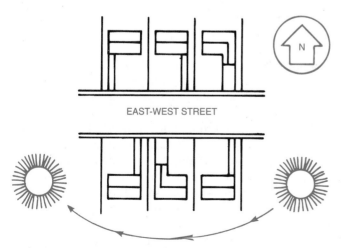

Figure 10–9   East-west street orientation.

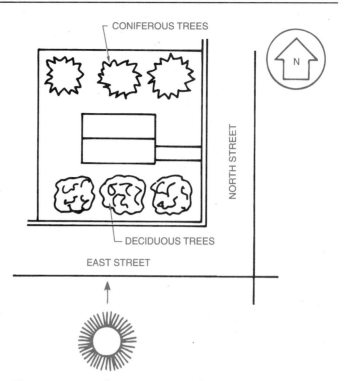

Figure 10–10   Solar orientation with trees.

▼ WIND ORIENTATION

The term *prevailing winds* refers to the direction from which the wind most frequently blows in a given area of the country. For example, if the prevailing winds are said to be southwesterly, that means that the winds in the area most generally flow from the southwest. There are some locations that may have southwesterly prevailing winds, but during certain times of the year, severe winds may blow from the northeast. The factors that influence these particular conditions may be mountains, large bodies of water, valleys, canyons, or river basins. The prevailing winds in the United States are from west to east, although some areas have wind patterns that differ from this.

Wind conditions should be taken into consideration in the orientation of a home or business. There is, in many cases, a conflict between the different aspects of site orientation. For example, the best solar orientation may be in conflict with the best wind orientaion. The best view may be out over the ocean, but the winter winds may also come from that direction.

One of the factors used to evaluate orientation may outweigh another. Personal judgment may be the final ruling factor. A good combination may be achieved if careful planning is used to take all of the environmental factors into consideration.

Site Location

Conditions that influence the site location may be found in almanacs or in the local library. Look for topics such as climate, microclimate, prevailing wind, or wind conditions.

Evaluate the direction of the prevailing winds in an area by calling the local weather bureau or by a discussion with local residents. Select an area where winter winds are at a minimum or where there is protection from these winds. Within a 25-mile (40-km) radius of a given area, there may be certain locations where wind is more of a problem than in others. A hill, mountain, or forest can protect the building site from general wind conditions. Figure 10–11 shows some examples of site locations that may be protected from prevailing winds.

It may be very difficult to avoid completely the negative effects of winds in a given location. When a site has been selected and wind continues to be a concern, construction or landscaping techniques can contribute to a satisfactory environment. A subterranean structure may totally reduce the effects of wind. These underground homes, when placed with the below-grade portion against the wind, can create a substantial wind control, as seen in Figure 10–12.

When subterranean housing is not considered a design alternative, then some other techniques can be used. Building the structure partly into the ground with a basement or using berming may be a successful alternative. Proper foundation drainage is a factor to consider when these methods are used. Figure 10–13 shows how wind protection can be achieved with proper use of the earth. Landscaping can also add an effective break between the cold winter wind and the home site. Coniferous trees or

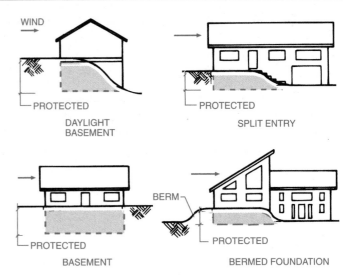
Figure 10–13   Wind protection using earth.

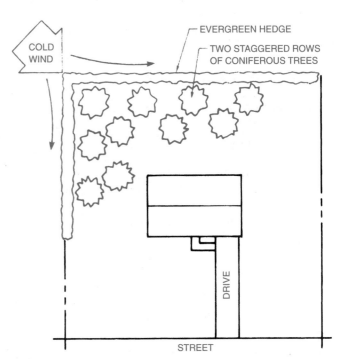
Figure 10–14   Wind protection with landscaping.

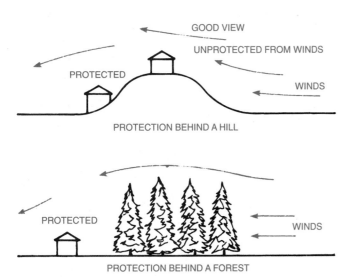
Figure 10–11   Locations protected from wind.

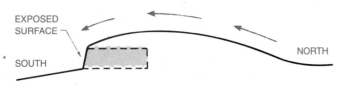
Figure 10–12   Subterranean wind protection.

other evergreen landscaping materials can provide an excellent windbreak. These trees should be planted in staggered rows. Two rows is a suggested minimum; three rows is better. Some hedge plants could be planted in one row to provide wind protection. Figure 10–14 shows how landscaping can protect the structure from cold wind.

Room and construction design can also influence wind protection. Place rooms that do not require a great deal of glass for view or solar use on the north side or the side toward severe winter winds. A garage, even though unheated, is an insulator and can provide an excellent break between the cold winter winds and the living areas of the home. Bedrooms with fairly small windows may be

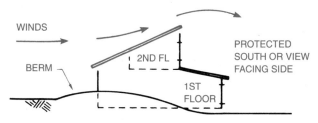

Figure 10–15 Roof slope wind protection.

placed to provide a barrier between the wind and the balance of the home. A common method of exterior construction design that helps deflect wind is a long, sloping roof. A flat two-story surface causes a great deal of wind resistance. A better alternative is achieved when a long, sloping roof is used to reduce wind resistance, resulting in more energy efficient construction. Figure 10–15 shows how roof construction can effectively deflect prevailing winds.

Summer Cooling Winds

Summer winds may be mild and help provide for a more comfortable living environment. Comfort can be achieved through design considerations for natural ventilation and through landscape design. Effective natural ventilation can be achieved when a structure has openings in opposite walls that allow for cross ventilation. In a two- or multiple-story structure, the openings can be effectively tied into the stairwells to provide for continuous ventilation. Figure 10–16 shows an example of how good natural ventilation can be achieved.

Landscaping can also help provide summer cooling. The discussion of wind protection showed how coniferous trees can protect against cold winter winds. These evergreen trees can also be used in conjunction with deciduous trees to help funnel summer winds into the

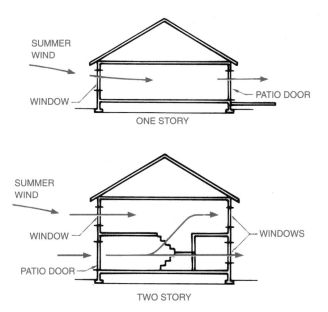

Figure 10–16 Natural ventilation.

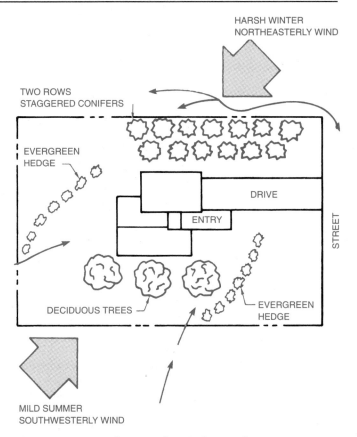

Figure 10–17 Landscaping for wind control.

home site and help provide natural summer cooling. Fences or other buildings can also help create a wind funnel. Deciduous trees serve a triple purpose. In the summer they provide needed shade and act as a filter to cool heated wind. In the winter they lose their leaves, thus allowing the sun's rays to help warm the house. Figure 10–17 shows how an effective landscape plan can be used to provide a total environment.

▼ SOUND ORIENTATION

If your construction site is located in the country near the great outdoors, sound orientation may not be a concern. The sounds that you will contend with are singing birds, croaking frogs, chirping crickets, and a few road noises. It is very difficult to eliminate road sounds from many other locations. A site that is within a mile of a major freeway may be plagued by the droning sounds of excessive road noise. A building site that is level with or slightly below a road may have less noise than a site that is above and overlooking the sound source.

A few landscaping designs can contribute to a quieter living environment. Berms, trees, hedges, and fences can all be helpful. Some landscape materials deflect sounds; others absorb sounds. The density of the sound barrier has an influence on sound reduction, although even a single hedge can reduce a sound problem somewhat. A mixture of materials can most effectively reduce sound.

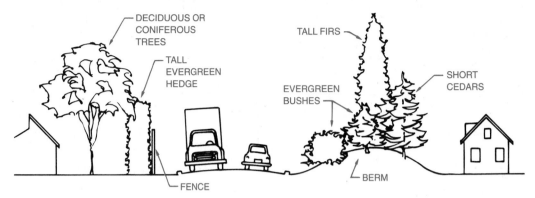

Figure 10–18 Sound insulation landscaping.

Keep in mind that while deciduous trees and plants help reduce noise problems during the summer, they are a poor sound insulation in the winter. See Figure 10–18.

Trees may provide sound insulation. The greater the width of the plantings for sound insulation, the better is the control. Trees planted in staggered rows provide the best design. Figure 10–19 shows the plan view of a site with sound barrier plants. Notice that the garage is placed between the living areas of the home and the street to help reduce sound problems substantially.

A number of factors influence site orientation. Solar orientation may be important to one builder; another homeowner may have view orientation as the main priority. A perfect construction site has elements of each site orientation design feature. When a perfect site is not available, the challenge for the designer or architect is to orient the structure to take the best advantage of each potential design feature. As you have seen, it is possible

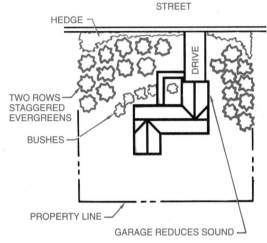

Figure 10–19 Sound reduction: Home and landscape plan.

to achieve some elements of good orientation with excavation and landscaping techniques. Always take advantage of the natural conditions whenever possible.

CHAPTER
10 Site Orientation Test

DIRECTIONS

Answer the questions with short complete statements or drawings as needed on an 8½ × 11 sheet of notebook paper, as follows:

1. Letter your name, Chapter 10 Test, and the date at the top of the sheet.
2. Letter the question number and provide the answer. You do not need to write out the question.

Answers may be prepared on a word processor if appropriate with course guidelines.

QUESTIONS

Question 10–1 Define site orientation.

Question 10–2 Describe in short, complete statements five factors that influence site orientation.

Question 10–3 Define magnetic declination.

Question 10–4 What is the magnetic declination of the area where you live?

Question 10–5 Name and describe two types of obstacles that can block the sun in an otherwise good solar site.

Question 10–6 Describe how trees can be an asset in solar orientation.

Question 10–7 What are prevailing winds?

Question 10–8   Name two features that can protect the structure from wind.

Question 10–9   Show in a sketch how landscaping can be used for winter wind protection and summer cooling.

Question 10–10   Show in a sketch how landscaping can be an effective sound control.

Question 10–11   List at least two factors that must be considered in subterranean construction.

Question 10–12   List at least two disadvantages of a two-story house on a narrow, treeless lot.

Question 10–13   List at least five advantages of considering site orientation in the planning of a home.

Question 10–14   Identify the type of homes that are a natural for placement on a level site.

Question 10–15   Sloped sites are a natural for what type of homes?

SECTION III

SITE PLANS

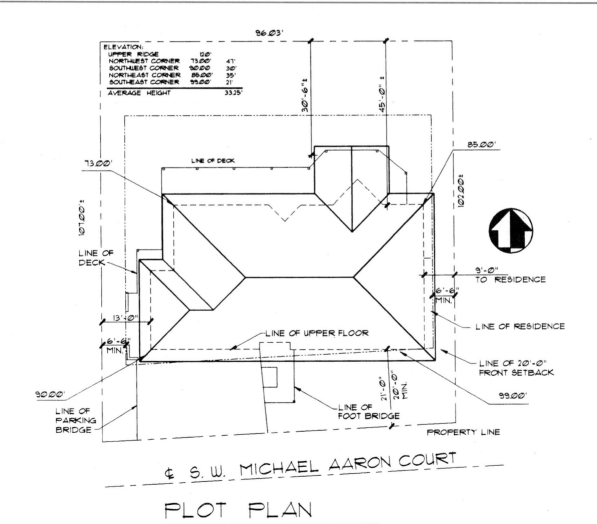

ELEVATION:
UPPER RIDGE	120'	
NORTHWEST CORNER	73.00'	47'
SOUTHWEST CORNER	90.00'	30'
NORTHEAST CORNER	85.00'	35'
SOUTHEAST CORNER	99.00'	21'
AVERAGE HEIGHT	33.25'	

96.03'

30'-6"±

45'-0"±

85.00'

102.00'±

LINE OF DECK

73.00'

101.00'±

LINE OF
DECK

9'-0"
TO RESIDENCE

6'-6"
MIN.

LINE OF RESIDENCE

13'-0"

LINE OF UPPER FLOOR

6'-6"
MIN.

LINE OF 20'-0"
FRONT SETBACK

21'-0"
20'-0"
MIN.

90.00'

99.00'

LINE OF
FOOT BRIDGE

LINE OF
PARKING
BRIDGE

PROPERTY LINE

¢ S. W. MICHAEL AARON COURT

PLOT PLAN

1" ———— 10'-0"

CHAPTER 11

Legal Descriptions and Plot Plan Requirements

▼ LEGAL DESCRIPTIONS

Virtually every piece of property in the United States is described for legal purposes. Legal descriptions of properties are filed in local jurisdictions, generally the county or parish courthouse. Legal descriptions are public records and may be reviewed at any time.

This section deals with plot plan characteristics and requirements. A *plot* is an area of land generally one lot or construction site in size. The term *plot* is synonymous with *lot*. A *plat* is a map of part of a city or township showing some specific area, such as a subdivision made up of several individual lots. There are usually many plots in a plat. Some dictionary definitions, however, do not differentiate between *plot* and *plat*.

There are three basic types of legal descriptions: metes and bounds, rectangular system, and lot and block.

Metes and Bounds System

Metes, or measurements, and *bounds,* or boundaries, may be used to identify the perimeters of any property. The metes are measured in feet, yards, rods (rd), or surveyor's chains (ch). There are 3 feet (914 mm) in 1 yard, 5.5 yards or 16.5 feet (5029 mm) in one rod, and 66 feet (20,117 mm) in one surveyor's chain. The boundaries may be a street, fence, or river. Boundaries are also established as *bearings,* directions with reference to one quadrant of the compass. There are 360° in a circle or compass, and each quadrant has 90°. Degrees are divided into minutes and seconds. There are 60 minutes (60') in 1° and 60 seconds (60") in 1 minute. Bearings are measured clockwise or counterclockwise from north or south. For example, a reading 45° from north to west is labeled N 45° W. See Figure 11–1. If a bearing reading requires great accuracy, fractions of a degree are used. For example, S 30° 20' 10" E reads from south 30 degrees 20 minutes 10 seconds to east.

The metes and bounds land survey begins with a *monument,* known as the *point-of-beginning.* This point is a fixed location and in times past has been a pile of rocks, a large tree, or an iron rod driven into the ground. Figure 11–2 shows an example of a plot plan that is laid out using metes and bounds, and the legal description for the plot is given.

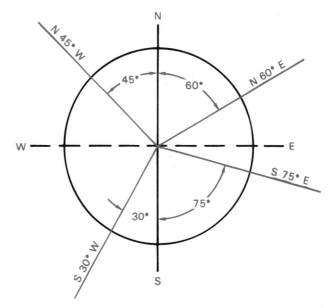

Figure 11–1 Bearings.

Rectangular System

The states in an area of the United States starting with the western boundary of Ohio, and including some southeastern states, to the Pacific Ocean, were described as public land states. Within this area the U.S. Bureau of Land Management devised a system for describing land known as the *rectangular system.*

Parallels of latitude and meridians of longitude were used to establish areas known as *great land surveys.* The point of beginning of each great land survey is where two basic reference lines cross. The lines of latitude, or parallels, are termed the *base lines,* and the lines of longitude, or meridians, are called *principal meridians.* There are 31 sets of these lines in the continental United States, with 3 in Alaska. At the beginning, the principal meridians were numbered, and the numbering system ended with the sixth principal meridian passing through Nebraska, Kansas, and Oklahoma. The remaining principal meridians were given local names. The meridian through one of the last great land surveys near the West Coast is named the Willamette Meridian because of its

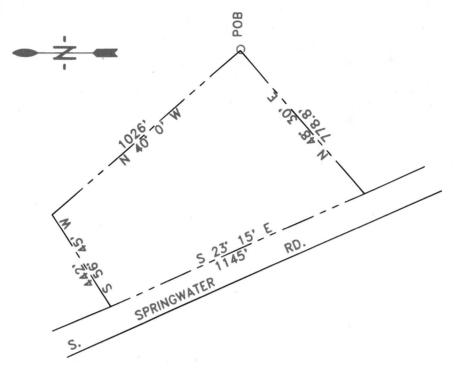

Figure 11–2 Metes and bounds plot plan and legal description for the plot. Beginning at a point 1200 ft. north 40° 0′ west from the southeast corner of the Asa Stone Donation Land Claim No. 49, thence north 40° 0′ west 1026 ft. to a pipe, thence south 56° 45′ west 442 ft. chains to center of road, thence south 23° 15′ east 1145 ft., thence north 48° 30′ east 778.8 ft. to place of beginning.

location in the Willamette Valley of Oregon. The principal meridians and base lines of the great land surveys are shown in Figure 11–3.

Townships. The great land surveys were broken down into smaller surveys known as *townships* and *sections*. The base lines and meridians were divided into blocks called *townships*. Each township measures 6 miles square. The townships are numbered by tiers running east-west. The tier numbering system is established either north or south of a principal base line. For example, the fourth tier south of the base line is labeled Township Number 4 South, abbreviated T. 4 S. Townships are also numbered according to vertical meridians, known as *ranges*. Ranges are established either east or west of a principal meridian. The third range east of the principal meridian is called Range Number 3 East, abbreviated R. 3 E. Now combine T. 4 S. and R. 3 E. to locate a township or a piece of land 6 miles by 6 miles or a total of 36 square miles. Figure 11–4 shows the township just described.

Sections. To further define the land within a 6 mile square township, the area was divided into units 1 mile square, *sections*. Sections in a township are numbered from 1 to 36. Section 1 always begins in the upper right corner, and consecutive numbers are arranged as shown in Figure 11–5. The legal descriptions of land can be carried one stage further. For example, Section 10 in the township given would be described as Sec. 10, T. 4 S.,

R. 3 E. This is an area of land 1 mile square. Sections are divided into acres. One acre equals 43,560 sq. ft. and one section of land contains 640 acres.

In addition to dividing sections into acres, sections are divided into quarters, as shown in Figure 11–6. The northeast one-quarter of Section 10 is a 160-acre piece of land described as NE ¼, Sec. 10, T. 4 S., R. 3 E. When this section is keyed to a specific meridian, it can only be one specific 160-acre area. The section can be broken further by dividing each quarter into quarters, as shown in Figure 11–7. If the SW ¼ of the NE ¼ of Section 10 were the desired property, then you would have 40 acres known as SW ¼, NE ¼, Sec. 10, T. 4 S., R. 3 E. The complete rectangular system legal description of a 2.5 acre piece of land in Section 10 reads: SW ¼, SE ¼, SE ¼, SE ¼ Sec. 10, T. 4 N., R. 8 W. of the San Bernardino Meridian, in the County of Los Angeles, State of California. See Figure 11–8.

The rectangular system of land survey may be used to describe very small properties by continuing to divide the section of a township. Often the township section legal description may be used to describe the location of the point of beginning of a metes and bounds legal description, especially when the surveyed land is an irregular plot within the rectangular system.

Lot and Block System

The *lot and block* legal description system can be derived from either the metes and bounds or the rectangular

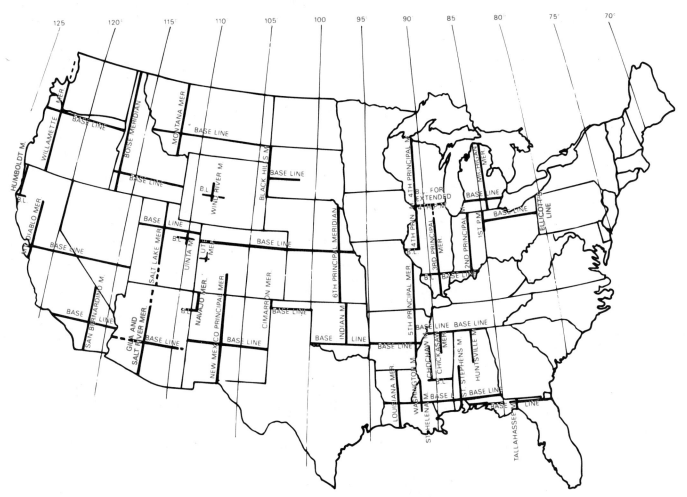

Figure 11–3   The principal meridians and base lines of the great land survey (not including Alaska).

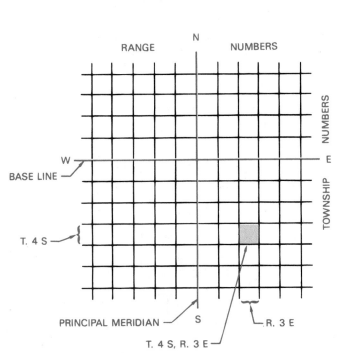

Figure 11–4   Townships.

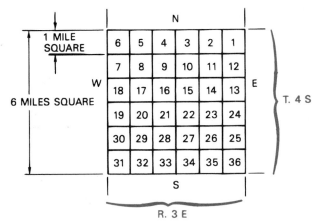

Figure 11–5   Sections.

systems. Generally when a portion of land is subdivided into individual building sites, the subdivision must be established as a legal plot and recorded as such in the local county records. The subdivision is given a name and broken into blocks of lots. A subdivision may have several blocks, each divided into a series of lots. Each lot may be 50′ × 100′, for example, depending on the zoning

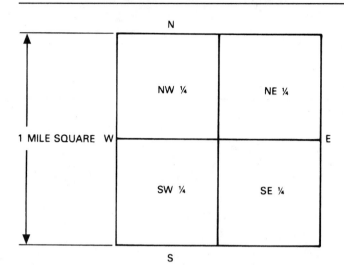

Figure 11–6 Section quarters.

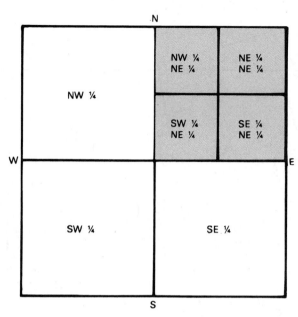

Figure 11–7 Dividing a quarter section.

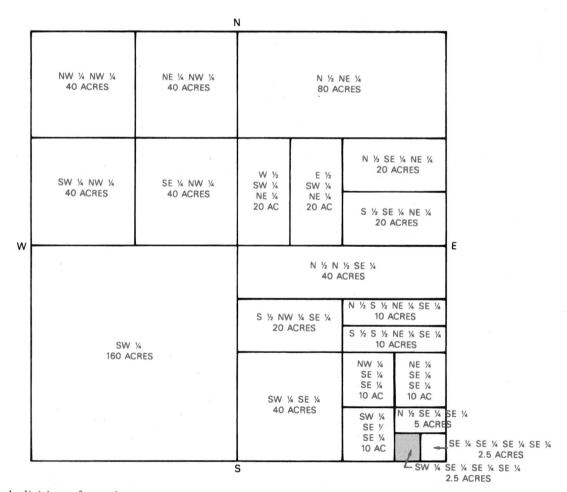

Figure 11–8 Sample divisions of a section.

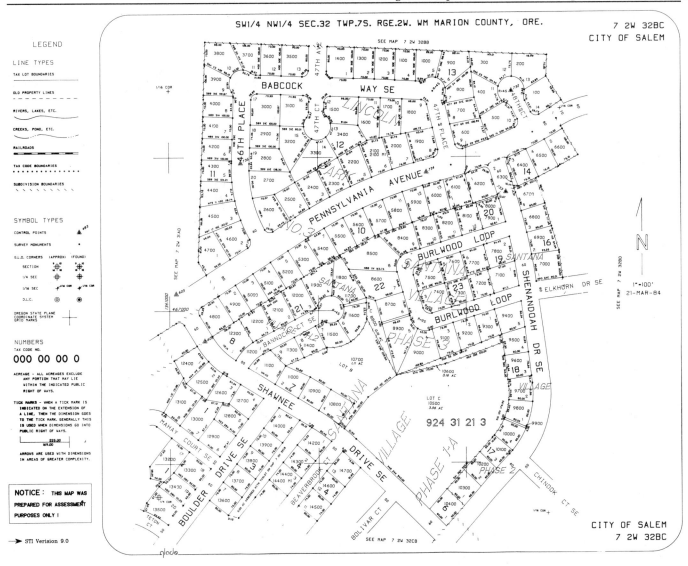

Figure 11–9 A computer-generated plat of a lot and block subdivision. *Courtesy GLADS Program.*

requirements of the specific area. Figure 11–9 shows an example of a typical lot and block system. A typical lot and block legal description might read: LOT 14, BLOCK 12, LINCOLN PARK NO. 3, CITY OF SALEM, STATE.

▼ PLAT AND PLOT PLAN REQUIREMENTS

A *plot plan,* also known as a *lot plan,* is a map of a piece of land that may be used for any number of purposes. Plot plans may show a proposed construction site for a specific property. Plots may show topography with contour lines, or the numerical value of land elevations may be given at certain locations. Plot plans are also used to show how a construction site will be excavated and are then known as a *grading plan.* Plot plans can be drawn to serve any number of required functions; all have some similar characteristics, which include showing the following:

○ A legal description of the property based on a survey
○ Property line bearings and directions

○ North direction
○ Roads and easements
○ Utilities
○ Elevations
○ Map scale

Plats are maps that are used to show an area of a town or township. They show several or many lots and may be used by a builder to show a proposed subdivision of land.

▼ TOPOGRAPHY

Topography is a physical description of land surface showing its variation in elevation, known as relief, and locating other features. Surface relief can be shown with graphic symbols that use shading methods to accentuate the character of land, or the differences in elevations can be shown with contour lines. Plot plans that require surface relief identification generally use *contour lines.*

These lines connect points of equal elevation and help show the general lay of the land.

A good way to visualize the meaning of contour lines is to look at a lake or ocean shore line. When the water is high during the winter or at high tide, a high-water line establishes a contour at that level. As the water recedes during the summer or at low tide, a new lower level line is obtained. This new line represents another contour. The high-water line goes all around the lake at one level, and the low-water line goes all around the lake at another level. These two lines represent *contours,* or lines of equal elevation. The vertical distance between contour lines is known as *contour interval.* When the contour lines are far apart, the contour interval shows relatively flat or gently sloping land. When the contour lines are close together, the contour interval shows a land that is much steeper. Contour lines are broken periodically, and the numerical value of the contour elevation above sea level is inserted. Figure 11–10 shows sample contour lines. Figure 11–11 shows a graphic example of land relief in pictorial form and contour lines of the same area.

Plot plans do not always require contour lines showing topography. Verify the requirements with the local building codes. In most instances the only contour-related information required is property corner elevations, street elevation at a driveway, and the elevation of the finished floor levels of the structure. Additionally, slope may be defined and labeled with an arrow.

▼ PLOT PLANS

Plot plan requirements vary by jurisdiction, although there are elements of plot plans that are similar around the country. Guidelines for plot plans can be obtained from the local building official or building permit department. Some agencies, for example, require that the plot plan be drawn on specific size paper such as 8½″ × 14″. Typical plot plan items include the following:

○ Plot plan scale.
○ Legal description of the property.

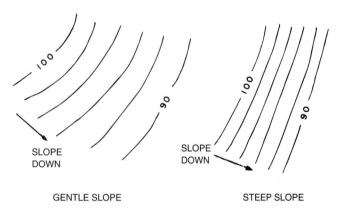

Figure 11–10 Contour lines showing both gentle and steep slopes.

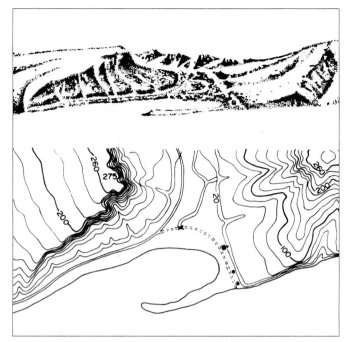

Figure 11–11 Land relief pictorial and contour lines. *Courtesy U.S. Department of the Interior, Geological Survey.*

○ Property line bearings and dimensions.
○ North direction.
○ Existing and proposed roads.
○ Driveways, patios, walks, and parking areas.
○ Existing and proposed structures.
○ Public or private water supply.
○ Public or private sewage disposal.
○ Location of utilities.
○ Rain and footing drains, and storm sewers or drainage.
○ Topography, including contour lines or land elevations at lot corners, street centerline, driveways, and floor elevations.
○ Setbacks—front, rear, and sides.
○ Specific items on adjacent properties may be required.
○ Existing and proposed trees may be required.

Figure 11–12 shows a plot plan layout that is used as an example at a local building department. Figure 11–13 shows a basic plot plan for a proposed residential addition.

The method of sewage disposal is generally an important item shown on a plot plan drawing. There are a number of alternative methods of sewage disposal, including public sewers and private systems. Chapter 17 gives more details of sewage disposal methods. The plot plan representation of a public sewer connection is shown in Figure 11–14. A private septic sewage disposal system is shown in a plot plan example in Figure 11–15.

▼ GRADING PLAN

Grading plans are plots of construction sites that generally show existing and proposed topography. The outline

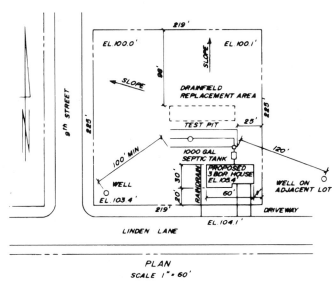

Figure 11–12 Recommended typical plot plan layout.

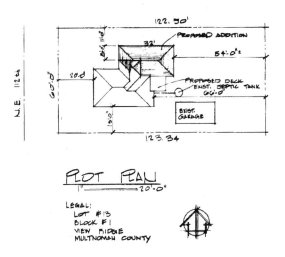

Figure 11–13 Sample plot plan showing existing home and proposed addition.

▼ SITE ANALYSIS PLAN

In areas or conditions where zoning and building permit applications require a design review, a site analysis plan may be required. The site analysis provides the basis for the proper design relationship of the proposed development to the site and to adjacent properties. The degree of detail of the site analysis is generally appropriate to the scale of the proposed project. A site analysis plan, shown in Figure 11–17, often includes the following:

○ A vicinity map showing the location of the property in relationship to adjacent properties, roads, and utilities.

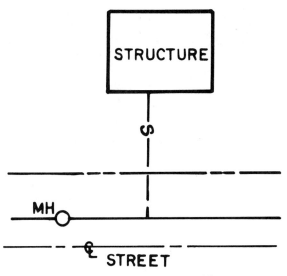

Figure 11–14 Plot plan showing a public sewer connection.

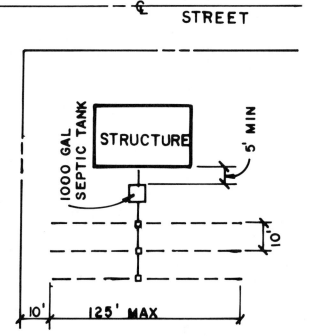

Figure 11–15 Plot plan showing a private septic sewage system.

of the structure may be shown with elevations at each building corner and the elevation given for each floor level. Figure 11–16 shows a detailed grading plan for a residential construction site. Notice that the legend identifies symbols for existing and finished contour lines. This particular grading plan provides retaining walls and graded slopes to accommodate a fairly level construction site from the front of the structure to the extent of the rear yard. The finished slope represents an embankment that establishes the relationship of the proposed contour to the existing contour. This particular grading plan also shows a proposed irrigation and landscaping layout.

Grading plan requirements may differ from one location to the next. Some grading plans may also show a cross section through the site at specified intervals or locations to evaluate the contour more fully.

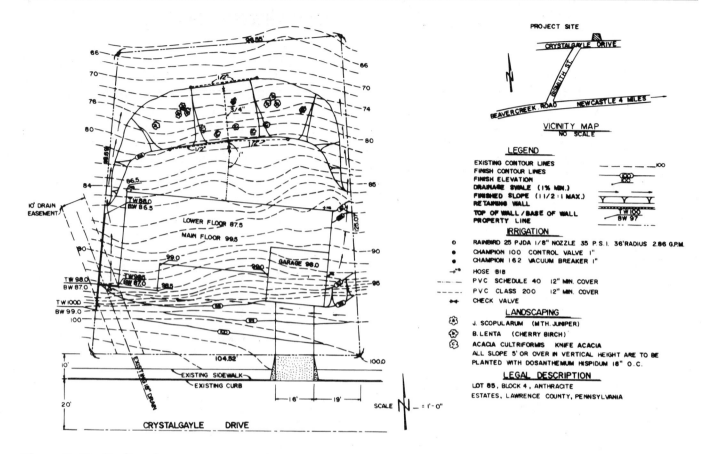

Figure 11–16 Grading plan.

○ Site features, such as existing structures and plants on the property and adjacent property.
○ The scale.
○ North direction.
○ Property boundaries.
○ Slope shown by contour lines or cross sections, or both.
○ Plan legend.
○ Traffic patterns.
○ Solar site information if solar application is intended.
○ Pedestrian patterns.

▼ SUBDIVISION PLANS

Local requirements for subdivisions should be confirmed because of the variety of procedures depending on local guidelines and zoning rules.

Some areas have guidelines for small subdivisions that differ from large subdivisions such as dividing a property into three or fewer parcels. The plat required for a minor subdivision may include the following:

1. Legal description.
2. Name, address, and telephone number of applicant.
3. Parcel layout with dimensions.
4. Direction of north.
5. All existing roads and road widths.

6. Number identification of parcels (e.g., Parcel 1, Parcel 2).
7. Location of well or proposed well, or name of water district.
8. Type of sewage disposal (septic tank or public sanitary sewers). Name of sewer district.
9. Zoning designation.
10. Size of parcel(s) in square feet or acres.
11. Slope of ground. (Arrows pointing downslope.)
12. Setbacks of all existing buildings, septic tanks, and drainfields from new property lines.
13. All utility and drainage easements.
14. Any natural drainage channels. Indicate direction of flow and whether drainage is seasonal or year-round.
15. Map scale.
16. Date.
17. Building permit application number, if any.

Figure 11–18 (page 152) shows a typical subdivision of land with three proposed parcels.

A major subdivision may require plats that are more detailed than a minor subdivision. Some of the items that may be included on the plat or in separate cover are the following:

1. The name, address, and telephone number of the property owner, applicant, and engineer or surveyor.

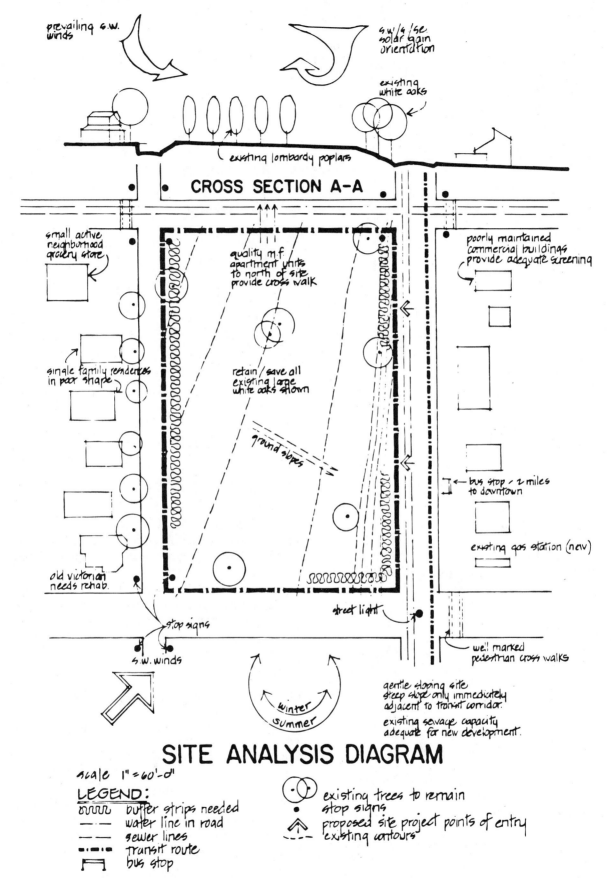

Figure 11–17 Site analysis plan. *Courtesy Planning Department, Clackamas County, Oregon.*

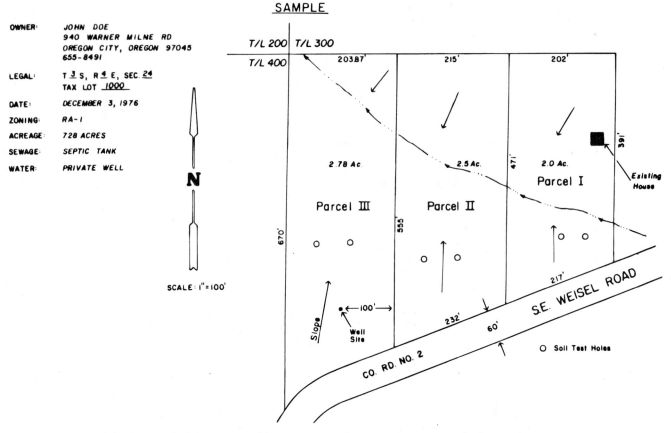

Figure 11–18 Subdivision with three proposed lots. *Courtesy Planning Department, Clackamas County, Oregon.*

2. Source of water.
3. Method of sewage disposal.
4. Existing zoning.
5. Proposed utilities.
6. Calculations justifying the proposed density.
7. Name of the major partitions or subdivision.
8. Date the drawing was made.
9. Legal description.
10. North arrow.
11. Vicinity sketch showing location of the subdivision.
12. Identification of each lot or parcel and block by number.
13. Gross acreage of property being subdivided or partitioned.
14. Dimensions and acreage of each lot or parcel.
15. Streets abutting the plat, including name, direction of drainage, and approximate grade.
16. Streets proposed, including names, approximate grades, and radius of curves.
17. Legal access to subdivision or partition other than public road.
18. Contour lines at 2′ interval for slopes of 10 percent or less, 5′ interval if exceeding a 10 percent slope.
19. Drainage channels, including width, depth, and direction of flow.
20. Existing and proposed easements locations.

21. Location of all existing structures, driveways, and pedestrian walkways.
22. All areas to be offered for public dedication.
23. Contiguous property under the same ownership, if any.
24. Boundaries of restricted areas, if any.
25. Significant vegetative areas such as major wooded areas or specimen trees.

Figure 11–19 shows an example of a small major subdivision plat.

▼ PLANNED UNIT DEVELOPMENT

A creative and flexible approach to land development is a *planned unit development.* Planned unit developments may include such uses as residential areas, recreational areas, open spaces, schools, libraries, churches, or convenient shopping facilities. Developers involved in these projects must pay particular attention to the impact on local existing developments. Generally the plats for these developments must include all of the same information shown on a subdivision plat, plus:

1. A detailed vicinity map, as shown in Figure 11–20.
2. Land use summary.
3. Symbol legend.

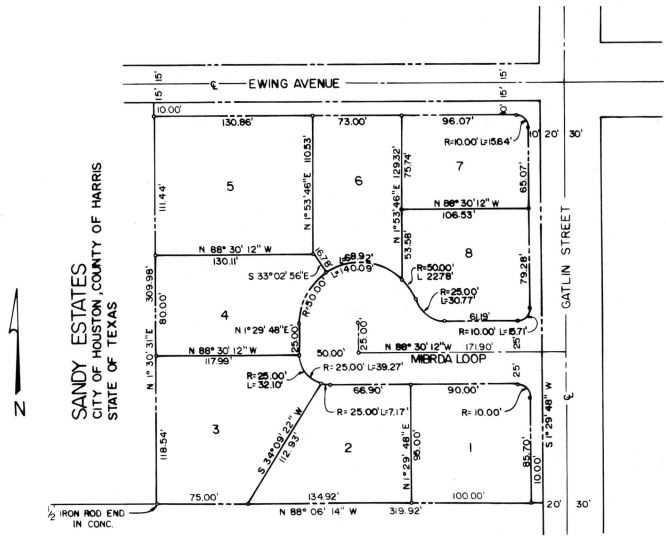

Figure 11–19 Major subdivision plat.

4. Special spaces such as recreational and open spaces or other unique characteristics.

Figure 11–21 shows a typical planned unit development plan. These plans, as in any proposed plat plan, may require changes before the final drawings are approved for development.

There are several specific applications for a plot or plat plan. The applied purpose of each is different, although the characteristics of each type of plot plan may be similar. Local districts have guidelines for the type of plot plan required. Be sure to evaluate local guidelines before preparing a plot plan for a specific purpose. Prepare the plot plan in strict accordance with the requirements in order to receive acceptance.

A variety of plot plan-related templates and CADD programs are available that can help make the preparation of these plans a little easier.

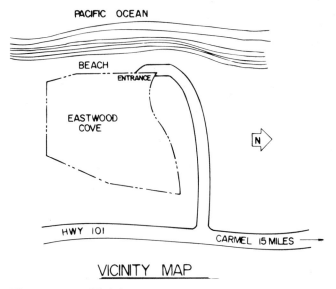

Figure 11–20 Vicinity map.

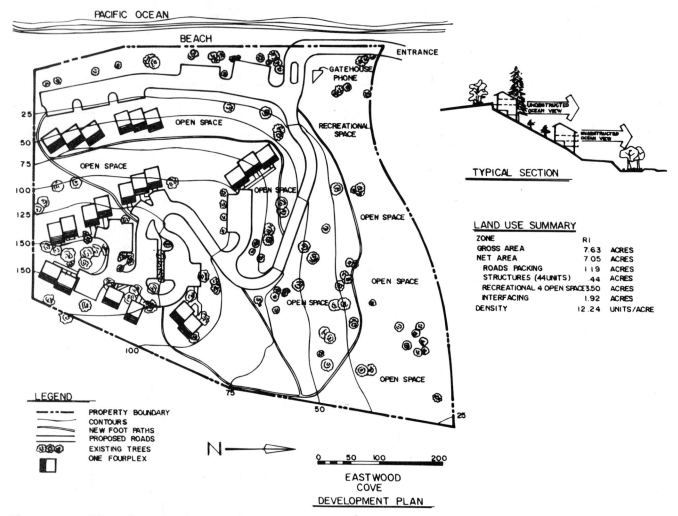

Figure 11–21 Planned unit development plan.

Legal Descriptions and Plot Plan Requirements Test

DIRECTIONS

Answer the questions with short complete statements or drawings as needed on an 8½ × 11 sheet of notebook paper, as follows:

1. Letter your name, Chapter 11 Test, and the date at the top of the sheet.
2. Letter the question number and provide the answer. You do not need to write out the question.
3. Answers may be prepared on a word processor if appropriate with course guidelines.

QUESTIONS

Question 11–1 Define plot.

Question 11–2 Name the three basic types of legal descriptions.

Question 11–3 How many total degrees are in a compass?

Question 11–4 How many minutes are in 1 degree?

Question 11–5 How many seconds are in 1 minute?

Question 11–6 Define bearing.

Question 11–7 Give the bearing of a property line positioned 60° from north toward west.

Question 11–8 Give the bearing of a property line positioned 20° from south toward east.

Question 11–9 Define base lines.

Question 11–10 Define principal meridians.

Question 11–11   Give the dimensions of a township.

Question 11–12   How many acres are in a section?

Question 11–13   How many square feet are in 1 acre?

Question 11–14   How are sections in a township numbered?

Question 11–15   How many acres are in a quarter section?

Question 11–16   Give the partial legal description of 2.5 acres of land in the farthest corner of the northwest corner of section 16, township six north, range four west.

CHAPTER 12

Plot Plan Layout

Plot plans may be drawn on media (bond paper, vellum, polyester film, or computer plot paper) ranging in size from 8½" × 11" up to 34" × 44" depending on the purpose of the plan and the guidelines of the local government agency requiring the plot plan. Many local jurisdictions recommend that plot plans be drawn on an 8½" × 14" size sheet.

Before you begin the plot plan layout there is some important information that you need. This information can often be obtained from the legal documents for the property, the surveyor's map, the local assessor's office, or the local zoning department. Figure 12–1 is a plat from a surveyor's map that can be used as a guide to prepare the plot plan. The scale of the surveyor's plat may vary, although in this case it is 1" = 200' (1:1000 metric). The plot plan to be drawn may have a scale ranging from 1" = 10' (1:50 metric) to 1" = 200'. The factors that influence the scale include the following:

○ Sheet size.
○ Plat size.
○ Amount of information required.
○ Amount of detail required.

Additional information that should be determined before the plat can be completed usually includes the following:

○ Legal description.
○ North direction.
○ All existing roads, utilities, water, sewage disposal, drainage, and slope of land.
○ Zoning information, including front, rear, and side yard setbacks.
○ Size of proposed structures.
○ Elevations at property corners, driveway at street, or contour elevations.

▼ STEPS IN PLOT PLAN LAYOUT

Follow these steps to draw a plot plan.

Step 1. Select the paper size. In this case the size is 8½" × 11". Evaluate the plot to be drawn. Lot 2

of Sandy Estates shown in Figure 12–1 will be used. Determine the scale to use by considering how the longest dimension (134.92') fits on the sheet. Always try to leave at least a ½" margin around the sheet.

Step 2. Use the given plat as an example to lay out the proposed plot plan. If a plat is not available, then the plot plan can be laid out from the legal description by establishing the boundaries using the bearings and dimensions in feet. Lay out the entire plot plan using construction lines. If errors are made, the construction lines are very easy to erase, or use CADD layers. See Figure 12–2.

Step 3. Lay out the proposed structure using construction lines. The proposed structure in this example is 54' long and 26' wide. The front setback is 25', and the east side is 20' Lay out all roads, driveways, walks, and utilities. Be sure the structure is inside or on the minimum setback requirements. *Setbacks* are imaginary boundaries within which the structure may not be placed. Think of them as property line offsets that are established by local regulations. Setback minimums may be confirmed with local zoning regulations. The minimum setbacks for this property are 25' front, 10' sides, and 35' back. Look at Figure 12–3.

Step 4. Darken all property lines, structures, roads, driveways, walks, and utilities, as shown in Figure 12–4. Some drafters use a thick line or shading for the structure.

Step 5. Add dimensions and contour lines (if any) or elevations. Add all labels, including the road name, property dimensions and bearings (if used), utility names, walks, and driveways, as shown in Figure 12–5.

Step 6. Complete the plot plan by adding the north indicator, the legal description, title, scale, client's name, and other title block information. Figure 12–6 shows a complete plot plan.

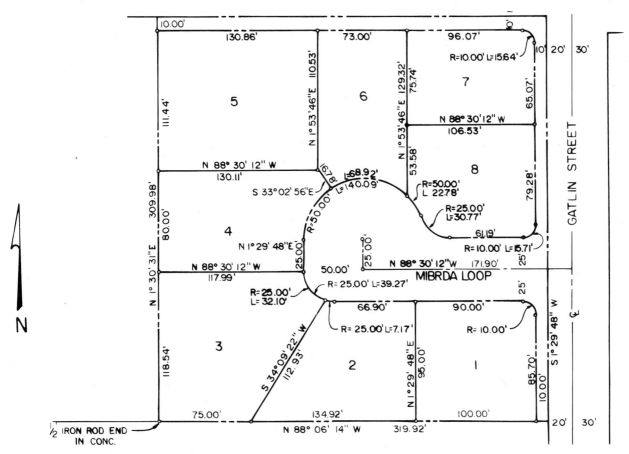

Figure 12–1 Plat from a surveyor's map.

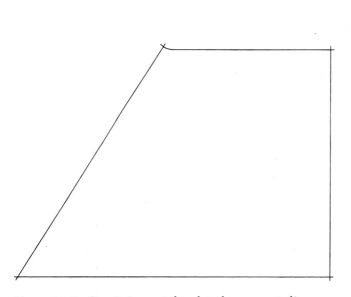

Figure 12–2 Step 2: Lay out the plot plan property lines.

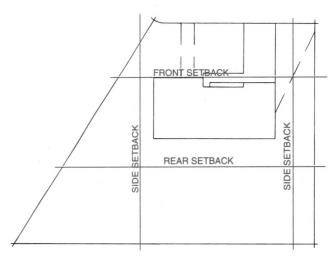

Figure 12–3 Step 3: Lay out the structures, roads, driveways, walks, and utilities. Be sure the structure is on or within the minimum required setbacks.

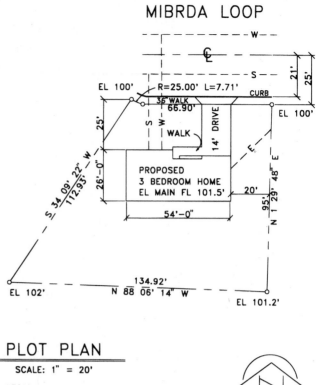

PLOT PLAN

SCALE: 1" = 20'

LEGAL:
LOT 2 SANDY ESTATES
CITY OF HUSTON,
COUNTY OF HARRIS,
STATE OF TEXAS

Figure 12–4 Step 4: Darken boundary lines, structures, roads, driveways, walks, and utilities.

Figure 12–6 Step 6: Complete the plot plan. Add title, scale, north arrow, legal description, and other necessary information, such as owner's name if required.

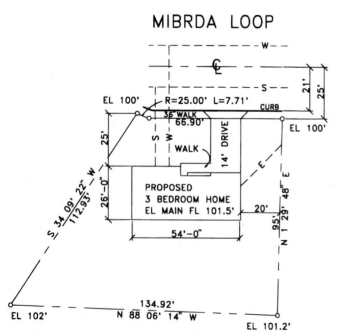

Figure 12–5 Step 5: Add dimensions and elevations; then label all roads, driveways, walks, and utilities.

CADD Applications

USING CADD TO DRAW PLOT PLANS

There are CADD software packages that may easily be customized to assist in drawing plot plans. Also available are complete CADD mapping packages that allow you to draw topographic maps, cut and fill layouts, grading plans, and land profiles. It all depends on the nature of your business and how much power you need in the CADD mapping program. One of the benefits of CADD over manual drafting is accuracy. For example, you can draw a property boundary line by giving the length and bearing. The computer automatically draws the line, then labels the length and bearing. Continue by entering information from the surveyor's notes to draw the entire property boundary in just a few minutes. Such features increase the speed and accuracy of drawing plot plans, but there are tablet menu overlays available that provide powerful plot plan capabilities, including standard symbols, scales, titles, north arrows, and landscaping features, as shown in Figure 12–7. A residential site plan, drawn using CADD mapping software, is shown in Figure 12–8.

The needs of the commercial site plan are a little different from the residential requirements. The commercial CADD site plan package should use the same features as the residential application and, additionally, have the ability to design street and parking lot layouts. A CADD template overlay with this capability is shown in Figure 12–9. The commercial CADD drafter uses features from symbol libraries, including utility symbols, street, curb and gutter designs, landscaping, parking lot layouts, titles, and scales. A commercial site plan is shown in Figure 12–10.

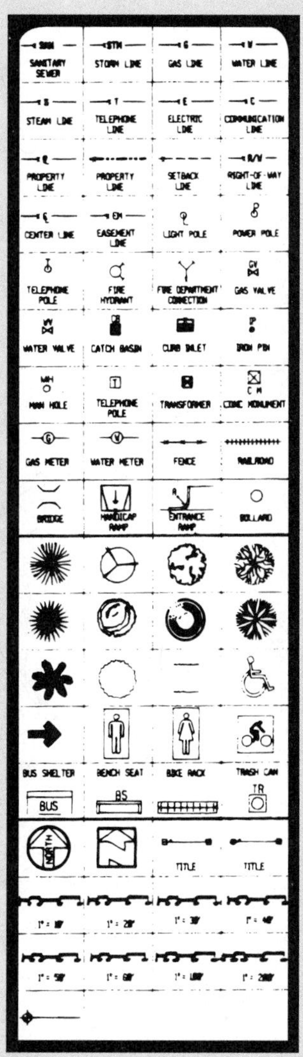

Figure 12–7 CADD tablet menu overlay showing plot plan symbols. *Courtesy Chase Systems.*

Figure 12–9 CADD tablet menu overlay for commercial site planning. *Courtesy Chase Systems.*

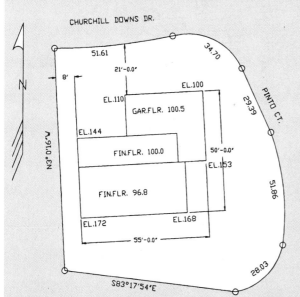

SITE PLAN

SCALE: 1" = 20'-0"

THE RUE RESIDENCE
LOT 11 BLOCK 17
HIDDEN SPRINGS RANCH #8 PHASE
CLACKAMAS COUNTY, OREGON 4/89

Figure 12–8 CADD-drawn site plan. *Courtesy Henry Fitzgibbon.*

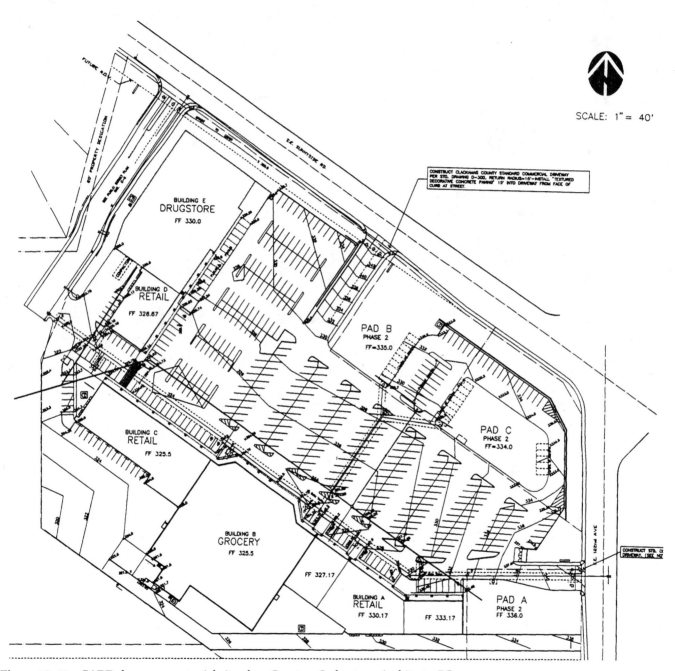

Figure 12–10 CADD-drawn commercial site plan. *Courtesy Soderstrom Architects, PC.*

CHAPTER

12 Plot Plan Layout Test

DIRECTIONS

Answer the questions with short complete statements or drawings as needed on an 8½ × 11 sheet of notebook paper, as follows:

1. Letter your name, Chapter 12 Test, and the date at the top of the sheet.
2. Letter the question number and provide the answer. You do not need to write out the question.

Answers may be prepared on a word processor if appropriate with course guidelines.

QUESTIONS

Question 12–1 What influences the size of the drawing sheet recommended for plot plans?

Question 12–2 What four factors influence scale selection when drawing a plot plan?

Question 12–3 List five elements of information that should be determined before starting a plot plan drawing.

Question 12–4 Why are construction lines helpful in plot plan layout?

Question 12–5 Information used to prepare a plot plan may come from one or more of several available sources. List four possible sources.

PROBLEMS

DIRECTIONS

1. Use an 8½ × 14 drawing sheet unless otherwise specified by your instructor.
2. Select an appropriate scale.
3. The minimum front setback is 25′–0″.
4. The minimum rear yard setback is 25′–0″.
5. The minimum side yard setback is 7′–0″.
6. Select, or have your instructor assign, one or more of the plot plan sketches or drawings that follow. The problems may be drawn using CADD if appropriate with course guidelines.

Problems 12–1 through 12–7 Begin the selected plot plan problem(s) by drawing the given information. The plot plan will then be completed after you have selected a residential design project(s) in Chapter 15. Two evaluations are recommended:
○ Plot plan without structures, drives, and utilities.
○ Complete plot plan after selection and placement of structures, drives, and utilities. Structures selected in Chapter 15.

Problem 12–8 Draw the complete plot plan shown.

Problem 12–9 Continue the drawing started in Problem 12–8 by drawing the complete grading plan using the layout for this problem as a guide.

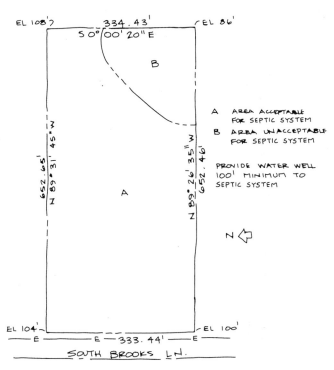

Problem 12–1

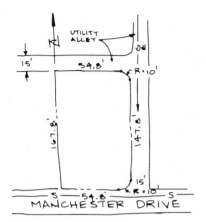

UTILITY ALLEY

15'

54.8'

R·10'

167.8'

147.8'

15'

S 54.8' S R·10'

MANCHESTER DRIVE

LOT 17, BLOCK 3, PLAT OF
GARTHWICK, YOUR CITY,
COUNTY, STATE

Problem 12–2

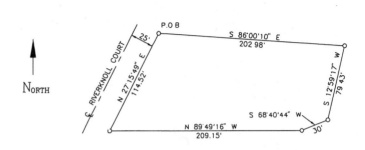

NORTH

RIVERKNOLL COURT

25'

P.O B

S 86°00'10" E
202 98'

N 27°15'49" E
114.52'

S 12°59'17" W
79 43'

S 68°40'44" W

N 89°49'16" W
209.15'

30'

LOT 15
BLOCK 3
BARRINGTON HEIGHTS
YOUR CITY, COUNTY
STATE

Problem 12–4

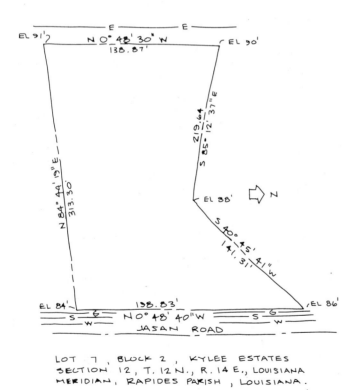

E E

EL 91'

N 0°48'30" W
138.87'

EL 90'

219.64
S 85°12'37" E

N 84°40'19" E
313.30

EL 88'

N

S 40°45'
141.31'
41" W

EL 84'

138.83'

EL 86'

S 6 N 0°48'40" W S 6
W W

JASAN ROAD

LOT 7, BLOCK 2, KYLEE ESTATES
SECTION 12, T. 12 N., R. 14 E., LOUISIANA
MERIDIAN, RAPIDES PARISH, LOUISIANA.

Problem 12–3

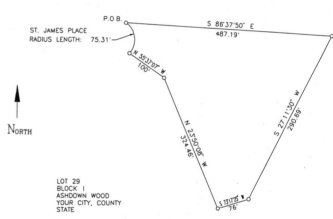

P.O.B.

ST. JAMES PLACE
RADIUS LENGTH: 75.31'

S 86°37'50" E
487.19'

N 55°37'07" W
100'

NORTH

N 23°50'08" W
324.46'

S 27°11'30" W
290.69'

S 73°17'25" W
76'

LOT 29
BLOCK 1
ASHDOWN WOOD
YOUR CITY, COUNTY
STATE

Problem 12–5

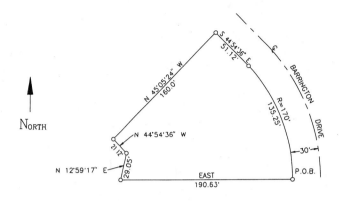

LOT 25
BLOCK 4
BARRIINGTON HEIGHTS
YOUR CITY, COUNTY
STATE

Problem 12–6

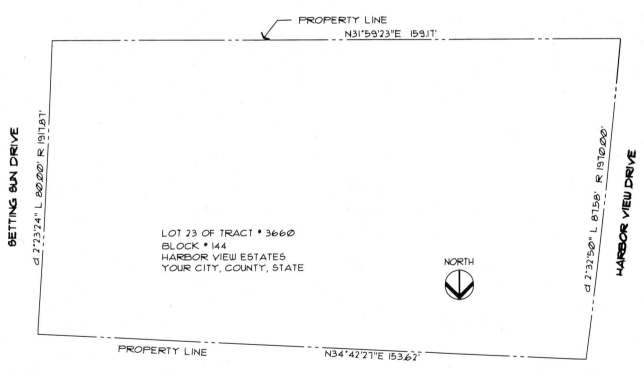

Problem 12–7

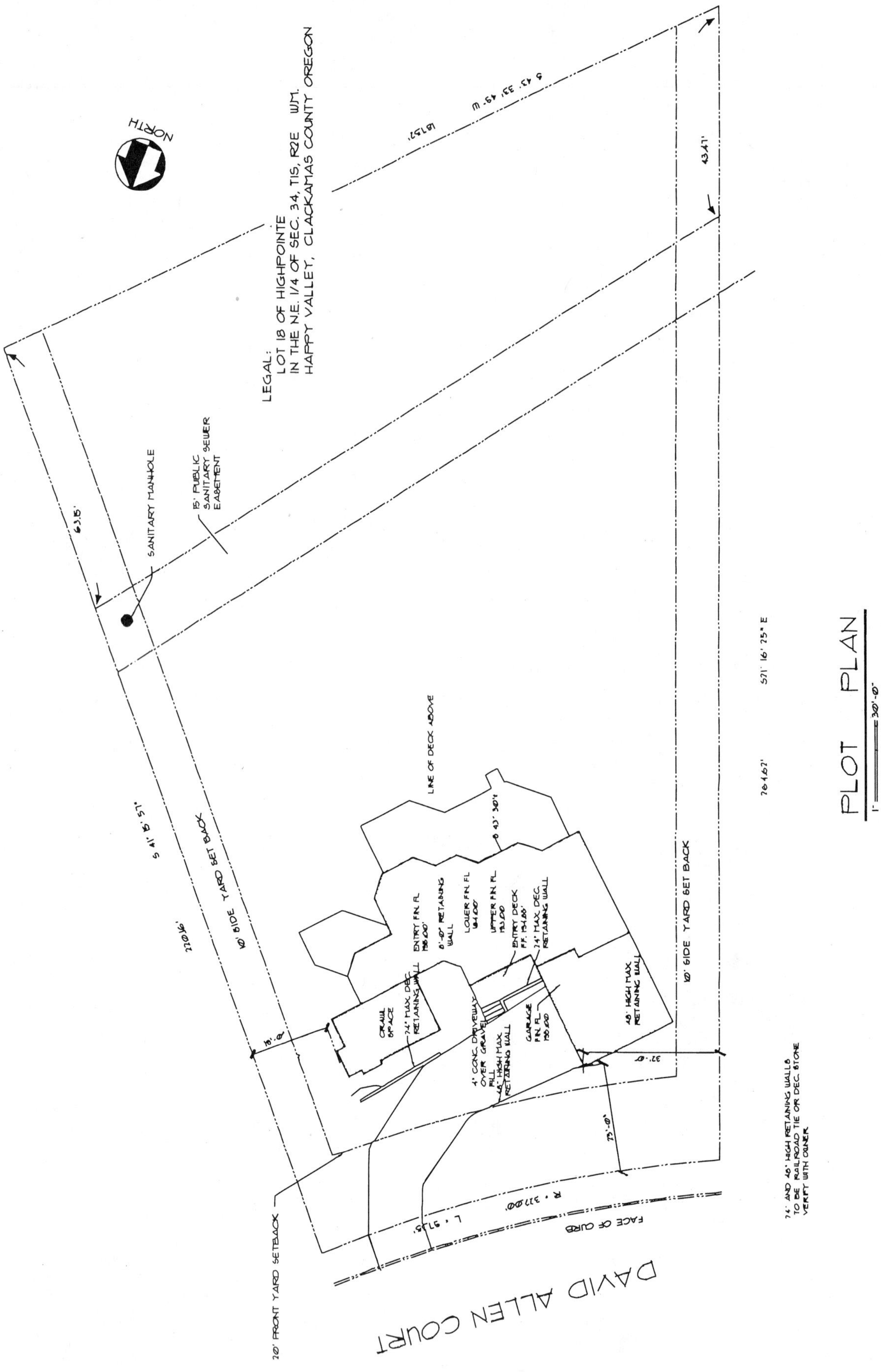

LEGAL:
LOT 18 OF HIGHPOINTE
IN THE N.E. 1/4 OF SEC. 34, T1S, R2E W.M.
HAPPY VALLEY, CLACKAMAS COUNTY OREGON

NORTH

PLOT PLAN
1" = 30'-0"

Problem 12-8

SANITARY MANHOLE

15' PUBLIC SANITARY SEWER EASEMENT

10' SIDE YARD SET BACK

20' FRONT YARD SETBACK

DAVID ALLEN COURT

FACE OF CURB

7½' AND 40" HIGH RETAINING WALLS
TO BE RAILROAD TIE OR DEC. STONE
VERIFY WITH OWNER.

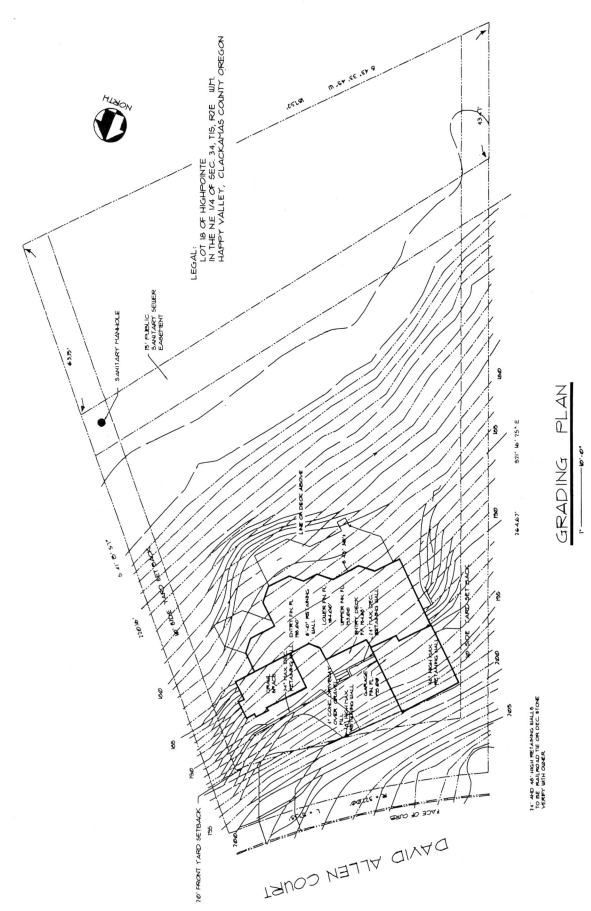

GRADING PLAN

LEGAL:
LOT 18 OF HIGHPOINTE
IN THE NE 1/4 OF SEC. 34, T1S, R2E W.M.
HAPPY VALLEY, CLACKAMAS COUNTY OREGON

Problem 12–9

SECTION IV

FLOOR PLANS

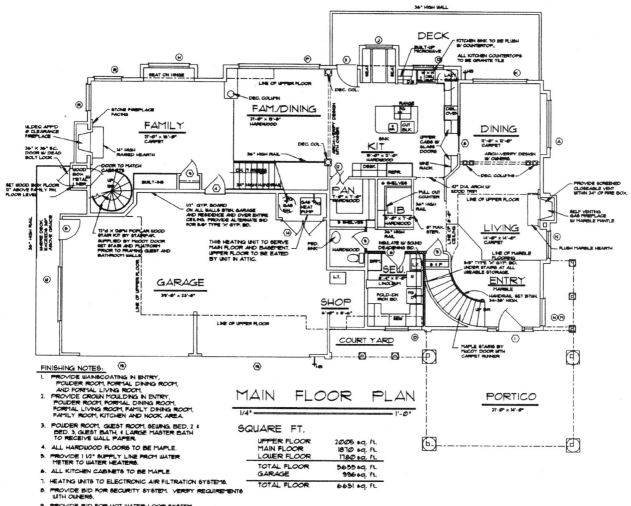

FINISHING NOTES:

1. PROVIDE WAINSCOATING IN ENTRY, POWDER ROOM, FORMAL DINING ROOM, AND FORMAL LIVING ROOM.
2. PROVIDE CROWN MOULDING IN ENTRY, POWDER ROOM, FORMAL DINING ROOM, FORMAL LIVING ROOM, FAMILY DINING ROOM, FAMILY ROOM, KITCHEN AND NOOK AREA.
3. POWDER ROOM, GUEST ROOM, SEWING, BED. 2 & BED. 3, GUEST BATH, & LARGE MASTER BATH TO RECEIVE WALL PAPER.
4. ALL HARDWOOD FLOORS TO BE MAPLE.
5. PROVIDE 1 1/2" SUPPLY LINE FROM WATER METER TO WATER HEATERS.
6. ALL KITCHEN CABINETS TO BE MAPLE.
7. HEATING UNITS TO ELECTRONIC AIR FILTRATION SYSTEMS.
8. PROVIDE BID FOR SECURITY SYSTEM. VERIFY REQUIREMENTS WITH OWNERS.
9. PROVIDE BID FOR HOT WATER LOOP SYSTEM.
10. ALL SHEET ROCK CORNERS TO BE BULLNOSE WITH NO CORNER BEAD.

MAIN FLOOR PLAN

1/4" = 1'-0"

SQUARE FT.

UPPER FLOOR	2005 sq. ft.
MAIN FLOOR	1870 sq. ft.
LOWER FLOOR	1780 sq. ft.
TOTAL FLOOR	5655 sq. ft.
GARAGE	996 sq. ft.
TOTAL FLOOR	6651 sq. ft.

CHAPTER 13

Floor-Plan Symbols

The floor plans for a proposed home provide the future homeowner with the opportunity to evaluate the design in terms of livability and suitability for the needs of the family. In addition, the floor plans quickly communicate the overall construction requirements to the builder. Symbols are used on floor plans to describe items that are associated with living in the home, such as doors, windows, cabinets, and plumbing fixtures. Other symbols are more closely related to the construction of the home, such as electrical circuits and material sizes and spacing. One of the most important concerns of the drafter is to combine carefully all of the symbols, notes, and dimensions on the floor plan so the plan is easy to read and uncluttered.

Floor plans that are easy to read are also easy to build since there is less of a possibility of construction errors than with unorganized, cluttered plans. Figure 13–1 shows a typical complete floor plan.

Figure 13–2 shows that floor plans are the representation of an imaginary horizontal cut made approximately 4′ (1220 mm) above the floor line. Residential floor plans are generally drawn at a scale of ¼″ = 1′–0″ (1:50 metric). Architectural templates are available with a wide variety of floor-plan symbols at the proper scale.

In addition to knowing the proper symbols to use, architects, designers, and architectural drafters should be familiar with standard products that the symbols represent. Products used in the structures they design such as plumbing fixtures, appliances, windows, and doors are usually available from local vendors. Occasionally some special items must be ordered far enough in advance to ensure delivery to the job site at the time needed. By becoming familiar with the standard products, you will become familiar with their cost also. Knowing costs allows you to know if the product you are considering using is affordable within the owner's budget.

One of the best sources of standard product information is Sweet's catalogs. Sweet's publishes several sets of catalogs in different categories, and the set you use depends on the kind of building you are designing. For residential designs, you would most likely use Sweet's *Home Building and Remodeling* catalog. Sweet's, as it is known, is a set of books containing building products manufacturers' catalogs arranged by category.

▼ WALL SYMBOLS

Exterior Walls

Exterior wood-frame walls are generally drawn 6″ thick at a ¼″ = 1′–0″ (1:50 metric) scale, as seen in Figure 13–3. Six-inch exterior walls are common when 2 × 4 (40 × 90 mm) studs are used. The 6″ (180 mm) is approximately equal to the thickness of wall studs plus interior and exterior construction materials. When 2 × 6 (40 × 150 mm) studs are used, the exterior walls are drawn thicker. The wall thickness depends on the type of construction. If the exterior walls are to be concrete or masonry construction with wood framing to finish the inside surface, then they are drawn substantially thicker, as shown in Figure 13–4. Exterior frame walls with masonry veneer construction applied to the outside surface are drawn an additional 4″ (approx. 100 mm) thick with the masonry veneer represented as 4″ (100 mm) thick over the 6″ (150 mm) wood frame walls, as seen in Figure 13–5.

Interior Partitions

Interior walls, known as *partitions*, are frequently drawn 5½″ (114 mm) thick (3½″ stud + (2) ½″ dry wall) when 2 × 4 (40 × 90 mm) studs are used with drywall applied to each side. Walls with 2 × 6 (40 × 150 mm) studs are generally used where additioanl plumbing is required within the wall cavity. This additional plumbing may occur behind a toilet to accommodate the soil pipe. Occasionally masonry veneer is used on interior walls and is drawn in a manner similar to the exterior application shown in Figure 13–5. In many architectural offices wood-frame exterior and interior walls are drawn the same thickness to save time. The result is a wall representation that clearly communicates the intent.

Wall Shading and Material Indications

Several methods are used to shade walls so the walls stand out clearly from the rest of the drawing. Wall shading is also referred to as *poché*, a French word that means the art of outlining large letters or areas and filling the centers. Wall shading should be the last drafting task performed.

Wall shading should be done on the back of the drawing, although some drafters shade on the front, a practice

169

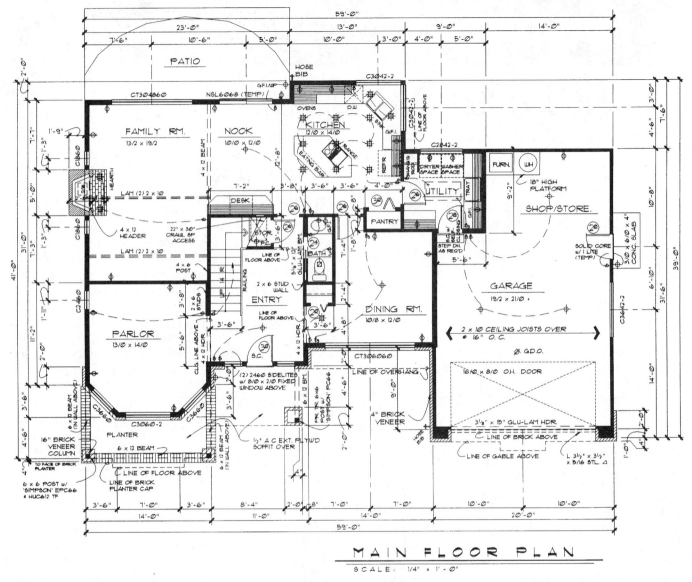

Figure 13–1 Complete floor plan. *Courtesy Piercy & Barclay Designers, Inc.*

that could cause the drawing to become smudged easily. When shading is done on the back of the sheet, blank paper should be placed on the drawing table to avoid transferring lines from the drawing image to the drafting table. Some drafters shade walls very dark for accent, while others shade the walls lightly for a more subtle effect, as shown in Figure 13–6. Wall shading should be done with a soft (F or HB) pencil lead. Other wall shading techniques include closely spaced thin lines, wood grain effect, or the use of colored pencils.

Office practice that requires light wall shading may also use thick wall lines to help accent the walls and partitions so they stand out from other floor-plan features. Thick wall lines may be achieved with an H or F lead in a 0.7-mm automatic pencil. Figure 13–7 shows how walls and partitions appear when outlined with thick lines. This method is not used when dark wall

shading is usedsince the darker shading results in walls and partitions that clearly stand out from other floor-plan features. Some architectural drafters prefer to draw walls and partitions unshaded with a thick outline.

Partial Walls

Partial walls are used as room dividers in situations where an open environment is desired, as guard rails at balconies, or next to a flight of stairs. Partial walls require a minimum height above the floor of 36″ and are often capped with wood or may have decorative spindles that connect to the ceiling. Partial walls are differentiated from other walls by wood grain or very light shading and should be defined with a note that specifies the height, as shown in Figure 13–8.

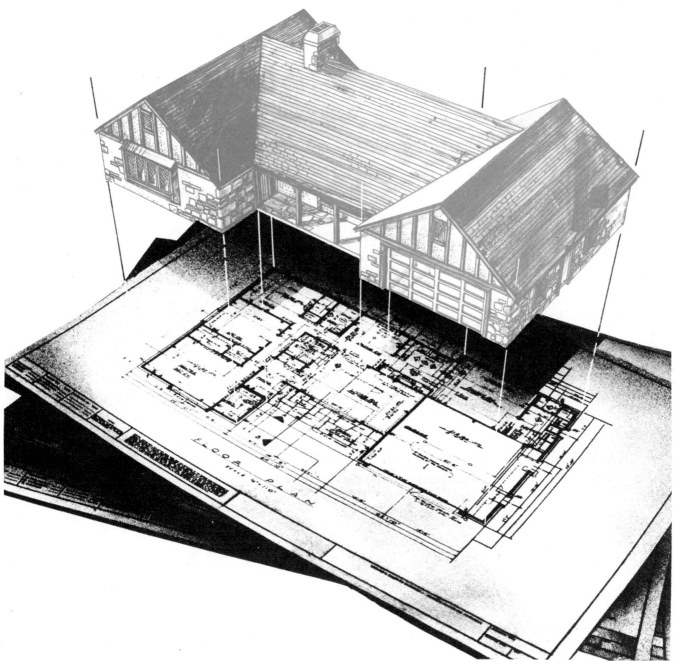

Figure 13–2 Floor plan representation.

Guardrails

Guardrails are used for safety at balconies, lofts, stairs, and decks over 30″ (762 mm) above the next lower level. Residential guardrails, according to CABO and UBC, should be noted on floor plans as being at least 36″ (915 mm) above the floor and may include another note specifying that the intermediate rails should not have more than a 6″ (1152 mm) open space. The minimum space-

helps ensure that small children will not fall through. Decorative guardrails are also used as room dividers—especially at a sunken area or to create an open effect between two rooms. Guardrails are commonly made of either wood or wrought iron. Creatively designed guardrails add a great deal to the aesthetics of the interior design. Figure 13–9 shows guardrail designs and how they are drawn on a floor plan.

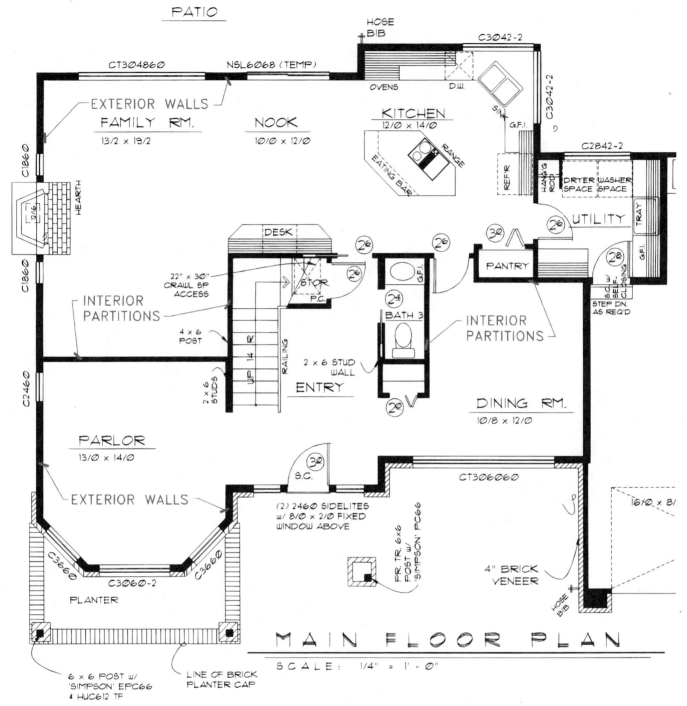

PATIO

CT304860   NSL6068 (TEMP)

HOSE BIB

C3042-2

OVENS   D.W.   C3042-2

EXTERIOR WALLS

KITCHEN
12/0 × 14/0

SINK   G.F.I.

FAMILY RM.
13/2 × 19/2

NOOK
10/0 × 12/0

RANGE

EATING BAR

C1860

HEARTH

12/6

REFR   HANG'G ROD   DRYER SPACE   WASHER SPACE   C2842-2

UTILITY   TRAY   G.F.I.

C1860

DESK

S.C. w/ SELF-CLOSING

PANTRY

STEP DN. AS REQ'D

22" × 30" CRAWL SP ACCESS

STOR.

P.C.

G.F.I.

INTERIOR PARTITIONS

4 × 6 POST

2 × 6 STUDS

UP 14   RAILING

BATH 3

2 × 6 STUD WALL

ENTRY

INTERIOR PARTITIONS

C2460

PARLOR
13/0 × 14/0

DINING RM.
10/8 × 12/0

S.C.

CT306060

EXTERIOR WALLS

(2) 2460 SIDELITES w/ 8/0 × 2/0 FIXED WINDOW ABOVE

16/0 × 8/

4" BRICK VENEER

C3660   C3660

C3060-2

PLANTER

PR. TR. 6×6 POST w/ 'SIMPSON' PC66

HOSE BIB

6 × 6 POST w/ 'SIMPSON' EPC66 & HUC612 TF

LINE OF BRICK PLANTER CAP

MAIN FLOOR PLAN

SCALE: 1/4" = 1'- 0"

Figure 13–3   A partial floor plan showing exterior walls and interior partitions. *Courtesy Piercy & Barclay Designers, Inc. CPBD.*

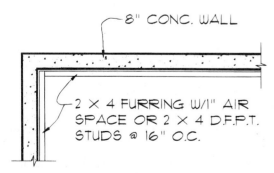

8" CONC. WALL

2 × 4 FURRING W/1" AIR SPACE OR 2 × 4 D.F.P.T. STUDS @ 16" O.C.

Figure 13–4   Concrete exterior wall.

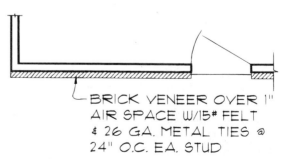

BRICK VENEER OVER 1" AIR SPACE W/15# FELT & 26 GA. METAL TIES @ 24" O.C. EA. STUD

Figure 13–5   Exterior masonry veneer.

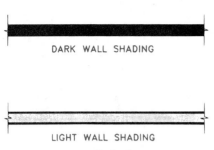

DARK WALL SHADING

LIGHT WALL SHADING

Figure 13-6 Wall shading.

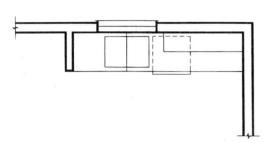

Figure 13-7 Thick lines used on walls and partitions.

Figure 13-8 Partial wall used as a partition wall or room divider.

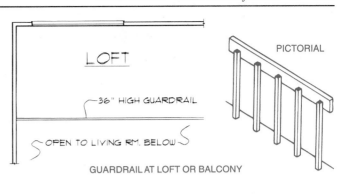

LOFT

36" HIGH GUARDRAIL

OPEN TO LIVING RM. BELOW

PICTORIAL

GUARDRAIL AT LOFT OR BALCONY

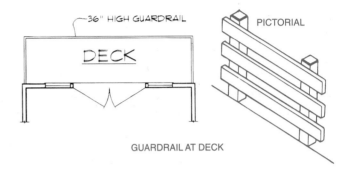

36" HIGH GUARDRAIL

DECK

PICTORIAL

GUARDRAIL AT DECK

Figure 13-9 Guardrail representations.

CADD Applications

DRAWING FLOOR PLAN WALLS USING CADD

There are a number of ways to draw walls using CADD. Custom architectural software packages are available that allow you to draw and edit walls in any desired layout. You can also use the standard drafting features of CADD software programs such as AutoCAD. For example, the AutoCAD LINE or PLINE commands may be used to draw the outside wall line, followed by using the OFFSET command to place the inside wall line a desired wall thickness away from the first line. The PLINE has the advantage of providing width variation for thick wall outlines. Use the BREAK, TRIM, or EXTEND commands to edit walls as needed. Figure 13–10 shows how walls may be drawn using the PLINE and OFFSET commands.

If you are an AutoCAD user, the MLINE command is available for drawing double lines. The double lines may be used to draw walls by specifying the wall thickness at the Width option. This command may be used to draw a variety of wall layouts, including intersecting walls and capped walls. Figure 13–11 shows some of the options using the MLINE command to draw straight wall elements. Figure 13–12 shows how to draw curved walls with the MLINE command.

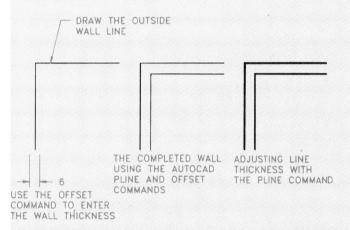

Figure 13–10 Drawing walls with the AutoCAD PLINE and OFFSET commands.

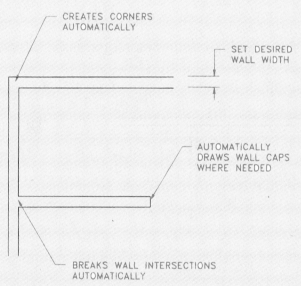

DRAWING STRAIGHT WALLS WITH THE DLINE COMMAND

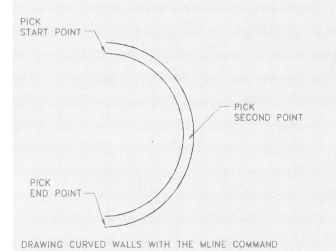

DRAWING CURVED WALLS WITH THE MLINE COMMAND

Figure 13–12 Using AutoCAD's MLINE command to draw curved walls.

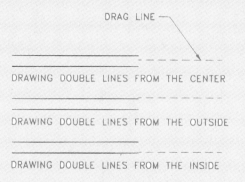

DRAWING DOUBLE LINES FROM THE CENTER

DRAWING DOUBLE LINES FROM THE OUTSIDE

DRAWING DOUBLE LINES FROM THE INSIDE

THE DRAGLINE OPTIONS OF THE MLINE COMMAND ALLOW YOU TO CREATE WALLS FROM THE CENTER FOR INTERIOR PARTITIONS OR FROM THE EDGE FOR EXTERIOR WALLS

Figure 13–11 Using AutoCAD's MLINE command to draw straight wall elements.

▼ DOOR SYMBOLS

Exterior doors are drawn on the floor plan with the sill shown on the outside of the house. The sill is commonly drawn projected about 1/16", although it may be drawn flush depending on individual company standards. See Figure 13–13. The main entry door is usually 3'–0" (915 mm) wide. An exterior door from a garage or utility room is usually 2'–8" (813 mm) wide. Exterior doors are typically solid wood or hollow metal with insulation. Doors may be either smooth, slab, or have a decorative pattern on the surface. Doors are typically 6'–8" (2032 mm) high, although 8'–0" (2438 mm) doors are available. Refer to Chapter 7 for a review of code requirements for exterior doors.

The interior door symbol, as shown in Figure 13–14, is drawn without a sill. Interior doors should swing into the room being entered and against a wall. Interior doors are typically placed within 3" (80 mm) of a corner, or at least 2' from the corner. This allows for furniture placement behind the door. Common interior door sizes are:

2'–8": Utility rooms

2'–8" to 2'–6": Bedrooms, dens, family, and dining rooms

2'–6" to 2'–4": Bathrooms

2'–4" to 2'–0": Closets (or larger for bifold and bipass doors)

The Americans with Disabilities Act (ADA) specifies a 32" (813 mm) wide opening for wheelchair access. This is a clear, unobstructive dimension measured to the edge of the door at 90°, if it does not open 180°. The larger-size door normally used on more expensive homes with wider

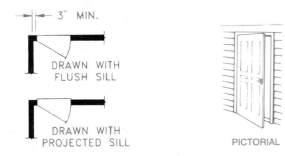

FLOOR PLAN REPRESENTATION

Figure 13–13 Exterior door.

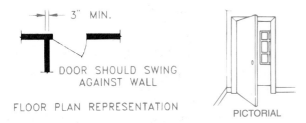

FLOOR PLAN REPRESENTATION

Figure 13–14 Interior door.

FLOOR PLAN REPRESENTATION

Figure 13–15 Pocket door.

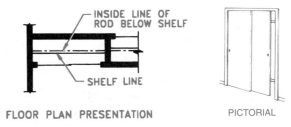

FLOOR PLAN PRESENTATION

Figure 13–16 Bipass door.

halls; smaller-size doors are usually used on homes where space is at a premium. Interior doors are usually flush but may be the raised-panel type or have glass panels.

Pocket doors are commonly used when space for a door swing is limited, as in a small room. Look at Figure 13–15. Pocket doors should not be placed where the pocket is in an exterior wall or there is interference with plumbing or electrical wiring. The size of pocket doors should follow the same guidelines as interior swinging doors. Pocket doors are more expensive to purchase and install than standard interior doors because the pocket door frame must be built while the house is being framed. The additional cost is often worth the results, however.

A common economical wardrobe is the bipass door, as shown in Figure 13–16. Bipass doors normally range in size from 4'–0" (1200 mm) through 12'–0" wide (3660 mm) in 1' intervals. Pairs of doors are usually adequate for widths up to about 8'. Three panels are usually provided for doors between 8 and 10' wide, and four panels are normal for doors wider than 10'. Bifold wardrobe or closet doors are used when complete access to the closet is required. Sometimes bifold doors are used on a utility closet that houses a washer and dryer or other utilities. See Figure 13–17. Bifold wardrobe doors often range in size from 4'–0" through 9'–0" wide, in 6" intervals.

Double-entry doors are common where a large formal foyer design requires a more elaborate entry than can be achieved by one door. The floor-plan symbol for double-entry and French doors is the same; therefore, the door schedule should clearly identify the type of doors to be installed. See Figure 13–18.

Glass sliding doors are made with wood or metal frames and tempered glass for safety. Figure 13–19 shows the floor-plan symbol for both a flush and projected exterior sill representation. These doors are used to provide glass areas and are excellent for access to a patio or deck. Glass sliding doors typically range in size from 5'–0" through 12'–0" wide at 1' intervals. 6'–0" and 8'–0" are common sizes.

French doors are used in place of glass sliding doors when a more traditional door design is required. Glass sliding doors are associated with contemporary design and do not take up as much floor space as French doors. French doors may be purchased with wood mullions and muntins (the upright and bar partitions respectively) between the glass panes or with one large glass pane and a removable decorative grill for easy cleaning. See Figure 13–18. French doors range in size from 2'–4" through 3'–6" wide in 2" increments. Doors may be used individually, in pairs, or in groups of threes or fours. The three- and four-panel doors typically have one or more fixed panels, which should be specified.

Double-acting doors are often used between a kitchen and eating area so the doors swing in either direction for easy passage. See Figure 13–20. Common size for pairs of doors range from 2'–6" through 4'–0" wide in 2" increments.

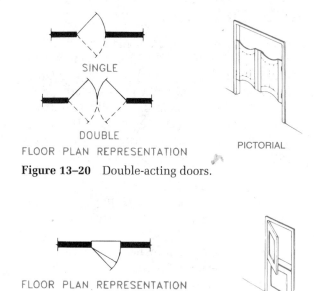

Figure 13–20 Double-acting doors.

Figure 13–21 Dutch doors.

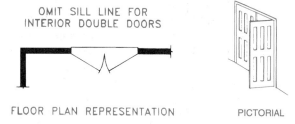

Figure 13–17 Bifold door.

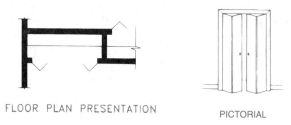

Figure 13–18 Double-entry or French doors.

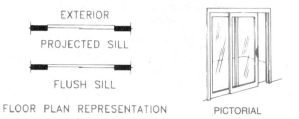

Figure 13–19 Glass sliding door.

Figure 13–22 Accordion door.

Dutch doors are used when it is desirable to have a door that may be half open and half closed. The top portion may be opened and used as a pass-through. Look at Figure 13–21. Dutch doors range in size from 2'–6" through 3'–6" wide in 2" increments.

Accordion doors may be used at closets or wardrobes, or they are often used as room dividers where an openable partition is needed. Figure 13–22 shows the accordion door floor-plan symbol. Accordion doors range in size from 4'–0" through 12'–0" in 1' increments.

The floor-plan symbol for an overhead garage door is shown in Figure 13–23. The dashed lines show the size and extent of the garage door when open. The extent of the garage door is typically shown when the door interferes with something on the ceiling.

Garage doors range in size from 8'–0" through 18'–0" wide. An 8'–0" door is common for a single car width. A door 9'–0" or 10'–0" is common for a single door and accommodates a pickup truck or large van. A door 16'–0" is common for a double car door. Door heights are typically 7'–0" high, although 8'–0" or 10'–0" high doors are common for campers or recreational vehicles. Metric sizes vary depending on hard or soft conversion. See Chapter 14.

▼ WINDOW SYMBOLS

Window symbols are drawn with a sill on the outside and inside. The sliding window, as shown in Figure 13–24, is a popular 50 percent openable window. Notice that windows may be drawn with exterior sills projected or flush, and the glass pane may be drawn with single or double lines. The method used should be consistent throughout the plan and should be determined by the preference of the specific architectural office. Many offices draw all windows with a projected sill and one line to represent the glass; they then specify the type of window in the window schedule.

Windows typically come in sizes that range from 2'–0" through 12'–0" wide at intervals of about 6".

The location of the window in the house and the way the window opens has an effect on the size. A window between 6'–0" and 12'–0" wide is common for a living, dining, or family room. To take advantage of a view while sitting, windows in these rooms are normally between 4'–0" and 5'–0" tall.

Windows in bedrooms are typically 3'–0" to 6'–0" wide. The depth often ranges from 3'–6" to 4'–0". The

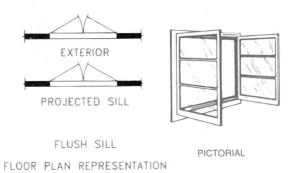

FLOOR PLAN REPRESENTATION

Figure 13–25 Casement window.

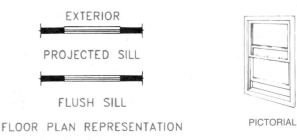

FLOOR PLAN REPRESENTATION

Figure 13–26 Double-hung window.

type of window used in the bedroom is important because of emergency egress requirements of most codes. See Chapter 7 for a review.

Kitchen windows are often between 3'–0" and 5'–0" wide and between 3'–0" and 3'–6" tall. Wide windows are nice to have in a kitchen for the added light they provide, but some of the upper cabinets may be eliminated. If the tops of windows are at the normal 6'–8" height, windows deeper than 3'–6" interfere with the countertops.

Bathroom windows often range between 2'–0" and 3'–0" wide, with an equal depth. A longer window with less depth is often specified if the window is to be located in a shower area. Most bathroom windows have obscure glass.

Casement windows may be 100 percent openable and are best used where extreme weather conditions require a tight seal when the window is closed, although these windows are in common use everywhere. See Figure 13–25. Casement windows may be more expensive than sliding units.

The traditional double-hung window has a bottom panel that slides upward, as shown in Figure 13–26. Double-hung wood-frame windows are designed for energy efficiency and are commonly used in traditional as opposed to contemporary architectural designs. Double-hung windows are typically taller than they are wide. Single windows usually range in width from 1'–6" through 4'–0". It is very common to have double-hung windows grouped together in pairs of two or more.

Awning windows often are used in basements or below a fixed window to provide ventilation. Another common use places awning windows between two different roof levels; this provides additional ventilation in vaulted

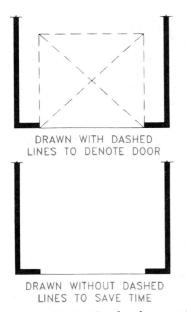

DRAWN WITH DASHED
LINES TO DENOTE DOOR

DRAWN WITHOUT DASHED
LINES TO SAVE TIME

Figure 13–23 Overhead garage door.

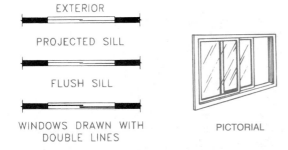

FLOOR PLAN REPRESENTATION

Figure 13–24 Horizontal sliding window.

rooms. These windows would then be opened with a long pole connected to the opening device. These windows are hinged at the top and swing outward, as shown in Figure 13–27. Hopper windows are drawn in the same manner; however, they hinge at the bottom and swing inward. Jalousie windows are used when a louvered effect is desired, as seen in Figure 13–28.

Fixed windows are popular when a large, unobstructed area of glass is required to take advantage of a view or to allow for solar heat gain. Figure 13–29 shows the floor plan symbol of a fixed window.

Bay windows are often used when a traditional style is desired. Figure 13–30 shows the representation of a bay. Usually the sides are drawn at a 45° or 30°. The depth of the bay is often between 18 and 24″. The total width of a bay is limited by the size of the center window, which is typically either a fixed panel, double-hung, or casement window. Bays can be either premanufactured or built at the job site.

A garden window, as shown in Figure 13–31, is a popular style for utility rooms or kitchens. Garden windows usually project between 12 and 18″ from the residence. Depending on the manufacturer, either the side or top panels open.

Skylights

When additional daylight is desirable in a room or for natural light to enter an interior room, consider using a skylight. Skylights are available fixed or openable. They are made of plastic in a dome shape or flat tempered glass. Tempered double-pane insulated skylights are energy

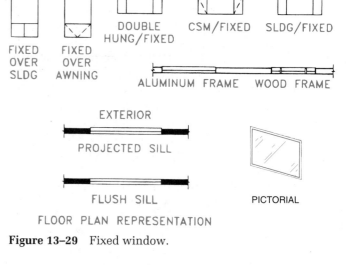

Figure 13–29 Fixed window.

Figure 13–30 Bay window.

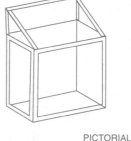

Figure 13–31 Garden window.

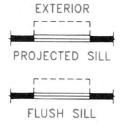

Figure 13–27 Awning window.

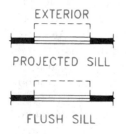

Figure 13–28 Jalousie window.

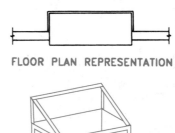

efficient, do not cause any distortion of view, and generally are not more expensive than plastic skylights. Figure 13–32 shows how a skylight is represented in the floor plan.

Metric sizes vary depending on hard or soft conversion. See Chapter 14.

▼ SCHEDULES

Numbered symbols used on the floor plan key specific items to charts known as *schedules*. Schedules are used to describe items such as doors, windows, appliances, materials, fixtures, hardware, and finishes. See Figure 13–33. Schedules help keep drawings clear of unnecessary notes since the details of the item are off the drawing or on another sheet. There are many different ways to set up a schedule, but it may include any or all of the following information about the product:

○ Vendor's name
○ Product name
○ Model number
○ Type

○ Quantity
○ Size
○ Rough opening size
○ Color

Placement of quantities on the schedules varies among offices. Some companies do not include quantities on the schedules, leaving this up to the contractor. During the early stages of plan development the drafter or designer refers to manufacturers' catalogs for products to be used in the design and begins to establish the schedule data.

Schedule Key

When doors and windows are described in a schedule, they must be keyed from the drawing to the schedule. The key may be to label doors with a number and windows with a letter. You can enclose the letters or numbers in different geometric figures, such as the following:

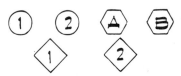

Figure 13–34 shows how the symbols key to the schedules. As an alternative method, you may consider using a divided circle for the key. Place the letter D for door or W for window above the dividing line and the number of the door or window, using consecutive numbers, below the line, as shown:

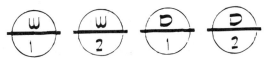

The exact method of representation depends on individual company standards.

Schedules are also used to identify finish materials used in different areas of the structure. Figure 13–35 shows a typical interior finish schedule.

Door and window sizes may be placed on the floor plan next to their symbols. This method, shown in Figure 13–36, is used by some companies to save time. While

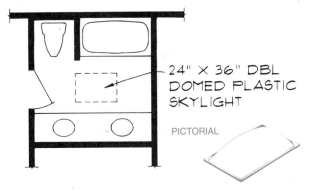

24" X 36" DBL DOMED PLASTIC SKYLIGHT

PICTORIAL

Figure 13–32   Skylight representation.

WINDOW SCHEDULE

SYM	SIZE	MODEL	ROUGH OPEN	QUAN.
A	1' x 5'	JOB BUILT	VERIFY	2
B	8' x 5'	W 4 N 5 CSM.	8'-0¾ x 5'-0⅛	1
C	4' x 5'	W 2 N 5 CSM.	4'-0¾ x 5'-0⅛	2
D	4' x 3⁶	W 2 N 3 CSM	4'-0¾ x 3'-6½	2
E	3⁶ x 3⁶	2 N 3 CSM	3'-6½ x 3'-6½	2
F	6' x 4'	G 64 SLDG.	6'-0½ x 4'-0½	1
G	5' x 3⁶	G 536 SLDG.	5'-0½ x 3'-6½	4
H	4' x 3⁶	G 436 SLDG.	4'-0½ x 3'-6½	1
J	4' x 2'	A 41 ALUN.	4'-0½ x 2'-0⅛	3

(a)

DOOR SCHEDULE

SYM.	SIZE	TYPE	QUAN
1	3' x 6⁸	S.C. R.P. METAL INSULATED	1
2	3' x 6⁸	S.C. FLUSH METAL INSULATED	2
3	2⁸ x 6⁸	S.C. SELF CLOSING	2
4	2⁸ x 6⁸	HOLLOW CORE	5
5	2⁶ x 6⁸	HOLLOW CORE	5
6	2⁶ x 6⁸	POCKET SLDG.	2

(b)

Figure 13–33   Window and door schedule.

this method is easy to do, it may not be used when specific data must be identified. The sizes are the numbers next to the windows and doors in Figure 13–36. The first

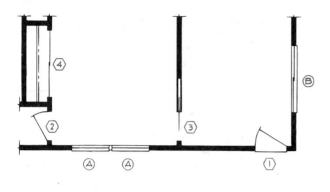

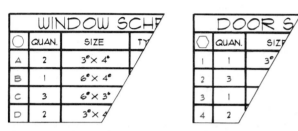

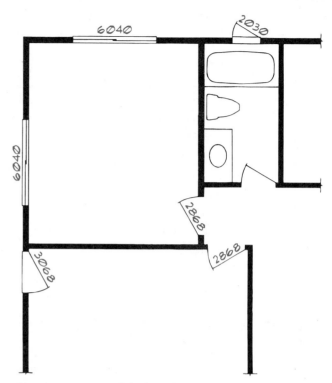

Figure 13-34 Method of keying windows and doors from the floor plan to the schedules.

Figure 13-36 Simplified method of labeling doors and windows.

INTERIOR FINISH SCHEDULE											
ROOM	FLOOR					WALLS			CIEL.		
	VINYL	CARPET	TILE	HARDWOOD	CONCRETE	PAINT	PAPER	TEXTURE	SPRAY	SMOOTH	BROCADE	PAINT
ENTRY					●							
FOYER			●			●			●		●	●
KITCHEN			●				●			●		●
DINING				●		●			●		●	●
FAMILY		●				●			●		●	●
LIVING		●				●		●			●	●
MSTR. BATH			●				●			●		●
BATH #2			●			●			●	●		●
MSTR. BED		●				●		●			●	●
BED #2		●				●			●		●	●
BED #3		●				●			●		●	●
UTILITY	●					●			●	●		●

Figure 13-35 Finish schedule.

two numbers indicate the width. For example, 30 means 3′-0″. The second two numbers indicate the height; 68 means 6′-8″. A 6040 window means 6′-0″ wide by 4′-0″ high, and a 2868 door means 2′-8″ wide by 6′-8″ high. Standard door height is 6′-8″. Some drafters use the same system in a slightly different manner by presenting the sizes as $2^8 \times 6^8$, meaning 2′-8″ wide by 6′-8″ high or with actual dimensions given, such as 2′-8″ × 6′-8″. Metric sizes vary depending on hard or soft conversion, as dis-

cussed in Chapter 14. This simplified method of identification is better suited for development housing in which doors, windows, and other items are not specified on the plans as to a given manufacturer. These details are included in a description of materials or specification sheet for each individual house. The principal reason for using this system is to reduce drafting time by omitting schedules. The building contractor may be required to submit alternate specifications to the client so actual items may be clarified.

When preparing a set of residential plans, give specific details about doors and windows regarding standard size, material, type of finish, energy efficiency, and style. This information is available through vendors' specifications. As an architectural drafter, you should obtain copies of vendors' specifications from manufacturers that have products available in your local area. Figure 13-37 is a typical page from a door catalog that shows styles and available sizes. Figure 13-38 is a page from a window catalog that describes the sizes of a particular product. Notice that specific information is given for the rough opening (framing size), finish size, and amount of glass. Vendors' catalogs also provide construction details and actual specifications for each product, as shown in Figure 13-39. (See page 182.)

▼ CABINETS, FIXTURES, AND APPLIANCES

Cabinets are found in kitchens, baths, dressing areas, utility rooms, bars, and workshops. Specialized cabinets

Doors with Beveled, Leaded Glass

Widths: 3'0"—single; 6'0"—double.
Heights: 6'8", 6'10", 7'0".

Embossed Doors

Widths: 2'8", 2'10", 3'0"—single; 5'4", 5'8", 6'0"—double.
Heights: 6'6", 6'8", 6'10",7'0".
NA: Certain styles not available in sizes as marked.

*Deadlocks on 2'8" 6-panel Embossed Doors must have a 2⅜" backset.
**Deadlocks on 2'8" 6-panel Embossed Doors with half-glass lites must have a 2⅜" backset and a 2" rose.

Figure 13–37 Partial door catalog page. *Courtesy Ceco Entry Systems, a United Dominion Company.*

Mas. Opg.	2-0 1/2	2-4 1/2	2-8 1/2	3-0 1/2	3-4 1/2	3-8 1/2	4-0 1/2
Rgh. Opg.	1-10 3/8	2-2 3/8	2-6 3/8	2-10 3/8	3-2 3/8	3-6 3/8	3-9 3/8
Frame Size	1-9 3/8	2-1 3/8	2-5 3/8	2-9 3/8	3-1 3/8	3-5 3/8	3-10 3/8
Sash Opg.	1-8	2-0	2-4	2-8	3-0	3-4	3-6
Glass Size	16"	20"	24"	28"	32"	36"	40"

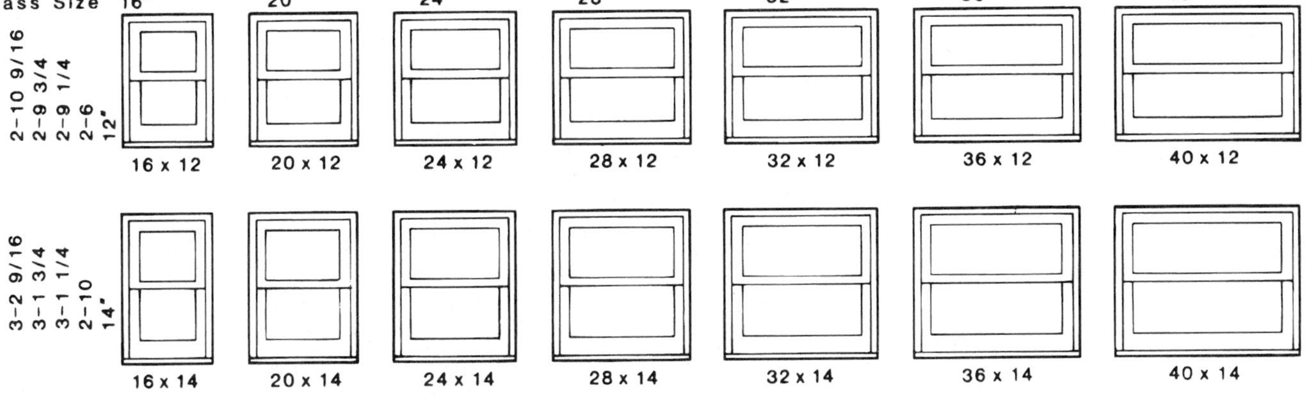

Figure 13–38 Partial window catalog page. *Courtesy Marvin Windows.*

such as desks or built-in dressers may be found in bedrooms. In general, the cabinets that are drawn on the floor plans are built-in units.

Kitchens

In conjunction with cabinets, locate fixtures such as sinks, butcher block cutting board countertops, and lighting, and appliances such as the range, refrigerator, dishwasher, trash compactor, and garbage disposal. Figure 13–40 shows the floor-plan view of the kitchen cabinets, fixtures, and appliances. Some drafters prefer to show upper cabinets, as in Figure 13–41.

The range should have a hood with a light and fan. You should note how the fan vents. Some ranges do not require a hood because they vent through the floor or wall. Vent direction should be specified on the plan. Some homes may have a range top with a built-in oven or an oven and microwave unit. There are many options. The size of the range and refrigerator may vary between 30 and 54" wide. Decide on the size you want from a vendor's catalog. Label the garbage disposal with a note

2x4 Frame Construction,
1/2" Drywall, 1/2" Sheathing,
Insulated Glass, Brick Veneer

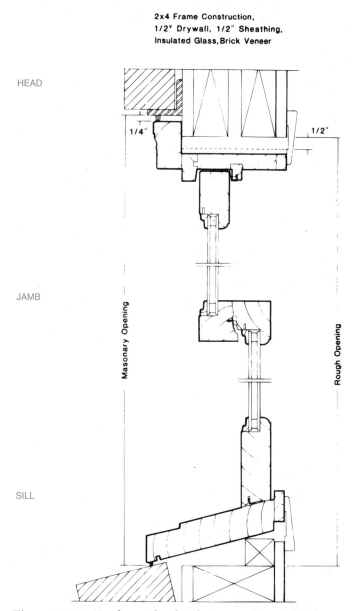

Figure 13–39 Vendor catalog detail. *Courtesy Marvin Windows.*

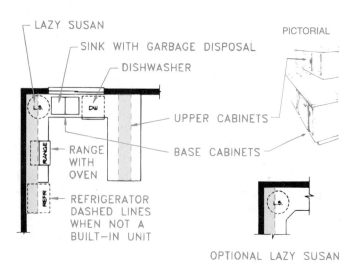

Figure 13–40 Kitchen cabinets, fixtures, and appliances.

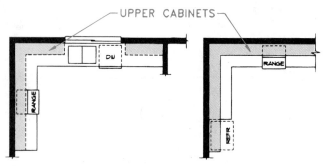

Figure 13–41 Alternate upper cabinet floor plan symbols.

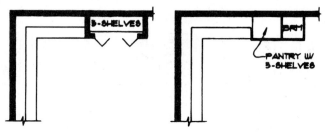

Figure 13–42 Kitchen pantry.

or with the abbreviation GD. Houses with septic tanks should not have garbage disposals.

Pantries are popular and may be drawn as shown in Figure 13–42. A pantry may also be designed to be part broom closet and part shelves for storage. Figure 13–43 shows a typical floor plan representation and cabinet sizes for a kitchen layout.

The *Manual of Acts and Relevant Regulations*, published for the Americans with Disabilities Act (ADA), specifies dimensions and design for people with disabilities.

Bathrooms

Bathroom cabinets and fixtures are shown in several typical floor plan layouts in Figure 13–44 (page 184). The vanity may be any length depending on the space available. The shower may be smaller or larger than the one shown depending on the vendor. Verify the size for a prefabricated shower unit before you draw one. Showers that are built on the job may be any size or design. They are usually lined with tile, marble, or other materials. The common size of tubs is 30" × 60", but larger and smaller ones are available. Sinks can be round, oval, or other shapes.

Figure 13–45 (page 185) shows a variety of products that are available for use when something special is required in a bathroom design. Verify the availability of products and vendors' specifications as you design and draw the floor plan. Figure 13–46 (page 185) shows the floor-plan representation of a 60" × 72" raised bathtub in a solarium. Often, due to cost considerations, the space available for bathrooms is minimal. Figure 13–47 (page 186) shows some minimum sizes to consider. The *Manual of Acts and Relevant Regulations*, published for the ADA,

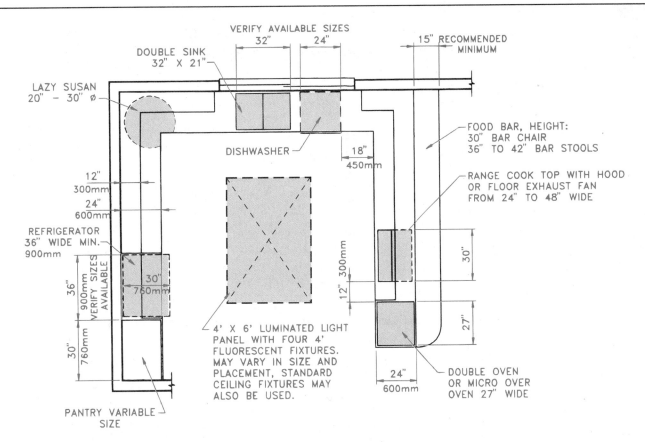

Figure 13–43   Standard kitchen cabinet, fixture, and appliance sizes.

specifies design and dimensions for people with disabilities. For example, wheelchair access requires up to 60″ (1524 mm) minimum width, depending on the design used.

Utility Rooms

The symbols for the clothes washer, dryer, and laundry tray are shown in Figure 13–48 (page 187). The clothes washer and dryer symbols may be drawn with dashed lines if these items are not part of the construction contract. Laundry utilities may be placed in a closet when only minimum space is available, as shown in Figure 13–49 (page 187). Ironing boards may be built into the laundry room wall or attached to the wall surface, as shown in Figure 13–50 (page 187). Provide a note to the electrician if power is required to the unit, or provide an adjacent outlet.

The furnace and hot water heater are sometimes placed together in a location central to the house. They may be placed in a closet, as shown in Figure 13–51 (page 187). These utilities may also be placed in a separate room, the basement, or the garage.

Wardrobes and closets, used for clothes and storage, are utilitarian in nature. Bedroom closets should be labeled WARDROBE. Other closets may be labeled ENTRY

CLOSET, LINEN CLOSET, BROOM CLOSET, and so-forth. Wardrobe or guest closets should be provided with a shelf and pole. The shelf and pole may be drawn with a thin line to represent the shelf and a centerline to show the pole as shown in Figure 13–52 (page 187). Some drafters used two dashed lines to symbolize the shelf and pole.

When a laundry room is below the bedroom area, provide a chute from a convenient area near the bedrooms through the ceiling and into a cabinet directly in the utility room. The cabinet should be above or next to the clothes washer. Figure 13–53 (page 188) shows a laundry chute noted in the floor plan.

▼ FLOOR-PLAN MATERIALS
Finish Materials

Finish materials used in construction may be identified on the floor plan with notes, characteristic symbols, or key symbols that relate to a finish schedule. The key symbol is also placed on the finish schedule next to the identification of the type of finish needed at the given location on the floor plan. Finish schedules may also be set up on a room-by-room basis. When flooring finish is

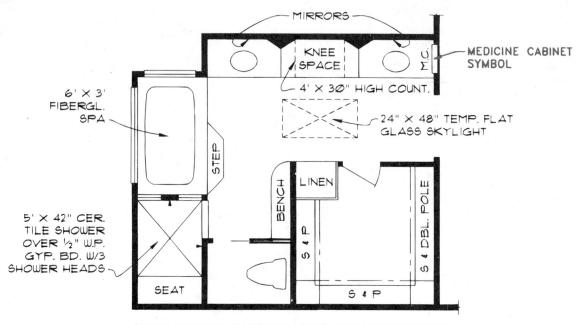

MIRRORS

KNEE SPACE

6' X 3' FIBERGL. SPA

4' X 30" HIGH COUNT.

MEDICINE CABINET SYMBOL

M.C.

STEP

24" X 48" TEMP. FLAT GLASS SKYLIGHT

5' X 42" CER. TILE SHOWER OVER ½" W.P. GYP. BD. W/3 SHOWER HEADS

BENCH

LINEN

S & DBL. POLE

S & P

SEAT

S & P

FULLY EQUIPPED BATHROOM WITH COMPARTMENTALIZED WATER CLOSET, TUB AND SHOWER, DOUBLE VANITY, BENCH, AND LINEN CLOSET

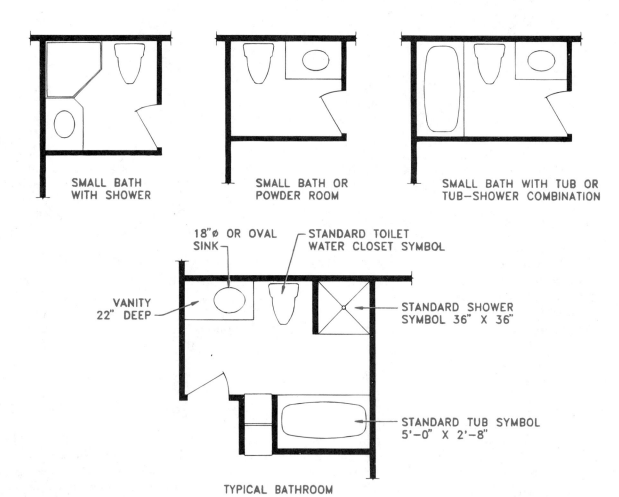

SMALL BATH WITH SHOWER

SMALL BATH OR POWDER ROOM

SMALL BATH WITH TUB OR TUB–SHOWER COMBINATION

18"ø OR OVAL SINK

STANDARD TOILET WATER CLOSET SYMBOL

VANITY 22" DEEP

STANDARD SHOWER SYMBOL 36" X 36"

STANDARD TUB SYMBOL 5'-0" X 2'-8"

TYPICAL BATHROOM

Figure 13–44 Layout of bathroom cabinet and fixture floor-plan symbols.

Figure 13–45 Unique bathroom products. *Courtesy Jacuzzi Whirlpool Bath.*

identified, the easiest method is to label the material directly under room designation as shown in Figure 13–54 (page 188). Other methods include using representative material symbols and a note to describe the finish materials as shown in Figure 13–55 (page 188).

Drafters with an artistic flair may draw floor-plan symbols freehand. Other drafters may use adhesive material symbols that are available in sheet form. You use this product by removing a section of the sheet that is slightly larger than the area to be covered. Then, with a sharp knife, cut away the extra material. Be careful to cut only the symbol material, not your drawing original. Remove the symbol material from areas where the pattern is not needed; then firmly press the pattern onto the desired area. Figure 13–56 (page 188) shows some patterns that are available.

Structural Materials

Structural materials are identified on floor plans with notes and symbols or in framing plans. This is discussed in detail in Chapter 30.

▼ STAIRS

Stair Planning

The design of a multistory home is often made complex by the need to plan stairs. Figure 13–57 (page 189) shows several common flights of stairs. Stairs should be conveniently located for easy access. Properly designed stairs have several important characteristics based on CABO requirements:

○ Minimum stair width is 30″ but 36 to 48″ is preferred.
○ Stair tread length should be 10 to 12″ with 9″ minimum.

○ Individual risers may range between 7 and 8¼″ (178–210 mm) in height. Risers should not vary more than ³⁄₈″ (10 mm). A comfortable riser height is 7¼″.
○ Landings at the top and bottom of stairs should be at least equal in size to the width of the stairs.
○ A clear height of 6′–8″ is the minimum amount of headroom required for the length of the stairs. A height of 7′ is preferred.

These characteristics along with stair calculation and construction are discussed in Chapter 37.

Handrails are placed at a blank wall along a flight of stairs to be used for support. Handrails should run the whole length of the stairs, even though the stairs are enclosed between two walls. Stairs with over three risers require handrails. Guardrails are placed where there is no protective wall. They serve the dual purpose of protecting people from falling off the edge of the stairs and providing a rail for support. Stair wells, or stair openings in the floor, should be enclosed with strong railings extending 6″ past the edge of the first step.

Stairs on Floor Plans

Stairs are shown on floor plans by the width of the tread, the direction and number of risers, and the lengths of handrails or guardrails. Abbreviations used are DN (down) UP (up) and R (risers). A note of 14R means there are 14 risers in the flight of stairs.

Stair layouts depend on the amount of space available. There is always one fewer run than rise. (*Run* is the individual tread; *rise* is each individual riser.) Consider the average set of stairs to have 14 risers. To figure the length of stairs, multiply the length of a tread by the number of runs. For example, stairs with 14 risers have 13 runs. If each individual run is 10″ then 10 × 13″ = 130″, or 10′–10″. Keep in mind that landings take up more space, and they should be as long as the stairs are

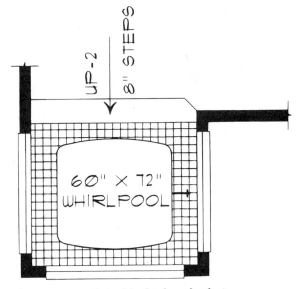

Figure 13–46 Raised bathtub and solarium.

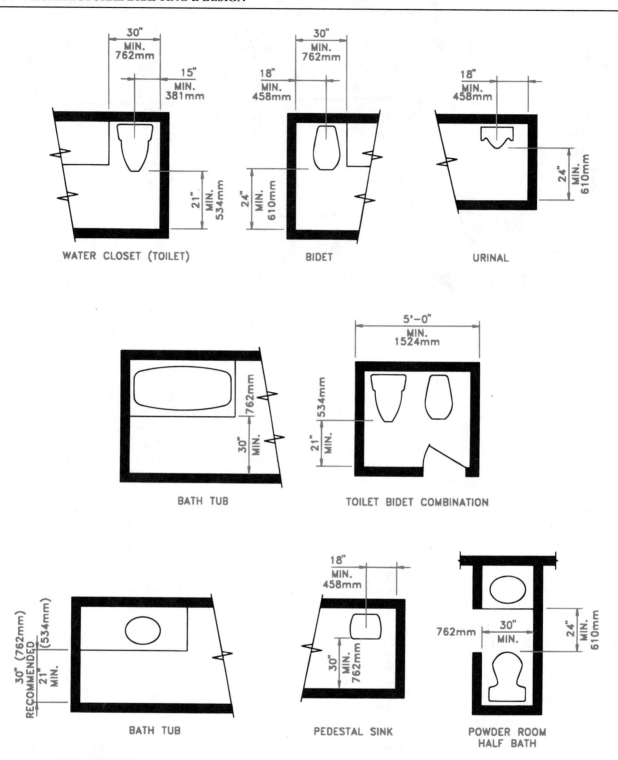

WATER CLOSET (TOILET)

BIDET

URINAL

BATH TUB

TOILET BIDET COMBINATION

BATH TUB

PEDESTAL SINK

POWDER ROOM
HALF BATH

INCH DIMENSIONS ARE BOCA MINIMUM REQUIREMENTS.
METRIC DIMENSIONS ARE PROVIDED FOR DESIGN REFERENCE.

VERIFY FIXTURE SIZES WITH PRODUCT SPECIFICATIONS.

VERIFY DESIGN AND DIMENSIONS FOR DISABLED ACCESS WITH THE MANUAL OF ACTS
AND RELEVANT REGULATIONS FOR THE AMERICAN'S WITH DISBILITIES ACT.
AND LOCAL CODE REQUIREMENTS.

Figure 13–47 Minimum bath spaces. All dimensions are CABO minimums.

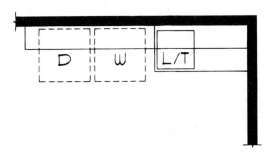

Figure 13–48   Washer, dryer, and laundry tray.

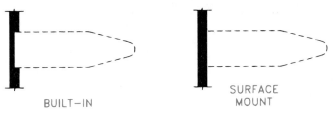

BUILT-IN

SURFACE MOUNT

Figure 13–50   Ironing boards.

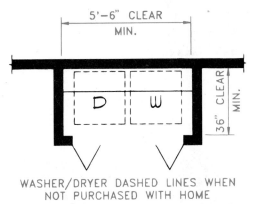

WASHER/DRYER DASHED LINES WHEN NOT PURCHASED WITH HOME

Figure 13–49   Minimum washer and dryer space.

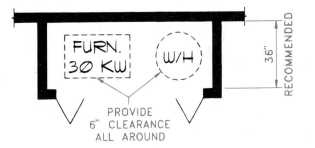

PROVIDE 6" CLEARANCE ALL AROUND

Figure 13–51   Furnace and water heater.

wide. If a 36″ wide stairway must have a landing, the landing should measure 36 × 36″ minimum.

Figure 13–58 (page 189) shows a common straight stair layout with a wall on one side and a guardrail on the other. Figure 13–59 (page 189) shows a flight of stairs with guardrails all around at the top level and a handrail running down the stairs. Figure 13–60 (page 189) shows stairs between two walls. Notice also that the stairs are drawn broken with a long break line at approximately mid-height. Figure 13–61 (page 189) shows stacked stairs with one flight going up and the other down. This situation is

common when access from the main floor to both the second floor and basement is designed for the same area.

Stairs with winders and spiral stairs may be used to conserve space. Winders are used to turn a corner instead of a landing. They must not be any smaller than 6″ at the smallest dimension. Spiral stairs should be at least 30″ wide from the center post to the outside, and the center post should have a 6″ minimum diameter. Spiral stairs and custom winding stairs may be manufactured in several designs. Figure 13–62 (page 190) shows the plan view of winder and spiral stairs.

A sunken living room or family room is a popular design feature. When a room is either sunken or raised, there is at least one step into the room. The steps are noted with an arrow, as shown in Figure 13–63 (page 190). The few steps up or down do not require a handrail unless there are over three risers, but provide a handrail next to the

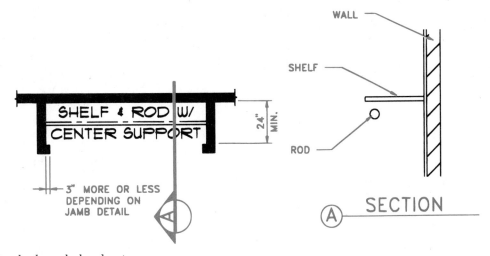

Figure 13–52   Standard wardrobe closet.

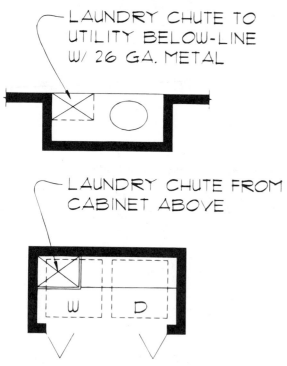

Figure 13–53 Laundry chute.

Figure 13–54 Room labels with floor finish material noted.

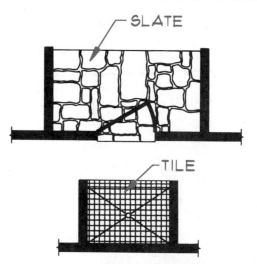

Figure 13–55 Finish material symbols.

steps when there are four or more. The guardrail shown in Figure 13–63 (page 190) is for decoration only.

The ADA specifies the use of a ramp for wheelchair access. The ramp should have a slope of 1:12 for 30″ (762 mm) maximum rise and 30′ (9 m) of horizontal run. Hallways shall be 36″ (915 mm) minimum width. (Metric dimensions vary depending on hard or soft conversion. See Chapter 14.)

▼ FIREPLACES

Floor Plan

The dimensions shown on the fireplace floor plan symbol in Figure 13–64 (page 190) are minimums. Figure 13–65 (page 190) shows several typical fireplace floor plan representations. Common fireplace opening sizes in inches are shown in the following table:

Opening Width	Opening Height	Unit Depth
36	24	22
40	27	22
48	30	25
60	33	25

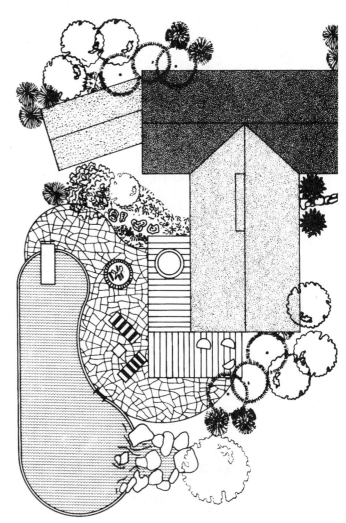

Figure 13–56 Adhesive patterns for finish materials. *Courtesy Chartpak.*

Steel Fireplaces

Fireplace fireboxes made of steel are available from various manufacturers. The fireplace you draw on the

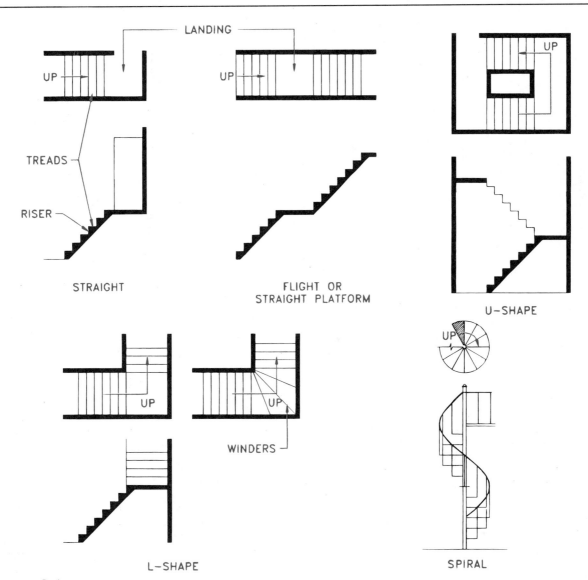

Figure 13-57 Stair types.

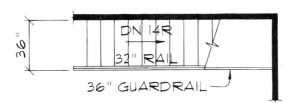

Figure 13-58 Stairs with a wall on one side and a guardrail on the other.

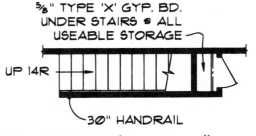

Figure 13-60 Stairs between two walls.

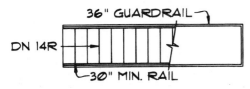

Figure 13-59 Stairs with guardrails all around.

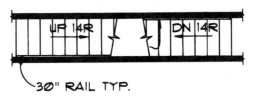

Figure 13-61 Stairs up and down in the same area.

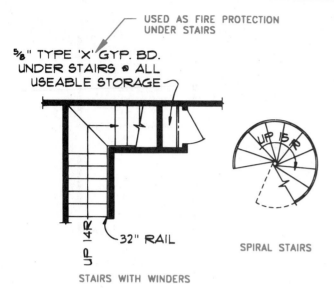

Figure 13–62 Winder and spiral stairs.

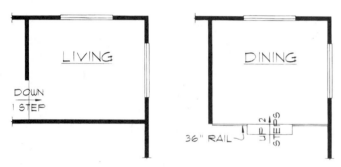

Figure 13–63 Sunken or raised rooms.

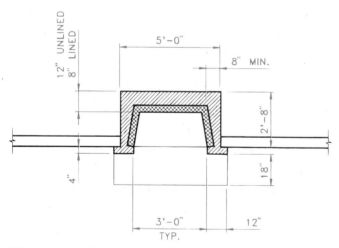

Figure 13–64 Single masonry minimum fireplace dimensions not generally shown on the plan.

floor plan will look the same whether the firebox is prefabricated of steel or constructed of masonry materials. See Figure 13–66. These prefabricated fireplaces are popular because of their increased efficiency. Some units add to their efficiency with fans that are used to circulate air around the firebox and into the room.

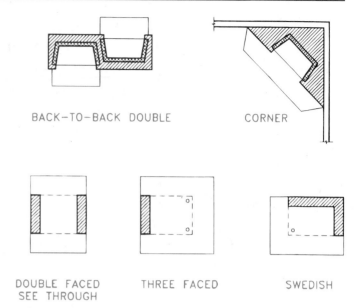

Figure 13–65 Floor plan representations of typical masonry fireplace.

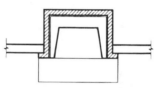

Figure 13–66 Manufactured firebox framed in masonry.

There are also insulated fireplace units made of steel that can be used with wood-framed chimneys. These units are often referred to as *zero-clearance units*: the insulated metal fireplace unit and flue can be placed next to a wood-framed structure. Verify building code clearances and vendors' specifications before calling for its use. The popularity of these fireplaces has increased due to their low construction cost as compared to masonry fireplace and chimney construction. See Figure 13–67. Figure 13–68 shows some common sizes of fireplace units that are available.

Wood Storage

Fireplace wood may be stored near the fireplace. A special room or an area in a garage adjacent to the fireplace may be ideal. A wood compartment built into the masonry next to a fireplace opening may also be provided for storing a small amount of wood. The floor-plan representation of such a wood storage box is shown in Figure 13–69.

Fireplace Cleanout

When you can provide easy access to the base of the fireplace, provide a *cleanout* (CO), a small door in the floor of the fireplace firebox that allows ashes to be dumped into a hollow cavity built into the fireplace below the floor. Access to the fireplace cavity to remove the

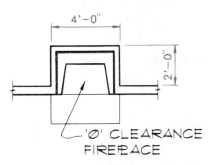

Figure 13–67   Manufactured firebox in wood frame.

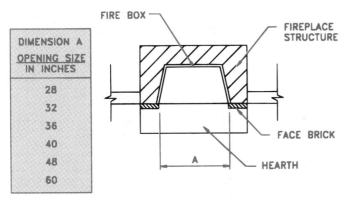

Figure 13–68   Prefabricated fireplace opening sizes. *Courtesy Heatilator, Inc.*

DIMENSION A OPENING SIZE IN INCHES
28
32
36
40
48
60

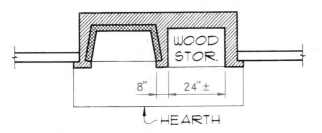

Figure 13–69   Wood storage.

stored ashes is provided from the basement or outside the house. Figure 13–70 shows a fireplace cleanout noted in the plan view.

Combustion Air

Building codes require that *combustion air* be provided to the fireplace. Combustion air is outside air supplied in sufficient quantity for fuel combustion. The air is supplied through a screened duct that is built by masons from the outside into the fireplace combustion chamber. By providing outside air, this venting device prevents the fireplace from using the heated air from the room, thus maintaining indoor oxygen levels and keeping heated air from going up the chimney. A note should be placed on the floor plan next to the fireplace that states the following: PROVIDE OUTSIDE COMBUSTION AIR TO WITHIN 24″, or PROVIDE SCREENED CLOSABLE VENT TO OUTSIDE AIR WITHIN 24″ OF FIREBOX.

Gas-burning Fireplace

Natural gas may be provided to the fireplace, either for starting the wood fire or for fuel to provide flames on artificial logs. The gas supply should be noted on the floor plan, as shown in Figure 13–71.

Built-in Masonry Barbecue

When a home has a fireplace in a room next to the dining room, nook, or kitchen, the masonry structure may also incorporate a built-in barbecue. The floor-plan representation for a barbecue is shown in Figure 13–72. A built-in barbecue may be purchased as a prefabricated unit that is set into the masonry structure surrounding the fireplace. There may be a gas or electricity supply to the barbecue as a source of heat for cooking. As an alternative, the barbecue unit may be built into the exterior structure of a fireplace for outdoor cooking. The barbecue may also be installed separately from a fireplace.

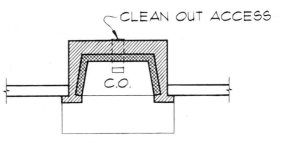

Figure 13–70   Fireplace cleanout.

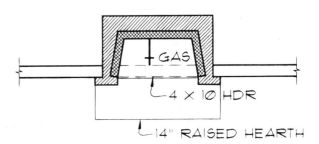

Figure 13–71   Gas supply to fireplace.

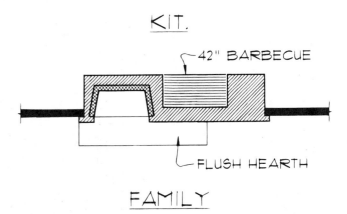

Figure 13–72   Barbecue in fireplace.

▼ SOLID FUEL-BURNING APPLIANCES

Solid fuel-burning appliances are such items as airtight stoves, free-standing fireplaces, fireplace stoves, room heaters, zero-clearance fireplaces, antique and home-made stoves, and fireplace inserts for existing masonry fireplaces. This discussion shows the floor-plan representation and minimum distance requirements for the typical installations of some approved appliances.

Appliances that comply with nationally recognized safety standards may be noted on the floor plan with a note such as: "ICBO APPROVED WOOD STOVE AND INSTALLATION REQUIRED [ICBO = International Conference of Building Officials] VERIFY ACTUAL INSTALLATION REQUIREMENTS WITH VENDORS' SPECIFICATIONS AND LOCAL FIRE MARSHAL OR BUILDING CODE GUIDELINES."

General Rules

Verify these rules with the local fire marshal or building code guidelines.

Floor Protection. Combustible floors must be protected. Floor protection material shall be noncombustible, with no cracks or holes, and strong enough not to crack, tear, or puncture with normal use. Materials commonly used are brick, stone, tile, or metal.

Wall Protection. Wall protection is critical whenever solid fuel-burning units are designed into a structure. Direct application of noncombustible materials will not provide adequate protection. When solid fuel-burning appliances are installed at recommended distances to combustible walls, a 7″ air space is necessary between the

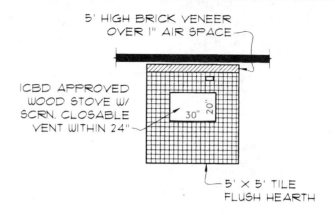

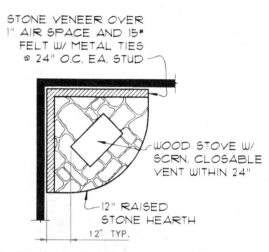

Figure 13–74 Wood stove installation.

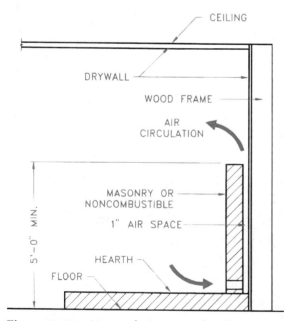

Figure 13–73 Air circulation around wall protection.

wall and floor to the noncombustible material, plus a bottom opening for air intake and a top opening for air exhaust to provide positive air change behind the structure. This helps reduce superheated air next to combustible material (Figure 13–73). Noncombustible materials include brick, stone, or tile over cement asbestos board. Minimum distances to walls should be verified in regard to vendors' specifications and local requirements.

Combustion Air. Combustion air is generally required as a screened closable vent installed within 24″ of the solid fuel–burning appliance.

Air Pollution. Some local areas have initiated guidelines that help control air pollution from solid fuel-burning appliances. The installation of a catalytic converter or other devices may be required. Check with local regulations.

Figure 13–74 shows floor-plan representations of common wood stove installations.

A current trend in housing is to construct a masonry alcove within which a solid fuel unit is installed. The floor-plan layout for a typical masonry alcove is shown in Figure 13–75.

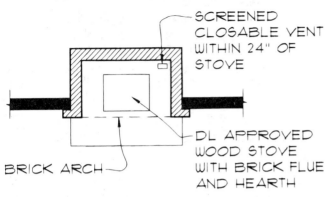

Figure 13–75   Masonry alcove for solid-fuel burning stove.

▼ DIRECT-VENT FIREPLACES

Many builders are now installing direct-vent gas fireplaces. Ease of installation allows them to be placed almost anywhere. These fireplaces give a gas fire around realistic-looking ceramic logs. The 4″ vent usually exits the wall a few feet above the fireplace, eliminating the need for expensive chimneys. The gas fire is controlled by a wall switch, remote control, or thermostat. The fire gives the appearance and warmth of a real wood fire without any of the mess.

Direct-vent fireplaces are sealed-combustion heaters. The doors open only for maintenance. The vent pipe is designed with an outer shell. The exhaust gases exit through the center, while combustion air is drawn into the fireplace through the outer chamber. The air is heated in passages behind the metal firebox, and the room air does not come in contact with the flames. Heat radiates through the glass doors and is forced by convection or fans around the firebox and into the room.

Direct-vent fireplaces do not affect indoor air quality, which makes them a good choice for energy-efficient construction. With restrictions on wood burning in some areas, direct-vent fireplaces are a good option to consider. Manufacturers' installation instructions and local codes must be followed to ensure proper clearances and venting specifications. Improper installation can result in fire or explosions, which will shatter the glass fireplace enclosure. Automatic shutoff switches may be considered to help keep the unit from overheating during extended use.

▼ ROOM TITLES

Rooms are labeled with a name on the floor plan. Generally the lettering is larger than that used for other notes and dimensions. Room titles may be lettered using ³⁄₁₆ or ¼″ (5 or 7 mm) high letters, as shown in Figure 13–76. The interior dimensions of the room may be lettered under the title, as shown in Figure 13–77.

▼ OTHER FLOOR-PLAN SYMBOLS
Hose Bibb

A hose bibb is an outdoor water faucet to which a garden hose may be attached. Hose bibbs should be placed at locations convenient for watering lawns or gardens and for washing a car. The floor-plan symbol for a hose bibb is shown in Figure 13–78. This symbol may be shown on a separate plumbing plan as discussed in Chapter 17. This often depends on the complexity of the structure.

Concrete Slab

Concrete slabs used for patio walks, garages, or driveways may be noted on the floor plan. A typical example would be 4″-THICK CONCRETE WALK. Concrete slabs used for the floor of a garage should slope toward the front, or to a floor drain in cold climates to allow water to drain. The amount of slope is ⅛″/ft minimum. A garage slab is noted in Figure 13–79. Calling for a slight slope on patios and driveway aprons is also common.

Figure 13–76   Room titles.

Figure 13–77   Room titles with room sizes noted.

Figure 13–78   Hose bibb symbol.

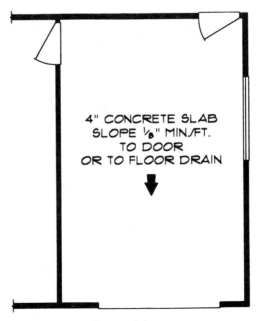

Figure 13–79   Garage slab note.

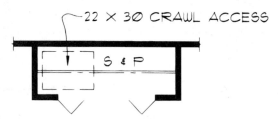

Figure 13–80   Crawl space access note.

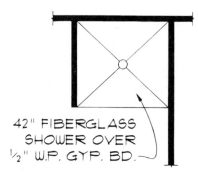

Figure 13–81   Shower symbol showing floor drain.

Attic and Crawl Space Access

Access is necessary to attics and crawl spaces. The crawl access may be placed in any convenient location that has 30″ minimum headroom, such as a closet or hallway. The minimum size of the attic access is 22 × 30″ (560 × 762 mm). The crawl access should be 22 × 30″ (560 × 762 mm) if located in the floor, as shown in Figure 13–80. The attic access may include a fold-down ladder if the attic is to be used for storage. A minimum of 30″ (762 mm) must be provided above the attic access. Crawl space access may be shown on the foundation plan when it is constructed through the foundation wall. When in the foundation wall, the access can be 24 × 18″ (610 × 460 mm).

Floor Drains

A floor drain is shown on the floor-plan symbol for a shower seen in Figure 13–81. Floor drains should be used in any location where water could accumulate on the floor, such as the laundry room, bathroom, or garage. The easiest application f a floor drain is in a concrete slab floor, although drains may be designed in any type of floor construction. Figure 13–82 shows a floor drain in a utility room.

Cross-section Symbol

The location on the floor plan where a cross section is taken is identified with symbols known as *cutting-plane lines*. These symbols are discussed in detail in Chapter 34.

Floor plans are a key element in a complete set of architectural drawings. Clients are interested in floor plans so they can see how their future home or business is laid out. As you have seen in this chapter, there are a large variety of symbols and drawing techniques that go into preparing a floor plan. The challenge to the architectural drafter is to be sure to include all the necessary symbols and notes needed for construction and yet make the plan easy to read.

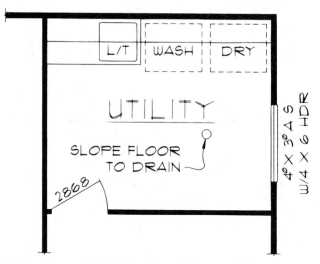

Figure 13–82   Floor drain symbol (symbol may also be square).

CADD Applications

CADD FLOOR-PLAN SYMBOLS

A variety of architectural CADD software packages is available that provides complete floor-plan symbol library templates. These architectural CADD programs usually include applications for windows, doors, structural symbols, electrical, furniture, appliance, stairs, and site-plan symbols. Also available are titles, drawing and editing functions, complete dimensioning capabilities, and pictorial drawing applications. The CADD commands provide (1) residential and commercial symbols for single insertion and (2) commercial symbols for multiple insertion. The multiple-insertion symbols are based on parametric design where you provide specifications for several variables and the CADD system automatically draws the units. For example, if you design a restroom for a hotel, all you do is select vanity sinks for multiple insertion, and you are given several sink types to choose from. You continue by specifying the sink size, the number of units, and the distance between walls. You also specify the cabinet width, height, and the backsplash dimensions. Figure 13–83 shows the multiple sink arrangement previously described.

Drawing stairs is also easy with the CADD system. Pick the start point of the stairs, specify the number of steps to the landing, give the stair width and direction, and give the total rise; the program automatically calculates the rise of each step. The program asks you to provide handrails with or without ballisters and provides several options of handrail ends. After you have provided all the required information, the stair is automatically drawn, as shown in Figure 13–84.

In most cases, the symbol you select is "dragged" onto the screen cursor at the symbol-insertion point. The term *drag* defines the CADD command that allows you to move the symbol to a desired location, where it is fixed in place on the drawing. If you decide at a later time that you do not like the selected position, you can use commands such as MOVE and ROTATE to reposi-

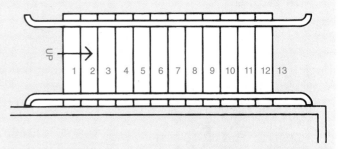

Figure 13–84   The stairs are automatically drawn from information provided.

tion the symbol; you might also use ERASE to get rid of it all together. Figure 13–85 shows a symbol being dragged into position and displays the insertion point of several common symbols. The insertion point refers to a convenient point on the symbol that is used when placing the symbol on the drawing.

Manufacturers' Symbols Libraries. Manufacturers of most architectural products such as doors, windows, plumbing fixtures, and appliances have CADD symbols libraries available for design and drafting. *Symbols libraries* are a collection of related shapes, views, symbols, and attributes that may be used repeatedly. *Attributes*, also known as *tags*, go along with the architectural symbols to provide identification and numerical values for the products. Attributes are also referred to as notes. There is more than the simple insertion of a symbol. The symbols also carry information that may be used as a bill of materials or compiled from the drawing into a bill of materials. Any part or all of the attributes may be displayed on the drawing or kept invisible. Manufacturers' symbol libraries are usually free to architects, designers, and builders. They are available to help the user select a product, make a specification, prepare drawings, and tabulate bills of material. Most of these products allow you to create drawings quickly in plan view, elevation, pictorial, and construction details. *Elevations* are drawings looking at the front and sides of the product. *Pictorial drawings* display the product in a three-dimensional representation. Figure 13–86 shows what the computer screen looks like when displaying a window from a manufacturer's symbol library. The left side of the screen shows a pictorial image, the upper right display is an elevation, and the lower right shows the plan view. Figure 13–87 displays the computer screen image of a detail drawing from a manufacturer's symbol library.

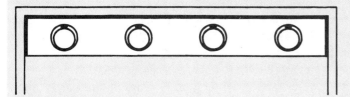

Figure 13–83   Multiple sink arrangement automatically placed with CADD.

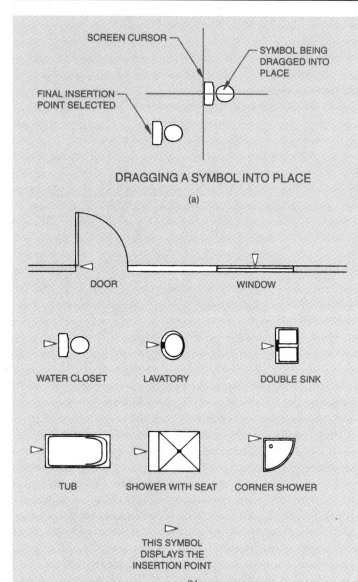

DRAGGING A SYMBOL INTO PLACE

(a)

DOOR   WINDOW

WATER CLOSET   LAVATORY   DOUBLE SINK

TUB   SHOWER WITH SEAT   CORNER SHOWER

THIS SYMBOL
DISPLAYS THE
INSERTION POINT

(b)

Figure 13–85 (a) Dragging symbol into place; (b) insertion points for common symbols.

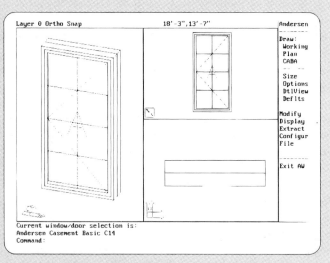

Figure 13–86 Computer screen displaying a window in 3D, elevation, and plan from a manufacturers' symbols library. *Courtesy Andersen Windows, Inc.*

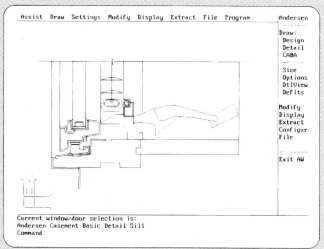

Figure 13–87 Computer screen image of a detail drawing from a manufacturers' symbol library. *Courtesy of Andersen Windows, Inc.*

In addition to the CADD symbol library, manufacturers provide a printed reference library. This library allows you to see what is available, and it provides the block name and default attributes. A *block* is a symbol that can be inserted into a drawing full size, or scaled, and rotated if necessary. A *default* is a value that is maintained by the computer until you change it. Figure 13–88 shows an example of several items found in the symbol library reference manual from a plumbing fixture manufacturer.

Creating Your Own CADD Symbols. You can create your own CADD symbols if a customized architectural menu is not available. The symbols that you make are called *blocks* or *wblocks* in AutoCAD. You can design an entire group of symbols for insertion into your architectural drawings. As you create these symbols, prepare a symbols library or catalog where each symbol is represented for reference. The symbol catalog should contain the symbol name, insertion point, and other important information about the symbol. To design your symbols follow these steps:

1. Draw the symbol just as you would draw any drawing.
2. Draw each symbol within the boundary of one unit square, so you can scale the symbol as needed when inserted in the drawing.
3. Use the BLOCK or WBLOCK command to save the symbol with a name such as DOOR.
4. Pick an insertion point—a point on the symbol where it would be convenient for you to insert into your drawing, such as the center of a door.
5. Use the INSERT command to place the symbol in your drawing. Figure 13–89 shows a symbol being created.

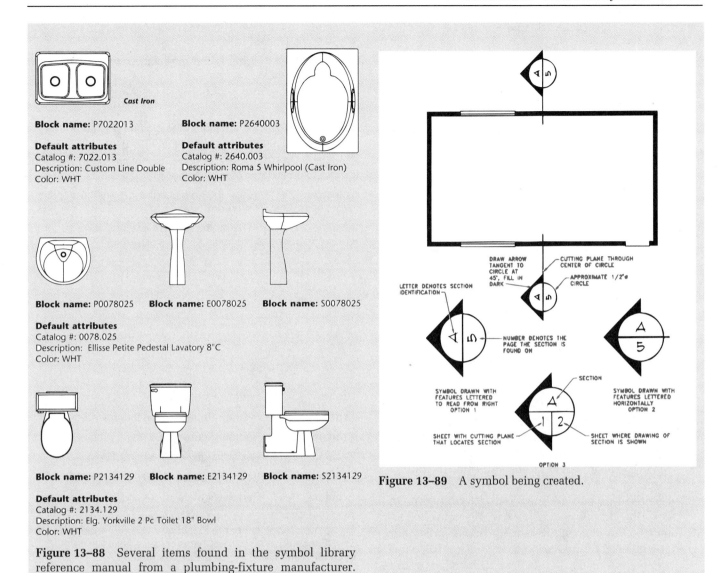

Cast Iron

Block name: P7022013

Default attributes
Catalog #: 7022.013
Description: Custom Line Double
Color: WHT

Block name: P2640003

Default attributes
Catalog #: 2640.003
Description: Roma 5 Whirlpool (Cast Iron)
Color: WHT

Block name: P0078025   **Block name:** E0078025   **Block name:** S0078025

Default attributes
Catalog #: 0078.025
Description: Ellisse Petite Pedestal Lavatory 8"C
Color: WHT

Block name: P2134129   **Block name:** E2134129   **Block name:** S2134129

Default attributes
Catalog #: 2134.129
Description: Elg. Yorkville 2 Pc Toilet 18" Bowl
Color: WHT

Figure 13–88   Several items found in the symbol library reference manual from a plumbing-fixture manufacturer. *Courtesy American Standard, Inc.*

Figure 13–89   A symbol being created.

CHAPTER 13

Floor-Plan Symbols Test

Answer the questions with short, complete statements or drawings as needed on an 8½ × 11″ sheet of notebook paper as follows:

1. Letter your name, Floor-Plans Symbol Test, and the date at the top of the sheet.
2. Letter the question number and provide the answer. You do not need to write out the question.
3. Answers may be prepared on a word processor if appropriate with course guidelines.

QUESTIONS

Question 13–1   Exterior walls for a wood-frame residence are usually drawn how thick?

Question 13–2   Interior walls are commonly drawn how thick?

Question 13–3   How does the floor-plan symbol for an interior door differ from the symbol for an exterior door?

Question 13–4   What are the recommended spaces required for the following?
a. wardrobe closet depth
b. water closet compartment
c. stair width
d. fireplace hearth depth

Question 13–5   Describe an advantage of using schedules for windows and doors.

Question 13–6 Sketch the following floor-plan symbols:
a. pocket door
b. bifold closet door
c. casement window
d. sliding window
e. skylight
f. ceiling beam
g. single-run stairs, up
h. hose bibb

Question 13–7 Letter the note that would properly identify a 48″ high wall.

Question 13–8 What is the required minimum height of guardrails?

Question 13–9 Provide the appropriate note used to identify a guardrail at a balcony overlooking the living room.

Question 13–10 What door would you recommend for use when space for a door swing is not available?

Question 13–11 Name the window that you would recommend when a 100 percent openable unit is required.

Question 13–12 What does the note 6040 next to a window on the floor plan mean?

Question 13–13 What is the minimum crawl space or attic access required?

Question 13–14 What is the abbreviation for garbage disposal?

Question 13–15 When should the clothes washer and dryer be shown with dashed lines on the floor plan?

Question 13–16 Identify in short, complete statements four factors to consider in stair planning.

Question 13–17 Show the note used to identify the steps at a two-step sunken living room.

Question 13–18 What is the amount of slope for a concrete garage floor and why should it have a slope?

PROBLEMS

DIRECTIONS

1. Draw the following problems manually or with CADD depending on your course guidelines. Use a ¼″ = 1′–0″ scale. Draw on 8½ × 11 vellum.
2. Make exterior walls 6″ thick and interior walls 5″ thick.
3. Use the line and architectural lettering methods described in this text. Letter only the notes that are part of the drawing. Estimate dimensions when not given and increase sizes to help fill paper when appropriate. Do not draw dimensions or pictorial illustrations.

Problem 13–1 Redraw Figures 13–4, 13–5, 13–7, and 13–8.

Problem 13–2 Redraw Figure 13–9.

Problem 13–3 Redraw Figures 13–13 through 13–18.

Problem 13–4 Redraw Figures 13–19 through 13–22.

Problem 13–5 Redraw Figure 13–23.

Problem 13–6 Redraw Figures 13–24 through 13–28 and 13–30.

Problem 13–7 Redraw Figure 13–32.

Problem 13–8 Redraw Figure 13–34. Draw the partial floor plan and door/window key symbols. Do not draw the door/window schedules.

Problem 13–9 Redraw Figure 13–36.

Problem 13–10 Redraw Figure 13–43.

Problem 13–11 Redraw Figure 13–44.

Problem 13–12 Redraw Figures 13–49, 13–51, and 13–55.

Problem 13–13 Redraw Figures 13–58 through 13–62.

Problem 13–14 Redraw Figure 13–63.

Problem 13–15 Redraw Figures 13–64 and 13–65.

Problem 13–16 Redraw Figures 13–66 through 13–72.

Problem 13–17 Redraw Figures 13–74 and 13–75.

Problem 13–18 Use your CADD system to redraw the following floor plan with either the PLINE and OFFSET method, or the MLINE technique discussed in this chapter. Use edit commands such as TRIM, EXTEND, or BREAK as needed.

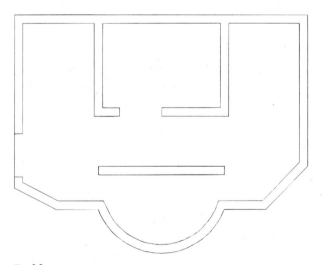

Problem 13–18

CHAPTER 14

Floor-Plan Dimensions

▼ ALIGNED DIMENSIONS

The dimensioning system most commonly used in architectural drafting is known as *aligned dimensioning*. With this system, dimensions are placed in line with the dimension lines and read from the bottom or right side of the sheet. Dimension numerals are centered on and placed above the solid dimension lines. Figure 14–1 shows a floor plan that has been dimensioned using the aligned dimensioning system.

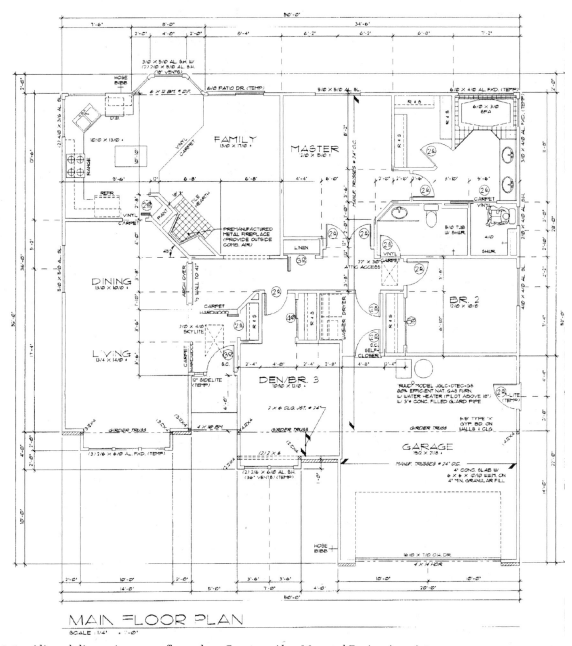

Figure 14–1 Aligned dimensions on a floor plan. *Courtesy Alan Mascord Design Associates.*

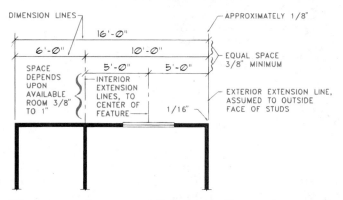

Figure 14–2 Recommended dimension line spacing.

▼ FLOOR-PLAN DIMENSIONS
Basic Dimensioning Concepts

You should place dimensions so the drawing does not appear crowded. Doing so is often difficult because of the great number of dimensions that must be placed on an architectural drawing. When placing dimensions, space dimension lines a minimum of ⅜″ (10 mm) from the object and from each other. If there is room, ½″ (13 mm) is preferred. Some drafters use 1″ (26 mm) or more if space is available. Other drafters place the first dimension line 1″ (26 mm) away from the plan and space additional dimensions equally at ½″ (13 mm) or ¾″ (20 mm) apart. The minimum recommended spacing of dimensions is shown in Figure 14–2. Regardless of distance chosen, be consistent so dimension lines are evenly spaced.

Extension and dimension lines are drawn thin and dark. They are drawn in this manner so they do not distract from the overall appearance and balance of the drawing. Dimension lines terminate at extension lines

with dots, arrowheads, or slash marks that are each drawn in the same direction. See Figure 14–3.

Dimension numerals are drawn ⅛″ (3 mm) high with the aid of guidelines. The dimension units used are feet and inches for all lengths over 12″. Inches and fractions are used for units less than 12″. Foot units are followed by the symbol (′) and inch units are shown with the symbol (″). Some drawing dimensions may not use the foot and inch symbols if company standards are to omit them.

Metric dimensions are given in millimeters (mm), as discussed later in this chapter; however, metric sizes vary depending on hard or soft conversion.

Exterior Dimensions

The overall dimensions on frame construction are understood to be given to the outside of the stud frame of the exterior walls. The reason for locating dimensions on the outside of the stud frame is that the frame is established first, and windows, doors, and partitions are usually put in place before sheathing and other wall-covering material is applied. The first line of dimensions on the plan is the smallest distance from the exterior wall to the center of windows, doors, and partition walls. The second line of dimensions generally gives the distance from the outside walls to partition centers. The third line of dimensions is usually the overall distance between two exterior walls. See Figure 14–4. This method of applying dimensions eliminates the need for workers to add dimensions at the job site and reduces the possibility of making an error.

This textbook shows dimensions placed from the outside face of exterior stud walls to the center of interior walls and partitions to the center of doors and windows. Many architectural offices, however, use a different dimensioning system. They place dimensions from

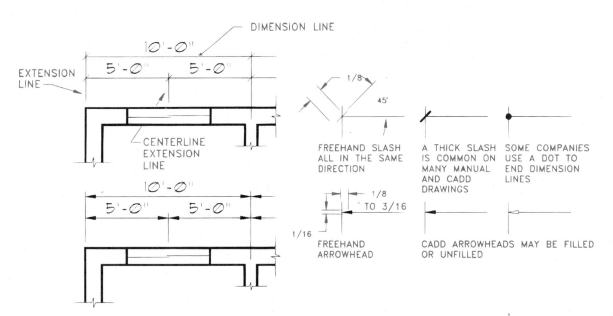

Figure 14–3 Methods for terminating dimension lines.

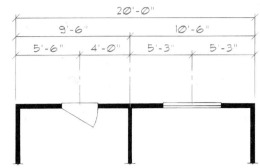

DIMENSIONING FROM OUTSIDE FACE OF EXTERIOR STUDS
TO THE CENTER OF INTERIOR WALLS, DOORS, AND WINDOWS

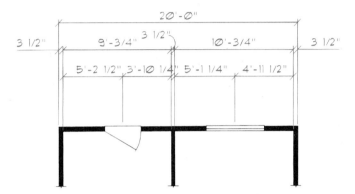

DIMENSIONING FROM THE OUTSIDE AND INSIDE FACE OF STUDS
BETWEEN EXTERIOR AND INTERIOR WALLS, FOLLOWED BY DIMENSIONS
FROM THE OUTSIDE FACE OF EXTERIOR STUDS TO THE CENTER
OF DOORS AND WINDOWS

Figure 14–4   Placing exterior dimensions.

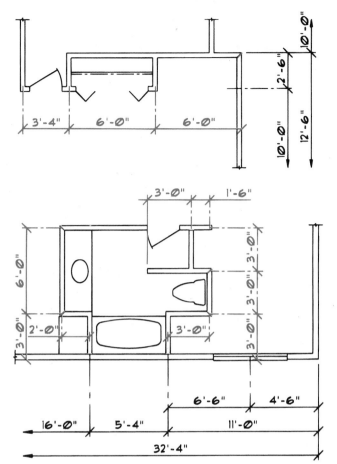

Figure 14–5   Placing interior dimensions.

the outside and inside face of studs between exterior and interior walls, followed by dimensions from the outside face of exterior studs to the center of doors and windows, as shown in Figure 14–4. Both methods provide an overall dimension as the last dimension.

Interior Dimensions

Interior dimensions locate all interior partitions and features in relationship to exterior walls. Figure 14–5 shows some common interior dimensions. Notice how they relate to the outside walls. When dimensioning interior features, ask yourself if you have provided workers with enough dimensions to build the house. The contractor should not have to guess where a wall or feature should be located, nor should workers have to use a scale to try to locate items on a plan.

Dimensions are generally given by locating interior walls and partitions from the outside face of exterior studs to their centers. Dimensions between interior walls are from center to center. *Stub walls*—walls that do not go all the way across the room—are dimensioned by giving their length. Windows and doors are dimensioned to their centers, unless the location is assumed. Some offices practice dimensioning all features to the face of studs rather than to the centers, as previously discussed. Figure 14–5 shows examples of dimensioning interior features.

Standard Features

Some interior features that are considered to be standard sizes may not require dimensions. Figure 14–6 shows a situation where the drafter elected to dimension the depth of the pantry even though it is directly next to a refrigerator, which has an assumed depth of 30″ (762 mm). Notice that the refrigerator width is given because there are many different widths available. The base cabinet is not dimensioned; such cabinets are typically 24″ (600 mm) deep. When there is any doubt, it is better to apply a dimension. The refrigerator and other appliances may be drawn with dashed lines or marked with the abbreviation NIC (not in contract) if the unit will not be supplied according to the contract for the home being constructed. Follow the practice of the architect's office in which you work.

Other situations where dimensions may be assumed are when a door is centered between two walls as at the end of a hallway, or when a door enters a room and the minimum distance from the wall to the door is assumed. See the examples in Figure 14–7. Some dimensions may be provided in the form of a note for standard features, as seen in Figure 14–8. The walls around a shower need not be dimensioned when the note, 36″ square shower,

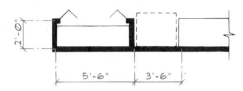

Figure 14–6 Assumed dimensions.

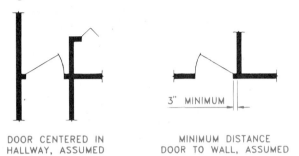

DOOR CENTERED IN HALLWAY, ASSUMED

MINIMUM DISTANCE DOOR TO WALL, ASSUMED

Figure 14–7 Assumed location of features without dimensions given.

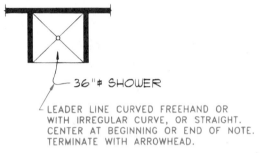

— 36"# SHOWER

LEADER LINE CURVED FREEHAND OR WITH IRREGULAR CURVE, OR STRAIGHT. CENTER AT BEGINNING OR END OF NOTE. TERMINATE WITH ARROWHEAD.

Figure 14–8 Standard features dimensioned with a specific note.

defines the inside dimensions. The shower must be located, and actual product dimensions verified during construction.

One of the best ways to learn how experienced drafters lay out dimensions is to study and evaluate existing plans. Metric dimensions may vary depending on hard or soft conversion, as discussed later in this chapter.

▼ COMMON SIZES OF ARCHITECTURAL FEATURES

All walls, edges of brick, and brick fireplaces are thick lines. All other lines are thin. The refrigerator, water heater, washer and dryer, dishwasher, and trash compactor are dashed lines. Closet poles are 15" from the back wall and are centerlines. Closet shelves are solid lines 12" from the back wall.

Room Components

- Exterior heated walls: 6".
- Exterior unheated walls: 4" (garage/shop).
- Interior: 4".
- Interior plumbing (toilet): 6".

- Hallways: 36" minimum clearance.
- Entry hallways: 42–60".
- Bedroom closets: 24" minimum depth; 48" length.
- Linen closets: 14–24" deep (not over 30").
- Base cabinets: 24" wide; 15–18" wide bar; 24" wide island with 12" minimum at each side of cooktop; 36" minimum from island to cabinet (for passage, 48" is better on sides with an appliance).
- Upper cabinets: 12" wide.
- Washer/dryer space: 36" deep, 5' × 6' long minimum.
- Stairways: 36" minimum wide; 10.5" tread (typical); 11'–4" total run.
- Fireplace: See pages 190–193.

Plumbing

- Kitchen sink: double, 32" × 21"; triple, 42" × 21".
- Vegetable sink and/or bar sink: 16" × 16" or 16" × 21".
- Laundry sink: 21" × 21".
- Bathroom sink: 19" × 16", oval; provide 9" from edge to wall and about 12" between two sinks; 36" minimum length.
- Toilet space: 30" wide (minimum); 24" clearance in front.
- Tub: 32" × 60" or 32" × 72".
- Shower: 36" square; 42" square; or combinations of 36", 42", 48", and 60" for fiberglass; any size for ceramic tile.
- Washer/dryer: 2'–4" square (approximately).

Appliances

- Forced air unit: gas, 18" square (minimum) with 6" space all around; cannot go under stair; electric, 24" × 30" (same space requirements as gas).
- Water heater: gas, 18–24" diameter; cannot go under stair.
- Refrigerator: 36" wide space; approximately 27" deep; 4" from wall; 4" from base cabinet.
- Stove/cooktop: 30" × 21" deep.
- Built-in oven: 27" × 24" deep.
- Dishwasher: 24" × 24"; place near upper cabinet to ease putting away dishes.
- Trash compactor: 15" × 24" deep; near sink, away from stove.
- Broom/pantry: 12" minimum × 24" deep, increasing by 3" increments.
- Desk: 30" × 24" deep (minimum); out of main work triangle.
- Built-in vacuum: 24–30" diameter.

Doors

- Entry: 36" × 6'–8"; 42" × 8'.
- Sliders or French: 5', 6', 8' (double); 9', 10' (triple); 12' (four panel).

○ Garage, utility, kitchen, and bedrooms on custom houses: 2'–8".
○ Bedrooms and bathrooms of nice houses: 2'–6".
○ Bathrooms closets: 2'–4".
○ Garage 8' × 7', 9' × 8', 16' × 7', and 18' × 7'.

Windows

○ Living, family: 8–10'.
○ Dining, den: 6–8'.
○ Bedrooms: 4–6'.
○ Kitchens: 3–5'.
○ Bathrooms: 2–3'.
○ Sliding: 4', 5', 6', 8', 10', 12'.
○ Single hung: 24", 30", 36", 42".
○ Casement: same as sliding.
○ Fixed/awning: 24", 30", 36", 42", 48".
○ Fixed/sliding: 24", 30", 36", 42", 48".
○ Picture 4', 5', 6', 8'.
○ Bay: 8–10' total; sides, 18–24" wide.

▼ DIMENSIONING FLOOR-PLAN FEATURES

Masonry Veneer

The dimensioning discussion thus far has provided examples of floor-plan dimensioning for wood-framed construction. Other methods of residential construction include masonry veneer, concrete block, and solid concrete. Masonry veneer construction is the application of thin (4") masonry, such as stone or brick, to the exterior of a wood-framed structure. This kind of construction provides a long-lasting, attractive exterior. Masonry veneer may also be applied to interior frame partitions where the appearance of stone or brick is desired. Brick or stone may cover an entire wall that contains a fireplace, for example. Occasionally either material may be applied extending to the lower half of a wall, with another material extending to the ceiling for a contrasting decorative effect. Masonry veneer construction is dimensioned on the floor plan in the same way as wood framing except the veneer is dimensioned and labeled with a note describing the product. See Figure 14–9. This construction requires foundation support, discussed in Chapters 31–33.

Concrete-block Construction

Concrete blocks are made in standard sizes and may be solid or have hollow cavities. Concrete block may be used to construct exterior or interior walls of residential or commercial structures. Some concrete blocks have a textured or sculptured surface to provide a pleasant exterior appearance. Most concrete-block construction must be covered, such as by masonry veneer for a finished look. Some structures use concrete block for the exterior bearing walls and wood-framed construction for interior partitions.

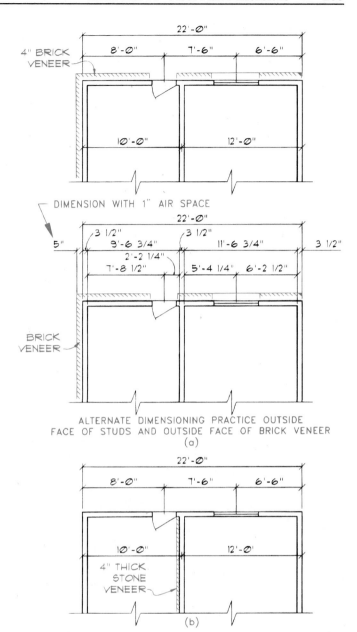

Figure 14–9 Dimensioning (a) brick exterior veneer; (b) interior stone veneer.

Dimensioning concrete-block construction is different from wood-framed construction in that each wall, partition, and window and door opening is dimensioned to the edge of the feature, as shown in Figure 14–10. While this method is most common, some drafters do prefer to dimension the concrete-block structure in the same way as wood-framed construction. To do so, they must add specific information in notes or section views about wall thicknesses and openings.

Solid Concrete Construction

Solid concrete construction is used in residential and commercial structures. In residences it is mostly limited to basements and subterranean homes. Concrete is poured into forms that mold the mixture to the shape

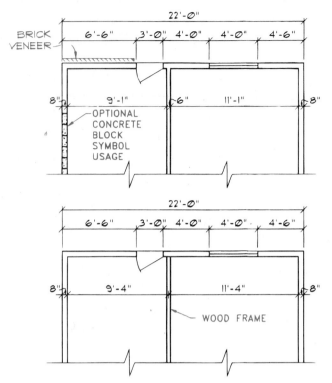

Figure 14–10 Dimensioning concrete-block construction.

desired. When the concrete cures (hardens), the forms are removed and the structure is complete, except for any surface preparation that may be needed.

Masonry veneer may be placed on either side of concrete walls for appearance. Wood framing and typical interior finish materials may also be added to the inside of concrete walls. The wood framing attached to the concrete walls is in the form of furring, which is usually nailed in place with special nails. (*Furring* is wood strips fastened to the studs or interior walls and ceilings and on which wall materials are attached. Furring is used on masonry walls to provide air space and a nailing surface for attaching finish wall materials. Furring attached to masonry is pressure-treated or foundation-grade wood or galvanized steel.) Standard interior finish materials such as drywall or open paneling may be added to the furring. Solid concrete construction is typically dimensioned the same as concrete-block construction. See Figure 14–11.

More information about masonry and concrete construction in a residence is found in Chapters 31 through 33.

▼ NOTES AND SPECIFICATIONS

Notes on plans are either specific or general. *Specific notes* relate to specific features within the floor plan, such as the header size over a window opening. The specific note is often connected to the feature with a leader line. Specific notes are also called *local notes* since they identify isolated items. Common information that may be specified in the form of local notes, is:

1. Window schedule. Place all windows in a schedule. Group by type (sliding, fixed, awning, etc.) from largest to smallest. Window size and type can also be placed directly on the floor plan.
2. Door schedule. Place all doors in schedule. Group by type (solid core, slab, sliding, bifold, etc.) from largest to smallest. Door size and type can be placed directly on the floor plan.
3. Place room names in the center of all habitable rooms (1/4" [6 mm] lettering) with the interior size below (1/8" [3 mm] lettering).
4. Label appliances such as furnace, water heater, dishwasher, compactor, stove, and refrigerator. Items that can be distinguished by shape such as toilets and sinks do not need to be identified.
5. Label all tubs, showers, or spas, giving size, type and material.
6. Label fireplace or solid-fuel-burning appliance. Specify vent within 24" (610 mm), hearth, U.L. approved materials, and wood box.
7. Label stairs, giving direction of travel, number of risers, and rail height.
8. Label all closets with shelves, S&P (shelf and pole), linen, broom, or pantry.
9. Specify attic and crawl access openings.
10. Designate 1-hour firewall between garage and residence with "5/8" (16 mm) type "X" gypsum board from floor to ceiling.
11. On multilevel structures, call out "line of upper floor," balcony above, line of lower floor, or other projections of one level beyond another.
12. Verify codes and construction methods for additional local notes.

General notes apply to the drawing overall rather than to specific items. General notes are commonly lettered in the *field* of the drawing. The field is any open area that surrounds the main views. A common location for general notes is the lower right or left corner of the sheet. Notes should not be placed closer than 1/2" from the drawing border. Some typical general notes are seen in Figure 14–12.

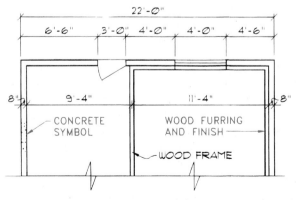

Figure 14–11 Dimensioning solid concrete construction.

GENERAL   NOTES:

1. ALL PENETRATIONS IN TOP OR BOTTOM
   PLATES FOR PLUMBING OR ELECTRICAL RUNS
   TO BE SEALED. SEE ELECTRICAL PLANS
   FOR ADDITIONAL SPECIFICATIONS.
2. PROVIDE 1/2" WATER PROOF GYPSUM BOARD
   AROND ALL TUBS, SHOWERS, AND SPAS.
3. VENT DRYER AND ALL FANS TO OUTSIDE AIR
   THRU VENT WITH DAMPER.
4. INSULATE WATER HEATER TO R-II. W.H. IN GARAGE
   TO BE ON 18" HIGH PLATFORM.

Figure 14–12   Some typical general notes.

Some specific notes that are too complex or take up too much space may be lettered with the general notes and keyed to the floor plan with a short identification such as with the phrase: see NOTE #1 or SEE NOTE 1, or with a number within a symbol such as ①.

Written specifications are separate notes that specifically identify the quality, quantity, or type of materials and fixtures that are used in the entire project. Specifications for construction are prepared in a format different from drawing sheets. Specifications may be printed in a format that categorizes each phase of the construction and indicates the precise methods and materials to be used. Architects and designers may publish specifications for a house so that the client knows exactly what the home will contain, even including the color and type of paint. This information also sets a standard that allows contractors to prepare construction estimates on an equal basis.

Lending institutions require material specifications to be submitted with plans when builders apply for financing. Generally each lender has a form to be used that supplies the description of all construction materials and methods along with a cost analysis of the structure. More information about specifications is found in Chapters 42 and 43.

▼ USING METRIC

The unit of measure commonly used is the millimeter (mm). Meters (m) are used for large site plans and civil engineering drawings. It is based on the International System of Units (SI). Canada is one country that uses metric dimensioning.

When materials are purchased from the United States, it is often necessary to make a *hard conversion* to metric units. This means that the typical inch units are converted directly to metric. For example, a 2 × 4 that is milled to 1½ × 3½" converts to 38 × 89 mm. When making the conversion from the imperial inch to millimeters, use the formula 25.4 × inch = millimeters.

The preferred method of metric dimensioning is called a *soft conversion*. This means that the lumber is milled directly to metric units. The 2 × 4 lumber is 40 × 90 mm using the soft conversion. This method is much more convenient to use when drawing plans and measuring in construction. When plywood thickness is measured in metric units, ⅝" thick equals 17 mm, and ¾" thick equals 20 mm. The length and width of plywood also changes from 48 × 96" to 1200 × 2400 mm. Modules for architectural design and construction in the United States are typically 12, 16 or 24". In countries using metric measurement, the dimensioning module is 100 mm. For example, construction members may be spaced 24" on center (OC) in the United States, while the spacing in Canada is 600 mm OC. Or the spacing between studs at 16" OC in the United States is 400 mm OC in Canada. These metric modules allow the 1200 × 2400 mm plywood to fit exactly on center with the construction members. Interior dimensions are also designed in 100-mm increments. For example, the kitchen base cabinet measures 600 mm deep.

Expressing Metric Units on a Drawing

When placing metric dimensions on a drawing, all dimensions within dimension lines are in millimeters and the millimeter symbol (mm) is omitted. When more than one dimension is quoted, the millimeter symbol (mm) is placed only after the last dimension. For example, the size of a plywood sheet reads 1200 × 2400 mm, or the size and length of a wood stud reads 40 × 90 × 600 mm. The millimeter symbol is omitted in the notes associated with a drawing, except when referring to a single dimension, such as the thickness of material or the spacing of members. For example, a note might read 90 × 1200 BEAM, while the reference to material thickness of 12 mm gypsum or the spacing of joists of 400 mm OC places the millimeter symbol after the size.

Rules for Writing Metric Symbols and Names

○ Unit names are given in lowercase, even those derived from proper names (e.g., millimeter, meter, kilogram, kelvin, newton, and pascal).

○ Use vertical text for unit symbols. Use lowercase text, such as mm (millimeter), m (meter), and kg (kilogram), unless the unit name is derived from a proper name as in K (kelvin), N (newton), or Pa (pascal).

○ Use lowercase for values less than 10^3, and use uppercase for values greater than 10^3.

○ Leave a space between a numeral and symbol, 55 kg, 24 m, 38° C. Do not close up the space like this: 55kg, 24m, 38° C.

○ Do not leave a space between a unit symbol and its prefix—for example, use kg not k g.

○ Do not use the plural of unit symbols—for example, use 55 kg, not 55 kgs.

○ The plural of a metric name should be used, such as 125 meters.

○ Do not mix unit names and symbols; use one or the other. Symbols are preferred on drawings where necessary. Millimeters (mm) are assumed on architectural drawings unless otherwise specified.

Metric Scales

In Chapter 2 metric scales were introduced with respect to the use of the metric scale as a drafting tool. When drawings are produced in metric, the floor plans, elevations, and foundation plans are generally drawn at a scale of 1:50 rather than the ¼″ = 1′–0″ scale used in the imperial system. Larger-scale drawings, such as those showing construction details, are often drawn at a scale of 1:5. Small-scale drawings, such as plot plans, may be drawn at a scale of 1:500. Figure 14–13 shows a floor plan completely drawn using metric dimensioning. The preferred method of soft conversion is used in the figure, and the metric scale is 1:50.

Metric Standards and Specifications

Refer to Appendix A for a complete list and examples of metric standards and specifications for architectural design and construction materials.

Metric Construction Dimensions

The following chart gives a comparison of metric and inch construction modules. These are not a direct conversion of metric and inches, but a relationship between the metric and inch standard based on soft conversion:

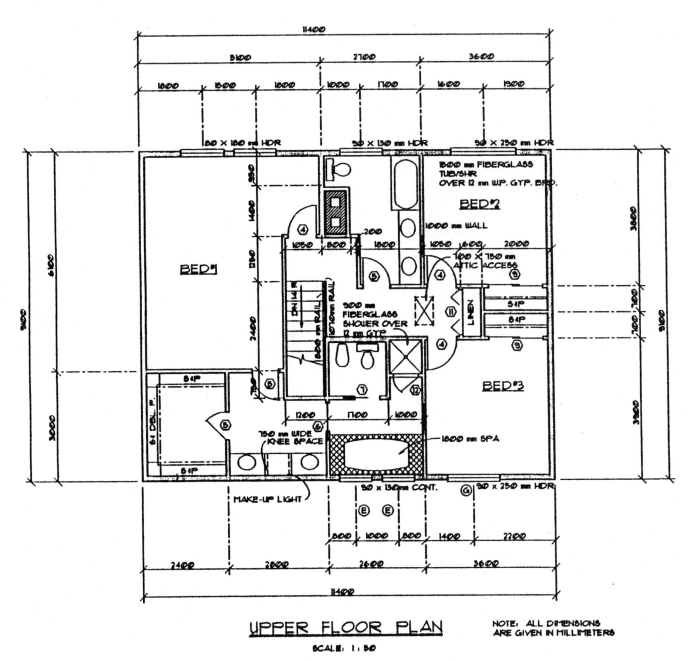

Figure 14–13 Floor plan drawn using metric dimensioning.

	Metric (mm)	Inches or Feet	Comments
Construction modules	100	4″	
	600	24″ or 2′	
Standard brick	90 × 57 × 190	4″ × 2″ × 8″	
Mortar joints	10	⅜″ and ½″	
Concrete block	200 × 200 × 400	8″ × 8″ × 16″	
Sheet metal	tenths of a mm	gage	
Drywall, plywood, and rigid insulation	1200 × 2400 1200 × 3000	4′ × 8′ 4′ × 10′	Thicknesses are in inches, so fire, acoustic, and thermal ratings will not have to be recalculated
Batt insulation	400 600	16″ 24″	Thicknesses are in inches, so thermal values will not have to be recalculated
Door height	2050 or 2100	6′–8″	
	2100	7′–0″	
Door width	750	2′–6″	
	800	2′–8″	
	850	2′–10″	
	900 or 950	3′–0″	
	1000	3′–4″	
Cut glass	mm	feet and inches, thickness is already in mm	

According to the Brick Institute of America (BIA), the American Society for Testing Materials (ASTM E835/E835M, *Standard Guide for Modular Construction of Clay and Concrete Masonry Units*) establishes dimensions for these products based on the basic 100 mm building module. Many common brick sizes are within a couple of millimeters of metric modular sizes. Nearly all common bricks can fit within a 100 mm vertical module by using a 10 mm joint. Appendix A displays common brick sizes in inches and in metric conversion. Due to the size of concrete blocks, however, they must be resized for metric applications to a 200- × -200- × -400 mm module.

CADD Applications

PLACING ARCHITECTURAL DIMENSIONS WITH CADD

Most CADD architectural dimensions automatically terminate with tick marks, which are popular in architectural drafting. Many architectural drafters prefer to use the tick mark for all dimensions and then change to another format, such as an arrow option, for all leader-line terminators. Figure 14–14 shows an example of each type of dimension-line terminator, and dimension text placement options.

Architectural dimensions are commonly placed using the continuous, or point-to-point, dimensioning method. Most architectural CADD programs allow this to happen automatically. When you pick the DIMENSION command, the computer prompts you to select vertical, horizontal, or aligned dimensioning. Aligned dimensions are placed in alignment with an angled wall or feature and read from the bottom or right side of the sheet. After you make this selection, you are asked for the first extension-line location. All you have to do is pick the origin of the first extension line. The program automatically places the traditional small gap between the building and the beginning of the extension line. Next you are asked to pick the second extension line location followed by the dimension line location. Now the computer automatically draws the extension lines, dimension line, and tick marks, and places the dimension text. After the first dimension is placed, the computer asks you to pick the next extension-line location. The second dimension is automatically placed, and this procedure continues until all dimensions are drawn, as shown in Figure 14–15. The CADD drawing is extremely accurate. If you draw the walls or other features where they are supposed to be, then the dimensions are perfect every time.

Notes, specifications, and titles are easily drawn with CADD. All you have to do is select the TEXT command, pick the location for the text, and type the desired information. Some examples of CADD general notes are shown in Figure 14–12.

A partial floor plan with dimensions and notes is shown in Figure 14–16.

Leader lines are drawn to specific notes. The leader may have an arrowhead, dot, or slash depending on the company practice or application. Leaders are generally drawn in CADD by picking the start point and the end of the leader if curved. Next, you are normally asked to enter the text, after which the text is automatically placed on the drawing. When drawing straight leaders, first pick the end point, followed by the first shoulder point and then the last shoulder point. You again enter the text for automatic placement. See Figure 14–17. Figure 14–18 shows an architectural detail with leaders and specific notes.

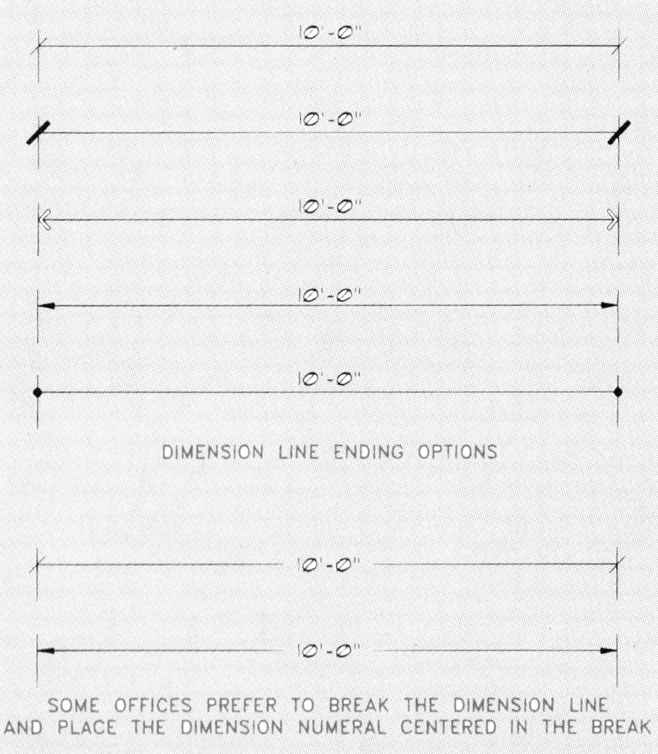

10'-0"

10'-0"

10'-0"

10'-0"

10'-0"

DIMENSION LINE ENDING OPTIONS

10'-0"

10'-0"

SOME OFFICES PREFER TO BREAK THE DIMENSION LINE AND PLACE THE DIMENSION NUMERAL CENTERED IN THE BREAK

Figure 14–14 Examples of dimension-line terminators.

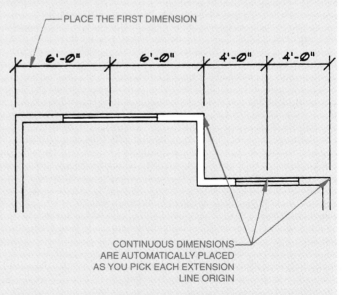

PLACE THE FIRST DIMENSION

6'-0" 6'-0" 4'-0" 4'-0"

CONTINUOUS DIMENSIONS ARE AUTOMATICALLY PLACED AS YOU PICK EACH EXTENSION LINE ORIGIN

Figure 14–15 Automatically placing continuous lines with CADD.

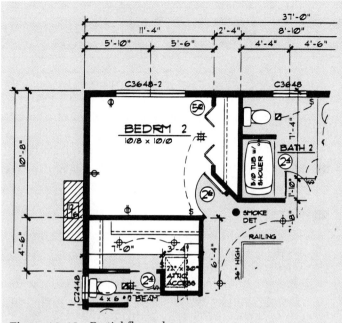

Figure 14–16   Partial floor plan.

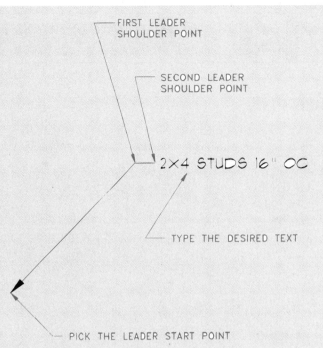

Figure 14–17   Text for automatic placement.

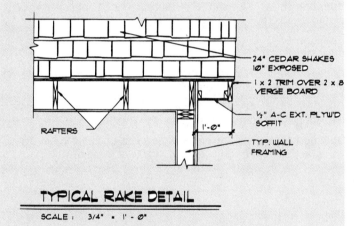

Figure 14–18   Drawing with leaders and notes.

14 Floor-Plan Dimensions Test

DIRECTIONS

Answer the questions with short complete statements or drawings as needed on an 8½ × 11 sheet of notebook paper, as follows:

1. Letter your name, Chapter 14 Test, and the date at the top of the sheet.
2. Letter the question number and provide the answer. You do not need to write out the question.

Answers may be prepared on a word processor if appropriate with course guidelines.

QUESTIONS

Question 14–1 Define aligned dimensioning.

Question 14–2 Why should dimension lines be spaced evenly and to avoid crowding?

Question 14–3 Should extension and dimension lines be drawn thick or thin?

Question 14–4 Shown an example of how dimension numerals that are less than 1 ft. are lettered.

Question 14–5 Show an example of how dimension numerals that are greater than 1 ft. are lettered.

Question 14–6 Are the overall dimensions on frame construction given to the outside of the stud frame at exterior walls?

Question 14–7 When three lines of dimensions are used, describe the dimensional information provided in each line.

Question 14–8 Describe and give an example of a specific note.

Question 14–9 What is another name for a specific note?

Question 14–10 Describe and give an example of a general note.

PROBLEMS

1. Draw the following problems using pencil on 8½ × 11 vellum or with CADD.
2. Use architectural lettering.
3. Prepare lines and dimensions as specified in your course guidelines, or select your own preferred technique as discussed in this text.

Problem 14–1 Draw an example of dimensioning a wood-framed structure on a floor plan. Show at least one interior partition, one exterior door, and one window.

Problem 14–2 Show an example of how masonry veneer is dimensioned on a floor plan when used with wood-framed construction.

Problem 14–3 Show an example of a floor plan constructed with concrete block. Show at least one woodframe interior partition, one exterior door, and one window.

Problem 14–4 Show an example of a floor plan constructed with solid concrete exterior walls. Show at least one interior wood-frame partition, one wall with interior wood furring, one exterior door, and one window.

CHAPTER 15

Floor-Plan Layout

Residential plans are commonly drawn on 17″ × 22″, 18″ × 24″, 22″ × 34″, or 24″ × 36″ drawing sheets. This discussion explains floor-plan layout techniques using 17″ × 22″ vellum. The layout methods may be used with any sheet size. A complete floor plan is shown in Figure 15–1. This floor plan is shown without the electrical layout and the structural elements. Designing and drawing the electrical plan is discussed in Chapter 16, and calculating and drawing the structural members is covered in Chapters 24 through 30. The step-by-step method used to draw this floor plan provides you with a method for laying out any floor plan. As you progress through the step-by-step discussion of the floor-plan layout, refer back to previous chapters to review specific floor-plan symbols and dimensioning techniques. Keep in mind that these layout techniques represent a suggested typical method used to establish a complete floor plan. You may alter these layout steps to suit your individual preference as your skills develop and knowledge increases.

▼ LAYING OUT A FLOOR PLAN

Before you begin the floor-plan layout, be sure that your hands and equipment are clean, tape your drawing sheet down very tightly, and align and square the scales of your drafting machine if you use one.

Step 1. Determine the working area of the paper to be used. This is the distance between borders. In this exercise, the 17″ × 22″ vellum has a 16″ vertical and 19½″ horizontal distance between borders.

Step 2. Determine the drawing area, which is approximately the area that the floor plan plus dimensions requires when completely drawn. Residential floor plans are drawn to a scale of ¼″ = 1′–0″. The house to be drawn is 60′ long and 36′ wide. At a scale of ¼″ = 1′–0″ the house will be drawn 15″ × 9″ (60′ ÷ 4 = 15″ by 36′ ÷ 4 = 9″). This size does not take into consideration any area needed for dimensions. However, it appears in this case there is enough space for dimensions because the house size is much less than the working area. You should consider at least 2″ on each side of the house for dimensions. When the plan does not fit within the

working area, you need to increase the sheet size or reduce the scale.

Reducing the floor-plan scale is not normally done, as floor plans are usually drawn at ¼″ to 1′–0″, except for some commercial applications where ⅛″ = 1′–0″ or ³⁄₁₆″ = 1′–0″ may be used.

Step 3. Center the drawing within the working area using these calculations:

$$
\begin{aligned}
\text{Working area length} &= 19.5'' \\
\text{Drawing area length} &= -15.0'' \\
\hline
&\;\;4.5''
\end{aligned}
$$

4.5″ ÷ 2 = 2.25″

2.25″ = Space on each side of floor plan

$$
\begin{aligned}
\text{Working area height} &= 16'' \\
\text{Drawing area height} &= -9'' \\
\hline
&\;\;7''
\end{aligned}
$$

7″ ÷ 2 = 3.5″

3.5″ = Space above and below floor plan

Figure 15–2 (page 213) shows the drawing area centered within the working area. Use construction lines drawn very lightly with a 6H or 4H lead to outline the drawing area. This method of centering a drawing is important for beginning drafters. It can be quite frustrating to start a drawing and find that there is not enough room to complete it. More experienced drafters may not use this process to this extent, although they make some quick calculations to determine where the drawing should be located. In actual practice, the drawing does not have to be perfectly centered, but a well-balanced drawing is important. Several factors contribute to a well-balanced drawing:

○ Actual size of the required drawing
○ Scale of the drawing
○ Amount of detail
○ Size of drawing sheet
○ Dimensions needed
○ Amount of general and local notes and schedules required

Step 4. Lay out all exterior walls 6″ thick within the drawing area using construction lines. Construction lines are very lightly drawn with a 6H or 4H lead and easy to erase if you make an error. If

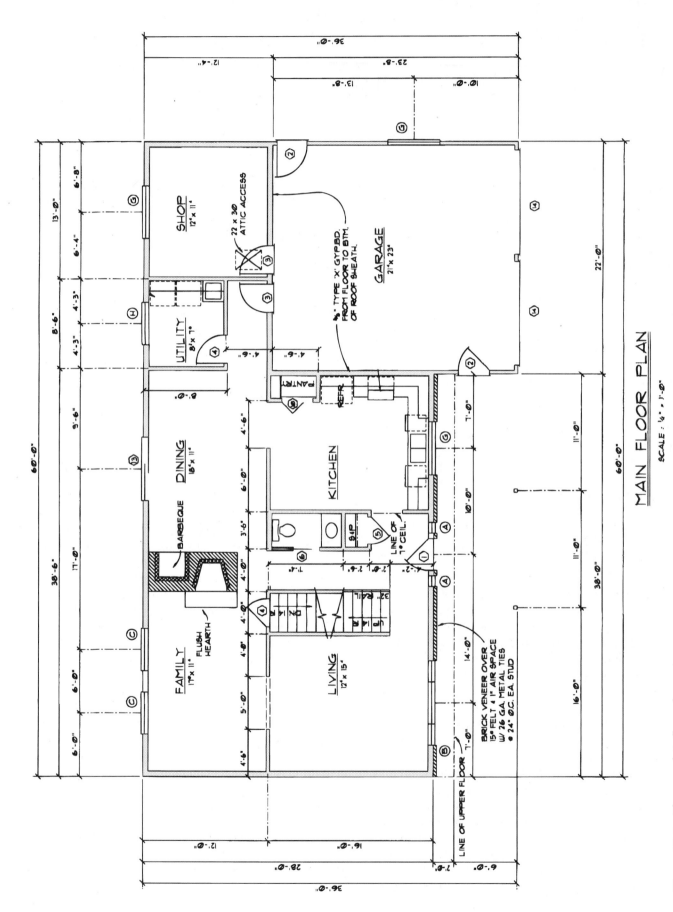

Figure 15–1 The complete floor plan.

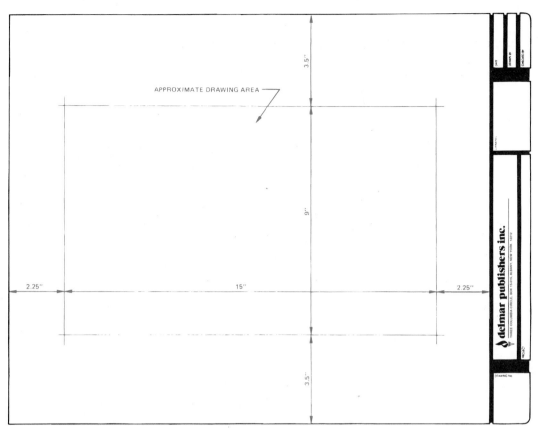

Figure 15–2 The approximate drawing area centered within the working area.

properly drawn, however, construction lines need not be erased. Properly drawn construction lines are too light to reproduce using a diazo print, although they may show as very light lines on a photocopy reproduction. Some drafters use a light-blue pencil lead for preliminary layout work because the light blue does not reproduce. Use your ¼″ = 1′–0″ architect's scale. Look at Figure 15–3.

Step 5. Lay out all interior walls 5″ thick within the floor plan using construction lines. See Figure 15–3.

Step 6. Block out all doors and windows in their proper locations using construction lines, as shown in Figure 15–4. Be sure that doors and windows are centered within the desired areas.

Step 7. Using construction lines, draw all cabinets. Base kitchen cabinets 5′ = wide are 24″, upper cabinets 12″, and bath vanities 22″ wide as shown in Figure 15–5.

Step 8. Draw all appliances and utilities using construction lines. Draw the refrigerator 36″ × 30″, range 30″ wide, dishwasher 24″ wide, furnace 24″ × 30″, water heater 24″ in diameter, and the washer and dryer 30″ × 30″ each. See Figure 15–5.

Step 9. Draw all plumbing fixtures, sinks, and toilets, as seen in Figure 15–5.

Step 10. Use construction lines to draw a 5′-wide single-face fireplace with a 36″-wide opening and an 18″ × 5′ hearth. Include a 30″ × 18″ barbecue. Add 4″ brick veneer to the front wall, as shown in Figure 15–5 (page 215).

Step 11. Lightly draw the stairs representation providing several 12″-wide treads to a long break line with necessary handrails or guardrails. See Figure 15–5.

Step 12. Check your accuracy; then darken all the construction lines that you have drawn. Add all door and window floor-plan symbols. Draw lines and symbols from top to bottom and from left to right if you are right handed or from right to left if left handed. Try to avoid passing your hands and equipment over darkened lines to help minimize smudging. Darken walls with a 0.7- or 0.9-mm pencil with H or F lead. For inked drawings or CADD, use a 0.7-mm pen to plot walls. All cabinets, appliances, fixtures, and other items are drawn with a 0.5-mm pencil or plotted line width. Lead grades are only suggestions. Look at Figure 15–6 (page 215).

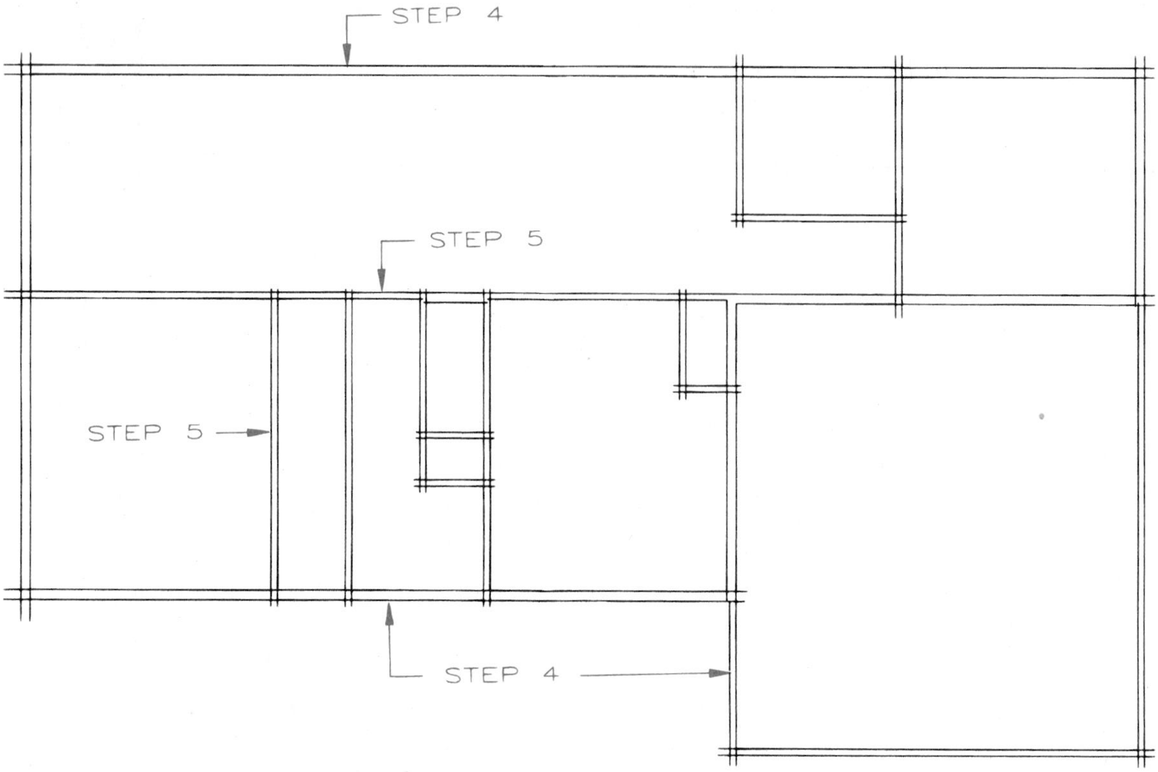

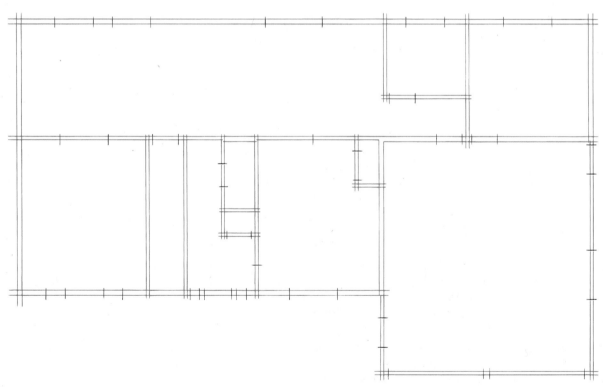

Figure 15–3 Steps 4 and 5: Lay out exterior walls and interior partitions.

Figure 15–4 Step 6: Block out all doors and windows.

At this point, the floor plan is ready to use as a preliminary drawing for client approval.

Step 13. Lay out all exterior dimension lines using construction lines. Place an overall dimension line on all sides of the residence. The second row of dimensions should go from exterior face to exterior face of major jogs. A third line should be used to go from the exterior face of exterior

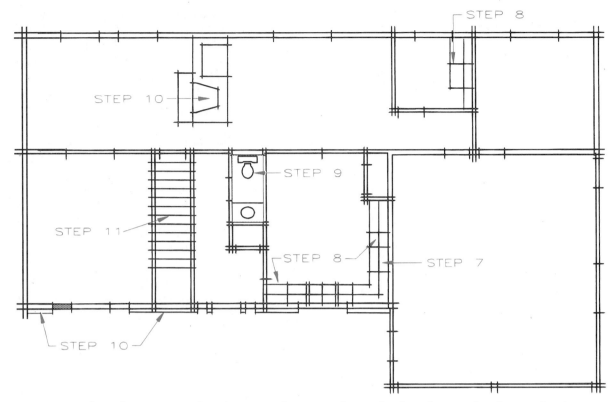

Figure 15–5 Steps 7 through 11: Lay out the cabinets, appliances, utilities, plumbing fixtures, fireplace, and stairs.

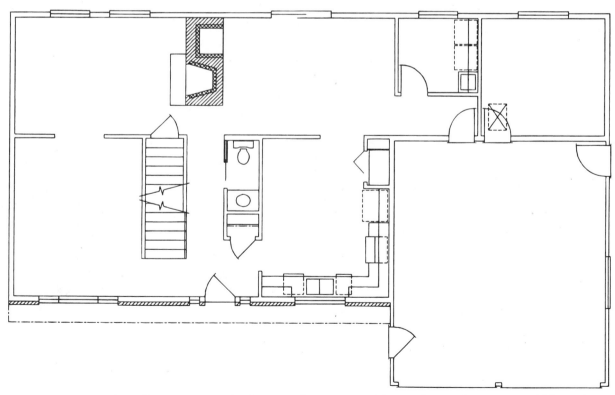

Figure 15–6 Step 12: Darken all items previously drawn with construction lines. This is where you would be in Step 11 with CADD. Next, add door and window symbols.

walls to the center of interior walls. There is a common exception: Dimension to the exterior face of the walls between the garage and the residence because of the change in foundation materials. A fourth line should be used to go from each wall to the center of each door and window. There is also a common exception: Do not locate the doors in the garage on the floor plan. These are dimensioned on the foundation plan. See Chapter 14 for a complete review of common dimensioning practices and alternatives. Do not place dimension numerals on at this time. Add symbols for door and window schedules. See Figure 15–7.

Step 14. Lay out all interior dimensions as needed. Locate interior dimensions from the exterior dimensions. Do not place dimension numerals at this time. See Figure 15–7. Darken all extension and dimension lines using 2H or H lead in a 0.5-mm pencil. Use a 0.35-mm technical pen for inked lines and CADD plots. *Experienced drafters prefer to darken dimension and extension lines as they are drawn.*

If the electrical layout is part of the floor plan, refer to Chapter 16, and then do the following:

○ Letter all switch locations and light fixture locations. Use 1/8″ diameter circle for light fixtures. If electrical symbols are too large, they distract from the total drawing. Draw

electrical circuits or switch legs from switches to fixtures using an irregular curve or a dash-line arc in CADD. Figure 16–13 shows the switches, lights, and relay lines added to the floor plan.

○ Add all outlets to the floor plan. Verify maximum spacing of outlets and proper symbols from Chapter 6. (See Figure 15–8).

○ Add notes for specific appliances, such as dishwasher (DW), water heater (WH), hood with fan over range, and oven. All exhaust fans must be noted to vent to the outside, and fireplace specifications must be added.

If the electrical plan is a separate drawing, then go directly to Step 15.

Step 15. Letter all dimension numerals centered above each dimension line. Dimension numerals should read from the bottom on horizontal dimension lines and from the right side of the sheet on vertical dimension lines. Prepare very light, 1/8″-high guidelines within 1/16″ from the dimension lines for all numerals. Do not letter without guidelines. Place a clean piece of paper under your hand as you letter to help avoid smudging. Use an H or F lead in a 0.5-mm pencil for lettering. For ink work or CADD plots, use a 0.35-mm pen width. Add arrowheads or slashes as preferred by your instructor or office. Provide window and door sizes. See Figure 15–9.

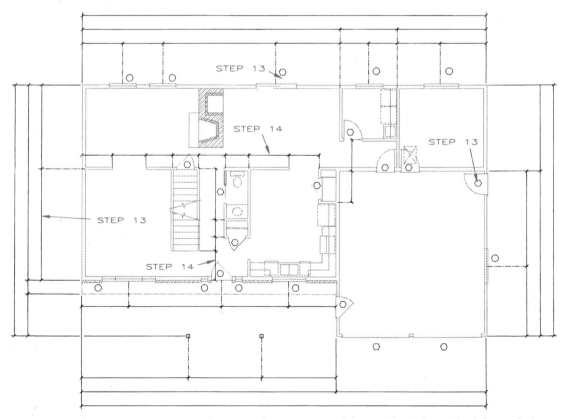

Figure 15–7 Steps 13 and 14: Lay out exterior and interior dimensions, and door and window schedule symbols.

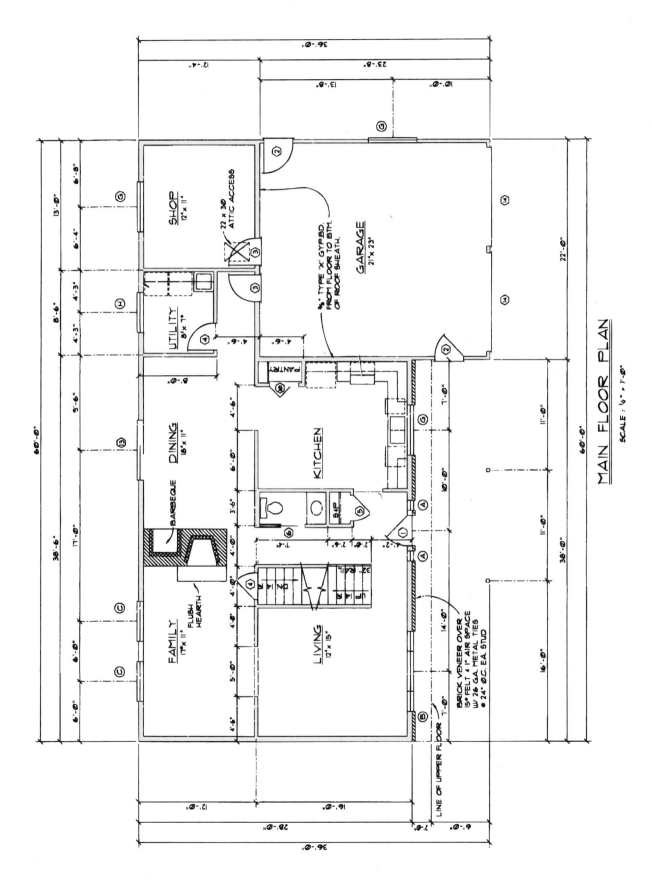

Figure 15–8 Composite CADD floor plan.

If the structural layout is part of the floor plan, refer to Chapters 24 through 30, and then do the following. Add all beam, header, joist, and rafter symbols and specifications. Use ⅛″ guidelines for all lettering, and add cross section indicators, as shown in Figure 15–8. If the structural plan is a separate drawing, then go directly to step 16.

Step 16. Draw ⅛″ guidelines and add all other local notes, such as those for a cantilever second floor or concrete slabs (see Figure 15–8).

Step 17. Label all rooms using 3⁄16″-high guidelines for lettering. The interior room dimensions may be placed under the room designation if required by your instructor. Use ⅛″ guidelines for room sizes if included, as in the following example:

LIVING ROOM
12'-6" X 14'-0"

Letter the title and scale of the drawing directly below and centered on the floor plan. Use ¼″-high guidelines for the title. Underline the title and letter the scale using ⅛″ guidelines. For example:

MAIN FLOOR PLAN
SCALE: 1/4" = 1'-0"

The scale may be omitted if it is located in the title block. Add all other general notes as required in the project or as specified by your instructor or office. Letter all title block information. Look at Figure 15–8.

Step 18. Add all material symbols such as tile or stone, if any. Shade all walls and handrails as needed. Wood-grain representation may be used on handrails, guardrails, and partial walls. Exterior and interior walls may be shaded on the back of the drawing using a soft F or HB lead. Shade on the back of the sheet to avoid smudging lines on the front. Turn the drawing over onto a blank sheet of scratch paper and carefully shade all walls. Do not shade without scratch paper between the drawing and table because lines from the drawing may transfer to the table. Figure 15–1 shows the completed floor plan.

▼ SECOND-FLOOR PLAN LAYOUT

The second-floor plan must be scaled so it fits over the main-floor plan. The exterior walls (unless cantilevered or supported beyond the main floor), the continued interior bearing partitions, and stairs must line up directly with the main floor. Be sure to check the scale of the print for accuracy, as prints may expand or contract slightly. An easy way to do this is to run a print of the main-floor plan and tape it to your table. Place your new drawing sheet for the second-floor plan over the print, and adjust the new sheet so the drawing area for the second floor is over the first-floor outline. On the second sheet take into consideration enough area for dimensions, notes, and possible schedules. If the second-floor plan is smaller than the main-floor plan, there may be additional space on the sheet for door and window schedules. These schedules should be near the floor plans when practical.

Tape the second drawing sheet to the board after adjusting it to the desired position. Lay out the second-floor plan using Steps 1 through 18 previously described. Figure 15–10 shows a complete second-floor plan.

▼ BASEMENT-PLAN LAYOUT

When a home has a basement, the basement-plan layout is accomplished in the same way as described in the second-floor plan layout. When a basement plan is very simple or possibly when left unfinished, it may be drawn in conjunction with the foundation plan. Use good judgment since you do not want to create a plan that results in confusion to the contractors. Figure 15–11 shows a partial basement floor plan and the door and window schedules for this house. A full basement would be under the entire main floor.

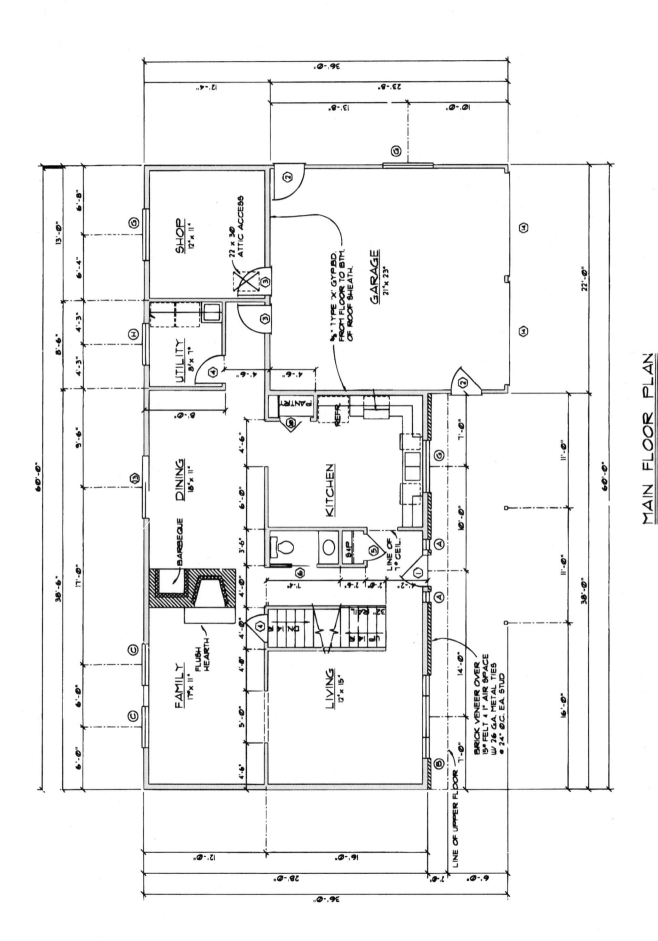

MAIN FLOOR PLAN
SCALE : ¼" : 1'-0"

Figure 15–9 Steps 15 through 18: Lay out all dimension numerals and window and door sizes. Place local notes, room labels, drawing title and scale, and material symbols.

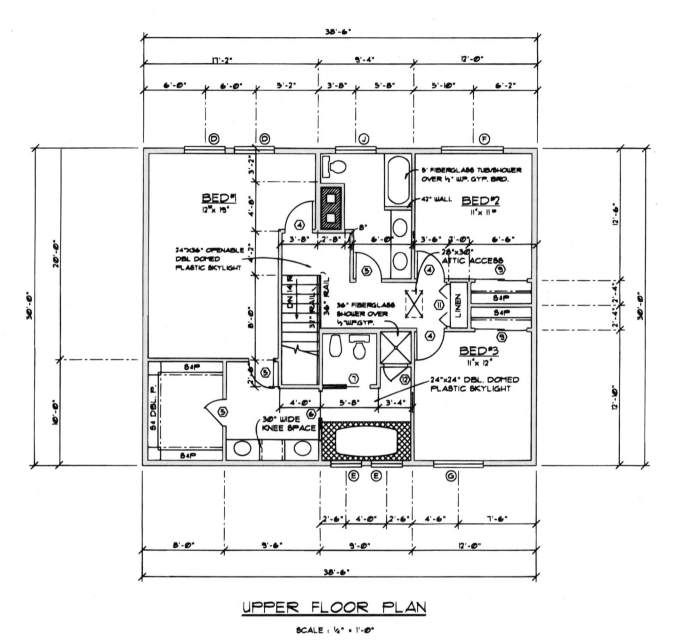

UPPER FLOOR PLAN

SCALE : ¼" = 1'-0"

Figure 15–10 Complete second-floor plan.

WINDOW SCHEDULE

SYM.	SIZE	MODEL	ROUGH OPEN	QUAN
A	$1^{0} \times 5^{0}$	JOB-BUILT		2
B	$8^{0} \times 5^{0}$	W4N5 CSM	8'-2¾"x5'-5⅛"	1
C	$4^{0} \times 5^{0}$	W2N5 CSM	4'-2¼"x5'-5⅛"	2
D	$4^{0} \times 3^{6}$	W2N3 CSM	4'-2¼"x3'-5½"	2
E	$3^{6} \times 3^{6}$	2N3 CSM	3'-5¼"x3'-5½"	2
F	$6^{0} \times 4^{0}$	G64 SLDG	6'-0½"x4'-0½"	1
G	$5^{0} \times 3^{6}$	G536 SLDG	5'-0½"x3'-6½"	4
H	$4^{0} \times 3^{6}$	G436 SLDG	4'-0½"x3'-6½"	1
J	$4^{0} \times 2^{0}$	A41 AWN	4-0½"x2'-0⅝"	1
K	$4^{0} \times 2^{0}$	G42 SLDG	4'-0½"x2'-0½"	3

DOOR SCHEDULE

SYM.	SIZE	TYPE	QUAN
1	$3^{0} \times 6^{8}$	S.C. R.P. METAL INSULATED	1
2	$3^{0} \times 6^{8}$	S.C.-FLUSH-METAL INSUL	2
3	$2^{8} \times 6^{8}$	S.C-SELF CLOSING	2
4	$2^{8} \times 6^{8}$	H.C.	5
5	$2^{6} \times 6^{8}$	H.C.	5
6	$2^{6} \times 6^{8}$	POCKET	2
7	$2^{4} \times 6^{8}$	POCKET	1
8	PR $2^{6} \times 6^{8}$	H.C.	1
9	$5^{0} \times 6^{8}$	BI-PASS	2
10	$3^{0} \times 6^{8}$	BI-FOLD	1
11	$4^{0} \times 6^{8}$	BI-FOLD	1
12	$2^{8} \times 6^{8}$	SHATTER PROOF	1
13	$6^{0} \times 6^{8}$	WOOD FRAME-TEMP. SLDG. GL.	1
14	$9^{0} \times 7^{0}$	OVERHEAD GARAGE	2

WALL AREAS *	2275 SQ. FT.
WINDOWS	250 SQ. FT.
SKYLITES	10 SQ. FT.
DOORS	77 SQ. FT.
TOTAL OPENINGS	337 SQ. FT.
% OPENINGS	15%

* BASEMENT EXCLUDED.

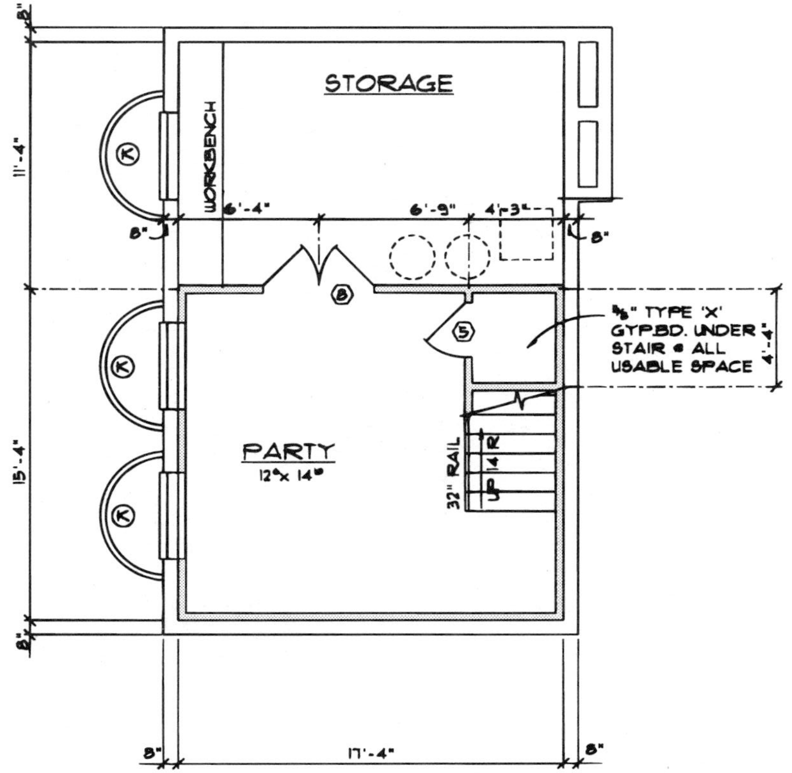
STORAGE
WORKBENCH
PARTY
12⁰x 14⁰
32" RAIL
UP 14 R
⅝" TYPE 'X' GYP.BD. UNDER STAIR @ ALL USABLE SPACE
11'-4"
15'-4"
17'-4"
6'-4"
6'-9"
4'-3"
4'-4"
8"

LOWER FLOOR PLAN

SCALE : ¼" = 1'-0"

Figure 15–11 Basement plan with the door and window schedules.

CADD Applications

USING CADD TO DRAW FLOOR PLANS

Procedures for laying out floor plans with CADD are similar to manual drafting. However, you do not need to lay out the walls and interior partitions with construction lines. The floor plan begins with walls and interior partitions drawn in place. If an error is made, it is easy to erase from the screen and redraw. After the walls are drawn, you may begin placing the symbols for doors, windows, plumbing fixtures, cabinets, and appliances. The drawing at this stage should look like Figure 15–6.

The next step is to place dimensions on the drawing. Dimensions are easily placed with CADD, and each dimension is accurate. As you draw individual dimensions, the computer gives you the exact dimension numeral so you can instantly see if the drawing is correct or requires alteration. With most architectural CADD systems, the drawing elements are placed on individual layers, so you can control the display of these items. For example, if you want the floor plan shown without dimensions, turn off the dimension layer. Refer to Chapter 6 and the CADD applications in each chapter, for discussions of drawing with CADD layers.

Some residential architects choose to display the electrical layout on the floor plan, while others create the electrical layout on *background drawings*—a subset of the information contained in the floor plan. For example, the only information shown on the electrical background drawing is the electrical layout. This drawing is sent to the electrical contractor for bidding and construction purposes. Sending the electrical layout without unnecessary detail often reduces bidding and construction errors, and simplifies the bidding process. With CADD, the drafter has the flexibility to display the floor plan with or without the background drawings. A composite drawing exists when the layer and background drawings are displayed together. A composite CADD floor plan drawing is shown in Figure 15–11.

In commercial applications, the background drawings become an even more important feature of the CADD capabilities. Background drawings are created for areas such as plumbing, mechanical (heating, ventilating, and air conditioning), structural, and interior design.

▼ OVERLAY DRAFTING

Overlay drafting, or *pin-registered drafting* as it is often called, is a process of making drawings in layers, each perfectly aligned with the other. Each layer contains its own independent information and all of the overlays may be reproduced as one composite drawing. Overlay drafting is frequently used in commercial architectural drafting, although this technique is also used in residential architecture.

In overlay drafting there are two main types of drawings: *base sheets* and *overlay sheets.* The base sheet contains general information that is common to the entire project. For example, a base sheet might show the main exterior structure with the permanent bearing walls and partitions. Each overlay sheet contains independent information related to the base sheet, but the base sheet information will not be redrawn each time. There may be a separate overlay for the electrical, mechanical, HVAC (heating, ventilating, and air-conditioning), and plumbing plan. There may be over a hundred overlays for a large commercial structure. Time is saved since the base information does not have to be redrawn each time. The basic components of overlay drafting are shown in Figure 15–12. The pin bar is a manufactured plastic or metal bar with precisely positioned pins that extend from its surface. This pin bar is fastened with tape to the drafting table and in careful alignment with the drafting machine or parallel bar. The base and overlay sheets are generally prepunched (some companies punch their own) so that each sheet will register perfectly over the pin bar. Polyester drafting film, at least 3 mils (.003″) thick, is usually used for base and overlay sheets because of its durability, dimensional stability, and reproduction quality when used in conjunction with ink or polyester lead.

The overlay drafting process is described as follows:

Step 1.   When using the overlay technique the base sheet drawing is prepared on prepunched polyester film. The information on this drawing will

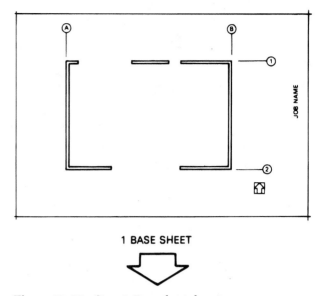

Figure 15–13   Step 1: Base sheet drawing.

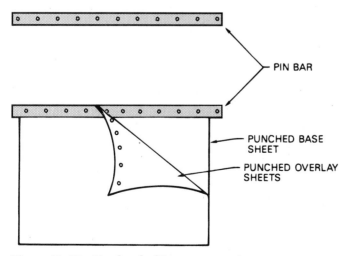

Figure 15–12   Overlay drafting components.

be unchanging from one overlay to the next. See Figure 15–13.

Step 2.   After the base sheet drawing is complete, diazo copies of the base sheet are made on prepunched clear polyester diazo reproduction material. These sheets, called *slicks,* are used as secondary base sheets. A slick is given to each drafter or subcontractor for use as the base sheet in preparation of the specific overlays.

Step 3.   The individual drafter or engineer places a clean overlay sheet over the slick and adds information by drawing on the overlay. Figure 15–14 shows interior partitions with related dimensions and specifications that have been drawn on an overlay covering the floor-plan base. The overlay is shown highlighted in color over the base, or slick, shown in black. The process continues with the foundation drawing shown in Figure 15–15. The slick is shown in black under the foundation with related dimensions and specifications. Figure 15–16 shows the roof framing plan over the slick. Some companies have several different departments so that one overlay may be prepared by the mechanical group, another by the plumbing department, and so on. Even small companies that do all engineering and drafting in one place by one or a few people can benefit from this technique. In some cases a slick may be sent to a subcontractor in another state for preparation of related overlays.

Step 4.   The base sheet and the completed overlay sheets can be registered together with registration clips and composite diazo or photocopy

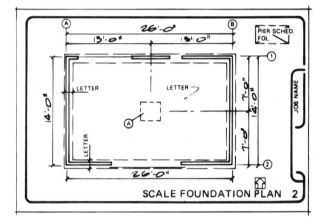

2 FLOOR PLAN OVERLAY

Figure 15–14 Step 3: Overlay of interior partitions. *Courtesy Structureform.*

3 FOUNDATION PLAN OVERLAY

Figure 15–15 Foundation overlay.

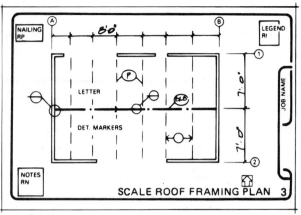

4 ROOF FRAMING PLAN OVERLAY

Figure 15–16 Roof framing play overlay.

made that integrates the components of all or any specific combination of sheets.

The advantages of overlay drafting include the following:

1. Time saving, which directly relates to cost saving.
2. It is easy to make interior designs using a separate overlay for each floor.
3. Subcontractor work will be easy and fast. With each working on his or her own base and overlays, the information required by one subcontractor will not get mixed up with another, and quality control is improved. Each subcontractor can be working simultaneously with the others, thus speeding progress on the complete set of drawings.
4. Composite prints may be made by combining the base sheet and any combination of overlays.

CADD Applications

CADD LAYERING COMPARED TO OVERLAY DRAFTING

CADD can be used together to achieve even more productivity. The CADD system is used to generate a base sheet, and then overlay sheets are developed that are registered to the base. Some CADD systems software is designed to make drawings in layers.

CADD layers must perform two basic functions: they must differentiate independent information and differentiate pen weights for plotting. In addition to the layer identification system established by the CADD software, many companies set up other layer systems based on their own project needs. For example, layers may be identified with a layer identification system, where the elements are divided into three groups: major group, minor group, and basic group. The major group contains the type of drawing, such as Floor Plan. The minor group identifies the element on the drawing, such as electrical, and the basic group labels the type of information, such as lines or text. An example of this method

of layer identification is FL1 ELEC L, which means first-floor plan, electrical plan, and lines. A typical layout of the three groups follows:

Major Group	Minor Group	Basic Group
BSMT	HVAC	LINE
FL1	ELEC	TXT
FL2	PLUMB	DIM
GAR	WALLS	INSERT
SITE	DOORS	HATCH
SCHED	WINDOWS	
NOTES	APPL	
FDN		
ELEV		
ROOF		
REFLECT		

The layer naming system recommended by the American Institute of Architects (AIA) is introduced in Chapter 6 and displayed in Appendix B.

CHAPTER

Floor-Plan Layout Test

DIRECTIONS

Answer the questions with short complete statements or drawings as needed on an 8½ × 11 sheet of notebook paper as follows:

1. Letter your name, Chapter 16 Test, and the date at the top of the sheet.
2. Letter the question number and provide the answer. You do not need to write out the question.

Answers may be prepared on a word processor if appropriate with course guidelines.

QUESTIONS

Question 15–1 Describe the drawing working area.

Question 15–2 What size sheets are generally used for plan drawings?

Question 15–3 Residential floor plans are generally drawn at what scale?

Question 15–4 Identify two alternative actions that could be considered when the drawing space is not adequate to accommodate the drawing.

Question 15–5 Define construction lines.

Question 15–6 Identify four factors that contribute to a well-balanced drawing.

Question 15–7 Describe a method to help determine the layout for a second-floor plan quickly.

Question 15–8 Why should layout work be drawn very lightly or with blue lead?

Question 15–9 Discuss why formulas for centering a drawing in the working area may not always be the way to lay out a drawing.

Question 15–10 Why should hands and equipment be clean before beginning a drawing?

PROBLEMS

DIRECTIONS

1. Select or your instructor will assign one or more of the following residential projects. The basic design is given. You need to complete the floor-plan drawing(s) by adding the missing and unidentified items. Try to keep dimensions in 2'–0" increments as much as possible for economic construction.
2. Completely label and dimension the floor plan.
3. Use one of the following layout methods for your selected or assigned floor-plan problem(s):
   ○ Draw a base sheet made up of walls, windows, cabinets, stairs, fireplaces, and other permanent items; room labels are optional. Use overlay drafting to create a dimension overlay with notes and room labels if they were not placed on the base sheet. The base sheet will be used again later when you design the electrical, plumbing, heating, and structural plans.
   ○ Draw a base sheet of walls, windows, doors, cabinets, stairs, fireplaces, and other permanent items; room labels are optional. Make a reverse sepia reproduction (see Chapter 3). Complete the floor plan on the sepia by adding the dimensions, notes, and room labels if they were omitted from the base sheet. The base sheet will be used again when you design the electrical, plumbing, heating, and structural plans.
   ○ Prepare one composite drawing with all information on it: walls, windows, doors, cabinets, stairs, fireplaces, other permanent items, room labels, dimensions, notes, and titles. If you use this technique, it means that you will add the electrical layout when you complete Chapter 16 and the structural layout after you study Chapters 24 through 30.
   ○ Use CADD to draw the floor plan. Establish layers as identified in the American Institute of Architects (AIA) layer naming system found in Appendix B.

Problem 15–1 This problem may be drawn manually or using CADD. Given the Cape Cod floor plan with approximate interior dimensions, do the following unless otherwise specified by your instructor:

1. Draw the main-floor plan with complete dimensions and notes using the scale ¼" = 1'-0". Use a 17" × 22" or 18" × 24" drawing sheet. This is sheet 2 of 5.

2. Draw the second-floor plan with complete dimensions using the scale ¼" = 1'-0". In the blank space available draw a complete door and window schedule. See Chapter 13 for size guidelines.
3. Establish all dimensions as per Chapter 14.
   Illustration courtesy Home Building Plan Service, Inc.

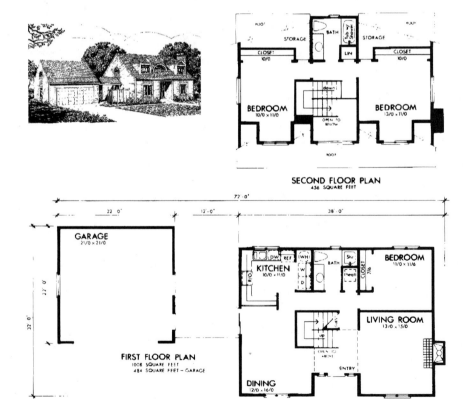

Problem 15–1

Problem 15–2 This problem may be drawn manually or with CADD. Given the cabin floor plan with approximate interior dimensions, do the following unless otherwise specified by your instructor:
1. Draw the main-floor plan with complete dimensions and notes using the scale ¼″ = 1′–0″. Use a 17″ × 22″ or 18″ × 24″ drawing sheet. This is sheet 2 of 3.
2. Draw a complete door and window schedule.
   Illustration courtesy Home Building Plan Service, Inc.

Problem 15–3 This problem may be drawn manually or with CADD. Given the ranch floor plan with approximate interior dimensions, do the following unless otherwise specified by your instructor:
1. Draw the main-floor plan with complete dimensions and notes using the scale ¼″ = 1′–0″.
2. In the blank space available draw a complete door and window schedule.
   Illustration courtesy Home Building Plan Service, Inc.

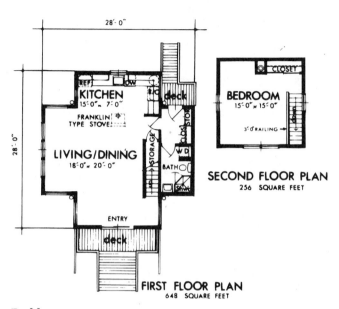

Problem 15–2

Problem 15–3

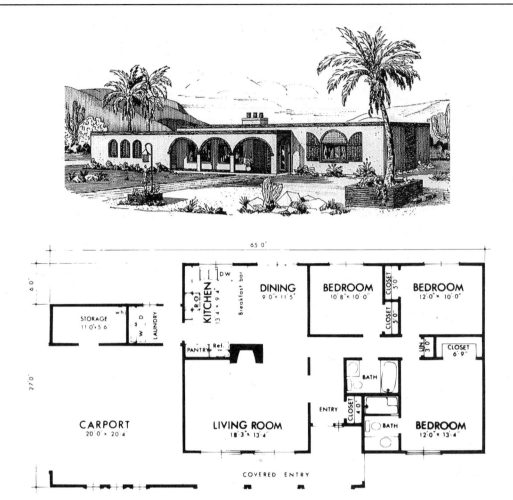

Problem 15–4

Problem 15–4 This problem may be drawn manually or with CADD. Given the Spanish floor plan with approximate interior dimensions, do the following unless otherwise specified by your instructor:

1. Draw the main-floor plan with complete dimensions and notes using the scale ¼″ = 1′–0″.
2. In the blank space available draw a complete door and window schedule.
3. If the weather patterns of your area require it, design a pitch roof with clay tiles. Verify design with your instructor prior to completion of drawings.
   Illustration courtesy Home Building Plan Service, Inc.

Problem 15–5 This problem may be drawn manually or with CADD. Given the early American floor plan with approximate interior dimensions, do the following unless otherwise specified by your instructor:

1. Draw the main-floor plan with complete dimensions using the scale ¼″ = 1′–0″.
2. Draw the second-floor plan with complete dimensions and notes using the scale ¼″ = 1′–0″. In the blank space available draw a complete door and window schedule.
3. Design and draw the basement-floor plan with complete dimensions using the scale ¼″ = 1′–0″. Design the basement so it may be used for one bedroom, a recreation room, and a bathroom. There will be a wall directly under the main-floor plan wall between the dining room and kitchen, entry and lavatory.
   Illustration courtesy Home Building Plan Service, Inc.

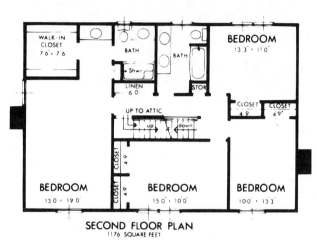

SECOND FLOOR PLAN
1176 SQUARE FEET

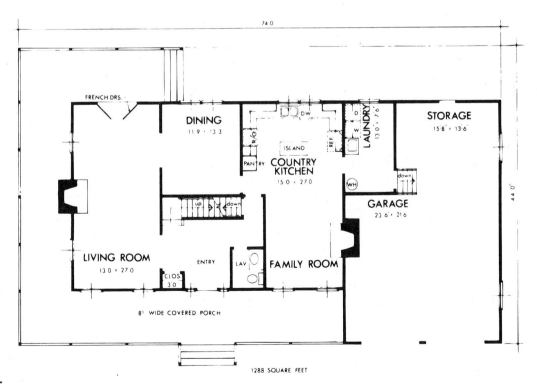

1288 SQUARE FEET

Problem 15–5

Problem 15–6 This problem may be drawn manually or with CADD. Given the side-to-rear sloped-lot (contemporary) floor plan with approximate interior dimensions, do the following unless otherwise specified by your instructor:

1. Draw the main-floor plan with complete dimensions using the scale ¼″ = 1′–0″.

2. Draw the second-floor plan with complete dimensions and notes using the scale ¼″ = 1′–0″. In the blank space available draw a complete door and window schedule. This is sheet 3 of 7.

3. Draw the basement-floor plan with complete dimensions using the scale ¼″ = 1′–0″. This is sheet 4 of 7.
   Illustration courtesy Home Building Plan Service, Inc.

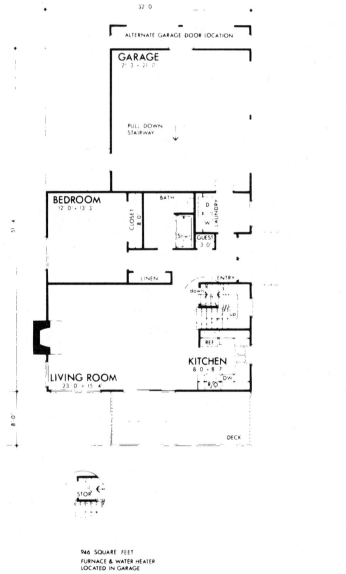

946 SQUARE FEET
FURNACE & WATER HEATER
LOCATED IN GARAGE

Problem 15–6

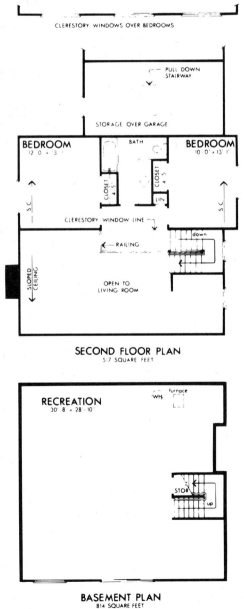

SECOND FLOOR PLAN
5/7 SQUARE FEET

BASEMENT PLAN
814 SQUARE FEET

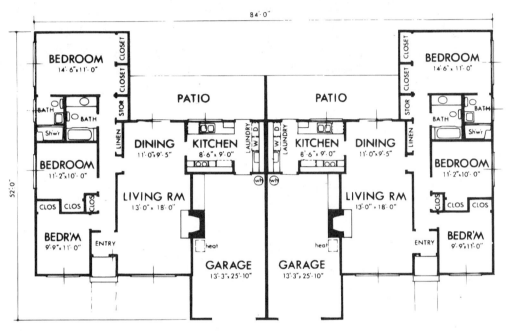

2 THREE-BEDROOM UNITS
2386 SQUARE FEET

Problem 15–7

Problem 15–7 This problem may be drawn manually or with CADD. Given the ranch duplex floor plan with approximate interior dimensions, do the following unless otherwise specified by your instructor:

1. Draw the main-floor plan with complete dimensions and notes using the scale ¼″ = 1′–0″. In the blank space available, draw a complete door and window schedule.
Illustration courtesy Home Building Plan Service, Inc.

Problems 15–8 through 15–14 Given the preliminary drawings and general information, draw the floor plan(s) using manual or computer-aided drafting as required by your course guidelines. Preliminary drawings for Problems 15–8 through 15–12: *Courtesy Computer Aided Designs, Inc.* Problem 15–13: *Courtesy Cynthia Weaver, Square Corner Designs.* Problem 15–14: *Courtesy April Muilenburg.* Problem 15–15: *Courtesy Paul Masterson.*

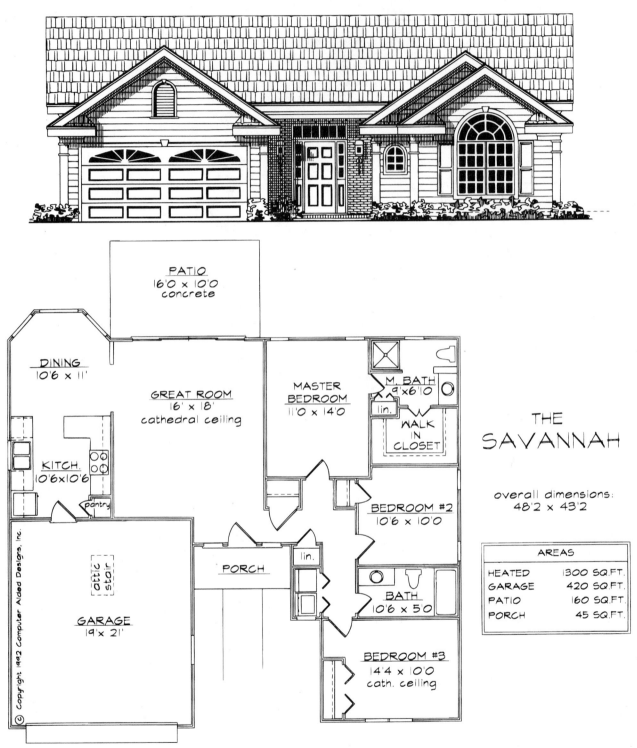

THE
SAVANNAH

overall dimensions:
48'2 x 43'2

AREAS	
HEATED	1300 SQ.FT.
GARAGE	420 SQ.FT.
PATIO	160 SQ.FT.
PORCH	45 SQ.FT.

Problem 15–8

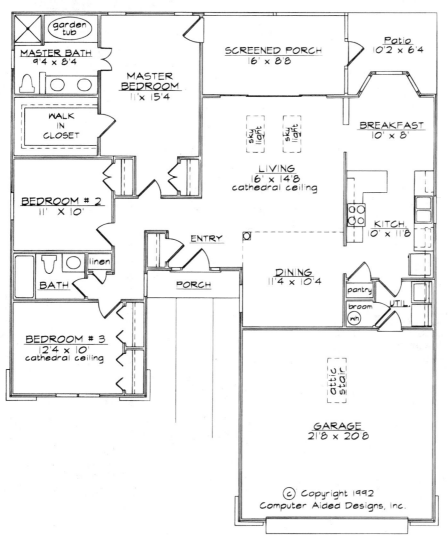

THE
CHARLESTON

overall dimensions:
48'-0" x 55'-4"

AREAS	
HEATED	1,485 SQ.FT.
GARAGE	467 SQ.FT.
SCREENED PORCH	147 SQ.FT.
PATIO	78 SQ.FT.
PORCH	31 SQ.FT.

Problem 15–9

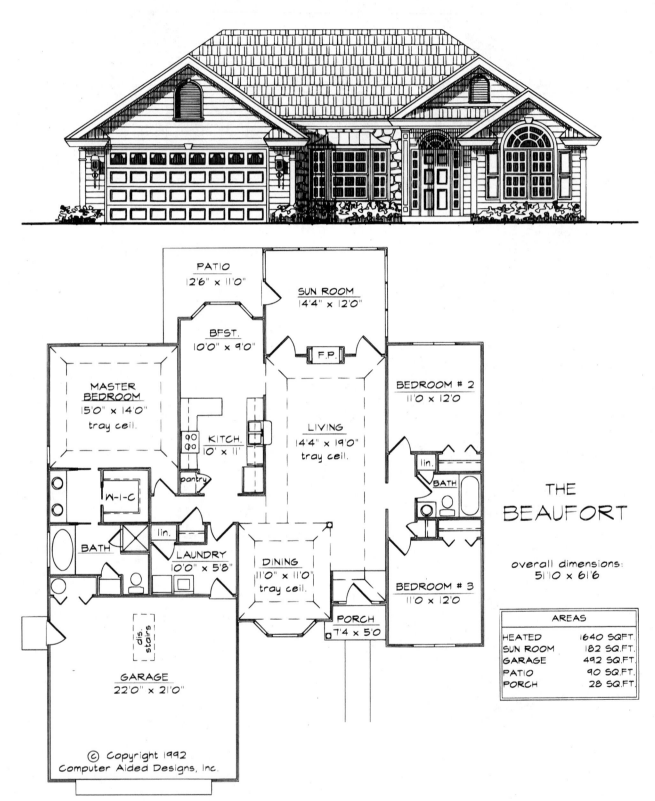

PATIO
12'6" x 11'0"

SUN ROOM
14'4" x 12'0"

BFST.
10'0" x 9'0"

F.P.

MASTER
BEDROOM
15'0" x 14'0"
tray ceil.

BEDROOM # 2
11'0 x 12'0

KITCH.
10' x 11'

LIVING
14'4" x 19'0"
tray ceil.

pantry

W-I-C

lin.

lin.

BATH

BATH

LAUNDRY
10'0" x 5'8"

DINING
11'0" x 11'0"
tray ceil.

BEDROOM # 3
11'0 x 12'0

dis.
stairs

PORCH
7'4 x 5'0

GARAGE
22'0" x 21'0"

© Copyright 1992
Computer Aided Designs, Inc.

THE
BEAUFORT

overall dimensions:
51'10 x 61'6

AREAS	
HEATED	1640 SQFT.
SUN ROOM	182 SQ.FT.
GARAGE	492 SQ.FT.
PATIO	90 SQ.FT.
PORCH	28 SQ.FT.

Problem 15–10

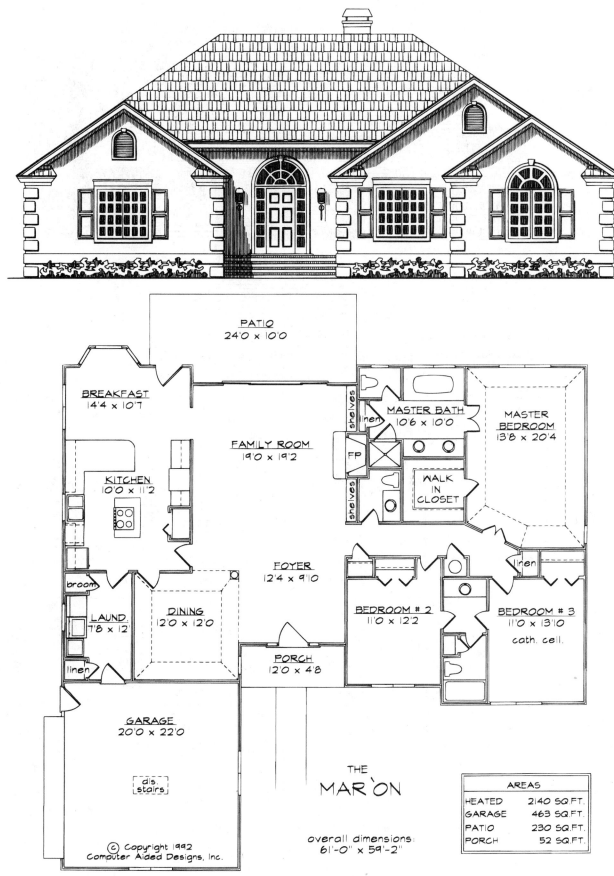

PATIO
24'0 x 10'0

BREAKFAST
14'4 x 10'7

MASTER BATH
10'6 x 10'0

MASTER
BEDROOM
13'8 x 20'4

shelves

linen

FAMILY ROOM
19'0 x 19'2

KITCHEN
10'0 x 11'2

FP

shelves

WALK
IN
CLOSET

broom

FOYER
12'4 x 9'10

linen

LAUND.
7'8 x 12'

DINING
12'0 x 12'0

BEDROOM # 2
11'0 x 12'2

BEDROOM # 3
11'0 x 13'10

cath. ceil.

linen

PORCH
12'0 x 4'8

GARAGE
20'0 x 22'0

dis.
stairs

THE
MAR'ON

© Copyright 1992
Computer Aided Designs, Inc.

overall dimensions:
61'-0" x 59'-2"

AREAS	
HEATED	2140 SQ.FT.
GARAGE	463 SQ.FT.
PATIO	230 SQ.FT.
PORCH	52 SQ.FT.

Problem 15–11

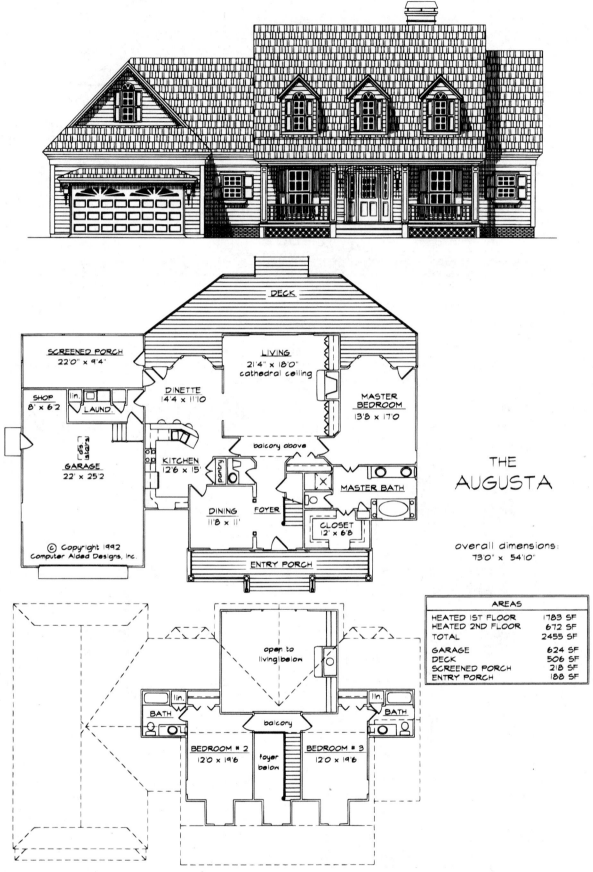

THE
AUGUSTA

overall dimensions:
73'0" x 54'10"

AREAS	
HEATED 1ST FLOOR	1783 SF
HEATED 2ND FLOOR	672 SF
TOTAL	2455 SF
GARAGE	624 SF
DECK	506 SF
SCREENED PORCH	218 SF
ENTRY PORCH	188 SF

Problem 15–12

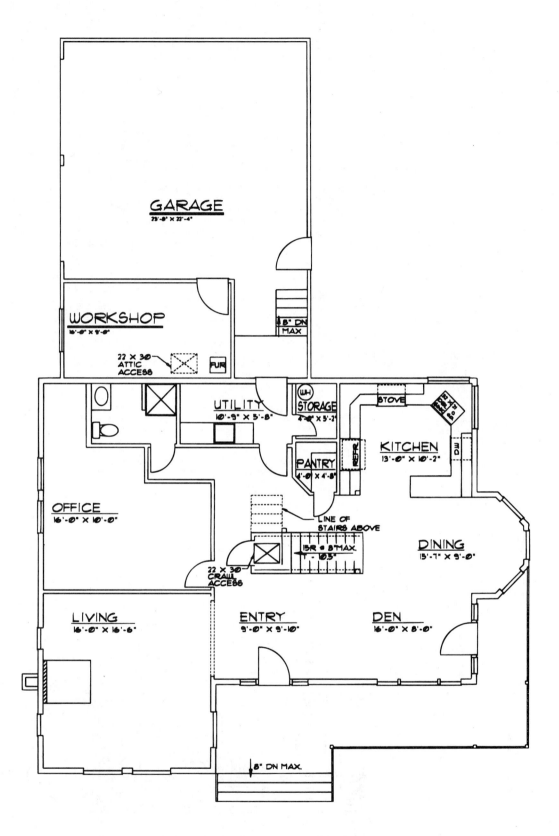

GARAGE
23'-8" × 22'-4"

WORKSHOP
15'-0" × 9'-0"

8" DN
MAX

22 × 30
ATTIC
ACCESS

FUR

UTILITY
10'-9" × 5'-8"

W.H.
STORAGE
4'-0" × 5'-2"

STOVE

KITCHEN
13'-0" × 10'-2"

PANTRY
4'-0" × 4'-8"

REFR

OFFICE
16'-0" × 10'-0"

LINE OF
STAIRS ABOVE

22 × 30
CRAWL
ACCESS

15R @ 8"MAX.
T - 10.5"

DINING
15'-7" × 9'-0"

LIVING
16'-0" × 16'-6"

ENTRY
9'-0" × 9'-10"

DEN
16'-0" × 8'-0"

8" DN MAX.

MAIN FLOOR PLAN

Problem 15–13a

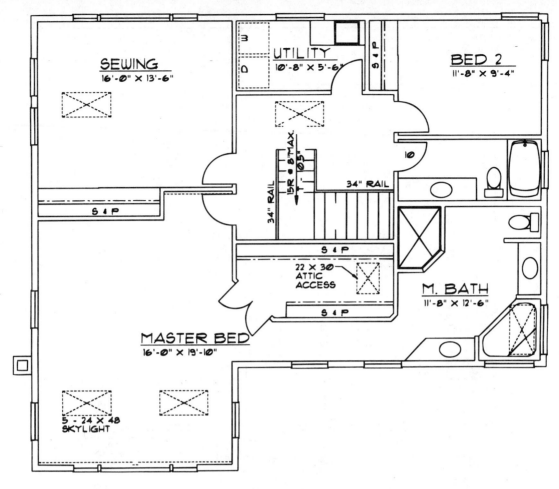

SEWING
16'-0" X 13'-6"

UTILITY
10'-8" X 5'-6"

BED 2
11'-8" X 9'-4"

34" RAIL

34" RAIL

15R @ 8"MAX.
@5"

34" RAIL

10

22 X 30
ATTIC
ACCESS

M. BATH
11'-8" X 12'-6"

S 4 P

S 4 P

S 4 P

MASTER BED
16'-0" X 19'-10"

5 - 24 X 48
SKYLIGHT

UPPER FLOOR PLAN

Problem 15–13b

9 | 12

FIN. CEIL.

5 | 12

FIN. FLOOR
FIN. CEIL.

FIN. FLOOR

FRONT ELEVATION

Problem 15–13c

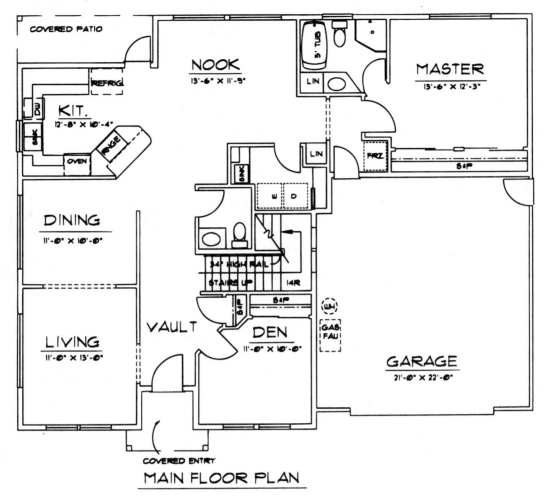

COVERED PATIO

REFRIG

NOOK
13'-6" X 11'-9"

5' TUB

LIN

MASTER
13'-6" X 12'-3"

KIT.
12'-8" X 10'-4"

DW

SINK

RANGE

OVEN

LIN

FRZ

SLP

DINING
11'-0" X 10'-0"

SINK

W   D

LIVING
11'-0" X 13'-0"

VAULT

34" HIGH RAIL

STAIRS UP   14R

SLP   SLP

DEN
11'-0" X 10'-0"

WH

GAS
FAU

GARAGE
21'-0" X 22'-0"

COVERED ENTRY

MAIN FLOOR PLAN

Problem 15–14a

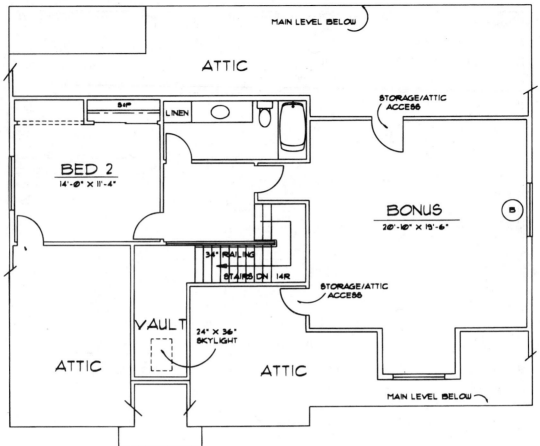

MAIN LEVEL BELOW

ATTIC

STORAGE/ATTIC ACCESS

54P

LINEN

BED 2
14'-0" X 11'-4"

BONUS
20'-10" X 19'-6"

34" RAILING
STAIRS DN 14R

STORAGE/ATTIC ACCESS

VAULT

24" X 36" SKYLIGHT

ATTIC

ATTIC

MAIN LEVEL BELOW

UPPER FLOOR PLAN

Problem 15–14b

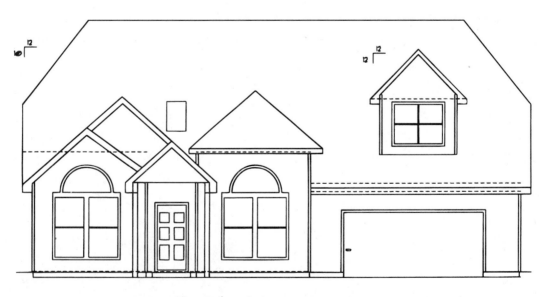

10 | 12

12 | 12

FRONT ELEVATION

Problem 15–14c

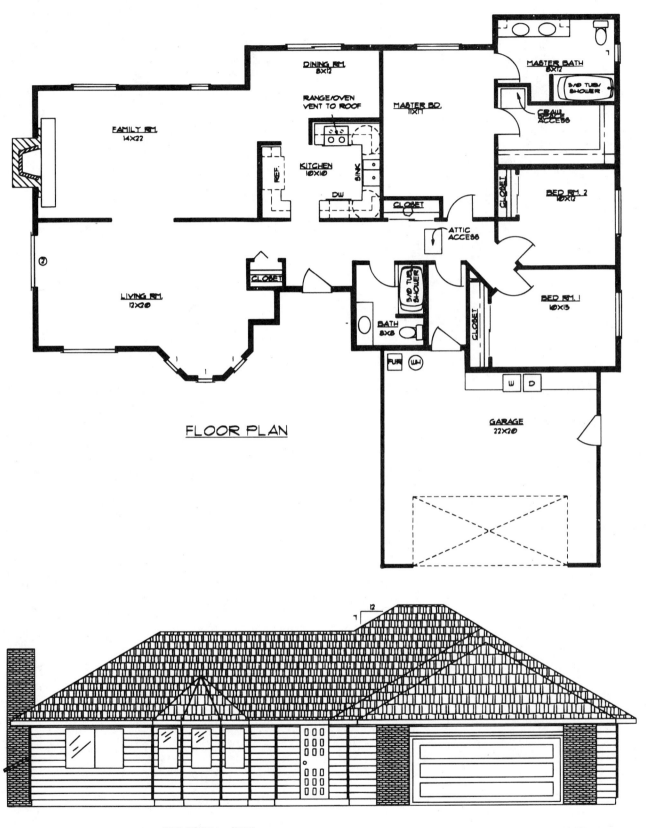

FLOOR PLAN

FRONT VIEW

Problem 15–15

SECTION V

SUPPLEMENTAL FLOOR PLAN DRAWINGS

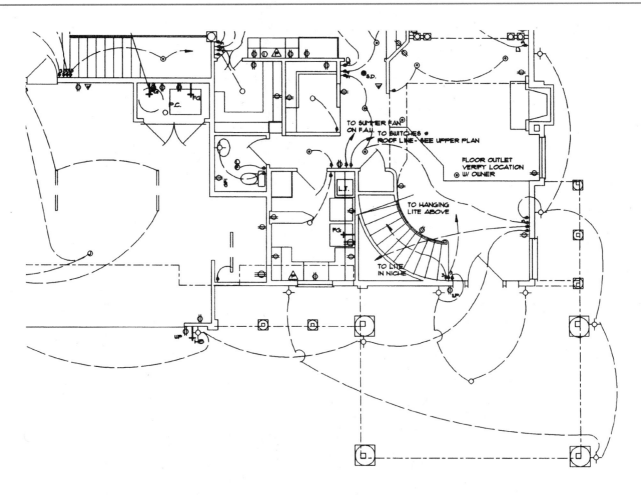

MAIN FLOOR ELECTRICAL PLAN

1/4" = 1'-0"

Electrical Plans

The electrical plans display all of the circuits and systems to be used by the electrical contractor during installation. Electrical installation for new construction occurs in these three phases.

○ Temporary—the installation of a temporary underground or overhead electrical service near the construction site and close to the final meter location.

○ Rough in—when the electrical boxes and wiring are installed. Rough-in happens after the structure is framed and covered with roofing. The electrical meter and permanent service is hooked up.

○ Finish—the installation of the light fixtures, outlets and covers, and appliances. This is one of the last construction phases.

Electrical plans may be placed on the floor plan with all the other symbols, information, and dimensions (as discussed in Chapters 13 through 15). This is a common practice on simple floor plans where the addition of the electrical symbols do not overcomplicate the drawing. Electrical plans are also drawn on a separate sheet, which displays the floor-plan walls and key symbols, such as doors, windows, stairs, fireplaces, cabinets, and room labels. This chapter takes the latter approach but provides discussion on how both methods might be used. CADD provides an excellent tool for creating electrical plans after the floor plans have been drawn. Floor plan layers that are not needed may be turned off or frozen while the electrical layer is turned on to create the electrical plan. It is easy to create a separate drawing from the key elements of the base drawing in this manner.

Following are definitions of typical terms related to electrical plans and construction. A basic understanding of terminology is important before beginning the electrical drawing.

▼ ELECTRICAL TERMS AND DEFINITIONS

Ampere: A measurement of electrical current flow. Referred to by its abbreviation: *amp* or *amps.*

Boxes: A box equipped with clamps, used to terminate a conduit. Also known as an outlet box. Connections are made in the box, and a variety of covers are available for finish electrical. A premanufactured box or casing is installed during electrical rough-in to house the switches, outlets, and fixture mounting. See *electrical work.*

Breaker: An electric safety switch that automatically opens the circuit when excessive amperage occurs in the circuit. Also referred to as a circuit breaker.

Circuit: The various conductors, connections, and devices found in the path of electrical flow from the source through the components and back to the source.

Conductor: A material that permits the free motion of electricity. Copper is a common conductor in architectural wiring.

Conduit: A metal or fiber pipe or tube used to enclose one or more electrical conductors.

Distribution panel: Where the conductor from the meter base is connected to individual circuit breakers, which are connected to separate circuits for distribution to various locations throughout the structure. Also known as a panel.

Electrical work: The installation of the wiring and fixtures for a complete residential or commercial electrical system. The installation of the wiring is referred to as the *rough-in.* Rough-in takes place after the framing is completed and the structure is *dried-in.* Dried-in refers to installing the roof or otherwise making the building dry. The light fixtures, outlets, and all other final electrical work is done when the construction is nearly complete. This is referred to as the *finish electrical.* The rough-in and finish electrical comprise the electrical work.

Ground: An electrical connection to the earth by means of a rod.

Meter: An instrument used to measure electrical quantities. The electrical meter for a building is where the power enters and is monitored for the electrical utility.

Meter base: The mounting base on which the electrical meter is attached. It contains all of the connections and clamps.

Outlet: An electrical connector used to plug in devices. A *duplex outlet,* with two outlets, is the typical wall plug.

Switch leg: The electrical conductor from a switch to the electrical device being controlled.

Volt: The unit of measure for electrical force.

Watt: A unit measure of power.

▼ ELECTRICAL CIRCUIT DESIGN

The design of the electrical circuits in a home is important because of the number of electrical appliances that make modern living enjoyable. Discuss with the client any anticipated needs for electric energy in the home. Such a discussion includes the intended use of each of the rooms and the potential placement of furniture in them. Try to design the electrical circuits so there are enough outlets and switches at convenient locations. National Electrical Code requirements dictate the size of some circuits and the placement of certain outlets and switches within the home.

In addition to convenience and code requirements, you have to consider cost. A client may want outlets 4′ (1220 mm) apart in each room, but the construction budget may not allow such a luxury. Try to establish budget guidelines and then work closely with the client to achieve desired results. While switches, convenience outlets, and electrical wire and boxes may not seem to cost very much individually, their installation in great quantity could be costly. In addition, attractive and useful light fixtures required in most rooms can also be very expensive.

Some common items to consider when designing electrical circuits based on convenience and national and local codes include the following:

○ Duplex convenience outlets (wall plugs) should be a maximum of 12′ (3657 mm) apart. Closer spacing is desirable, although economy is a factor.

○ Duplex outlets should be no more than 6′ (1820 mm) from a corner, and there should be a duplex outlet in any wall over 2′ (610 mm).

○ Each wall over 2′ (610 mm) in length must be within 2′ of an outlet.

○ Consider furniture layout, if possible, so duplex outlets do not become inaccessible behind large pieces of furniture.

○ Place a duplex outlet next to or behind a desk or table where light is needed or near a probable reading area.

○ Place a duplex outlet near a fireplace, for an adjacent table light, for a vacuum, or for fireplace maintenance.

○ Place a duplex outlet in a hallway, for a vacuum.

○ Duplex outlets should be placed close together in kitchens and installed where fixed and portable appliances will be located for use. Outlets are required every 2′ of counter space. Some designs include an outlet in a pantry for use with portable appliances.

○ Kitchen, bathroom, laundry, and outdoor circuits are required to be protected by ground-fault circuit interrupters (GFI of GFCI), which trips a circuit breaker when there is any unbalance in the circuit current. The standard breaker condition is .005 amp fault in ¹⁄₄₀ second. This protection is required for any electric fixture within 6′ of water. Fixtures and appliances produce a potential hazard for electrocution.

○ Each bath sink or makeup table should have a duplex outlet.

○ Each enclosed bath or laundry (utility) room should have an exhaust fan. Rooms with openable windows do not require fans, although they are desirable.

○ A lighting outlet is necessary outside all entries and exterior doors.

○ A waterproof duplex outlet should be placed at a convenient exterior location, such as a patio. This exterior outlet must have a GFCI if within direct grade level access.

○ Ceiling lights are common in children's bedrooms for easy illumination. Master bedrooms may have a switched duplex outlet for a bedside lamp.

○ The kitchen may have a light over the sink, although in a very well-lighted kitchen, this may be an extra light.

○ Bathroom lights are commonly placed above the mirror at the vanity and located in any area that requires additional light. In a large bath, a recessed light should be placed in a shower or above a tub.

○ Place switches in a manner that provides the easiest control of the lights.

○ Locate lights in large closets or any alcove or pantry that requires light.

○ Light stairways well, and provide a switch at each level.

○ Place lights and outlets in garages or shops in relationship to their use. For example, a welder or shop equipment requires 220-volt outlets.

○ Place exterior lights properly to illuminate walks, drives, patios, decks, and other high-use areas.

▼ ENERGY CONSERVATION

Energy efficiency is a very important consideration in today's home design and construction market. There are a number of steps that can easily help contribute to energy savings for the homeowner at little additional cost. Some of these items will have to be explained by specific or general notes or included in construction specifications; other methods must be defined with detail drawings. Energy-efficient considerations related to the electrical design include the following:

○ Keep electrical outlets and recessed appliances or panels to a minimum at exterior walls. Any units recessed into an exterior wall eliminate or severely compress the insulation and reduce its insulating value.

○ Fans or other exhaust systems exhausting air from the building should be provided with back draft or automatic dampers to limit air leakage.

○ Timed switches or humidistats should be installed on exhaust fans to control unnecessary operation.

○ Electrical wiring is often located in exterior walls at a level convenient for electricians to run wires. This practice causes insulation to become compacted, and a loss of insulating value results. Wires could be run along the bottom of the studs at the bottom plate. This is easily accomplished during framing by cutting a V groove in the stud bottoms while the studs are piled at the site after delivery. Figure 16–1 shows a comparison of standard and energy-saving installation methods.

○ Select energy-efficient appliances such as a self-heating dishwasher or a high-insulation water heater. Evaluate the vendor's energy statement before purchase.

○ Use energy-saving fluorescent lighting fixtures where practical, such as in the kitchen, laundry, utility, garage, or shop.

○ Fully insulate above and around recessed lighting fixtures. Verify code and vendor specifications for this practice.

○ Carefully caulk and seal around all light and convenience outlets. Also, caulk and seal where electrical wires penetrate the top and bottom plates. Figure 16–2 shows some methods that could be used to increase energy efficiency.

○ Use recessed lights that are IC (insulation cover) rated. This allows you to insulate around and over the recessed light to help avoid heat loss.

▼ HOME AUTOMATION

Many modern homes are being built with home automation systems. This technology is rapidly changing and requires that you continuously research the available products. The installation of home automation systems allows the builder to sell the same-square-footage home for more and increases the marketability. *Automation* is a method or process of controlling and operating mechanical devices by other than human power. Such operations include the computerized control of the heating,

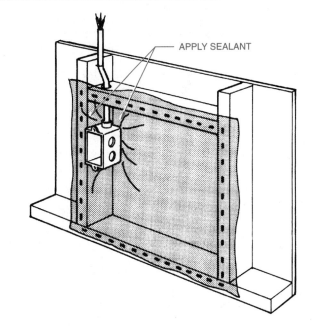

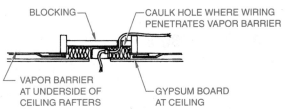

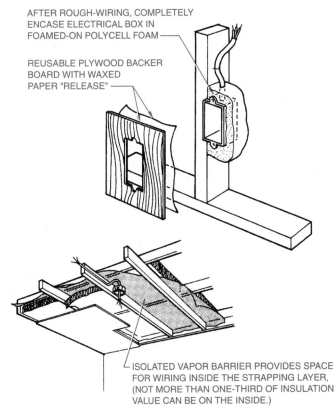

Figure 16–2 Energy conservation construction methods.

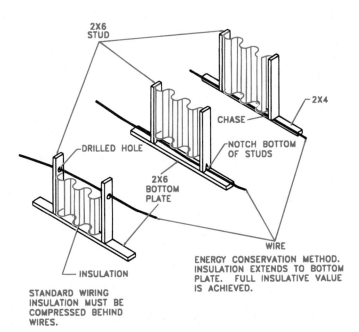

Figure 16–1 Wiring techniques for energy conservation.

ventilating, air conditioning, landscape sprinkling systems, lighting, and security systems. Among the many home automation systems available are the following:

○ Entertainment centers, to provde an in-home theater. These are fully automatic and contain the best technology in picture and sound performance. For example, pressing one button automatically dims the lights, opens the curtain, and begins the movie.

○ Computerized programming of house functions from a personal computer. These systems allow the computer user to set and monitor a variety of electrical circuits throughout the home, including security, sound, yard watering, cooking, and lighting.

○ Residential elevators, to transport people who require a wheelchair between floors.

The design and drafting of the home automation system may be done by the product supplier or the architect. Electrical symbols and specific notes are placed on the floor plan or on separate drawings used for construction purposes.

▼ ELECTRICAL SYMBOLS

Electrical symbols are used to show the lighting arrangement desired in the home. This includes all switches, fixtures, and outlets. Lighting fixture templates are available with a variety of fixture shapes for drafting convenience. Switch symbols are generally lettered freehand unless CADD is used.

All electrical symbols should be drawn with ⅛″ (3 mm) diameter circles, as seen in Figure 16–3. The electrical layout should be subordinate to the plan and not clutter or distract from the other information. All lettering for switches and other notes should be ⅛″ (3 mm) high depending on space requirements and office practice.

Switch symbols are generally drawn perpendicular to the wall and are placed to read from the right side or bottom of the sheet. Look at Figure 16–4. Also, notice that the switch relay should intersect the symbol at right angles to the wall, or the relay may begin next to the symbol. Verify the preference of your instructor or employer, and do not mix methods.

Figure 16–5 shows several typical electrical installations with switches to light outlets. The switch leg or electrical circuit line may be drawn with an irregular (French) curve.

When special characteristics are required, such as a specific size fixture, a location requirement, or any other specification, a local note that briefly describes the situation may be applied next to the outlet, as shown in Figure 16–6.

Figure 16–7 (page 250) shows some common errors related to the placement and practice of electrical floor plan layout.

Figure 16–8 (page 251) shows some examples of maximum spacing recommended for installation of wall outlets.

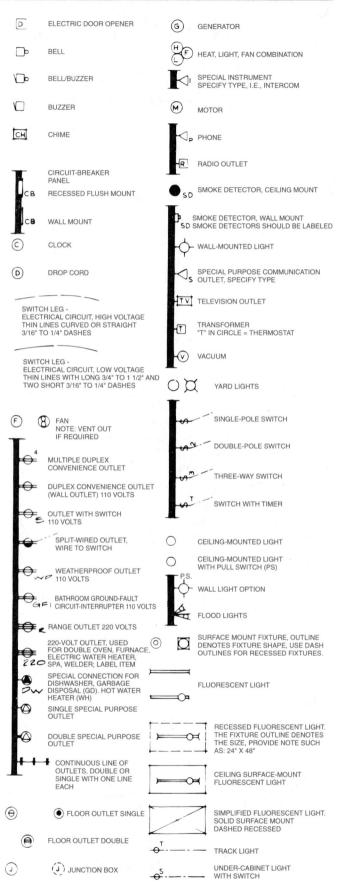

Figure 16–3 Common electrical symbols.

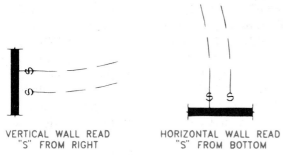

VERTICAL WALL READ
"S" FROM RIGHT

HORIZONTAL WALL READ
"S" FROM BOTTOM

Figure 16-4 Switch symbol placement.

Figure 16-9 (see page 251) shows some examples of typical electrical layouts. Figure 16-10 (see page 252) shows a bath layout. Figure 16-11 (see page 252) shows a typical kitchen electrical layout.

▼ ELECTRICAL WIRING SPECIFICATIONS

The following specifications describing the installation of the service entrance and meter base for a residence have been adapted from information furnished by Home Building Plan Service.

Service Entrance and Meter Base Installation

Before proceeding with the installation of the service entrance and meter base, the following questions must be resolved:

1. *What is the service capacity to be installed?* For capacity of service, the average single-family residence should be equipped with a 200-amp service entrance. If heating, cooking, water heating, and similar heavy loads are supplied by energy sources other than electricity, the service entrance conductors may be sized for 100 amps, provided local codes are not violated. Some local codes require a minimum capacity of 200 amps for each single dwelling. Local inspection authorities and the power company serving the area can be of assistance if difficulty is encountered in sizing the service entrance.

2. *Where is the service entrance to be located?* The service entrance location depends on whether the power company serves the houses in the area from an overhead or underground distribution system. The underground service entrance is the most desirable from both an aesthetic and reliability standpoint. However, if the power company services the area from an overhead pole line, an underground service entrance may be impractical, prohibitively expensive, or even impossible.

3. *Where will the meter base be located?* The meter base must always be mounted so the meter socket is on the exterior of the house and preferably on the side wall of a garage. The meter base should not be installed on the front of the house for aesthetic reasons. If the house does not have a garage, the meter base should

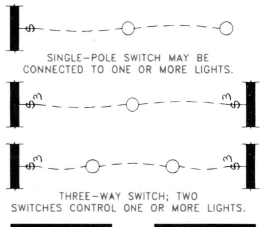

SINGLE-POLE SWITCH MAY BE
CONNECTED TO ONE OR MORE LIGHTS.

THREE-WAY SWITCH; TWO
SWITCHES CONTROL ONE OR MORE LIGHTS.

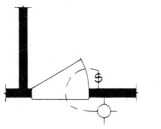

FOUR-WAY SWITCH; THREE
SWITCHES CONTROL ONE OR MORE LIGHTS.

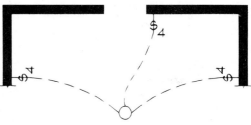

SINGLE-POLE SWITCH TO WALL-MOUNTED
LIGHT. TYPICAL INSTALLATION AT AN
ENTRY OR PORCH.

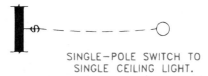

SINGLE-POLE SWITCH TO
SINGLE CEILING LIGHT.

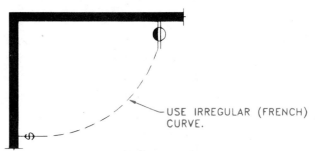

USE IRREGULAR (FRENCH)
CURVE.

SINGLE-POLE SWITCH TO SPLIT-WIRED
OUTLET. COMMON APPLICATION IN A
ROOM WITHOUT A CEILING LIGHT.
ALLOWS SWITCHING A TABLE LAMP.

Figure 16-5 Typical electrical installations.

Figure 16–6 Special notes for electrical fixtures.

be located on an exterior side wall as near to the distribution panel as practical.

4. *Where will the distribution panel be located?* The preferred location for the distribution panel is on an inside garage wall, close to the heaviest electrical loads, such as a range, clothes dryer, or electric furnace. The panel should be mounted flush unless the location does not permit this. If the house does not

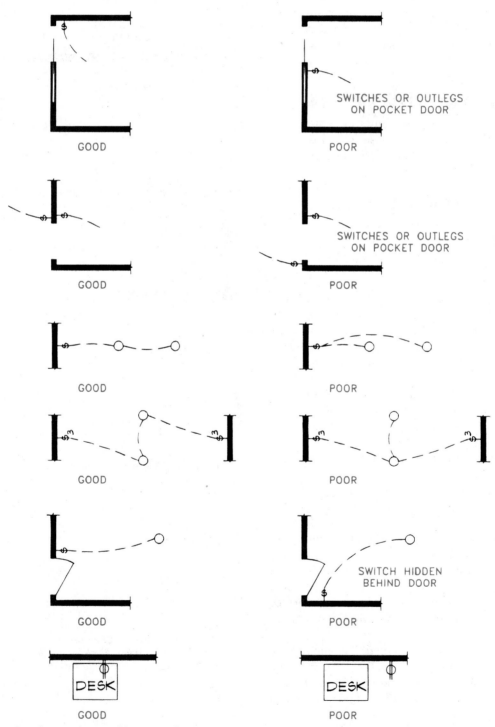

Figure 16–7 Good and poor electrical layout techniques.

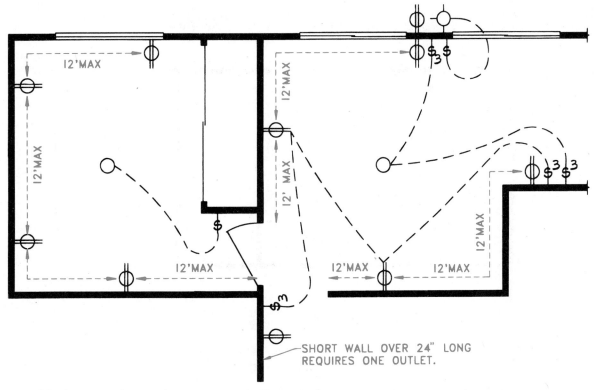

Figure 16–8 Maximum spacing requirements.

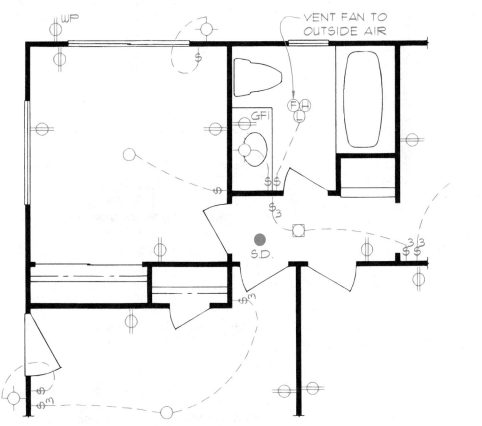

Figure 16–9 Typical electrical layouts.

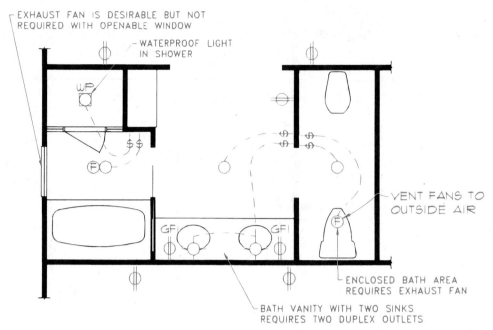

Figure 16–10 Bath electrical layout example.

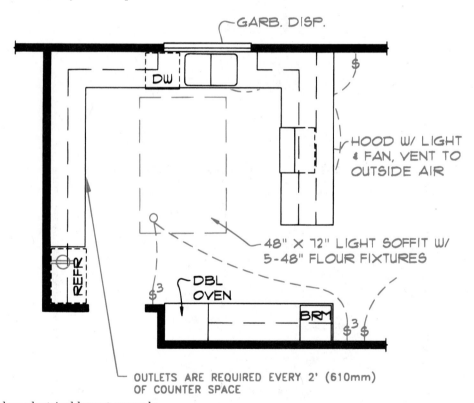

Figure 16–11 Kitchen electrical layout example.

have a garage, the panel should be located in a readily accessible area such as a utility room, or kitchen wall.

In some localities the electrical power billing structure requires two meters. For example, lighting and general use energy may be billed at a rate different from energy used for space and water heating. In such a situation,

provide two meter bases and two distribution panels to suit requirements.

▼ **METRICS IN ELECTRICAL INSTALLATIONS**

Electrical conduit designations are expressed in millimeters. *Electrical conduit* is a metal or fiber pipe or tube used to enclose a single or several electrical conductors. Electrical conduit is produced in decimal inch dimensions and is identified in nominal inch sizes. *Nominal size* is referred to as the conventional size; for example, a 0.500″ pipe has a ½″ nominal size. The actual size of conduit will remain in inches but will be labeled in metric:

Inch	Metric (mm)	Inch	Metric (mm)
½	16	2½	63
¾	21	3	78
1	27	3½	91
1¼	35	4	103
1½	41	5	129
2	53	6	155

Existing American Wire Gage (AWG) sizes will remain the same without a metric conversion. The diameter of wires conform to various gaging systems. The AWS is one system for the designation of wire sizes.

▼ STEPS IN DRAWING THE ELECTRICAL FLOOR PLAN

The electrical plan is prepared as a separate floor-plan drawing, labeled ELECTRICAL PLAN, for complex homes, or combined with all other floor-plan information for basic one-story homes. When drawing the electrical plan with all other floor-plan information, follow the next steps after placing everything on the drawing, as explained in Chapter 15. If you are creating a separate electrical plan, use overlay drafting techniques or the CADD layering system. To do this, first begin with a base drawing of the floor plan, with all of the walls, doors, windows, stairs, cabinets, and the fireplace; the basic room titles are optional. A base drawing for the residence that was first used in Chapter 15 is shown in Figure 16–12.

Step 1. Letter all switch locations and draw all light fixture locations. Use a ⅛″ diameter circle (6″ at a scale of ¼″ = 1′–0″) for the light fixture symbols. If electrical symbols are too large, they distract from the appearance of the drawing. If they are too small, they can be difficult to read. CADD electrical symbol libraries provide quick insertion of these symbols. Look at Figure 16–13.

Step 2. Draw electrical circuits or switch legs from switches to fixtures using a dashed arc line. For

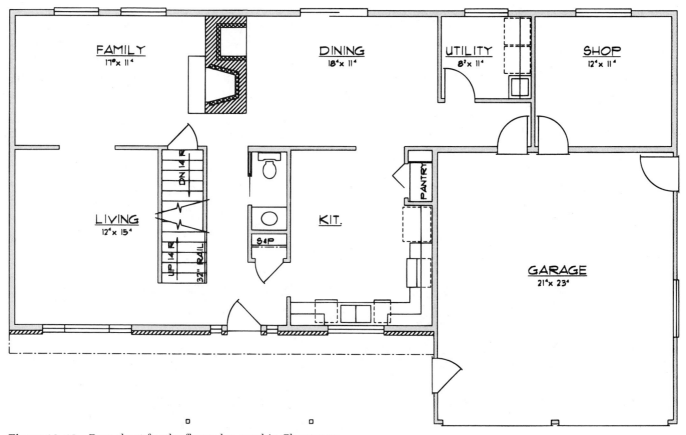

Figure 16–12 Base sheet for the floor plan used in Chapter 15.

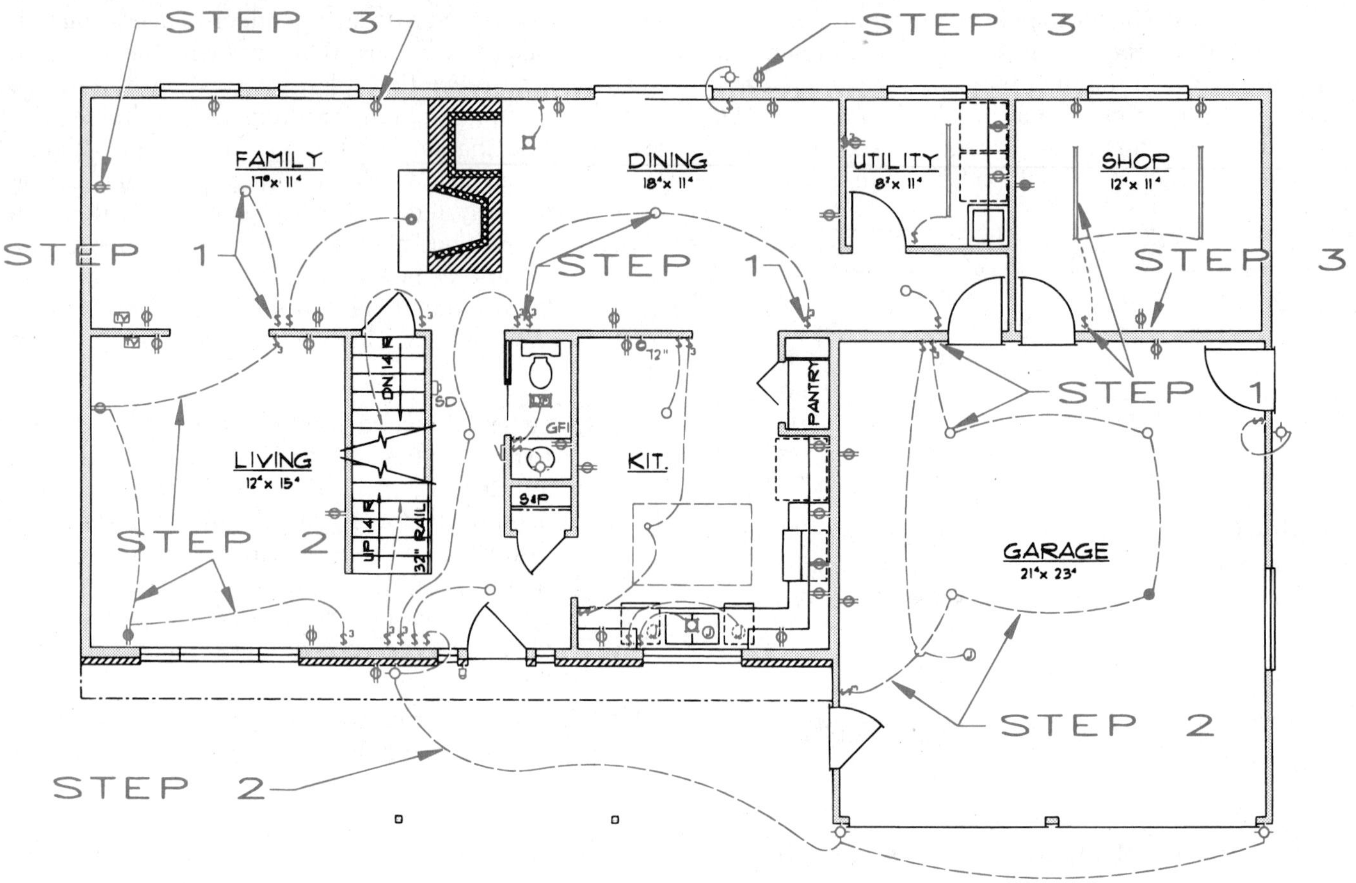

Figure 16–13 Step 1: Placing the switches and lighting outlets. Step 2: Drawing the switch legs. Step 3: Placing the outlets.

manual drafting, use an irregular curve and a 0.5-mm 2H or H automatic pencil. For CADD applications, use either a dashed or hidden line type and the arc command. Figure 16–13 shows Step 2.

Step 3. Place all electrical outlets, such as duplex convenience, range, and television. Figure 16–13 shows the placement of items for Step 3.

Step 4. Letter all notes, and the drawing title and scale, as shown in Figure 16–14.

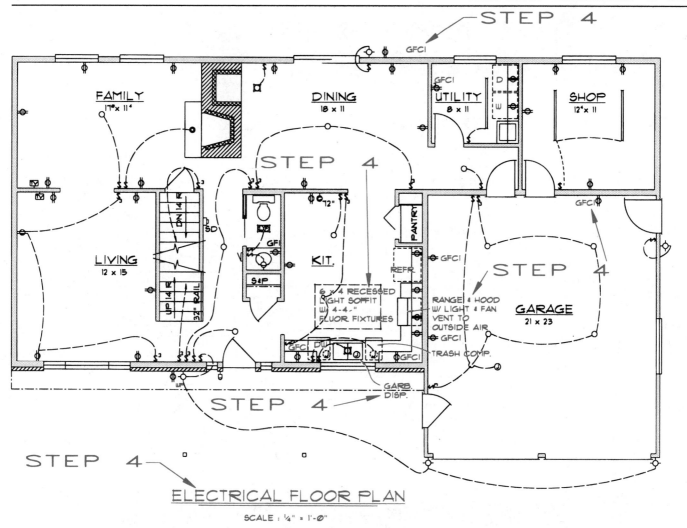

Figure 16–14 Lettering the notes and drawing title.

▼ CADD Applications

DRAWING ELECTRICAL SYMBOLS WITH CADD

In most cases, electrical symbols are drawn on a separate CADD layer. The floor plan serves as the base sheet. So, by turning the electrical layer ON you have a composite drawing with the floor plan, electrical symbols, and electrical layout. You also have the flexibility to turn the electrical layer OFF when you want to display the floor plan without electrical. Many companies provide two separate drawings for commercial projects: the floor plan completely dimensioned and the floor plan with only the walls and electrical layout. All you have to do is turn ON or OFF specific layers as needed to display what you want. Most architectural CADD programs are tablet menu driven with a symbols library for electrical applications, as shown in Figure 16–15. A drafter simply selects the desired symbol from the tablet menu and then drags the symbol into position on the drawing. Parametric design plays an important role in applications such as ceiling lighting grids. You specify the room width, length, and grid size; the computer then auto-

Figure 16–15 CADD symbols library for architectural applications. Notice the ELEC, LIGHT/CLG, and APP symbols for electrical applications. *Courtesy Autodesk, Inc.*

matically draws the entire ceiling grid. Figure 16–16 shows a complete floor plan with electrical symbols.

CHAPTER

16 Electrical Plans Test

DIRECTIONS

Answer the questions with short, complete statements or drawings as needed on an 8½ × 11 sheet of notebook paper, as follows:

1. Letter your name, Chapter 16 Test, and the date at the top of the sheet.
2. Letter the question number and provide the answer. You do not need to write out the question.

Answers may be prepared on a word processor if appropriate with course guidelines.

QUESTIONS

Question 16–1 Duplex convenience outlets should be a maximum of how many feet apart?

Question 16–2 Duplex convenience outlets should be no more than how many feet from a corner?

Question 16–3 Describe at least four energy-efficient considerations related to electrical design.

Question 16–4 Draw the proper floor-plan symbol for:
a. Duplex convenience outlet
b. Range outlet
c. Recessed circuit breaker panel
d. Phone
e. Light
f. Wall-mounted light
g. Single-pole switch
h. Simplified fluorescent light fixture
i. Fan

Question 16–5 Why is it poor practice to place a switch behind a door swing?

Question 16–6 What is a GFI duplex convenience outlet?

Question 16–7 The average single-family residence should be equipped with how many amps of electrical service?

PROBLEMS

Problem 16–1 Draw a typical floor-plan representation of the following on 8½ × 11 vellum in pencil or using CADD:
a. Three-way switch controlling three ceiling-mounted lighting fixtures.
b. Four-way switch to one ceiling-mounted lighting fixture.
c. Single-pole switch to wall-mounted light fixture.
d. Single-pole switch to split-wired duplex convenience outlet.

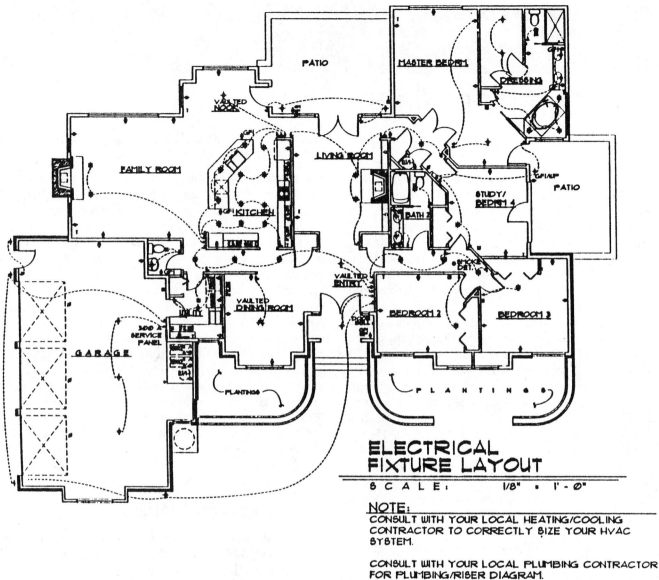

Figure 16–16 Complete floor plan with electrical symbols. *Courtesy Piercy & Barclay Designers, Inc. CPBD.*

Problem 16–2 Typical small bathroom layout with tub, water closet, one sink vanity, one wall-mounted light over sink with single-pole switch at door, GFCI duplex convenience outlet next to sink, ceiling-mounted light-fan-heat unit with timed switch at door, proper vent note for fan.

Problem 16–3 Typical kitchen layout with double light above sink, dishwasher, range with oven below and proper vent note, refrigerator, centered ceiling-recessed fluorescent lighting fixture, adequate GFI duplex convenience outlets, and garbage disposal.

Problem 16–4 Use the drawing of the house that you started in the floor-plan problem from Chapter 15 and do one of the following:
a. If you created a base sheet for overlay drafting, design an electrical floor-plan overlay for the house.
b. If you are using CADD, create an ELECT layer, and design the electrical plan on this layer.
c. If you drew the entire floor plan with the intention of entering the electrical plan on the same drawing, do so now.

CHAPTER 17

Plumbing Plans

There are two classifications of piping: industrial and residential. Industrial piping is used to carry liquids and gases used in the manufacture of products. Steel pipe with welded or threaded connections and fittings is used in heavy construction.

Residential piping is called plumbing and carries fresh water, gas, or liquid and solid waste. The pipe used in plumbing may be made of copper, plastic, galvanized steel, or cast iron.

Copper pipes have soldered joints and fittings, and are used for carrying hot or cold water. Plastic pipes have glued joints and fittings, and are used for vents and for carrying fresh water or solid waste. Many contractors are replacing copper pipe with plastic piping for both hot and cold water applications. A plastic pipe with the chemical name of polybutylene (PB), also known as *poly pipe*, has been used for both hot and cold water applications. Some concerns have been addressed about the strength and life expectancy of PB pipe in construction. Plastic polyvinyl chloride (PVC) pipe has been used effectively for cold water installations. Corrosion-resistant plastic piping is available in a thermoplastic with the chemical name of postchlorinated polyvinyl chloride (CPVC). While metal piping loses heat, the CPVC pipe retains insulation and saves energy. CPVC piping lasts longer than copper pipe because it is corrosion resistant, it maintains water purity even under severe conditions, and it does not cause condensation, as copper does. Plastic pipe is considered to be quieter than copper pipe, and it costs less to buy and install.

Very little galvanized steel pipe is used except where conditions require such installation for corrosion resistance. Steel pipe has threaded joints and fittings and is used to carry natural gas. Cast-iron pipe is commonly used to carry solid and liquid waste as the sewer pipe that connects the structure with the local or regional sewer system. Cast-iron pipe may also be used for the drain system throughout the structure to help reduce water flow noise substantially in the pipes. It is more expensive than plastic pipe but may be worth the price if a quiet plumbing system is desired.

Residential plans may not require a complete plumbing plan. The need for a complete plumbing plan should be verified with the local building code. In most cases, the plumbing requirements can be clearly provided on the floor plan in the form of symbols for fixtures and notes for specific applications or conditions. The plumbing fixtures are drawn in their proper locations on the floor plans at a scale of ¼″ = 1′–0″. Templates with a large variety of floor plan plumbing symbols are available.

Other plumbing items to be added to the floor plan include floor drains, vent pipes, and sewer or water connections. Floor drains are shown in their approximate location and are identified with a note identifying size, type, and slope to drain. Vent pipes are shown in the wall where they are to be located and labeled by size. Sewer and water service lines are located in relationship to the position in which these utilities enter the home. The service lines are commonly found on the plot plan. In the situation described here, where a very detailed plumbing layout is not provided, the plumbing contractor is required to install quality plumbing in a manner that meets local code requirements and is economical. Figure 17–1 shows plumbing fixture symbols in plan, frontal, and profile view.

▼ ENERGY CONSERVATION

Energy conservation methods of construction can be applied to the plumbing installation. If these types of methods are used, they can be applied to the drawing in the form of specific or general notes on detailed drawings or in construction specifications. Some energy conservation methods include:

○ Insulating all exposed hot water pipes. Cold water pipes should be insulated in climates where freezing is a problem.
○ Running water pipes in insulated spaces where possible.
○ Keeping water pipes out of exterior walls where practical.
○ Placing thermosiphon traps in hot water pipes to reduce heat loss from excess hot water in the pipes.
○ Locating the water heater in a heated space and insulating it well.
○ Selecting low-flow shower heads.
○ Putting flow restrictors in faucets.
○ Completely caulking plumbing pipes where they pass through the plates.
○ Sealing and covering drain penetrations in floors.
○ Sealing all wall penetrations.

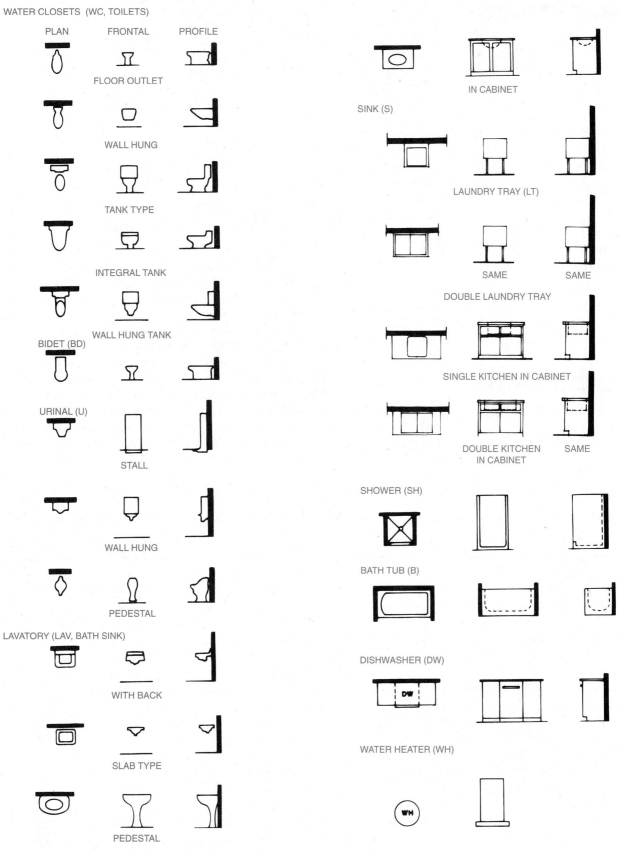

Figure 17–1 Some common plumbing fixture symbols.

▼ PLUMBING SCHEDULES

Plumbing schedules are similar to door, window, and finish schedules. Schedules provide specific information regarding plumbing equipment, fixtures, and supplies. The information is condensed in a chart so the floor plan is not unnecessarily crowded. Figure 17–2 shows a typical plumbing fixture schedule.

Other schedules may include specific information regarding floor drains, water heaters, pumps, boilers, or radiators. These schedules generally key specific items to the floor plan with complete information describing size, manufacturer, type and other specifications as appropriate.

Description of Plumbing Fixtures and Materials

Mortgage lenders may require a complete description of materials for the structure. Part of the description often includes a plumbing section in which certain plumbing specifications are described, as shown in Figure 17–3.

▼ PLUMBING DRAWINGS

Plumbing drawings are usually not drawn on the same sheet as the complete floor plan. The only plumbing items shown on the floor plans are fixtures, as previously shown. The drafter or designer prepares an accurate outline of the floor plan showing all walls, partitions, doors, windows, plumbing fixtures, and utilities. This outline can be used as a preliminary drawing for client approval and can also be used to prepare a sepia or photocopy secondary original. This secondary original allows the drafter to save the original drawing for the complete floor plan. The secondary original saves drafting time so the drafter will not have to prepare another outline for the plumbing drawing. When the drawings are done on CADD, the base drawing is the floor plan, and layers are set up to display the plumbing symbols and piping layout when needed.

PLUMBING FIXTURE SCHEDULE

LOCATION	ITEM	MANUFACTURER	REMARKS
MASTER BATH	36" F.G. SHR. COLOR BIDET C.I. SR. COLOR LAV. COLOR W.C.	HYTEC K-4868 K-2904 K-3402-PBR	M2620 BRASS K1940 BRASS M4625 BRASS PLAS. SEAT
BATH #2	KEG STYLE TUB C.I. COLOR PED. LAV. COLOR W.C.	KOHLER KOHLER K3402-PBR	M2850 BRASS M4625 BRASS PLAS. SEAT
BATH #3	URINAL F.G. SHOWER C.I. LAVS.	K-4980 HYTEC K2904	M2620 BRASS M4625 BRASS
KITCHEN	C.I. 3 HOLE SINK	K5960	M7531 BRASS
WATER HTR.	82 GAL. ELEC.	MORFIO	P & T VALVE
UTILITY	F.G. LAUN. TRAY	24 ×21	D2121 BRASS

Figure 17–2 Plumbing fixture schedule.

PLUMBING:

Fixture	Number	Location	Make	Mfr's Fixture Identification No.	Size	Color
Sink						
Lavatory						
Water closet						
Bathtub						
Shower over tub △						
Stall shower △						
Laundry trays						

△□ Curtain rod   △□ Door   □ Shower pan: material _____

Water supply: □ public; □ community system; □ individual (private) system. ★

Sewage disposal: □ public; □ community system; □ individual (private) system. ★

★Show and describe individual system in complete detail in separate drawings and specifications according to requirements.

House drain (inside): □ cast iron; □ tile; □ other _____ House sewer (outside): □ cast iron; □ tile; □ other _____

Water piping: □ galvanized steel; □ copper tubing; □ other _____ Sill cocks, number _____

Domestic water heater: type _____ ; make and model _____ ; heating capacity _____

_____ gph 100° rise.   Storage tank: material _____ ; capacity _____ gallons.

Gas service: □ utility company; □ liq. pet. gas; □ other _____ Gas piping: □ cooking; □ house heating.

Footing drains connected to: □ storm sewer; □ sanitary sewer; □ dry well.   Sump pump; make and model _____

_____ ; capacity _____ ; discharges into _____

Figure 17–3 Plumbing description of materials form.

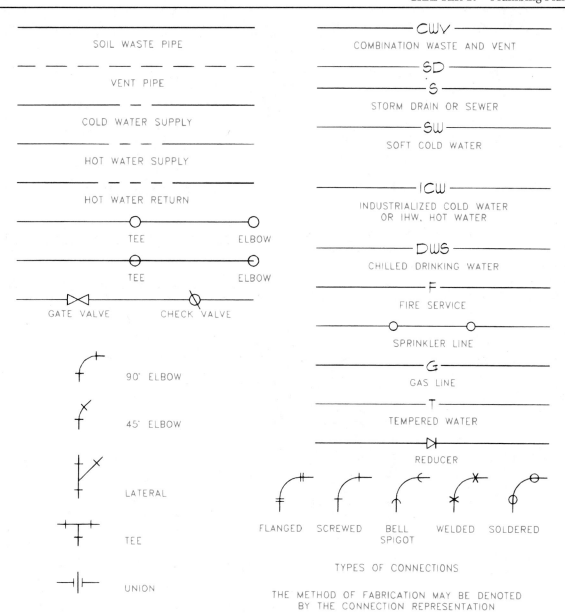

Figure 17–4 Typical plumbing pipe symbols.

Plumbing drawings may be prepared by a drafter in an architectural office or in conjunction with a plumbing contractor. In some situations, when necessary, plumbing contractors or mechanical engineers work up their own rough sketches or field drawings. Plumbing drawings are made up of lines and symbols that show how liquids, gases, or solids are transported to various locations in the structure. Plumbing lines and features are drawn thicker than wall lines so they are clearly distinguishable. Symbols identify types of pipes, fittings, valves, and other components of the system. Sizes and specifications are provided in local or general notes. Figure 17–4 shows some typical plumbing symbols. Certain abbreviations are commonly used in plumbing drawings as shown in Figure 17–5.

CURRENT		MCS	
CW	Cold-Water Supply	WC	Water Closet (Toilet)
HW	Hot-Water Supply	LA	Lavatory (Bath Sink)
HWR	Hot-Water Return	B	Bathtub
HB	Hose Bibb	S	Sink
CO	Clean Out	U	Urinal
DS	Downspout	SH	Shower
RD	Rain Drain	DF	Drinking Fountain
FD	Floor Drain	WH	Water Heater
SD	Shower Drain	DW	Dishwasher
CB	Catch Basin	BD	Bidet
MH	Manhole	GD	Garbage Disposal
VTR	Vent Thru Roof		

Figure 17–5 Some typical plumbing abbreviations.

▼ WATER SYSTEMS

Water supply to a structure begins at a water meter for public systems or from a water storage tank for private well systems. The water supply to the home or business, known as the *main line*, is generally 1″ plastic pipe. This size may vary in relationship to the service needed. The plastic main line joins a copper line within a few feet of the structure. The rest of the water system piping is usually copper pipe, although plastic pipe is increasing in popularity for cold water applications. The 1″ main supply often changes to ¾″ pipe where a junction is made to distribute water to various specific locations. From the ¾″ distribution lines, ½″ pipe usually supplies water to specific fixtures, for example, the kitchen sink. Figure 17–6 shows a typical installation from the water meter of a house with distribution to a kitchen. The water meter location and main line representation are generally shown on the plot plan. Verify local codes regarding the use of plastic pipe.

There is some advantage to placing plumbing fixtures back-to-back when possible. This practice saves materials and labor costs. Also, when designing a two-story structure, placing plumbing fixtures one above the other aids in an economical installation. If the functional design of the floor plan clearly does not allow for such economy measures, then good judgment should be used in the placement of plumbing fixtures so the installation of the plumbing is physically possible. Figure 17–7 shows a back-to-back bath situation. If the design allows, another common installation may be a bath and laundry room next to each other to provide a common plumbing wall. See Figure 17–8. If a specific water temperature is required, that specification can be applied to the hot water line, as shown in Figure 17–9.

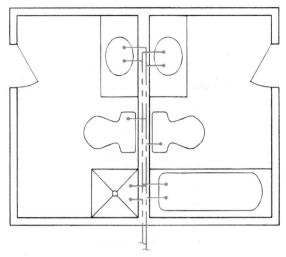

Figure 17–7 Common plumbing wall with back-to-back baths.

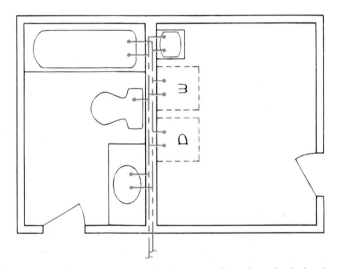

Figure 17–8 Common plumbing wall with a bath back to laundry.

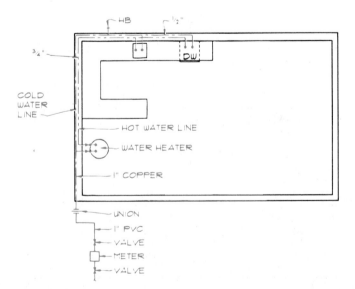

Figure 17–6 A partial water supply system.

Figure 17–9 Hot water temperature.

▼ DRAINAGE AND VENT SYSTEMS

The *drainage system* provides for the distribution of solid and liquid waste to the sewer line. The *vent system* allows for a continuous flow of air through the system so gases and odors may dissipate and bacteria do not have an opportunity to develop. These pipes throughout the house are generally made of PVC plastic, although the pipe from the house to the concrete sewer pipe is commonly 3″ or 4″ cast iron. Drainage and vent systems, as with water systems, are drawn with thick lines using symbol, abbreviations, and notes. Figure 17–10 shows a sample drainage vent system. Figure 17–11 shows a house plumbing plan.

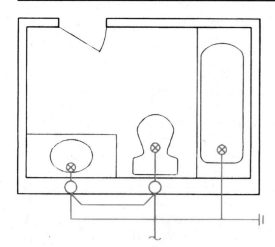

Figure 17–10   Drainage and vent drawing.

▼ ISOMETRIC PLUMBING DRAWINGS

Isometric drawings may be used to provide a three-dimensional representation of a plumbing layout, sometimes called a plumbing riser diagram. Isometric plumbing templates and CADD programs are available to make the job easier. Especially for a two-story structure, an isometric drawing provides an easy-to-understand pictorial drawing. Figure 17–12 shows an isometric drawing of the system shown in plan view in Figure 17–10. Figure 17–13 shows a detailed isometric drawing of a typical

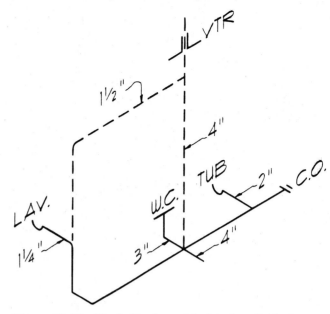

Figure 17–12   Single-line isometric drawing of a drainage and vent system.

drain, waste, and vent system. Figure 17–14 shows a single-line isometric drawing of the detailed isometric drawing shown in Figure 17–13. Figure 17–15 shows a detailed isometric drawing of a hot and cold water supply system. Figure 17–16 shows a single-line isometric drawing of that same system.

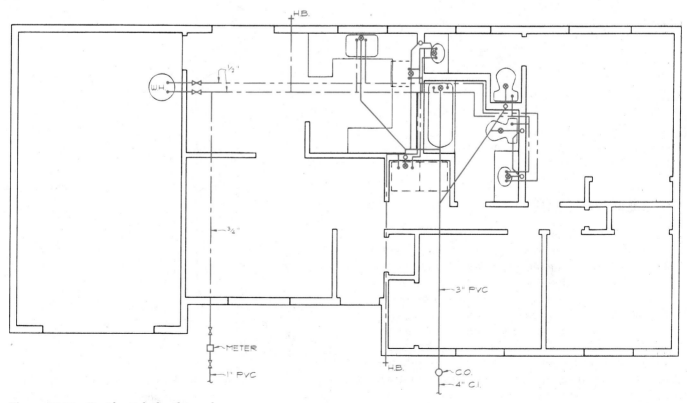

Figure 17-11   Residential plumbing plan.

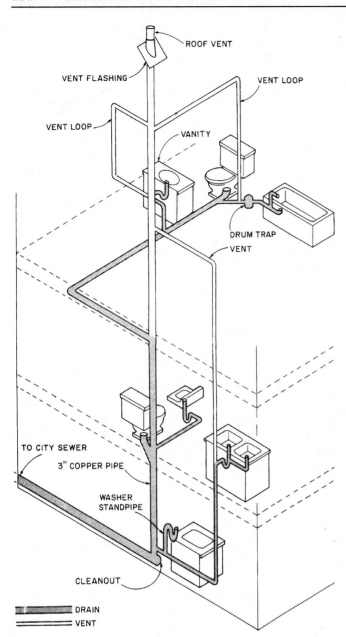

Figure 17–13 Detailed isometric drawing of drain, waste, and vent system. *From Huth,* Understanding Construction Drawings, *Delmar Publishers Inc.*

▼ SOLAR HOT WATER

Solar hot water collectors are available for new or existing residential, commercial, or industrial installations. Solar collectors may be located on a roof, wall, or on the ground. Figure 17–17 shows a typical application of solar collectors. Solar systems vary in efficiency. The number of collectors needed to provide heat to a structure depends on the size of the structure and the volume of heat needed.

The flat-plate collector is the heart of a solar system. Its main parts are the transparent glass cover, absorber plate, flow tubes, and insulated enclosure. Piping and wiring drawings for a complete solar hot water heating system are available through the manufacturer.

▼ SEWAGE DISPOSAL

Public Sewers

Public sewers are available in and near most cities and towns. The public sewers are generally located under the street or an easement next to the construction site. In some situations the sewer line may have to be extended to accommodate the addition of another home or business in a newly developed area. The cost of this extension may be the responsibility of the developer. Usually the public sewer line is under the street so the new construction has a line from the structure to the sewer. The cost of this construction usually includes installation expenses, street repair, sewer tap, and permit fees. Figure 17–18

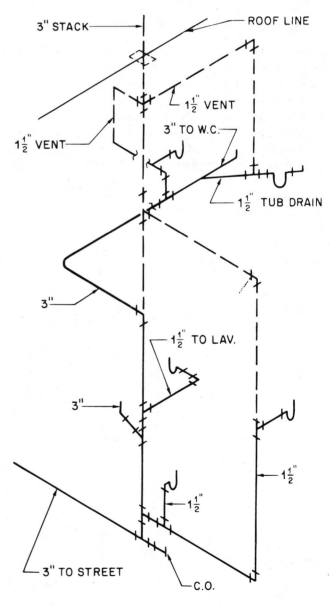

Figure 17–14 Single-line isometric drawing of drain, waste, and vent system. *From Huth,* Understanding Construction Design, *Delmar Publishers Inc.*

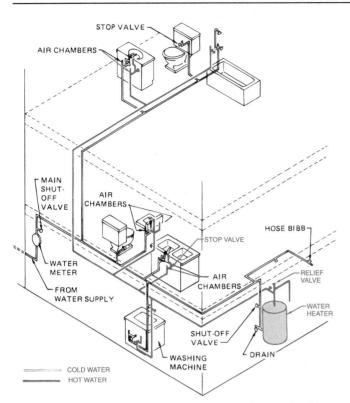

Figure 17–15 Detailed isometric drawing of hot and cold water system.

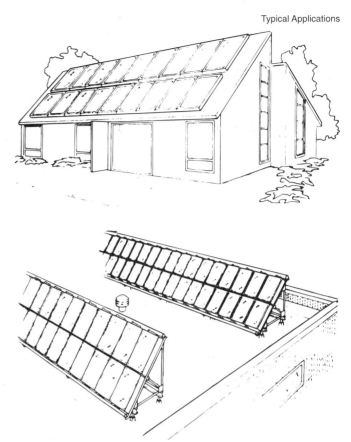

Typical Applications

Figure 17–17 Typical application of solar hot water collectors. *Courtesy Lennox Industries, Inc.*

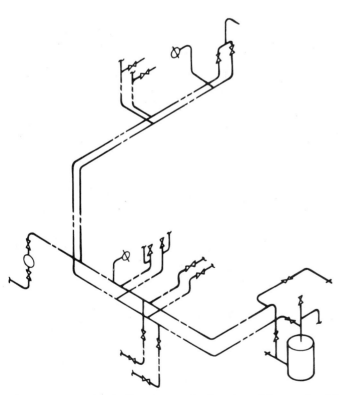

Figure 17–16 Single-line isometric drawing of hot and cold water system.

shows an illustration of a sewer connection and the plan view usually found on the plot plan.

Private Sewage Disposal: A Septic System

The *septic system* consists of a storage tank and an absorption field and operates as follows. Solid and liquid waste enters the septic tank, where it is stored and begins decomposition into sludge. Liquid material, or effluent, flows from the tank outlet and is dispersed into a soil-absorption field, or drain field (also known as *leach lines*). When the solid waste has effectively decomposed, it also dissipates into the soil absorption field. The owner should use a recommended chemical to work as a catalyst for complete decomposition of solid waste. Septic tanks may become overloaded in a period of up to ten years and may require pumping.

The characteristics of the soil must be verified for suitability for a septic system by a soil feasibility test, also known as a *percolation test*. This test, performed by a soil scientist or someone from the local government, determines if the soil will accommodate a septic system. The test should also identify certain specifications that should be followed for system installation. The Veterans Administration (VA) and the Federal Housing Administration (FHA) require a minimum of 240′ of field

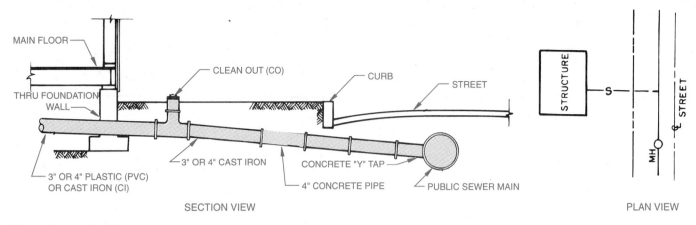

Figure 17–18 Public sewer system.

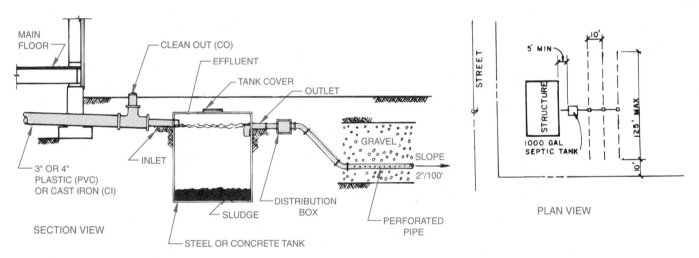

Figure 17–19 Septic sewer system.

line or more if the soil feasibility test shows it necessary. Verify these dimensions with local building officials. When the soil characteristics do not allow a conventional system, they may be some alternatives such as a sand filter system, which filters the effluent through a specially designed sand filter before it enters the soil absorption field. Check with your local code officials before calling for such a system. Figure 17–19 shows a typical serial septic system. The serial system allows for one drain field line to fill before the next line is used. The drain field lines must be level and must follow the contour of the land perpendicular to the slope. The drain field should be at least 100′ from a water well, but verify the distance with local codes. There is usually no minimum distance to a public water supply.

▼ METRICS IN PLUMBING

Pipe is made of a wide variety of materials identified by trade names; the nominal sizes are related only loosely to actual dimensions. For example, a 2″ galvanized pipe has an inside diameter of about 2 ⅛″, but is called 2″ pipe for convenience. Since few pipe products have even inch dimensions that match their specifications, there is no

reason to establish even metric sizes. Metric values established by the International Organization for Standardization relate nominal pipe sizes (NPS) in inches to metric equivalents, referred to as *diameter nominal* (DN). The following equivalents relate to all plumbing, natural gas, heating, oil, drainage, and miscellaneous piping used in buildings and civil works projects:

NPS (in.)	DN (mm)	NPS (in.)	DN (mm)
⅛	6	8	200
³⁄₁₆	7	10	250
¼	8	12	300
⅜	10	14	350
½	15	16	400
⅝	18	18	450
¾	20	20	500
1	25	24	600
1¼	32	28	700
1½	40	30	750
2	50	42	800
2½	65	36	900
3	80	40	1000
3½	90	44	1100

NPS (in.)	DN (mm)	NPS (in.)	DN (mm)
4	100	48	1200
4½	115	52	1300
5	125	56	1400
6	150	60	1500

When giving a metric pipe size, it is recommended that DN precede the metric value. For example, the conversion of a 2½″ pipe to metric is DN65.

The standard thread for thread pipe is the National Standard Taper Pipe Threads (NPT). The thread on ½″ pipe reads ½–14NPT, where the 14 is threads per inch. The metric conversion affects only the nominal pipe size—½ in this case. The conversion of the ½–14NPT pipe thread to metric is DN15–14NPT.

▼ ADDING PLUMBING INFORMATION TO THE FLOOR PLAN

Most residential floor plan drawings prepared by the architect or designer have the plumbing symbols and notes placed with all other information. Plumbing symbols such as the sinks and water closets aid in reading and understanding the drawing. Plumbing notes are generally minimal. The plumbing symbols and notes may also

be placed on a drawing overlay if overlay drafting is used or on plumbing layers if CADD is used. Figure 17–20 shows the base floor plan drawing with the plumbing symbols and notes displayed in color.

▼ COMMERCIAL PLUMBING DRAWINGS

In most cases the plumbing drawings are found as an individual component of the complete set of plans for the commercial building. The architect or mechanical engineer prepares the plumbing drawings over the floor-plan layer. This method keeps the drawing clear of any other unwanted information and makes it easier for the plumbing contractor to read the print. An example is shown in Figure 17–21. Notice the floor-plan outline with unnecessary detail and information omitted. The plumbing plan shown in color is the only other information provided. As you can see, the plumbing drawing may be fairly complex.

Common Commercial Plumbing Symbols

Displaying the fixture is the most important aspect of a residential plumbing drawing. In fact, usually the actual piping is left off the residential plan. Commercial plumbing drawings are different; they must clearly show the

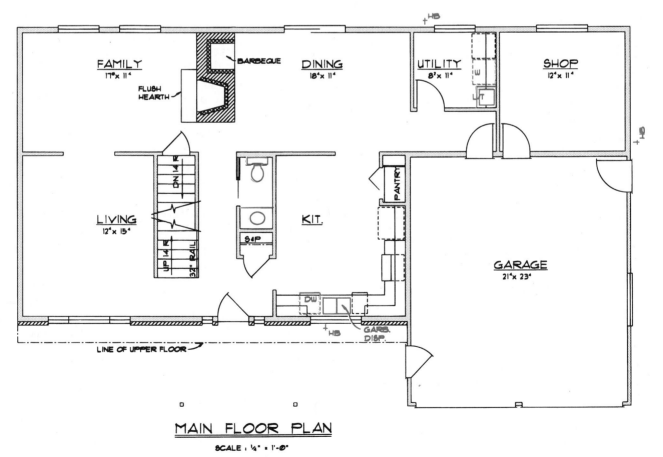

Figure 17–20 Base floor plan with plumbing layer shown in color.

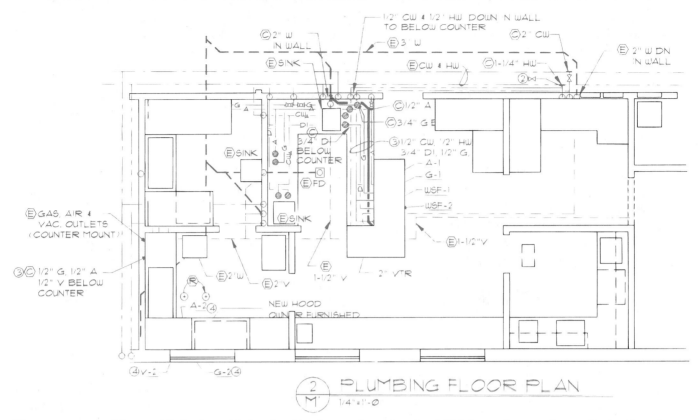

Figure 17–21 Architectural floor plan with the plumbing layout shown in color CADD layer. *Courtesy System Design Consultants.*

pipe lines and fittings as symbols, with related information given in specific and general notes. Some of the common plumbing symbols are shown in Figure 17–22. A variety of valve symbols may be used in a plumbing print. A *valve* is a device used to regulate the flow of a gas or liquid. The symbols for common valves are shown in Figure 17–23. The direction in which the pipe is running is also important to plumbing drawings. When the pipe is rising, the end of the pipe is shown as if it were cut through, and the inside is cross-hatched with section lines. When the pipe is turned down, or running away from you, the symbol appears as if you are looking at the outside of the pipe. Figure 17–24 shows several common symbols in side view, in a rising pipe view, and in a view with the pipe turning down.

Pipe Size and Elevation Shown on the Drawing

The size of the pipe may be shown with a leader and a note, or with an area of the pipe expanded to provide room for the size to be given as shown in Figure 17–25.

Some plumbing installations provide the elevation of a run of pipe or the elevation of a pipe fixture when a specific location is required. *Elevation* is the height of a feature from a known base, usually given as 0 (zero elevation). The elevation of a pipe or fitting is given in feet and inches from the base—for example, EL 24′–6″, where EL is the abbreviation of elevation and 24′–6″ is the height from the zero elevation base reference. The elevation may be noted on the print in relationship to a construction member such as TOC ("top of concrete") or TOS ("top of steel"). The elevation may also be related to a location on the pipe, such as BOP ("bottom of pipe") or CL ("center line of pipe").

Special Symbols, Schedules, and Notes on a Plumbing Drawing

As with any other drawing, plumbing information may be provided in the form of special symbols, schedules, and general notes. Figure 17–22 shows special symbols in a LEGEND, and an example of a plumbing connection schedule and a general notes. Architectural and mechanical engineering offices often place a *legend* of symbols on the drawing for reference. The legend displays symbols used on the drawing. The legend and many general notes are often saved as a BLOCK for insertion in any drawing when needed.

LEGEND

————————	W	SANITARY WASTE ABOVE SLAB
— — — — —	W	SANITARY WASTE BELOW SLAB
———— — ————	HW	DOMESTIC HOT WATER
— — — — —	V	VENT
———— — ————	CW	DOMESTIC COLD WATER
————G————	G	GAS (NATURAL)
————A————	A	COMPRESSED AIR
————VA————	VA	LABORATORY VACUUM
————CW_A————	CW_A	COLD WATER (ASPIRATOR)
————DI————	DI	DEIONIZED WATER
—⋈——⋈—		GATE VALVE/GLOBE VALVE
——▷—		PIPE REDUCER
——⊕—		SPRINKLER HEAD
ⓒ		CONNECT TO EXISTING
ⓔ		EXISTING TO REMAIN
ⓡ		RELOCATE EXISTING
⊗		REMOVE EXISTING

CONNECTION SCHEDULE Ⓐ

SYMBOL	DESCRIPTION	CW	HW	W	V	REMARKS
S-1	SINK	1/2"	1/2"	2"	1-1/2"	STAINLESS
S-2	SINK	1/2"	1/2"	2"	1-1/2"	STAINLESS
WSF-1	WATER SUPPLY FITTING	1/2"	1/2"	—	—	ASPIRATOR
WSF-2	WATER SUPPLY FITTING	1/2"	1/2"	—	—	DEIONIZED WATER

Ⓐ PLUMBING FIXTURES ONLY - G, VC AND A
BRANCH SIZES NOTED ON 2/M1. VERIFY
CONNECTION REQUIREMENTS WITH FIXTURES
FURNISHED.

NOTES

② PROVIDE GATE VALVE IN BRANCH BEFORE TAKEOFF.

③ MOUNT ALL PIPING BELOW COUNTER TIGHT TO WALL

④ PROVIDE HARD CONNECTION FROM GAS OUTLET
TO 3/8" NPT HOOD GAS COCKS. PROVIDE MISC.
FITTINGS AS REQUIRED. VERIFY EXACT CONNECTION
LOCATIONS WITH ACTUAL HOOD FURNISHED BY OWNER.

Figure 17–22 Some common plumbing symbols, schedule, and general notes. *Courtesy System Design Consultants.*

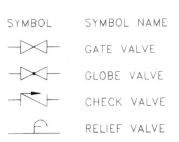

SYMBOL	SYMBOL NAME
—⋈—	GATE VALVE
—◁▷—	GLOBE VALVE
—◁|—	CHECK VALVE
—∩—	RELIEF VALVE

Figure 17–23 Common valve symbols.

PIPE FITTING FITTING SYMBOL

PIPE FITTING	PIPE RISING	SIDE VIEW	PIPE TURNING DOWN
90° ELBOW			
45° ELBOW			
TEE			
REDUCER			
CAP			
LATERAL			
CROSS			
UNION			

Figure 17–24 Several common piping symbols in side view, in a rising pipe view, and in a view with the pipe turning down.

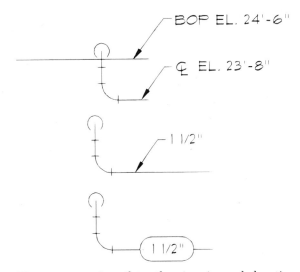

BOP EL. 24'-6"

℄ EL. 23'-8"

1 1/2"

1 1/2"

Figure 17–25 Specifying the pipe size and elevation.

CADD Applications

DRAWING PLUMBING SYMBOLS WITH CADD

In residential architecture, the plumbing drawings usually amount to placing floor-plan symbols such as sinks, water closets, and tubs. In commercial architecture applications, a separate plumbing drawing shows the complete plumbing layout with symbols and pipes on a separate CADD layer. The floor plan serves as the base sheet. So, by turning the plumbing layer ON, you have a composite drawing with the floor plan, plumbing symbols, and plumbing layout shown. You also have the flexibility to turn the plumbing layer OFF when you want to display the floor plan without plumbing. Many companies show two separate drawings on commercial projects: the (1) floor plan completely dimensioned and (2) the floor-plan walls and plumbing layout. This is easy to do, because all you have to do is turn ON or OFF specific layers as needed to display what you want. Most architectural CADD programs are tablet menu driven with a symbols library for plumbing applications, as shown in Figure 17–26. A drafter simply selects the desired symbol from the tablet menu and then drags the symbol into position on the drawing. Parametric design plays an important role in applications such as multiple water closet stalls in a commercial building rest room. You specify the distance between walls or the width of each stall, the number of stalls, the stall length, and the door-swing direction; the computer then automatically draws the entire series of rest room stalls, as shown in Figure 17–27.

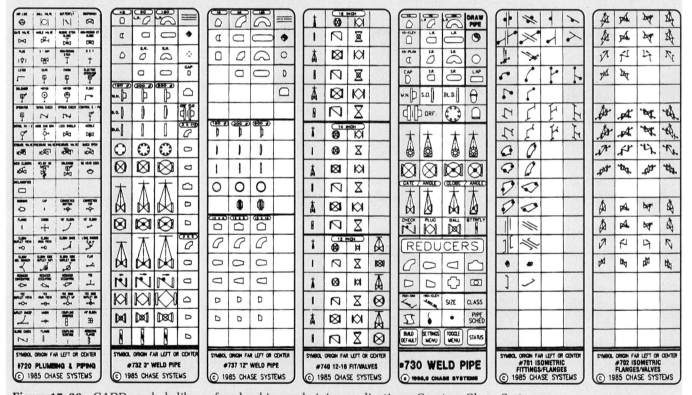

Figure 17–26 CADD symbols library for plumbing and piping applications. *Courtesy Chase Systems.*

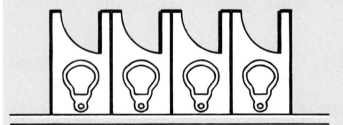

Figure 17–27 A series of stalls automatically drawn from given specifications using CADD.

17 Plumbing Plans Test

DIRECTIONS

Answer the questions with short, complete statements or drawings as needed on an 8½ × 11 sheet of notebook paper, as follows:

1. Letter your name, Chapter 17 Test, and the date at the top of the sheet.
2. Letter the question number and provide the answer. You do not need to write out the question.

Answers may be prepared on a word processor if appropriate with course guidelines.

QUESTIONS

Question 17–1 Floor-plan plumbing symbols are generally drawn at what scale?

Question 17–2 Identify four methods that can contribute to energy efficient plumbing.

Question 17–3 Define the following plumbing abbreviations:
a. CO
b. FD
c. VTR
d. WC
e. WH
f. SH

Question 17–4 What information is required on a plumbing schedule?

Question 17–5 Are plumbing drawings required by all contractors? Why?

Question 17–6 List at least one advantage and one disadvantage of solar hot water systems.

Question 17–7 Briefly explain how public sewer and private septic systems differ.

PROBLEMS

Use 8½ × 11 vellum and pencil or a CADD system to prepare the following drawings as specified.

Problem 17–1 Draw the following plumbing fixture symbols in plan view:
a. Water closet
b. 5'–0" long × 2'–6" wide bathtub
c. Shower 3'–0" × 3'–0"
d. Urinal

Problem 17–2 Draw the following plumbing piping symbols:
a. Hot water supply
b. Cold water supply
c. Soil waste pipe
d. Gate valve
e. 90° elbow
f. Tee

Problem 17–3 Make a single-line isometric drawing of a typical kitchen sink piping diagram.

Problem 17–4 Make a single-line isometric drawing of a typical drain, waste, and vent system.

Problem 17–5 Make a floor-plan plumbing drawing of a typical back-to-back bath arrangement.

Problem 17–6 Draw a partial single line water supply system (see Figure 17–6).

Problem 17–7 Draw a typical bathroom drainage and vent system. See Figure 17–10.

Problem 17–8 Continue the floor plan(s) that you started in Chapter 15 by adding any additional symbols and notes related to the plumbing, such as hose bibbs, washer, laundry treys, sinks, and water heater.

CHAPTER 18

Heating, Ventilating, and Air-Conditioning

▼ THE BOCA NATIONAL ENERGY CONSERVATION CODE

The Building Officials and Code Administrators (BOCA) International updates the National Energy Conservation Code every three years. The purpose of the code is to regulate the design and construction of the exterior envelope and selection of heating, ventilating, and air-conditioning (HVAC), service water heating, electrical distribution and illuminating systems and equipment required for effective use of energy in buildings for human occupancy. The *exterior envelope* is made up of elements of a building that enclose conditioned (heated and cooled) spaces through which thermal energy transfers to or from the exterior.

▼ CENTRAL FORCED-AIR SYSTEMS

One of the most common systems for heating and air-conditioning circulates the air from the living spaces through or around heating or cooling devices. A fan forces the air into sheet metal or plastic pipes called *ducts*, which connect to openings called *diffusers*, or *air supply registers*. Warm air or cold air passes through the ducts and registers to enter the rooms, heating or cooling them as needed.

Air then flows from the room through another opening into the return duct, or return air register. The return duct directs the air from the rooms over the heating or cooling device. If warm air is required, the return air is either passed over the surface of a combustion chamber (the part of a furnace where fuel is burned) or a heating coil. If cool air is required, the return air passes over the surface of a cooling coil. Finally, the conditioned air is picked up again by the fan and the air cycle is repeated. Figure 18–1 shows the air cycle in a forced-air system.

Heating Cycle

If the air cycle just described is used for heating, the heat is generated in a furnace. Furnaces for residential heating produce heat by burning fuel oil or natural gas, from electric heating coils, or through the use of heat pumps. If the heat comes from burning fuel oil or natural

gas, the combustion (burning) takes place inside a combustion chamber. The air to be heated does not enter the combustion chamber but absorbs heat from the outer

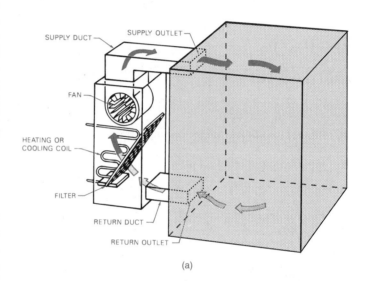

(a)

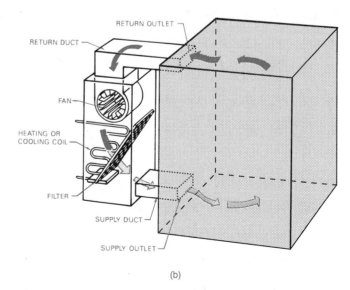

(b)

Figure 18–1   (a) Downdraft forced-air system heated air cycle. *From Huth*, Understanding Construction Drawings, *Delmar Publishers Inc.*; (b) updraft forced-air system heated air cycle.

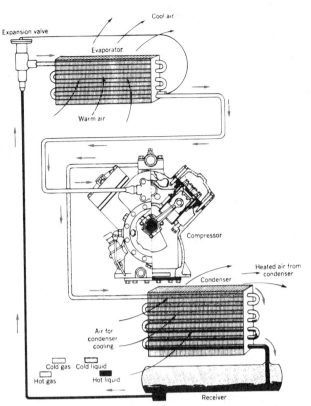

Figure 18–2 Schematic diagram of refrigeration cycle. From *Lang*, Principles of Air Conditioning, *3d ed., Delmar Publishers Inc.*

surface of the chamber. The gases given off by combustion are vented through a chimney. In an electric furnace, the air to be heated is passed directly over the heating coils. This type of furnace does not require a chimney.

Cooling Cycle

If the air from the room is to be cooled, it is passed over a cooling coil. The most common type of residential cooling system is based on two principles:

1. As liquid changes to vapor, it absorbs large amounts of heat.
2. The boiling point of a liquid can be changed by changing the pressure applied to the liquid. This is the same as saying that the temperature of a liquid can be raised by increasing its pressure and lowered by reducing its pressure.

Common refrigerants boil (change to a vapor) at very low temperatures, some as low as 21° F(6° C) below zero.

The principal parts of a refrigeration system are the cooling coil (evaporator), the compressor, the condenser, and the expansion valve. Figure 18–2 shows a diagram of the cooling cycle. The cooling cycle operates as follows. Warm air from the ducts is passed over the evaporator. As the cold liquid refrigerant moves through the evaporator coil, it picks up heat from the warm air. As the liquid picks up heat, it changes to a vapor. The heated refriger-

ant vapor is then drawn into the compressor, where it is put under high pressure. This causes the temperature of the vapor to rise even more.

Next, the high-temperature, high-pressure vapor passes to the condenser, where the heat is removed. This is done by blowing air over the coils of the condenser. As the condenser removes heat, the vapor changes to a liquid. It is still under high pressure, however. From the condenser, the refrigerant flows to the expansion valve. As the liquid refrigerant passes through the valve, the pressure is reduced, which lowers the temperature of the liquid still further so that it is ready to pick up more heat.

The cold low-pressure liquid then moves to the evaporator. The pressure in the evaporator is low enough to allow the refrigerant to boil again and absorb more heat from the air passing over the coil of the evaporator.

Forced-Air Heating Plans

Complete plans for the heating system may be needed when applying for a building permit or a mortgage, depending on the requirements of the local building jurisdiction or the lending agency. If a complete heating layout is required, it can be prepared by the architectural drafter, a heating contractor or a mechanical engineer.

When forced-air electric, gas, or oil heating systems are used, the warm-air outlets and return air locations may be shown, as in Figure 18–3. Notice the registers are normally placed in front of a window so warm air is circulated next to the coldest part of the room. As the warm air rises, a circulation action is created as the air goes through the complete heating cycle. Cold-air returns are often placed in the ceiling or floor of a central location.

A complete forced-air heating plan shows the size, location and number of Btu dispersed to the rooms from the warm-air supplies. Btu stands for *British thermal unit*, which is a measure of heat. The location and size of the cold-air return and the location, type, and output of the furnace are also shown.

The warm-air registers are sized in inches—for example, 4″ × 12″. The size of the duct is also given, as shown in Figure 18–4. The note 20/24 means a 20″ × 24″ register while a number 8 next to a duct means an 8″ diameter duct. This same system may be used as a central cooling system when cool air is forced from an air conditioner through the ducts and into the rooms. WA means warm air, and RA is return air. CFM is cubic feet per minute, the rate of air flow.

▼ HOT WATER SYSTEMS

In a hot water system, the water is heated in an oil- or gas-fired boiler and then circulated through pipes to radiators or convectors in the rooms. The boiler is supplied with water from the fresh water supply for the house. The

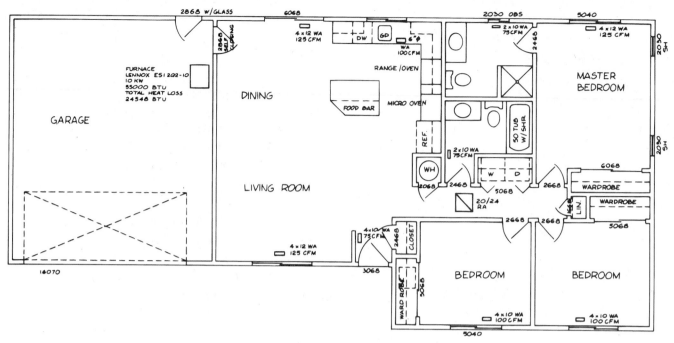

Figure 18–3 Simplified forced-air plan.

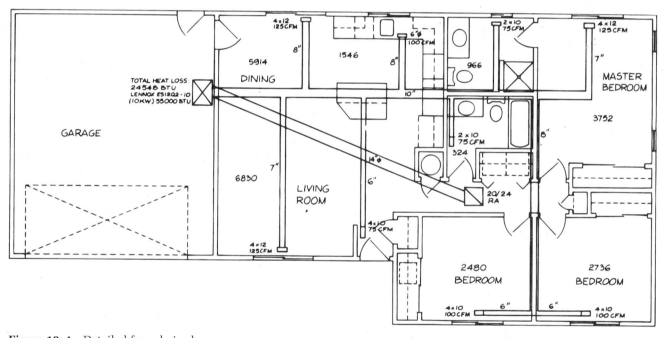

Figure 18–4 Detailed forced-air plan.

water is circulated around the combustion chamber, where it absorbs heat.

In one kind of system, one pipe leaves the boiler and runs through the rooms of the building and back to the boiler. In this one-pipe system, the heated water leaves the supply, is circulated through the outlet, and is returned to the same pipe, as shown in Figure 18–5. In a two-pipe system, two pipes run throughout the building. One pipe supplies heated water to all of the outlets. The other is a return pipe, which carries the water back to the boiler for reheating, as seen in Figure 18–6.

Hot water systems use a pump, called a *circulator*, to move the water through the system. The water is kept at a temperature between 150° F and 180° F. in the boiler. When heat is needed, the thermostat starts the circulator, which supplies hot water to the convectors in the rooms.

▼ HVAC SYMBOLS

There are over a hundred HVAC symbols that may be used in residential and commercial heating plans. Only a

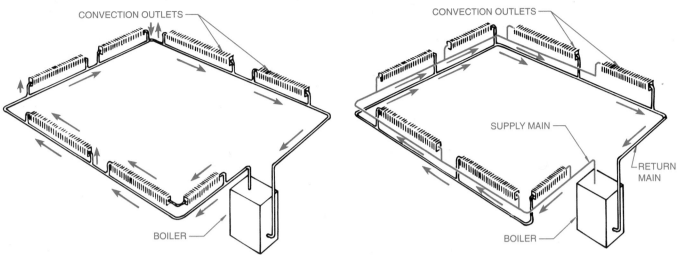

Figure 18–5   One-pipe hot-water system. *From Huth*, Understanding Construction Drawings, *Delmar Publishers Inc.*

Figure 18–6   Two-pipe hot-water system. *From Huth*, Understanding Construction Drawings, *Delmar Publishers Inc.*

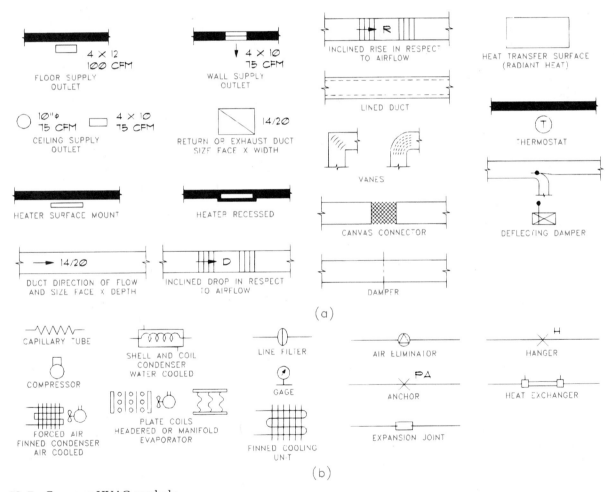

Figure 18–7   Common HVAC symbols.

few of the symbols are typically used in residential HVAC drawings. Figure 18–7 shows some common HVAC symbols. Sheet metal conduit template and heating and air-conditioning templates are time-saving devices that are used to help improve the quality of drafting for HVAC systems. Custom HVAC CADD programs are also available to help improve drafting productivity. CADD applications are discussed later.

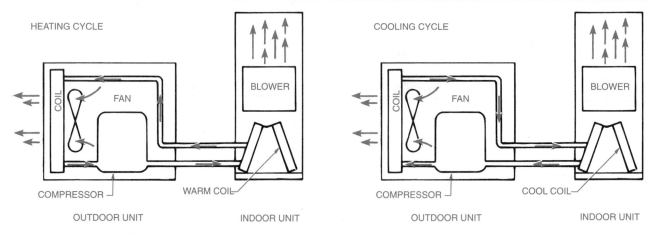

Figure 18–8   Heat pump heating and cooling cycle. *Courtesy Lennox Industries, Inc.*

▼ HEAT PUMP SYSTEMS

The *heat pump* is a forced-air central heating and cooling system. It operates using a compressor and a circulating refrigerant system. Heat is extracted from the outside air and pumped inside the structure. The heat pump supplies up to three times as much heat per year for the same amount of electrical consumption as a standard electric forced-air heating system. In comparison, this can result in a 30 to 50 percent annual energy savings. In the summer the cycle is reversed, and the unit operates as an air conditioner. In this mode, the heat is extracted from the inside air and pumped outside. On the cooling cycle the heat pump also acts as a dehumidifier. Figure 18–8 shows a graphic example of how a heat pump works.

Residential heat pumps vary in size from 2 to 5 tons. Some vendors carry two-stage heat pumps that alternate—between 3 and 5 tons, for example. During minimal demand, the more efficient 3-ton phase is used, while the 5-ton phase is operable during peak demand. Each ton of rating removes approximately 12,000 Btu per hour (Btuh) of heat.

Residential or commercial structures that, either due to size or design, cannot be uniformly heated or cooled from one compressor may require a split system with two or more compressors. The advantage to such an arrangement may be that two smaller units can more effectively control the needs of two zones within the structure.

Other features, such as an air cleaner, humidifier, or air freshener, can be added to the system. In general, the initial cost of the heat pump will be over twice as much as a conventional forced-air electric heat system. However, the advantages and long-range energy savings make it a significant option for heating. Keep in mind that if cooling is a requirement, the cost difference is less.

The total heat pump system uses an outside compressor, an inside blower to circulate air, a backup heating coil, and complete duct system. Heat pump systems move a large volume of air; therefore, the return air and supply ducts must be adequately sized. Figure 18–9 shows the relationship between the compressor and the balance of the system. The compressor should be placed in a location where some noise will not cause a problem. The compressor should be placed on a concrete slab about 36″ × 48″ × 6″ in size in a location that will allow adequate service access. Do not connect the slab to the structure to avoid transmitting vibration. Show or note the concrete slab on the foundation plan. Verify the vendor's specifications. In some areas of the country, heat pumps may not be as efficient as in other areas because of annual low or high temperatures. Verify the product efficiency with local vendors.

▼ ZONE CONTROL SYSTEMS

A zoned heating system requires one heater and one thermostat per room. No ductwork is required, and only the heaters in occupied rooms need be turned on.

One of the major differences between a zonal and a central system is flexibility. A zonal heating system allows the home occupant to determine how many rooms are heated, how much energy is used, and how much money is spent on heat. A zonal system allows the home to be heated to the family's needs; a central system requires

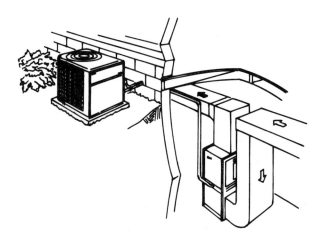

Figure 18–9   Heat pump compressor. *Courtesy Lennox Industries, Inc.*

using all the heat produced. If the air flow is restricted, the efficiency of the central system is reduced. There is also a 10 to 15 percent heat loss through ductwork in central systems.

Regardless of the square footage in a house, its occupants normally use less than 40 percent of the entire area on a regular basis. A zoned system is very adaptable to heating the 40 percent of the home that is occupied using automatic controls that allow for night setback, day setback, and non-heated areas. The homeowner can save as much as 60 percent on energy costs through controlled heating systems.

There are typically two types of zone heaters: baseboard and fan. Baseboard heaters have been the most popular type for zoned heating system for the past several decades. They are used in many different climates and under various operating conditions. There are no ducts, motors, or fans required. Baseboard units have an electric heating element, which causes a convection current as the air around the unit is heated. The heated air rises into the room and is replaced by cooler air that falls to the floor. Baseboard heaters should be placed on exterior walls under or next to windows or other openings. These units do project a few inches into the room at floor level. Some homeowners do not care for this obstruction in comparison to the floor duct vent of a central system or the recessed wall-mounted fan heater. Furniture arrangements should be a factor in locating any heating element.

Fan heaters are generally mounted in a wall recess. A resistance heater is used to generate the heat, and a fan circulates the heat into the room. These units should be placed to circulate the warmed air in each room adequately. Avoid placing the heaters on exterior walls, as the recessed unit reduces or eliminates the insulation in that area. The fan in these units does cause some noise.

Split systems are also possible using zonal heat in part of the home and central heat in the rest of the structure. This is also an option for additions to homes that have a central system. Zoned heat can be used effectively in an addition so the central system is not overloaded or required to be replaced.

▼ RADIANT HEAT

Radiant heating and cooling systems provide a comfortable environment by means of controlling surface temperatures and minimizing excessive air motion within the space. Warm ceiling panels are effective for winter heating because they warm the floor surfaces and glass surfaces by direct transfer of radiant energy. The surface temperature of well-constructed and properly insulated floors is 2° F to 3° F ($-16.4°$ C) above the ambient air temperature, and the inside surface temperature of glass is increased significantly. *Ambient temperature* is the temperature of the air surrounding a heating or cooling device. As a result of these heated surfaces, downdrafts are minimized to the point where no discomfort is felt.

Radiant heat systems generate operating cost savings of 20 to 50 percent annually compared to conventional convective systems. This savings is accomplished through lower thermostat settings. Savings are also due to the superior, cost-effective design inherent in radiant heating products.

It is generally accepted that 3 to 4 percent of the energy cost is saved for each degree the thermostat is lowered. Users of surface-mounted radiant panels take advantage of this fact in two ways. First, they are able to achieve comfort at a lower ambient air temperature, normally 60° F to 64° F (16.6° C) as compared with convection heating air temperatures of 68° F to 72° F (21° C). Second, they are able to practice comfortable day and night temperature setback of about 58° F in areas used frequently, 55° F (12.8° C) in areas used occasionally, and 50° F (10° C) in those areas seldom used.

Radiant heat may be achieved with oil- or gas-heated hot water piping in the floor or ceiling, to electric coils, wiring, or elements either in or above the ceiling gypsum board, and transferred to metal radiator panels generally mounted by means of a bracket about an inch below the ceiling surface.

Zone radiant heating panels are an alternative when space is limited or the location of built-in zone heaters may be a factor.

▼ THERMOSTATS

The *thermostat* is an automatic mechanism for controlling the amount of heating or cooling given by a central or zonal heating or cooling system. The thermostat floor-plan symbol is shown in Figure 18–10.

The location of the thermostat is an important consideration to the proper functioning of the system. For zoned heating or cooling units thermostats may be placed in each room, or a central thermostat panel that controls each room may be placed in a convenient location. For central heating and cooling systems there may be one or more thermostats depending on the layout of the system or the number of units required to service the structure. For example, a very large home or office building may have a split system that divides the structure into two or more zones, each individual zone with its own thermostat.

Several factors contribute to the effective placement of the thermostat for a central system. A good common location is near the center of the structure and close to a return air duct for a central forced-air system. The air entering the return air duct is usually temperate, thus causing little variation in temperature on the thermostat. A key to successful thermostat placement is to find a stable location where an average temperature reading can

Figure 18–10   Thermostat floor-plan symbol.

be achieved. There should be no drafts that would adversely affect temperature settings. The thermostat should not be placed in a location where sunlight or a heat register causes an unreliable reading. The same rationale suggests that a thermostat not be placed close to an exterior door where temperatures can change quickly. Thermostats should be placed on inside partitions rather than on outside walls where a false temperature reading could also be obtained. Avoid placing the thermostat near stairs or a similar traffic area where significant bouncing or shaking could cause the mechanism to alter the actual reading.

Energy consumption can be reduced by controlling thermostats in individual rooms. Central panels are available that make it easy to lower or raise temperatures in any room where each panel switch controls one remote thermostat.

Programmable microcomputer thermostats effectively help reduce the cost of heating or cooling. Some units automatically switch from heat to cool while minimizing temperature deviation from the setting under varying load conditions. These computers can be used to alter heating and cooling temperature settings automatically for different days of the week or different months of the year.

▼ HEAT RECOVERY AND VENTILATION
Sources of Pollutants

Air pollution in a structure is a principal reason for installing a heat recovery and ventilation system. A number of sources contribute to an unhealthy environment within a home or business:

○ Moisture in the form of relative humidity can cause structural damage as well as health problems, such as respiratory problems. The source of relative humidity is the atmosphere, steam from cooking and showers, and individuals, who can produce up to 1 gal. (3.8 liters) of water vapor per day.

○ Incomplete combustion from gas-fired appliances or wood-burning stoves and fireplaces can generate a variety of pollutants, including carbon monoxide, aldehydes, and soot.

○ Humans and pets can transmit bacterial and viral diseases through the air.

○ Tobacco smoke contributes chemical compounds to the air, which can affect both smokers and nonsmokers alike.

○ Formaldehyde, when present, is considered a factor in the cause of eye irritation, certain diseases, and respiratory problems. Formaldehyde is found in carpets, furniture, and the glue used in construction materials, such as plywood and particle board, as well as some insulation products.

○ Radon is a naturally occurring radioactive gas that breaks down into compounds that are carcinogenic (cancer causing) when large quantities are inhaled over a long period of time. Radon may be more apparent in a structure that contains a large amount of concrete or in certain areas of the country. Radon can be monitored scientifically at a nominal cost, and barriers can be built that help reduce concern about radon contamination.

○ Household products such as those available in aerosol spray cans and craft materials such as glues and paints can contribute a number of toxic pollutants.

Air-to-Air Heat Exchangers

Government energy agencies, architects, designers, and contractors around the country have been evaluating construction methods that are designed to reduce energy consumption. Some of the tests have produced superinsulated, vapor-barrier-lined, airtight structures. The result has been a dramatic reduction in heating and cooling costs; however, the air quality in these houses has been significantly reduced and may even be harmful to health. In essence, the structure does not breathe, and the stale air and pollutants have no place to go. A technology has emerged from this dilemma in the form of an air-to-air heat exchanger. In the past the air in a structure was exchanged due to leakage through walls, floors, ceilings, and around openings. Although this random leakage was no insurance that the building was properly ventilated, it did ensure a certain amount of heat loss. Now, with the concern of energy conservation, it is clear the internal air quality of a home or business cannot be left to chance.

An *air-to-air heat exchanger* is a heat recovery and ventilation device that pulls polluted, stale warm air from the living space and transfers the heat in that air to fresh, cold air being pulled into the house. Heat exchangers do not produce heat; they only exchange heat from one airstream to another. The heat transfer takes place in the core of the heat exchanger, which is designed to avoid mixing the two airstreams to ensure that indoor pollutants are expelled. Moisture in the stale air condenses in the core and is drained from the unit. Figure 18–11 shows the function and basic components of an air-to-air heat exchanger.

The recommended minimum effective air change rate is 0.5 air changes per hour (ACH). Codes in some areas of the country have established a rate of 0.7 ACH. The American Society of Heating, Refrigeration, and Air Conditioning Engineers, Inc. (ASHRAE) recommends ventilation levels based on the amount of air entering a room. The recommended amount of air entering most rooms is 10 cubic ft. per minute (CFM). The rate for kitchens is 100 CFM and for bathrooms 50 CFM. Mechanical exhaust devices vented to outside air should be added to kitchens and baths to maintain the recommended air exchange rate.

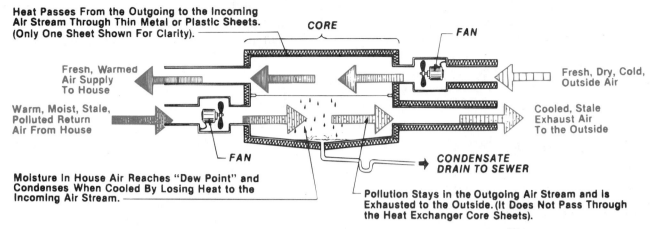

Figure 18-11   Components and function of air-to-air heat exchanger. *Courtesy U.S. Department of Energy.*

The minimum heat exchanger capacity needed for a structure can easily be determined. Assume a 0.5 ACH rate in a 1,500 sq. ft. single-level, energy-efficient house and follow these steps:

Step 1.   Determine the total floor area in sq. ft. Use the outside dimensions of the living area only.

$$30' \times 50' = 1,500 \text{ sq. ft.}$$

Step 2.   Determine the total volume within the house in cubic ft. by multiplying the total floor area by the ceiling height.

$$1,500 \text{ sq. ft.} \times 8' = 12,000 \text{ cu. ft.}$$

Step 3.   Determine the minimum exchanger capacity in CFM by first finding the capacity in cubic feet per hour (CFH). To do this, multiply the house volume in cubic feet (CUFT) by the ventilation rate in air changes per hour (ACH).

$$12,000 \text{ CU FT} \times 0.5 \text{ ACH} = 6,000 \text{ CFH}$$

where AC = air change, hr = hour.

Step 4.   Convert the CFH rate to cubic feet per minute (CFM) by dividing the CFH rate by 60 min.

$$6,000 \text{ CFH} \div 60 \text{ min} = 100 \text{ CFM}$$

There is a percentage of capacity loss due to mechanical resistance that should be considered by the system designer.

Most homeowners do not have the knowledge of potential air pollution and how internal air may be adequately controlled by ventilation. Therefore, the architect designing an energy-efficient home must plan the proper ventilation. The ducts, intake, and exhaust for such a system should be located on the advice of a ventilation consultant or a knowledgeable HVAC contractor.

▼ CENTRAL VACUUM SYSTEMS

Central vacuum systems have a number of advantages over portable vacuum cleaners.

1.   Some systems cost no more than a major appliance.
2.   Increased resale value of home.
3.   Removes dirt too heavy for most portable units.
4.   Exhausts dirt and dust out of the house or office.
5.   No motor unit or electric cords; however, there are often long hoses.
6.   No noise.
7.   Saves cleaning time.
8.   Vacuum pressure may be varied.

A well-designed system requires only a few inlets to cover the entire home or business, including exterior use. The hose plugs into a wall outlet and the vacuum is ready for use, as shown in Figure 18–12. The hose is generally

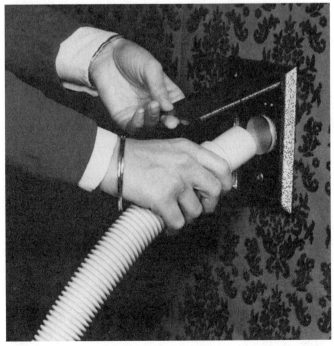

Figure 18–12   Vacuum wall outlet. *Courtesy Vacu-maid, Inc.*

lightweight and without a portable unit to carry, it is often easier to vacuum in hard-to-reach places. The central canister empties easily to remove the dust and debris from the house or business. The floor plan symbol for vacuum cleaner outlets is shown in Figure 18–13. The central unit may be shown in the garage or storage area as a circle that is labeled CENTRAL VACUUM SYSTEM.

▼ THERMAL CALCULATIONS FOR HEATING/COOLING
Development of Heat Loss Estimation Methodology

The basic methodology used to estimate residential heat loss today has been in use since the early 1900s. Historically, its primary use was to calculate the design heat load of houses in order to estimate the size of gas and oil heating systems required. Because home designers, homebuilders, and heating system installers did not want to receive complaints from cold homeowners, they commonly designed the heating system for worst-case weather conditions with a bias toward overestimating the design heat load to ensure the furnace would never be too small. Consequently, many gas and oil furnaces were oversized.

As early as 1915 an engineer for a gas utility began modifying the design heat load with a degree-day method to estimate annual energy consumption. Oil companies also began using this method to predict when to refill their customers' tanks. Data from this period, for houses with little or no insulation, indicated that the proportionality between annual heating energy requirements and average outside temperature began at 65° F (18° C) in residential buildings. This was the beginning of the 65° F base for degree days. Studies made by the American Gas Association up to 1932 and by the National District Heating Association in 1932 also indicated a 65° F base was appropriate for houses of that period.

Experience in the 1950s and early 1960s with electrically heated homes indicated the traditional degree-day procedure was overestimating annual heating loads. This was due primarily to tighter, better insulated houses with balance temperatures below 65° F, and to more appliances. These later studies led to use of the modified degree-day procedure incorporating a modifying factor, C. This C factor compensates for such things as higher insulation levels and more heat-producing appliances in the house. The high insulation levels found in homes built to current codes, and those that have been retrofitted with insulation, cause even lower balance tempera-

tures than those of houses built in the 1950s and 1960s. (This history has been adapted from *Standard Heat Loss Methodology of July, 1983*, Bonneville Power Administration.)

Terminology

Btu (British thermal unit): A unit of measure determined by the amount of heat required to raise one pound of water one degree Fahrenheit. Heat loss is calculated in Btus per hour (Btuh).

Compass point: The relationship of the structure to compass orientation. When referring to cooling calculations it is important to evaluate the amount of glass in each wall as related to compass orientation. This is due to the differences effected by solar gain.

Duct loss: Heat loss through duct work in an unheated space, which has an effect on total heat loss. Insulating those ducts helps increase efficiency.

Grains: The amount of moisture in the air (grains of moisture in one cubic foot). A grain is a unit of weight. Air at different temperatures and humidities holds different amounts of moisture. The effects of this moisture content in the southern and midwestern states becomes more of a factor than in other parts of the country.

Heat transfer multiplier: The amount of heat that flows through 1 sq. ft. of the building surface. The product depends on the type of surface and whether it is applied to heating or cooling.

Indoor temperature: The indoor design temperature is generally 60 to 70° F.

Indoor wet bulb: Relates to the use of the wet-bulb thermometer inside. A wet-bulb thermometer is one in which the bulb is kept moistened and is used to determine humidity level.

Infiltration: The inward flow of air through a porous wall or crack. In a loosely constructed home, infiltration substantially increases heat loss. Infiltration around windows and doors is calculated in CFM per linear foot of crack. Window infiltration greater than 0.5 CFM per linear foot of crack is excessive.

Internal heat gain: Heat gain associated with factors such as heat transmitted from appliances, lights, other equipment, and occupants.

Latent load: The effects of moisture entering the structure from the outside by humidity infiltration or from the inside produced by people and plants, and daily activities such as cooking, showers, and laundry.

Mechanical ventilation: Amount of heat loss through mechanical ventilators such as range hood fans, or bathroom exhaust fans.

Outdoor temperature: Related to average winter and summer temperatures for a local area. If the outdoor winter design temperature is 20°, this means that the temperature during the winter is 20° or higher 97½ percent of the

Figure 18–13 Vacuum outlet floor-plan symbol.

time. If the outdoor summer design temperature in an area is 100° this means that the temperature during the summer is 100° or less 97½ percent of the time. Figures for each area of the country have been established by ASHRAE. Verify the outdoor temperature with your local building department or heating contractor.

Outdoor wet bulb: Relates to the use of a wet-bulb thermometer outside. A wet-bulb thermometer is one in which the bulb is kept moistened and is used to determine humidity level.

R factor: Resistance to heat flow. The more resistance to heat flow, the higher the R value. For example, 3½″ of mineral wool insulation has a value of R-11; 6″ of the same insulation has a value of R-19. R is equal to the reciprocal of the U factor, 1/U = R.

Sensible load calculations: Load calculations associated with temperature change that occurs when a structure loses or gains heat.

Temperature difference: The indoor temperature less the outdoor temperature.

U factor: The coefficient of heat transfer expressed to Btuh sq. ft./°F of surface area.

Steps in Filling Out the Residential Heating Data Sheet

Figure 18–14 is a completed Residential Heating Data Sheet. The two-page data sheet is divided into several categories with calculations resulting in total heat loss for the structure shown in Figure 18–15. The large numbers by each category refer to the following steps used in completing the form. Notice that the calculations are rounded off to the nearest whole unit.

Step 1. *Outdoor temperature.* Use the recommended outdoor design temperature for your area. The area selected for this problem is Dallas, Texas, with an outdoor design temperature of 22° F, which has been rounded off to 20° F to make calculations simple to understand. The proper calculations would interpolate the tables for a 22° F outdoor design temperature.

Step 2. *Indoor temperature:* 70° F.

Step 3. *Temperature difference:*

$$70° - 20° = 50° F.$$

Step 4. *Movable glass windows:* Select double glass; find area of each window (frame length × width); then combine for total: 114 sq. ft.

Step 5. *Btuh heat loss:* Using 50° design temperature difference, find 46 approximate heat transfer multiplier:

$$46 \times 114 = 5244 \text{ Btuh.}$$

Step 6. *Sliding glass doors:* Select double glass; find total area of 34 sq. ft.

Step 7. *Btuh heat loss:*

$$34 \times 48 = 1632 \text{ Btuh}$$

Step 8. *Doors:* Weatherstripped solid wood; 2 doors at 39 sq. ft.

Step 9. *Btuh heat loss:*

$$39 \times 30 = 1170 \text{ Btuh}$$

Step 10. *Walls:* Excluding garage; perimeter in running feet 132 × ceiling height 8′ = Gross wall area 1056 sq. ft. Subtract window and door areas − 187 sq. ft. = net wall area 869 sq. ft.

Step 11. *Frame wall:* No masonry wall above or below grade. When there is masonry wall, subtract the square footage of masonry wall from total wall for the net frame wall. Fill in net amount on approximate insulation value; 869 sq. ft. frame wall, R-13 (given).

Step 12. *Btuh heat loss:*

$$869 \times 3.5 = 3042 \text{ Btuh.}$$

Steps 13–14. No masonry above grade in this structure.

Steps 15–16. No masonry below grade in this structure.

Step 17. Heat loss subtotal: Add together items 5, 7, 9, and 12; 11,088 Btuh. Transfer the amount to the top of page 2 of the form.

Step 18. Ceilings: R-30 insulation (given), same square footage as floor plan 976 sq. ft. (given).

Step 19. *Btuh heat loss:*

$$976 \times 1.6 = 1562 \text{ Btuh.}$$

Step 20. *Floor over an unconditioned space:* R-19 insulation (given), 97 sq. ft.

Step 21. *Btuh heat loss:*

$$976 \times 2.6 = 2538 \text{ Btuh.}$$

Steps 22–23. *Basement floor:* Does not apply to this house.

Steps 24–25. *Concrete slab without perimeter system:* Does not apply to this house.

Steps 26–27. *Concrete slab with perimeter system:* Does not apply to this house.

Step 28. *Infiltration:* 976 sq. ft. floor × 8′ ceiling height = 7808 cu. ft.

$$0.40 \times 7808 \text{ cu. ft.} \div 60 = 52 \text{ CFM}$$

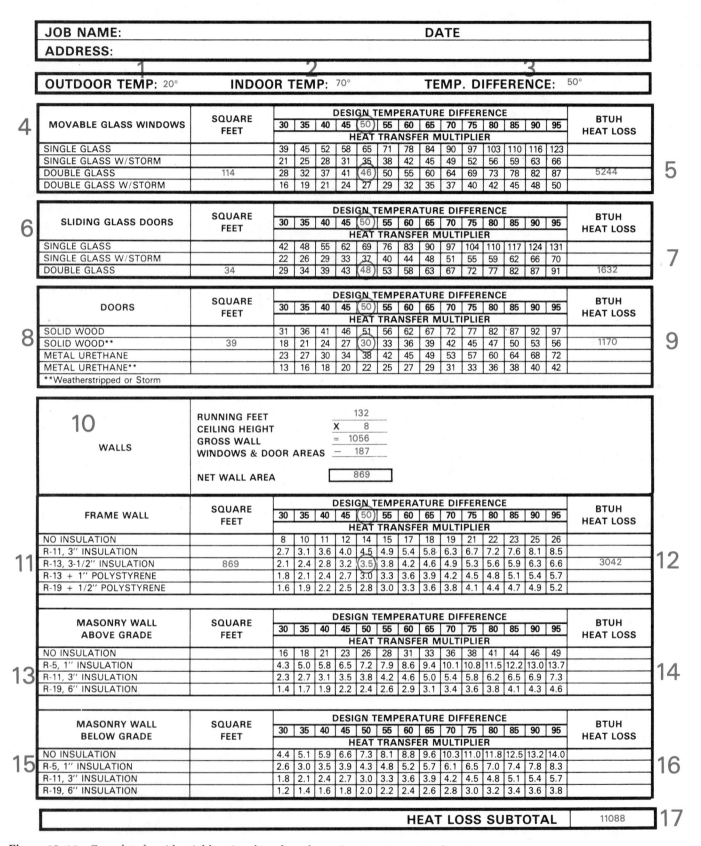

JOB NAME:		DATE	
ADDRESS:			

1 OUTDOOR TEMP: 20°   **2** INDOOR TEMP: 70°   **3** TEMP. DIFFERENCE: 50°

4

MOVABLE GLASS WINDOWS	SQUARE FEET	DESIGN TEMPERATURE DIFFERENCE														BTUH HEAT LOSS
		30	35	40	45	(50)	55	60	65	70	75	80	85	90	95	
		HEAT TRANSFER MULTIPLIER														
SINGLE GLASS		39	45	52	58	65	71	78	84	90	97	103	110	116	123	
SINGLE GLASS W/STORM		21	25	28	31	35	38	42	45	49	52	56	59	63	66	
DOUBLE GLASS	114	28	32	37	41	(46)	50	55	60	64	69	73	78	82	87	5244
DOUBLE GLASS W/STORM		16	19	21	24	27	29	32	35	37	40	42	45	48	50	

5

6

SLIDING GLASS DOORS	SQUARE FEET	DESIGN TEMPERATURE DIFFERENCE														BTUH HEAT LOSS
		30	35	40	45	(50)	55	60	65	70	75	80	85	90	95	
		HEAT TRANSFER MULTIPLIER														
SINGLE GLASS		42	48	55	62	69	76	83	90	97	104	110	117	124	131	
SINGLE GLASS W/STORM		22	26	29	33	37	40	44	48	51	55	59	62	66	70	
DOUBLE GLASS	34	29	34	39	43	(48)	53	58	63	67	72	77	82	87	91	1632

7

8

DOORS	SQUARE FEET	DESIGN TEMPERATURE DIFFERENCE														BTUH HEAT LOSS
		30	35	40	45	(50)	55	60	65	70	75	80	85	90	95	
		HEAT TRANSFER MULTIPLIER														
SOLID WOOD		31	36	41	46	51	56	62	67	72	77	82	87	92	97	
SOLID WOOD**	39	18	21	24	27	(30)	33	36	39	42	45	47	50	53	56	1170
METAL URETHANE		23	27	30	34	38	42	45	49	53	57	60	64	68	72	
METAL URETHANE**		13	16	18	20	22	25	27	29	31	33	36	38	40	42	
**Weatherstripped or Storm																

9

10 WALLS

RUNNING FEET	132
CEILING HEIGHT	X 8
GROSS WALL	= 1056
WINDOWS & DOOR AREAS	− 187
NET WALL AREA	869

11

FRAME WALL	SQUARE FEET	DESIGN TEMPERATURE DIFFERENCE														BTUH HEAT LOSS
		30	35	40	45	(50)	55	60	65	70	75	80	85	90	95	
		HEAT TRANSFER MULTIPLIER														
NO INSULATION		8	10	11	12	14	15	17	18	19	21	22	23	25	26	
R-11, 3" INSULATION		2.7	3.1	3.6	4.0	4.5	4.9	5.4	5.8	6.3	6.7	7.2	7.6	8.1	8.5	
R-13, 3-1/2" INSULATION	869	2.1	2.4	2.8	3.2	(3.5)	3.8	4.2	4.6	4.9	5.3	5.6	5.9	6.3	6.6	3042
R-13 + 1" POLYSTYRENE		1.8	2.1	2.4	2.7	3.0	3.3	3.6	3.9	4.2	4.5	4.8	5.1	5.4	5.7	
R-19 + 1/2" POLYSTYRENE		1.6	1.9	2.2	2.5	2.8	3.0	3.3	3.6	3.8	4.1	4.4	4.7	4.9	5.2	

12

13

MASONRY WALL ABOVE GRADE	SQUARE FEET	DESIGN TEMPERATURE DIFFERENCE														BTUH HEAT LOSS
		30	35	40	45	50	55	60	65	70	75	80	85	90	95	
		HEAT TRANSFER MULTIPLIER														
NO INSULATION		16	18	21	23	26	28	31	33	36	38	41	44	46	49	
R-5, 1" INSULATION		4.3	5.0	5.8	6.5	7.2	7.9	8.6	9.4	10.1	10.8	11.5	12.2	13.0	13.7	
R-11, 3" INSULATION		2.3	2.7	3.1	3.5	3.8	4.2	4.6	5.0	5.4	5.8	6.2	6.5	6.9	7.3	
R-19, 6" INSULATION		1.4	1.7	1.9	2.2	2.4	2.6	2.9	3.1	3.4	3.6	3.8	4.1	4.3	4.6	

14

15

MASONRY WALL BELOW GRADE	SQUARE FEET	DESIGN TEMPERATURE DIFFERENCE														BTUH HEAT LOSS
		30	35	40	45	50	55	60	65	70	75	80	85	90	95	
		HEAT TRANSFER MULTIPLIER														
NO INSULATION		4.4	5.1	5.9	6.6	7.3	8.1	8.8	9.6	10.3	11.0	11.8	12.5	13.2	14.0	
R-5, 1" INSULATION		2.6	3.0	3.5	3.9	4.3	4.8	5.2	5.7	6.1	6.5	7.0	7.4	7.8	8.3	
R-11, 3" INSULATION		1.8	2.1	2.4	2.7	3.0	3.3	3.6	3.9	4.2	4.5	4.8	5.1	5.4	5.7	
R-19, 6" INSULATION		1.2	1.4	1.6	1.8	2.0	2.2	2.4	2.6	2.8	3.0	3.2	3.4	3.6	3.8	

16

HEAT LOSS SUBTOTAL	11088

17

Figure 18–14 Completed residential heating data sheet form. *Courtesy Lennox Industries, Inc.*

| | | Heat Loss Subtotal from Page 1 | 11088 | 17 |

18 **19**

CEILING	SQUARE FEET	DESIGN TEMPERATURE DIFFERENCE													BTUH HEAT LOSS
		30	35	40	45	50	55	60	65	70	75	80	85	90	95	
		HEAT TRANSFER MULTIPLIER														
NO INSULATION		18	21	24	27	30	33	36	39	42	45	48	51	54	57	
R-11, 3″ INSULATION		2.6	3.1	3.5	4.0	4.4	4.8	5.3	5.7	6.2	6.6	7.0	7.5	7.9	8.4	
R-19, 6″ INSULATION		1.6	1.9	2.1	2.4	2.6	2.9	3.2	3.4	3.7	4.0	4.2	4.5	4.8	5.0	
R-30, 10″ INSULATION	976	1.0	1.2	1.3	1.5	1.6	1.8	2.0	2.1	2.3	2.5	2.6	2.8	3.0	3.1	1562
R-38, 12″ INSULATION		0.8	0.9	1.0	1.2	1.3	1.4	1.6	1.7	1.8	2.0	2.1	2.2	2.3	2.5	

20 **21**

FLOOR OVER AN UNCONDITIONED SPACE	SQUARE FEET	DESIGN TEMPERATURE DIFFERENCE													BTUH HEAT LOSS
		30	35	40	45	50	55	60	65	70	75	80	85	90	95	
		HEAT TRANSFER MULTIPLIER														
NO INSULATION		10	11	13	14	16	17	19	21	22	24	25	27	28	30	
R-11, 3″ INSULATION		2.4	2.8	3.2	3.6	4.0	4.4	4.8	5.2	5.6	6.0	6.4	6.8	7.2	7.6	
R-19, 6″INSULATION	976	1.6	1.8	2.1	2.3	2.6	2.9	3.1	3.4	3.6	3.9	4.2	4.4	4.7	4.9	2538
R-30, 10″INSULATION		1.1	1.3	1.5	1.7	1.8	2.0	2.2	2.4	2.6	2.8	3.0	3.1	3.3	3.5	

22 **23**

BASEMENT FLOOR	SQUARE FEET	DESIGN TEMPERATURE DIFFERENCE													BTUH HEAT LOSS
		30	35	40	45	50	55	60	65	70	75	80	85	90	95	
		HEAT TRANSFER MULTIPLIER														
BASEMENT FLOOR		0.8	1.0	1.1	1.3	1.4	1.5	1.7	1.8	2.0	2.1	2.2	2.4	2.5	2.7	

24 **25**

CONCRETE SLAB WITHOUT PERIMETER SYSTEM	LINEAR FOOT	DESIGN TEMPERATURE DIFFERENCE													BTUH HEAT LOSS
		30	35	40	45	50	55	60	65	70	75	80	85	90	95	
		HEAT TRANSFER MULTIPLIER														
NO EDGE INSULATION		25	29	33	37	41	45	49	53	57	61	65	69	73	77	
1″ EDGE INSULATION		13	15	17	19	21	23	25	27	29	31	33	35	37	39	
2″ INSULATION		6.3	7.4	8.4	9.4	10.5	11.5	12.6	13.6	14.7	15.8	16.8	17.8	18.9	20.0	

26 **27**

CONCRETE SLAB WITH PERIMETER SYSTEM	LINEAR FOOT	DESIGN TEMPERATURE DIFFERENCE													BTUH HEAT LOSS
		30	35	40	45	50	55	60	65	70	75	80	85	90	95	
		HEAT TRANSFER MULTIPLIER														
NO EDGE INSULATION		57	67	76	86	95	105	114	124	133	143	152	162	171	181	
1″ EDGE INSULATION		34	40	46	52	57	63	69	74	80	86	91	97	103	109	
2″ EDGE INSULATION		28	33	37	42	47	51	56	61	65	70	75	79	84	89	

An additional infiltration load is calculated **only if** the home is loosely constructed or when window infiltration is greater than .5 CFM per linear foot of crack.

28 INFILTRATION/ VENTILATION

_976_ FLOOR SQ. FT. x _8_ CEILING HEIGHT = _7808_ CUBIC FT

0.40 x _7808_ CUBIC FT ÷ 60 = _52_ CFM

MECHANICAL VENTILATION CFM = FRESH AIR INTAKE

	CFM	DESIGN TEMPERATURE DIFFERENCE													BTUH HEAT LOSS
		30	35	40	45	50	55	60	65	70	75	80	85	90	95	
		HEAT TRANSFER MULTIPLIER														
INFILTRATION	52	33	39	44	50	55	61	66	72	77	83	88	94	99	105	2860
MECHANICAL VENTILATION	52	33	39	44	50	55	61	66	72	77	83	88	94	99	105	2860

29

| HEAT LOSS SUBTOTAL | 20908 | 30 |

31

DUCT LOSS	BTUH HEAT LOSS
R-4, 1″ Flexible Blanket Insulation: ADD 15% (.15)	3138
R-7, 2″ Flexible Blanket Insulation: ADD 10% (.10)	

31

| TOTAL HEAT LOSS | 24046 | 32 |

NOTE: All Heat Transfer Multipliers from ACCA Manual "J" Sixth Edition.

HL-841-L7   002344   Litho U.S.A.

Figure 18–14   Continued.

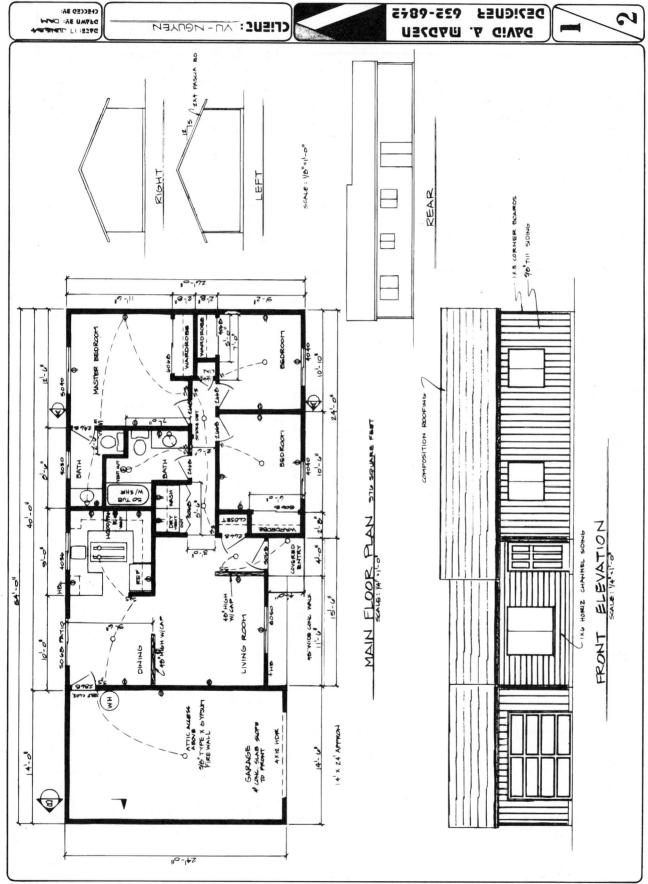

Figure 18–15 Sample floor plan for heat loss and heat gain calculations. *Courtesy Madsen Designs.*

Infiltration = mechanical ventilation CFM = Fresh air intake.

Step 29. *Btuh heat loss:*

$$52 \times 55 = 2860 \text{ Btuh.}$$

Step 30. *Heat loss subtotal:* Add items 17, 19, 21, and 29; 20,908 Btuh.

Step 31. *Duct loss:* R-4 insulation (given); add 15 percent to item 30:

$$.15 \times 20,923 = 3138.$$

When no ducts are used, exclude this item.

Step 32. *Total heat loss:* Add items 30 and 31; 24,046 Btuh.

Steps in Filling Out the Residential Cooling Data Sheet

Figure 18–16 is a completed residential cooling data sheet. The two-page data sheet is divided into several categories, with calculations resulting in total sensible and latent heat gain for the structure, shown in Figure 18–15, used for heat loss calculations. The large numbers by each category refer to the following steps used in completing the form.

Step 1. *Outdoor temperature:* 100° F.

Step 2. *Indoor temperature:* 70° F.

Step 3. *Temperature difference.* 100° − 70° = 30° F.

Steps 4–13. Glass, no shade, double glazed.

Step 4. *North glass:* Including sliding glass door; 81 sq. ft.

Step 5. *Btuh heat gain:*

$$81 \times 28 = 2268 \text{ Btuh.}$$

Steps 6–7. *NE and NW glass:* None; house faces N, E, W, S.

Steps 8–9. *East and West glass:* None in this plan.

Steps 10–11. *SE and SW glass:* None; house faces N, E, W, S.

Step 12. South glass: 67 sq. ft.

Step 13. *Btuh heat loss:*

$$67 \times 43 = 2881 \text{ Btuh.}$$

Steps 14–23. Applies to glass with inside shade. This house is sized without the consideration of inside shade. If items 14–23 are used, omit items 4–13.

Step 24. *Doors:* Solid wood weatherstripped, 2 doors; 39 sq. ft.

Step 25. *Btuh heat gain:*

$$39 \times 9.6 = 374 \text{ Btuh.}$$

Step 26. Walls (excluding garage walls):

$$132' \times 8' = 1056 \text{ sq. ft.} - 187 \text{ sq. ft.} = 869 \text{ sq. ft. net wall area.}$$

Step 27. *Frame wall:* 869 sq. ft.

Step 28. *Btuh heat gain:*

$$869 \times 2.5 = 2173 \text{ Btuh.}$$

Steps 29–30. *Masonry wall above grade:* None in this house.

Step 31. *Sensible heat gain subtotal:* Add together items 5, 13, 25, 28. 7696 Btuh heat gain. Transfer the amount to the top of page 2 of the form.

Step 32. *Ceiling:* R-30 insulation; 976 sq. ft.

Step 33. *Btuh heat gain:*

$$976 \times 1.7 = 1659 \text{ Btuh.}$$

Step 34. *Floor over unconditioned space:* R-19 insulation; 976 sq. ft.

Step 35. *Btuh heat gain:*

$$976 \times 1.2 = 1171 \text{ Btuh.}$$

Step 36. *Infiltration/ventilation:*

$$976 \text{ sq. ft.} \times 8' = 7808 \text{ cu. ft.}$$
$$0.40 \times 7808 \text{ cu. ft.} \div 60 = 52 \text{ CFM.}$$

Step 37. *Btuh heat gain:*

$$52 \times 32 = 1664 \text{ Btuh.}$$

Step 38. *Internal heat gain:* Number of people (assume 4 for this house) 4 × 300 Btuh per person = 1200 Btuh. Kitchen allowance given 1200 Btuh.

Step 39. *Sensible heat gain subtotal:* Add together items 31, 33, 35, 37, and 38; 15,054 Btuh.

Step 40. *Duct gain:*

$$15,054 \times .15 = 2258 \text{ Btuh.}$$

Step 41. *Total sensible heat gain:* Add together items 39 and 40; 17,312 Btuh.

Steps 42–45. Latent load calculations.

RESIDENTIAL COOLING DATA SHEET

JOB NAME:	DATE
ADDRESS:	

OUTDOOR TEMP: 100°   **INDOOR TEMP:** 70°   **TEMP DIFFERENCE:** 30°

SENSIBLE LOAD CALCULATIONS

GLASS NO SHADE		SINGLE						DOUBLE						TRIPLE						BTUH HEAT GAIN
COMPASS POINT	GLASS AREA SQ. FEET	DESIGN TEMPERATURE DIFFERENCE																		BTUH HEAT GAIN
		10	15	20	25	30	35	10	15	20	25	30	35	10	15	20	25	30	35	
		HEAT TRANSFER MULTIPLIER																		
N	81	25	29	33	37	41	45	20	22	24	26	28	30	15	16	18	19	20	21	2268
NE & NW		55	60	65	70	75	80	50	52	54	56	58	60	37	38	40	41	42	44	
E & W		80	85	90	95	100	105	70	72	74	76	78	80	55	56	58	59	60	62	
SE & SW		70	74	78	82	86	90	60	62	64	66	68	70	47	49	51	52	53	54	
S	67	40	44	48	52	56	60	35	37	39	41	43	45	26	27	29	31	32	33	2881

GLASS INSIDE SHADE		SINGLE						DOUBLE						TRIPLE						BTUH HEAT GAIN
COMPASS POINT	GLASS AREA SQ. FEET	DESIGN TEMPERATURE DIFFERENCE																		BTUH HEAT GAIN
		10	15	20	25	30	35	10	15	20	25	30	35	10	15	20	25	30	35	
		HEAT TRANSFER MULTIPLIER																		
N		15	19	23	27	31	35	15	17	19	21	23	25	10	12	14	16	17	19	
NE & NW		35	39	43	47	51	55	30	32	34	36	38	40	22	24	26	28	30	31	
E & W		50	54	58	62	66	70	45	47	49	51	53	55	35	36	38	40	42	44	
SE & SW		40	44	48	52	56	60	35	37	39	41	43	45	29	30	32	34	36	38	
S		25	29	33	37	41	45	20	22	24	26	28	30	16	18	20	22	24	26	

DOORS	SQUARE FEET	DESIGN TEMPERATURE DIFFERENCE						BTUH HEAT GAIN
		10	15	20	25	30	35	
		HEAT TRANSFER MULTIPLIER						
SOLID WOOD		6.3	8.6	10.9	13.2	14.4	15.5	
SOLID WOOD **	39	4.2	5.7	7.3	8.8	9.6	10.4	374
METAL URETHANE		2.6	3.5	4.5	5.4	5.9	6.4	
METAL URETHANE **		2.2	3.0	3.8	4.6	5.0	5.4	

** Weatherstripped or Storm

26 WALLS

RUNNING FEET	132
CEILING HEIGHT	X 8
GROSS WALL	= 1056
WINDOWS & DOOR AREAS	− 187
NET WALL AREA	869

FRAME WALL	SQUARE FEET	DESIGN TEMPERATURE DIFFERENCE						BTUH HEAT GAIN
		10	15	20	25	30	35	
		HEAT TRANSFER MULTIPLIER						
NO INSULATION		3.7	5.0	6.4	7.8	8.5	9.1	
R-11, 3" INSULATION		1.2	1.7	2.1	2.6	2.8	3.0	
R-13, 3-1/2" INSULATION	869	1.1	1.5	1.9	2.3	2.5	2.7	2173
R-13 + 1" POLYSTYRENE		0.8	1.1	1.4	1.7	1.8	2.0	
R-19 + 1/2" POLYSTYRENE		0.7	1.0	1.3	1.6	1.7	1.8	

MASONRY WALL ABOVE GRADE	SQUARE FEET	DESIGN TEMPERATURE DIFFERENCE						BTUH HEAT GAIN
		10	15	20	25	30	35	
		HEAT TRANSFER MULTIPLIER						
NO INSULATION		3.2	5.8	8.3	10.9	12.2	13.4	
R-5, 1" INSULATION		0.9	1.6	2.3	3.1	3.5	3.8	
R-11, 3" INSULATION		0.5	0.9	1.3	1.6	1.8	2.0	
R-19, 6" INSULATION		0.3	0.5	0.8	1.0	1.2	1.3	

SENSIBLE HEAT GAIN SUBTOTAL	7696

Figure 18–16 Completed residential cooling data sheet form. *Courtesy Lennox Industries, Inc.*

| | | Sensible Heat Gain Subtotal from Page 1 | | 7696 | **31** |

CEILING	**SQUARE FEET**	**DESIGN TEMPERATURE DIFFERENCE**						**BTUH HEAT GAIN**
		10	15	20	25	30	35	
		HEAT TRANSFER MULTIPLIER						
No Insulation		14.9	17.0	19.2	21.4	22.5	23.6	
R-11, 3" Insulation		2.8	3.2	3.7	4.1	4.3	4.5	
R-19, 6" Insulation		1.8	2.1	2.3	2.6	2.8	2.9	
R-30, 10" Insulation	976	1.1	1.3	1.5	1.6	(1.7)	1.8	1659
R-38, 12" Insulation		0.9	1.0	1.1	1.3	1.3	1.4	

32 **33**

FLOOR OVER UNCONDITIONED SPACE	**SQUARE FEET**	**DESIGN TEMPERATURE DIFFERENCE**						**BTUH HEAT GAIN**
		10	15	20	25	30	35	
		HEAT TRANSFER MULTIPLIER						
No Insulation		1.9	3.9	5.8	7.7	8.7	9.6	
CARPET FLOOR-NO INSULATION		1.3	2.5	3.8	5.1	5.7	6.3	
R-11, 3" INSULATION		0.4	0.8	1.3	1.7	1.9	2.1	
R-19, 6" INSULATION	976	0.3	0.5	0.8	1.1	(1.2)	1.3	1171
R-30 10" INSULATION		0.2	0.4	0.6	0.7	0.8	0.9	

34 **35**

36	**INFILTRATION/ VENTILATION**	_976_ FLOOR SQ. FT. x _8_ CEILING HEIGHT = _7808_ CUBIC FT
		0.40 x _7808_ CUBIC FT ÷ 60 = _52_ CFM
		MECHANICAL VENTILATION CFM — FRESH AIR INTAKE

	CFM	**DESIGN TEMPERATURE DIFFERENCE**						**BTUH HEAT GAIN**
		10	15	20	25	30	35	
		HEAT TRANSFER MULTIPLIER						
INFILTRATION	52	11.0	16.5	22.0	27.0	(32.0)	38.0	1664
MECHANICAL VENTILATION	52	11.0	16.5	22.0	27.0	(32.0)	38.0	1664

37

INTERNAL HEAT GAIN		**BTUH HEAT GAIN**
Number of People _4_ × 300		1200
Kitchen Allowance		1,200

38 **38**

| **SENSIBLE HEAT GAIN SUBTOTAL** | 15054 |

39

DUCT GAIN		**BTUH HEAT GAIN**
R-4, 1" Flexible Blanket Insulation: ADD 15% (.15)		2258
R-7, 2" Flexible Blanket Insulation: ADD 10% (.10)		

40 **40**

| **TOTAL SENSIBLE HEAT GAIN** | 17312 |

41

LATENT LOAD CALCULATIONS

Conditions	**Outdoor Wet Bulb**	**Indoor Wet Bulb**	**Grains**
Wet	80	62.5	50
Medium	75	62.5	(35)
Medium Dry	70	62.5	20
Dry	65	62.5	0

42

Based on 75°F Indoor Dry Bulb at 50% RH.

| **LATENT LOAD-INFILTRATION** | |
| 0.68 × _35_ Grains × _52_ Infiltration CFM | 1238 |

43 **43**

| **LATENT LOAD-VENTILATION** | |
| 0.68 × _35_ Grains × _52_ Ventilation CFM | 1238 |

44 **44**

| **LATENT LOAD-PEOPLE** | |
| Number of People _4_ × 230 | 920 |

45 **45**

| **TOTAL LATENT HEAT GAIN** | 3396 |

46

| **TOTAL SENSIBLE AND LATENT HEAT GAIN** | 20716 |

47

NOTE: All Heat Transfer Multipliers from ACCA Manual "J" Sixth Edition and for a medium outdoor daily temperature range.

241-17 002345

Litho U.S.A.

Figure 18–16 Continued.

Step 42. Determine the local relative humidity conditions, either wet, medium, medium dry, or dry. Our selected location is medium, which is 35 grains.

Step 43. *Latent load infiltration:*

0.68 × 35 grains = 52 CFM (from item 36) = 1238 Btuh.

Step 44. *Latent load ventilation:*

0.68 × 35 grains × 52 CFM = 1238 Btuh.

Step 45. *Latent load people:*

4 people × 230 Btuh = 920 Btuh.

Step 46. *Total latent heat gain:* Add together items 43, 44, and 45; 3396 Btuh.

Step 47. *Total sensible and latent heat gain:* Add together items 41 and 46; 20,708 Btuh.

▼ HVAC DRAWINGS

Drawings for the HVAC system show the size and location of all equipment, duct work, and components with accurate symbols, specifications, notes, and schedules that form the basis of contract requirements for construction. *Specifications* are documents that accompany the drawings and contain all pertinent written information related to the HVAC system.

HVAC drawings for residential structures may or may not be necessary, depending on the requirements of the local building jurisdiction or the lending agency. Drawings may be prepared by the architect, the architectural drafter, or the heating contractor when a complete HVAC layout is necessary.

For commercial structures the HVAC plan may be prepared by an HVAC engineer as a consultant for the architect. The consulting engineer is responsible for the HVAC design and installation. The engineer determines the placement of all equipment and the location of all duct runs and components. He or she also determines all of the specifications for unit and duct size based on calculations of structure volume, exterior surface areas and construction materials, rate of air flow, and pressure. The engineer may prepare single-line sketches or submit data and calculations to a design drafter, who prepares design sketches or final drawings. Drafters without design experience work from engineering or design sketches to prepare formal drawings. A single-line engineer's sketch is shown in Figure 18–17. The next step in the HVAC design is for the drafter to convert the rough sketch into a preliminary drawing. This preliminary drawing goes back to the engineer and architect for verification and corrections or changes. The final step in the design process is for the drafter to implement the design changes on the preliminary drawing to establish the final HVAC drawing. The final HVAC drawing is shown in Figure 18–18.

When the drafter converts an engineering sketch to a formal drawing, the easiest-to-read format should be used. For example:

1. Draw duct runs using thick (0.7- or 0.9-mm) line widths.
2. Label duct sizes within the duct when appropriate, or use a note with a leader to the duct in other situations.
3. Duct sizes may be noted as 22 × 12 or 22/12, where the first number, 22, is the duct width and the second numeral, 12, is the duct depth.
4. Place notes on the drawing to avoid crowding. Aligned techniques may be used, where horizontal notes read from the bottom of the sheet and vertical notes read from the right side of the sheet. Make notes clear and concise.
5. Refer to schedules to get specific drawing information that may not otherwise be available on the sketch. Schedule information is found on the following discussion.
6. Label equipment to stand out clearly from other information on the drawing—either blocked out or bold.

Several examples of duct system elements are shown comparing the engineering sketch and formal drawing in Figure 18–19.

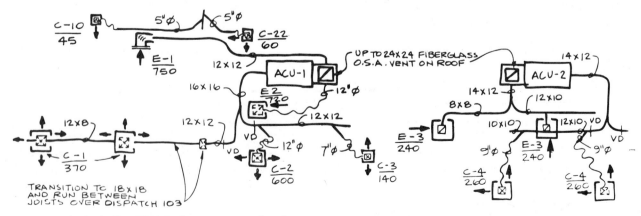

Figure 18–17 A single-line HVAC plan engineer's sketch.

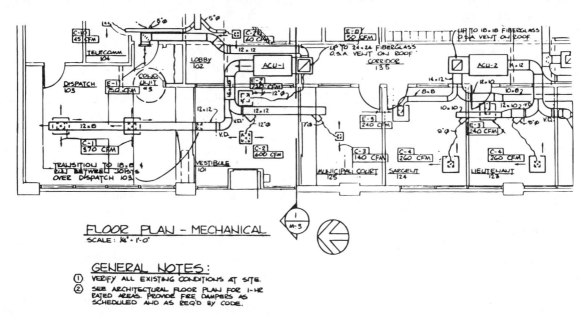

Figure 18–18 HVAC plan for the engineer's sketch shown in Figure 18–17. *Courtesy W. Alan Gold Consulting Mechanical Engineer and Robert Evenson Associates AIA Architects.*

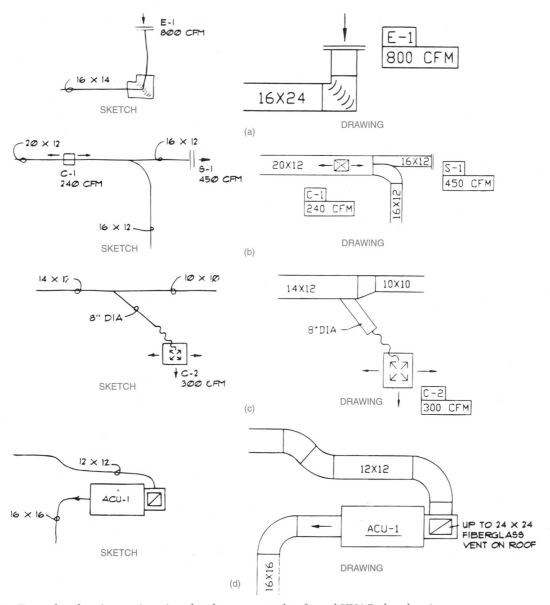

Figure 18–19 Examples showing engineering sketches converted to formal HVAC plan drawings.

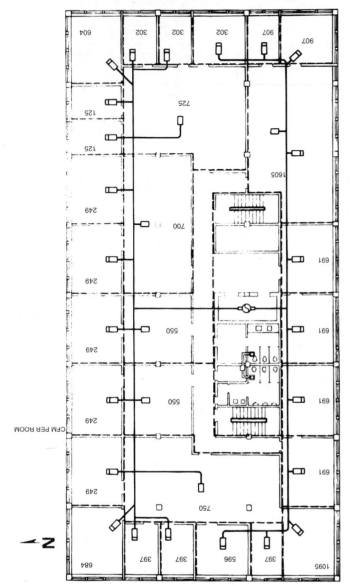

Figure 18–20 Single-line ducted system showing a layout of the proposed trunk and runout ductwork. *Courtesy The Trane Company, LaCrosse, Wisconsin.*

Single- and Double-line HVAC Plans

HVAC plans are drawn over the floor plan as a layer. The floor-plan layout is drawn first using thin lines as a base sheet for the HVAC layout and other layers. The HVAC plan is then drawn or plotted using thick lines and notes for contrast with the floor plan. The HVAC plan shows the placement of equipment and duct work. The size (in inches) and shape (with symbols, ∅ = round, □ = square or rectangular) of duct work and system component labeling are placed on the drawing or keyed to schedules. Drawings may be either single or double line, depending on the needs of the client or how much detail must be shown. Single-line drawings are easier and faster to draw. In many situations they are adequate to provide

the equipment placement and duct routing, as shown in Figure 18–20. Double-line drawings take up more space and are more time-consuming than single line, but they are often necessary when complex systems require more detail, as shown in Figure 18–18.

Detail Drawings

Detail drawings are used to clarify specific features of the HVAC plan. Single- and double-line drawings are intended to establish the general arrangement of the system; they do not always provide enough information to fabricate specific components. When further clarification of features is required, detail drawings are made. A *detail drawing* is an enlarged view(s) of equipment, equipment installations, duct components, or any feature that is not defined on the plan. Detail drawings may be scaled or unscaled and provide adequate views and dimensions for sheet metal shops to prepare fabrication patterns, as shown in Figure 18–21. When scaled, details are drawn at ½″ = 1′-0″ to 1½″ = 1′-0″ scales.

Section Drawings

Sections or sectional views are used to show and describe the interior portions of an object or structure that would otherwise be difficult to visualize. Section drawings provide a clear representation of construction details or a profile of the HVAC plan as taken through one or more locations in the building. There are two basic types of section drawings used in HVAC. One method is used to show the construction of the HVAC system in relationship to the structure. In this case the building is sectioned, and the duct system is shown unsectioned. This drawing provides a profile of the HVAC system. There may be one or more sections taken through the structure, depending on the complexity of the project. The building

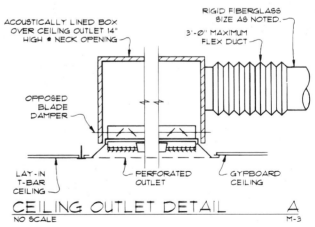

Figure 18–21 A sample detail drawing. *Courtesy W. Alan Gold Consulting Mechanical Engineer and Robert Evenson Associates AIA Architects.*

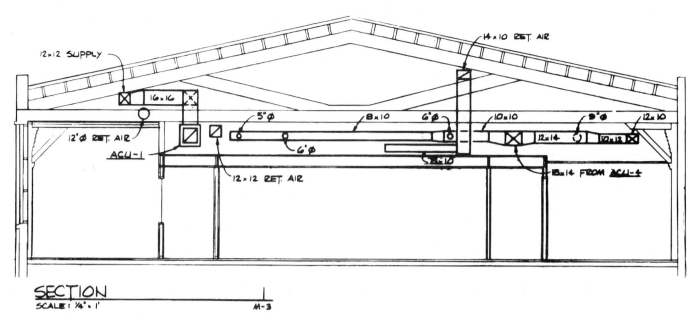

Figure 18–22 Section drawing. *Courtesy W. Alan Gold Consulting Mechanical Engineer and Robert Evenson Associates AIA Architects.*

structure may be drawn using thin lines as shown in Figure 18–22. Figure 18–22 shows a section through the HVAC plan first shown in Figure 18–18. The other sectioning method is used to show detail of equipment or to show how parts of an assembly might fit together (Figure 18–23).

Schedules

Numbered symbols that are used on the HVAC plan to key specific items to charts are known as *schedules*. Schedules describe items such as ceiling outlets, supply and exhaust grills, hardware, and equipment. Schedules are charts of materials or products that include size, description, quantity used, capacity, location, vendor's specification, and any other information needed to construct or finish the system. Schedules aid the drawing by keeping it clear of unnecessary notes. They are generally placed in any convenient area of the drawing field or on a separate sheet.

Items on the plan may be keyed to schedules by using a letter and number combination, such as C-1 for CEILING OUTLET NO. 1, E-1 for EXHAUST GRILL NO. 1, or ACU-1 for EQUIPMENT UNIT NO. 1. The exhaust grill schedule keyed to the HVAC plan that is in Figure 18–18 may be set up as a chart, as demonstrated in Figure 18–24.

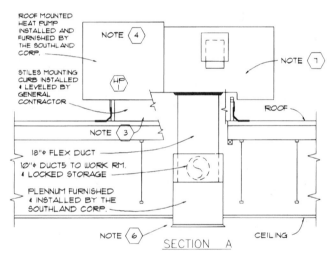

Figure 18–23 Detailed section showing HVAC equipment installation. *Courtesy The Southland Corporation.*

Pictorial Drawings

Pictoral drawings may be isometric or oblique as shown in Figure 18–25. Pictorial drawings are usually not drawn to scale. They are used in HVAC for a number of applications, such as assisting in visualization of the duct system and when the plan and sectional views are not adequate to show difficult duct routing.

EXHAUST GRILL SCHEDULE

SYMBOL	SIZE	CFM	LOCATION	DAMPER TYPE				TYPE	REMARKS
				FIRE DPR.	KEY OP. OPPBLD	KEY OP. EXTR	NO. DPR.		
E-1	24x12	750	HIGH WALL	X	X			4	
E-2	18x18	720	CEILING	X	X			2	
E-3	10x10	240	CEILING		X			1	24x24 PANEL
E-4	10x10	350	CEILING		X			1	
E-5	12x12	280	CEILING		X			1	
E-6	10x10	500	CEILING	X	X			1	
E-7	6x6	350			X			2	
E-8	12x6	50			X			3	
E-9	12x6	200			X			3	
E-10	12x8	150			X			3	
E-11	10x10	290			X			3	
E-12	9x4	160			X			1	24x24 PANEL
E-13	9x4	75	HIGH WALL	X	X			4	

TYPE 1: KRUGER 1190 SERIES STEEL PERFORATED FRAME 23 FOR LAY-IN TILE

TYPE 2: KRUGER 1190 SERIES STEEL PERFORATED FRAME 22 FOR SURFACE MOUNT

TYPE 3: KRUGER EGC-5:1/2"X1/2"X1/2" ALUMINUM GRID.

TYPE 4: KRUGER S80H:35° HORIZ. BLADES 3/4"O.C.

Figure 18–24 Exhaust grill schedule.

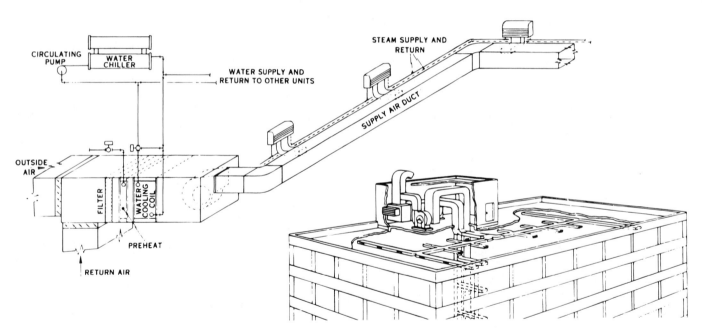

Figure 18–25 Pictorial drawings. *Courtesy The Trane Company, LaCrosse, Wisconsin.*

CADD Applications

CADD USES IN THE HVAC INDUSTRY

HVAC CADD software is available that allows you to place duct fittings and then automatically sizes ducts in accordance with common mechanical equipment supplier's specifications. The floor plan is commonly used as a reference layer, and the HVAC plan is a separate layer. The CADD drafter performs the following simple steps:

Step 1. For the preliminary layout, draw the duct center lines, as shown in Figure 18–26.

Step 2. Select supply and return registers from a template menu symbols library, and add the symbols to the end of the center lines where appropriate, as shown in Figure 18–27.

Step 3. The program then automatically identifies and records the lengths of individual duct runs, and tags each run. Fittings are located and identified by the type of intersection, as shown in Figure 18–28. While all of this drawing information is added to the layout, the computer automatically gathers design information into a file for duct sizing based on a specific mechanical manufacturer's specifications that you select.

Step 4. After both the fitting location and sizes are determined, the program transforms each fitting into accurate double-line symbols exactly to the ANSI Y32.2.4 standard, as shown in Figure 18–29.

Step 5. When the fittings are in place, the program calculates and draws the connecting ducts and adds couplings automatically at the maximum duct lengths. If a transition is needed in a duct run, the program recommends the location; all you have to do is pick a transition fitting from the menu library. See Figure 18–30 for the complete HVAC layout.

The increase in productivity over manual methods is excellent. Plus, the program automatically records information while you draw, generating a complete bill of materials. The systems that offer the greatest flexibility and productivity are designed as a parametric package. This program type allows you to set the design parameters that you want to use. Then the computer automatically draws and details according to these settings. As you draw, information is placed with each fitting, including type of fitting, CFM, and gauge. A complete HVAC plan and related schedules are shown in Figure 18–31 (page 295).

CADD HVAC packages provide a variety of tablet menu overlays that assist in the rapid selection of symbols. Some of these tablet menu overlays are shown in Figure 18–32.

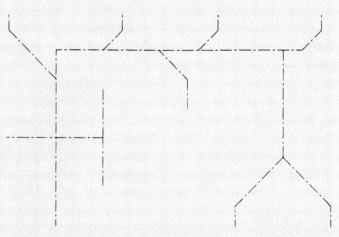

Figure 18–26 Preliminary layout of duct center lines. *Courtesy Chase Systems.*

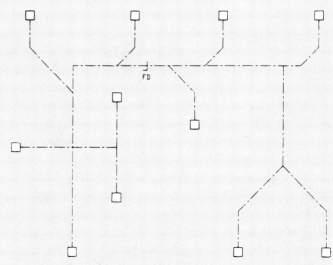

Figure 18–27 Automatically placing supply and return registers. *Courtesy Chase Systems.*

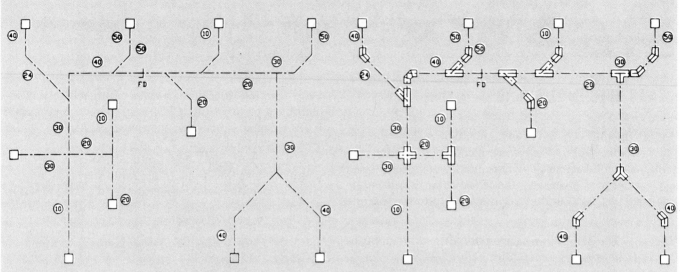

Figure 18–28 Fittings located and identified. *Courtesy Chase Systems.*

Figure 18–29 Fitting symbols automatically drawn. *Courtesy Chase Systems.*

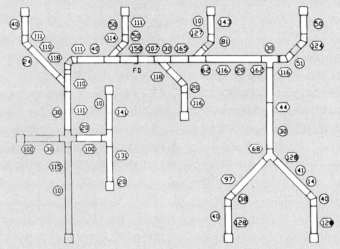

Figure 18–30 Complete HVAC CADD layout. *Courtesy Chase Systems.*

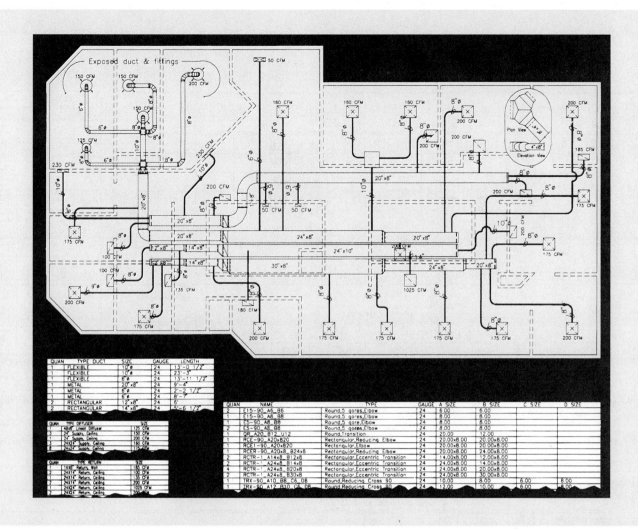

Figure 18–31 Complete HVAC plan, related symbols, and schedules automatically set up and drawn with the CADD systems. *Courtesy Chase Systems.*

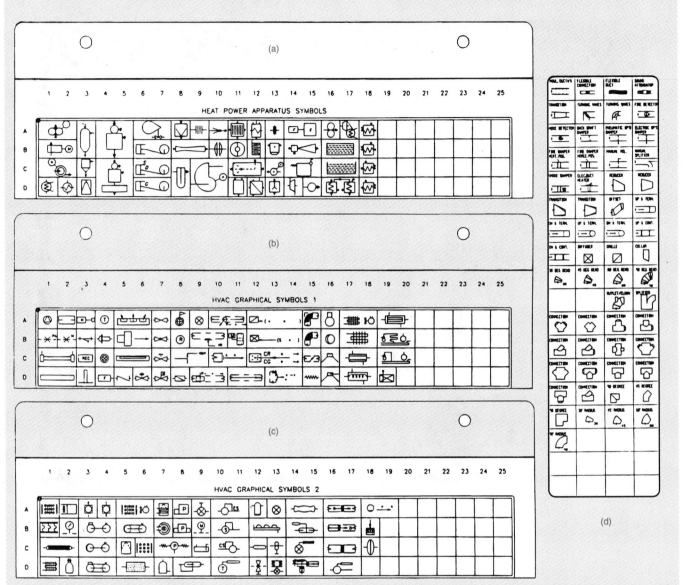

Figure 18–32 HVAC tablet menu overlays. (a)–(c) *Courtesy Drafting Technology Services, Inc., Bartlesville, Oklahoma;* (d) *Courtesy Chase Systems.*

CADD Applications

HVAC PICTORIALS

CADD applications often make pictorial representations much easier to implement, especially when the HVAC program allows direct conversion from the plan view to the pictorial. The CADD pictorials, known as *graphic models*, may be used to view the HVAC system from any angle or orientation. Some CADD programs automatically analyze the layout for obstacles where an error in design may result in a duct that does not have a clear path. One of the biggest advantages of a CADD system occurs when changes are made to the HVAC plan. These changes are simultaneously corrected on all drawings, schedules, and lists of materials. Figure 18–33 shows a CADD-generated perspective of HVAC duct routing.

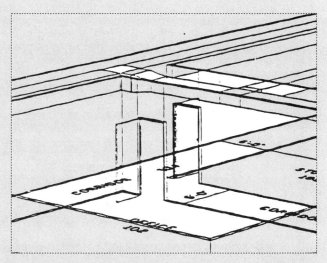

Figure 18–33   CADD-generated perspective of HVAC duct routing. *Courtesy Computervision Corporation.*

CHAPTER 18

Heating, Ventilating, and Air-Conditioning Test

DIRECTIONS

Answer the questions with short, complete statements or drawings as needed on an 8½ × 11 sheet of notebook paper, as follows:

1. Letter your name, Chapter 18 Test, and the date at the top of the sheet.
2. Letter the question number and provide the answer. You do not need to write out the question.

Answers may be prepared on a word processor if appropriate with course guidelines.

QUESTIONS

Question 18–1   Describe the following heating and cooling systems:
a. Central forced air
b. Hot water
c. Heat pump
d. Zoned heating

Question 18–2   Describe a type of zoned heating called radiant heat.

Question 18–3   List two advantages and two disadvantages of zonal heat as compared to central forced-air heating.

Question 18–4   A heat pump may supply up to how many times as much heat per year for the same amount of electrical consumption as a standard electric forced-air system?

Question 18–5   Discuss four factors that influence the placement of a thermostat.

Question 18–6   Describe five sources that can contribute to an unhealthy living environment.

Question 18–7   Discuss the function of an air-to-air heat exchanger.

Question 18–8   List five advantages of a central vacuum system.

Question 18–9   Define the following terms related to thermal performance calculations:
a. Outdoor temperature
b. Indoor temperature
c. Temperature difference
d. Infiltration
e. Mechanical ventilation
f. Duct loss
g. R factor
h. U factor
i. Btu
j. Compass point
k. Internal heat gain
l. Latent load
m. Grains

PROBLEMS

Problem 18–1 Prepare heat loss and gain calculations for one of the floor plans found at the end of Chapter 15 Floor Plan Problems. You may use the floor plan of your choice, or your instructor will assign a floor plan. Use the residential heating data sheets and the residential cooling data sheets (found in the Instructor's Guide) provided by your instructor, or prepare your own data sheets by copying the proper headings from Figures 18–14 and 18–16. Complete the data sheets according to the method used in this chapter to solve for the total winter heat loss and summer heat gain. Use the following criteria unless otherwise specified by your instructor:

1. Outdoor temperature as recommended for your area.
2. Indoor temperature: 70° F.
3. Double glass windows and glass doors.
4. Urethane insulated metal exterior doors, weatherstripped.
5. Ceiling height: 8′–0″.
6. Ceiling insulation: R-30.
7. Frame wall insulation: R-19 + ½″ polystyrene.
8. Insulation in floors over unconditioned space: R-19.
9. Duct insulation: R-4.
10. Masonry above and below grade as required in plan used.
11. Basement floor or concrete slab construction as required in plan used. Verify use with your instructor.
12. Assume entry foyer door faces south to establish window orientation.
13. Assume glass with inside shade.
14. Number of occupants is optional. Select the family size you desire or count the bedrooms and add one. For example, a three-bedroom home plus one equals four people.
15. Use latent load conditions typical for your local area. Select wet, medium, medium dry, or dry as appropriate. Consult your instructor.

Problem 18–2 Calculate the minimum heat exchanger capacity for one of the floor plans found in Floor Plan Problems at the end of Chapter 15. Use the formulas established earlier in the Air-to-Air Heat Exchangers section of this chapter.

Problem 18–3 Using this problem's residential heating engineering sketch of main floor plan and basement, do the following on appropriately sized vellum:

1. Make a formal double-line HVAC floor plan layout at a ¼″ = 1′-0″ scale.
2. Approximate the location of undimensioned items such as windows.
3. Use thin lines for the floor plan and thick lines for the heating equipment and duct runs.

Problem 18–4 Using this problem's residential air-to-air heat exchanger ducting engineering sketch of a basement floor plan, do the following on appropriately sized vellum (one B or C size sheet is recommended):

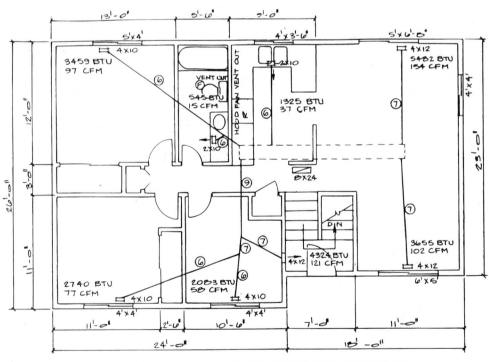

TOTAL HEAT LOSS 23,885 BTU
675 CFM

Problem 18–3a

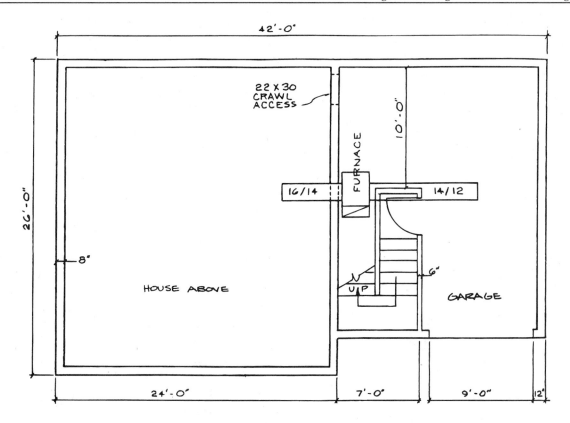

BASEMENT FURNACE HEATING PLAN

Problem 18–3b

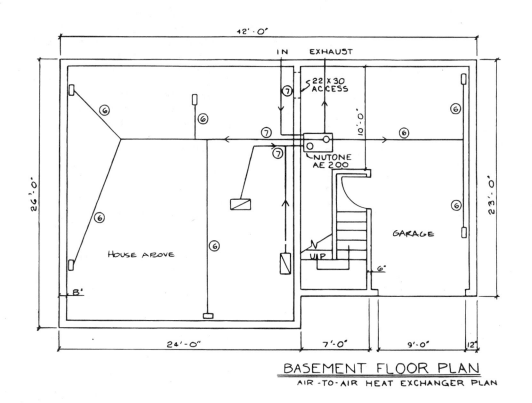

BASEMENT FLOOR PLAN

AIR-TO-AIR HEAT EXCHANGER PLAN

Problem 18–4

1. Make a formal single-line air-to-air heat exchanger floor plan layout at a ¼″ = 1′–0″ scale.
2. Approximate the location of undimensioned items such as doors.
3. Use thin lines for the floor plan and use thick lines for the air-to-air heat exchanger equipment and duct runs.

Problem 18–5 Given the following:
1. HVAC floor plan engineering layout at approximately ¹⁄₁₆″ = 1′–0″. The engineer's layout is rough, so round off dimensions to the nearest convenient units. For example, if the

dimension you scale reads 24′–3″, round off to 24′–0″. The floor plan will not require dimensioning; therefore, the representation is more important than the specific dimensions.
2. Related schedules.
3. Engineer's sketch for exhaust hood.

Do the following on appropriately sized vellum (C or D size is recommended; all required items will fit on one sheet with careful planning):

SCHEDULES				
CEILING OUTLET SCHEDULE				
SYMBOL	SIZE	CFM	DAMPER TYPE	PANEL SIZE
C-10	9 × 9	230	Key Operated	12 × 12
C-11	8 × 8	185	Key Operated	12 × 12
C-12	6 × 6	40	Key Operated	12 × 12
C-13	6 × 6	45	Key Operated	12 × 12
C-14	6 × 18	300	Fire Damper	24 × 24

SUPPLY GRILL SCHEDULE				
SYMBOL	SIZE	CFM	LOCATION	DAMPER TYPE
S-1	20 × 8	450	High Wall	Key Operation
S-2	20 × 8	450	High Wall	External Operation

EXHAUST GRILL SCHEDULE				
SYMBOL	SIZE	CFM	LOCATION	DAMPER TYPE
E-5	18 × 24	1000	Low Wall	No Damper

ROOF EXHAUST FAN SCHEDULE			
SYMBOL	AREA SERVED	CFM	FAN SPECIFICATIONS
REF-1	Solvent Tank	900	¼ HP, 12″ Non-spark Wheel, 1050 Max. Outlet Velocity

1. Make a formal double-line HVAC floor plan layout a ¼″ = 1′–0″ scale. (Note: You measured the given engineer's sketch at ¹⁄₁₆″ = 1′–0″.) Now you convert the established dimensions to a formal drawing at ¼″ = 1′–0″. Approximate the location of the HVAC duct runs and equipment in proportion to the presentation on the sketch.
2. Prepare correlated schedules in the space available. Set up the schedules in a manner similar to the examples below.
3. Make a detail drawing of the exhaust hood, either scaled or unscaled. Make the detail large enough to clearly show the features. Refer to the detail drawings below for examples.

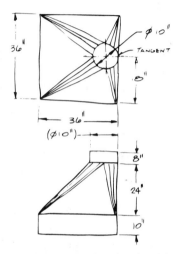

EXHAUST HOOD DETAIL

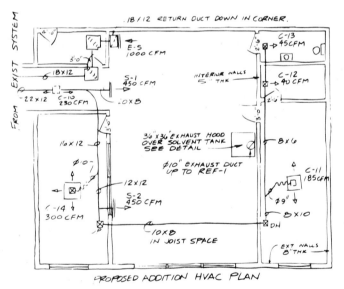

PROPOSED ADDITION HVAC PLAN

Problem 18–5

SECTION VI

ROOF PLANS

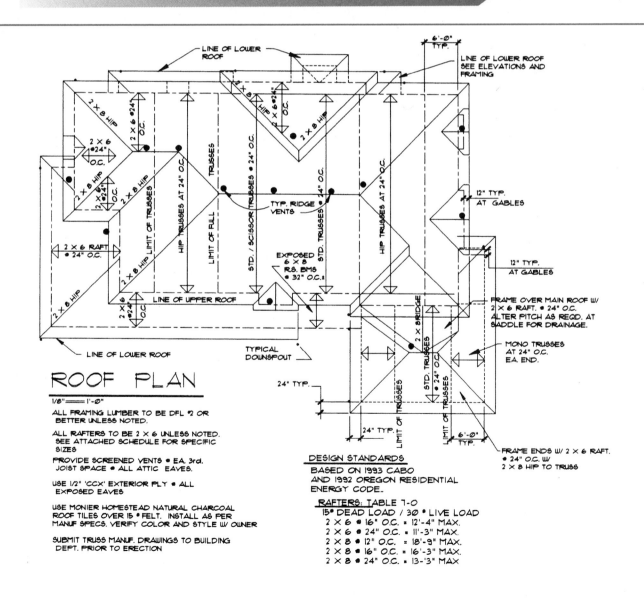

LINE OF LOWER ROOF

LINE OF LOWER ROOF
SEE ELEVATIONS AND FRAMING

6'-0" TYP.

2 X 8 HIP

2 X 6 24" O.C.

2 X 6 24" O.C.

2 X 8 HIP

2 X 6 24" O.C.

2 X 8 HIP

2 X 8 HIP

2 X 6 24" O.C.

12" TYP. AT GABLES

HIP TRUSSES AT 24" O.C.

LIMIT OF FULL TRUSSES

STD. / SCISSOR TRUSSES 24" O.C.

STD. TRUSSES 24" O.C.

HIP TRUSSES AT 24" O.C.

TYP. RIDGE VENTS

12" TYP. AT GABLES

2 X 6 RAFT. 24" O.C.

2 X 8 HIP LIMIT OF TRUSSES

2 X 6 24" O.C.

LINE OF UPPER ROOF

EXPOSED 6 X 8 R.S. BMS 32" O.C.±

2 X 8 HIP

FRAME OVER MAIN ROOF W/ 2 X 6 RAFT. 24" O.C. ALTER PITCH AS REQD. AT SADDLE FOR DRAINAGE.

MONO TRUSSES AT 24" O.C. EA. END.

2 X BRIDGE

LINE OF LOWER ROOF

TYPICAL DOWNSPOUT

STD. TRUSSES 24" O.C.

LIMIT OF TRUSSES

LIMIT OF TRUSSES

24" TYP.

24" TYP.

6'-0" TYP.

FRAME ENDS W/ 2 X 6 RAFT. 24" O.C. W/ 2 X 8 HIP TO TRUSS

ROOF PLAN

1/8" = 1'-0"

ALL FRAMING LUMBER TO BE DFL #2 OR BETTER UNLESS NOTED.

ALL RAFTERS TO BE 2 X 6 UNLESS NOTED. SEE ATTACHED SCHEDULE FOR SPECIFIC SIZES

PROVIDE SCREENED VENTS EA. 3rd. JOIST SPACE ALL ATTIC EAVES.

USE 1/2" 'CCX' EXTERIOR PLY ALL EXPOSED EAVES

USE MONIER HOMESTEAD NATURAL CHARCOAL ROOF TILES OVER 15 # FELT. INSTALL AS PER MANUF SPECS. VERIFY COLOR AND STYLE W/ OWNER

SUBMIT TRUSS MANUF. DRAWINGS TO BUILDING DEPT. PRIOR TO ERECTION

DESIGN STANDARDS

BASED ON 1993 CABO AND 1992 OREGON RESIDENTIAL ENERGY CODE.

RAFTERS: TABLE 7-0
15# DEAD LOAD / 30 # LIVE LOAD
2 X 6 16" O.C. = 12'-4" MAX.
2 X 6 24" O.C. = 11'-3" MAX.
2 X 8 12" O.C. = 18'-9" MAX.
2 X 8 16" O.C. = 16'-3" MAX.
2 X 8 24" O.C. = 13'-3" MAX

CHAPTER 19

Roof Plan Components

The design of the roof must be considered long before the roof plan is drawn. The architect or designer will typically design the basic shape of the roof as the floor plan and elevations are drawn in the preliminary design stage. This does not mean that the designer plans the entire structural system for the roof during the initial stages, but the general shape and type of roofing material to be used will be planned. By examining the structure in Figure 19–1, you can easily see how much impact the roof design has on the structure. Often the roof can present a larger visible surface area than the walls. In addition to aesthetic considerations, the roof can also be used to provide rigidity in a structure when wall areas are filled with glass, as seen in Figure 19–2. To insure that the roof will meet the designer's criteria, a roof plan is usually drawn by the drafter to provide construction information. In order to draw the roof plan, a drafter should understand types of roof plans, various pitches, common roof shapes, and common roof materials.

▼ TYPES OF ROOF PLANS

The plan that is drawn of the roof area may be either a *roof plan* or a *roof framing plan*. For some types of roofs a roof drainage plan may also be drawn. Roof framing and drainage plans will be discussed in Chapter 30.

Figure 19–1 The shape of the roof can play an important role in the design of the structure as seen in this home by architect Robert Roloson. *Photo courtesy Red Cedar Shingle & Handsplit Shake Bureau.*

Figure 19–2 In addition to aesthetic considerations, the roof is often used to resist wind and seismic forces when walls of the structure contain large amounts of glass, as seen in this home by architect Claude Miquelle. *Photo courtesy Red Cedar Shingle & Handsplit Shake Bureau.*

Roof Plans

A roof plan is used to show the shape of the roof. Materials such as the roofing material, vents and their location, and the type of underlayment are also typically specified on the roof plan, as seen in Figure 19–3. Roof plans are typically drawn at a scale smaller than the scale used for the floor plan. A scale of ⅛″ = 1′–0″ or ¹⁄₁₆″ = 1′–0″ is commonly used when drawing a roof plan. A roof plan is typically drawn on the same sheet as the exterior elevations.

Roof Framing Plans

Roof framing plans are usually required on complicated residential roof shapes and with most commercial projects. A roof framing plan shows the size and direction of the construction members that are required to frame the roof. Figure 19–4 shows an example of a roof framing plan. On very complex projects, every framing member is shown, as seen in Figure 19–5. Framing plans will be further discussed in Chapter 30.

▼ ROOF PITCHES

Roof pitch, or *slope,* is a description of the angle of the roof that compares the horizontal run to the vertical rise. The slope is shown when the elevations and sections are drawn and will be shown in Chapters 22 and 25. The

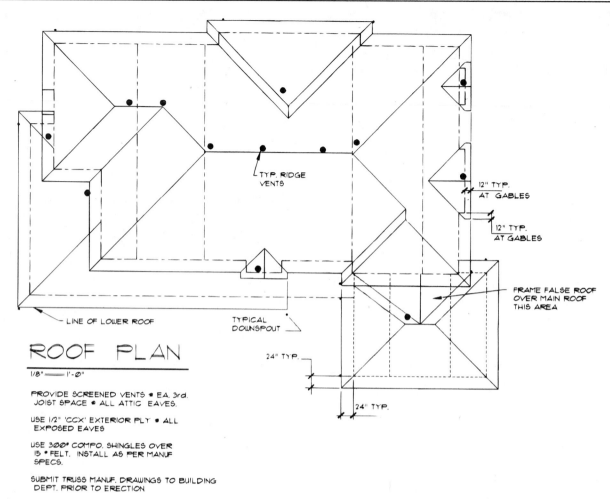

TYP. RIDGE VENTS

12" TYP. AT GABLES

12" TYP. AT GABLES

FRAME FALSE ROOF OVER MAIN ROOF THIS AREA

LINE OF LOWER ROOF

TYPICAL DOWNSPOUT

24" TYP.

24" TYP.

ROOF PLAN
1/8" = 1'-0"

PROVIDE SCREENED VENTS @ EA. 3rd. JOIST SPACE @ ALL ATTIC EAVES.

USE 1/2" 'CCX' EXTERIOR PLY @ ALL EXPOSED EAVES

USE 300# COMPO. SHINGLES OVER 15 # FELT. INSTALL AS PER MANUF SPECS.

SUBMIT TRUSS MANUF. DRAWINGS TO BUILDING DEPT. PRIOR TO ERECTION

Figure 19–3 A roof plan is drawn to show the shape of the roof.

intersections that result from various roof pitches must be shown on the roof plan. In order to plot the intersection between two roof surfaces correctly, the drafter must understand how various roof pitches are drawn. Figure 19–6 shows how the pitch can be visualized. The drafter can plot the roof shape using this method for any pitch. Adjustable triangles for plotting roof angles are available and save the time of having to measure the rise and run of a roof. The roof pitch can also be drawn if the drafter knows the proper angle that a certain pitch represents. Knowing that a 4/12 roof equals 18½° allows the drafter to plot the correct angle without having to plot the layout. Figure 19–7 shows angles for common roof pitches.

▼ ROOF SHAPES

By changing the roof pitch, the shape of the roof may be changed. Common roof shapes include flat, shed, gable, A-frame, gambrel, hip, Dutch hip, and mansard. See Chapter 25 for a complete discussion of roof framing terms.

Flat Roofs

The flat roof is a very common style of roof in areas with little rain or snow. In addition to being used in residential construction, the flat roof is typically used on commercial structures to provide a platform for heating and other mechanical equipment. The flat roof is economical to construct because ceiling joists are eliminated and rafters are used to support both the roof and ceiling loads. Figure 19–8 shows the common materials used to frame a flat roof. Figure 19–9 shows how a flat roof could be represented on the roof plan.

Often the flat roof has a slight pitch in the rafters. A pitch of ¼" per ft is often used to help prevent water from ponding on the roof. As water flows to the edge, a metal diverter is usually placed at the eave to prevent dripping at walkways. A flat roof will often have a *parapet*, or false wall, surrounding the perimeter of the roof. Figure 19–10 provides an example of a parapet wall. This wall can be used for decoration or for protection of mechanical equipment. When used it must be shown on the roof plan.

Shed Roofs

The shed roof, as seen in Figure 19–11, offers the same simplicity and economical construction methods of a flat roof but does not have the drainage problems associated with a flat roof. Figure 19–12 shows the construction

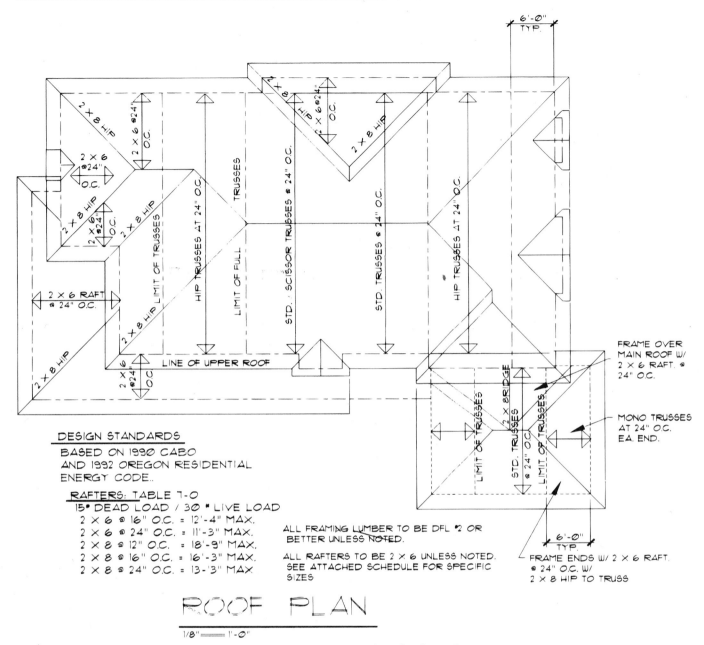

Figure 19–4 A roof framing plan is used to show the framing members for the roof.

methods of shed roofs. The shed roof may be constructed at any pitch. The roofing material and the desired aesthetic considerations are the only limiting factors of the pitch. When drawn in plan view, the shed roof will resemble the flat roof, as seen in Figure 19–13.

Gable Roofs

A gable roof, one of the most common types of roof used in residential construction, uses two shed roofs that meet to form a ridge between the support walls. Figure 19–14 shows the construction of a gable roof system. The gable can be constructed at any pitch, with the choice of pitch limited only by the roofing material and the effect desired. A gable roof is often used on designs seeking to

aintain a traditional appearance with formal balance. Figure 19–15 shows how a gable roof is typically represented in plan view. Many plans use two or more gables at 90° angles to each other. The intersections of gable surfaces are called either hips or valleys. Typically the valley and hip are specified on the roof plan.

A-Frame Roofs

An A-frame is a method of framing walls, as well as a system of framing roofs. An A-frame structure uses rafters to form its supporting walls, as shown in Figure 19–16. The structure gets its name from the letter A that is formed by the roof and floor systems. The roof plan for an A-frame is very similar to the plan for a gable roof.

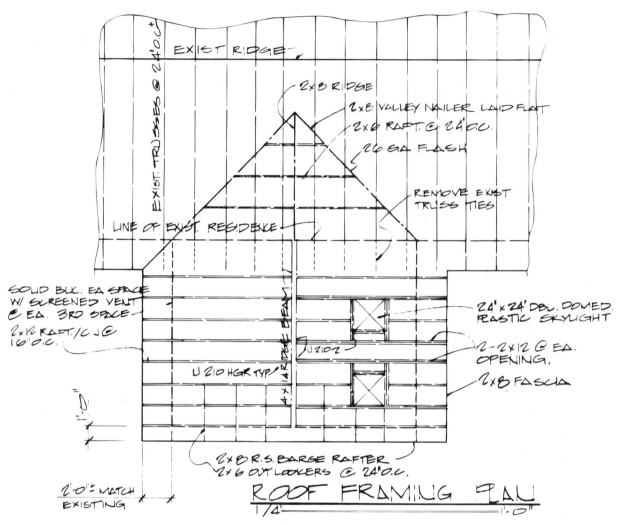

Figure 19–5 A roof framing plan may be drawn for complicated roofs to show the size and location of every structural member.

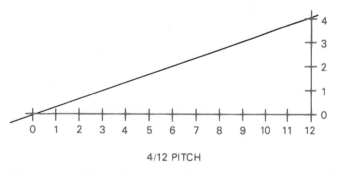

4/12 PITCH

Figure 19–6 When plotting a roof slope, the angle is expressed as a comparison of equal units. Units may be inches, feet, meters, etc., as long as the horizontal and vertical units are of equal length.

COMMON ANGLES FOR DRAWING ROOF PITCHES	
Roof Pitch	Angle
1/12	4°–30′
2/12	9°–30′
3/12	14°–0′
4/12	18°–30′
5/12	22°–30′
6/12	26°–30′
7/12	30°–0′
8/12	33°–45′
9/12	37°–0′
10/12	40°–0′
11/12	42°–30′
12/12	45°–0′

Figure 19–7 Common roof pitches and angles. Angles shown are approximate and are to be used for drawing purposes only.

However, the framing materials are usually quite different. Figure 19–17 shows how an A-frame can be represented on the roof plan.

Gambrel Roofs

A gambrel roof can be seen in Figure 19–18. The gambrel roof is a very traditional roof shape that dates back to

the colonial period. Figure 19–19 shows the construction methods of a gambrel roof. Typically used on two-level structures, the upper level is covered with a steep roof surface, which connects into a roof system with a slighter pitch. By covering the upper level with roofing material rather than siding, the height of the structure will appear shorter than it is. This roof system can also be used to

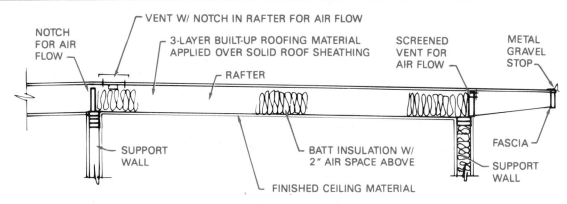

Figure 19–8 Common construction components of a flat roof.

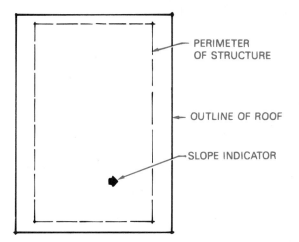

Figure 19–9 A flat roof in plan view.

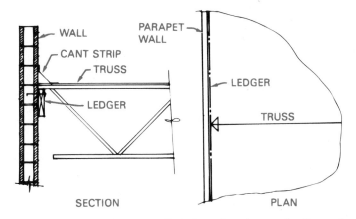

Figure 19–10 A parapet wall is often placed around a flat roof to hide mechanical roof equipment. The thickness of the wall should be represented on the roof plan.

reduce the cost of siding materials with less expensive roofing materials. Figure 19–20 shows a plan view of a gambrel roof.

Hip Roofs

The hip roof of Figure 19–21 is a traditional roof shape. This shape can be used to help eliminate some of the roof

Figure 19–11 A shed roof is simple, economical, and attractive. *Courtesy American Plywood Association.*

mass and create a structure with a smaller appearance. A hip roof has many similarities to a gable roof but has four surfaces instead of two. The intersection between each surface is called a *hip*. If built on a square structure, the hips will come together to form a point. If built on a rectangular structure, the hips will form two points with a ridge spanning between them. When hips are placed over an L- or T-shaped structure, an interior intersection will be formed that is called a valley. The valley of a hip roof is the same as the valley for a gable roof. Hip roofs can be seen in plan view as shown in Figure 19–22.

Dutch Hip Roofs

The Dutch hip roof is a combination of a hip and a gable roof. The center section of the roof is framed in a method similar to a gable roof. The ends of the roof are framed with a partial hip that blends into the gable. A small wall is formed between the hip and the gable roofs, as seen in Figure 19–23. On the roof plan, the shape, distance, and wall location must be shown similar to the plan in Figure 19–24.

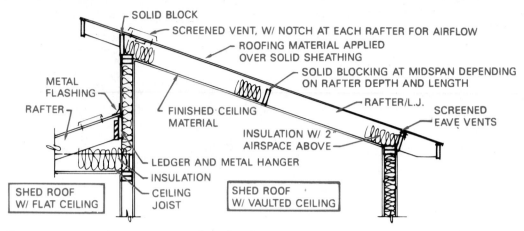

Figure 19–12 Common construction components of shed roofs.

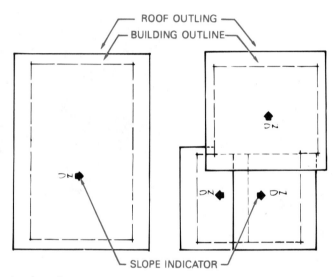

Figure 19–13 Shed roof shapes in plan view.

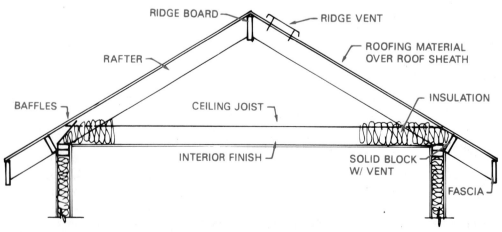

Figure 19–14 Common construction components of a gable roof.

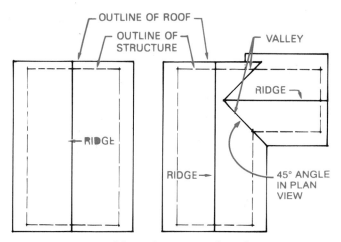

Figure 19–15   A gable roof represented in plan view.

Figure 19–18   The gambrel roof is often used to enhance the traditional appearance of a residence. *Designed by architect Russell Swinton Oatman, R.A., Princeton. Photo courtesy Cabot's Stains.*

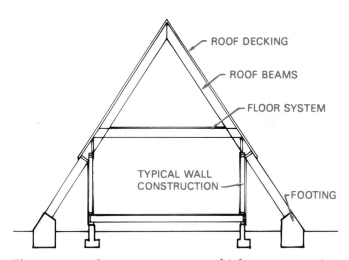

Figure 19–16   Common components of A-frame construction.

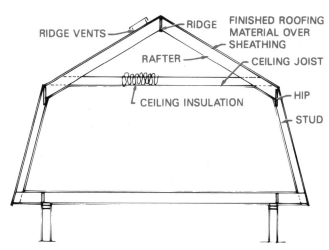

Figure 19–19   Common construction components of a gambrel roof.

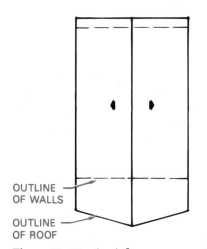

Figure 19–17   An A-frame represented in plan view.

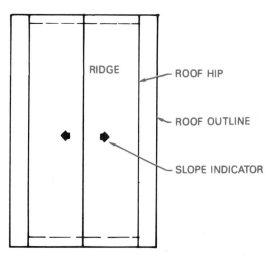

Figure 19–20   A gambrel roof represented in plan view.

Figure 19–21 A hip roof is often used to decrease the amount of roof that is visible. *Courtesy Owens Corning Fiberglas.*

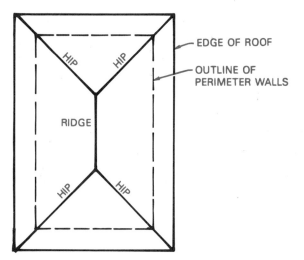

Figure 19–22 A hip roof in plan view.

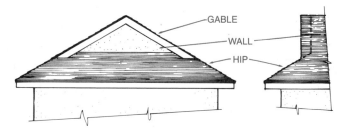

Figure 19–23 A wall is formed between the hip and gable roof.

Mansard Roofs

The mansard roof is similar to a gambrel roof but has the angled lower roof on all four sides rather than just two. The mansard roof is often used as a parapet wall to hide mechanical equipment on the roof or to help hide the height of the upper level of a structure. Examples of each can be seen in Figure 19–25. Mansard roofs can be constructed in many different ways. Figure 19–26 shows

two common methods of constructing a mansard roof. The roof plan for a mansard roof will resemble the plans shown in Figure 19–27.

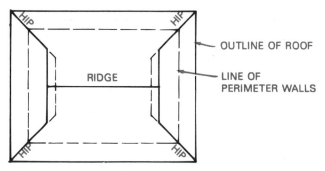

Figure 19–24 A Dutch hip roof represented in plan view.

Figure 19–25 Mansard roofs are often used to hide equipment on the roof or to help disguise the height of the structure. *Landow and Landow, AIA, Architects. Smithtown, NY; photo courtesy Follansbee Steel Corporation.*

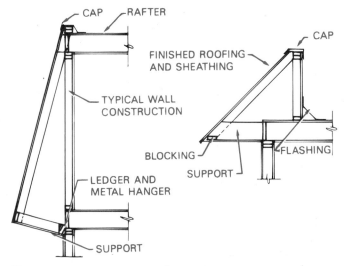

Figure 19–26 Common methods of constructing a mansard roof.

Dormers

A dormer is an opening framed in the roof to allow for window placement. Figure 19–28 shows an example of dormers that have been added to provide light and ventilation to rooms in what would have been attic space. Dormers are most frequently used on traditional style roofs such as the gable or hip. Figure 19–29 shows one of the many ways that dormers can be constructed. Dormers are usually shown on the roof plan as seen in Figure 19–30.

▼ ROOFING MATERIALS

The material to be used on the roof is dependent on the pitch, exterior style, weather, and cost of the structure. Common roofing materials include built-up roofing, composition and wood shingles, clay and cement tiles, and metal panels. When ordering or specifying these materials, the term *square* is often used. A square is used to describe an area of roofing that covers 100 sq. ft. (9.3 m²). The drafter will need to be aware of the weight per square and the required pitch as the plan is being drawn. The

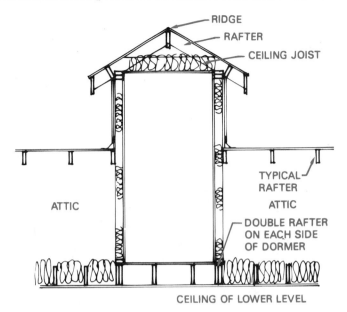

Figure 19–29   Typical components of dormer construction.

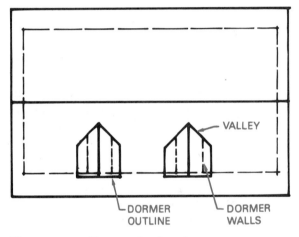

Figure 19–30   Dormers in plan view.

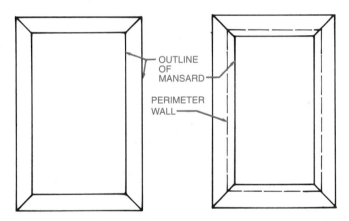

Figure 19–27   Mansard in plan view.

Figure 19–28   Dormers are added to allow windows to be added to attic areas. *Courtesy John Black.*

weight of the roofing material will affect the size of the framing members all the way down to the foundation level. The material will also affect the required pitch and the appearance that results from the selected pitch.

Built-up Roofing

Built-up roofing of felt and asphalt is typically used on flat or low-sloped roofs below a ³⁄₁₂ pitch. When the roof has a low pitch, water will either pond or drain very slowly. To prevent water from leaking into a structure, built-up roofing is used because of its lack of seams. On a residence, a built-up roof may consist of three alternate layers of felt and hot asphalt placed over solid roof decking. The decking is usually plywood. In commercial uses, a four- or five-layer roof is used to provide added durability. Gravel is often used as a finishing layer to help cover the felt. On roofs with a pitch over 2/12, course rocks 2″ or 3″ (50–75 mm) in diameter are used for protecting the roof and for appearance. When built-up roofs are to be specified on the roof plan, the note should include the number of layers, the material to be used, and the size of the finishing material. A typical note would be: 3 LAYER BUILT UP ROOF WITH HOT ASPHALTIC EMULSION BETWEEN EACH LAYER WITH ¼″ (6 mm) PEA GRAVEL.

Shingles

Asphalt, asbestos-cement, fiberglass, and wood are the most typical types of shingles used as roofing materials. Most building codes and manufacturers require a minimum roof pitch of 4/12 with an underlayment of one layer of 15 lb. felt. Asphalt and fiberglass shingles can be laid on roofs as low as 2/12 if two layers of 15 lb. felt are laid under the shingles and if the shingles are sealed. Asbestos-cement and wood shingles must usually be installed on roofs having a pitch of at least 3/12. Asphalt, asbestos-cement, and fiberglass are similar in appearance and application. Each comes in many different patterns and colors.

Asphalt shingles come in a variety of colors and patterns. Also known as composition shingles, they are typically made of fiberglass backing and covered with asphalt and a filler with a coating of finely crushed particles of stone. The asphalt waterproofs the shingle, and the filler provides fire protection. The standard shingle is a three tab rectangular strip weighing 235 lb. per square. The upper portion of the strip is coated with self-sealing adhesive and is covered by the next row of shingles. The lower portion of a three tab shingle is divided into three flaps that are exposed to the weather. See Figure 19–31.

Composition shingles are also available in random width and thickness to give the appearance of cedar

Figure 19–31 235-lb. composition shingles are made of three equal-sized tabs. *Courtesy Owens Corning Fiberglas.*

Figure 19–32 300-lb. composition shingles are made with tabs of random width and length. *Courtesy Owens Corning Fiberglas.*

shakes. These shingles weigh approximately 300 lb. per square. Both types of shingles can be used in a variety of conditions on roofs having a minimum slope of 2/12. The lifetime of shingles varies from 20- to 40-year guarantees. See Figure 19–32. Asbestos cement shingles are also available; they weigh approximately 560 lb. per square, depending on the manufacturer and the pattern used.

Shingles are typically specified on drawings in note form listing the material, the weight, and the underlayment. The color and manufacture may also be specified. This information is often omitted in residential construction to allow the contractor to purchase a suitable brand at the best cost. A typical call-out would be:

○ 235 lb. composition shingles over 15 lb. felt.
○ 300 lb. composition shingles over 15 lb. felt.
○ Architect 80 'Driftwood' class A fiberglass shingles by Genstar with 5⅝″ exposure over 15 lb. felt underlayment with 30-year warranty.

Wood is also used for shakes and shingles. Wood shakes are thicker than shingles and are also more irregular in their texture. Wood shakes and shingles are generally installed on roofs with a minimum pitch of ⁴⁄₁₂ using a base layer of 15 lb. felt. An additional layer of 15 lb. by 18″ (457 mm) wide felt is also placed between each course or layer of shingles. Wood shakes and shingles can be installed over solid or spaced sheathing. The weather, material availability, and labor practices affect the type of underlayment used.

Depending on the area of the country, shakes and shingles are usually made of cedar, redwood, or cypress. They are also produced in various lengths. When shakes or shingles are specified on the roof plan, the note should usually include the thickness, the material, the exposure, the underlayment, and the type of sheathing. Other materials such as Masonite and metal are also used to simulate shakes. These materials are typically specified on plans in note form listing the material, underlayment, and amount of shingle exposed to the weather. A typical specification for wood shakes would be: MEDIUM CEDAR SHAKES OVER 15# FELT W/ 15# × 18″ WIDE FELT BETWEEN EACH COURSE. LAY WITH 10½″ EXPOSURE.

Metal is sometimes used for roof shingles. Metal shingles provide a durable, fire-resistant roofing material, but they are generally not used in residential construction because of their high cost. Metal shingles are usually installed using the same precautions applied to asphalt shingles. Metal shingles are typically specified on the roof plan in a note listing the manufacturer, the type of shingle, and the underlayment.

Clay and Cement Tiles

Tile is the material most often used for homes on the high end of the price scale or where the risk of fire is extreme. Although tile may cost twice as much as the better grades of asphalt shingle, it offers a lifetime guarantee. Tile is available in a variety of colors, materials, and patterns. Clay, concrete, and metal are the most common materials. See Figure 19–33.

Roof tiles are manufactured in both curved and flat shapes. Curved tiles are often called Spanish tiles and come in a variety of curved shapes and colors. The flat, or barr, tiles are also produced in many different colors and shapes.

Tiles are typically installed on roofs having a roof pitch of 4/12 or greater. Interlocking roof tiles may be placed on roofs as low as 3/12 pitch if a 15-lb. minimum underlayment is used. Tiles can be placed over either spaced or solid sheathing. If solid sheathing is used, wood strips are generally added on top of the sheathing to support the tiles.

When tile is to be used, special precautions must be taken with the design of the structure. Tile roofs weigh between 850 and 1000 lb. per square. These weights require rafters, headers, and other supporting members to be larger than normally required for other types of roofing material. Tiles are generally specified on the roof plan in a note, which lists the manufacturer, style, color, weight,

fastening method, and underlayment. A typical note on the roof plan might be: MONIER BURNT TERRA COTTA MISSION S ROOF TILE OVER 15# FELT AND 1 × 3 SKIP SHEATHING. USE A 3″ MIN. HEAD LAP AND INSTALL AS PER MANUF. SPECIFICATIONS.

Metal Panels

Metal roofing panels often provide savings because of the speed and ease of installation. Metal roof panels provide a water- and fireproof material that comes with a warranty for a protected period that can range from 20 to 50 years. Panels are typically produced in either 22- or 24-gage metal in widths of either 18″ or 24″. The length of the panel can be specified to meet the needs of the roof in lengths up to 40′. Metal roofing panels typically weigh between 50 and 100 lb. per square. Metal roofs are manufactured in many colors and patterns and can be used to blend with almost any material. Steel, stainless steel, aluminum, copper, and zinc alloys are most typically used for metal roofing. Steel panels are heavier and more durable than other metals but must be covered with a protective coating to provide protection from rust and corrosion. A baked-on acrylic coating typically provides both color and weather protection. Stainless steel does not rust or corrode but is more expensive than steel. Stainless steel weathers to a natural matte-gray finish. Aluminum is extremely lightweight and does not rust. Finish coatings are typical to those used for steel. Copper has been used for centuries as a roofing material. Copper roofs do not rust and weather to a blue-green color. When specifying metal roofing on the roof plan, the note should include the manufacturer, the pattern, the material, the underlayment, and the trim and flashing. A typical note would be: AMER-X-9 GA. 36″ WIDE, KODIAK BROWN METAL ROOFING BY AMERICAN BUILDING PRODUCTS OR EQUAL. INSTALL OVER 15# FELT AS PER MANUFACTURER'S SPECIFICATIONS.

▼ ROOF VENTILATION AND ACCESS

As the roof plan is drawn, the drafter must determine the size of the attic space. The attic is the space formed between the ceiling joists and the rafters. The UBC, SBC, and BOCA all require that the attic space be provided with vents that are covered with ¼″ screen mesh. These vents must have an area equal to 1/150 of the attic area. This area can be reduced to 1/300 of the attic area if a vapor barrier is provided on the heated side of the attic floor and half of the required vents are placed in the upper half of the roof area.

The method used to provide the required vents varies throughout the country. Vents may be placed in the gabled end walls near the ridge. This allows the roof surface to remain vent free. In other areas, a continuous vent is placed in the eaves or a vent may be placed in each third rafter

Figure 19–33 Flat tiles provide a durable, fireproof roof material. *Courtesy Saundra Powell.*

space. The drafter needs to specify the proper area of vents that are required and the area in which they are to be placed.

The drafter must also specify how to get into the attic space. The actual opening into the attic is usually shown on the floor plan, but its location must be considered while drawing the roof plan. The size of the access opening depends on the code that you are using. The UBC requires an opening that is 22″ × 30″ (560 × 760 mm) with 30″ (760 mm) minimum of headroom. The SBC requires an access opening of 22″ × 36″ (560 × 900 mm)

with 24″ (610 mm) minimum of headroom. While planning the roof shape, the drafter must find a suitable location for the attic access that meets both code and aesthetic requirements. The access should be placed where it can be easily reached but should not be placed where it will visually dominate a space. Avoid placing the access in areas such as the garage, and areas with high moisture content, such as bathrooms and utility rooms, or in bedrooms that will be used by young children. Hallways usually provide an area to place the access that is both easily accessible but not in a focal point of the structure.

CHAPTER

Roof Plan Components Test

DIRECTIONS

1. Letter your name, Chapter 19 Test, and the date at the top of the sheet.
2. Letter the question number and provide the answer. You do not need to write out the question.
3. Do all lettering with vertical uppercase architectural letters. If the answer requires line work, use proper drafting tools and technique.

Answers may be prepared on a word processor if appropriate with course guidelines.

QUESTIONS

Question 19–1 List and describe three different types of roof plans.

Question 19–2 When describing roof pitch, what do the numbers 4/12 represent?

Question 19–3 What angle represents a 6/12 pitch?

Question 19–4 Is a surface built at a 28° angle from vertical a wall or a roof?

Question 19–5 What are two advantages of using a flat roof?

Question 19–6 What is the major disadvantage of using a flat roof?

Question 19–7 List three traditional roof shapes.

Question 19–8 Sketch and define the difference between a hip and a Dutch hip roof.

Question 19–9 What are the two uses for a mansard roof?

Question 19–10 List two common weights for asphalt or fiberglass shingles.

Question 19–11 What are two common shapes of clay roof tiles?

Question 19–12 What advantage do metal roofing panels have over other roofing materials?

Question 19–13 According to the SBC, what is the minimum headroom required at the attic access?

Question 19–14 What is the minimum size of an attic access opening according to the UBC?

Question 19–15 What type of roof is both a roof system and a framing system?

CHAPTER 20

Roof Plan Layout

▼ LINES AND SYMBOLS

Several different types of lines and symbols are required when drawing the roof plan. These include lines used to represent the roof shape, nonstructural materials, dimensions, and written specifications.

Roof Shape

The overhang of the roof that forms the outline of the roof is usually drawn with bold solid lines. The size of the overhang varies depending on the pitch of the roof and the amount of shade desired. As seen in Figure 20–1, the steeper the roof pitch, the smaller the overhang that is required to shade an area. If you are drawing a home for a southern location, an overhang large enough to protect glazing from direct sunlight is usually desirable. In northern areas, the overhang is usually restricted to maximize the amount of sunlight received during winter months. Figure 20–2 compares the effect of an overhang at different times of the year. Another important consideration regarding the size of the overhang is how the view from windows will be affected. As the angle of the roof is increased, the size of the overhang may need to be decreased so that the eave will not extend into the line of sight from a window. The designer will need to size the overhang based on the roof pitch so a view is not hindered.

Changes in the shape of the roof such as ridges, hips, Dutch hips, and valleys are also drawn with solid lines. Each can be seen in Figure 20–3. These changes in shape can be drawn easily when the pitches are of equal angle. Because the ridge is centered between the two bearing walls, it can be located by measuring the distance between the two supporting walls and dividing that distance in half. A faster method of locating the ridge is to

draw two 45° angles, as shown in Figure 20–4. The ridge extends through the intersection of the two angles. When the roof pitches are unequal, the drafter must plot the proper pitches to determine the location of the ridge. The ridge location can be determined by drawing a partial section to reflect the angle of the roof. See Figure 20–5.

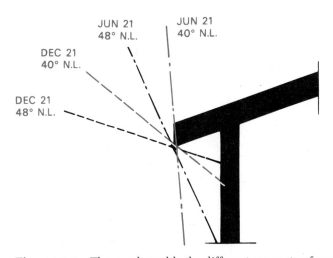

Figure 20–2 The overhang blocks different amounts of sunlight at different times of the year. Notice the difference in the sun's angles at 40° and 48° north latitudes.

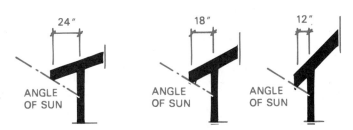

Figure 20–1 The steeper the roof pitch, the more shadow will be cast. Typically the overhang is decreased as the pitch is increased.

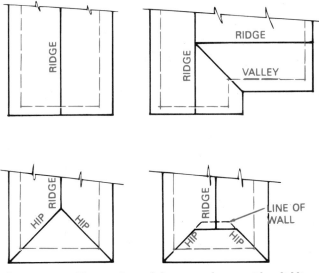

Figure 20–3 Changes in roof shape are shown with solid lines.

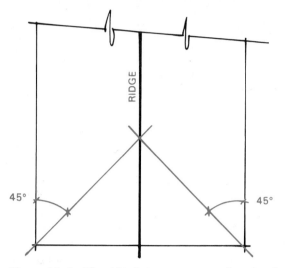

Figure 20–4   The ridge between two equal roof surfaces can be plotted by drawing two 45° angles. The ridge extends through the intersection.

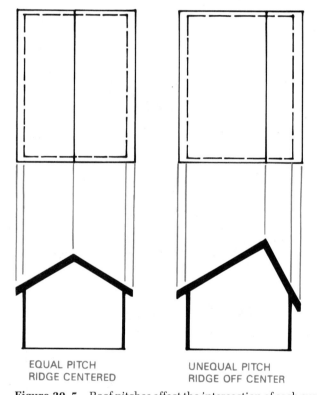

EQUAL PITCH
RIDGE CENTERED

UNEQUAL PITCH
RIDGE OFF CENTER

Figure 20–5   Roof pitches affect the intersection of each surface.

Hip Roofs.   If a hip roof is to be drawn, the hips, or external corners of the roof, can be drawn in a manner similar to that used to locate the ridge. The hips represent the intersection between two roof planes and are represented by a line drawn at an angle that is one-half of the angle formed between the two supporting walls. For walls that are perpendicular, an internal corner or valley is formed. The valley will be drawn at an angle of 45° between the two intersecting walls, as seen in Figure 20–6. Keep in mind that the 45° line represents the inter-

section between two equally pitched roofs in plan view only. This angle has nothing to do with the slope of the roof. The angle that represents the slope or pitch of the roof can be shown only in an elevation of the structure. Drawing elevations will be introduced in Chapters 21 and 22.

Dutch Hip Roofs.   A Dutch hip is drawn by first lightly drawing a hip, as shown in Figure 20–7. Once the hip is drawn, determine the location of the gable wall. The wall is usually located over a framing member, spaced at 24″ (600 mm) on center. The wall is typically located 48 or 72″ (1200–1800 mm) from the exterior wall. With the wall located, the overhang can be drawn in a similar manner to a gable roof. The overhang line will intersect the hip lines to form the outline of the Dutch hip.

When two perpendicular roofs intersect each other, a valley will be formed at 45° to the walls. The drafter must consider both the distance between supporting walls and the pitch to determine how the roof intersections are represented on the plan. Notice in Figure 20–8 that the shape of the roof changes dramatically as the width between the walls is changed. Remember that when the pitches are equal, the wider the distances are between the walls, the higher the roof will be.

When the walls are equally spaced but not perpendicular, the hip, valley, and ridge intersection will occur along a line, as shown in Figure 20–9. With unequal distances between the supporting walls, the hips, valleys, and ridge can be drawn, as shown in Figure 20–10. Such a drawing is often a difficult procedure for an inexperienced drafter. The process is simplified if you draw partial sections by the roof plan, as shown in Figure 20–11. By comparing the heights and distances in the sections, the intersections often can be visualized better.

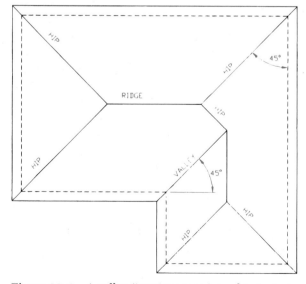

Figure 20–6   A valley (interior corner) or a hip (exterior corner) is formed between two intersecting roof planes. When each plane is framed from walls that are perpendicular to each other and the slope of each roof is identical, the valley is drawn at a 45° angle.

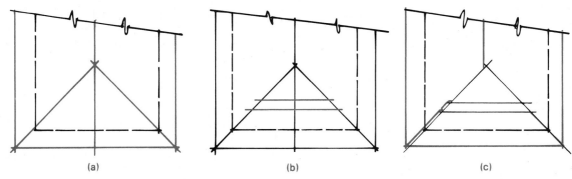

Figure 20–7 Layout of a Dutch hip roof: (a) Establish the ridge and hip locations; (b) locate the gable end wall; (c) use bold lines to complete the roof.

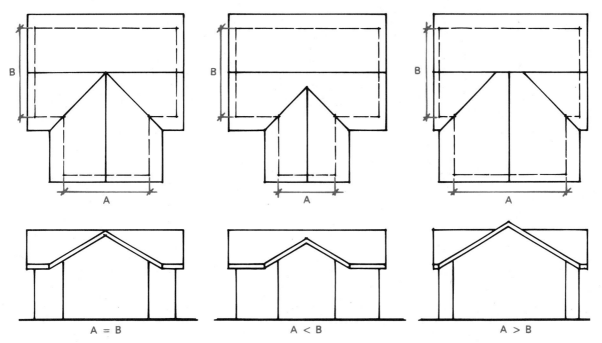

Figure 20–8 Although the basic shape remains the same, the roof plan and elevations change as the distance between walls varies.

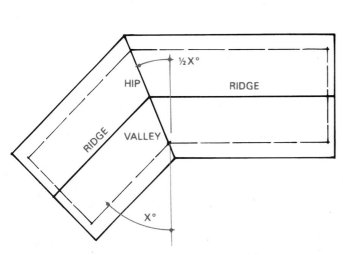

Figure 20–9 When walls are equally spaced but not perpendicular, the roof surfaces occur along a line, as shown.

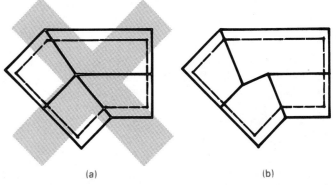

Figure 20–10 When walls are unequally spaced, a drafter may be tempted to draw the roof as shown at (a). The actual intersection can be seen at (b).

Bay Projections. If a bay is to be included on the floor plan, special consideration will be required to draw the roof covering it. Figure 20–12 shows the steps required to lay out the bay roof. The layout of the roof can be eased by

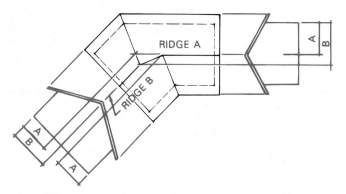

Figure 20–11 The true roof shape can be seen by drawing sections of the roof. Project the heights represented on the sections onto the roof.

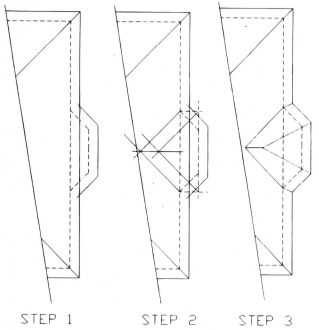

STEP 1   STEP 2   STEP 3

Figure 20–12 The roof to cover a bay can be drawn by drawing the outline of the bay and the desired overhangs parallel to the bay walls (Step 1). Draw construction lines to represent the roof that would be created if the bay were a rectangle. This will establish the ridge and the intersection of the ridge with the main roof (Step 2). Drawing the hips created by a rectangular bay will also establish the point where the true hips intersect the ridge. The true hips will start at the intersection of the overhangs, pass through the intersection of the bay walls, and end at the ridge (Step 3).

squaring the walls of the bay. Lay out the line of the ridge from the hips and valleys created by the intersecting rectangles. The true hips over the bay are drawn with a line from the intersection of the roof overhangs, which extends up to the end of the ridge.

Roof Intersections with Varied Wall Heights. Another feature common to roof plans is the intersection between two roof sections of different heights. Figure 20–13 shows

how an entry portico framed with a 10′ (3000 mm) wall would intersect with a residence framed with walls 8′ (2400 mm) high. The roof pitch must be known to determine where the intersection between the two roofs will occur. In this example a 6/12 pitch was used. At this pitch, the lower roof must extend 4′ (1200 mm) before it will be 10′ (3000 mm) high. A line was drawn 48″ (1200 mm) from the wall representing the horizontal distance. The valley between the two roofs will occur where the lines representing the higher walls intersect the line representing the horizontal distance.

Structural Material

The type of plan to be drawn will affect the method used to show the structural material. Many offices draw a separate roof framing plan. (See Figure 19–4). Chapter 27 explains in detail how to represent structural members on a roof framing plan. Chapter 30 will discuss drawing framing plans. If a framing roof is to be drawn on the roof plan, the rafters or trusses can be represented as shown in Figure 20–14.

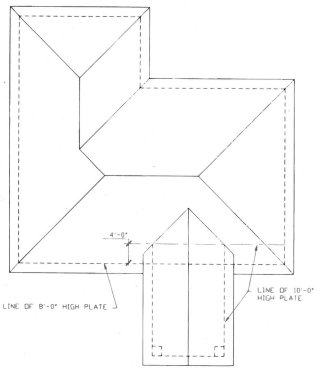

4′-0″

LINE OF 8′-0″ HIGH PLATE

LINE OF 10′-0″ HIGH PLATE

Figure 20–13 The slope of the roof must be known when drawing the intersection between roofs built on walls of differing heights. With a 6/12 pitch, a horizontal distance of 48″ (1200 mm) would be required to reach the starting height of the upper roof (6″ per ft. × 48″ = 24″ in vertical rise). By projecting the line of the upper walls to a line 48″ in from the lower wall, the intersection of the two roof planes can be determined. A valley will be formed between the two intersecting roofs starting at the intersection of the wall and the line representing the vertical rise.

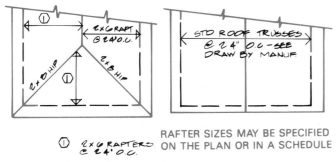

① 2×6 RAFTERS
@ 24' O.C.

RAFTER SIZES MAY BE SPECIFIED
ON THE PLAN OR IN A SCHEDULE

Figure 20–14 Rafters or trusses are typically represented on the roof framing plan by a thin line showing the proper direction and span.

Nonstructural Materials

Vents, chimneys, skylights, solar panels, diverters, cant strips, slope indicators, and downspouts are the most common materials that will need to be shown on the roof plan. Vents are typically placed as close to the ridge as possible, usually on the side of the roof that is the least visible. The size of the vent varies with each manufacturer, but an area equal to 1/300 of the attic area must be provided for ventilation. Vents can typically be represented by a circle 12″ (300 mm) in diameter or a square of 12″ (300 mm) placed at approximately 10′–0″ (3000 mm). Figure 20–15 shows how ridge vents are represented.

The method used to represent the chimney depends on the chimney material. Two common methods of representing a chimney can be seen in Figure 20–15. The metal chimney can be represented by a circle 14″ (300 mm) in diameter. The masonry chimney can range from a minimum of 16″ (400 mm) square, up to matching the size used to represent the fireplace on the floor plan. The location of skylights can usually be determined from their location on the floor plan. As seen in Figure 20–16, the opening in the roof for the skylight does not have to align directly with the opening in the ceiling. The skylight is connected to the ceiling by an enclosed area, called a *chase.* By adjusting the angle of the chase, the size and location of the skylight can be adjusted. If solar panels are to be represented, the size, angle, and manufacturer should be specified.

The amount of rainfall determines the need for gutters and downspouts. In semiarid regions, a metal strip called a *diverter* can be placed on the roof to route runoff on the roof from doorways. The runoff should be diverted to an area of the roof where it can drop to the ground and be adequately diverted from the foundation. In wetter regions, gutters should be provided to collect and divert the water collected by the roof. Local codes will specify if the drains must be connected to storm sewers, private drywell, or a splash block. The downspouts that bring the roof runoff to ground level can be represented by approximately a 3″ (70 mm) circle or square. Care should be taken to keep downspouts out of major lines of sight. Each downspout can typically drain approximately 20′

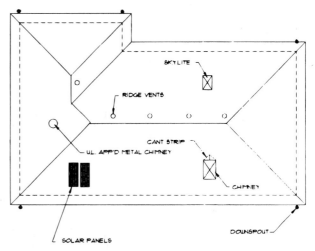

Figure 20–15 Nonstructural materials shown on the roof plan.

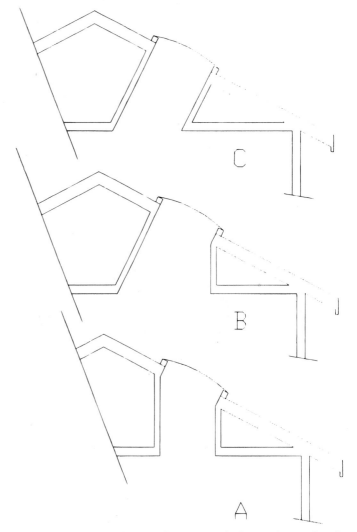

Figure 20–16 By altering the angle of chase (the enclosed space connecting the skylight to the ceiling), the location of the opening in the ceiling can be altered. Walls can be framed perpendicular to the ceiling (A), with one side perpendicular to both the roof and the ceiling (B), perpendicular to the roof framing (C).

(6100 mm) of roof on each side of the downspout, allowing the downspouts to be placed at approximately 40′ (12,200 mm) spacing along the eave. This spacing can be seen in Figure 20–17. The distance between downspouts will vary depending on the amount and rate of rainfall and should be verified with local manufacturers. Common methods of showing downspouts can be seen in Figure 20–15. The cant strip or saddle is a small gable built behind the chimney to divert water away from the chimney, as seen in Figure 20–18.

Dimensions

The roof plan requires very few dimensions. Typically only the overhangs and openings are dimensioned. These may even be specified in note form rather than with dimensions. Figure 20–19 shows how dimensions are placed on roof and framing plans. When drawing framing plans for commercial projects, beam locations are often dimensioned to help in estimating the materials required.

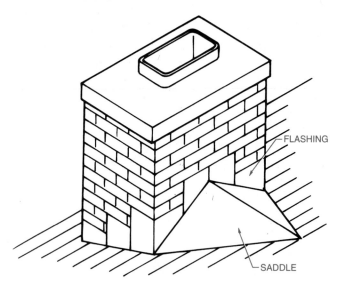

Figure 20–18 A saddle or cant strip is used to divert water around the chimney.

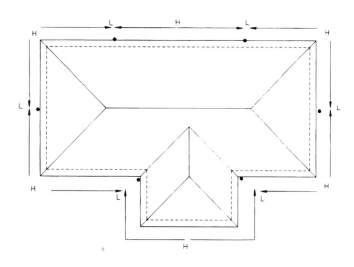

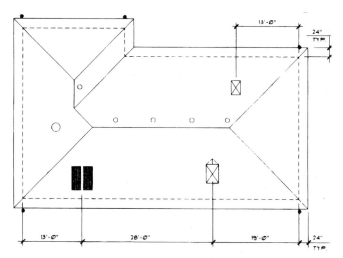

Figure 20–19 Dimensions should be placed by using leader and extension lines or in a note.

Notes

As with the other drawings, notes on the roof or framing plans can be divided into general and local notes. General notes might include the following:

○ Vent notes
○ Sheathing information
○ Roof covering
○ Eave sheathing
○ Pitch

Material that should be specified in local notes includes the following:

○ Skylight type, size, and material
○ Chimney caps
○ Solar panel type and size
○ Cant strips

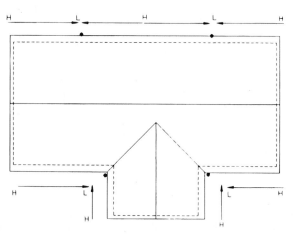

Figure 20–17 The location of gutters and downspouts will be determined by the type of roof and amount of rainfall to be drained. Gutters must be sloped to the downspouts, which should be placed so that no view is blocked.

In addition to these notes, the drafter should also place a title and scale on the drawing.

▼ DRAWING THE ROOF PLAN

The following instructions are for the roof plan to accompany the residence that was used in Chapter 15. The plan will be drawn using a gable roof. Use construction lines for Steps 1 through 6. Each step can be seen in Figure 20–20. Use the dimensions on the floor plan to determine all sizes.

Step 1.   Draw the perimeter walls.

Step 2.   Locate any supports required for covered porches.

Step 3.   Draw the limits of the overhangs. Unless your instructor provides different instructions, use 1'–0" (305 mm) overhangs for the gable end walls and 2'–0" (610 mm) overhangs for the eaves.

Step 4.   Locate and draw the ridge or ridges.

Step 5.   Locate and draw any hips and valleys required by design.

Step 6.   Locate the chimney.

Step 7.   Using the line quality described in this chapter, draw the materials of Steps 1 through 6. Draw the outline of the upper roof, and then work down to lower levels. Your drawing should now resemble Figure 20–21.

Use the line quality described in this chapter to draw Steps 8 through 13. Each step can be seen in Figure 20–22.

Step 8.   Draw any skylights that are specified on the floor plans.

Step 9.   Calculate the area of the attic and determine the required number of vents. Assume that 12" (305 mm) round vents will be used. Draw the

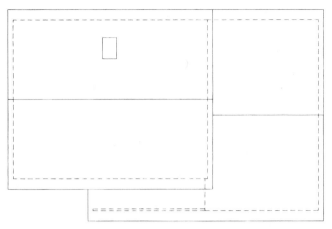

Figure 20–21   Drawing the roof plan using finished quality lines.

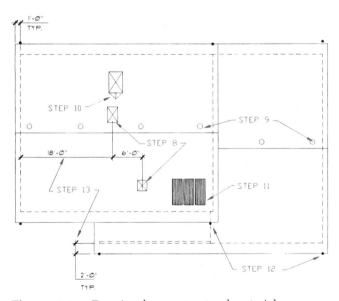

Figure 20–22   Drawing the nonstructural material.

required vents on a surface of the roof that will make the vents least visible.

Step 10.   Draw the cant strip by the chimney if a masonry chimney or wood chase was used.

Step 11.   Draw solar panels.

Step 12.   Draw the downspouts.

Step 13.   Add dimensions for the overhangs and skylights.

Step 14.   Label all materials using local and general notes, a title, and a scale. Your drawing should now resemble Figure 20–23.

Step 15.   Evaluate your drawing for completeness, and make any minor revisions required before giving your drawing to your instructor.

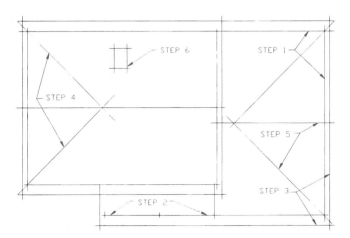

Figure 20–20   The layout of the roof plan should be done using construction lines.

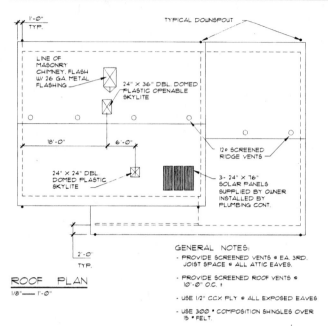

ROOF PLAN
1/8"——1'-0"

Figure 20–23 The size of the overhangs and the location of all openings in the roof should be dimensioned. The roof plan is completed by adding the required local and general notes to specify all roofing materials.

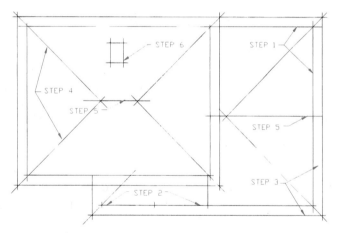

Figure 20–24 The layout of a hip roof plan should be done with construction lines.

▼ DRAWING HIP ROOF PLANS

The step-by-step process can be used for other styles of roofs. A hip roof can be drawn by completing the following steps. Although drawing roof plans may prove frustrating, remember that lines will always be vertical, horizontal, or at a 45° angle. Another helpful hint for drawing a hip roof is to remember that three lines are always required to represent an intersection of hips, valleys, and ridges. Use construction lines for Steps 1–6. Each step can be seen in Figure 20–24. Use the dimensions on the floor plan to determine all sizes.

Step 1. Draw the perimeter walls.

Step 2. Locate all supports required for any covered porches.

Step 3. Draw the limits of the overhangs. Unless your instructor provides different instructions, assume 24″ (600 mm) overhangs.

Step 4. Locate and draw all hips and valleys.

Step 5. Locate and draw all ridges.

Step 6. Locate the chimney if required.

Step 7. Using the line quality described in this chapter, draw the materials of Steps 1 through 6. Draw the upper roof, and then work down to lower roof levels. When complete, your drawing should resemble Figure 20–25.

Use the line quality described in this chapter to draw Steps 8 through 13. When complete, your drawing should resemble Figure 20–26.

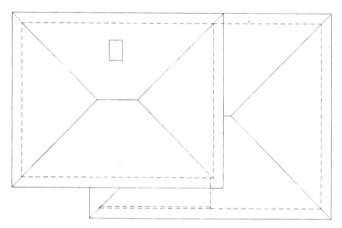

Figure 20–25 A hip roof plan drawn with finished line quality.

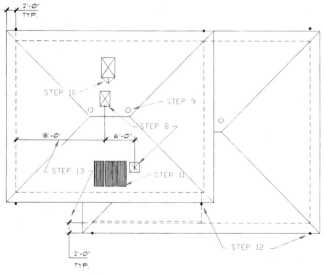

Figure 20–26 All openings in the roof, such as the chimney and skylights, and materials to be mounted on the roof should be identified. The size of the overhangs and the location of all openings in the roof should be dimensioned.

Step 8. Draw any skylights that are specified on the floor plan.

Step 9. Calculate the area of the attic, and determine the required number of vents. Assume 12″ (300 mm) round vents will be used.

Step 10. Draw a cant strip or saddle for any chimneys.

Step 11. Draw solar panels if required.

Step 12. Draw downspouts.

Step 13. Provide dimensions to locate any roof openings and all overhangs.

Step 14. Label all materials using local and general notes, a title, and a scale. Your drawing should now resemble Figure 20–27.

Step 15. Evaluate your drawing for completeness, and make any minor revisions required before giving your drawing to your instructor.

▼ DRAWING DUTCH HIP ROOF PLANS

The step-by-step process can be used for other styles of roofs. A Dutch hip roof can be drawn by completing the following steps. Use construction lines for Steps 1 through 6. Each step can be seen in Figure 20–28. Use the dimensions on the floor plan to determine all sizes.

Step 1. Draw the perimeter walls.

Step 2. Locate supports required for any covered porches.

Step 3. Draw the limits of the overhangs. Unless your instructor provides different instructions, as-

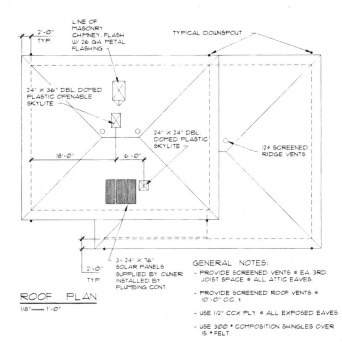

Figure 20–27 The roof plan is completed by adding the required local and general notes to specify all roofing materials.

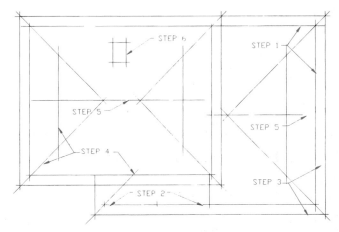

Figure 20–28 The layout of a Dutch hip roof plan should be done with construction lines.

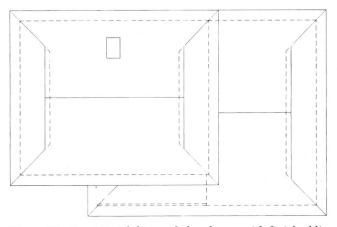

Figure 20–29 A Dutch hip roof plan drawn with finished line quality.

sume 24″ (600 mm) overhangs for eaves and 12″ (300 mm) at gable end walls.

Step 4. Locate and draw all hips, valleys, and Dutch hip locations. Hips and valleys are located following methods used to draw a hip roof. The location of a Dutch hip is determined by the framing method. Typically the gable end wall, which is formed between the hip and the gable roof, is located approximately 6′-0″ (1800 mm) from the end wall. Chapter 30 will explain framing considerations that can affect the location of the Dutch hip gable end walls.

Step 5. Locate and draw all ridges.

Step 6. Locate the chimney if required.

Step 7. Using line quality described in this chapter, draw the materials of Steps 1 through 6. Draw the upper roof, and then work down to lower roof levels. When complete, your drawing should resemble Figure 20–29.

Use the line quality described in this chapter to draw Steps 8 through 13. When complete, your drawing should resemble Figure 20–30.

Step 8. Draw any skylights that are specified on the floor plan.

Step 9. Calculate the area of the attic and determine the required number of vents. Assume 12″ (300 mm) round vents will be used.

Step 10. Draw a cant strip or saddle for any chimneys.

Step 11. Draw solar panels if required.

Step 12. Draw downspouts.

Step 13. Provide dimensions to locate any roof openings and all overhangs.

Step 14. Label all materials using local and general notes, a title, and a scale. Your drawing should now resemble Figure 20–31.

Step 15. Evaluate your drawing for completeness and make any minor revisions required before giving your drawing to your supervisor.

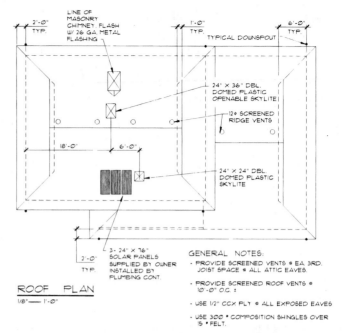

Figure 20–31 The roof plan is completed by adding the required local and general notes to specify all roofing materials.

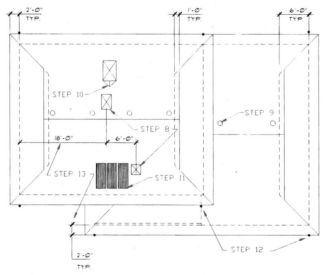

Figure 20–30 All openings in the roof, such as the chimney and skylights, and materials to be mounted on the roof should be identified. The size of the overhangs and the location of all openings in the roof should be dimensioned.

CADD Applications

DRAWING ROOF PLANS WITH CADD

A roof plan can be drawn using a CADD program following similar procedures that are used for a manual drawing. The drawing can be created in the file containing the floor plan.

Start by freezing all floor-plan material except the wall layout. Create layers to contain the roof plan using the ROOF prefix. Layers such as ROOFWALL, ROOFLINE, ROOFTEXT, and ROOFDIMEN can be used to keep the roof information separate from the floor-plan information and to ease plotting. Draw the outline of the residence on the ROOFWALL layer, and then freeze the walls of the floor plan. The plan can now be completed using the appropriate step-by-step process for the required roof type. Figure 20–32 shows an example of a roof plan drawn by CADD. When the roof plan is completed, a copy can be saved as a WBLOCK with a title ROOF. This drawing eventually can be inserted into and plotted with the drawing file that will contain the elevations. To ease development of the elevations of a structure, a copy of the roof plan is usually left as an overlay of the FLOOR drawing file.

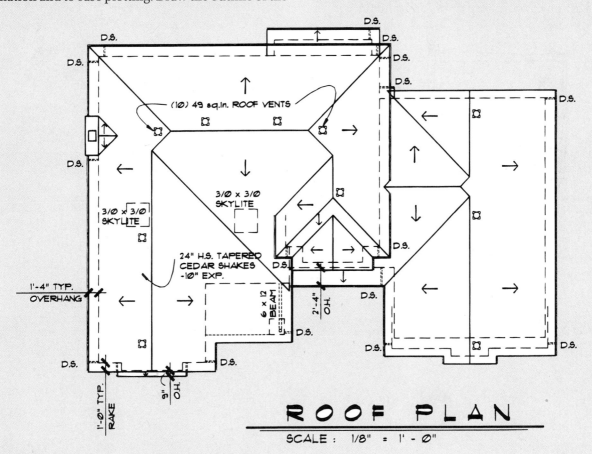

Figure 20–32   CADD-drawn plans showing a roof plan.

20 Roof Plan Layout Test

DIRECTIONS

1. Using a print of your floor plan drawn for problems from Chapter 15, draw a roof plan for your house. If no sketch has been given for the plan you have drawn, design a roof system appropriate for your area.

2. Place the plan on the same sheet as the elevations if possible. If a new sheet is required, place the drawing so that other drawings can be put on the same sheet.

3. When you have completed your drawing, turn in a diazo copy to your instructor for evaluation.

PROBLEMS

Problem 20–1 Cape Cod.

Problem 20–2 Cabin.

Problem 20–3 Ranch.

Problem 20–4 Early American.

Problem 20–5 Contemporary.

Problem 20–6 to 20–15 Without the use of a sketch, design a roof plan for the project you started in Chapter 15. Sketch the layout you will use and have it approved by your instructor prior to starting your drawing.

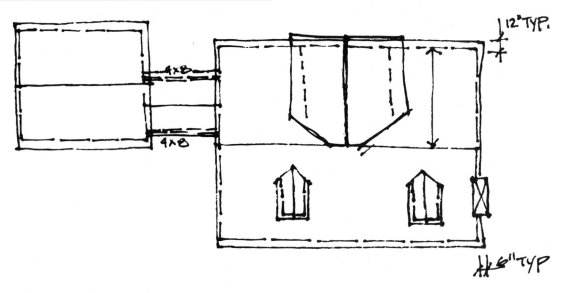

Problem 20–1 Cape Cod.

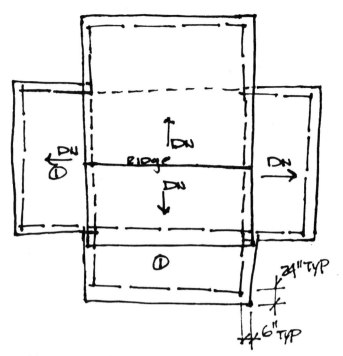

Problem 20–2   Cabin.

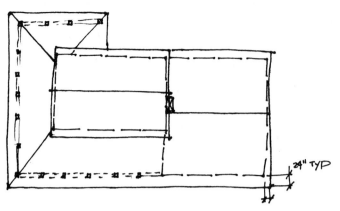

Problem 20–4   Early American.

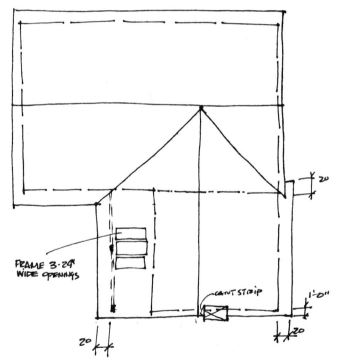

Problem 20–3   Ranch.

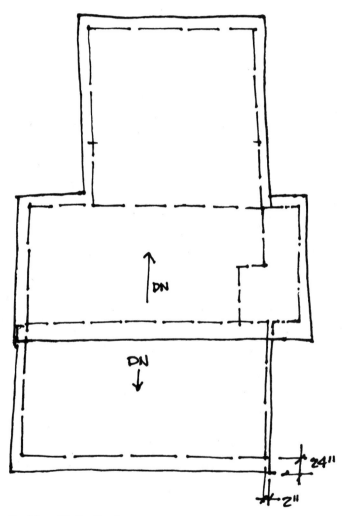

Problem 20–5   Contemporary.

SECTION VII

ELEVATIONS

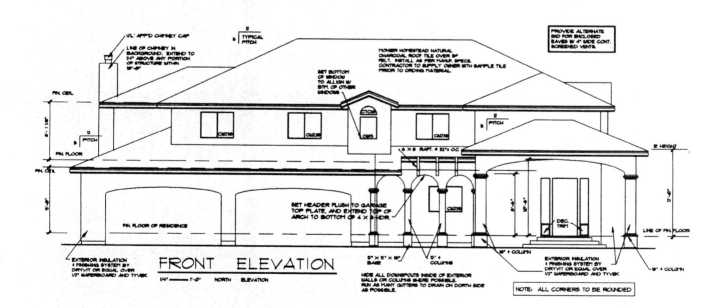

FRONT ELEVATION

1/4" ——— 1'-0"   NORTH ELEVATION

CHAPTER 21

Introduction to Elevations

Elevations are an essential part of the design and drafting process. The elevations are a group of drawings that show the exterior of a building. An elevation is an orthographic drawing that shows one side of a building. Elevations are projected as shown in Figure 21–1. Elevations are drawn to show exterior shapes and finishes, as well as the vertical relationships of the building levels. By using the elevations, sections, and floor plans, the exterior shape of a building can be determined.

▼ REQUIRED ELEVATIONS

Typically, four elevations will be required to show the features of a building. On a simple building, only three elevations will be needed, as seen in Figure 21–2. When drawing a building with an irregular shape, parts of the house may be hidden. An elevation of each different

surface should be drawn as shown in Figure 21–3. If a building has walls that are not at 90° to each other, a true orthographic drawing could be very confusing. In the orthographic projection, part of the elevation will be distorted, as can be seen in Figure 21–4. Elevations of this type of building are usually expanded so that a separate elevation of each different face is drawn, similar to Figure 21–5.

▼ TYPES OF ELEVATIONS

Elevations can be drawn as either presentation drawings or working drawings. Presentation drawings were introduced in Chapter 1 and will be covered in depth in Chapter 39. These are drawings that are part of the initial design process and may range from sketches to very detailed drawings intended to help the owner and lending institution understand the basic design concepts. See Figure 21–6.

Working elevations are part of the working drawings and are drawn to provide information for the building team. They include information on roofing, siding, openings, chimneys, land shape, and sometimes even the depth of footings, as shown in Figure 21–7. The floor plans and elevations provide the information for the contractor to determine surface areas. Once surface areas are known, exact quantities of material can be determined. The elevations are also necessary when making heat loss calculations, as described in Chapter 15. The elevations are used to determine the surface area of walls and wall openings for the required heat loss formulas.

▼ ELEVATION SCALES

Elevations are typically drawn at the same scale as the floor plan. For most plans, this means a scale of ¼″ = 1′–0″ will be used. This allows for the elevations to be projected directly from the floor plans. Some floor plans for multifamily and commercial projects may be laid out at a scale of ¹⁄₁₆″ = 1′–0″ or even as small as ¹⁄₃₂″ = 1′–0″. When a scale of ⅛″ = 1′–0″ or less is used, generally very little detail is placed on the drawings, as seen in Figure 21–8. Depending on the complexity of the project or the amount of space on a page the front elevation may be drawn at ¼″ = 1′–0″ scale and the balance of the elevations drawn at a smaller scale.

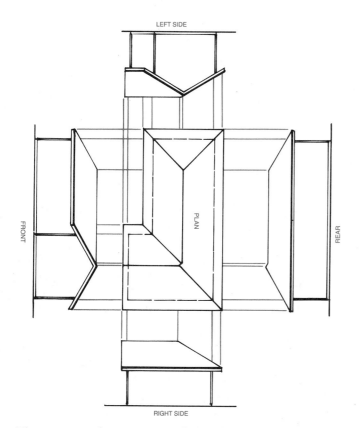

Figure 21–1 Elevations are orthographic projections showing each side of a structure.

LEFT SIDE

FRONT

PLAN

REAR

RIGHT SIDE

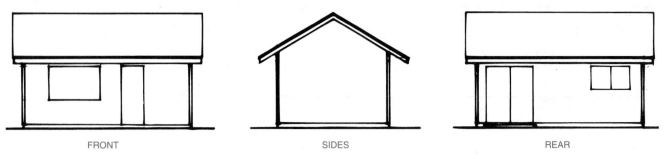

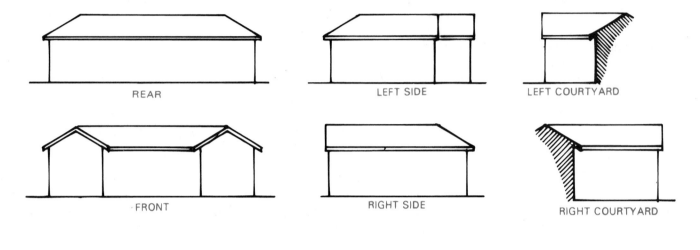

Figure 21–2 Elevations are used to show the exterior shape and material of a building. On a simple structure, only three views are required.

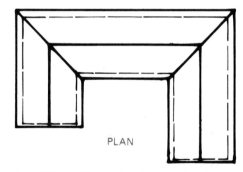

Figure 21–3 Plans of irregular shapes often require an elevation of each surface.

▼ ELEVATION PLACEMENT

It is usually the drafter's responsibility to plan the layout for drawing the elevations. The layout will depend on the scale to be used, size of drawing sheet, and number of drawings required. Because of size limitations, the elevations are not usually laid out in the true orthographic projection of Front, Side, Rear, Side. A common method of layout for four elevations can be seen in Figure 21–9. This layout places a side elevation by both the front and rear elevations so that true vertical heights may be projected from one view to another. If the drawing paper is not long enough for this placement, the layout of Figure 21–10 may be used. This arrangement often is used when

the elevations are placed next to the floor plan to conserve space. A drawback to this layout is that the heights established on the side elevations cannot be quickly transferred. A print of one elevation will have to be taped next to the location of the next elevation so that heights can be transferred.

▼ SURFACE MATERIALS IN ELEVATION

The materials that are used to protect the building from the weather need to be shown on the elevations. This information will be considered in the categories of roofing, wall coverings, doors, and windows.

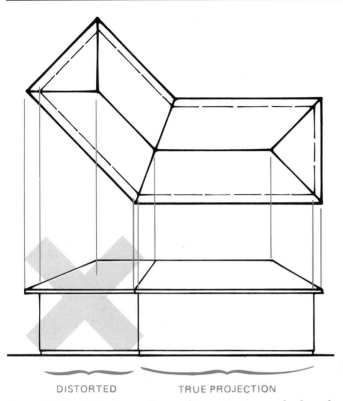

Figure 21–4 Using true orthographic projection methods with a plan of irregular shapes will result in a distortion of part of the view.

Roofing Materials

Several common materials are used to protect the roof. Among the most frequently used are asphalt shingles, wood shakes and shingles, clay and concrete tiles, metal sheets, and built-up roofing materials, which were introduced in Chapter 19. It is important that the architectural drafter have an idea of what each material looks like so that it can be drawn in a realistic style. It is also important to remember that the purpose of the elevations is for the framing crew. The framer's job will not be made easier by seeing every shingle drawn or other techniques that are used on presentation drawings and renderings. The elevations need to have materials represented clearly and quickly.

Asphalt Shingles. Asphalt shingles come in many colors and patterns. Figure 21–11 shows a roof covered with the random pattern of 300# asphalt shingles. Asphalt shingles are typically drawn using the method seen in Figure 21–12.

Wood Shakes and Shingles. Figure 21–13 shows a roof protected with wood shakes. Other materials such as Masonite are used to simulate wood shakes and can be seen in Figure 21–14. Shakes and Masonite create a jagged surface at the ridge and the edge. These types of materials are often represented as shown in Figure 21–15.

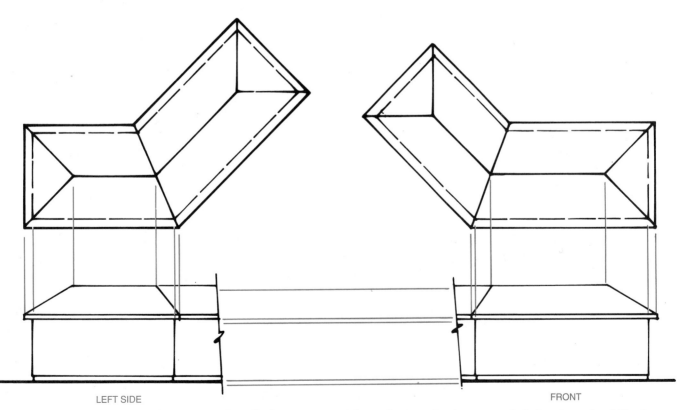

Figure 21–5 When drawing elevations with walls that are not at right angles to each other, expanded elevations should be drawn.

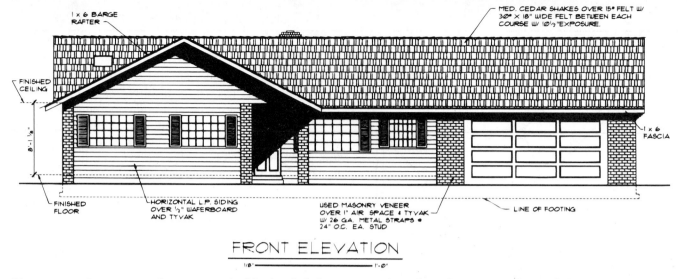

FRONT ELEVATION

Figure 21–6 Presentation drawings are highly detailed drawings used to show the exterior shape of the structure. *Courtesy Michelle Cartwright.*

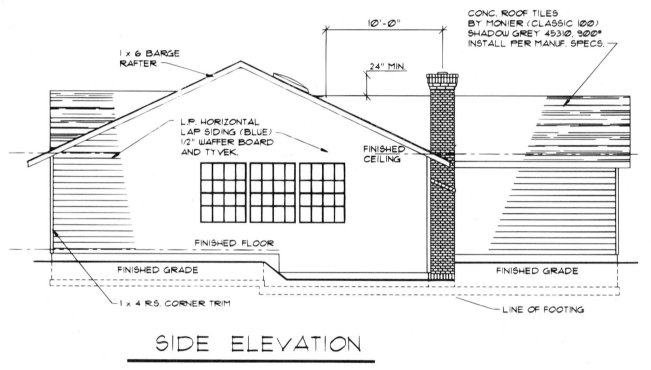

SIDE ELEVATION

Figure 21–7 Working elevations contain less finish detail but still accurately show the shape of a structure. *Courtesy April Muilenburg.*

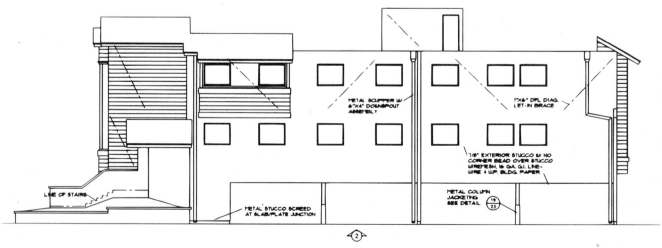

Figure 21–8 Elevations of large structures are typically drawn at a small scale such as $\frac{1}{16}'' = 1'-0''$ with very little detail shown. *Courtesy Structureform Masters, Inc.*

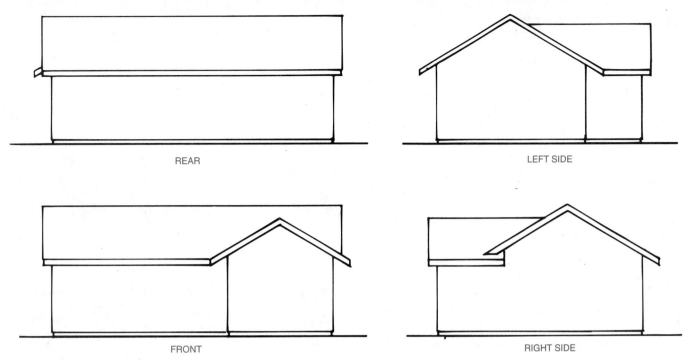

Figure 21–9 A common method of elevation layout is to place a side elevation beside both the front and the rear elevations. This allows for heights to be directly transferred from one view to the other.

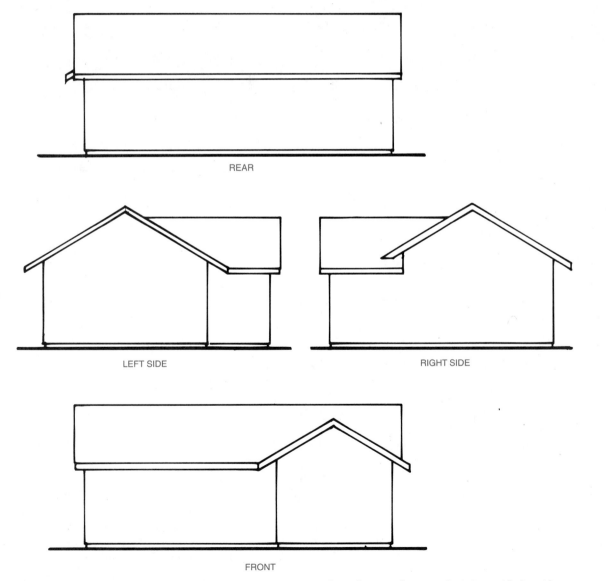

Figure 21–10 An alternative elevation arrangement is to place the two shortest elevations side by side.

Figure 21–11 A common alternative to the uniform size and texture of 235# composition shingles are 300# shingles. The random pattern is typified in this photo of "Architect 80." *Courtesy Genstar Roofing Products Company.*

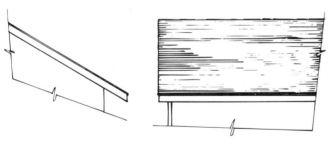

Figure 21–12 Drawing asphalt shingles.

Figure 21–13 Wood shakes and shingles have much more texture than asphalt shingles.

Figure 21–14 Shingles are made to resemble wood shakes. These shingles are Masonite. *Courtesy Masonite Corporation.*

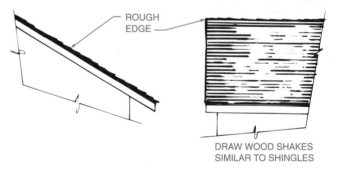

Figure 21–15 Drawing wood shakes and shingles.

Tile. Concrete, clay, or a lightweight simulated tile material present a very rugged surface at the ridge and edge, as well as many shadows throughout the roof. Figure 21–16 shows flat tile. This type of tile is usually drawn in a manner similar to the drawing in Figure 21–17. Figure 21–18 shows a roof with Spanish tile. This type of tile is often drawn as shown in Figure 21–19. Rub-on films are available for some types of tile roofs. Figure 21–20 shows an elevation where the drafter used a rub-on film to represent the tile rather than drawing each tile.

Metal. Metal shingles are usually drawn in a manner similar to asphalt shingles.

Built-up Roofs. Because of the low pitch and the lack of surface texture, built-up roofs are usually outlined and left blank. Occasionally a built-up roof will be covered with 2″ or 3″ (50–75 mm) diameter rock, as seen in Figure 21–21. The drawing technique for this roof can be seen in Figure 21–22.

Skylights. Skylights may be made of either flat glass or domed plastic. Although they come in a variety of shapes and styles, skylights usually resemble those shown in Figure 21–23. Depending on the pitch of the roof, skylights may or may not be drawn. On very low-pitched roofs a skylight may be unrecognizable. On roofs over

Figure 21–16 Flat tiles are a common roofing material in many parts of the country.

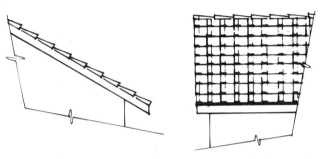

Figure 21–17 Flat tiles are drawn using methods similar to drawing shakes.

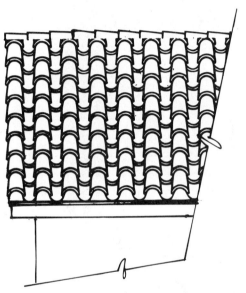

Figure 21–20 Tiles may be drawn in elevation by using rub-on film. *Courtesy Chartpak.*

Figure 21–18 Curved, or Spanish, tiles are a traditional roofing material in many areas of the country.

Figure 21–21 Built-up roofing with a gravel cover is common in many areas where the weather is mild all year.

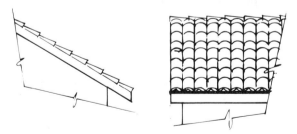

Figure 21–19 Drawing curved tiles.

3/12 pitch the shape of the skylight can usually be drawn without creating confusion. Unless the roof is very steep, a rectangular skylight will appear almost square. The flatter the roof, the more distortion there will be in the size of the skylight. Figure 21–24 shows common methods of drawing both flat glass and domed skylights.

Wall Coverings

Exterior wall coverings are usually made of wood, wood substitutes, masonry, metal, or plaster or stucco. Each has its own distinctive look in elevation.

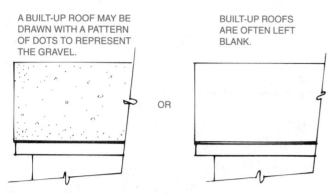

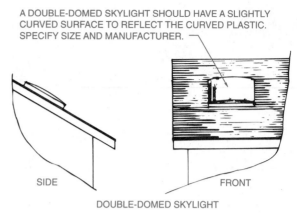

A DOUBLE-DOMED SKYLIGHT SHOULD HAVE A SLIGHTLY CURVED SURFACE TO REFLECT THE CURVED PLASTIC. SPECIFY SIZE AND MANUFACTURER.

DOUBLE-DOMED SKYLIGHT

Figure 21–22 Drawing built-up roofs.

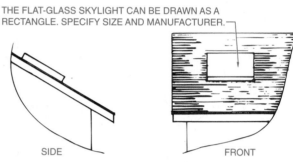

THE FLAT-GLASS SKYLIGHT CAN BE DRAWN AS A RECTANGLE. SPECIFY SIZE AND MANUFACTURER.

FLAT-GLASS SKYLIGHT

Figure 21–24 Common methods of drawing skylights.

(a)

(b)

Figure 21–23 (a) Domed and (b) flat glass skylights are a common feature of many roofs. (a) *Courtesy Duralite;* (b) *courtesy VELUX-AMERICA, INC.*

Wood. Wood siding can be installed in large sheets or in individual pieces. Plywood sheets are a popular wood siding because of the low cost and ease of installation. Individual pieces of wood provide an attractive finish but usually cost more than plywood. This higher cost results from differences in material and the labor to install each individual piece.

Plywood siding can have many textures, finishes, and patterns. Textures and finishes are not shown on the elevations but may be specified in a general note. Patterns in the plywood are usually shown. The most common patterns in plywood are T-1-11 (Figure 21–25), reverse board and batten shown (Figure 21–26), and plain or rough-cut ply (Figure 21–27). Figure 21–28 shows methods for drawing each type of siding.

Lumber siding comes in several types and is laid in many patterns. Among the most common lumber for sidings are cedar, redwood, pine, fir, spruce, and hemlock. Common styles of lumber siding are tongue and groove, bevel, and channel siding. Each kind can be seen in Figures 21–29 through 21–31. Figure 21–32 shows common shape of wood siding. Each of these materials can be installed in a vertical, horizontal, or diagonal position. The material and type of siding must be specified in a general note on the elevations. The pattern in which the siding is to be installed must be shown on the elevations in a manner similar to Figure 21–33. The type of siding and the position in which it is laid will affect how the siding appears at a corner. Figure 21–34 shows two common methods of corner treatment.

Figure 21–25 Plywood T-1-11 is a common siding. *Courtesy American Plywood Association.*

Figure 21–27 Plywood siding can be seen on this home designed by the Santa Rosa, California, architectural firm of Roland/Miller/Associate. *Courtesy American Plywood Association.*

Figure 21-26 Batt-on-board siding. *Courtesy Weather Shield Mfg., Inc.*

Wood shingles can either be installed individually or in panels. Shingles are often drawn as shown in Figure 21–35.

Wood Substitutes. Hardboard, aluminum, and vinyl siding can be produced to resemble lumber siding. Figure 21–36 shows a home finished with hardboard siding. Hardboard siding is generally installed in large sheets similar to plywood but often has more detail than plywood or lumber siding. It is typically drawn using methods similar to those used for drawing lumber sidings. Each of the major national wood distributors has also

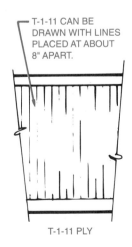
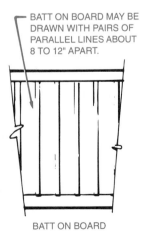
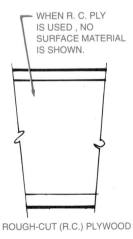

T-1-11 CAN BE DRAWN WITH LINES PLACED AT ABOUT 8" APART.

BATT ON BOARD MAY BE DRAWN WITH PAIRS OF PARALLEL LINES ABOUT 8 TO 12" APART.

WHEN R. C. PLY IS USED , NO SURFACE MATERIAL IS SHOWN.

T-1-11 PLY

BATT ON BOARD

ROUGH-CUT (R.C.) PLYWOOD

Figure 21–28 Drawing plywood siding in elevation.

Figure 21–29 Tongue and groove siding. *Courtesy Western Wood Products Association.*

Figure 21–31 Channel siding.

Figure 21–30 Bevel siding can be seen on this residence in Idyllwild, California, designed by Outback Design & Construction. *Courtesy California Redwood Association.*

developed siding products made from wood by-products that resemble individual pieces of beveled siding. Strands of wood created during the milling process are saturated with a water-resistant resin binder and compressed under extreme heat and pressure. The exterior surface typically has an embossed finish to resemble the natural surface of cedar. Most engineered lap sidings are primed to provide protection from moisture prior to installation.

Aluminum and vinyl sidings also resemble lumber siding in appearance, as shown in Figure 21–37. Aluminum and vinyl sidings are drawn similar to their lumber counterpart.

Masonry. Masonry finishes may be made of brick, concrete block, or stone.

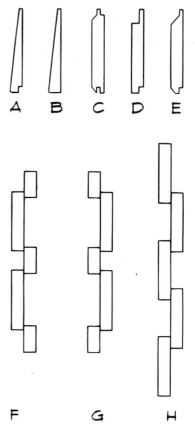

Figure 21–32 Common types of siding: (a) bevel; (b) rabbeted; (c) tongue and groove; (d) channel shiplap; (e) V shiplap; (f–h) these types can have a variety of appearances depending on the width of the boards and battens being used.

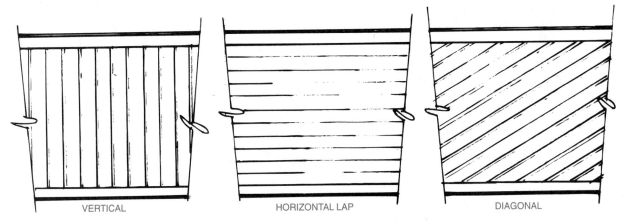

Figure 21–33   The type of siding and the position in which it is to be installed must be shown on the elevations.

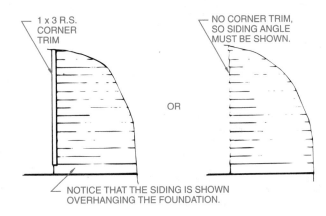

Figure 21–34   Common methods of corner treatment to be shown on elevations.

Figure 21–35   Drawing shingles. Notice the contrast in line weights to create shadows and texture. *Courtesy Cedar Shake and Shingle, Bureau.*

Bricks come in a variety of sizes, patterns, and textures. When drawing elevations, the drafter must represent the pattern of the bricks on the drawing and the material and texture in the written specifications. A common method for drawing bricks is shown in Figure 21–38. Although bricks are not usually drawn exactly to scale, the proportions of the brick must be maintained. Figure 21–39 shows common patterns available for rub-on bricks.

A drafter will often be required show size, pattern and texture on the elevation when drawing concrete blocks. Figure 21–40 shows methods of representing some of the various concrete blocks.

Figure 21–36 Hardboard siding. *Courtesy Georgia-Pacific Corporation.*

Figure 21–37 Vinyl and aluminum siding match the shape and size of their wood counterparts and offer the strength and protection from weathering by the use of lightweight metal. *Courtesy Alside.*

Figure 21–39 Rub-on brick patterns. *Courtesy Chartpak.*

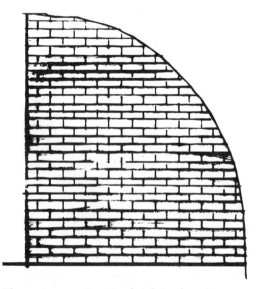

Figure 21–38 Drawing brick in elevation.

Figure 21–40 Methods of representing concrete blocks in elevation.

Stone or rock finishes also come in a wide variety of sizes and shapes, and are laid in a variety of patterns, as shown in Figure 21–41. Stone or rock may be natural or artificial. Both appear the same when drawn in elevation. When representing stone and rock surfaces in elevation, the drafter must be careful to represent the irregular shape, as shown in Figure 21–42.

Metal. Although primarily a roofing material, metal can be used as an attractive wall covering. Drawing metal in elevation uses a method similar to drawing lumber siding.

Plaster or Stucco. Although primarily used in areas with little rainfall, plaster or stucco can be found throughout the country. Figure 21–43 shows an example

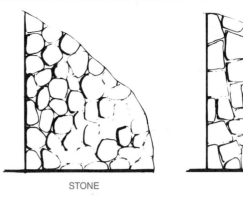

STONE   ROCK

Figure 21–42 Drawing stone and rock surfaces in elevation.

Figure 21–43 Stucco and plaster can be installed in many different patterns and colors to provide a durable finish. *Courtesy Western Red Cedar Lumber Association.*

Figure 21–41 Stone is used in a variety of shapes and patterns. *Courtesy Stucco Stone Products, Inc., Napa.*

of a common stucco pattern. No matter what the pattern, stucco is typically drawn as shown in Figure 21–44.

Similar to the appearance of stucco or plaster are exterior insulation and finishing systems (EIFS) such as Dryvit. Typically a rigid insulation board is used as a base for a fiberglass-reinforced base coat. A weather-resistant colored finish coat is then applied by trowel to seal the structure. Shapes made out of insulation board can be added wherever three-dimensional details are desired. Figure 21–45 shows a home sealed with Dryvit.

Doors

On elevations, doors are drawn to resemble the type of door specified in the door schedule. The drafter should be careful not to try to reproduce an exact likeness of a door. This is especially true on entry doors and garage doors. Typically these doors have a decorative pattern on them. It is important to show this pattern, but do not

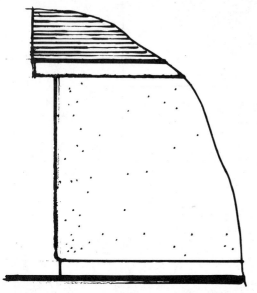

Figure 21–44   Drawing stucco or plaster in elevation.

Figure 21–45   EIFS offers the look of stucco while providing added insulation and durability. *Courtesy Dryvit.*

spend time trying to reproduce the exact pattern. Since the door is manufactured, the drafter is wasting time on details that add nothing to the plan. Figure 21–46 shows the layout of a raised-panel door, and Figure 21–47 shows how other common types of doors can be drawn.

Windows.   The same precautions about drawing needless details for doors should be taken when drawing windows. Care must be given to the frame material when drawing windows. Wooden frames are wider than metal frames. When the elevations are started in the prelimi-

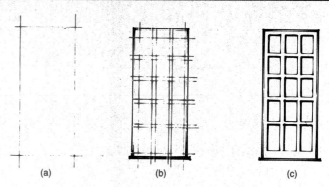

Figure 21–46   Layout steps for drawing a raised-panel door.

nary stages of design, the drafter may not know what type of frames will be used. In this case, the drafter should draw the windows in the most typical usage for the area. Figure 21–48 shows the layout steps for an aluminum sliding window. Figure 21–49 shows how other common types of windows can be drawn.

Rails

Rails can be solid to match the wall material or open. Open rails are typically made of wood or wrought iron. Although spacing varies, vertical rails are usually required to be no more than 6" (152 mm) clear. Rails are often drawn as shown in Figure 21–50. Although it is not necessary to measure the location of each vertical, care must be taken to space the rails evenly.

Shutters

Shutters are sometimes used as part of the exterior design and must be shown on the elevations. Figure 21–51 shows a typical shutter and how it could be drawn. Although the spacing should be uniform, the drafter should try to lay out the shutter by eye rather than measuring each louver.

Eave Vents

Drawing eave vents is similar to drawing shutters. Figure 21–52 shows common methods of drawing eave vents.

Chimney

Several different methods can be used to represent a chimney. Figure 21–53 shows examples of wood and masonry chimneys.

▼ CADD ELEVATION SYMBOLS

Drawing exterior and interior elevation features with CADD is easy. Most architectural CADD programs have a variety of elevation symbols representing doors, windows, and materials. For example, when selecting symbols, pick a DOOR or WINDOW symbol from a tablet menu library; a screen menu will display your options.

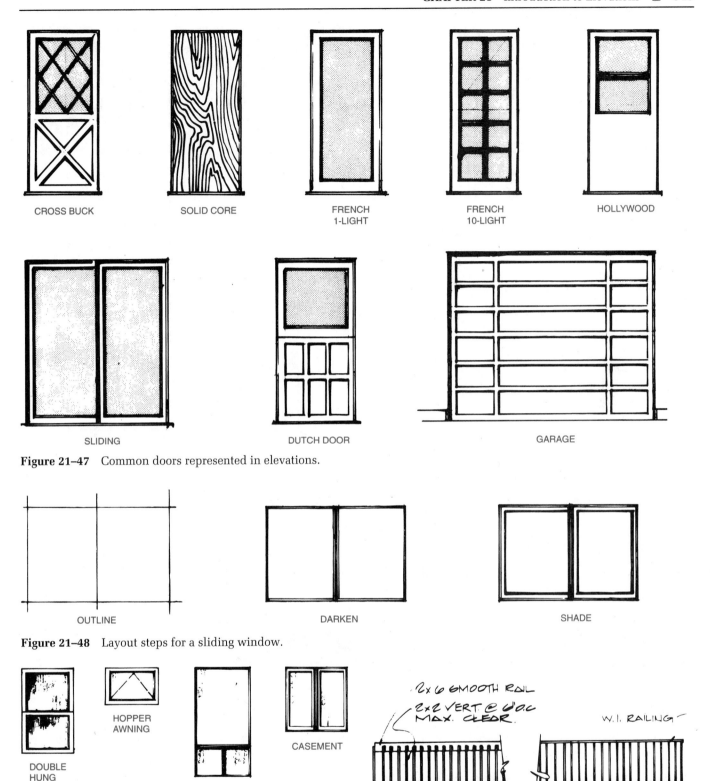

Figure 21–47 Common doors represented in elevations.

Figure 21–48 Layout steps for a sliding window.

Figure 21–49 Representing common window shapes on elevations.

Figure 21–50 Common types of railings.

You then pick doors such as entrance, flush, or French. You can also add trim and door knobs. When you pick WINDOW from the tablet menu library, you get options such as sliders, casement, double hung, awning, and fixed. Figure 21–54 shows door and window tablet menu overlays. You can customize your window symbols with divided lights, arrows, reflection marks, solid black glass, or hinge marks.

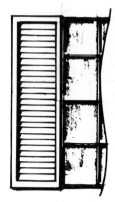

Figure 21–51 Representing shutters in elevation.

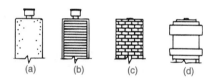

Figure 21–52 Common methods of representing attic vents on elevations.

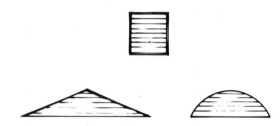

Figure 21–53 Common methods of drawing chimneys: (a) stucco chase with metal cap; (b) horizontal wood siding with metal cap; (c) common brick; (d) decorative masonry.

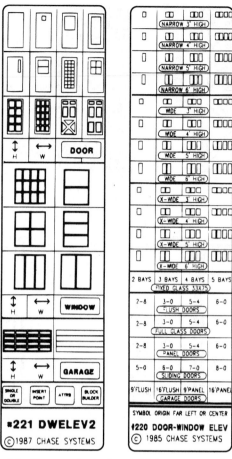

Figure 21–54 Door and window tablet menu overlays. *Courtesy Chase Systems.*

CHAPTER 21 — Introduction to Elevations Test

DIRECTIONS

Answer the questions with short, complete statements or drawings as needed.

1. Letter your name and the date at the top of the sheet.
2. Letter the question number, and provide the answer. You do not need to write out the question.

QUESTIONS

Question 21–1 Under what circumstances could a drafter be required to draw only three elevations?

Question 21–2 When are more than four elevations required?

Question 21–3 What are the goals of an elevation?

Question 21–4 What is the most common scale for drawing elevations?

Question 21–5 Would an elevation that has been drawn at a scale of $\frac{1}{16}'' = 1'-0''$ require the same methods to draw the finishing materials as an elevation that has been drawn at some larger scale?

Question 21–6 Describe two methods of transferring the heights of one elevation to another.

Question 21–7 Describe two different methods of showing concrete tile roofs.

Question 21–8 What are the two major types of wood siding?

Question 21–9   When drawing a home with plywood siding, how should the texture of the wood be expressed?

Question 21–10   Sketch the most typically used pattern for brick work.

Question 21–11   What problems is the drafter likely to encounter when drawing stone?

Question 21–12   Sketch the way wood shingles appear when looking at the gable end of a roof.

Question 21–13   Give the major consideration when drawing doors in elevation.

Question 21–14   What are the two most common materials used for rails?

Question 21–15   When drawing stucco, how should the pattern be expressed?

CHAPTER 22

Elevation Layout and Drawing Techniques

Drawing elevations is one of the most enjoyable projects in architecture. As a student, though, you should keep in mind that a beginning drafter very rarely gets to draw the main elevations. The architect or designer usually draws the main elevations in the preliminary stage of design, and then a senior drafter gets to finish the elevations. As a beginner you may be introduced to elevations at your first job by making corrections resulting from design changes. Not a lot of creativity is involved in corrections, but it is a start. As you display your ability, you'll most likely be involved in earlier stages of the drawing process. Elevations are drawn in the three steps of layout, drawing, and lettering. Usually you will be given the preliminary design elevation or the designer's sketch to use as a guide.

▼ LAYOUT

The size of drawing sheet you'll be using will affect the scale, placement, and layout of the elevations. Because C-sized paper is so commonly used in educational settings, these instructions will assume the drawing is to be made on that size sheet. Usually two elevations cannot be placed side by side on C-sized material. Thus, you will not be able to use a side elevation for projecting heights onto the front or rear elevations. The following instructions will be given for drawing the front elevation at a scale of ¼" = 1'-0" and the rear and both side elevations at a scale of ⅛" = 1'-0". Using two scales will require more steps than when all of the elevations are drawn at one scale.

The layout process can be divided into the stages of overall shape layout, roof layout, and layout of openings. Use construction lines for each stage.

Overall-shape Layout

Step 1.   With a clean sheet of drawing paper in place, tape down a print of the floor and roof plans to use as a guide for all horizontal measurements. If a separate roof plan has not been drawn, draw the outline of the roof shape on the print of the floor plan. Assume 24" (600 mm) overhangs and

a 12" (300 mm) gable end wall overhang. Each horizontal measurement can be projected as shown in Figure 22–1.

Step 2.   See Figure 22–2 for steps 2 through 8. Lay out a baseline to represent the finish grade.

Step 3.   Project construction lines down from the floor plan to represent the edges of the house.

Step 4.   Establish the finish floor line. See Figure 22–3 for help in determining minimum measurements.

Step 5.   Measure up from the finished floor line, and draw a line to represent the ceiling level, (typically 8'-0" (2440 mm).

Step 6.   Measure up the depth of the floor joist to establish the finished floor level of the second floor. If the floor joist size has not been determined, assume 8" (200 mm) depth (2 × 8s) will be used.

Step 7.   Measure up 8'-0" (2440 mm) from the second finished floor line, and draw a line to represent the upper ceiling level.

Step 8.   Measure up from each finished floor line 6'-8" (2030 mm), and draw a line to represent the tops of the windows and doors.

Roof Layout

Because a side elevation is not being drawn beside the front elevation, there is no easy way to determine the true roof height. If the sections have been drawn before the elevations, the height from the ceiling to the ridge could be measured on one of the section drawings and then transferred to the elevations. Although this might save time, it also might duplicate an error. If the roof was incorrectly drawn on the section drawing, the error will be repeated if dimensions are simply transferred from section to elevation.

The best procedure to determine the height of the roof in the front view is to lay out a side view lightly. This can be done in the middle of what will become the front view. An example of this can be seen in Figure 22–4. On this home, three different roofs must be drawn: the roof over the two-story area, garage, and porch. It would be best to lay out the porch roof last since it is a projection of the

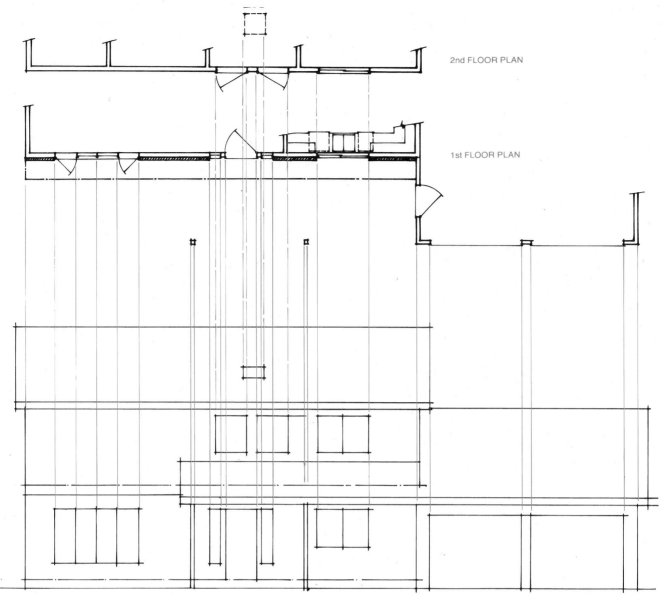

2nd FLOOR PLAN

1st FLOOR PLAN

Figure 22–1 Layout of an elevation. With a print of the floor plans taped above the area where the elevations will be drawn, horizontal distances can be projected down to the elevation.

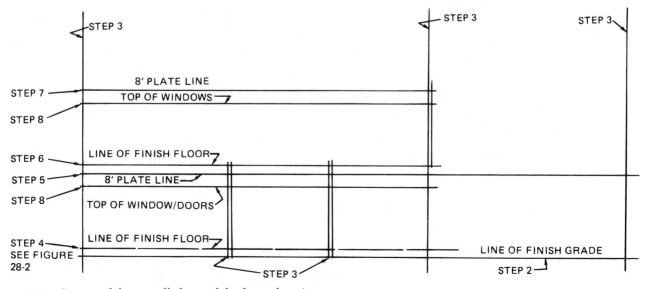

STEP 3

STEP 3

STEP 3

8' PLATE LINE

STEP 7 → TOP OF WINDOWS

STEP 8

STEP 6 → LINE OF FINISH FLOOR

STEP 5 → 8' PLATE LINE

STEP 8 → TOP OF WINDOW/DOORS

STEP 4 → LINE OF FINISH FLOOR
SEE FIGURE
28-2

LINE OF FINISH GRADE

STEP 3

STEP 2

Figure 22–2 Layout of the overall shape of the front elevation.

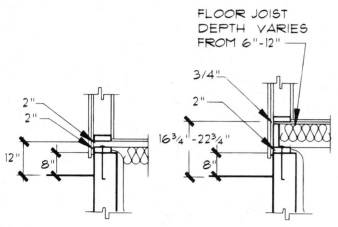

FLOOR JOIST
DEPTH VARIES
FROM 6"-12"

3/4"

2"

2"

2"

12"

8"

16¾"-22¾"

8"

Figure 22-3 The height of the finish floor above grade will be determined by the type of the floor framing system. Exact framing sizes are often rounded up to the nearest inch when drawing elevations.

garage roof. The upper or garage roof can be drawn by following steps 9 through 19 in Figure 22–5.

Step 9. Lay out a line 1'-0" (300 mm) from the end of each wall to represent the edge of the roof for each gable end wall overhang.

Step 10. Lay out a line 2'-0" (600 mm) from one of the end walls to represent the overhang projection.

Step 11. Lay out a line to represent the ridge. Because the lower floor is 36' wide (11,000 mm), the ridge will be 18' (5500 mm) from the front wall. For the upper roof, the ridge will be 15' (4570 mm) from the front wall.

Step 12. Determine the roof slope. For this home, draw the roof at a ⁵⁄₁₂ pitch. A ⁵⁄₁₂ slope equals 22½ degrees. See Section V for a complete explanation of pitches.

Step 13. At the intersection of the wall and ceiling lines (lines drawn in steps 3 and 7), draw a line at the desired angle to represent the roof. This line will represent the bottom of the truss or rafter.

Step 14. Measure up 6" (150 mm), and draw a line parallel to the line just drawn, which will represent the top of the rafter. On the elevations, it is not important that the exact depth of the rafters be represented.

Step 15. Establish the top of the roof. The true height of the roof can be measured where the line from step 14 intersects the line from step 11.

Step 16. Establish the bottom of the eave. This can be determined from the intersection of the line from step 10 with the lines of steps 13 and 14.

Step 17. Lay out the upper roof using the same procedure of drawing a partial side elevation and projecting exact heights.

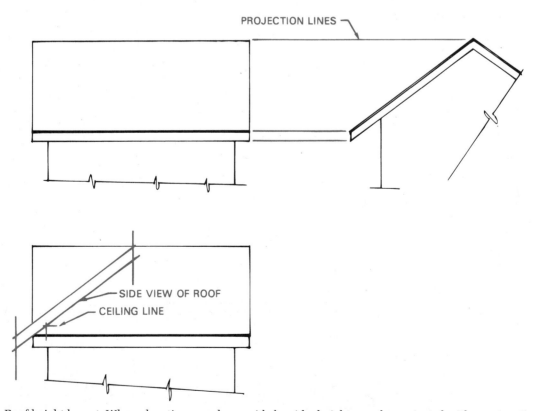

PROJECTION LINES

SIDE VIEW OF ROOF
CEILING LINE

Figure 22-4 Roof height layout. When elevations are drawn side by side, heights can be projected with construction lines. When only one elevation is drawn, a side elevation will need to be drawn in the middle of what will become the front elevation.

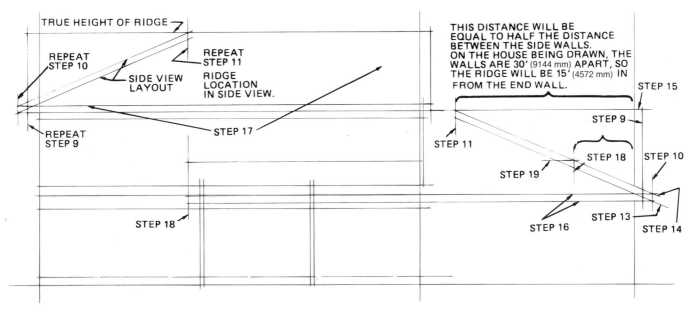

Figure 22–5 Layout of the roof shapes. When looking at a roof from the front, the true height cannot be determined without drawing a side elevation. If the roof pitch and the ridge location are known, the true height can be quickly and accurately drawn.

The porch roof will be an extension of the garage roof, so much of the layout work will have been done. Only the left end and the upper limits of the roof must be determined. Because the ends of the upper and garage roof extend 12″ (300 mm), this distance will be maintained. The upper edge of the roof will be formed by the intersection of the roof and the wall for the upper floor. This height can be determined by following steps 18 and 19.

Step 18. Place a line on the slope lines (the lines from steps 13 and 14) to represent the distance from the garage wall to the front wall on the upper floor plan.

Step 19. At the intersection of the lines from steps 14 and 18, project a line to represent the top of the roof.

You now have the basic shape of the walls and roof blocked out. The only items to be drawn now are the door and window openings.

Layout of Openings

See Figure 22–6 for steps 20 through 23.

Step 20. Lay out the width of all doors, windows, and skylights by projecting their location from the floor plan.

Step 21. Lay out the bottom of the windows and doors. The doors should extend to the line representing the finished floor. Determine the depth of the windows from the window schedule, and draw a line to represent the correct depth.

Step 22. Determine the height of the skylights. This must be done by measuring the true size of the

skylights on the line from step 14, and then projecting this height to the proper location.

Step 23. Draw the shape of the garage doors. Remember that the height of these doors is measured from the line representing the ground, not from the finished floor level.

You now have the front elevation blocked out. The same steps can be used to draw the right and left side and rear elevations. The major difference in laying out these elevations is that the widths of walls, windows, and doors cannot be projected from the floor plan. To lay out these items, you will need to rely on the dimensions of the floor plan.

Start by drawing either side elevation. The side elevation can be drawn by following the procedure for the layout of the front elevation. Figure 22–7 shows a completed side elevation. Once this is drawn, use the heights of the side elevation to establish the heights for the other side and the rear elevation, as shown in Figure 22–8.

▼ DRAWING FINISHED-QUALITY LINES

With all of the elevations drawn with construction lines, they must now be completed with the finished-quality lines. An H and 2H lead will provide good contrast for many of the materials. The use of ink and graphite will also produce a very nice effect. Methods for drawing items required in elevations were discussed in Chapter 21. Line quality cannot be overstressed. By using varied line widths and density, a more realistic elevation can be created.

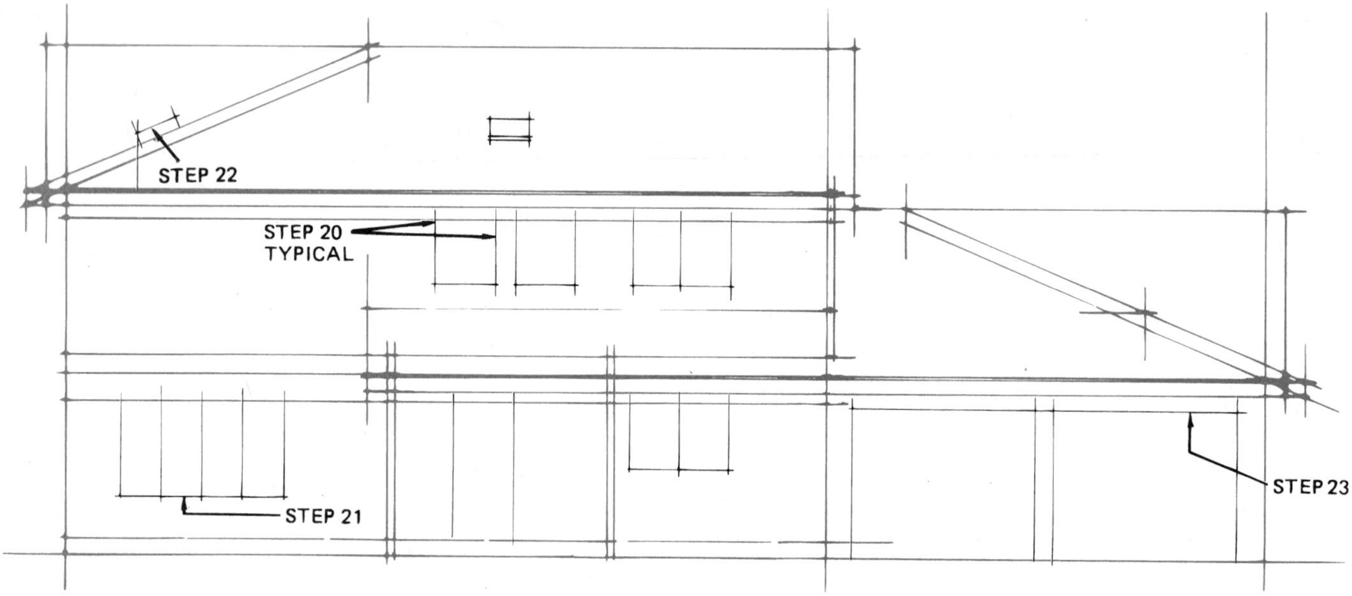

Figure 22–6 Layout of the doors, windows, and skylights. The widths can be projected down from the floor plan. Vertical heights for windows came from the window schedule.

Figure 22–7 The layout of a side elevation is similar to the layout of the front elevation.

A helpful procedure for completing the elevations is to begin to darken the lines at the top of the page first and work down the page to the bottom. Because you will be using a soft lead, it will tend to smear easily. If the lower elevations are drawn first, cover them with a sheet of paper when they are complete to prevent smearing.

Another helpful procedure for completing the elevations is to start at the top of one elevation and work toward the ground level. Also start by drawing surfaces that are in the foreground and work toward the background of the elevation. This procedure will help cut down on erasing background items for items that are in the foreground. Figure 22–9 shows a completed front elevation. It can be drawn by following steps 24 through 28.

Step 24. Draw the outline of all roofs.

Step 25. Draw the roofing material.

Step 26. Draw all windows and doors.

Step 27. Draw all corner trim and posts.

Step 28. Draw the siding.

At this point your front elevation is completely drawn. The same procedures can be used to finish the rear and side elevations, shown in Figure 22–10. Note that only a small portion of the siding has been shown in these elevations. It is common practice to have the front elevation highly detailed and have the other elevations show just the minimum information for the construction crew.

▼ LETTERING

With the elevations completely drawn, the material must now be specified. Chapter 25 will discuss each of the materials to be specified in the notes that must be placed on the exterior elevations. If the elevations are all on one page, a material only needs to be specified on one elevation. Use Figures 22–9 and 22–10 as guides for placing notes. The notes needed include the following:

1. Siding. Generally the kind of siding, its thickness, and the backing material must be specified. Examples might include:
    - ○ Horiz. siding over ⅜″ (9.5 mm) plywood and 15# felt.
    - ○ Horiz. siding over ½″ (12.7 mm) waferboard and Tyvek.
    - ○ L. P. siding over ½″ (12.7) waferboard and Tyvek.
    - ○ Fishscale siding by Shakertown or equal over ½″ (12.7 mm) waferboard and Tyvek.
    - ○ ⅝″ (15.8 mm) T-1-11 siding w/grooves @ 4″ (100 mm) o.c. over Tyvek.
    - ○ 1″ (25 mm) ext. stucco over 15# felt w/ 26 ga. linewire at 24″ (600 mm) o.c. w/stucco wire mesh, and no corner bead.

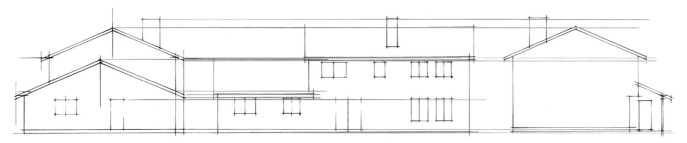

Figure 22–8 With a side elevation drawn, heights for the rear and other side can be projected quickly.

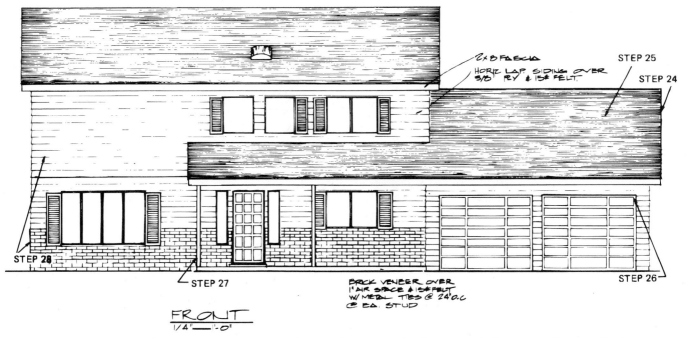

Figure 22–9 Drawing the front elevation using finished-quality lines. Note that features in the background such as the chimney are often omitted for clarity.

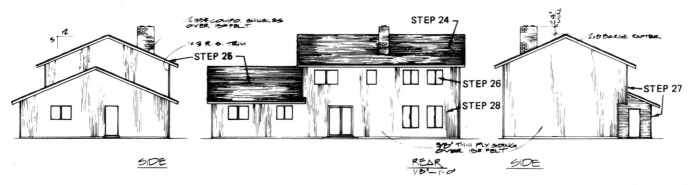

Figure 22–10 Drawing small-scale elevations does not require drawing all of the finishing materials. The same procedures used on the front elevation can be used on the side and rear elevations.

○ Exterior insulation and finishing system by Dryvit. Install as per manuf. specifications.

○ Monterey Sand, Nova IV Vinyl siding by Alside or equal installed over insulation board as per manuf. instructions.

○ Brick veneer over 15# felt over 1" (25 mm) air space with 26 ga. metal ties @ 24" (600 mm) o.c. @ ea. stud.

2. Corner and decorative trims. Common trim might include:

○ 1 × 4 (25 × 100 mm) R.S. corner trim.

○ Provide corner bead [for stucco or plaster applications].

○ Omit corner bead [for stucco or plaster applications].

3. Chimney height (2′ minimum above any roof within 10′).
4. Flatwork that would include any decks or concrete patios. Include rail material, vertical material, and spacing such as:
   ○ 2 × 6 (50 × 150 mm) dfl. smooth rail w/ 2 × 2 (50 × 50 mm) vert. @ 6″ (150 mm) max. clear.
   ○ 6 × 6 (150 × 150 mm) decorative post.
   ○ 2 × 6 (50 × 150 mm) cedar decking laid flat w/ ¼″ (6 mm) gap between.
   ○ W.I. railing w/ vert. @ 6″ (150 mm) clear max.
   ○ 4″ (100 mm) concrete flatwork.
5. Posts and headers.
6. Fascias and barge rafters.
7. Roof material. For example:
   ○ 235# composition shingles over 15# felt.
   ○ 300# composition shingles over 15# felt.
   ○ Med. cedar shakes over 15# felt w/ 30# × 18″ (450 mm) wide felt btwn. ea. course w/ 10½″ (270 mm) exposure.
   ○ Concrete tile roof [give manufacturer's name, style, color, and weight of tile; e.g., Chestnut brown, Classic 100 concrete roof tile (950# per sq.) by Monier over 15# felt]. Install as per manuf. specs.
8. Roof pitch.
9. Dimensions. The use of dimensions on elevations varies greatly from office to office. Some companies place no dimensions on elevations, using the sec-

tions to show all vertical heights. Other companies show the finished floor lines on the elevations and dimension floor-to-ceiling levels. Unless your instructor tells you otherwise, no dimensions will be placed on the elevations.

▼ ALTERNATIVE ELEVATIONS

Many homes such as those that will be constructed in a subdivision are designed to be built more than once, and they require alternative facades. Figures 22–11 and 22–12 show examples of alternative elevations that were drawn using the same step-by-step process as the plan in Figure 22–9. These plans could be completed by using the information for the floor and roof plans. If drawn using manual drafting, much of the information such as wall and opening locations could be traced from the original elevation. By using CADD, information common to all three drawings can be placed on a base layer, with information specific to each roof plan placed on separate layers. The elevations for the gable, hip, and Dutch hip roofs can be stored in one drawing file or separated for ease of plotting.

▼ DRAWING IRREGULAR SHAPES

Not all plans that will be drawn have walls constructed at 90° to each other. When the house has an irregular shape, a floor plan and a roof plan will be required to draw the

Figure 22–11 An elevation with a hip roof can be drawn using the process described in Chapters 20 and 21. Locate the ridge height, assuming a gable roof will be used. Once the ridge height is determined, the guideline used to locate the ridge can be used to form the hips.

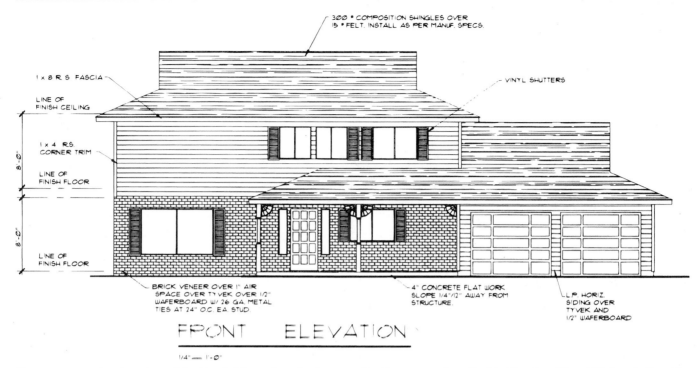

300 # COMPOSITION SHINGLES OVER
15 # FELT. INSTALL AS PER MANUF. SPECS.

1 x 8 R.S. FASCIA

LINE OF
FINISH CEILING

1 x 4 R.S.
CORNER TRIM

LINE OF
FINISH FLOOR

8'-0"

8'-0"

LINE OF
FINISH FLOOR

VINYL SHUTTERS

BRICK VENEER OVER 1" AIR
SPACE OVER TYVEK OVER 1/2"
WAFERBOARD W/ 26 GA. METAL
TIES AT 24" O.C. EA. STUD.

4" CONCRETE FLAT WORK
SLOPE 1/4"/12" AWAY FROM
STRUCTURE.

L.P. HORIZ.
SIDING OVER
TYVEK AND
1/2" WAFERBOARD

FRONT   ELEVATION

1/4" = 1'-0"

Figure 22-12   An elevation with a Dutch hip roof can be drawn using the process described in Chapters 20 and 21. Locate the ridge height, assuming a gable roof will be used. Once the ridge height is determined, the guideline used to locate the ridge can be used to form the hips. The location of the Dutch hip can be projected from the roof plan.

elevations. The process is similar to the process just described but will require more patience in projecting all of the roof and wall intersections. The projection of an irregularly shaped home can be seen in Figure 22-13. See Section V for an explanation of how the roof plan is drawn. Once the roof plan has been determined, it can be drawn on a print of the floor plan, and the intersections of the roof can be projected to the elevation.

▼ PROJECTING GRADES

Grade is used to describe the shape of the finish ground. Projecting ground levels is required on elevations when a

structure is being constructed on a hillside. Figure 22-14 shows the layout of a hillside residence. To complete this elevation a plot, floor, and roof plan were used to project locations. Usually the grading plan and the floor plan are drawn at different scales. The easiest method of projecting the ground level onto the elevation is to measure the distane on the plot plan from one contour line to the next. This distance can then be marked on the floor plan and projected onto the elevation.

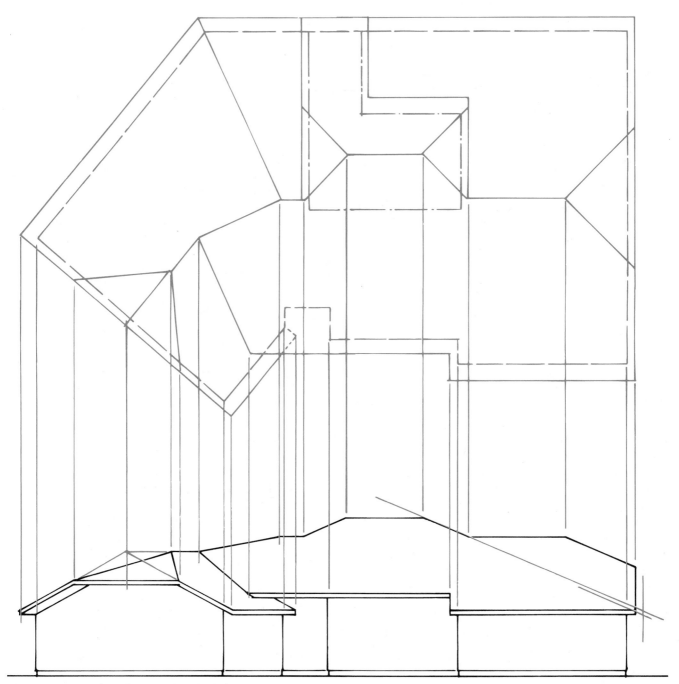

Figure 22–13 When drawing the elevations of a structure with an irregular shape, great care must be taken in the projection methods. This elevation was prepared by drawing the outline of the roof on a print of the floor plan and then projecting the needed intersections down to the drawing area. Notice on the right side of the elevation a line at the required angle has been drawn to represent the proper roof angle for establishing the true heights.

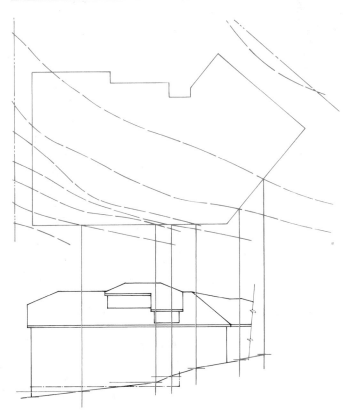

Figure 22–14 Projecting grades is similar to the method used to project roof lines. This elevation was drawn by using the information on the roof, floor, and grading plans. Only the projection lines for the grades are shown for clarity.

CADD Applications

DRAWING ELEVATIONS WITH CADD

Drawing exterior elevations with CADD is similar to techniques used to draw elevations manually. With manual drafting, you project lines from the floor plan to create lines for the elevation drawing. This was shown in Figure 22–1. With CADD, the same thing is done using a computer.

Create layers for the elevations with prefixes that clearly identify the contents. Layer titles such as ELEVWALLS, ELEVLAYOUT, ELEVROOF, ELEVWIN, ELEVSIDING, or ELEVTEXT will be useful for keeping track of features as the elevations are created.

After the floor plan is drawn, freeze all layers that are not relevant to the exterior of the residence. Make a *block* of the floor plan. A block is a custom symbol or drawing that can be transferred between drawings and displayed at any time. Set the drawing limits large enough so that the block of the floor can be inserted and rotated; in this way, each elevation can be projected from the floor plan using techniques similar to manual methods (Figure 22–15).

Once the roof and wall locations have been projected, the locations of doors and windows can be determined. Windows and doors representing specific styles can be created and stored as blocks in a library and then inserted as needed. Trees and bushes can also be created as blocks and inserted as needed. Once the design is completed, the elevations can be moved into the desired position for presentation and plotting. Figure 22–16 shows an example of elevations created using AutoCAD.

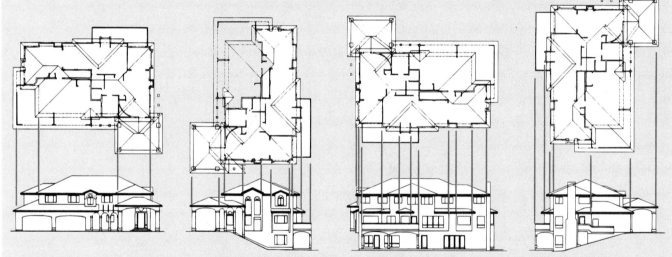

Figure 22–15 Use a block of the relevant parts of the floor and roof plans to line out each of the exterior elevations. Once tthe shapes have been projected, doors, windows, and other elements can be added. The completed elevations can be moved into any position for plotting.

Figure 22–16 A complete CADD elevation. *Courtesy of Piercy & Barclay Designers, Inc.*

CHAPTER

Elevation Layout and Drawing Techniques Test

DIRECTIONS

Answer the questions with short, complete statements or drawings as needed.

1. Letter your name and the date at the top of the sheet.
2. Letter the question number and provide the answer. You do not need to write out the question.

QUESTIONS

Question 22–1 Who typically draws the main elevations of a structure? Explain.

Question 22–2 List three methods of projecting heights from one elevation to another.

Question 22–3 What scales are typically used to draw elevations?

Question 22–4 What is the best method of determining the height of a roof when it is being viewed in a manner that does not show the pitch?

Question 22–5 What lines should be drawn first when using finished-line quality?

Question 22–6 What drawings are required to project grades onto an elevation? Briefly explain the process.

Question 22–7 What drawings are required to draw a structure with an irregular shape?

Question 22–8 Sketch the method of showing the chimney height, and provide the minimum dimensions that must be shown.

Question 22–9 What dimensions are typically shown on elevations?

Question 22–10 What angle is used to represent a 4/12 pitch?

Problems

DIRECTIONS

1. Select from the following sketches the one that corresponds to your floor-plan problem from Chapter 15, and draw the required elevations.
2. Draw the required elevations using the same type and size of drawing material that you used to draw your floor plan.
3. See each sketch to determine the scale and method of layout to be used. If no sketch has been given for the elevations of your plan, design the structure using materials suitable for your area.

4. Refer to your floor plan to determine all sizes and locations. If you find material that cannot be located using your floor plan, make additions to your floor plan as required.
5. Use the sketch as a reference only. Refer to the text of this chapter and notes from class lectures to complete the drawings.
6. When your drawing is complete, turn in a print for evaluation to your instructor.

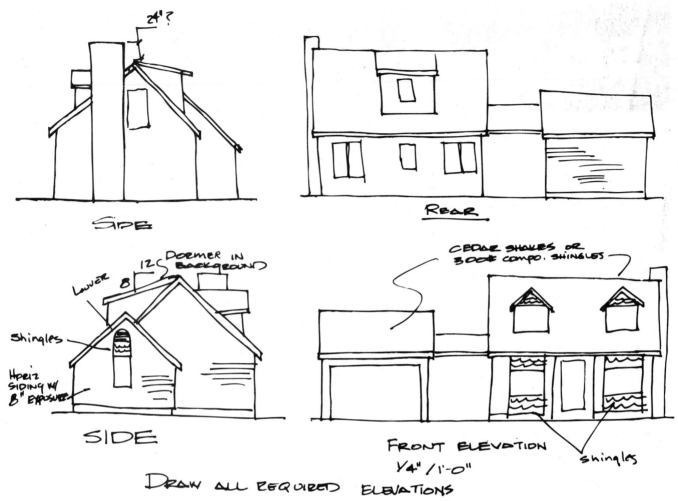

Problem 22–1   Cape Cod.

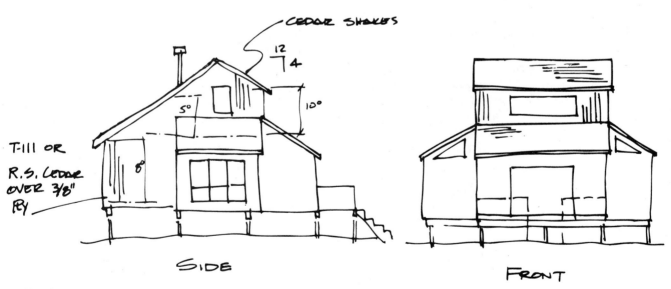

Problem 22–2   Cabin.

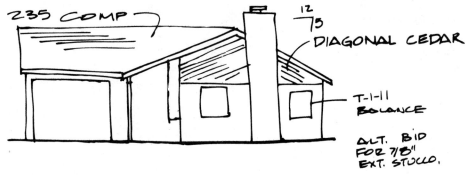

235 COMP

12
5

DIAGONAL CEDAR

T-1-11
BALANCE

ALT. BID
FOR 7/8"
EXT. STUCCO.

Problem 22–3 Ranch.

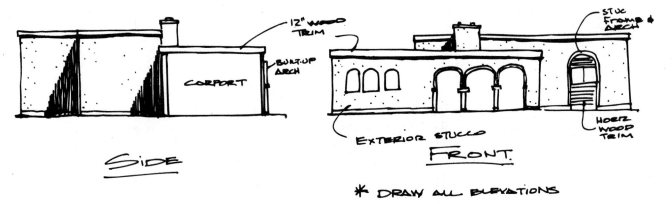

12" WOOD TRIM

BUILT-UP ARCH

CORPORT

SIDE

STUC FRAME & ARCH

HORIZ WOOD TRIM

EXTERIOR STUCCO

FRONT.

* DRAW ALL ELEVATIONS

Problem 22–4 Spanish.

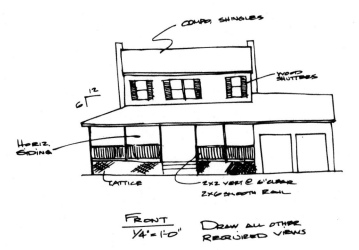

COMPO, SHINGLES

WOOD SHUTTERS

12
6

HORIZ. SIDING

LATTICE

2X2 VERT @ 6" CLEAR
2X6 SMOOTH RAIL

FRONT
1/4" = 1'-0"

DRAW ALL OTHER REQUIRED VIEWS

Problem 22–5 Early American.

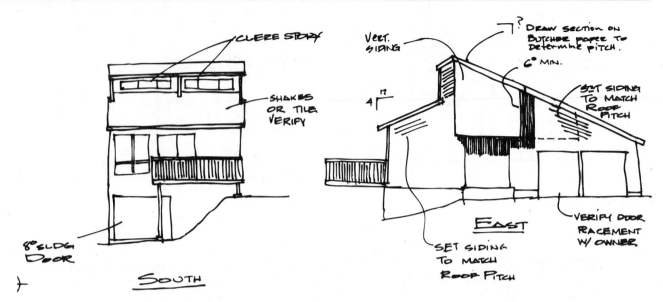

Problem 22–6 Contemporary.

Problem 22–7 to 22–15 Design the elevations for the structures started in Chapter 15 using suitable roof pitch and materials for your area. Sketch an elevation for each surface that will need to be drawn, and have it approved by your instructor prior to starting your drawing.

CHAPTER 23

Millwork and Cabinet Technology, Cabinet Elevations, and Layout

Cabinet construction is an element of the final details of a structure known as millwork. *Millwork* is any item that is considered finish trim or finish woodwork. Architects, designers, and drafters do not always draw complete details of millwork items. The practice of drawing millwork representations depends on the specific requirements of the project. For example, the custom plans for a residential or commercial structure may show very detailed and specific drawings of the finish woodwork in the form of plan views, elevations, construction details, and written specifications. Plans of a house may show only floor-plan views of cabinetry. Factors that determine the extent of millwork drawings on a set of plans are the requirements of specific lenders and local code jurisdictions. In most situations when a set of plans is ready for construction, it must be submitted to a lender for loan approval and to code officials for a building permit. The approval of local building officials is always necessary in areas where building permits are required. Before a set of plans can be completed, verify the requirements of the client, the lender, and the local building officials.

▼ TYPES OF MILLWORK

Millwork may be designed for appearance or for function. When designed for appearance, ornate and decorative millwork may be created with a group of shaped wooden forms placed together to capture a style of architecture. Molded millwork may be reproduced in as many styles as the designer can imagine. There are also vendors that manufacture a wide variety of prefabricated millwork moldings that are available at less cost than custom designs. See Figure 23–1.

Millwork that is used for function may be very plain in appearance. This type of millwork is also less expensive than standard sculptured forms. In some situations, wood millwork may be replaced with plastic or ceramic products. For example, in a public restroom or in a home laundry room a plastic or rubber base strip may be used around the wall at the floor to protect the wall. This material stands up to abuse better than wood. In some cases, drafters are involved in drawing details for specific millwork applications. There are as many possible de-

Figure 23–1 Prefabricated millwork. *Courtesy Cumberland Woodcraft Co., Inc., Carlisle, Pennsylvania.*

tails as there are design ideas. The following discussion provides a general example of the items to help define the terms.

Baseboards

Baseboards are placed at the intersection of walls and floors and are generally used to protect the wall from damage, as shown in Figure 23–2. Baseboards can be as ornate or as plain as the specific design or location dictates. In some designs, baseboards are the same shape as other millwork members, such as trim around doors and windows. Figure 23–3 shows some standard molded baseboards.

Wainscots

A *wainscot* is any wall finish where the material on the bottom portion of the wall is different from the upper portion. The lower portion is called the wainscot, and the material used is called wainscoting. Wainscots may be used on the interior or exterior of the structure. Exterior wainscoting is often brick veneer. Interior wainscoting may be any material that is used to divide walls into two visual sections. For example, wood paneling, plaster tex-ture, ceramic tile, wallpaper, or masonry may be used as wainscoting. Figure 23–4 shows the detail of wood wain-scot with plywood panels. The plywood panel may have an oak or other hardwood outer veneer to match the surrounding hardwood material. This is a less expensive method of constructing an attractive wood wainscot than with the hardwood panels shown in Figure 23–5.

Chair Rail

The *chair rail* has traditionally been placed horizon-tally on the wall at a height where chair backs would otherwise damage the wall. See Figure 23–6. Chair rails, when used, are usually found in the dining room, den, office, or in other areas where chairs are frequently

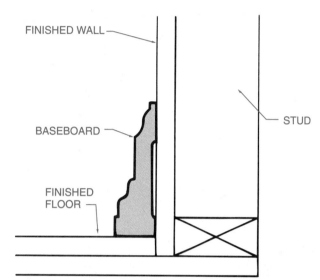

Figure 23–2 Baseboard.

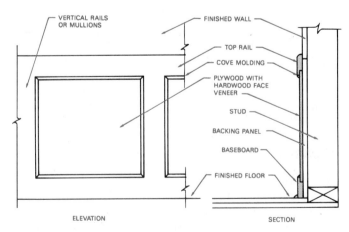

Figure 23–3 Standard baseboards. *Courtesy Hillsdale Sash and Door Co.*

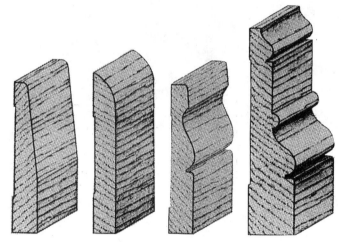

Figure 23–4 Plywood panel wood wainscot.

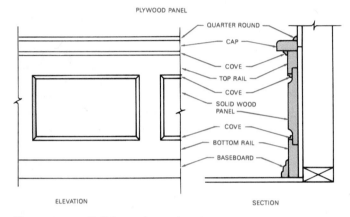

Figure 23–5 Solid wood-panel wainscot.

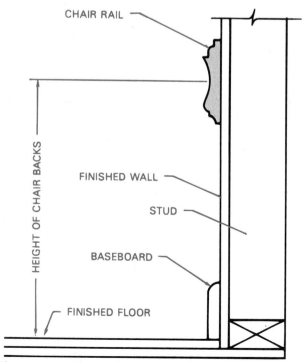

Figure 23–6 Chair rail.

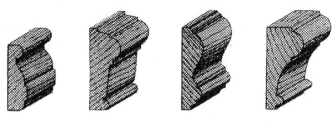

Figure 23–7 Standard chair rails. *Courtesy Hillsdale Pozzi Co.*

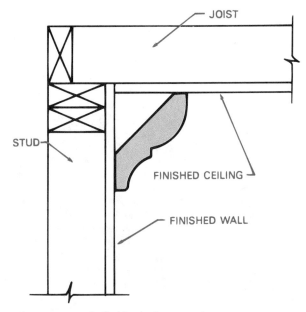

Figure 23–8 Individual piece cornice.

moved against a wall. Chair rails may be used in conjunction with wainscoting. In some applications the chair rail is an excellent division between two different materials or wall textures. Figure 23–7 shows sample chair rail moldings.

Cornice

The *cornice* is decorative trim placed in the corner where the wall meets the ceiling. A cornice may be a single shaped wood member called *cove* or *crown* molding, as seen in Figure 23–8, or the cornice may be a more elaborate structure made up of several individual wood members, as shown in Figure 23–9. Cornice boards are not commonly used; they traditionally fit into specific types of architectural styles, such as English Tudor, Victorian, or colonial. Figure 23–10 shows some standard cornice moldings. In most construction, where contemporary architecture or cost saving is important, wall-to-ceiling corners are left square.

Casings

Casings are the members that are used to trim around windows and doors. Casings are attached to the window or door jamb (frame) and to the adjacent wall, as shown in Figure 23–11. Casings may be decorative to match other

moldings or plain to serve the functional purpose of covering the space between the door or window jamb and the wall. Figure 23–12 shows a variety of standard casings.

Mantels

The mantel is an ornamental shelf or structure that is built above a fireplace opening, as seen in Figure 23–13. Mantel designs vary with individual preference. Mantels may be made of masonry as part of the fireplace structure, or ornate decorative wood moldings, or even a rough-sawn length of lumber bolted to the fireplace face. Figure 23–14 shows a traditional mantel application.

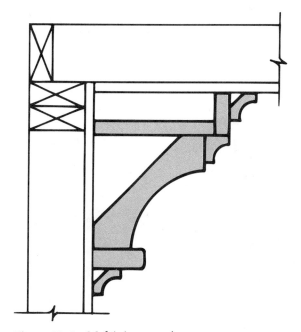

Figure 23–9 Multipiece cornice.

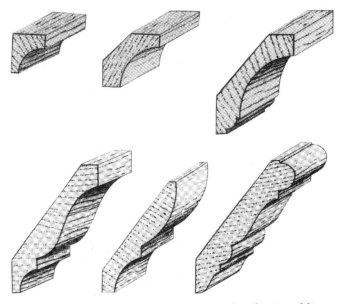

Figure 23–10 Standard cornice (cove and ceiling) moldings. *Courtesy Hillsdale Sash and Door Co.*

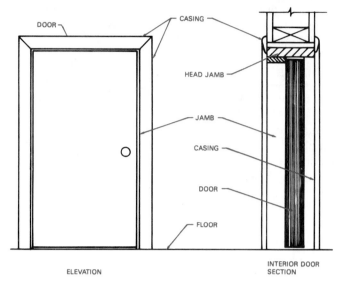

ELEVATION INTERIOR DOOR SECTION

Figure 23–11 Casings.

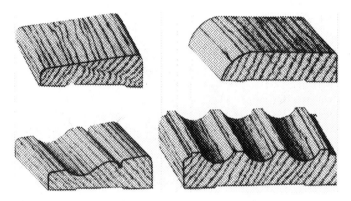

Figure 23–12 Standard casings. *Courtesy Hillsdale Sash and Door Co.*

Book Shelves

Book and display shelves may have simple construction with metal brackets and metal or wood shelving, or they may be built detailed as fine furniture. Book shelves are commonly found in the den, library, office, or living room. Shelves that are designed to display items other than books are also found in almost any room of the house. A common application is placing shelves on each side of a fireplace, as in Figure 23–13. Shelves are also used for functional purposes in storage rooms, linen closets, laundry rooms, and any other location where additional storage is needed.

Railings

Railings are used for safety at stairs, landings, decks, and open balconies where people could fall. Rails are recommended at any rise that measures 24″ (610 mm) or is three or more stair risers. Verify the size with local codes. Railings may also be used as decorative room dividers or for special accents. Railings may be built en-

closed or open and constructed of wood or metal. Enclosed rails are often the least expensive to build because they require less detailed labor than open rails. A decorative wood cap is typically used to trim the top of enclosed railings. See Figure 23–15.

Open railings can be one of the most attractive elements of interior design. Open railings may be as detailed as the designer's or craftsperson's imagination. Detailed open railings built of exotic hardwoods can be one of the most expensive items in the structure. Figure 23–16 shows some standard railing components. *Newels* are the posts that support the rail at the ends and at landings. *Balusters* are placed between the newels.

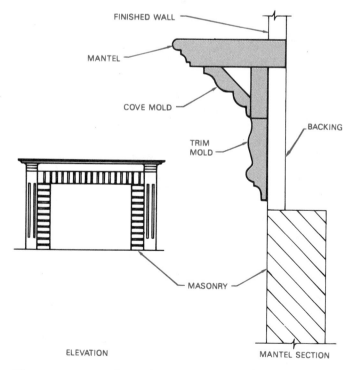

Figure 23–13 Mantel. *Photo by Bruce Davies, courtesy Park Place Wood Products, Inc.*

ELEVATION MANTEL SECTION

Figure 23–14 Traditional mantel.

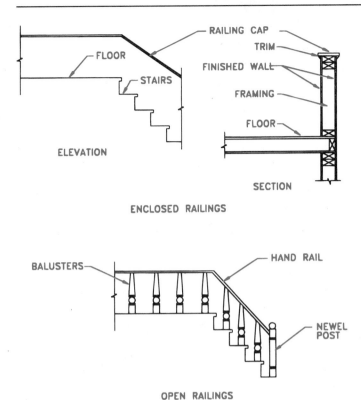

Figure 23–15 Railings.

▼ CABINETS

One of the most important items for buyers of a new home is the cabinets. The quality of cabinetry can vary greatly. Cabinet designs may reflect individual tastes, and a variety of styles are available for selection. Cabinets are used for storage and as furniture. The most obvious locations for cabinets are the kitchen and bath.

The design and arrangement of kitchen cabinets has been the object of numerous studies. Over the years design ideas have resulted in attractive and functional kitchens. Kitchen cabinets have two basic elements: the base cabinet and the upper cabinet. See Figure 23–17. Drawers, shelves, cutting boards, pantries, and appliance locations must be carefully considered.

Bathroom cabinets are called vanities, linen cabinets, and medicine cabinets. See Figure 23–18. Other cabinetry may be found throughout a house such as in utility or laundry rooms for storage. For example, a storage cabinet above a washer and dryer is common.

Cabinet Types

There are as many cabinet styles and designs as the individual can imagine. However, there are only two general types of cabinets based on their method of construction: (1) modular or prefabricated cabinets and (2) custom cabinets.

Modular Cabinets. The term *modular* refers to prefabricated cabinets because they are constructed in specific sizes called modules. The best use of modular cabinets is

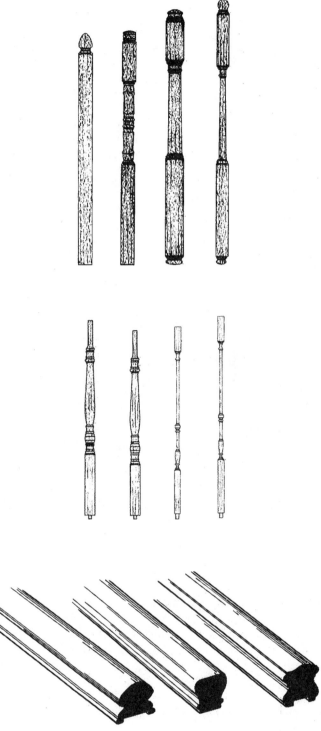

Figure 23–16 Examples of standard railing components. *Courtesy Hillsdale Pozzi Co.*

when a group of modules can be placed side by side in a given space. If a little more space is available, then pieces of wood, called *filler*, are spliced between modules. Many brands of modular cabinets are available that are well crafted and very attractive. Most modular cabinet vendors offer different door styles, wood species, or finish colors on their cabinets. Modular cabinets are sized in relationship to standard or typical applications, although

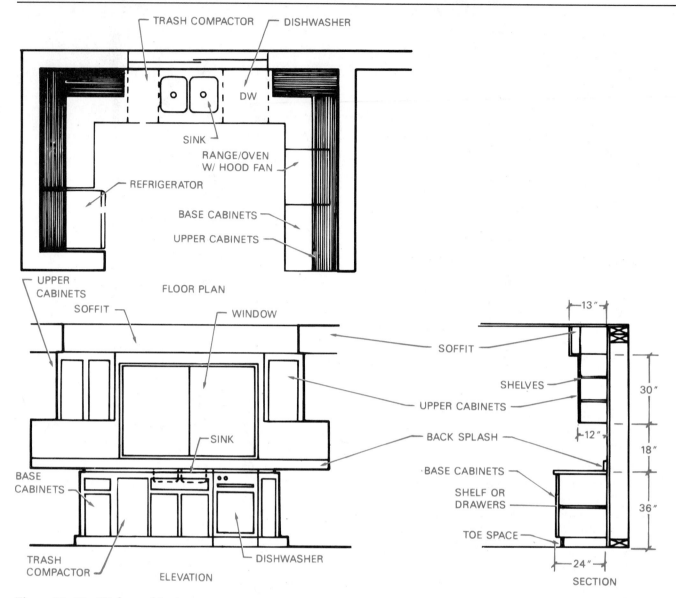

Figure 23-17 Kitchen cabinet components.

many modular cabinet manufacturers can make cabinet components that will fit nearly every design situation. The modular cabinets are often manufactured nationally and delivered in modules. See Figure 23–19.

Custom Cabinets. The word *custom* means "made to order." Custom cabinets are generally fabricated locally in a shop to the specifications of an architect or designer, and after construction they are delivered to the job site and installed in large sections.

One of the advantages of custom cabinets is that their design is limited only by the imagination of the architect and the cabinet shop. Custom cabinets may be built for any situation, such as for any height, space, type of exotic hardwood, type of hardware, geometric shape, or other design criteria.

Custom cabinet shops estimate a price based on the architectural drawings. The actual cabinets are constructed from measurements taken at the job site after the

building has been framed. In most cases, if cabinets are ordered as soon as framing is completed, the cabinets should be delivered on time.

Cabinet Options

Some of the cabinet design alternatives that are available from either custom or modular cabinet manufacturers include the following:

○ Door styles, materials, and finishes.
○ Self-closing hinges.
○ A variety of drawer slides, rollers, and hardware.
○ Glass cabinet fronts for a more traditional appearance.
○ Wooden range hoods.
○ Specially designed pantries, appliance hutches, and a lazy susan or other corner cabinet design for efficient storage.
○ Bath, linen storage, and kitchen specialties.

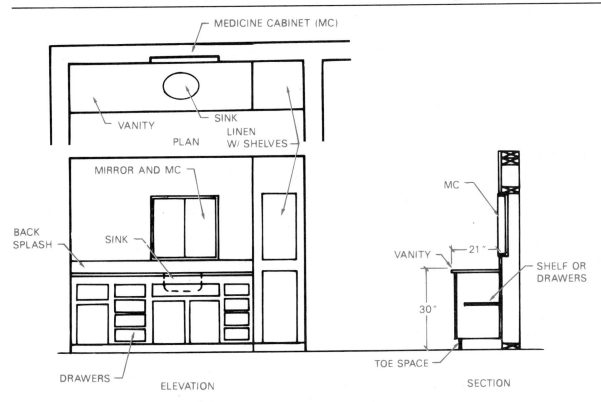

Figure 23–18   Bath cabinet components.

▼ CABINET ELEVATIONS AND LAYOUT

Cabinet floor plan layouts and symbols were discussed in Chapter 13. Cabinet elevations or exterior views are developed directly from the floor-plan drawings. The purpose of the elevations is to show how the exterior of the cabinets look when completed and to give general dimensions, notes, and specifications. Cabinet elevations may be as detailed as the architect or designer feels is necessary.

In Figure 23–20 the cabinet elevations are very clear and well done without detail or artwork that is not specifically necessary to construct the cabinets. In Figure 23–21 the cabinet elevations are more artistically drawn. Neither set of cabinet drawings is necessarily more functional than the other.

▼ KEYING CABINET ELEVATIONS TO FLOOR PLANS

Several methods may be used to key the cabinet elevations to the floor plan. In Figure 23–20, the designer keyed the cabinet elevations to the floor plans with room titles such as KITCHEN ELEVATION or BATH ELEVATION. In Figure 23–21, the designer used an arrow with a letter inside to correlate the elevation to the floor plan. For example, in Figure 23–22 the E and F arrows pointing to the vanities are keyed to letters E and F that appear below the vanity elevation in Figure 23–21.

Given the commercial drawing shown in Figure 23–23, notice that the floor plan has identification numbers within a combined circle and arrow. The symbols, similar to that shown in Figure 23–24, are pointing to various areas that correlate to the same symbols labeled below the related elevations. A similar method may be used to relate cabinet construction details to the elevations. The detail identification symbol as shown in Figure 23–25 is used to correlate a detail to the location from which it originated on the elevation. With some plans, small areas such as the women's and men's rest rooms have interior elevations without specific orientation to the floor plan. The reason is that the relationship of floor plan to elevations is relatively obvious.

There may be various degrees of detailed representation needed for millwork depending on its complexity and the design specifications. The sheet from a set of architectural working drawings shown in Figure 23–26 is an example of how millwork drawings can provide very clear and precise construction techniques.

▼ CABINET LAYOUT

Cabinet layout is possible after the floor-plan drawings are complete. The floor-plan cabinet drawings are used to establish the cabinet elevation lengths. Cabinet elevations may be projected directly from the floor plan, or dimensions may be transferred from the floor plan to the elevations with a scale or dividers.

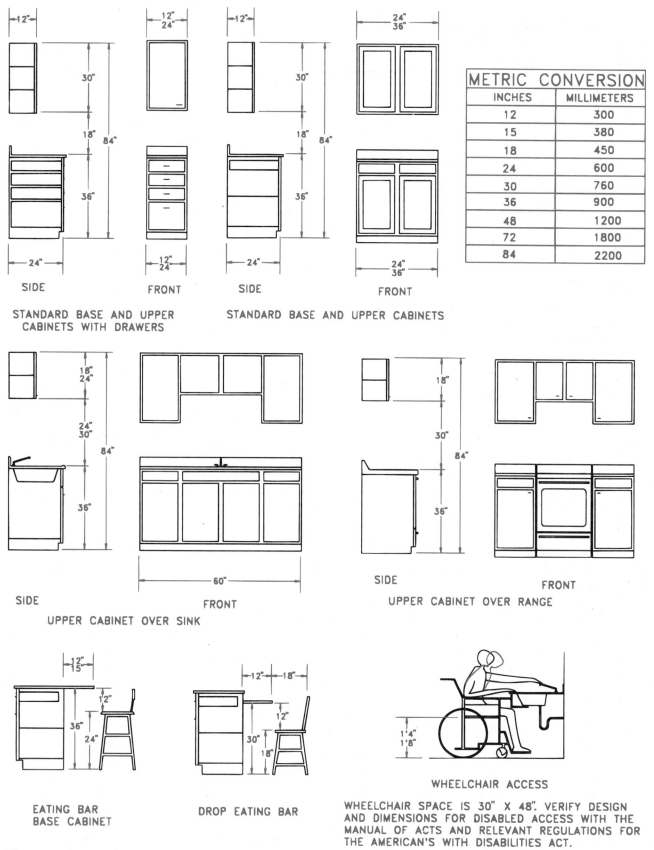

METRIC CONVERSION	
INCHES	MILLIMETERS
12	300
15	380
18	450
24	600
30	760
36	900
48	1200
72	1800
84	2200

STANDARD BASE AND UPPER CABINETS WITH DRAWERS

STANDARD BASE AND UPPER CABINETS

UPPER CABINET OVER SINK

UPPER CABINET OVER RANGE

EATING BAR BASE CABINET

DROP EATING BAR

WHEELCHAIR ACCESS

WHEELCHAIR SPACE IS 30" X 48". VERIFY DESIGN AND DIMENSIONS FOR DISABLED ACCESS WITH THE MANUAL OF ACTS AND RELEVANT REGULATIONS FOR THE AMERICAN'S WITH DISABILITIES ACT.

Figure 23–19 Standard cabinet dimensions. Note that all metric equivalents are soft conversions.

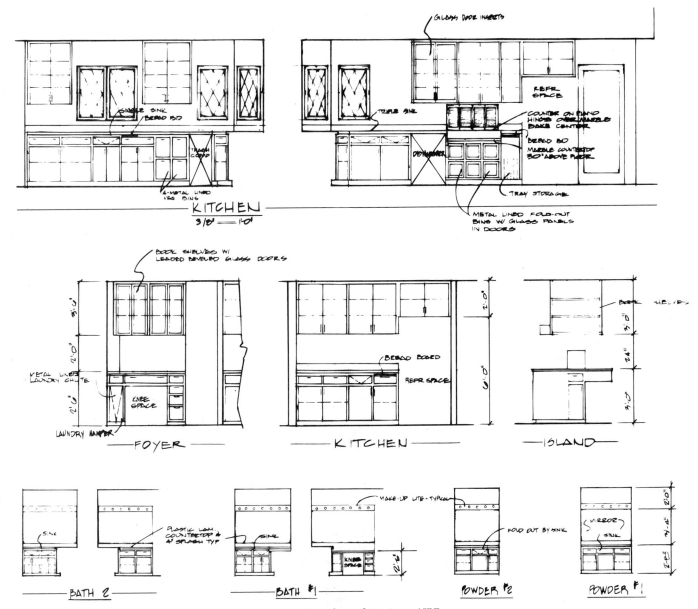

Figure 23–20 Simplified cabinet elevations. *Courtesy Residential Designs, AIBD.*

Kitchen Cabinet Layout

The cabinet elevations are drawn as though you were standing in the room looking directly at the cabinets. Cabinet elevations are two dimensional. Height and length are shown in an external view, and depth is shown where cabinets are cut with a cutting plane that is the line of sight. See Figure 23–27. For the kitchen shown in Figure 23–27 there are two elevations: one looking at the sink and dishwasher cabinet group, the other viewing the range, refrigerator, and adjacent cabinets.

Given the floor-plan drawing shown in Figure 23–27, draw the cabinet elevations using the following methods:

Step 1. Make a copy of the cabinet floor-plan area, and tape the copy to the drafting table above the area where the elevations are to be drawn. Transfer length dimensions from the floor plan onto the blank drafting surface in the location where the desired elevation is to be drawn using construction lines as shown in Figure 23–28. Some drafters prefer to transfer dimensions from the floor-plan drawings with a scale or dividers. As you gain layout experience, you will probably prefer to transfer dimensions. Notice when the elevations are transferred from the floor plan, the scale is the same as the floor plan, $\frac{1}{4}'' = 1'-0''$ (1:50 metric). Other scales are also used, such as $\frac{3}{8}'' = 1'-0''$ or $\frac{1}{2}'' = 1'-0''$ when additional clarity is needed to show small detail. When the cabinet elevation scale is different from the floor-plan scale, the dimensions may not be projected. In this case the dimensions are established from

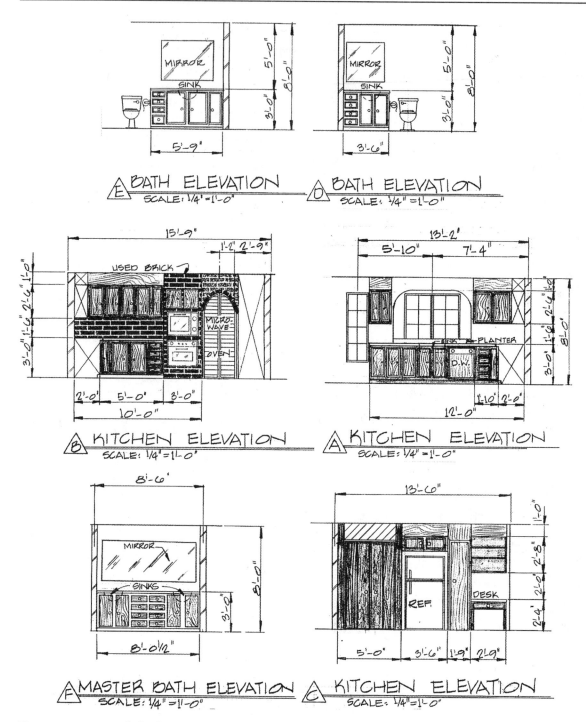

Figure 23–21 Detailed cabinet elevations. *Courtesy Madsen Designs.*

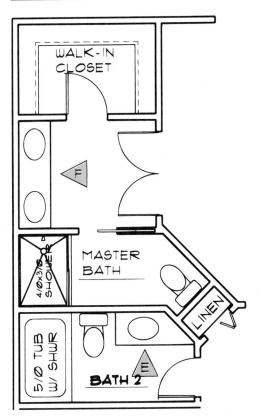

Figure 23–22 Using the floor plan and elevation cabinet location symbol. *Courtesy Piercy & Barclay Designers, Inc.* CPBD.

the floor plan and a scale change is implemented before transfer to the elevation drawing. When using CADD, the cabinets may be projected directly from the floor plan on their own layer. After drawing, the floor-plan layer may be turned off, leaving only the cabinet drawings on the screen. The cabinets may then be moved or scaled as needed.

Step 2. Establish the room height from floor to ceiling. If the floor to ceiling height is 8', then draw two lines 8' apart with construction lines using the ¼" = 1'-0" scale. See Figure 23–29.

Step 3. Establish the portion of the cabinetry where the line of sight cuts through base and upper cabinets, and soffits. Use construction lines to draw lightly in these areas or project on a construction layer in CADD, as shown in Figure 23–30.

Step 4. Use construction lines to draw cabinet elevation features such as doors, drawers, and appliances. See Figure 23–31. Also, draw background features such as backsplash, windows, and doors, if any. Notice the rectangular shape that is drawn to the left of the dishwasher and below the countertop where the sink would be. This is called a *false front*, designed to look like a drawer. An actual drawer cannot be placed here as it would run into the sink. A false drawer front is used to establish the appearance of a line of drawers just below the base cabinet top and before doors are used. Some designs use cabinet doors from the toe kick to the base cabinet top, although this method is not as common.

Step 5. Darken all cabinet lines. Some drafters use the technique of drawing the outline of the ends of the cabinets and surrounding areas with a thick line. Draw dashed lines that represent hidden features, such as sinks, and shelves, as seen in Figure 23–32.

Step 6. Add dimensions and notes. See Figure 23–32. The dimensions shown in Figure 23–32 are given in feet and inches; it is standard practice in many offices to dimension cabinetry in inches or millimeters only.

Follow the same process for the remaining cabinet elevations. See Figure 23–33. The same procedure may be used to draw bath vanity elevations as shown in Figure 23–34 or other special cabinet or millwork elevations and details.

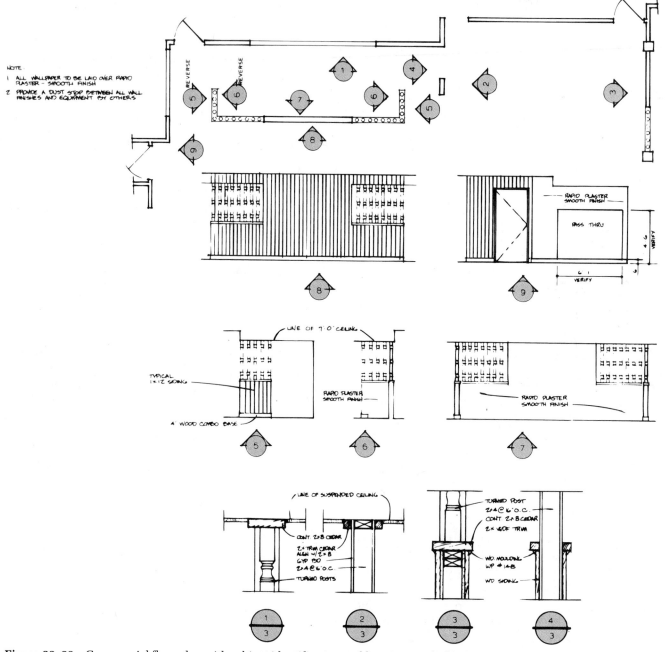

Figure 23–23 Commercial floor plan with cabinet identification and location symbols. *Courtesy Ken Smith Structureform Masters, Inc.*

Figure 23–24 Floor plan and elevation cabinet identification and location symbol.

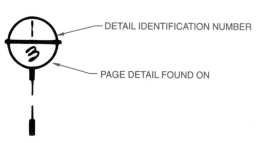

Figure 23–25 Detail identification symbol.

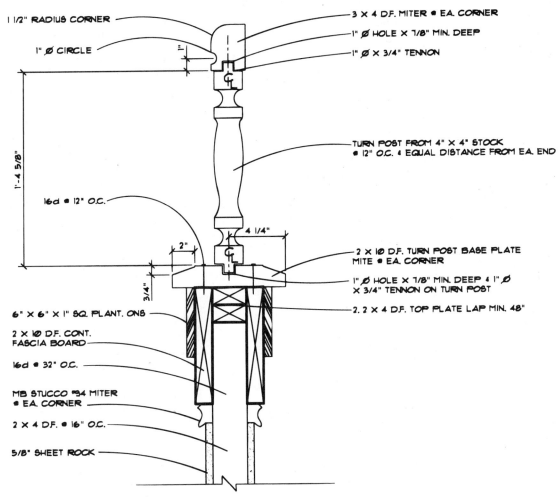

1 1/2" RADIUS CORNER

1" ∅ CIRCLE

3 X 4 D.F. MITER @ EA. CORNER

1" ∅ HOLE X 7/8" MIN. DEEP

1" ∅ X 3/4" TENNON

1'-4 5/8"

1"

C.L.

TURN POST FROM 4" X 4" STOCK
@ 12" O.C. & EQUAL DISTANCE FROM EA. END

16d @ 12" O.C.

4 1/4"

C.L.

2"

2 X 10 D.F. TURN POST BASE PLATE
MITE @ EA. CORNER

3/4"

1" ∅ HOLE X 7/8" MIN. DEEP & 1" ∅
X 3/4" TENNON ON TURN POST

6" X 6" X 1" SQ. PLANT. ONS

2 X 10 D.F. CONT.
FASCIA BOARD

16d @ 32" O.C.

MB STUCCO #34 MITER
@ EA. CORNER

2 X 4 D.F. @ 16" O.C.

5/8" SHEET ROCK

2. 2 X 4 D.F. TOP PLATE LAP MIN. 48"

Figure 23–26   Millwork and cabinet construction details. *Courtesy Structureform Masters, Inc.*

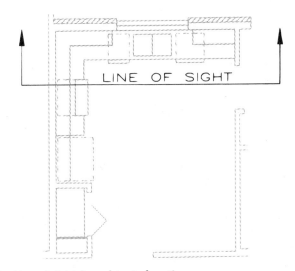

LINE OF SIGHT

Figure 23–27   Establishing the line of sight for cabinet elevations.

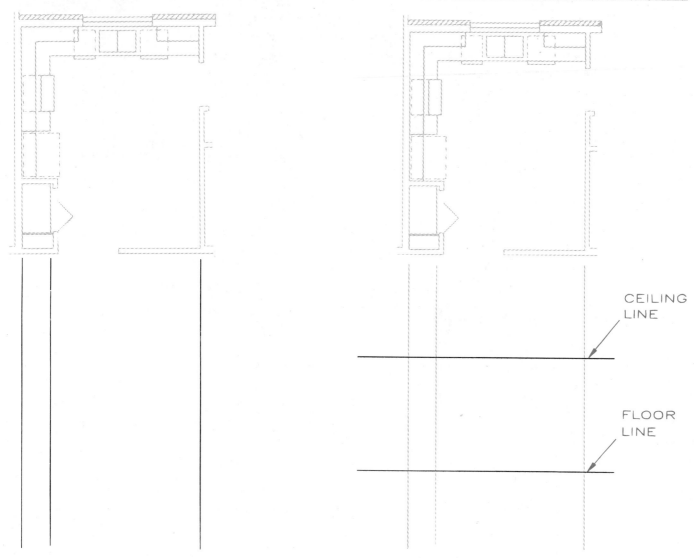

Figure 23–28 Step 1: Transfer length dimensions from the floor plan to the cabinet elevation using construction lines.

Figure 23–29 Step 2: Establish the floor and ceiling lines using construction lines.

CEILING
LINE

FLOOR
LINE

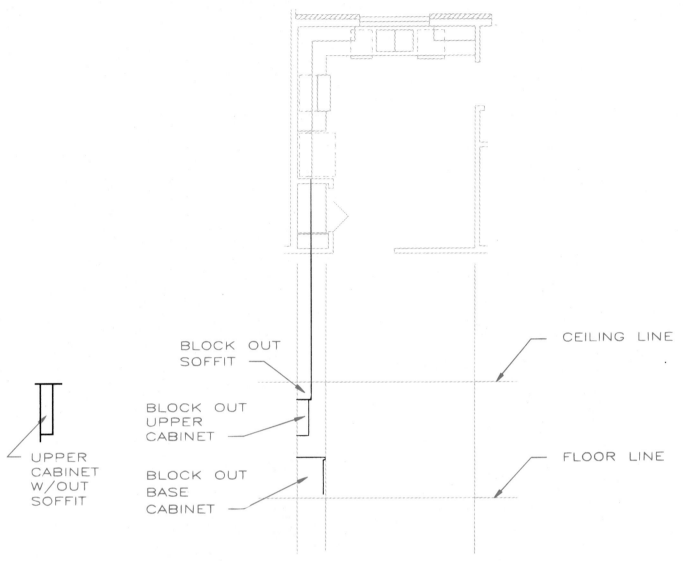

Figure 23–30   Step 3: Establish base cabinets, upper cabinets, and soffits using construction lines.

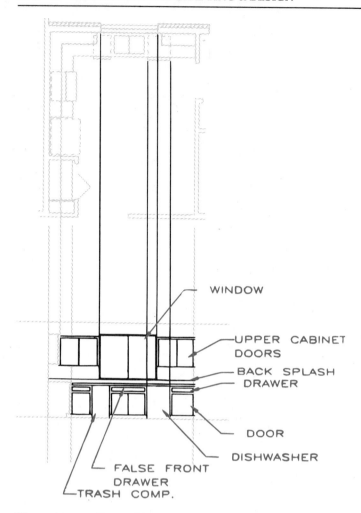

Figure labels:
- WINDOW
- UPPER CABINET DOORS
- BACK SPLASH
- DRAWER
- DOOR
- DISHWASHER
- FALSE FRONT DRAWER
- TRASH COMP.

Figure 23–31 Step 4: Use construction lines to establish cabinet elevation features.

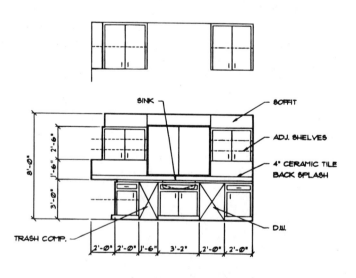

SINK
SOFFIT
ADJ. SHELVES
4" CERAMIC TILE BACK SPLASH
D.W.
TRASH COMP.

8'-0"   2'-6"   1'-6"   3'-0"

2'-0" 2'-0" 1'-6" 3'-2" 2'-0" 2'-0"

KITCHEN CABINET ELEV.

SCALE: ¼"=1'-0"

Figure 23–32 Steps 5 and 6: Darken all cabinet lines, and add dimensions.

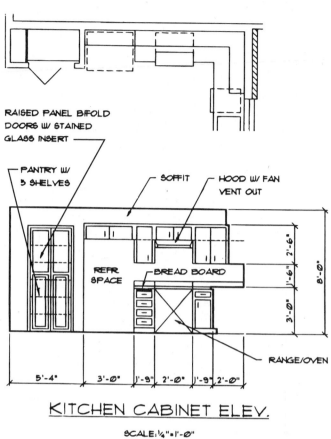

- RAISED PANEL BIFOLD DOORS W/ STAINED GLASS INSERT
- PANTRY W/ 3 SHELVES
- SOFFIT
- HOOD W/ FAN VENT OUT
- REFR SPACE
- BREAD BOARD
- RANGE/OVEN

2'-6" 1'-6" 3'-0" 8'-0"

5'-4"   3'-0"   1'-9" 2'-0" 1'-9" 2'-0"

KITCHEN CABINET ELEV.

SCALE: ¼"=1'-0"

Figure 23–33 Follow steps 1 through 6 to draw the other cabinet elevations.

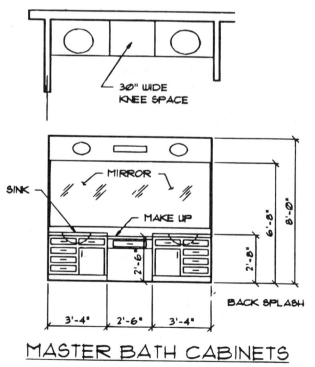

- 30" WIDE KNEE SPACE
- SINK
- MIRROR
- MAKE UP
- BACK SPLASH

6'-8" 8'-0" 2'-6" 2'-8"

3'-4"   2'-6"   3'-4"

MASTER BATH CABINETS

SCALE: ¼"=1'-0"

Figure 23–34 Use cabinet layout steps 1 through 6 to draw all bath cabinet elevations.

CADD Applications

DRAWING MILLWORK AND CABINET ELEVATIONS WITH CADD

CADD makes it easy to draw millwork details and sections with the use of tablet menu symbol libraries that contain millwork profiles, as shown in Figure 23–35.

CADD tablet menu commands for cabinet floor plan layouts place base and upper cabinets on the drawing, then add symbols for appliances and kitchen or bath features. Draw cabinet elevations with CADD by selecting cabinet module symbols from a template menu library; then organize them on the drawing. These cabinet modules are combinations of (1) a cabinet door and drawer or (2) banks of drawers for the base units and door modules for the upper units. There are also typical cabinet profiles that display the section of a cabinet through the end view. For bath cabinet elevations, there are modules with base units and mirror or medicine cabinets. Figure 23–36 shows a CADD tablet menu library used to draw cabinet elevations. A complete set of CADD cabinet elevation drawings is shown in Figure 23–37.

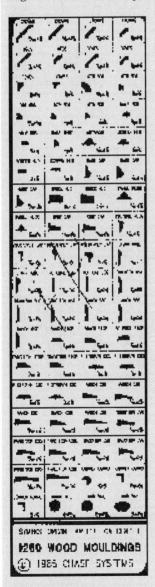

Figure 23–35 CADD tablet menu symbol library for millwork profiles. *Courtesy Chase Systems.*

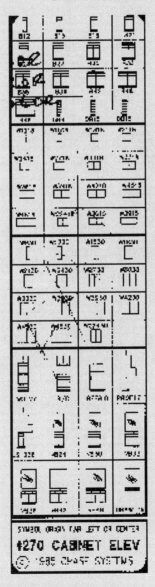

Figure 23–36 CADD tablet menu library for cabinet elevations. *Courtesy Chase Systems.*

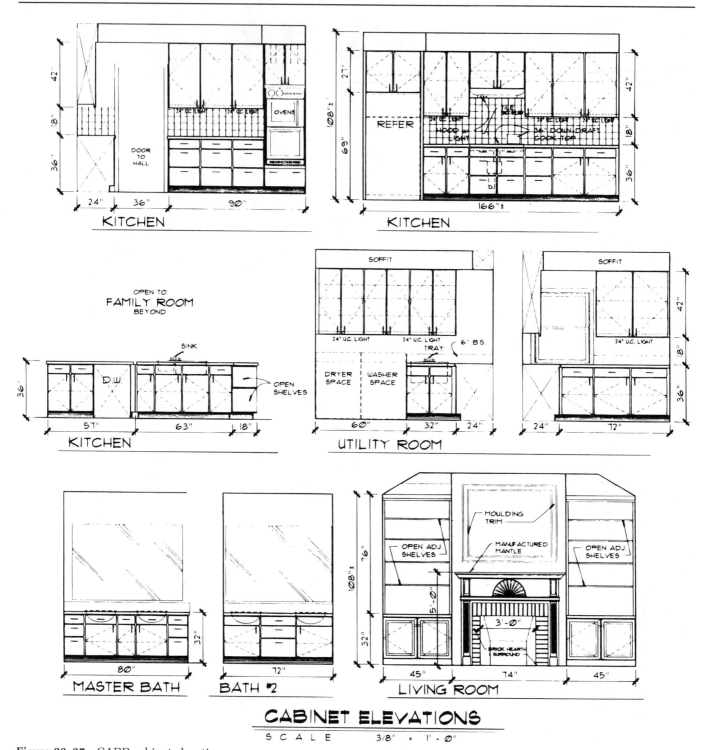

Figure 23–37 CADD cabinet elevations.

Millwork and Cabinet Technolgy, Cabinet Elevations, and Layout Test

DIRECTIONS

Answer the questions with short complete statements or drawings as needed on an 8½ × 11 sheet as follows:

1. Letter your name and the date at the top of the sheet.
2. Letter the question number and provide the answer. You do not need to write out the question. Answers may be prepared on a word processor if appropriate with course guidelines.

QUESTIONS

Question 23–1 Define millwork.

Question 23–2 Give an example of when plastic or ceramic products may be used rather than wood millwork.

Question 23–3 What are baseboards used for?

Question 23–4 Describe wainscots.

Question 23–5 Define cornice.

Question 23–6 When is the application of cornice millwork appropriate?

Question 23–7 Where are casings used?

Question 23–8 Where are mantels used?

Question 23–9 Give an example of where book shelves may be used effectively as millwork in the interior design of a structure.

Question 23–10 Give two examples of where railings may be used.

Question 23–11 Describe modular cabinets.

Question 23–12 Describe custom cabinets.

Question 23–13 What is another name for modular cabinets?

Question 23–14 What is the big difference between modular and custom cabinets?

Question 23–15 How is it possible to compare cabinetry and millwork to fine furniture?

Question 23–16 What is the purpose of cabinet elevations?

Question 23–17 Why is it possible that cabinet elevations may not be part of a complete set of architectural drawings?

Question 23–18 Show by sketch and/or explanation three methods of how cabinet elevations may be keyed to the floor plans.

Question 23–19 Make a sketch of a kitchen cabinet elevation that has the following components labeled or dimensioned:
a. Drawers
b. False front
c. Toe space
d. Backsplash
e. Base cabinet
f. Upper cabinet
g. Soffit
h. Range and hood
i. Dishwasher
j. Cutting board
k. Shelves
l. Base cabinet height
m. Dimension from top of base cabinet to bottom of upper cabinet
n. Upper cabinet height
o. Dimension from top of range to bottom of range hood
p. Cabinet doors

Question 23–20 Make a sketch of a bathroom vanity elevation that has the following components labeled or dimensioned:
a. Vanity cabinet
b. Backsplash
c. Toe space
d. Cabinet doors
e. Cabinet drawers
f. Mirror
g. Medicine cabinet
h. Height of vanity cabinet

PROBLEM

1. Using as a guide either the floor plan of the problem you have been drawing as a project or a floor plan that has been assigned to you by your instructor, draw the necessary elevations for all kitchen, bath, and specialty cabinets. The specific location of cabinet doors, drawers, and other features is flexible. Verify your cabinet designs with the contents of this chapter and your instructor's guidelines. The placement of appliances and fixtures is predetermined by the floor-plan drawing unless otherwise specified by your instructor.

2. Use a scale of ¼″ = 1'–0″ or ⅜″ = 1'–0″ for the cabinet elevations depending on the amount of space available on your drawing sheet. The cabinet elevations may be placed on a sheet with other drawings if convenient. Do not crowd the cabinet elevations on the sheet you select. If there is not enough area for the cabinet elevations on a sheet with other drawings, then place them on a separate sheet. Avoid placing cabinet elevations on a sheet with exterior elevations or the main floor plan.

3. These elevations may be prepared using CADD if appropriate with course guidelines.

SECTION VIII

FRAMING METHODS AND PLANS

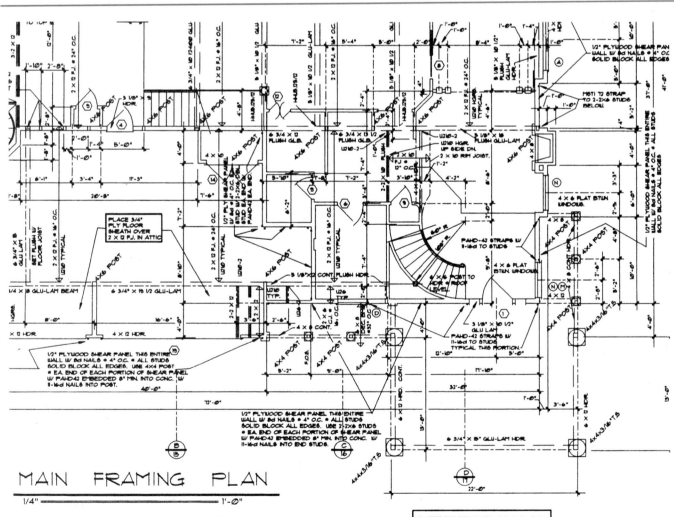

MAIN   FRAMING   PLAN

1/4" ================== 1'-0"

SET ALL DOOR AND WINDOW HEADERS AT 6'-10" EXCEPT FOR
WINDOW HEADERS IN FAMILY, FAMILY/DINING, KITCHEN, AND
DINING ROOMS. SET THESE HEADERS AT 7'-4". SEE ELEVATIONS
AND SECTIONS.

EXCEPT FOR DESIGNATED SHEAR PANELS,
BALANCE OF WALLS ON THIS FLOOR MAY
BE SHEATHED W/ 1/2" WAFERBOARD.

CHAPTER 24

Framing Methods

More so than any other drawing, the *structural drawings*—the framing plans, sections and details, and foundation plan—require a thorough understanding of the materials and process of construction. A drafter can successfully complete the architectural drawings with minimal understanding of the structural processes that are involved. To work on drawings that will show how a structure is assembled requires an understanding of basic construction principles and materials.

Wood, steel, masonry, and concrete are the most common materials used in the construction of residential and small office buildings. With each material several different framing methods can be used to assemble the components. Wood is the most widely used material for the framing of houses and apartments. The most common framing systems used with wood are balloon, platform, and post and beam. There have also been several variations of platform framing to provide for better energy efficiency. Components for each system will be introduced in Chapter 25.

▼ BALLOON FRAMING

Although balloon framing is not widely used, a knowledge of this system will prove helpful if an old building is being remodeled. With the *balloon* or *eastern framing* method, the exterior studs run from the top of the foundation to the top of the highest level as seen in Figure 24–1. This is one of the benefits of the system. Wood has a tendency to shrink as the moisture content decreases and shrinks more in width than in length. Because the wall members are continuous from foundation to the roof, fewer horizontal members are used, resulting in less shrinkage. Since brick veneer or stucco is often applied to the exterior face of the wall, the minimal shrinkage of the balloon system helps to keep the exterior finish from cracking.

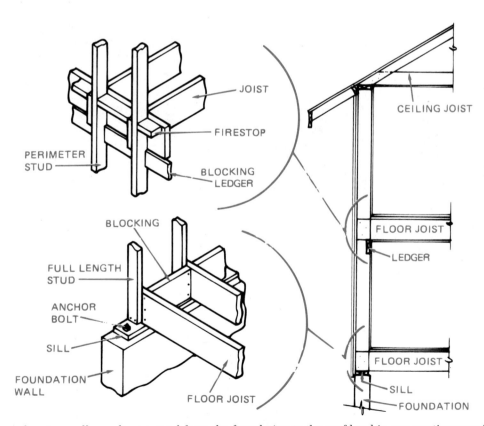

Figure 24–1   Balloon framing wall members extend from the foundation to the roof level in one continuous piece.

Because of the long pieces of lumber needed, a two-story structure was the maximum that could be built easily using balloon framing. Floor framing at the midlevel was supported by a ledger set into the studs. Structural members were usually spaced at 12″, 16″, or 24″ (300, 400 or 600 mm) on center (o.c.). Figure 24–2 shows typical construction methods used in the balloon framing system.

Although the length of the stud gave the building stability, it also caused the demise of the system. The major flaw with balloon framing is fire danger. A fire starting in the lower level could quickly race through the cavities formed in the wall or floor systems of the building. Blocking or smoke-activated dampers are now required by many building codes at all levels to resist the spread of fire. See Figure 24–3.

▼ PLATFORM FRAMING

Platform or *western platform framing* is the most common framing system now in use. The system gets its name from the platform created by each floor as the building is being framed. The framing crew is able to use the floor as a platform to assemble the walls for that level. See Figure 24–4. Platform framing grew out of the need for the fireblocks in the balloon framing system. The fireblocks that had been put in individually between the studs in bal-

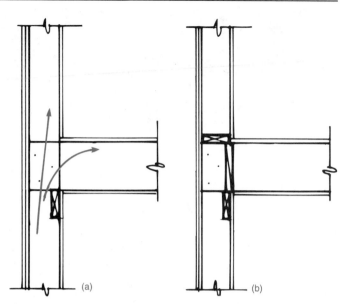

Figure 24–3 (a) The risk of fire spreading with balloon framing is very great because flames can race through the floor and wall systems. (b) Blocking, or fire blocking, is required by building codes to be added to the intersection of each floor and wall to resist the spread of fire and smoke through the floor and wall systems of a structure.

Figure 24–4 Western platform framing allows the framers to use the floor to construct the walls to support the next level. *Courtesy John Black.*

loon framing became continuous members placed over the studs to form a solid bearing surface for the floor or roof system. See Figure 24–5.

Building with the Platform System

Once the foundation is in place, the framing crew sets the girders and the floor members. The major components of platform construction can be seen in Figure 24–6. Floor joists ranging in size from 2″ × 6″ through 2″ × 14″ (50 × 150 – 50 × 350 mm) can be used depending on the distance they are required to span. Figure 24–7 shows the construction of a joist flow system. Plywood of ½″, ⅝″, or ¾″ (12.5, 15.5, or 18.5 mm) thickness can be

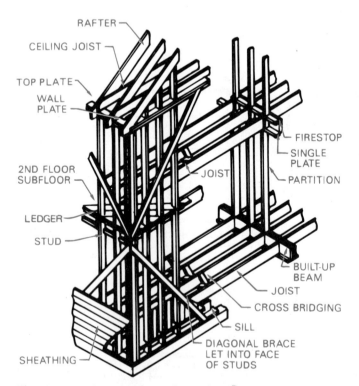

Figure 24–2 Structural members of the balloon framing system are typically spaced at 12, 16, or 24″ (300, 400, or 610 mm) o.c. Exterior walls are erected first, and then the upper floor is hung from the walls.

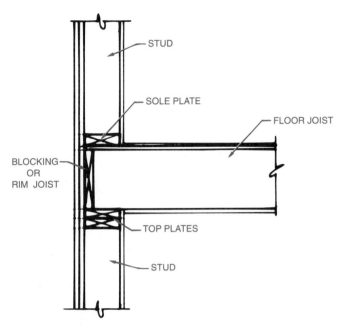

Figure 24–5 The fireblocks of the balloon system gave way to continuous supports at each floor and ceiling level.

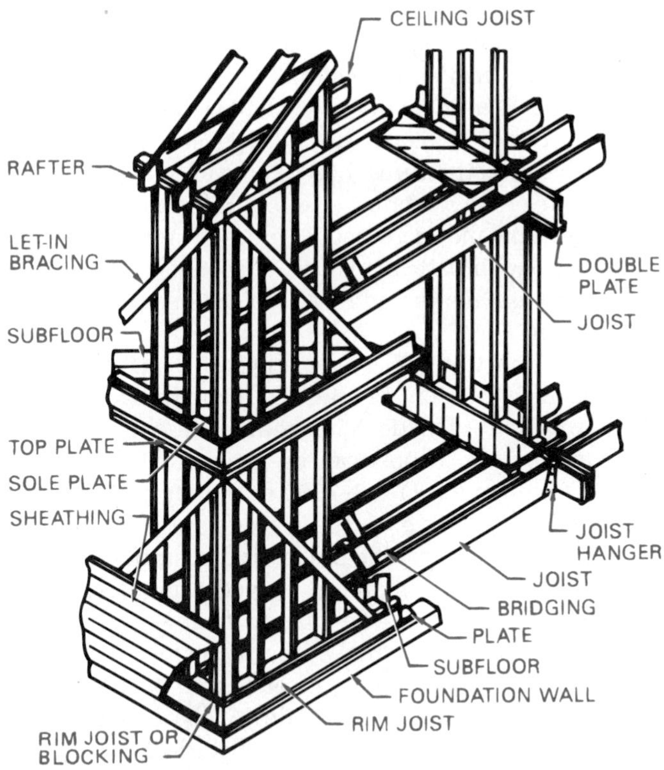

150 mm) will be laid diagonal to the floor joists to form the floor system, but plywood is primarily used because of the speed with which it can be installed.

With the floor in place, the walls are constructed using the floor as a clean, flat, layout surface. Walls are typically built flat on the floor using a bottom plate, studs, and two top plates. With the walls squared, sheathing can be nailed to the exterior face of the wall, and then the wall tilted up into place. When all of the bearing walls are in place the next level can be started. Similar to the balloon system, studs are typically placed at 12, 16, or 24″ (300, 400, or 600 mm), with 16″ (400 mm) o.c. the most common spacing. Figure 24–8 shows a building constructed with platform building methods.

Although in theory the height of a structure is limitless using the platform framing method, most codes do not allow a structure greater than three floor levels to be constructed. The height restriction is due to the risk of fire to the inhabitants because of the combustible nature of wood. Chapter 45 provides a further discussion of codes affecting wood construction.

Figure 24–6 Structural members of the platform framing system.

Figure 24–7 A joist floor system typically places support members at 16″ (400 mm) o.c. *Courtesy Louisiana Pacific Corporation.*

Figure 24–8 A residence framed with western platform framing methods. *Courtesy Southern Forest Products Association.*

used for the subfloor, which covers the floor joists. The plywood usually has a tongue-and-groove (T&G) pattern in the edge to help minimize floor squeaking. Gluing the plywood to the supporting members in addition to the normal nailing also helps to eliminate squeaks. Occasionally T&G lumber 1 × 4 or 1 × 6 (25 × 100 or 25 ×

▼ ENERGY-EFFICIENT FRAMING

Typically exterior walls have been framed with 2 × 4 (50 × 100 mm) studs at 16" (400 mm) spacing. Since the early 1980s, many framers have substituted 2 × 6 (50 × 150 mm) studs at 24" (600 mm) spacing to allow for added insulation in the wall cavity. Many long-held framing practices are being altered to allow for greater energy savings. The framing practices are usually referred to as *advanced framing techniques* (AFT) in the building codes.

AFT systems eliminate nonstructural wood from the building shell and replace it with insulation. Wood has an average resistive value for heat loss of R-1 per inch of wood compared to R-3.5 through R-8.3 per inch of insulation. By reducing the amount of wood in the shell, the energy efficiency of the structure is increased. Advanced framing methods usually include 24" (600 mm) stud spacing, insulated corners, insulation in exterior walls behind partition intersections, and insulation headers. Figure 24–9 compares standard and advanced framing methods. Many codes limit advanced framing methods to

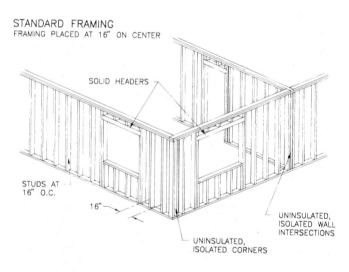

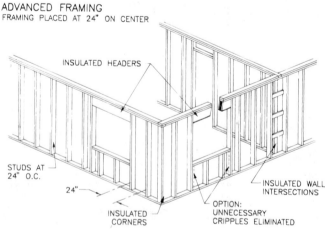

Figure 24–9 Advanced framing methods space studs at 24" (600 mm) o.c. and provide insulation at corners, wall intersections, and headers.

one-level construction. On a multilevel residence, advanced methods can be used on the upper level, and the spacing can be altered to 16" (400 mm) for the lower floor.

The insulation that is added to corners, wall intersections, and headers must be equal to the insulation value of the surrounding wall. (Chapter 26 introduces insulation requirements.) Figure 24–10 shows examples of framing intersections using advanced framing techniques. Figure 24–11 shows examplles of how insulated heaers can be framed. Advanced framing methods also affect how the roof will intersect the walls. Figure 24–12 compares standard roof and wall intersection with those of advanced methods.

Structural insulated panels (SIPs) have been used for decades in the construction of refrigerated buildings where thermal efficiency is important. SIPs have also been used in residential construction since the 1930s as part of research projects. Their popularity increased as energy costs rose through the 1980s, and they have become an economical method of construction in many parts of the country.

SIPs are composed of a continuous core of rigid foam insulation, which is laminated between two layers of structural board with an adhesive to form a single, solid panel. Typically, expanded polystyrene (EPS) foam is used as the insulation material, and oriented strand board (OSB) is used for the outer shell of the panel. Panels with an EPS core and drywall skin are available for timber-framed structures. EPS retains its shape throughout the life of the structure, to offer uniform resistance to air infiltration and increased R values over a similar-size wood wall with batt insulation. Although values range for each manufacturer, an R value of 4.35 per inch of EPS is common.

In addition to their energy efficiency, SIPs offer increased construction savings by combining framing, sheathing, and insulation procedures into one step during construction. Panels can be made in sizes ranging from 4' through 24' (1200–7300 mm) long and 8' (2400 mm) high, depending on the manufacturer and design requirements. SIPs can be precut and custom fabricated for use on the most elaborately shaped structures. Panels are quickly assembled on-site by fitting splines into routed grooves in the panel edges, as seen in Figure 24–13. Door and window openings can be precut during their assembling or cut at the job site. A chase is typically installed in the panels to allow for electrical wiring.

▼ POST-AND-BEAM FRAMING

Post-and-beam or timber framing construction places framing members at greater distances apart than platform methods do. In residential construction, posts and beams are usually spaced at 48" (1200 mm) on center. Although the system uses less lumber than other methods, the members required are larger. Sizes vary depending on the

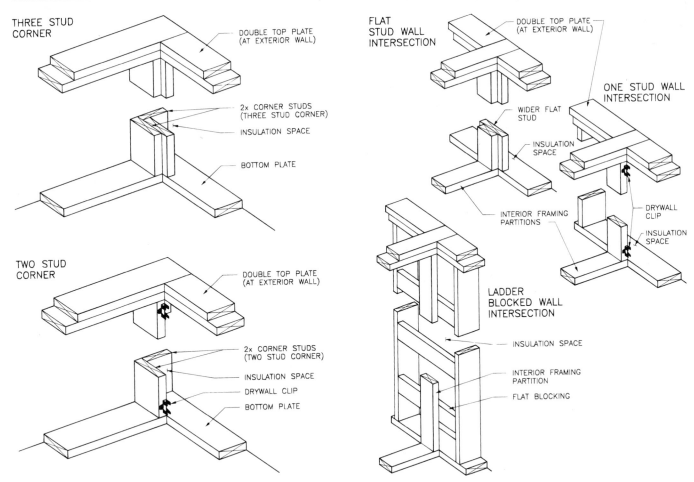

Figure 24–10 Reducing the amount of lumber used at wall intersections reduces the framing cost and allows added insulation to be used.

span, but beams of 4″ or 6″ (100–150 mm) wide are typically used. The subflooring and roofing over the beams is commonly 2″ × 6″, 2″ × 8″, or 1⅛ (50 × 150, 50 × 200 or 28 mm) T&G plywood. Figure 24–14 shows typical construction members of post-and-beam construction. Post-and-beam construction can offer great savings in both the lumber and nonstructural materials. This savings results from careful planning of the locations of the posts, and doors and windows that will be located between them. Savings also result by having the building conform to the modular dimensions of the material being used.

▼ TIMBER CONSTRUCTION

Although timber framing has been used for over 2000 years, the post-and-beam system was not widely used for the last 100 years. The development of balloon framing methods with its smaller materials greatly reduced the desire for timber methods. Many homeowners are now re

turning to timber framing methods for the warmth and cozy feelings timber homes tend to create. See Figure 24–15.

The length and availability of lumber affect the size of the frame. Although custom sizes are available for added cost, many mills no longer stock timbers longer than 16′ (4900 mm). If laminated timbers are used, spans will not be a consideration. The method for lifting the timbers into place may also affect the size of the frame. Beams are often lifted into place by brute force, by winch, by forklift, or by crane. The method of joining the beams at joints affects the frame size. Figure 24–16 shows common timber components.

Although an entire home can be framed with post-and-beam or timber methods, many contractors use the post-and-beam system for supporting (1) the lower floor (when no basement is required) and (2) roof systems, then use conventional framing methods for the walls and upper levels, as seen in Figure 24–17. Figure 24–18 shows plywood decking being installed on the beams of a post-and-beam floor system.

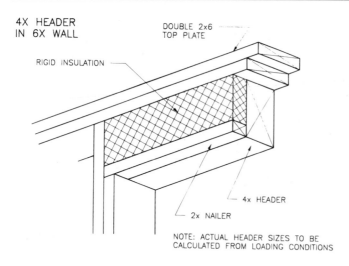

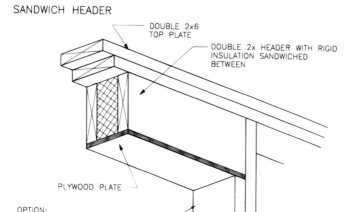

Figure 24–11 Rigid insulation placed behind or between headers increases the insulation value of a header from R-4 to R-11.

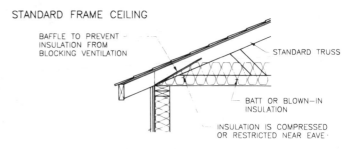

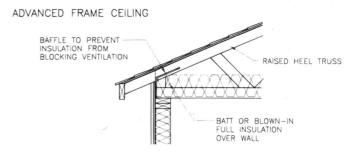

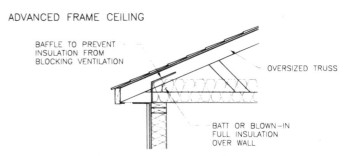

Figure 24–12 By altering the bearing point of a truss or rafter, the ceiling insulation can be carried to the outer edge of the exterior wall and reduce heat loss.

Figure 24–13 Structural insulated panels have a core of expanded polystyrene, which is laminated between two layers of oriented strand board. *Photo courtesy Insulspan.*

▼ STEEL FRAMING

Although primarily used for light commercial and high-rise construction, steel framing is becoming increasingly competitive with wood framing techniques in residential construction. Lower energy cost, higher strength, and insurance considerations are helping steel framing companies make inroads into residential markets. Because exterior walls are wider, insulation of R-30 can be used for exterior walls. Steel framing also has excellent properties for resisting stresses from snow, wind, and seismic forces, as well as termite and fire damage.

Many residential steel structures incorporate techniques similar to western platform construction methods. Steel studs, as seen in Figure 24–19, are used to frame walls. Walls are normally framed in a horizontal position on the floor and then tilted into place and bolted together. Steel trusses are typically used to provide design flexibility by requiring no interior bearing walls. Trusses can also be assembled on the ground and then lifted into place.

▼ CONCRETE MASONRY CONSTRUCTION

Concrete masonry units (CMUs) are a durable, economical building material that provides excellent structural

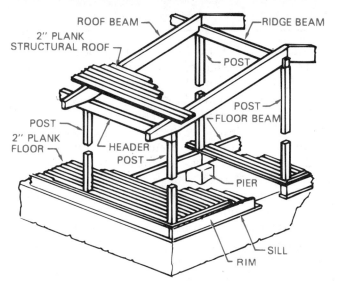

Figure 24–14 Structural members of a post-and-beam framing system. In residential construction, supporting members are usually placed at 48″ o.c. In commercial uses, spacing may be 20′ or greater.

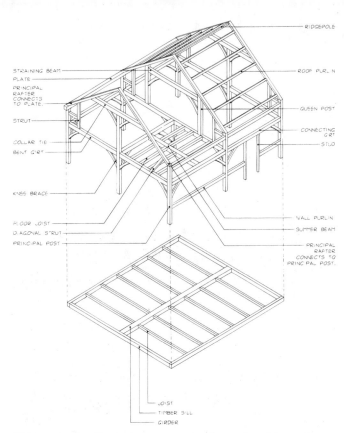

Figure 24–16 Typical components of a post-and-beam home. *Courtesy Timberpeg Post and Beam Homes.*

Figure 24–15 The use of exposed timbers throughout the residence creates a warm, cozy feeling. *Courtesy Ross Chapple, Timberpeg Post and Beam Homes.*

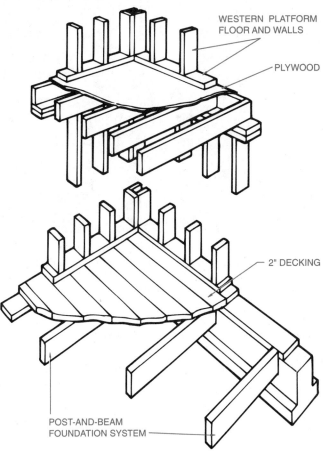

Figure 24–17 Post-and-beam construction (lower floor) is often mixed with platform construction for upper floors and roof framing.

Figure 24–18 A post-and-beam floor system being constructed. *Courtesy American Plywood Association.*

load bearing (ASTM C 145), and nonload-bearing blocks (ASTM C 129), which can be either solid or hollow blocks. Solid blocks must be 75 percent solid material in any cross-sectional plane. Blocks are also classed by their weight as normal-, medium- and lightweight. The weight of the block is affected by the type of aggregate used to form the unit. Normal aggregates such as crushed rock and gravel produce a block weight of between 40 and 50 lbs for an 8″ × 8″ × 16″ (200 × 200 × 400 mm) block. Lightweight aggregates include coal cinders, shale, clay, volcanic cinders, pumice, and vermiculite. Use of a lightweight aggregate will produce approximately a 50 percent savings in weight.

CMUs come in a wide variety of patterns and shapes. The most common sizes are 8″ × 8″ × 16″ (200 × 200 × 400 mm), 8″ × 4″ × 16″ (200 × 100 × 400 mm), and 8″ × 12″ × 16″ (200 × 300 × 400 mm). The nominal dimensions for block widths are 4″, 6″, 8″, 10″, and 12″ (100, 150, 200, 250, and 300 mm), and lengths are available in 6″, 8″, 12″, 16″, and 24″ (150, 200, 300, 400, 600 mm). Each dimension of a concrete block is actually 3/8″ (10 mm) smaller to allow for a 3/8″ (10 mm) mortar joint. In addition to the exterior patterns, concrete blocks come in a variety of shapes, as seen in Figure 22–20. These shapes allow for the placement of steel reinforcement bars, but steel reinforcing mesh can also be used. Chapter 31 will provide further information on concrete reinforcement. Reinforcing is typically required at approximately 48″

Figure 24–19 Lightweight steel is becoming increasingly popular in the framing of single- and multistory dwelling units. *Courtesy Steel Framing Systems, a unit of United States Gypsum Company.*

and insulation values. Concrete blocks are used primarily in warmer climates, from Florida to southern California, as an above-ground building material. CMUs can be waterproofed with cement-based paints and used as the exterior finish, or they can be covered with stucco. Waterproof wood furring strips are normally attached to the interior side of the block to support sheet rock. In areas with cooler climates, concrete blocks are often only used for below-grade construction.

Four classifications are used to define concrete blocks for construction: hollow, load bearing (ASTM C 90), solid

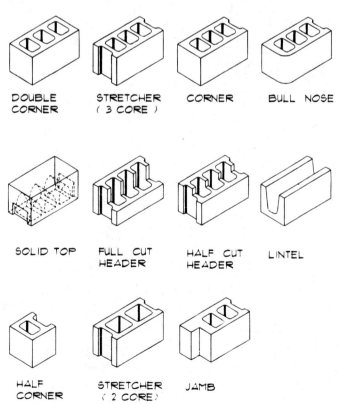

Figure 24–20 Common shapes of concrete masonry units.

(1200 mm) o.c. vertically. Horizontal reinforcement is approximately 16" (400 mm) o.c., but the exact spacing depends on the seismic or wind stresses to be resisted. Reinforcement is typically specified on the framing plan and also shown in a cross-section and detail.

▼ SOLID MASONRY CONSTRUCTION

One of the most popular features of brick is the wide variety of positions and patterns in which it can be placed. These patterns are achieved by placing the brick in various positions to each other. The position in which the brick is placed may alter what the brick is called. Figure 24–21 shows the names of common brick positions.

Bricks can be placed in various positions to form a variety of bonds and patterns. A *bond* is the connecting of two wythes to form stability within the wall. The pattern is the arrangement of the bricks within one wythe. The Flemish and English bonds in Figure 24–22 are the most common methods of bonding two wythes or vertical section of a wall that is one brick thick. The Flemish bond consists of alternating headers and stretchers in every course. A course of brick is one row in height. An English bond consists of alternating courses of headers and stretchers. The headers span between wythes to keep the wall from separating.

Masonry walls must be reinforced using similar methods to those used with concrete blocks. The loads to be supported and the stress from wind and seismic loads determine the size and spacing of the rebar to be used and are established by the architect or engineer. If a masonry wall is to be used to support a floor, a space one wythe wide will be left to support the joist. Joists are usually required to be strapped to the wall so that the wall and floor will move together under lateral stress. The end of the joist must be cut on an angle, called a *fire cut*. If the floor joist were to be damaged by fire, the fire cut will

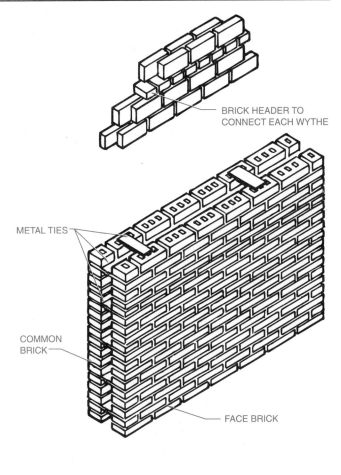

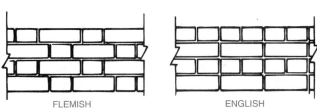

Figure 24–22 Brick walls can be strengthened by metal ties or bricks connecting each wythe. The most common types of brick bonds are the Flemish and English bonds.

allow the floor joist to fall out of the wall, without destroying the wall.

Because brick is very porous and absorbs moisture easily, some method must be provided to protect the end of the joist from absorbing moisture from the masonry. Typically the end of the joist is wrapped with 55# felt and set in a ½" (13 mm) air space. The interior of a masonry wall also must be protected from moisture. A layer of hot asphaltic emulsion can be applied to the inner side of the interior wythe and a furring strip attached to the wall. In addition to supporting sheetrock or plaster, the space between the furr strips can be used to hold batt or rigid insulation.

When a roof framing system is to be supported on masonry, a pressure-treated plate is usually bolted to the brick much the same way that a plate is attached to a

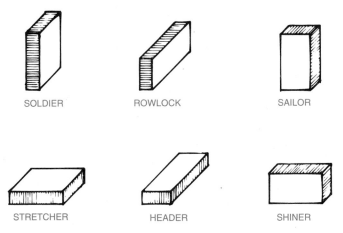

Figure 24–21 The position in which a brick is placed alters the name of the unit.

concrete foundation. Figure 24–23 shows how floor and roof members are typically attached to masonry.

Another method of using brick is to form a cavity between each wythe of brick. The cavity is typically 2″ (50 mm) wide and creates a wall approximately 10″ (250 mm) wide with masonry exposed on both the exterior and interior surfaces. The air space between wythes provides an effective barrier to moisture penetration to the interior wall. Weep holes in the lower course of the exterior wythe will allow moisture that collects to escape. Rigid insulation can be applied to the interior wythe to increase the insulation value of the air space in cold climates. Care must be taken to keep the insulation from touching the exterior wythe so that moisture is not transferred to the interior. Metal ties are typically embedded in mortar joints at approximately 16″ (400 mm) to tie each wythe together.

▼ MASONRY VENEER

A common method of using brick and stone in residential construction is as a *veneer*, a nonstructural covering material. Using brick as a veneer offers the charm and warmth of brick, with a lower construction cost than structural brick. If brick is used as a veneer over a wood bearing wall, for example, the amount of brick required is half that of structural brick. Care must be taken to protect the wood frame from moisture in the masonry. Typically brick is installed over a 1″ (25 mm) air space and a 15# layer of felt applied to the framing. The veneer is attached to the framing with 26-gage metal ties at 24″ (600 mm) o.c.

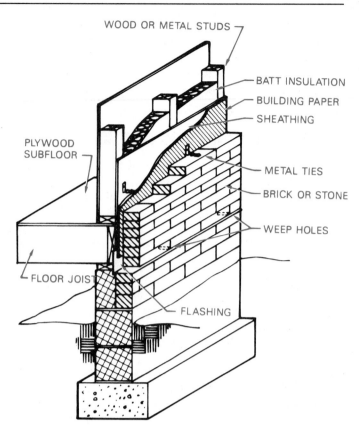

Figure 24–24 Brick is typically used as a nonload-bearing veneer attached to wood or concrete construction with metal ties.

along each stud. Figure 24–24 shows an example of how masonry veneer is attached to a wood frame wall.

▼ MODULAR FRAMING METHODS

Many people confuse the term *modular homes* with mobile or manufactured homes. Manufactured homes are built on a steel chassis and are built to meet codes other than the UBC, CABO, or SBC. Modular homes are governed by the same codes that govern site-built homes but are built off site.

Modular construction is a highly engineered method of constructing a building or building components in an efficient and cost-effective manner. Modular homes begin as components designed, engineered, and assembled in the controlled environment of a factory using assembly line techniques. Work is never slowed by weather, and materials are not subject to warping due to moisture absorption. A home travels to different workstations, where members of each building trade perform the required assembly. Rooms that require plumbing can be assembled in an area separate from other rooms and then assembled into a module. Modules can be shipped to the construction site, where they are assembled on a conventional foundation. Modules can be shipped complete or may require additional site constructions, depending on

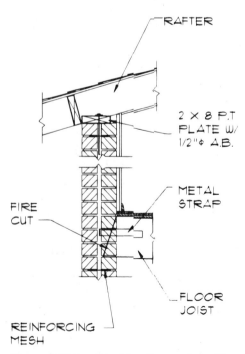

Figure 24–23 Attaching floor and roof framings members to a masonry wall.

the complexity of the module. Figure 24–25 shows the components of a modular home being assembled at the construction site.

Structures also can be constructed using preassembled components. Panelized walls, which include the windows, doors, and interior and exterior finish, can be assembled in a factory but installed at the job site. Floor and roof framing members can be precut in a factory, numbered for their exact location, and then shipped for assembly at the job site. This system of construction can offer economical construction in remote areas where materials are not readily available.

Figure 24–25 Modular construction allows as much as 90 percent of the construction work to be completed inside an environmentally controlled factory. *Courtesy J. Rouleau & Assoc.*

CHAPTER 24 Framing Methods Test

DIRECTIONS

Answer the questions with short, complete statements or drawings as needed on an 8½ × 11 drawing sheet as follows:

1. Letter your name and the date at the top of the sheet.
2. Letter the question number and provide the answer. You do not need to write out the question.

QUESTIONS

Question 24–1   Explain two advantages of platform framing.

Question 24–2   Sketch the platform construction methods for a one-level house.

Question 24–3   Why is balloon framing not commonly used today?

Question 24–4   List five framing methods that can be used to frame a residence.

Question 24–5   Sketch a section showing a post-and-beam foundation system.

Question 24–6   How is brick typically used in residential construction?

Question 24–7   What factor dictates the reinforcing in masonry walls?

Question 24–8   Why would a post-and-beam construction roof be used with platform construction walls.

Question 24–9   What would be the R value for a wood wall 4″ wide?

Question 24–10   What is the typical spacing of posts and beams in residential construction?

Question 24–11   List three qualities that make steel framing a popular residential method?

Question 24–12   List four classifications of CMUs.

Question 24–13   Sketch two methods of framing a header with advance framing technology.

Question 24–14   What is a firecut, and how is it used?

Question 24–15   How is moisture removed from a masonry cavity wall?

Question 24–16   What are two materials typically used to reinforce concrete block construction?

CHAPTER 25

Structural Components

As with every other phase of drafting, section drawing includes its own terminology. The terms referred to in this chapter are basic for structural components of sections. The terms will cover floors, walls, and roofs.

▼ FLOOR CONSTRUCTION

Two common methods of framing the floor system are conventional joist and post and beam. In some areas of the country it is common to use post-and-beam framing for the lower floor when a basement is not used and conventional framing for the upper floor. The project designer chooses the floor system or systems to be used. The size of the framing crew and the shape of the ground at the job site are the two main factors in the choice of framing methods.

Conventional Floor Framing

Conventional, or stick, framing involves the use of 2″ (50 mm) wide members placed one at a time in a repeti-

tive manner. Basic terms of the system are *mudsill, floor joist, girder,* and *rim joist.* Each can be seen in Figure 25–1 and throughout this chapter.

Floor joists can range in size from 2″ × 6″ (50 × 150 mm) through 2″ × 14″ (50 × 350 mm) and may be spaced at 12″, 16″, or 24″ (300, 400, or 600 mm) on center depending on the load, span and size of joist to be used. A spacing of 16″ (400 mm) is most common. See Chapter 28 for a complete explanation of sizing joist. Because of the decreasing supply and escalating price of sawn lumber, several alternatives to floor joist have been developed. Three of the most common substitutes to floor joist include open web trusses, I joist, and laminated veneer lumber.

The *mudsill,* or *base plate,* is the first of the structural lumber used in framing the home. The mudsill is the plate that rests on the masonry foundation and provides a base for all framing. Because of the moisture content of concrete and soil, the mudsill is required to be pressure treated or made of foundation-grade redwood or cedar. Mudsills are set along the entire perimeter of the foundation and attached to the foundation with anchor bolts.

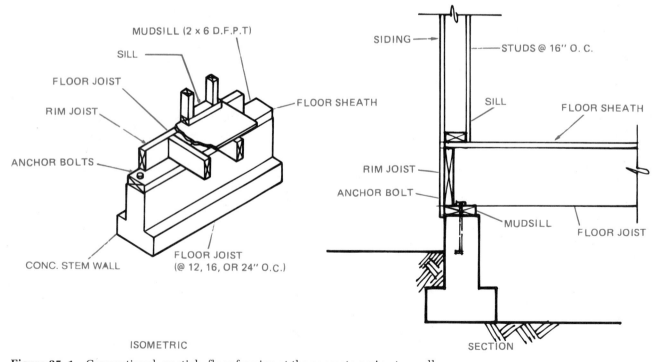

Figure 25–1 Conventional, or stick, floor framing at the concrete perimeter wall.

Figure 25–2 Glu-lam beams span longer distances than sawn lumber and eliminate problems with twisting, shrinking, and splitting. *Courtesy Georgia Pacific Corporation.*

The size of the mudsill is typically 2″ × 6″ (50 × 150 mm). Anchor bolts are typically ½ × 10″ (13 × 250 mm). Sizes of ⅝ and ¾″ (16 and 19 mm) diameter bolts are also common. Bolt spacing of 6′–0″ (1830 mm) is the maximum allowed by all codes and is common for most residential usage. Areas of a structure that are subject to lateral loads or uplifting will reflect bolts at a much closer spacing based on the engineers load calculations. Section IX will further discuss anchor bolts.

With the mudsills in place, the *girders* can be set to support the floor joists. A girder is another name for a beam. The girder is used to support the floor joists as they span across the foundation. Girders are typically sawn lumber, 4 or 6″ (100 or 150 mm) wide, with the depth determined by the load to be supported and the span to be covered. Sawn members 2″ (50 mm) wide can often be joined together to form a girder. If only two members are required to support the load, members can be nailed, based on the nailing schedules provided in each code. If more than three members must be joined together to support a load, most building departments require the members to be bolted together so the inspector can readily see that all members are attached. Another form of a built-up beam is a *flitch beam*, which consists of steel plates bolted between wood members. Because they are so labor intensive and heavy, and because of the availability of engineered wood beams, flitch beams are becoming less popular. Chapter 29 introduces formulas that designers use to determine beam sizes and if built-up beams can be used.

In areas such as basements where a large, open area is desirable, laminated wood beams called *glu-lam beams* are often used. They are made of sawn lumber that is glued together under pressure to form a beam that is stronger than its sawn lumber counterpart. Beam widths of 3⅛, 5⅛, and 6¾″ (80, 130, and 170 mm) are typical of residential construction. Although much larger sizes are available, depths typically range from 9 through 16½″ (229–419 mm) in 1½″ (38 mm) intervals. See Figure 25–2.

As the span of the beam increases, a *camber,* or curve, is built into the beam to help resist the tendency of the beam to sag when a load is applied. Glu-lams are often used where the beam will be left exposed because they do not drip sap and do not twist or crack, as a sawn beam does as it dries.

Engineered wood girders and beams are also becoming common throughout residential construction. Unlike glu-lam beams, which are laminated from sawn lumber such as a 2 × 4 or 2 × 6 (50 × 100 or 50 × 150 mm), parallel strand lumber (PSL) is laminated from veneer strips of fir and southern pine, which are coated with resin and then compressed and heated. Typical widths are 3½, 5¼, and 7″ (90, 130, and 180 mm). Depths range from 7 through 18″ (180–460 mm). PSL beams have no crown or camber.

Laminated veneer lumber (LVL) is made from ultrasonically graded douglas fir veneers, which are laminated with all grains parallel to each other with exterior grade adhesives under heat and pressure. LVL beams come in widths of 1¾″ (45 mm) and 3½″ (90 mm) and range in depths of 5½″ through 18″ (140–460 mm). LVL beams offer performance and durability superior to that of other engineered products and often offer the smallest and lightest wood girder solution.

Steel girders such as those seen in Figure 25–3 are often used where foundation supports must be kept to a minimum and offer a tremendous advantage over wood for total load that can be supported on long spans. Steel beams are often the only type of beam that can support a specified load and still be hidden within the depth of the floor framing. They also offer the advantage of no expansion or shrinkage due to moisture content. Steel beams are named for their shape, depth, and weight. A steel beam with a designation of W 16″ × 19″ would represent a wide-flange steel beam with the shape of an I. The number 16 represents the approximate depth of the beam

Figure 25–3 The use of steel girders allows for greater spans with fewer supporting columns for this residential foundation. *Courtesy Janice Jefferis.*

in inches, and the 19 represents the approximate weight of the beam in pounds per linear foot. Floor joists are usually set on top of the girder, as shown in Figure 25–4, but they also may be hung from the girder with joist hangers.

Posts are used to support the girders. As a general rule of thumb, a 4″ × 4″ (100 × 100 mm) post is typically used below a girder 4″ (100 mm) wide, and a post 6″ (150 mm) wide is typically used below a girder 6″ (150 mm) wide. Sizes can vary based on the load and the height of the post. LVL posts ranging in size from 3½″ through 7″ (90–180 mm) are increasingly being used for their ability to support loads. A minimum bearing surface of ½″ (38 mm) is required by code to support a girder resting on wood support, and a 3″ (75 mm) bearing surface is required if the girder is resting on concrete.

Steel columns may be used in place of a wooden post depending on the load to be transferred to the foundation. Because a wooden post will draw moisture out of the concrete foundation it must rest on 55 pound felt. Sometimes an asphalt roofing shingle is used between the post and the girder. If the post is subject to uplift or lateral forces, a metal post base or strap may be specified by the engineer to attach the post to the concrete firmly. Figure 25–5 shows the steel connectors used to keep a post from lifting off the foundation. Forces causing uplift will be further discussed in Chapter 27.

Once the framing crew has set the support system, the floor joists can be set in place. *Floor joists* are the structural members used to support the subfloor, or rough floor. Floor joists usually span between the foundation and a girder, but as shown in Figure 25–6, a joist may extend past its support. This extension is known as a *cantilever*.

Open web floor trusses are a common alternative to using 2× sawn lumber for floor joist. Open web trusses are typically spaced at 24″ (600 mm) o.c. for residential

Figure 25–5 Steel straps typically are used to resist stress from lateral and uplift by providing a connection between the residential frame and the foundation. *Courtesy Janice Jefferis.*

and 32″ (800 mm) for light commercial floors. Open web trusses are typically available for spans up to 38′ (11,590 mm) for residential floors. Figure 25–7 shows an example of an open web floor truss. The horizontal members of the truss are called *top* and *bottom chords*, respectively, and are typically made from 1.5″ × 3″ (40 × 75 mm) lumber laid flat. The diagonal members, called *webs*, are made of tubular steel approximately 1″ (25 mm) in diameter.

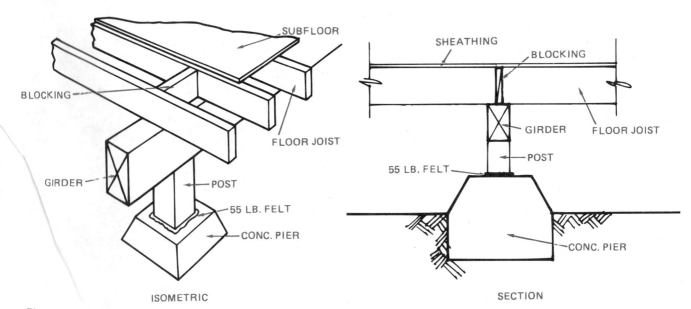

ISOMETRIC

SECTION

Figure 25–4 Floor joists supported by a girder.

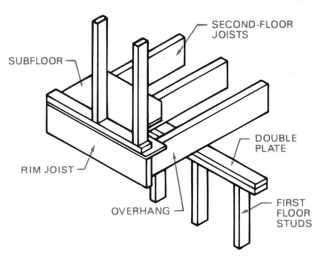

FRAMING A SECOND-FLOOR OVERHANG

Figure 25–6 A floor joist or beam that extends past its supporting member is cantilevered.

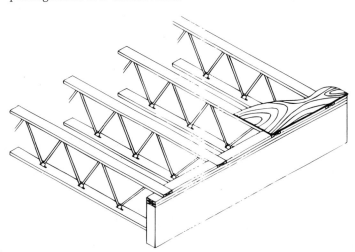

Figure 25–7 Open web floor joists allow for spans of up to 38′ for residential construction. *Courtesy Truss Joist MacMillan.*

When drawn in section or details, the webs are typically represented with a bold center line drawn at a 45° angle. The exact angle is determined by the truss manufacturer and is unimportant to the detail. As the span or load on the joist increases, the size and type of material used are increased. Pairs of 1.5″ × 2.3″ (40 × 60 mm) laminated veneer lumber are used for the chords of many residential floor trusses. When drawing sections showing floor trusses, it is important to work with the manufacturer's details to find exact sizes and truss depth. Most truss manufacturers supply disk copies of common truss connections. Figure 25–8 shows an example of a truss detail.

Trusses that resemble an I are often referred to as I-joist and are a high-strength, lightweight, cost-efficient alternative to sawn lumber. First developed in the late 1960s, I-joists form a uniform size, have no crown, and do not shrink. They come in depths of 9½″, 11⅞″, 14″, and 18″

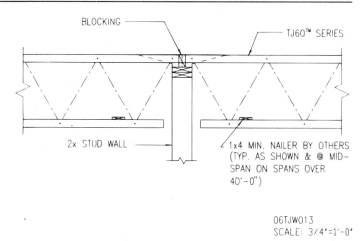

06TJW013
SCALE: 3/4″=1′-0′

Figure 25–8 Disk copies of truss details, supplied by most manufacturers, can be inserted into the working drawings. *Courtesy Truss Joist Macmillan.*

(240, 300, 360, and 460 mm). I-joists are able to span greater distances than comparable-sized sawn joists and are suitable for spans up to 40′ (12,200 mm) for residential uses. Figure 25–9 shows an example of a solid web truss with a plywood web.

Figure 25–9 Each of the national wood manufacturers makes trusses with solid webs made of plywood. Depths of 9½ and 11⅞″ (240 and 300 mm) are typical for residential construction. *Courtesy Georgia Pacific.*

Webs can be made from plywood or oriented strand board (OSB). OSB is increasingly being used to replace lumber products and is made from wood fibers arranged in a precisely controlled pattern; the fibers are coated with resin and then compressed and cured under heat. Holes can be placed in the web to allow for HVAC ducts and electrical requirements based on the manufacturer's specifications. Figure 25–10 shows a truss made with OSB.

Laminated veneer lumber (LVL) is made from ultrasonically graded douglas fir veneers, which are laminated just as LVL beams are. LVL joists are 1¾″ (45 mm) wide and range in depths of 5½″ through 18″ (140–460 mm). LVL joists offer superior performance and durability to other engineered products and are designed for single- or multi-span uses that must support heavy loads. Figure 25–11 shows an example of an LVL joist.

Floor Bracing

Because of the height-to-depth proportions of a joist, it will tend to roll over onto its side as loads are applied. To resist this tendency, a rim joist or blocking is used to support the joist. A *rim joist,* which is sometimes referred to as a band or header, is usually aligned with the outer face of the foundation and mudsill. Some framing crews set a rim joist around the entire perimeter and then end-nail the floor joists. An alternative to the rim joist is to use solid blocking at the sill between each floor joist. See Figure 25–12.

Blocking in the floor system is typically used at the center of the joist span, as shown in Figure 25–13. Spans longer than 10′ (250 mm) must be blocked at center span to help transfer lateral forces from one joist to another and then to the foundation system. Another use of blocking is at the end of the floor joists at their bearing point. See Figure 25–12. These blocks help keep the entire floor system rigid, and are used in place of the rim joist. Blocking is also used to provide added support to the floor

Figure 25–11 Laminated veneer lumber offers superior performance and durability over sawn lumber for supporting heavy loads. *Courtesy Louisiana Pacific.*

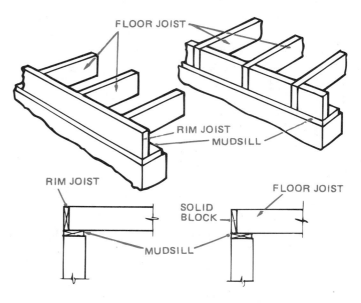

Figure 25–12 Floor joist blocking at the center and end of the span. At midspan, solid or cross blocking may be used. At the ends of the floor joists, solid blocking or a continuous rim joist may be used to provide stability.

sheathing. Often walls are reinforced to resist lateral loads. These loads are, in turn, transferred to the foundation through the floor system and are resisted by a *diaphragm,* a rigid plate that acts similar to a beam and can be found in the roof level, walls, or floor system. In a floor diaphragm, 2″ × 4″ (50 × 100 mm) blocking is typically laid flat between the floor joist to allow the edges of plywood panels to be supported. Nailing all edges of a plywood panel allows for approximately one-half to two

Figure 25–10 Truss joists are now being made from oriented strand board because of the shortage of quality old-growth timber. *Courtesy Louisiana Pacific.*

Figure 25–13 Blocking is used to provide stiffness to a floor system and to support plumbing and heating equipment. *Courtesy John Black.*

times the design load to be resisted over unblocked diaphragms. The architect or engineer determines the size and spacing of nails and blocking required. Blocking is also used to reduce the spread of fire and smoke through the floor system.

Floor sheathing is installed over the floor joists to form the subfloor. The sheathing provides a surface for the base plate of the wall to rest on. Until the mid-1940s, 1 × 4s or 1 × 6s (25 × 100 or 25 × 150 mm) laid perpendicularly or diagonally to the floor joists were used for floor sheathing. Plywood is now the most common floor sheathing. Depending on the spacing of the joist, 1/2″, 5/8″, or 3/4″ (13, 16, or 19 mm) with an APA grade of EXP 1 or 2, EXT, STRUCT I-EXP-1 or STRUCT 1-EXT plywood is used for sheathing because of its labor-saving advantages. EXT represents exterior grade, STRUCT represents structural, and EXP represents exposure. Plywood also is printed with a number to represent the *span rating*, which represents the maximum spacing from center to center of supports. The span rating is listed as two numbers, such as 32/16, separated by a slash. The first number represents the maximum recommended spacing of support-

ers if the panel is used for roof sheathing and the long dimension of the sheathing is placed across three or more supports. The second number represents the maximum recommended spacing of supports if the panel is used for floor sheathing and the long dimension of the sheathing is placed across three or more supports. Plywood is usually laid so that the face grain on each surface is perpendicular to the floor joist. This provides a rigid floor system without having to block the edges of the plywood. Floor sheathing that will support ceramic tile is typically 32/16″ APA span rated 15/31″ (12 mm) or span rated 40/20″ and 19/32″ (15 mm) thick.

Waferboard or oriented strand board has also become a popular choice to traditional plywood subfloors since the mid-1980s. OSB is made from three layers of small strips of wood that have been saturated with a binder and then compressed under heat and pressure. The exterior layers of strips are parallel to the length of the panel, and the core layer is laid in a random pattern.

Once the subfloor has been installed, an underlayment for the finish flooring is put down. The underlayment is not installed until the walls, windows, and roof are in place, making the house weathertight. The underlayment provides a smooth impact-resistant surface on which to install the finished flooring. Underlayment is usually 3/8″ or 1/2″ (9 or 13 mm) APA underlayment GROUP 1, EXPOSURE 1 plywood, hardboard, or waferboard. Hardboard is typically referred to as medium- or high-density fiberboard (MDF or HDF) and is made from wood particles of various sizes that are bonded together with a synthetic resin under heat and pressure. The underlayment may be omitted if the holes in the plywood are filled. APA STURD-I-FLOOR rated plywood of 19/32″ through 1 3/32″ (15–28 mm) thick can also be used to eliminate the underlayment.

Post-and-Beam Construction

Terms to be familiar with when working with post-and-beam floor systems include *girder, post, decking,* and *finished* floor. Notice there are no floor joists with this system. See Figure 25–14.

A mudsill is installed with post-and-beam construction just as with platform construction. Once set, the girders are also placed. With post-and-beam construction the girder is supporting the floor decking instead of floor joists. Girders are usually 4 × 6 (100 × 150 mm) beams spaced at 48″ (1200 mm) o.c., but the size and distance may vary depending on the loads to be supported. Similar to conventional methods, posts are used to support the girders. Typically a 4 × 4 (100 × 100 mm) is the minimum size used for a post, with a 4 × 6 (100 × 150 mm) post used to support joints in the girders. With the support system in place, the floor system can be installed. Figure 25–15 shows the components of a post-and-beam floor system.

Decking is the material laid over the girders to form the subfloor. Typically decking is 2 × 6 or 2 × 8 (50 ×

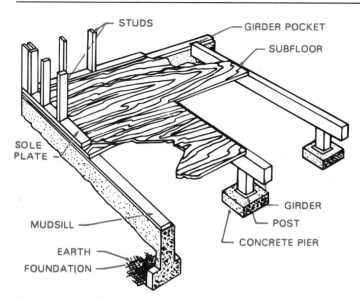

Figure 25–14 Post-and-beam construction components.

Figure 25–15 Floors built with a post-and-beam system require no floor joist to support the floor sheathing. *Courtesy Saundra Powell.*

150 or 50 × 200 mm) tongue and groove (T&G) boards or 1³⁄₃₂″ (28 mm) APA rated 2-4-1 STURD-I-FLOOR EXP-1 plywood T&G floor sheathing. The decking is usually finished in similar fashion to conventional decking with a hardboard overlay.

▼ FRAMED WALL CONSTRUCTION

As a drafter you will be concerned with two types of walls, bearing and nonbearing. A bearing wall not only supports itself but also the weight of the roof or other floors constructed above it. A bearing wall requires some type of support under it at the foundation or lower floor level in the form of a girder or another bearing wall. Nonbearing walls are sometimes called partitions. A nonbearing wall serves no structural purpose. It is a partition used to divide rooms and could be removed without causing damage to the building. Bearing and nonbearing walls can be seen in Figure 25–16. In post-and-beam construction, any exterior walls placed between posts are nonbearing walls.

Bearing and nonbearing walls are both constructed with similar materials using a sole plate, studs, and a top plate. Each can be seen in Figure 25–17. The *sole*, or bottom, plate is used to help disperse the loads from the wall studs to the floor system. The sole plate also holds the studs in position as the wall is being tilted into place. *Studs* are the vertical framing members used to transfer loads from the top of the wall to the floor system. Typically studs 2 × 4 (50 × 100 mm) are spaced at 16″ (400 mm) o.c. and provide a nailing surface for the wall sheathing on the exterior side and the sheetrock on the interior side. Studs 2 × 6 (50 × 150 mm) are often substituted for 2 × 4 studs (50 × 100 mm) to provide added resistance for lateral loads and to provide a wider area to install insulation. Engineered studs in 2 × 4 or 2 × 6 (50 × 100 or 50 × 150 mm) are also available in 8, 9, and 10′ (2400, 2700, and 3000 mm) lengths. Engineered studs are made from short sections of stud-grade lumber that have had the knots and splits removed. Sections of wood are joined together with ⁵⁄₈″ (16 mm) finger

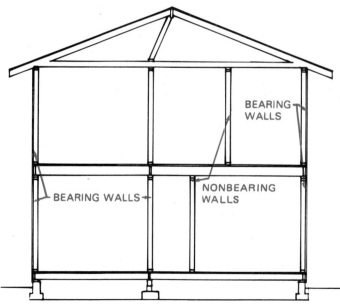

Figure 25–16 Bearing and nonbearing walls. A bearing wall supports its own weight and the weight of floor and roof members. A nonbearing wall supports only its own weight. Most building codes allow ceiling weight to be supported on a wall and still be considered a nonbearing wall.

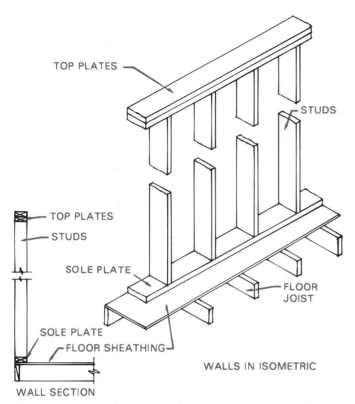

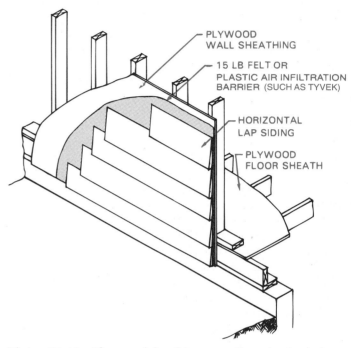

Figure 25–17 Standard wall construction uses a double top plate on the top of the wall, a sole plate at the bottom of the wall, and studs.

Plywood sheathing is primarily used as an insulator against the weather and also as a backing for the exterior siding, as shown in Figure 25–19. Sheathing may also be installed over TYVEK (see Chapter 26) as seen in Figure 25–20. Sheathing may be considered optional depending on your area of the country. When sheathing is used on exterior walls, it provides what is called *double-wall construction*. In *single-wall construction*, wall sheathing is

Figure 25–19 The use of sheathing on walls varies throughout the country. In some areas, siding is applied over the sheathing to provide double-wall construction.

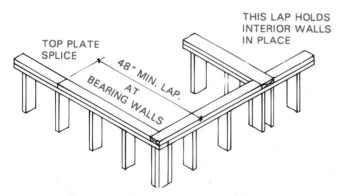

Figure 25–18 Top plate laps are required to be 48″ minimum in bearing walls or held together with steel straps.

joints. The top plate is located on top of the studs and is used to hold the wall together. See Figure 25–18. Two top plates are required on bearing walls, and each must lap the other a minimum of 48″ (1200 mm). This lap distance provides a continuous member on top of the wall to keep the studs from separating. An alternative to the double top plate is to use one plate with a steel strap at each joint in the plate. The top plate also provides a bearing surface for the floor joists from an upper level or for the roof members.

Figure 25–20 Wood siding can be placed directly over a vapor barrier and the studs, forming single-wall construction. Vapor barriers such as Tyvek allow moisture to escape but not enter the structure. *Courtesy Cynthia Weaver.*

not used, and the siding is attached over building paper to the studs, as in Figure 25–21. Plywood sheathing should be APA-rated EXP 1 or 2, EXT, STRUCT 1, EXT 1, or STRUCT 1 EXT. The cost of the home and its location will have a great influence on whether wall sheathing is to be used. In areas where double-wall construction is required for weather or structural reasons, 3/8" plywood is used on exterior walls as an underlayment. Many builders prefer to use 1/2" (13 mm) OSB for sheathing in place of plywood.

In areas where wood is neither inexpensive nor plentiful, plywood is used for its ability to resist the tendency of a wall to twist or rack. Racking can be caused from wind or seismic forces. Plywood used to resist these forces is called a *shear panel*. See Figure 25–22 for an example of racking and how plywood can be used to resist this motion.

An option to plywood for shear panels to keep the studs from racking is to use let-in braces. Figure 25–23 shows how a brace can be used to stiffen the studs. A notch is cut into the studs, and a 1 × 4 (25 × 100 mm) is laid flat in this notch at a 45° angle to the studs. If plywood siding rated APA Sturd-I-Wall is used, no underlayment or let-in braces are required.

Prior to the mid-1960s, blocking was common in walls to help provide stiffness. Blocking for structural or fire reasons is now no longer required unless a wall exceeds 10′ (3050 mm) in height. Blocking is often installed for a nailing surface for mounting cabinets and plumbing fixtures. Blocking is sometimes used to provide extra strength in some seismic zones.

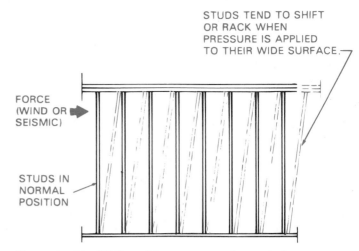

Figure 25–22 Wall racking occurs when wind or seismic forces push the studs out of their normal position. Plywood panels can be used to resist these forces and keep the studs perpendicular to the floor. These plywood panels are called *shear panels*.

Figure 25–23 Let-in braces can sometimes be used instead of shear panels to resist wall racking. *Courtesy Gang-Nail Systems, Inc.*

Figure 25–21 With single-wall construction, siding is applied directly over the building paper and studs. *Courtesy American Plywood Association.*

In addition to the wall components mentioned, there are several other terms that the drafter needs to be aware of. These are terms that are used to describe the parts used to frame around an opening in a wall: *headers, subsill, trimmers, king studs,* and *jack studs.* Each can be seen in Figure 25–24.

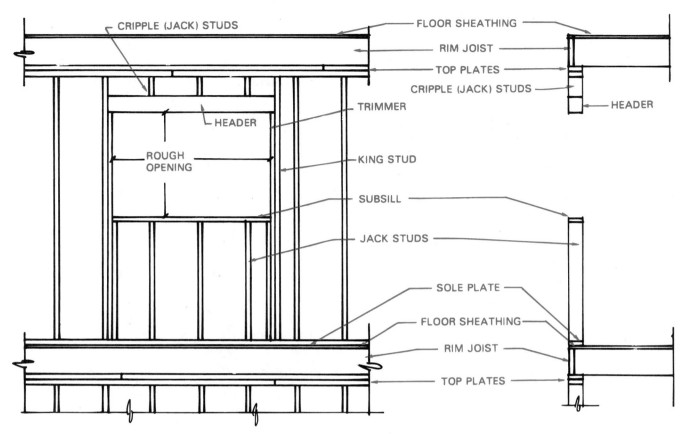

Figure 25–24   Construction components at an opening.

A *header* in a wall is used over an opening such as a door or window. When an opening is made in a wall, one or more studs must be omitted. A header is used to support the weight that the missing studs would have carried. The header is supported on each side by a *trimmer*. Depending on the weight the header is supporting, double trimmers may be required. The trimmers also provide a nailing surface for the window and the interior and exterior finishing materials. A *king stud* is placed beside each trimmer and extends from the sill to the top plates. It provides support for the trimmers so that the weight imposed from the header can only go downward, not sideways.

Between the trimmers is a *subsill* located on the bottom side of a window opening. It provides a nailing surface for the interior and exterior finishing materials. *Jack studs,* or *cripples,* are studs that are not full height. They are placed between the sill and the sole plate and between a header and the top plates.

▼ ROOF CONSTRUCTION

Roof framing includes both conventional and truss framing methods. Each has its own special terminology, but there are many terms that apply to both systems. These common terms will be described first, followed by the terms for conventional and truss framing methods.

Basic Roof Terms

Roof terms common to conventional and trussed roofs are *eave, cornice, eave blocking, fascia, ridge, sheathing, finishing roofing, flashing,* and *roof pitch dimensions.*

The *eave* is the portion of the roof that extends beyond the walls. The *cornice* is the covering that is applied to the eaves. Common methods for constructing the cornice can be seen in Figure 25–25. Eave, or *bird blocking,* is a spacer block placed between the rafters or truss tails at the eave. This block keeps the spacing of the rafters or trusses uniform and keeps small animals from entering the attic. It also provides a cap to the exterior siding, as seen in Figure 25–26.

A *fascia* is a trim board placed at the end of the rafters or truss tails and usually perpendicular to the building wall. It hides the truss or rafter tails from sight and also provides a surface where the gutters may be mounted. The fascia can be made from either 1x or 2x material, depending on the need to resist warping. See Figure 25–26. The fascia is typically 2″ (50 mm) deeper than the rafters or truss tails. At the opposite end of the rafter from the fascia is the ridge. The *ridge* is the highest point of a

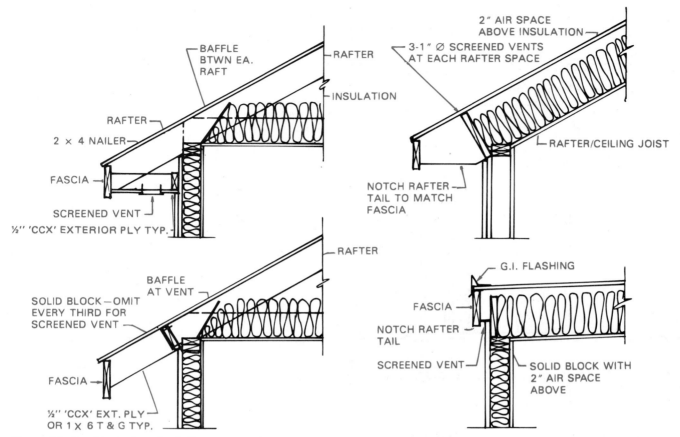

Figure 25–25 Typical methods for constructing a cornice.

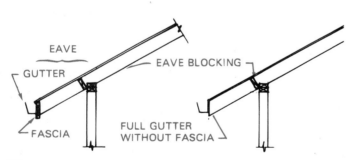

Figure 25–26 Eave components.

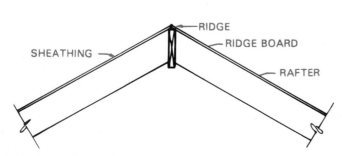

Figure 25–27 Ridge construction.

roof and formed by the intersection of the rafters or the top chords of a truss. See Figure 25–27.

Roof sheathing is similar to wall and floor sheathing in that it is used to cover the structural members. Roof sheathing may be either solid or skip. The area of the country and the finished roofing to be used will determine which type of sheathing is used. For solid sheathing, ½″ (13 mm) thick CDX plywood is generally used. CDX is the specification given by the American Plywood Association to designate standard-grade plywood. It provides an economical, strong covering for the framing, as well as an even base for installing the finished roofing. Common span ratings used for residential roofs are 24/16, 32/16, 40/20, and 48/24. Figure 25–28 shows plywood sheathing being applied to a conventionally framed roof.

Skip sheathing is generally used with either tile or shakes. Typically 1 × 4s (25 × 100 mm) are laid perpendicular to the rafters with a 4″ (100 mm) space between each piece of sheathing. See Figure 25–29. A water-resistant sheathing must be used when the eaves are exposed to weather. This usually consists of plywood rated CCX or 1″ (25 mm) T&G decking. CCX is the specification for exterior use plywood. It is designated for exterior use because of the glue and the type of veneers used to make the panel.

The *finished roofing* is the weather protection system. Typically roofing might include built-up roofing, asphalt shingles, fiberglass shingles, cedar, tile, or metal panels.

Figure 25–28 Plywood with a span rating of 24/16 is typically placed over rafters to support the finish roofing material. *Courtesy Louisiana Pacific.*

Figure 25–29 Skip or spaced sheathing is used under shakes and tile roofs. *Courtesy Saundra Powell.*

(See Chapter 19.) Flashing is generally 20 to 26 gage metal used at wall and roof intersections to keep water out.

Pitch, span, and overhang are dimensions that are needed to define the angle, or steepness, of the roof. Each can be seen in Figure 25–30. *Pitch*, used to describe the slope of the roof, is the ratio between the horizontal run and the vertical rise of the roof. The *run* is the horizontal measurement from the outside edge of the wall to the centerline of the ridge. The *rise* is the vertical distance from the top of the wall to the highest point of the rafter being measured. Review Chapter 19 for a complete discussion of roof pitch. The *span* is the horizontal measurement between the inside edges of the supporting walls. (Chapter 27 will further discuss span.) The *overhang* is the horizontal measurement between the exterior face of the wall and the end of the rafter tail.

Conventionally Framed Roof Terms

Conventional, or stick, framing methods involve the use of wood members placed in repetitive fashion. Stick framing involves the use of such members as a ridge board, rafter, and ceiling joists. The *ridge board* is the horizontal member at the ridge, which runs perpendicular to the rafters. The ridge board is centered between the exterior walls when the pitch on each side is equal. The ridge board resists the downward thrust resulting from gravity trying to force the rafters into a V shape between the walls. The ridge board does not support the rafters but is used to align the rafters so that their forces are pushing against each other.

Rafters are the sloping members used to support the roof sheathing and finished roofing. Rafters are typically spaced at 24" O.C. (600 mm), but 12 and 16" (300 and 400 mm) spacings are also used. There are various kinds of rafters: *common*, *hip, valley*, and *jack*. Each can be seen in Figure 25–31.

A *common rafter* is used to span and support the roof loads from the ridge to the top plate. Common rafters run perpendicular to both the ridge and the wall supporting them. The upper end rests squarely against the ridge board, and the lower end receives a bird's mouth notch and rests on the top plate of the wall. A *bird's mouth* is a notch cut in a rafter at the point where the rafter intersects a beam or bearing wall. This notch increases the contact area of the rafter by placing more rafter surface against the top of the wall, as shown in Figure 25–32.

Hip rafters are used when adjacent slopes of the roof meet to form an inclined ridge. The hip rafter extends

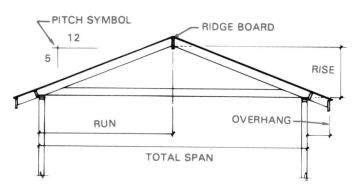

Figure 25–30 Roof dimensions needed for construction.

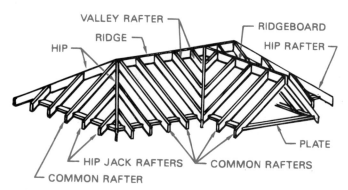

Figure 25–31 Roof members in conventional construction.

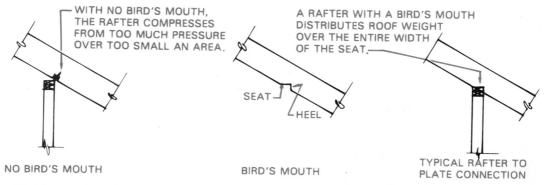

Figure 25–32   A bird's mouth is a notch cut into the rafter to increase the bearing surface.

Figure 25–33   A hip rafter is an inclined ridge board used to frame an exterior corner. *Courtesy Michelle Cartwright.*

diagonally across the common rafters and provides support to the upper end of the rafters. See Figure 25–33. The hip is inclined at the same pitch as the rafters. A *valley rafter* is similar to a hip rafter. It is inclined at the same pitch as the common rafters that it supports. Valley rafters get their name because they are located where adjacent roof slopes meet to form a valley. *Jack rafters* span from a wall to a hip or valley rafter. They are similar to a common rafter but span a shorter distance. Typically, a section will only show common rafters, with hip, valley, and jack rafters reserved for a very complex section.

Rafters tend to settle because of the weight of the roof and because of gravity. As the rafters settle, they push supporting walls outward. These two actions, downward and outward, require special members to resist these forces. These members are *ceiling joists, ridge bracing, collar ties, purlins,* and *purlin blocks and braces.* Each can be seen in Figure 25–34.

Ceiling joists span between the top plates of bearing walls to resist the outward force placed on the walls from the rafters. The ceiling joists also support the finished ceiling. *Collar ties* are also used to help resist the outward thrust of the rafters. They are usually the same cross-section size as the rafter and placed in the upper third of the roof.

Ridge braces are used to support the downward action of the ridge board. The brace is usually a 2 × 4 (50 × 100) spaced at 48″ (1200 mm) o.c. maximum. The brace must be set at 45° maximum to the ceiling joist. A *purlin* is a brace used to provide support for the rafters as they span between the ridge and the wall. The purlin is usually the same cross-section size as the rafter and is placed below the rafter to reduce the span. As the rafter span is reduced, the size of the rafter can be reduced. See Chapter 28 for a further explanation of rafter sizes. A *purlin brace* is used to support the purlins. They are typically 2 × 4s (50 × 100s) spaced at 48″ (1220 mm) o.c. along the purlin and transfer weight from the purlin to a supporting wall. The brace is supported by an interior wall, or a 2 × 4 (50 × 100) lying across the ceiling joist. They can be installed at no more than 45° from vertical. A scrap block of wood is used to keep the purlin from sliding down the brace. When there is no wall to support the ridge brace, a strong back is added. A *strong back* is a beam placed over the ceiling joist to support the ceiling and roof loads.

If a vaulted ceiling is to be represented, two additional terms will be used on the sections: *rafter/ceiling joist* and *ridge beam.* Both can be seen in Figure 25–35. A *rafter/ceiling joist,* or *rafter joist,* is a combination of rafter and ceiling joist. The rafter/ceiling joist is used to support both the roof loads and the finished ceiling. Typically a 2 × 12 (50 × 300 mm) rafter/ceiling joist is used to allow room for 10″ (250 mm) of insulation and 2″ (50 mm) of air space above the insulation. The size of the rafter/ceiling joist must be determined by the load and span.

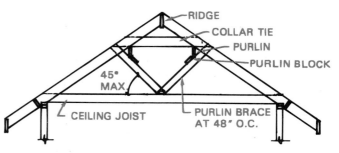

Figure 25–34   Common roof supports.

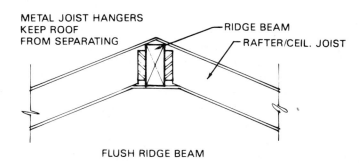

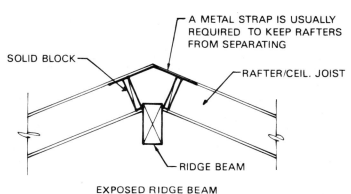

Figure 25–35 Common connections between the ridge beam and rafters. The ridge may be exposed or hidden.

A *ridge beam* is used to support the upper end of the rafter/ceiling joist. Since there are no horizontal ceiling joists, metal joist hangers must be used to keep the rafters from separating from the ridge beam.

The final terms that you will need to be familiar with to draw a stick roof are *header* and *trimmer*. Both are terms that are used in wall construction, and they have a similar function when used as roof members. See Figure 25–36. A *header* at the roof level consists of two members nailed together to support rafters around an opening, such as for a skylight or chimney. *Trimmers* are two rafters nailed together to support the roofing on the inclined edge of an opening.

Truss Roof Construction Terms

Truss construction is generally considered nonconventional construction. A *truss* is a component used to span large distances without intermediate supports. Residential trusses can be as short as 15′ (4570 mm) or as long as 50′ (15,200 mm). Trusses can be either prefabricated or job built. Prefabricated trusses are commonly used in residential construction. Assembled at the truss company and shipped to the job site, the truss roof can quickly be set in place. A roof that might take two or three days to frame using conventional framing can be set in place in two or three hours using trusses, which are set in place by crane. The size of material used to frame trusses is smaller than with conventional framing. Typically, truss members need only be 2 × 4s (50 × 100 mm) set at 24″ (600 mm) o.c. The exact size of the truss members will

be determined by an engineer working for the truss manufacturer. The drafter's responsibility when drawing a structure framed with trusses is to represent the span of the trusses on the framing plan and show the general truss shape and bearing points in the section drawings. Chapter 30 will introduce framing plans, and Chapter 35 will introduce sections.

A knowledge of truss terms is helpful in drawing these drawings. Terms tht must be understood are *top chord, bottom chord, webs, ridge block,* and *truss clips.* Each is shown in Figure 25–37. The *top chord* serves a similar function to a rafter. It is the upper member of the truss and supports the roof sheathing. The *bottom chord* serves a similar purpose as a ceiling joist. It resists the outward thrust of the top chord and supports the finished ceiling material. *Webs* are the interior members of the truss that span between the top and bottom chord. They are attached to the chords by the manufacturer using metal plate connectors. *Ridge blocks* are blocks of wood used to provide a nailing surface for the roof sheathing and as a

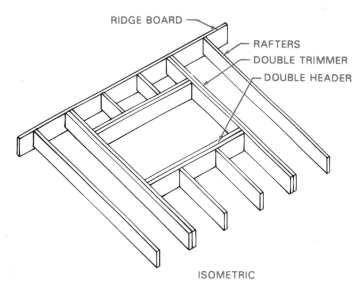

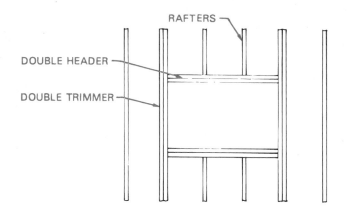

Figure 25–36 Typical construction at roof openings.

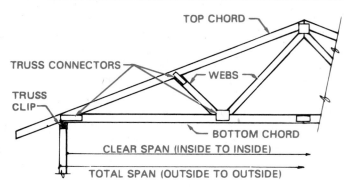

Figure 25–37 Truss construction members.

spacer when setting the trusses into position. *Truss clips*, also known as *hurricane ties*, are used to strengthen the connection between the truss and top plate or header, which is used to support the truss. The truss clips transfer wind forces applied to the roof, which cause uplift down through the wall framing into the foundation. A block is also used where the trusses intersect a support. Common truss intersections with blocking and hurricane ties are shown in Figure 25–38.

The truss gains its strength from triangles formed throughout the truss. The shape of each triangle cannot be changed unless the length of one of the three sides is altered. The entire truss will tend to bend under the roof loads as it spans between its bearing points. Most residential trusses can be supported by a bearing point at or near each end of the truss. As spans exceed 40' or as the shape of the bottom chord is altered, it may be more economical to provide a third bearing point at or near the center. For a two-point bearing truss, the top chords are in compression from the roof loads and tend to push out the heels and down at the center of the truss. The bottom chord is attached to the top chords and is in tension as it resists the outward thrust of the top chords. Webs closest

to the center are usually stressed by tension, and the outer webs are usually stressed by compression. Figure 25–39 shows how the tendency of the truss to bend under the roof loads is resisted and the loads transferred to the bearing points.

When they were introduced to the residential market, trusses were used primarily to frame gable roofs. Computers and sophisticated design software have enabled trusses to be easily manufactured in nearly any shape. Several of the common shapes and types of trusses available for residential roof construction can be seen in Figure 25–40. The most commonly used truss in residential construction is a *standard truss*, which is used to frame gable roofs. A *gable end wall truss* is used to form the exterior ends of the roof system and is aligned with the exterior side of the end walls of a structure. This truss is more like a wall than a truss, with the vertical supports typically spaced at 24" (600 mm) o.c. to support exterior finishing material. A *girder truss* is used on houses with an L- or U-shaped roof where the roofs intersect. A girder truss is typically formed by the manufacturer's bolting two or three standard trusses together. The truss manufacturer determines the size and method of constructing the girder truss.

The *cantilevered truss* is used where a truss must extend past its support to align with other roof members. Cantilevered trusses are typically used where walls jog to provide an interior courtyard or patio. A *stub truss* can be used where an opening will be provided in the roof or the roof must be interrupted. Rooms with glass skywalls as in Figure 25–41 can often be framed using stub trusses. The shortened end of the truss can be supported by either a bearing wall or beam or by a header truss. A *header truss* has a flat top, which is used to support stub trusses. The header truss has a depth to match that of the stub truss and is similar in function to a girder truss. The header truss spans between and is hung from two standard trusses. Stub trusses are hung from the header truss.

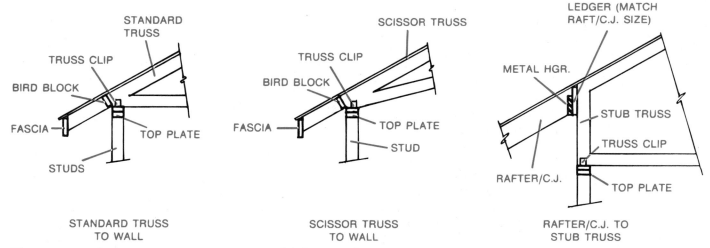

Figure 25–38 Common truss connections normally shown in the sections. Notice that the top chord aligns with the outer face of the top plate. When detailing a scissor truss, the bottom chord is typically drawn a minimum of two pitches less than the top chord.

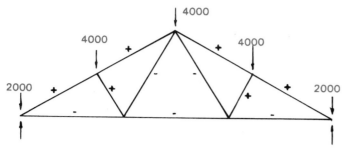

Figure 25–39 Trusses are designed so that the weight to be supported is spread to the outer walls. This is done by placing some members in tension and some in compression. A member in compression is indicated by a plus sign (+) and one in tension is represented by a minus sign (−).

Hip trusses are used to form hip roofs. Each succeeding truss increases in height until the full height of the roof is achieved and standard trusses can be used. The height of each hip truss decreases as they get closer to the exterior wall, which is perpendicular to the ridge. Typically hip trusses must be 6′ (1500 mm) from the exterior wall to achieve enough height to support the roof loads. Figure 25–41 shows where hip trusses could be used to frame a roof. The exact distance will be determined by the truss manufacturer and will be further explained in Chapter 30. A *mono truss* is a single pitched truss, which can often be used in conjunction with hip trusses to form the external 6′ (1500 mm) of the hip. A mono truss is also useful in blending a one-level structure with a two-level structure, as seen in Figure 25–42.

If a vaulted roof is desired, vaulted or scissor trusses can be used. *Vaulted* and *scissor trusses* have inclined bottom chords. Typically there must be at least a two-pitch difference between the top and bottom chord. If the top chord is set at a 6/12 pitch, the bottom chord usually cannot exceed a 4/12 pitch. A section will need to be drawn to give the exact requirement for the vaulted portion of the truss similar to Figure 25–43. If a portion of the truss needs to have a flat ceiling, it may be more economical to frame the lowered portion with conventional framing materials, a process often referred to as *scabbing on*. Although this process requires extra labor at the job site, it might eliminate a third bearing point near the center of the truss. Figure 25–44 shows an example of two areas with vaulted ceilings separated by a portion of a room with a flat ceiling.

▼ METAL HANGERS

Metal hangers are used on floor, ceiling, and roof members, typically to keep structural members from separating. Figure 25–45 shows several common types of connectors that are used in light construction. These connectors keep beams from lifting off posts, keep posts from lifting off foundations, or hold one beam or joist to another.

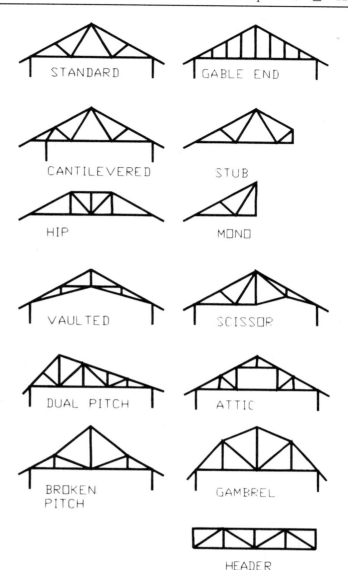

Figure 25–40 Common types of roof trusses for residential construction.

▼ ENVIRONMENTALLY FRIENDLY DESIGN AND CONSTRUCTION

Throughout this chapter, many building components have been discussed. Most of us live and work in structures that are constructed of these components and never even think about the building materials. The Environmental Protection Agency (EPA) list over 48,000 chemicals, with no information on the toxic effects of 79 percent of them. Of the chemicals that have been tested, many are found in the residential construction industry and can cause severe medical problems to individuals who are chemically sensitive. Allergies and diseases such as chronic fatigue syndrome are being linked to construction materials. Indoor air of new homes often contains as much as six times the acceptable outdoor levels of pollutants. The greatest indoor health risks

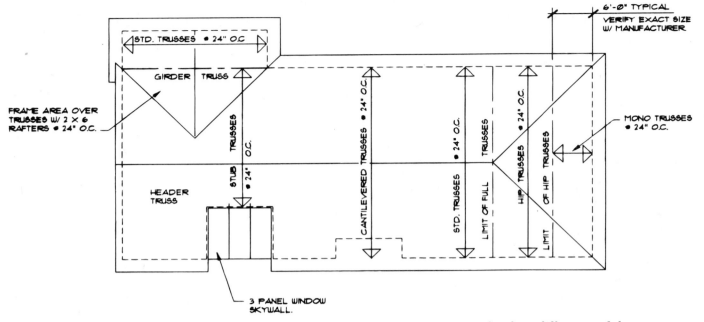

Figure 25–41 Standard, girder, stub, cantilever, hip, and mono trusses can each be used to form different roof shapes.

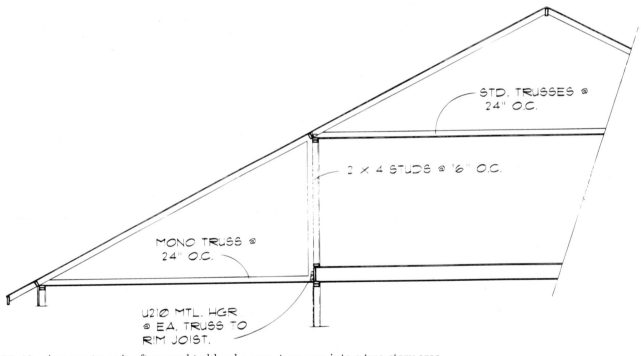

Figure 25–42 A mono truss is often used to blend a one-story area into a two-story area.

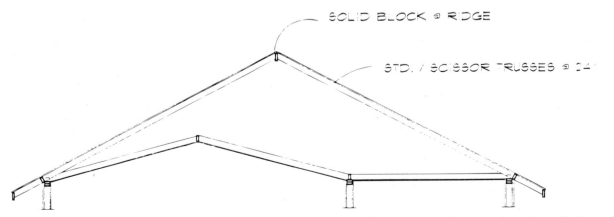

Figure 25–43 A drawing of a standard/scissor truss is usually provided to show the manufacturer where to vault the roof.

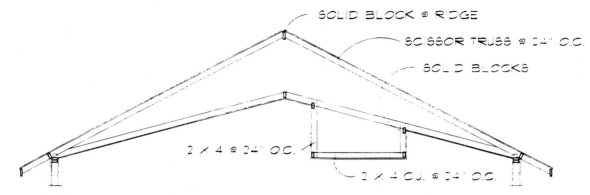

Figure 25–44 An alternative to a standard/scissor truss is to "scab on" a false ceiling at the job site.

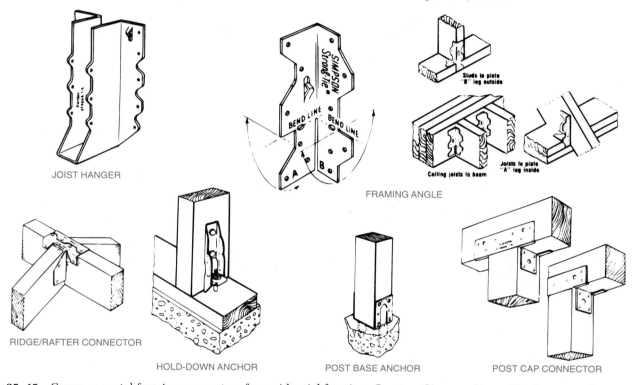

JOIST HANGER

FRAMING ANGLE

RIDGE/RAFTER CONNECTOR

HOLD-DOWN ANCHOR

POST BASE ANCHOR

POST CAP CONNECTOR

Figure 25–45 Common metal framing connectors for residential framing. *Courtesy Simpson Strong-Tie Company, Inc.*

come from airborne pollutants from products containing formaldehyde-based resins and solvents used in the construction process containing volatile organic compounds.

Formaldehyde is a colorless gas compound composed of carbon, hydrogen, and oxygen and is found in most resin-based construction products. Resin is used in products such as plywood, HDF, MHF, PSL, OSB, LVL, linoleum, lacquer, gypsum board, paneling, wallpaper, caulking compounds, insulation, adhesives, upholstery, and carpet. The toxins contained in the products can be present in harmful levels to healthy adults for up to 20 years after insulation. Many household cleaning products also contain small amounts of formaldehyde. Exposure to low-level concentrations can cause irritation of the nose, throat, and eyes and headaches, coughing, and

fatigue. Higher levels of exposure can cause severe allergic reactions, skin rash, and possibly some types of cancer.

Alternatives to formaldehyde-based products are often expensive and difficult to find. The AIA is becoming a leader in providing education in environmentally friendly construction methods. Concern is now being expressed by many groups for the need to examine the effects of manufacturing, using, and disposing of specific products in relation to the environment.

Pressure-treated lumber can be removed from a residence by substituing products with a natural ability to repel moisture, such as cedar or redwood. Low-toxic products such as AM Penetrating Seal, Protek, and Boracare Timbor Impel can be used to protect lumber from moisture.

Sheets of grass-based boards such as Meadowboard or Medite can be used in place of plywood sheathing. These boards resemble OSB in appearance but contain no dangerous resins. If plywood must be used, exterior grade plywoods contain lower levels of formaldehyde than most interior grades. The sides, edges, and interior side of plywoods can be sealed with sealers such as AM Safeseal or Crystalaire to prevent the leaking of fumes to the home interior. Fiberglass insulation materials are also harmful to chemically sensitive people. Products such as AirKrete can be sprayed into the stud space to provide a toxin-free insulation. Foil-based insulation materials such as Kshield or Dennyfoil with taped joints can also be used to shield the interior of a structure from toxins contained in wall cavity material.

Adhesives, joint compound, strippers, paints, and sealants all contain volatile organic chemicals, which can be harmful for years. Such products as AM, Murco, Crystalaire, and Auro are nontoxic or very low toxic.

Carpeting is one of the worst causes of indoor pollution. Typically carpeting contains over 50 chemicals, which are used as bonding and protective agents. In addition to the chemical content, carpet harbors pollutants such as dust, mold, and pet dander, which affect many people with allergies. All-natural carpets of untreated wool, cotton, or other natural blends such as coir of seagrass are the best choices for allergy-sensitive people. Nylon is the most inert of the synthetic materials used for carpets. Stain treatments, foam pads, and the adhesives used for attaching the pad must be carefully considered because of their chemical content. AFM Carpet Guard seals noxious fumes into most carpets.

For a chemically sensitive person, hard-surface floors are a better alternative than carpet. Tile, hardwood, linoleum, and slate are safe choices. Hardwood floors made of natural wood are safe, but the protective coating should be carefully considered. Wood parquet flooring should not be used because of the resins in the products and the adhesives used to bond them to the subfloor. Vinyl flooring is typically chemically saturated. Natural linoleum made from linseed oil and wood floor with jute backing provide a nontoxic durable flooring.

In addition to the material used for construction, heating equipment, appliances, and furnishings must be considered if the toxins used in construction are to be reduced. Although gas and oil furnaces are cost-effective, the fumes created by the burning of fossil fuels cause allergies for many people. Electric forced air and radiant heating are nontoxic.

Many appliances within a residence should also be carefully considered. Gas appliances should be avoided for the same problems associated with gas heaters. Ovens should not be self-cleaning and should initially be heated outside the residence to burn off oils, paints, and plastic fumes. Dishwashers and clothes dryers also typically contain materials that affect chemically sensitive users.

CHAPTER

Structural Components Test

DIRECTIONS

Answer the questions with short complete statements or drawings as needed.

1. Letter your name and the date at the top of the sheet.
2. Letter the question number and provide the answer. You do not need to write out the question.

QUESTIONS

Question 25–1 List two different types of floor framing methods, and explain the differences. Provide sketches to illustrate your answer.

Question 25–2 What is the difference among a girder, header, and beam?

Question 25–3 List the differences between a rim joist and solid blocking at the sill.

Question 25–4 How wide are girders typically for a residence framed with conventional floor system?

Question 25–5 What are let-in braces and wall sheathing used for?

Question 25–6 A let-in brace must be at what maximum angle?

Question 25–7 Define the following abbreviations:

PSL	MDF	APA
OSB	EXP	EXT
LVL	HDF	STRUCT

Question 25–8 How is a foundation post protected from the moisture in the concrete support?

Question 25–9 Blocking is required in walls over how many feet high?

Question 25–10 List four common materials suitable for residential beams and girders.

Question 25–11 What purpose does the bird's mouth of a rafter serve?

Question 25–12 What do the numbers 4/12 mean when placed on a roof pitch symbol?

Question 25–13 Explain the difference among ridge, ridge blocking, and ridge board.

Question 25–14 List three types of truss web materials.

Question 25–15 Sketch, list, and define five types of rafters.

Question 25–16 List two functions of a ceiling joist.

Question 25–17 Define the four typical parts of a truss.

Question 25–18 What two elements are typically applied to wood to make an engineered wood product?

Question 25–19 Define the term mudsill.

Question 25–20 What function does a purlin serve?

Question 25–21 Explain the difference between a bearing and nonbearing wall.

Question 25–22 What is the minimum lap for top plates in a bearing wall?

Question 25–23 Sketch, label, and define the members supporting the loads around a window.

Question 25–24 What is the most common spacing for rafters?

Question 25–25 What is the function of the top plate of a wall?

Question 25–26 What is the advantage of providing blocking at the edge of a floor diaphragm?

Question 25–27 List the common sizes of engineered wood studs.

Question 25–28 What are the common span ratings for plywood suitable for roof sheathing?

Question 25–29 What grades of plywood are typically used for floor sheathing?

Question 25–30 List eight different qualities typically given in an APA wood rating.

CHAPTER 26

Energy-Efficient Design and Construction

▼ ENERGY-EFFICIENT DESIGN

Energy-conscious individuals have been experimenting with energy-efficient construction for decades. Their goal has been to reduce home heating and cooling costs. In recent years several formal studies have been done around the country, sponsored by various private and governmental agencies with a commitment to conservation as a result of rising energy costs. The experimental programs have had the following goals:

○ Create a better living environment.
○ Meet consumer demand for more economical living.
○ Evaluate realistic material and construction alternatives that may be used to alter building codes in the future.

Today's home buyers are concerned about energy-efficient design. They expect energy-efficient design and construction to be a part of the total home package, and, in fact, such design and construction can be achieved without a great deal of additional cost. For example, consumers look for construction that uses air infiltration barriers because when air infiltration is reduced, the amount of unconditioned air that must be heated or cooled to the desired temperature is also reduced. Such construction increases comfort while reducing energy costs.

Home builders around the country have realized the advantages of energy-efficient construction for both consumers and themselves. An energy-efficient home is easier to sell and will sell sooner than ordinary construction. This is especially true during times when a consumer's ratio of income to purchasing power makes it difficult for a potential home buyer to qualify for a loan. The person applying for a mortgage loan on an energy-efficient home will qualify sooner than for ordinary construction because the estimated energy savings will reduce the purchaser's monthly expenses. The money saved on energy will be available for mortgage payments. Several nationwide programs assist lenders in identifying energy efficiency. These programs include the following:

○ National Association of Home Builders (NAHB) thermal performance guidelines.
○ Virginia and Maryland Homebuilders E-7 Program.

○ Owens-Corning Fiberglas Corporation Energy Performance Design System (EPDS).
○ Massachusetts Home Energy Rating System.
○ (VEPCO), Energy Saver Home Program.
○ Western Resources Center, Residential Energy Evaluation Program.
○ Tennessee Valley Authority Energy Saver Home Program.
○ Bonneville Power Administration and Oregon Department of Energy.

In several controlled situations where energy-efficient construction techniques were implemented to determine if there would be a significant reduction in energy consumption, the results clearly show that there is energy savings. There are a number of techniques that can be used that do not cost much more than standard construction to implement. To the other extreme, there are some energy conservation construction techniques that are complex and involve labor- and material-intensive installations.

▼ ENERGY CODES

The Model Energy Code was originally developed jointly by Building Officials and Code Administrators International, Inc. (BOCA), International Conference of Building Officials (ICBO), National Conference of States on Building Codes and Standards (NCSBCS), and Southern Building Code Congress International, Inc. (SBCCI) under a contract funded by the U.S. Department of Energy (DOE). The Model Energy Code sets minimum requirements for the design of new buildings and additions to existing buildings. BOCA has also established the National Energy Conservation Code, which is updated every three years. The purpose of the code is to regulate the design and construction of the exterior envelope and selection of HVAC, service water heating, electrical distribution and illuminating systems, and equipment required for effective use of energy in buildings for human occupancy. The *exterior envelope* is made up of the elements of a building that enclose conditioned (heated and cooled) spaces through which thermal energy transfers to or from the exterior.

▼ ENERGY-EFFICIENT CONSTRUCTION

No matter what system of construction is used, energy efficiency can be a part of the construction process. Some of the construction techniques presented in Chapter 24 may seem like excessive protection, and depending on the area of the country where you live, some of the methods are inappropriate. The examples given are simply some of the methods that have been used and found effective. The goal of energy-efficient construction is to decrease the dependency on the heating and cooling system. This is best done by the use of framing techniques, caulking, vapor retarders, and insulation.

Caulking

Caulking normally consists of filling small seams in the siding or the trim to reduce air drafts. In energy-efficient construction, caulking is added during construction at the following places:

○ Exterior joints around the window and door frames.
○ Joints between wall cavities and window and door frames.
○ Joints between the wall and foundation.
○ Joints between the wall and roof.
○ Joints between wall panels.
○ Penetrations or utility services through exterior walls, floors, and roofs.
○ All other openings that may cause air leakage. Check with local building officials for a complete list.

Additional energy-efficient construction techniques related to specific construction trades are discussed in Chapters 16, 17, and 18. Figure 26–1 shows typical areas where caulking can be added. These beads of caulk keep air from leaking between joints in construction materials. Figure 26–2 shows the general caulking notes that may be found on a drawing. It may seem like extra work for a small effect, but caulking is well worth the effort. Caulking involves minimal expense for material and labor.

Vapor Retarders

For an energy-efficient system, air tightness is critical. The ability to eliminate air infiltration through small cracks is imperative if heat loss is to be minimized. Vapor retarders are a very effective method of decreasing heat loss. Most building codes require 6-mil-thick plastic to be placed over the earth in the crawl space. Many energy-efficient construction methods add a continuous vapor retarder to the walls. This added vapor retarder is designed to keep exterior moisture from the walls and insulation. Figure 26–3 shows the effect of water vapor on insulated walls with and without a vapor retarder.

Under normal conditions, wall and ceiling insulation with a foil face on the interior side allows small amounts

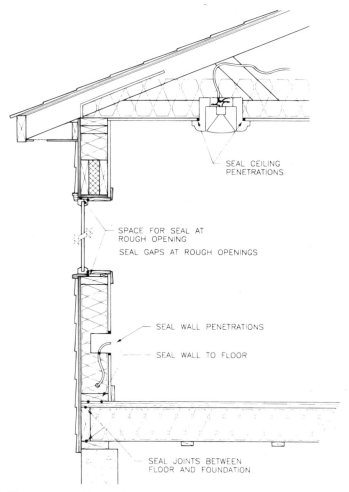

Figure 26–1 Typical areas where caulking should be added. *Courtesy Oregon Residential Energy Code.*

CAULKING NOTES:

CAULKING REQUIREMENTS BASED ON 1992 OREGON RESIDENTIAL ENERGY CODE

1. SEAL THE EXTERIOR SHEATHING @ CORNERS JOINTS DOOR AND WINDOW AND FOUNDATION SILLS WITH SILICONE CAULKING.

2. CAULK THE FOLLOWING OPENINGS W/ EXPANDED FOAM OR BACKER RODS. POLYURETHANE, ELASTOMERIC COPOLYMER SILCONIZED ACRYLIC LATTEX CAULKS MAY ALSO BE USED WHERE APPROPRIATE.

ANY SPACE BETWEEN WINDOW AND DOOR FRAMES

BETWEEN ALL EXTERIOR WALL SOLE PLATES AND PLY SHEATHING

ON TOP OF RIM JOIST PRIOR TO PLYWOOD FLOOR APPLICATION

WALL SHEATHING TO TOP PLATE.

JOINTS BETWEEN WALL AND FOUNDATION

JOINTS BETWEEN WALL AND ROOF

JOINTS BETWEEN WALL PANELS

AROUND OPENINGS FOR DUCTS, PLUMBING ELECTRICA TELEPHONE AND GAS LINES IN CEILINGS WALLS AND FLOORS. ALL VOIDS AROUND PIPING RUNNING THROUGH FRAMING OR SHEATHING TO BE PACKED W/ GASKETING OR OAKUM TO PROVIDE A DRAFT FREE BARRIER.

Figure 26–2 General caulking notes that may be found on a set of architectural drawings.

of air to leak in at each seam. To eliminate this leakage, a continuous vapor retarder can be installed. To be effective, the vapor retarder must be lapped and sealed to keep air from penetrating through the seams in the plastic.

The vapor retarder can be installed in the ceiling, walls, and floor system for effective air control. Figure 26–4 shows three different ceiling applications. Each is designed to help prevent small amounts of heated air from escaping to the attic. All the effort required to keep the vapor retarder intact at the seams must be continued wherever an opening in the wall or ceiling is required.

The cost of materials for a vapor retarder is low compared to the overall cost of the project. The expense of the vapor retarder comes in the labor to install and to maintain its seal. Great care is required by the entire construction crew to maintain the barrier.

Caution should be exercised when building a completely airtight structure with increased insulation and vapor retarders. This type of construction may cause inside air quality problems. These potential problems may be countered with the installation of an air-to-air ex-

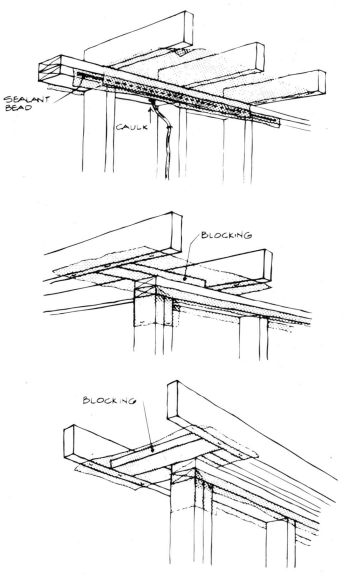

Figure 26–4 Ceiling application of continuous vapor retarders. *Courtesy Oregon Department of Energy.*

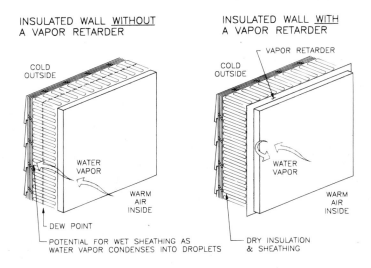

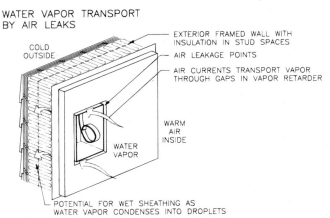

Figure 26–3 Effect of water vapor on insulated walls with and without a vapor retarder. *Courtesy Oregon Residential Energy Code.*

changer. A complete discussion of air contaminants and air-to-air exchangers is found in Chapter 18. A heating, ventilating, and air-conditioning (mechanical) engineer should be consulted when designing these structures. Garages need screened openings through exterior walls at or near the floor level with a clear area of screen not less than 60 square inches per motor vehicle to be accommodated.

Insulation

Many energy-efficient construction methods depend on added insulation to help reduce air infiltration and heat loss. Some of these systems require not only adding more insulation to the structure, but also adding more framing material to contain the insulation.

Building codes require a minimum of R-30 insulation in sloped ceilings and R-38 insulation in flat ceilings to help retain heat. In energy-efficient homes, the R value of the ceiling insulation may range from R-38 to R-80. Ceiling insulation of 16″ of cellulose insulation is often used to provide an R value of 60.

Some codes regulate the insulation based on the efficiency of the heating system and the amount of openings. For instance, a house with an 80 percent efficient furnace could have the following minimum insulation R values:

Flat ceiling: R–38

Vaulted ceiling: R–30

Walls: R–21

Floor: R–25

Wall openings, such as windows, doors, and skylights, cannot exceed 17 percent of the total heated wall area without increasing other energy-conservation methods.

Figure 26–5 shows the general insulation notes that may be found in a set of plans designed for energy-efficient construction.

INSULATION NOTES:

INSULATION BASED ON PATH #1
OF 1992 OREGON RESIDENTIAL
ENERGY CODE.

1. INSULATE ALL EXTERIOR HEATED WALLS W/ HIGH DENSITY FIBERGLASS BATT INSULATION R-21 MIN. W/ PAPER FACE. INSULATE EXTERIOR WALLS PRIOR TO INSULATION OF TUB / SHOWER UNITS.

2. INSULATE ALL FLAT CEILINGS W/ 12" R-38 FIBERGLASS BATT INSULATION (NO PAPER FACE REQUIRED).

3. INSULATE ALL VAULTED CEILINGS TO W/ 10 ¼" HIGH DENSITY PAPER FACED FIBERGLASS BATTS R-38 MIN. W. 2" MIN. AIR SPACE ABOVE.

4. INSULATE ALL WOOD FLOORS W. 8" FIBERGLASS BATTS R-25 MIN. W/ PAPER FACE OR 1 PERM FORMULATED VAPOR RETARDED INSTALLED ABOVE FLOOR DECKING PRIOR TO INSTALLING FINISHED FLOORING. INSTALL PLUMBING ON HEATED SIDE OF INSULATION.

5. INSULATE CONCRETE SLAB FLOORS BENEATH HEATED ROOMS W. 3" EXTRUDED POLYSTYRENE R-15 MIN. X 24" WIDE AT ALL SLAB EDGES IN CONTACT WITH UNHEATED EARTH.
   • PROVIDE ALT. BID FOR R-15 RIGID INSULATION ON EXTERIOR FACE OF STEM WALL FROM TOP OF SLAB TO FOUNDATION BTM. PROVIDE FLASHING AND PROTECTION BOARD AT EXPOSED INSULATION ABOVE GRADE.

6. INSULATE BASEMENT WALLS W/ R-21 HIGH DENSITY BATTS IN 2 X 4 FURRING WALL. •• PROVIDE ALTERNATE BID FOR R-21 RIGID INSULATION ON EXTERIOR FACE OF WALL W/ FLASHING AND PROTECTION BOARD FOR EXPOSED INSULATION ABOVE GRADE.

7. COVER THE EXTERIOR FACE OF ALL EXTERIOR HEATED WALLS W/ TYVEK VAPOR BARRIER. LAP ALL JOINTS 6" MIN. AND TAPE ALL JOINTS.

8. WEATHER STRIP THE ATTIC AND CRAWL ACCESS DOORS. INSULATE ATTIC ACCESS DOOR TO R-38.

9. SET ALL MUDSILLS FOR HEATED WALLS ON NON PORUS SILL SEAL.

10. INSULATE ALL HEATING DUCTS IN UNHEATED AREAS TO R-8. INSULATION TO HAVE A FLAME SPREAD RATING OF 50 MAX. W/ AND A SMOKE DEVELOPMENT RATING OF 100 MAX. ALL DUCT SEAMS TO BE SEALED.

Figure 26–5 General insulation notes that may be found on a set of architectural drawings. *Courtesy Residential Design, AIBD.*

▼ SOLAR ENERGY DESIGN

Sunlight becomes solar energy when it is transferred to a medium that has the capacity or ability to provide useful heat. This useful solar energy may be regulated in order to heat water or the inside of a building, or create power to run electrical utilities. Most areas of the earth receive about 60 percent direct sunlight each year, and in very clear areas, up to 80 percent of the annual sunlight is available for use as solar energy. When the sun's rays reach the earth, air and the things on the earth become heated. Certain dense materials such as concrete can absorb more heat than less dense materials such as wood. During the day the dense materials absorb and store solar energy. Then at night with the source of energy gone, the stored energy is released in the form of heat. Some substances, such as glass, absorb thermal radiation while transmitting light. This concept, in part, is what makes solar heating possible. Solar radiation enters a structure through a glass panel and warms the surfaces of the interior areas. The glass keeps the heat inside by absorbing the radiation.

Two basic residential and commercial uses for solar energy are heating spaces and hot water. Other uses are industrial and include drying materials such as lumber, masonry, or crops. Solar energy has also been used for nearly a century with desalinization plants to provide fresh water from either mineral or salt water.

The types of solar space heating systems available are either passive, active, or a combination of the two. *Passive,* or architectural, systems use no mechanical devices to retain, store, and radiate solar heat. *Active,* or mechanical, systems do use mechanical devices to absorb, store, and use solar heat.

The potential reduction of fossil fuel energy consumption can make solar heat an economical alternative. There are a number of factors that contribute to the effective use of solar energy. Among them are building a structure with energy-efficient construction techniques and fully insulating the building to reduce heat loss and air infiltration.

An auxiliary heating system is often used as a backup or supplemental system in conjunction with solar heat. The amount of heat needed from the auxiliary system depends on the effectiveness of the solar system. Both the primary and supplemental systems should be professionally engineered for optimum efficiency and comfort.

A southern exposure, when available, provides the best site orientation for solar construction. A perfect solar site allows the structure to have an unobstructed southern exposure. See Chapter 10.

Room placement is another factor to consider when taking advantage of solar heat. Living areas, such as the living and family rooms, should be on the south side of the house, while inactive rooms, such as bedrooms, laundries, and baths, should be located on the north side of the structure, where a cooler environment is desirable. If possible, the garage should be placed on the north, northeast, or northwest side of the home. A garage can act as an effective barrier for insulating the living areas from cold exterior elements.

Another energy-efficient element of the design is an *air-lock entry,* known as a *vestibule.* This is an entry that provides a hall or chamber between an exterior and interior door to the building. The vestibule should be designed so neither interior nor exterior door should be open at the same time. The distance between doors should be at least 7′ to help force the occupants to close one door before they reach the other. The main idea behind a vestibule entry is to provide a chamber that is always closed to the living area by a door. When the exterior door is opened, the air lock loses heat, but the heat loss is confined to the small space of the vestibule and the warm air of the living area is exposed to minimum heat loss.

Figure 26–6 shows a solar design with vestibule, and living areas and a solarium, at the southern exposure where extensive glass allows the rooms to be warmed by solar heat. The inactive rooms, including the bedrooms, baths, laundry, pantry, and garage, shelter the living area from the northern exposure. The use of masonry walls also allows the house to be built into a slope on its northern side or for berms to be used as shelter from the elements.

▼ LIVING WITH SOLAR ENERGY SYSTEMS

The use of solar energy and energy-efficient construction requires a commitment to energy conservation. Each individual must evaluate cost against potential savings and become aware of the responsibility of living with energy conservation.

Solar systems that provide some heat from the sun and require little or no involvement from the occupant to assist the process can be designed. Such a minimal energy-saving design is worth the effort in many cases.

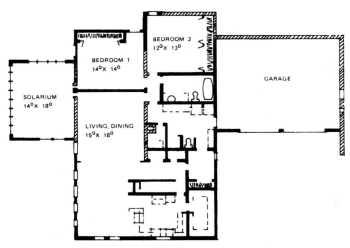

Figure 26–6 Solar design. Notice the solarium and the air-lock entry. *Courtesy Residential Designs, AIBD.*

Active solar space heating systems are available that can provide a substantial amount of needed heat energy. These systems are automatic and also do not generally require involvement from the homeowner, although he or she should be aware of operation procedures and maintenance schedules. On the other hand, some passive solar heating systems require much participation by the occupant. For example, a mechanical shade that must be maneuvered by the homeowner can be used to block the summer sun's rays from entering southern exposure windows. During the winter months when the sun is heating the living area, the heat should be retained as long as possible. In the morning the homeowner needs to open shutters or drapes to allow sunlight to enter and heat the rooms. In the early evening before the heat begins to radiate out of the house, the occupant should close the shutters or drapes to keep the heat within.

▼ CODES AND SOLAR RIGHTS

Building permits are generally required for the installation of active solar systems or the construction of passive solar systems. Some installations also require plumbing and electrical permits. Verify the exact requirements for solar installations with local building officials. During the initial planning process, always check the local zoning ordinances to determine the feasibility of the installation. For example, many areas have dwelling height restrictions. If the planned solar system encroaches on this zoning rule, then a different approach or a variance to the restriction should be considered.

The individual's rights to solar access are not always guaranteed. A solar home may be built in an area that has excellent solar orientation, and then a few years later, a tall structure may be built across the street that blocks the sun. The neighbor's trees may grow tall and reduce solar access. Determine the possibility of such a problem arising before construction begins. Some local zoning ordinances, laws, or even deed restrictions do protect the individual's right to *light.* In the past, laws generally provided the right to receive light from above the property but not from across neighboring land. This situation is changing in many areas of the country.

▼ ROOF OVERHANG

Since the sun's angle changes from season to season, lower on the horizon in the winter and higher in the summer, the addition of overhang can shield a major glass area from the heat of the summer sun and also allow the lower winter sun to help warm the home. Figure 26–7 shows an example of how a properly designed overhang can aid in the effective use of the sun's heat.

An overhang that provides about 100 percent of shading at noon on the longest day of the year can be calculated with a formula that divides the window

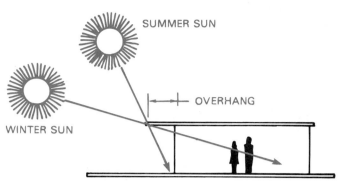

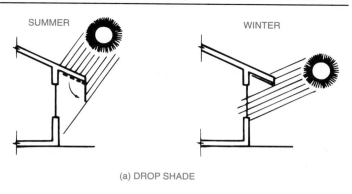

(a) DROP SHADE

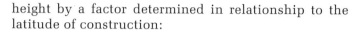

Figure 26–7 Effective overhang.

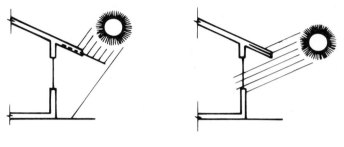

(b) SLIDING SHADE

height by a factor determined in relationship to the latitude of construction:

$$\text{Overhang} = \frac{\text{Windowsill height}}{F}$$

NORTH LATITUDE	F	NORTH LATITUDE	F
28°	8.4	44°	2.4
32°	5.2	48°	2.0
36°	3.8	52°	1.7
40°	3.0	56°	1.4

Calculate the recommended southern overhang for a location at 36° latitude and provide for a 6′–8″ window height.

$$\text{Overhang (OH)} = \frac{6'-8'' \text{ (Window height)}}{3.8 \text{ (F at 36° latitude)}}$$

$$= \frac{6.6}{3.8} = 1.8 = \text{approx. } 1'-10''$$

The overhang recommendation is greater for a more northerly latitude.

Overhang protection can be constructed in ways other than a continuation of the roof structure. An awning, porch cover, or trellis may be built to serve the same function. There is a great deal of flexibility in architectural design that provides the same function. Alternate methods of shading from summer heat and exposing the window areas to winter sun can be achieved with mechanical devices. These movable devices require that the occupant be aware of the need for shade or heat at different times of the year. Figure 26–8 shows some shading options.

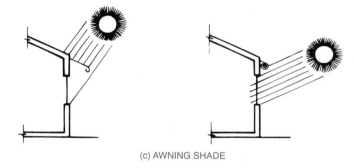

(c) AWNING SHADE

Figure 26–8 Optional shading devices.

▼ PASSIVE SOLAR SYSTEMS

In passive solar architecture, the structure is designed so the sun directly warms the interior. A passive solar system allows the sun to enter the structure and be absorbed into a structural mass. The stored heat then warms the living space. Shutters or drapes are used to control the

amount of sun entering the home, and vents help provide temperature control. In passive solar construction, the structure is the system. The amount of material needed to store heat depends on the amount of sun, the desired temperature within the structure, and the heat storage ability of the material used. Materials such as water, steel, concrete, and masonry have good heat capacity; wood does not. There are several passive solar architectural methods that have been used individually and together, including south-facing glass, thermal storage walls, roof ponds, solariums, and envelope construction.

South-facing Glass

Direct solar gain is a direct gain in heat created by the sun. This direct gain is readily created in a structure through south-facing windows. Large window areas facing south can provide up to 60 percent of a structure's heating needs when the windows are insulated at night with tight-fitting shutters or insulated curtains. When the window insulation is not used at night, the heat gain during the day is quickly lost. The sun's energy must heat a dense material in order to be retained. Floors and walls are often covered or made of materials other than traditional wood and plasterboard. Floors constructed of or covered with tile, brick, or concrete, and special walls made of concrete or masonry, or containing water tubes can provide heat storage for direct gain applications. Figure 26–9 shows a typical direct-gain south-facing glass application.

Clerestory windows can be used to provide light and direct solar gain to a second floor living area and increase the total solar heating capacity of a house. These clerestory units can also be used to help ventilate a structure during the summer months when the need for cooling may be greater than heating. Look at Figure 26–10.

Skylights can be used effectively for direct solar gain. When skylights are placed on a south-sloping roof, they provide needed direct gain during the winter. In the summer, these units can cause the area to overheat unless additional care is taken to provide for ventilation or a shade cover. Some manufacturers have skylights that open and may provide sufficient ventilation. Figure 26–11 shows an application of openable skylights.

Thermal Storage Walls

Thermal storage walls can be constructed of any good heat-absorbing material such as concrete, masonry, or water-filled cylinders. The storage wall can be built inside and next to a large southern-exposed window or group of windows. The wall receives and stores energy during the day and releases the heat slowly at night.

The *Trombe wall*, designed by French scientist Dr. Felix Trombe, is a commonly used thermal storage wall. The Trombe wall is a massive dark-painted masonry or concrete wall situated a few inches inside and

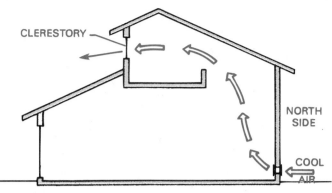

Figure 26–10   Clerestory circulation.

Figure 26–11   Skylights. *Courtesy VELUX-AMERICA, INC.*

next to south-facing glass. The sun heats the air between the wall and the glass. The heated air rises and enters the room through vents at the top of the wall. At the same time, cool air from the floor level of adjacent rooms is pulled in through vents at the bottom of the wall. The vents in the Trombe wall must be closable to avoid losing the warm air from within the structure through the windows at night. The heat absorbed in the wall during the day radiates back into the room during the night hours. The Trombe wall also acts to cool the structure during summer. This happens when warm air rises between the wall and glass and is vented to the outside. The air currents thus created work to pull cooler air from an open north-side window or vent. Figure 26–12 shows an example of the thermal storage wall.

Some passive solar structures use water as the storage mass where large vertical water-filled tubes or drums painted dark to absorb heat are installed as the Trombe wall. The water functions as a medium to store heat during the day and release heat at night.

Figure 26–9   South-facing glass provides direct gain. *Courtesy Residential Designs, AIBD.*

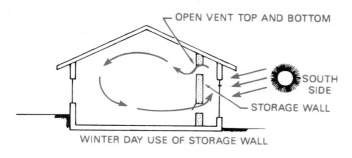

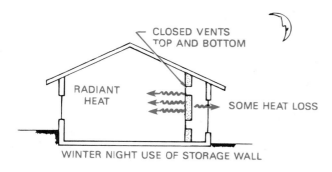

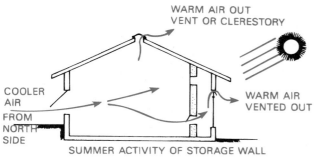

Figure 26–12 Thermal storage wall.

Roof Ponds

Roof ponds are used occasionally in residential architecture, although they are more common in commercial construction. A roof pond is usually constructed of containers filled with antifreeze and water on a flat roof. The water is heated during the winter days, and then at night the structure is covered with insulation, which allows the absorbed heat to radiate into the living space. This process functions in reverse during the summer. The water-filled units are covered with insulation during the day and uncovered at night to allow any stored heat to escape. In order to assist in the radiation of heat at night, the structure should be constructed of a good thermal-conducting material such as steel. Figure 26–13 shows an example of the roof pond system.

Solariums

A *solarium,* sometimes called a sun room or solar greenhouse, is designed to be built on the south side of a house next to the living area. Solariums are a nice part of the living area when designed into a home. The addi-

tional advantage of the solarium is the greenhouse effect. These rooms allow plants to grow well all year.

The theory behind the solarium is to absorb a great amount of solar energy and transmit it to the balance of the structure. A thermal mass may also be used in conjunction with the solarium. Heat from the solarium can be circulated throughout the entire house by natural convection or through a forced-air system. The circulation of hot air during the winter and cool air during the summer in the solar greenhouse functions similar to a Trombe wall.

The solarium can overheat during long, hot summer days. This potential problem can be reduced by mechanical ventilators or a mechanical humidifier. Exterior shading devices are often advantageous where a cover can be rolled down over the greenhouse glass area. The use of landscaping with southern deciduous trees, as discussed in Chapter 10, is a suggested alternative. Figure 26–14 shows how the solarium operates to provide solar heat and circulate summer cooling.

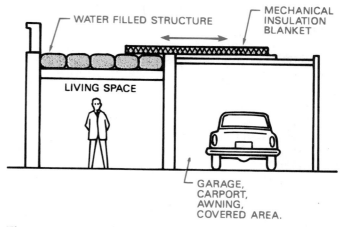

Figure 26–13 Roof ponds.

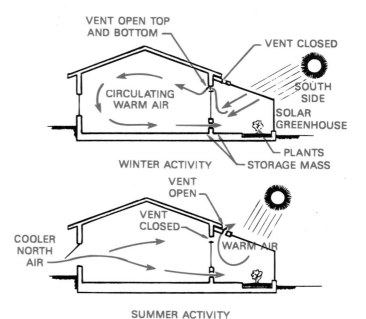

Figure 26–14 Solarium construction.

Envelope Design

The *envelope design* is based on the idea of constructing an envelope or continuous cavity around the perimeter of the structure. A solarium is built on the south side with insulating double-pane glass to act as a solar collector. During the winter when solar energy is required to heat the structure, the warm air that is created in the solarium rises, and the convection currents cause the heated air to flow around the structure through the envelope cavity. Look at Figure 26–15.

The envelope crawl space floor is a storage mass. The rock stores heat during the day and releases the energy slowly at night. This functions in reverse during the summer to help cool the cavity and the structure. The greenhouse floor acts as a vent, with floor decking spaced to allow air currents to pass through and provide the continuous current. The heated air in the envelope cavity helps keep the living space at a constant temperature. The properly designed envelope structure can completely eliminate the need for a backup heating system. Some envelope dwellings use a wood- or coal-burning appliance for backup heat, while others use small forced-air systems.

The summer heat is often a concern in solar design, although an envelope can effectively cool when the sun is high in the sky. Deciduous shade trees are an added advantage for summer cooling when planted on the south side. The function of the cooling process is a reverse of the heating operation. The clerestory windows are opened, and heated air is allowed to escape as it rises. Another design feature that assists in cooling is the underground pipe or tube that is constructed in connection with the cool crawl space. The underground temperature remains about 55° F, and when open to outside air ventilation, a convection current is again created. This time the flow of air is cool and flows under the cool crawl space and around the envelope. This same system can be

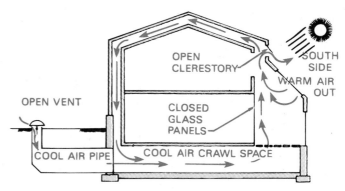

Figure 26–16 Summer envelope activity.

used in conjunction with the solar greenhouse. Figure 26–16 shows how the envelope design can cool during the summer.

Insulation is an important factor in the envelope design with both the outer shell and the inner shell being insulated. Generally R-19 insulation is adequate on the outer shell, and R-11 insulation on the inner shell is common. Insulative values in excess of these amounts generally do not add greater efficiency to the structure. Foundation walls at the crawl space perimeter and floor should be adequately insulated.

The passive envelope system is an effective method of reducing energy consumption and creating an excellent living environment. Standard construction techniques are generally used, although the cost of two walls and roof structures is a factor. Another problem is the framing and installation of windows and openings in the double walls of the envelope. Research local and national building codes when considering the envelope design. Factors that should be discussed include fire safety. Codes often require that specific safety precautions be taken to damper the cavity mechanically in case of fire. The same air currents that carry the air around the envelope also allow flames to engulf the structure quickly.

If proper precautions are not taken, a fire in the home can turn the envelope into a giant wooden flue, and the inside of the house will ultimately become the firebox and proliferate the flames into an all-fuel fireplace. National codes in general specify fire blocking at the floor and ceiling plates, and require the home designer to submit alternatives for the envelope design. Most alternatives include the addition of smoke-activated dampers in the cavity at the floor and ceiling lines plus lining the air chamber with drywall. The liner may improve the efficiency of the structure but add tremendously to the cost. Some envelope designs have proved better at cooling in the summer than heating in the winter, which may be a plus. However, the popularity of the envelope design is declining due to the added cost compared to other systems. The potential energy cost savings may not be much better than that achieved with excellent energy-efficient construction.

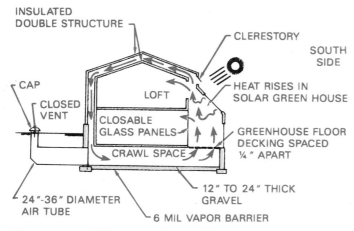

Figure 26–15 Winter envelope activity.

▼ ACTIVE SOLAR SYSTEMS

Active solar systems for space heating use collectors to gather heat from the sun, which is transferred to a fluid. Fans, pumps, valves, and thermostats move the heated fluid from the collectors to an area of heat storage. The heat collected and stored is then distributed to the structure. Heat in the system is transported to the living space by insulated ducts, which are similar to the duct work used in a conventional forced-air heating system. The active solar heating system generally requires a backup heating system that is capable of handling the entire heating needs of the building. Some backup systems are forced-air systems that use the same ducts as the solar system. The backup system may be heated by electricity, natural gas, or oil. Some people rely on wood- or coal-burning appliances for supplemental heat, but local codes should be checked. Active solar systems are commonly adapted to existing homes or businesses that normally use a forced-air heating system.

Solar Collectors

As the name implies, solar collectors catch sunlight and convert this light to heat. Well-designed solar collectors are nearly 100 percent efficient.

The number of solar collectors needed to provide heat to a given structure depends on the size of the structure and volume of heat needed. These are the same kinds of determinations made when sizing a conventional heating system. Collectors are commonly placed in rows on a roof or on the ground next to the structure, as shown in Figure 26–17. Architects and designers are integrating the solar panels into the total design in an effort to make them less obvious. The best placement requires an unobstructed southern exposure. The angle of the solar collectors should be in proper relationship to the angle of the sun during winter months when the demand for heat is greatest. Verify the collector angle that is best suited for the specific location. Some collectors are designed to be positioned at 60° from horizontal so that winter sun hits at

about 90° to the collector. While this is the ideal situation, a slight change from this angle of up to 15° may not alter the efficiency very much. Other collectors are designed to obtain solar heat by direct and reflected sunlight. Reflecting, or focusing, lenses help concentrate the light on the collector surface. Figure 26–18 shows an example of solar collector tilt.

Storage

During periods of sunlight, active solar collectors transfer the heat energy to a storage area and then to the living space. After the demand for heat is met, the storage

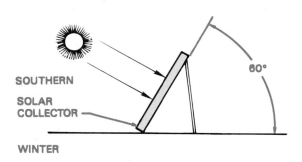
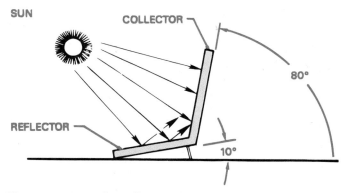

Figure 26–18 Solar collector tilt.

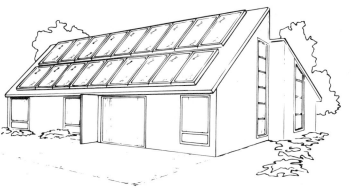
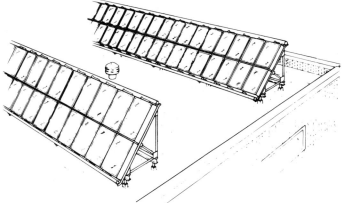

Figure 26–17 Typical applications of solar hot water collectors. *Courtesy Lennox Industries, Inc.*

facilities allow the heat to be contained for use when solar activity is reduced, such as at night or during cloud cover.

The kind of storage facilities used depends on the type of collector system used. There are water, rock, or chemical storage systems. Water storage is excellent since water has a high capacity for storing heat. Water in a storage tank is used to absorb heat from a collector. When the demand for heat exceeds the collector's output, the hot water from the storage tank is pumped into a radiator, or through a water-to-air heat exchanger for dispersement to the forced-air system. Domestic hot water may be provided by a water-to-water heat exchanger in the storage tank.

Rock storage is often used when air is the fluid used by the collectors rather than water. The heated air flows over a rock storage bed. The rock absorbs some heat from the air while the balance is distributed to the living space. When the solar gain is minimal, cooler air passes over the rock storage, where it absorbs heat and is distributed by fans to the living space. Domestic hot water may also be provided by an air-to-water heat exchanger situated in the rock storage or in the hot air duct leading from the collector.

Chemicals used in collectors and in storage facilities absorb large amounts of heat at low temperatures. Many of these systems claim to absorb heat on cloudy days or even when the collectors are snow covered. Chemicals with very low freezing temperatures can actually absorb heat during winter months when the outside temperature is low.

Solar Architectural Cement Products

Solar collectors may be built into the patio, driveway, tennis court, pool deck, or tile roof.

Solar architectural cement products make the most versatile solar collector on the market. It can be used as a driveway, sidewalk, patio, pool deck, roofing material, or the side surface of a building wall or fence. The finished surface can resemble cobblestone, brick, or roof tiles. Pigments are added to the specially formulated material,

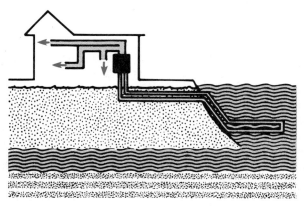

Figure 26–20 Geothermal freon exchanger in a lake. *Courtesy Solar Oriented Environmental Systems, Inc.*

which give it a lasting color that is pleasing to the eye. These solar products are completely different from all other solar collectors in that they add to the aesthetic value of the property. These products are manufactured from a specially formulated mixture that is strong and dense, allowing it to be very conductive and waterproof. Solar architectural cement products absorb and collect heat from the sun and outside air, and transfer the heat into water, glycol, or other heat transferring fluid passing through imbedded tubes.

Geothermal Systems

Geothermal heating and cooling equipment is designed to use the constant, moderate temperature of the ground to provide space heating and cooling, or domestic hot water, by placing a heat exchanger in the ground, or in wells, lakes, rivers, or streams.

A geothermal system operates by pumping ground water from a supply well and then circulating it through a heat exchanger, where either heat or cold is transferred by freon. The water, which has undergone only a temperature change, is then returned through a discharge well back to the strata, as shown in Figure 26–19. Lakes, rivers, ponds, streams, and swimming pools may be alternate sources of water. Rather than pumping the water up to the heat exchanger, a geothermal freon exchanger can be inserted into a lake, river, or other natural body of water to extract heat or cold as desired, as shown in Figure 26–20.

Another alternative is to insert a geothermal freon exchanger directly into a well, where the natural convection of the ground water temperature, assisted by the normal flow of underground water in the strata, is used to transfer the hot or cold temperature that is required. Sometimes the thermal transfer must be assisted by forcing the freon to circulate through tubes within the well by a low-horsepower pump. See Figure 26–21.

A geothermal system may assist a solar system that uses water as a heat storage and transfer medium. After the water heated by solar collectors has given up enough

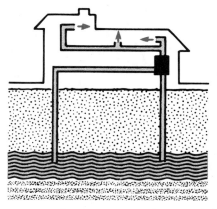

Figure 26–19 Ground water system. *Courtesy Solar Oriented Environmental Systems, Inc.*

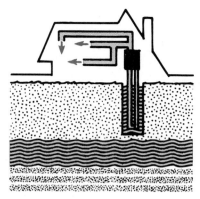

Figure 26–21 Geothermal freon exchanger in a well. *Courtesy Solar Oriented Environmental Systems, Inc.*

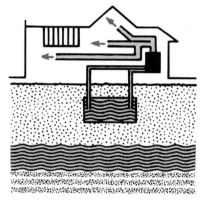

Figure 26–22 Solar-assisted system. *Courtesy Solar Oriented Environmental Systems, Inc.*

heat to reduce its temperature to below 100°F, this water becomes the source for operating the geothermal system. See Figure 26–22.

When an adequate supply of water is unavailable, the ground, which always maintains a constant temperature below the frost line, can be used to extract either heat or cold. The thermal extraction is done through the use of a closed-loop system consisting of polybutylene tubing filled with a glycol solution and circulated through the geothermal system. One alternative system is shown in Figure 26–23. Another method is to use a vertical dry-hole well which is sealed to enable it to function as a closed-loop system, as shown in Figure 26–24.

The geothermal system is mechanically similar to a conventional heat pump, except that it uses available water to cool the refrigerant or to extract heat as opposed to using 90° F air for cooling, or 20 to 40° F air for heating. Water can store great amounts of geothermal energy due to its high *specific heat* (the amount of energy required to raise the temperature of any substance 1° F). The specific heat of air is only 0.018, so it can absorb and release only ¹⁄₅₀ of the energy that water can. Fifty times more air by weight must pass through a heat pump to produce as much heat as the same amount of water.

Photovoltaic Modules

Photovoltaic technology has been developed and refined, and photovoltaic modules are now powering thousands of installations worldwide. Solar photovoltaic systems are now producing millions of watts of electricity to supply power for remote cabins, homes, railroad signals, water pumps, telecommunications stations, and even utilities. Uses for this technology are continuing to expand, and new installations are being constructed.

Photovoltaic cells turn light into electricity. The word *photovoltaic* is derived from the Greek *photo*, meaning "light," and *voltaic*, meaning "to produce electricity by chemical action." Photons strike the surface of a silicon wafer, which is a semiconductor diode, to stimulate the release of mobile electric charges that can be guided into a circuit to become a useful electric current.

Photovoltaic modules produce direct current (DC) electricity. This type of power is useful for many applications and for charging storage batteries. When alternating current (AC) is required, DC can be changed to AC by an inverter. Some photovoltaic systems are designed to use immediately the energy produced, as is often the case

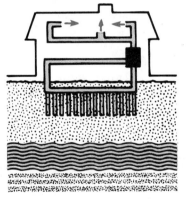

Figure 26–23 Ground loop system. *Courtesy Solar Oriented Environmental Systems, Inc.*

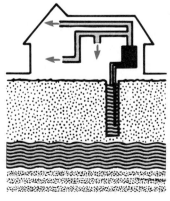

Figure 26–24 Vertical dry system. *Courtesy Solar Oriented Environmental Systems, Inc.*

in water-pumping installations. When the energy produced by solar electric systems is not to be used immediately, or when an energy reserve is required for use when sunlight is not available, the energy must be stored. The most common storage devices used are batteries.

Although both DC and AC systems can stand alone, AC systems can also be connected to a utility grid. During times of peak power usage, the system can draw on the grid for extra electricty if needed. At other times, such a system may actually return extra power to the grid. In most areas of the country the utility (grid) is required to purchase excess power.

Correct site selection is vitally important. The solar modules should be situated where they receive maximum exposure to direct sunlight for the longest period of time every day. Other considerations are distance from the load (appliance), shade from trees or buildings, which changes with the seasons, and accessibility.

The future looks bright for solar technologies. The costs and efficiencies of high-technology systems are improving. As homeowners watch their heating and cooling costs rise and become more concerned about shortages of oil and gas, solar heating becomes an attractive alternative. When preparing a preliminary design, solar alternatives should be considered. A qualified solar engineer should evaluate the site and recommend solar design alternatives. Additional assistance is usually available from state or national Departments of Energy.

CHAPTER

 Energy Efficient Design and Construction Test

DIRECTIONS

Answer the questions with short, complete statements or drawings as needed on an 8½ × 11 sheet as follows:

1. Letter your name, Chapter 26 Test, and the date at the top of the sheet.
2. Letter the question number and provide the answer. You do not need to write out the question.
3. Answers may be prepared on a word processor if appropriate with course guidelines.

QUESTIONS

Question 26–1 List three reasons that energy-efficient construction techniques are becoming important.

Question 26–2 Describe at least one factor that may be a disadvantage of energy-efficient construction.

Question 26–3 What is the primary goal of energy-efficient design and construction?

Question 26–4 Why might it be easier to qualify for a mortgage to purchase an energy-efficient home as opposed to a nonenergy-efficient home?

Question 26–5 Why is it possible to use energy-efficient design in any architectural style?

Question 26–6 Why is it possible to implement some energy-efficient design in every new home?

Question 26–7 List two site factors that contribute to energy-efficient design.

Question 26–8 List two room layout factors that contribute to energy-efficient design.

Question 26–9 List two window placement factors that contribute to energy-efficient design.

Question 26–10 List two factors that influence a reduction in air infiltration into a structure.

Question 26–11 What are two goals of an energy-efficient construction system?

Question 26–12 List three general materials for energy-efficient construction.

Question 26–13 What is caulking? Where is it typically used?

Question 26–14 What is the thickness for a vapor barrier in a crawl space?

Question 26–15 List the typical R value for the following uses:
a. flat ceilings
b. vaulted ceilings
c. energy-efficient ceilings
d. walls

Question 26–16 What are the two basic residential and commercial uses for solar heat?

Question 26–17 Describe passive and active solar heating.

Question 26–18 List and describe three factors that influence a good solar design.

Question 26–19 Discuss the concept of right to light.

Question 26–20 Define and show a sketch of how each of the following solar systems function:
a. south-facing glass
b. clerestory windows
c. thermal storage wall
d. roof ponds
e. solarium
f. envelope design

Question 26–21 Define solar collector.

Question 26–22 What are solar architectural cement products, and how do they function?

Question 26–23 Describe three applications of geothermal heating and cooling systems.

Question 26–24 Define the function of photovoltaic cells.

CHAPTER 27

Design Criteria for Structural Loading

As a beginning drafter you need not be concerned about how the structure to be drawn is supported. To advance in the field of architecture, however, requires a very thorough understanding of how the weight of the materials used for construction will be supported. This will require a knowledge of loads and determining how they are dispersed throughout a structure.

▼ TYPES OF LOADS

As a drafter you will need to be concerned with several types of loads acting upon a building: dead, live, and dynamic loads. Because of the complexity of determining these loads exactly, building codes have tables of conventional safe loads, which can be used to help determine the amount of weight or stress acting on any given member.

Dead Loads

Dead loads consist of the weight of the structure: walls, floors, and roofs, plus any permanently fixed loads such as fixed service equipment. Building codes typically require design values to be based on a minimum dead load of 10 lb per square foot (psf) (0.48kN/m^2) for floors and ceilings. A design value of between 7 and 15 psf (0.48–0.72kN/m^2) for rafters is used, depending on the weight of the finished roofing. The design load for dead loads can be verified in the design criteria section of each joist and rafter table of each major code. See Chapter 28. The symbol DL is often used to represent the values for dead loads.

Live Loads

Live loads are those superimposed on the building through its use. These loads include such things as people and furniture, and weather-related items such as ice, snow, and water (rain). The most commonly encountered live loads are floor, moving, roof, and snow loads. Live loads are represented by the symbol LL.

Floor Live Loads. Buildings are designed for a specific use or occupancy. Depending on the occupancy, the floor live load will vary greatly. For a residence, the UBC requires floor members to be designed to support a minimum live load of 50 psf (2.4kN/m^2). Exterior balconies must be able to support a minimum live load of 60 psf (2.88kN/m^2), and decks must be designed to support a minimum live load of 40 psf (1.92kN/m^2). CABO allows the live design load to be reduced to 30 psf (1.44kN/m^2) for sleeping areas and attic floors. Live loads for floors will be further discussed in Chapter 45.

Moving Live Loads. In residential construction, moving loads typically occur only in garage areas due to the weight of a car or truck. When the weight is being supported by a slab over soil, these loads do not usually cause concern. When moving weights are being supported by wood, the designer needs to take special care in the design. Consult with the local building department to determine what design weight should be used for moving loads.

Roof Live Loads. Roof live loads vary from 20 to 40 psf (0.96–1.92kN/m^2) depending on the pitch and the use of the roof. Some roofs also are used for sundecks and are designed in a manner similar to floors. Other roofs are so steep that they may be designed like a wall. Many building departments use 30 psf (1.44kN/m^2) as a safe live load for roofs. Consult the building department in your area for roof live load values.

Snow Loads. Snow loads may or may not be a problem in the area for which you are designing. You may be designing in an area where snow is something you dream about, not design for. If you are designing in an area where winter is something you shovel, it is also something you must allow for in your design. Because snow loads vary so greatly, the designer should consult the local building department to determine what amount of snow load should be considered. In addition to climatic variables, the elevation, wind frequency, duration of snowfall, and the exposure of the roof all influence the amount of live load design.

Dynamic Loads

Dynamic loads are those imposed on a structure from a sudden gust of wind or from an earthquake.

Wind Loads. Although wind design should be done only by a competent architect or engineer, a drafter must understand the areas of a structure that are subject to failure. Figure 27–1 shows a map of minimum basic wind speeds that should be designed for. Wind pressure creates wind loads on a structure. These loads vary greatly based on the location and the height above the ground of the structure. Because such wide variations in wind speed can be encountered, the designer of the structure must rely on the local building department to provide information.

Figure 27–2 shows a simplified explanation of how winds can affect a house. Wind affects a wall just as it would the sail of a boat. With a boat, the desired effect is to move the boat. With a structure, this tendency to move must be resisted. The walls resisting the wind will tend to bow under the force of the wind pressure. The tendency can be resisted by roof and foundation members and perpendicular support walls. These supporting walls will tend to become parallelograms and collapse. The designer of the structure determines the anticipated wind speed and designs walls, typically referred to as *shear walls*, to resist this pressure.

In planning for loads from wind, prevailing wind direction cannot be assumed. Winds are assumed to act in any horizontal direction and will create a positive pressure on the windward side of a structure. A negative

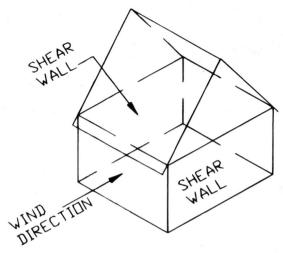

Figure 27–2 Wind pressure on a wall will be resisted by the bolts connecting the wall sill to the foundation and by the roof structure. The wind is also resisted by shear walls, which are perpendicular to the wall under pressure.

pressure on the leeward (downwind) side of the structure creates a partial vacuum. Design pressures for a structure are based on the total pressure a structure might be

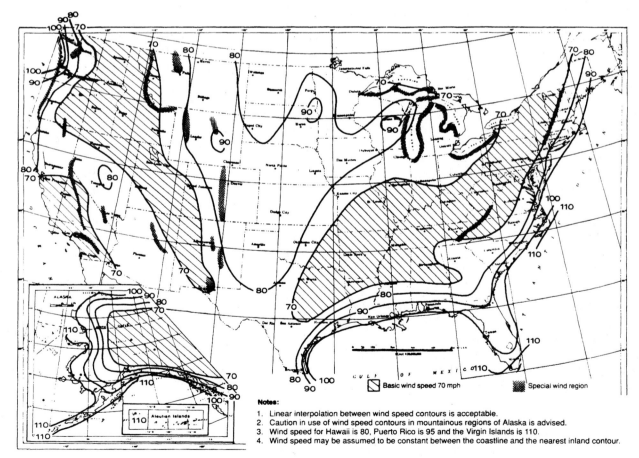

Notes:
1. Linear interpolation between wind speed contours is acceptable.
2. Caution in use of wind speed contours in mountainous regions of Alaska is advised.
3. Wind speed for Hawaii is 80, Puerto Rico is 95 and the Virgin Islands is 110.
4. Wind speed may be assumed to be constant between the coastline and the nearest inland contour.

Figure 27–1 Basic minimum wind speeds for the design of wind pressure on a residence. *Reproduced from the* Uniform Building Code, *1994 edition. Courtesy International Conference of Building Officials.*

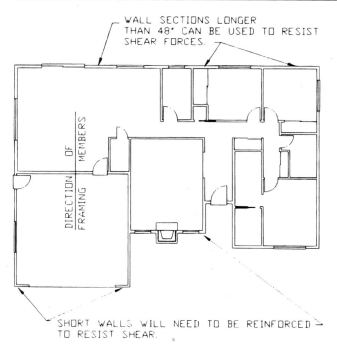

Figure 27–3 Special provisions for shear walls are generally required where openings break up normal wall construction. For residential construction, ⅜–½″ (9.5–12.7 mm) plywood is typically used to reinforce the framing. *Courtesy Doug Major.*

expected to encounter, which is equal to the sum of the positive and negative pressure. A design value of 30 psf (1.44kN/m²) is common for residential projects, but this value should be verified with local building departments. Thirty psf (1.44kN/m²) equals a wind speed of 108 miles per hour (mph). Other common values include:

psf	(kN/m²)	mph	km/h	psf	kN/m²	mph	km/h
15	0.72	76	122	50	2.40	140	225
20	0.96	88	142	55	2.64	147	237
25	1.20	99	159	60	2.88	153	246
30	1.44	108	174	70	3.36	165	266
35	1.68	117	188	80	3.84	177	285
40	1.92	125	201	90	4.32	188	303
45	2.16	133	214	100	4.80	198	319

Areas subject to high winds from hurricanes or tornados should be able to withstand winds of 125 mph (201 km/h). Design studies by the APA suggest wind speeds as high as 145 mph (233 km/h) with gusts as high as 160 mph (257 km/h) should become the design standard. Using these values, the designer can determine existing wall areas and the resulting wind pressure that must be resisted. This information is then used to determine the size and spacing of anchor bolts and any necessary metal straps or ties needed to reinforce the structure.

Wind pressure is also critical to the design and placement of doors and windows. The design wind pressure will affect the size of the glazing area and the method of framing the rough openings for doors and windows. Openings in shear walls will reduce the effectiveness of the wall. Framing around openings must be connected to the frame and foundation to resist forces of uplift from wind pressure. *Uplift*, the tendency of members to move upward due to wind or seismic pressure, is typically resisted by the use of steel straps or connectors that join the trimmer and king studs beside the window to the foundation or to framing members in the floor lever below. Wall areas with large areas of openings must also have design studies to determine the amount of wall area that will be required to resist the lateral pressure created by wind pressure. As a general rule of thumb, a section of wall 4′–0″ long is required per each 25 linear feet of the structure. Where the 48″ solid wall cannot be provided, the remaining wall areas will need to be reinforced to resist forces of shear. Figure 27–3 shows the relationship of walls to openings for shear walls.

Another problem created by wind pressure is the tendency of a structure to overturn. Structures built on pilings or other posted foundations allow wind pressure under the structure to exceed the pressure on the leeward side of the structure. Codes typically require that the resistance of the dead loads of a building to overturning be one and a half times the overturning effect of the wind. Typically metal ties are used to connect the walls to the floor, the floor joist to support beams, the support beams to supporting columns, and columns to the foundation. These ties are determined by an architect or engineer for each structure based on the area exposed to the wind and the wind pressure. Figure 27–4 shows an example of a detail designed by an engineer to resist the forces of uplift on a structure.

Seismic Loads.   Seismic loads result from earthquakes. Figure 27–5 shows a seismic map of the United States and the risk of each area to damage from earthquakes. The

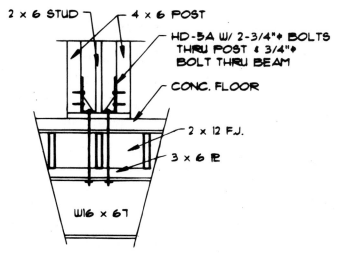

Figure 27–4 Metal straps are used to transfer the loads from one floor through another and down to the foundation.

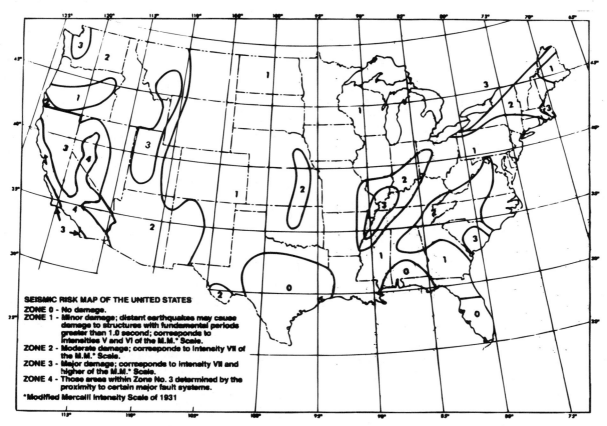

Figure 27–5 Seismic zone maps can be used to help determine the potential risk of damage from earthquakes. *Reproduced from the Uniform Building Code, 1994 edition. Courtesy International Conference of Building Officials.*

UBC, BOCA, and CABO all specify specific requirements for seismic design.

Stress which results from an earthquake is termed a *seismic* load and is usually treated as a lateral load that involves the entire structure. Lateral loads created by wind only effect certain parts of a structure while lateral forces created by seismic forces effect the entire structure. As the ground moves in varying directions at varying speeds, the entire structure is set in motion. Even after the ground comes to rest, the structure tends to wobble much like a jello structure. Typically, structures fall to seismic forces in much the same way as they do to wind pressure. Intersections of roofs-to-wall, wall-to-floor and floor-to-foundation are critical to the ability of a structure to resist seismic forces. An architect or engineer should carefully design these connections. Typically, a structure must be fluid enough to move with the shock wave but so connected that individual components move as a unit and all units of a structure move as one. Figure 27–6 shows a detail of an engineer's design for a connection of a garage-door king stud designed to resist seismic forces. The straps that are attached to the wall cause the wall and foundation to move as one unit.

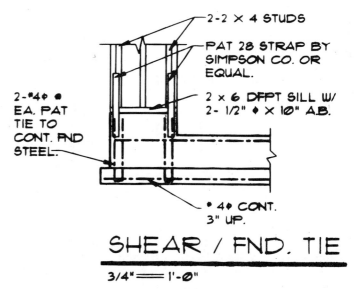

Figure 27–6 Hold-down anchors are typically used to resist seismic forces and form a stable intersection between the floor and foundation.

▼ LOAD DESIGN

Once the floor plan and elevations have been designed, the designer can start the process of determining how the

Member	DL	LL	Total
Floor	10	40	50
Decks	10	60	70
Ceiling	5	10	15
Roofs			
235 lb Comp.	8	30	38
Cedar shakes	7	30	37
Tile	25	30	55
Built-up	8	30	38

Figure 27–7 Typical loads for residential construction.

structure will resist the loads that will be imposed on it. To determine the sizes of material required, it is always best to start at the roof and work down to the foundation. By calculating from the top down, the loads will be accumulating, and when you work down to the foundation, you will have the total loads needed to size the footings.

As a beginning drafter you are not expected to be designing the structural components. In most offices, the structural design of even the simplest buildings is done by a designer or an engineer. The information of this chapter is a brief introduction into the size of structural members. The design of even simple beams can be very complex.

▼ LOAD DISTRIBUTION

A beam may be simple or complex. A *simple beam* has a uniform load evenly distributed over the entire length of the beam and is supported at each end. With a simple beam, the load on it is equally dispersed to each support. An individual floor joist, rafter, or truss or a wall may be thought of as a simple beam. For instance, the wall resisting the wind load in Figure 27–2 can be thought of as a simple beam because it spans between two supporting

walls and has a uniform load. If a beam is supporting a uniformly distributed load of 10,000 lb (4536 kg), each supporting post would be resisting 5000 lb (2268 kg). A *complex beam* has a nonuniform load at any point of the beam, or supports that are not located at the end of the beam. Chapter 29 will introduce methods of determining beam sizes when the loads are not evenly distributed.

Figure 27–7 shows a summary of typical building weights based on minimum design values from the Uniform Building Code. Figure 27–8 shows the bearing walls for a one-story structure framed with a truss roof system and a post-and-beam floor system. Remember that with a typical truss system, all interior walls are nonbearing. Half of the roof weight will be supported by the left wall, and half of the roof weight will be supported by the right wall. For a structure 32' wide with 2' overhangs, the entire roof will weigh 1440 lb (36' × 40#). Each linear foot of wall will hold 720 lb (18 × 40# psf), which is half the total roof weight.

If the walls are 8' tall, each wall will weigh 80 lb (8' × 10#) per linear feet of wall. The foundation will hold 100 lb of floor load (2' × 50#) per linear foot. Only 2' of floor will be supported since beams are typically placed at 48" o.c. Half of the floor weight (2' × 50#) is placed on the stem wall, and half of the weight (2' × 50#) will be supported by the girder parallel to the stem wall. Each interior girder will support 200 lb (4' × 50#) per linear foot of floor weight. The total weight on the stem wall per foot will be the sum of the roof, wall, and floor loads, which equals 900 lb.

Figure 27–9 shows the bearing walls of a two-level structure framed using western platform construction methods. This building has a bearing wall that is located approximately halfway between the exterior walls. For this type of building, one-half of the total building loads

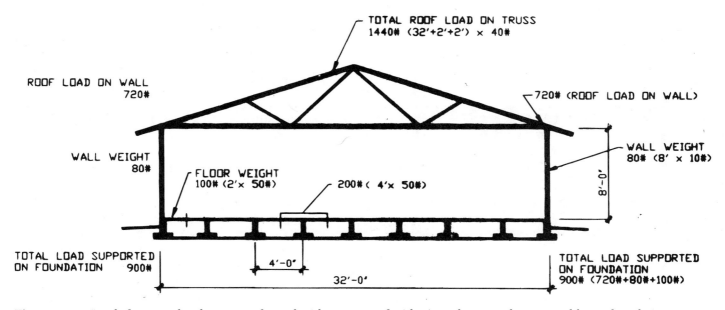

Figure 27–8 Loads for a one-level structure framed with a truss roof with 2' overhangs and a post-and-beam foundation system.

will be on the central bearing wall, and one-quarter of the total building loads will be on each exterior wall. Examine the building one floor at a time and see why.

The Upper Floor

At the upper floor level, the roof and ceiling loads are being supported. In a building 32′ × 15′, as shown in Figure 27–10, each rafter is spanning 16′. Of course the rafter is longer than 16′ but remember that the span is a horizontal measurement. If the loads are uniformly distributed throughout the roof, half of the weight of the roof will be supported at each end of the rafter. At the ridge, half of the total roof weight is being supported. At each exterior wall, one-quarter of the total roof load is being supported. At the ceiling and the other floor levels, the loading is the same. One-half of each joist is supported at the center wall and half at the outer wall.

Using the loads from Figure 27–7, the weights being supported can be determined. To determine the total weight that a wall supports is a matter of determining the area being supported and multiplying by the weight.

Roof

The area being supported at an exterior wall is 15′ long by 8′ wide, which is 120 sq ft. The roof load equals the sum of the live and dead loads. Assume a live load of 30

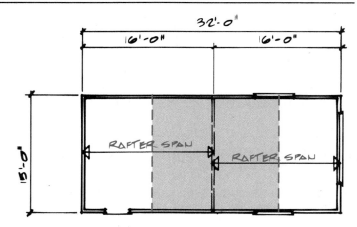

Figure 27–10 Load distribution on a simple beam.

psf. For the dead load assume the roof is built with asphalt shingles, ½″ ply, and 2 × 8 rafters for a dead load of approximately 8 psf. For simplicity 8 can be rounded up to 10 psf for the dead load. By adding the dead and live loads, you can determine that the total roof load is 40 psf. The total roof weight the wall is supporting is 4800 lb (120 × 40 lb = 4800 lb).

Because the center area in the example is twice as big, the weight will also be twice as big. But just to be sure, check the total weight on the center wall. You should be using 15 × 16 × 40 lb for your calculations. The wall is 15′ long and holds 8′ of rafters on each side of the wall, for a total of 16′. Using a LL of 40 lb, the wall is supporting 9600 pounds.

Ceiling

The procedure for calculating a ceiling is the same as for a roof, but the loads are different. A typical loading pattern for a ceiling is 15 lb. At the outer walls the formula would be 15 × 8 × 15 lb, or 1800 lb. The center wall would be holding 15 × 16 × 40 lb or 3600 lb.

The Lower Floor

Finding the weight of a floor is the same as finding the weight of a ceiling, but the loads are much greater. The LL for residential floors is usually 40 lb and the DL is 10 lb for a total of 50 lb per sq ft.

Walls

The only other weight left to be determined is the weight of the walls. Generally, walls average about 10 lb per sq ft. Determine the height of the wall and multiply by the length of the wall to find its area. Multiply the area by the weight per square foot, and you will have the total wall weight. Figure 27–11 shows the total weight that will be supported by the footings.

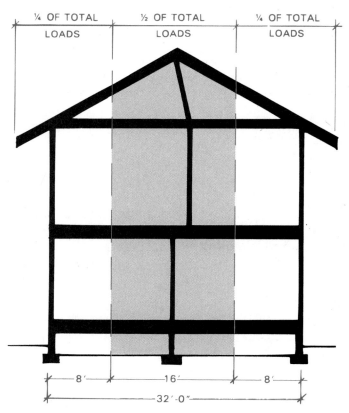

Figure 27–9 Bearing walls and load distribution of a typical home.

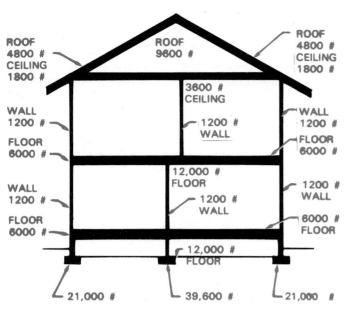

Figure 27–11 Total loads supported by footings. Using the assumed loads multiplied by 15 (the length of the building) produces the total loads.

CHAPTER 27

Design Criteria for Structural Loading Test

DIRECTIONS

Answer the questions with short complete statements. Show all work and provide a sketch of how loads will be supported.

1. Letter your name, and the date at the top of the sheet.
2. Letter the question number and provide the answer. You do not need to write out the question.

QUESTIONS

Question 27–1 What are the two major categories of loads that affect buildings?

Question 27–2 What is the safe design live load for a residential floor?

Question 27–3 What is the safe design total load for a residential floor?

Question 27–4 What factors cause snow loads to vary so widely within the same area?

Question 27–5 On a uniformly loaded joist, how will the weight be distributed?

Question 27–6 What load will a floor 15′ wide by 25′ long generate?

Question 27–7 If the floor in problem 6 had a girder to support it 7′–6″ in from the edge (centered), how much weight will the girder be supporting?

Question 27–8 A building is 20′ wide with a tile roof supported by trusses. The roof shape is a gable, and the building is 30′ long. What amount of roof weight are the walls supporting? What amount of weight is a footing at the bottom of the wall supporting?

Question 27–9 A one-story home with a post-and-beam floor needs a foundation design. Girders will be placed at 4′ o.c. with support every 8′. What weight will the girders be supporting? What weight will the stem wall at the end of the girder be supporting?

PROBLEMS

Problem 27–1 If a joist with a total uniform load of 500 lb is supported at each end, how much weight will each end be supporting?

Problem 27–2 If a joist with a total uniform load of 700 lb is supported at the midpoint and at each end, how much weight will be supported at each support?

Problem 27–3 A 16' rafter ceiling joist at a ⁶⁄₁₂ pitch will be supporting a cedar shake roof. What will be the total weight supported if the rafters are at 12" spacings? 16"? 24"?

Problem 27–4 A truss with 24" overhangs will span 28' over a residence. It will be supported by a wall at one end and a girder truss with a metal hanger at the other end. How much weight will the hanger need to support?

Problem 27–5 Steel columns will be used to support a girder 16' long with a weight of 1260 lb per linear foot. How much weight will the columns support?

Problem 27–6 A residence 24' wide will be built with a truss roof, with built-up roofing and 30" overhangs. How much weight would a header over a window 8' wide need to support?

Use the following information to complete problems 7 through 11. Answer the questions by providing the weight per linear foot. A one-level residence will be built using trusses with 36" overhangs. The residence will be 24' wide. The roof is 235 lb composition shingles.

Problem 27–7 How much weight will be supported by a header over a window 6' wide in an exterior wall?

Problem 27–8 A wall is to be built 11'–9" from the right bearing wall. If a 2'–6" pocket door is to be installed in the wall, how long will the header need to be, and how much weight will it be supporting?

Problem 27–9 Determine the load to be supported and specify the width of footing required if a wall is placed 14' from the left bearing wall.

Problem 27–10 What will be the total dead load that the bearing wall on the right side will be supporting?

Problem 27–11 What will be the total load that the bearing wall on the left side will be supporting?

CHAPTER 28

Sizing Joists and Rafters Using Span Tables

The complexity of the structure and the experience of the drafter determine who will size the framing members. Even if each framing member will be specified by an engineer, a drafter should have an understanding of span tables used to determine joists and rafters. Because joists and rafters are considered simple beams with uniform loads, their sizes can be determined from span tables contained in the building code.

Standard framing practice is to place structural members at 12, 16, or 24″ (305, 406, or 610 mm) on center. Because of this practice, standard tables have been developed for the sizing of repetitive members. These tables are found in most code books. In addition to building codes, many building departments publish their own standards for repetitive framing members. In order to use these tables you must understand a few basic facts, including typical loading reactions of framing members and the structural capabilities of various species and grades of wood.

▼ LOADING REACTIONS OF WOOD MEMBERS

For every action there is an equal and opposite reaction. This is a law of physics that affects every structure ever built. There are two actions, or stresses, that must be understood before beam reactions can be considered: fiber bending stress and modulus of elasticity. There are many other stresses that act on a structural member, but they do not need to be understood to use standard loading tables. You do not need to know how these stresses are generated, only that they exist, to determine the size of a framing member using standard loading tables. Each term will be covered in greater depth in the next chapter.

Fiber Bending Stress

Fiber bending stresses are represented on span tables by the symbol F_b. The figures that are listed on span tables are the safe, allowable fiber bending stresses for a specific load over a specific span. The designer must compare the calculated values of F_b that must be supported, with the F_b values listed for the species and grade of lumber used to support the load.

Modulus of Elasticity

The modulus of elasticity deals with the stiffness of a structural member. Represented by the symbol E, the modulus of elasticity, or deflection, is concerned with how much a structural member sags.

▼ DETERMINING WOOD CHARACTERISTICS

Before a span table can be used, the values for F_b and E must be determined. Before these can be determined, you must be familiar with the species and grade of wood being used for framing in your area. Some of the most common types of lumber in use for framing are douglas fir–larch (DFL #2), southern pine (SP #2), spruce–pine–fir (SPF #2), hemlock–fir (Hem-Fir #2), and hemlock (Hem #2). Notice that each species is followed by #2. This number refers to the grade value of the species. Usually only #1 or #2 grade lumber is used as structural lumber.

Once the type of framing lumber is known, the safe working values can be determined. Figure 28–1 shows the safe working stresses for joists and rafters. Notice that all values are given in terms of F_b and E. A useful procedure for using such a table would be as follows:

1. Determine the type of wood to be used—for example, DFL #2.
2. Determine the expected lumber size. Some lumber references may still show one value for all sizes of a given species of lumber. The 1994 edition of the *Uniform Building Code* (UBC) provides values based on each standard size of lumber per species.
3. When sizing rafters, determine if snow or no-snow loads will be used. If snow loads are used, verify the design loads with the building department.
4. Select the needed values based on the lumber size. For a DFL #2 under normal load duration, the F_b value would be: 2 × 6-1235, 2 × 8-1140, 2 × 10-1045, and 2 × 12-950.

The E value for each member is 1,600,000. Notice that F_b values are listed in three columns. Although the UBC allows for different loading conditions, only the values

Values for Joists and Rafters—Visually Graded Lumber

These "F_b" values are for use where repetitive members are spaced not more than 24 inches (457 mm). For wider spacing, the "F_b" values shall be reduced 13 percent. Values for surfaced dry or surfaced green lumber apply at 19 percent maximum moisture content in use.

SPECIES AND GRADE	SIZE (inches) × 25.4 for mm	DESIGN VALUE IN BENDING "F_b" psi — Normal Duration	Snow Loading	7-day Loading	MODULUS OF ELASTICITY "E" psi × 0.00689 for N/mm²
DOUGLAS FIR-LARCH (North)					
Select Structural		2,245	2,580	2,805	1,900,000
No. 1/No. 2		1,425	1,635	1,780	1,600,000
No. 3		820	940	1,025	1,400,000
Stud	2 × 4	820	945	1,030	1,400,000
Construction		1,095	1,255	1,365	1,500,000
Standard		605	695	755	1,400,000
Utility		290	330	360	1,300,000
Select Structural		1,945	2,235	2,430	1,900,000
No. 1/No. 2	2 × 6	1,235	1,420	1,540	1,600,000
No. 3		710	815	890	1,400,000
Stud		750	860	935	1,400,000
Select Structural		1,795	2,065	2,245	1,900,000
No. 1/No. 2	2 × 8	1,140	1,310	1,425	1,600,000
No. 3		655	755	820	1,400,000
Select Structural		1,645	1,890	2,055	1,900,000
No. 1/No. 2	2 × 10	1,045	1,200	1,305	1,600,000
No. 3		600	690	750	1,400,000
Select Structural		1,495	1,720	1,870	1,900,000
No. 1/No. 2	2 × 12	950	1,090	1,185	1,600,000
No. 3		545	630	685	1,400,000
HEM-FIR (North)					
Select Structural		2,245	2,580	2,805	1,700,000
No. 1/No. 2		1,725	1,985	2,155	1,600,000
No. 3		990	1,140	1,240	1,400,000
Stud	2 × 4	980	1,125	1,225	1,400,000
Construction		1,325	1,520	1,655	1,500,000
Standard		720	825	900	1,400,000
Utility		345	395	430	1,300,000
Select Structural		1,945	2,235	2,430	1,700,000
No. 1/No. 2	2 × 6	1,495	1,720	1,870	1,600,000
No. 3		860	990	1,075	1,400,000
Stud		890	1,025	1,115	1,400,000
Select Structural		1,795	2,065	2,245	1,700,000
No. 1/No. 2	2 × 8	1,380	1,585	1,725	1,600,000
No. 3		795	915	990	1,400,000
Select Structural		1,645	1,890	2,055	1,700,000
No. 1/No. 2	2 × 10	1,265	1,455	1,580	1,600,000
No. 3		725	835	910	1,400,000
Select Structural		1,495	1,720	1,870	1,700,000
No. 1/No. 2	2 × 12	1,150	1,325	1,440	1,600,000
No. 3		660	760	825	1,400,000
SPRUCE-PINE-FIR					
Select Structural		2,155	2,480	2,695	1,500,000
No. 1/No. 2		1,510	1,735	1,885	1,400,000
No. 3		865	990	1,080	1,200,000
Stud	2 × 4	855	980	1,065	1,200,000
Construction		1,120	1,290	1,400	1,300,000
Standard		635	725	790	1,200,000
Utility		290	330	360	1,100,000
Select Structural		1,870	2,150	2,335	1,500,000
No. 1/No. 2	2 × 6	1,310	1,505	1,635	1,400,000
No. 3		750	860	935	1,200,000
Sttud		7>2<75	895	970	1,200,000
Select Structural		1,725	1,985	2,155	1,500,000
No. 1/No. 2	2 × 8	1,210	1,390	1,510	1,400,000
No. 3		690	795	865	1,200,000
Select Structural		1,580	1,820	1,975	1,500,000
No. 1/No. 2	2 × 10	1,105	1,275	1,385	1,400,000
No. 3		635	725	790	1,200,000
Select Structural		1,440	1,655	1,795	1,500,000
No. 1/No. 2	2 × 12	1,005	1,155	1,260	1,400,000
No. 3		575	660	720	1,200,000

SPECIES AND GRADE	SIZE (inches) × 25.4 for mm	DESIGN VALUE IN BENDING "F_b" psi — Normal Duration	Snow Loading	7-day Loading	MODULUS OF ELASTICITY "E" psi × 0.00689 for N/mm²
SOUTHERN PINE					
Dense Select Structural		3,510	4,030	4,380	1,900,000
Select Structural		3,280	3,770	4,100	1,800,000
Non-Dense Select Structural		3,050	3,500	3,810	1,700,000
No. 1 Dense		2,300	2,650	2,880	1,800,000
No. 1		2,130	2,450	2,660	1,700,000
No. 1 Non-Dense		1,950	2,250	2,440	1,600,000
No. 2 Dense	2 × 4	1,960	2,250	2,440	1,700,000
No. 2		1,720	1,980	2,160	1,600,000
No. 2 Non-Dense		1,550	1,790	1,940	1,400,000
No. 3		980	1,120	1,220	1,400,000
Stud		1,010	1,160	1,260	1,400,000
Construction		1,270	1,450	1,580	1,500,000
Standard		720	825	900	1,300,000
Utility		345	395	430	1,300,000
Dense Select Structural		3,100	3,570	3,880	1,900,000
Select Structural		2,930	3,370	3,670	1,800,000
Non-Dense Select Structural		2,700	3,110	3,380	1,700,000
No. 1 Dense		2,010	2,310	2,520	1,800,000
No. 1	2 × 6	1,900	2,180	2,370	1,700,000
No. 1 Non-Dense		1,720	1,980	2,160	1,600,000
No. 2 Dense		1,670	1,920	2,080	1,700,000
No. 2		1,440	1,650	1,800	1,600,000
No. 2 Non-Dense		1,320	1,520	1,650	1,400,000
No. 3		865	990	1,080	1,400,000
Stud		890	1,020	1,110	1,400,000
Dense Select Structural		2,820	3,240	3,520	1,900,000
Select Structural		2,650	3,040	3,310	1,800,000
Non-Dense Select Structural		2,420	2,780	3,020	1,700,000
No. 1 Dense		1,900	2,180	2,370	1,800,000
No. 1	2 × 8	1,730	1,980	2,160	1,700,000
No. 1 Non-Dense		1,550	1,790	1,940	1,600,000
No. 2 Dense		1,610	1,850	2,010	1,700,000
No. 2		1,380	1,590	1,720	1,600,000
No. 2 Non-Dense		1,260	1,450	1,580	1,400,000
No. 3		805	925	1,010	1,400,000
Dense Select Structural		2,470	2,840	3,090	1,900,000
Select Structural		2,360	2,710	2,950	1,800,000
Non-Dense Select Structural		2,130	2,450	2,660	1,700,000
No. 1 Dense		1,670	1,920	2,080	1,800,000
No. 1	2 × 10	1,500	1,720	1,870	1,700,000
No. 1 Non-Dense		1,380	1,590	1,730	1,600,000
No. 2 Dense		1,380	1,590	1,730	1,700,000
No. 2		1,210	1,390	1,510	1,600,000
No. 2 Non-Dense		1,090	1,260	1,370	1,400,000
No. 3		690	795	865	1,400,000
Dense Select Structural		2,360	2,710	2,950	1,900,000
Select Structural		2,190	2,510	2,730	1,800,000
Non-Dense Select Structural		2,010	2,310	2,520	1,700,000
No. 1 Dense		1,550	1,790	1,940	1,800,000
No. 1	2 × 12	1,440	1,650	1,800	1,700,000
No. 1 Non-Dense		1,320	1,520	1,650	1,600,000
No. 2 Dense		1,320	1,520	1,650	1,700,000
No. 2		1,120	1,290	1,400	1,600,000
No. 2 Non-Dense		1,040	1,190	1,290	1,400,000
No. 3		660	760	825	1,400,000

Figure 28–1 Working stress values of common framing woods. Southern pine is graded according to the SPIB (Southern Pine Inspection Bureau). All other grades shown here are graded according to NLGA (National Lumber Grades Authority). Because wood values are being revised nationally by most lumber associations, it is important to know the standard used to rate the lumber. *Reproduced from the Structural Engineering Design Provision of the Uniform Building Code, 1994 edition. Courtesy International Conference of Building Officials.*

for **normal duration** should be used. Sizing members with snow loading or seven-day loading conditions should be left for an engineer. Values may vary slightly, depending on the code of lumber grading system being used. Values that have been industry standards since the mid 1960s are in a state of flux as the lumber industry revises its standards based on available timber supplies. Values for this chapter are taken from the 1994 UBC and list the safe working stress for each size of lumber. If a value for a span is determined that exceeds the safe values from Figure 28–1, the member will fail.

▼ DETERMINING SIZE AND SPAN

Once the safe working values of the framing lumber are known, the size and span can be determined. Each code provides allowable span tables for floor joists, ceiling joists, and rafters. Determine how the wood is to be used and proceed to the proper table.

A few simple headings must be located at the top of the table before any spans can be identified. See Figure 28–2 to become familiar with the basics of the span table. Some of the key information to be gleaned from this part of the table follows:

○ *The title.* Most codes include about ten different span charts and it is extremely easy to use the wrong table. Double-check the title to see that you have the right table.

○ *The loads.* Within some categories of tables, the values for the table are determined by the loads assumed to be supported. In Figure 28–2, the table is based on an assumed LL value of 40 lb per sq ft. and a dead load of 10 psf. Live loads of 30 or 50 lb and dead loads of 10 or 20 lb are also found in some codes. The values in this table will work fine for a residence, but not work for an office.

○ *Listed values.* Once you determine that you have the right table, determine which values are actually being given. The span is always governed by F_b or E. Notice in Figure 28–2 that this table gives spans that are governed by E. To use this table, the safe E working stress must be known. This value was given as 1.600,000 or 1.6 from Figure 28–2 or DFL #2 in Figure 28–1.

By starting at the bottom of the table with the **Fb** values, spans can also be determined. Given two beams supporting an equal load, the modulus of elasticity will tend to govern as the length increases. A

FLOOR JOISTS WITH L/360 DEFLECTION LIMITS
The allowable bending stress (F_b) and modulus of elasticity (E) used in this table shall be from Tables 23-I-X-1 and 23-I-X-2 only.

DESIGN CRITERIA:
Deflection—For 40 psf (1.92 kN/m^2)live load.
Limited to span in inches (mm) divided by 360.
Strength—Live load of 40 psf (1.92 kN/m^2) plus dead load of 10 psf (0.48 kN/m^2) determines the required bending design value.

Joist Size (in)	Spacing (in)	Modulus of Elasticity, E, in 1,000,000 psi																
		× 0.00689 for N/mm^2																
× 25.4 for mm		0.8	0.9	1.0	1.1	1.2	1.3	1.4	1.5	1.6	1.7	1.8	1.9	2.0	2.1	2.2	2.3	2.4
2 × 6	12.0	8-6	8-10	9-2	9-6	9-9	10-0	10-3	10-6	10-9	10-11	11-2	11-4	11-7	11-9	11-11	12-1	12-3
	16.0	7-9	8-0	8-4	8-7	8-10	9-1	9-4	9-6	9-9	9-11	10-2	10-4	10-6	10-8	10-10	11-0	11-2
	19.2	7-3	7-7	7-10	8-1	8-4	8-7	8-9	9-0	9-2	9-4	9-6	9-8	9-10	10-0	10-2	10-4	10-6
	24.0	6-9	7-0	7-3	7-6	7-9	7-11	8-2	8-4	8-6	8-8	8-10	9-0	9-2	9-4	9-6	9-7	9-9
2 × 8	12.0	11-3	11-8	12-1	12-6	12-10	13-2	13-6	13-10	14-2	14-5	14-8	15-0	15-3	15-6	15-9	15-11	16-2
	16.0	10-2	10-7	11-0	11-4	11-8	12-0	12-3	12-7	12-10	13-1	13-4	13-7	13-10	14-1	14-3	14-6	14-8
	19.2	9-7	10-0	10-4	10-8	11-0	11-3	11-7	11-10	12-1	12-4	12-7	12-10	13-0	13-3	13-5	13-8	13-10
	24.0	8-11	9-3	9-7	9-11	10-2	10-6	10-9	11-0	11-3	11-5	11-8	11-11	12-1	12-3	12-6	12-8	12-10
2 × 10	12.0	14-4	14-11	15-5	15-11	16-5	16-10	17-3	17-8	18-0	18-5	18-9	19-1	19-5	19-9	20-1	20-4	20-8
	16.0	13-0	13-6	14-0	14-6	14-11	15-3	15-8	16-0	16-5	16-9	17-0	17-4	17-8	17-11	18-3	18-6	18-9
	19.2	12-3	12-9	13-2	13-7	14-0	14-5	14-9	15-1	15-5	15-9	16-0	16-4	16-7	16-11	17-2	17-5	17-8
	24.0	11-4	11-10	12-3	12-8	13-0	13-4	13-8	14-0	14-4	14-7	14-11	15-2	15-5	15-8	15-11	16-2	16-5
2 × 12	12.0	17-5	18-1	18-9	19-4	19-11	20-6	21-0	21-6	21-11	22-5	22-10	23-3	23-3	24-0	24-5	24-9	25-1
	16.0	15-10	16-5	17-0	17-7	18-1	18-7	19-1	19-6	19-11	20-4	20-9	21-1	21-6	21-10	22-2	22-6	22-10
	19.2	14-11	15-6	16-0	16-7	17-0	17-6	17-11	18-4	18-9	19-2	19-6	19-10	20-2	20-6	20-10	21-2	21-6
	24.0	13-10	14-4	14-11	15-4	15-10	16-3	16-8	17-0	17-5	17-9	18-1	18-5	18-9	19-1	19-4	19-8	19-11
F_b	12.0	718	777	833	888	941	993	1,043	1,092	1,140	1,187	1,233	1,278	1,323	1,367	1,410	1,452	1,494
	16.0	790	855	917	977	1,036	1,093	1,148	1,202	1,255	1,306	1,357	1,407	1,456	1,504	1,551	1,598	1,644
	19.2	840	909	975	1,039	1,101	1,161	1,220	1,277	1,333	1,388	1,442	1,495	1,547	1,598	1,649	1,698	1,747
	24.0	905	979	1,050	1,119	1,186	1,251	1,314	1,376	1,436	1,496	1,554	1,611	1,667	1,722	1,776	1,829	1,882

NOTE: The required bending design value, F_b, in pounds per square inch (× 0.00689 for N/mm^2) is shown at the bottom of this table and is applicable to all lumber sizes shown. Spans are shown in feet-inches (1 foot = 304.8 mm, 1 inch = 25.4 mm) and are limited to 26 feet (7925 mm) and less.

Figure 28–2 This span table is suitable for determining floor joists with a live load of 40 psf and a dead load of 10 psf. *Reproduced from the Structural Engineering Design Provision of the* Uniform Building Code, *1994 edition. Courtesy International Conference of Building Officials.*

shorter beam, with an equal load will tend to be governed by fiber bending.

○ *Size and spacing of lumber.* The left side of most span tables is usually reserved for the size and spacing of the framing lumber. In Figure 28–2 you will notice that each size of structural member has a value for the spacings of 12″, 16″, 19.2″, or 24″ (305, 406, 488, or 610 mm) on center.

Sizing Floor Joist

To use Figure 28–2, find the column that represents proper E value for the lumber used in your area. For our example, douglas fir #2 for a floor will be used, which means that the 1.6 column will contain all the span information we need. The first listing in this column is 10–9. The 10–9 represents a maximum allowable span of 10′–9″. Drop down to the Fb portion at the bottom of the table, and the stress in fiber bending can be determined. Because the 10–9 value is derived from the 12″ spacing, the fiber bending value must also be taken from the 12″ spacing row. The corresponding Fb value is 1140. This number represents the stress that would be created if 2 × 6 DFL #2 floor joists were used to support a live load of 40 lb and a dead load of 10 lb and span 10′–9″. Because 1140 is less than the safe working stress of 1235 for a 2 × 6 from Table 28–1, a 2 × 6 is safe for spanning 10′–9″.

Using Figure 28–2, determine the distance a 2 × 8 Hem-Fir #2 floor joist will span at 16″ o.c. Use the 1.6 E column. Work down the left side until you come to 2 × 8. Now work to the right in the 16″ spacing row. The 1.6 E column and the 16″ spacing row will intersect at 12–10. The span for a 2 × 8 at 16″ o.c. is 12′–10″ with a fiber bending value of 1255. Because the safe limit is 1380, the 2 × 8 is safe to use.

Use the same procedure for determining the span for a 2 × 10 DFL #2 floor joist at 16″ o.c. The listed span is 18′–0″. The corresponding Fb value is 1255, which exceeds the safe fiber bending value of 1045. To determine the distance a 2 × 10 DFL #2 can span, enter the table through the Fb values rather than the E values. Because the desired spacing is 16″ o.c., use the 16 row. Move to the right in this row until the largest value without exceeding the safe working value (1045) is found. A value of 1036 is the largest value listed that does not exceed the safe working fiber bending value. The 1036 value is in the 1.2 column. Move up the 1.2 column to the 2 × 10 @ 16″ spacing row. The maximum span for a 2 × 10 DFL #2 floor joist at 16″ o.c. is 14′–11″. For this span, the tendency to fail is in fiber bending rather than modulus of elasticity.

Typically, the span will be known, and the drafter will need to determine the size of member. For instance, determine the size of floor joist needed for a room 14′ wide. Assume southern pine will be used. Using an E value of 1.6, a 2 × 10 floor joist with 16″ spacing will span 16′–5″, with an Fb value of 1255. The safe working value of Fb is exceeded. To determine the size using the Fb value, enter the 16 row, and go the maximum value that does not exceed 1210, which is the safe working value for a 2 × 10 SP. Using the 1202 value, now go up the 1.5 column to the 2 × 10 at 16″ o.c. row. A 2 × 10 SP #2 floor joist will safely span 16′–0″ and be adequate to support the floor of a room 14′ wide.

Sizing Ceiling Joists

To determine the size of a ceiling joist, use the same procedure that was used to size a floor joist using the proper table. Because they are both horizontal members, they will have similar loading patterns. A table for sizing ceiling joists can be seen in Figure 28–3. For ceiling joists the live load is only 10 lb. Using Figure 28–3, determine the smallest size DFL #2 ceiling joist that will span 16′ at 16 in o.c. By using the 1.6 column, you will find that a 2 × 6 at 16″ o.c. will span 17′–8″ with an F_b of 1243. The safe limit in fiber bending is 1235 so the member will fail at 17′–8″. Since the joist was safe in modules of elasticity but failed in fiber bending, enter the table through the Fb values at the bottom of the table. This will allow the stress in fiber bending to be determined. Enter the table in the 16 row of Fb, and proceed to the right to the highest value that does not exceed the safe value of 1235. Using the Fb value of 1191 in the 1.5 E column, a 2 × 6 DFL #2 can safely be used for spans up to but not exceeding 17′–4″.

Sizing Rafters

Although the selection of rafters is similar to the selection of joists, two major differences will be encountered. Rafter tables have columns based on Fb values rather than E values. The other noticeable difference is that there are many rafter tables to choose from. In the UBC, rafter tables are divided into tables for live loads of either 20 or 30 psf, dead loads of 10, 15, or 20 psf, and groups with deflection limits of either 1/240 or 1/180 of the rafter length.

The *deflection limit*, the amount a member is allowed to sag, is based on the length of the framing member in inches. While sizing joists, a deflection limit of 1/360 is used. If a joist is 10′ long, its length in inches (l) is 120, with a maximum allowable deflection limit of 1/360 (120/360) or 0.33″. A rafter 10′ long is allowed to sag 0.50″ with a deflection limit of 1/240 and 0.66″ with a deflection limit of 1/180. The UBC requires a deflection of 1/240 to be used for rafters, although some local municipalities may allow a limt of 1/180 to be used for unfinished attics.

Your local building department will determine which live load value is to be used. The dead load is determined by the construction materials to be used. The table in Figure 28–4 with a dead load of 10 psf would be suitable for rafters supporting most materials except tile, with no interior finish. The table in Figure 28–5 with a load of 15 psf is suitable for rafters supporting roofing materials other than tile, and finished with sheet rock or some other

CEILING JOISTS WITH L/240 DEFLECTION LIMITS
The allowable bending stress (F_b) and modulus of elasticity (E) used in this table shall be from Tables 23-I-X-1 and 23-I-X-2 only.

DESIGN CRITERIA:
Deflection—For 10 (0.48 kN/m²) psf live load.
Limited to span in inches (mm) divided by 240.
Strength—Live load of 10 psf (0.48 kN/mm²) plus dead load of 5 psf (0.24 kN/m²) determines the required fiber stress value.

Joist Size (in)	Spacing (in)	Modulus of Elasticity, E, in 1,000,000 psi																
× 25.4 for mm		0.8	0.9	1.0	1.1	1.2	1.3	1.4	1.5	1.6	1.7	1.8	1.9	2.0	2.1	2.2	2.3	2.4
2 × 4	12.0	9-10	10-3	10-7	10-11	11-3	11-7	11-10	12-2	12-5	12-8	12-11	13-2	13-4	13-7	13-9	14-0	14-2
	16.0	8-11	9-4	9-8	9-11	10-3	10-6	10-9	11-0	11-3	11-6	11-9	11-11	12-2	12-4	12-6	12-9	12-11
	19.2	8-5	8-9	9-1	9-4	9-8	9-11	10-2	10-4	10-7	10-10	11-0	11-3	11-5	11-7	11-9	12-0	12-2
	24.0	7-10	8-1	8-5	8-8	8-11	9-2	9-5	9-8	9-10	10-0	10-3	10-5	10-7	10-9	10-11	11-1	11-3
2 × 6	12.0	15-6	16-1	16-8	17-2	17-8	18-2	18-8	19-1	19-6	19-11	20-3	20-8	21-0	21-4	21-8	22-0	22-4
	16.0	14-1	14-7	15-2	15-7	16-1	16-6	16-11	17-4	17-8	18-1	18-5	18-9	19-1	19-5	19-8	20-0	20-3
	19.2	13-3	13-9	14-3	14-8	15-2	15-7	15-11	16-4	16-8	17-0	17-4	17-8	17-11	18-3	18-6	18-10	19-1
	24.0	12-3	12-9	13-3	13-8	14-1	14-5	14-9	15-2	15-6	15-9	16-1	16-4	16-8	16-11	17-2	17-5	17-8
2 × 8	12.0	20-5	21-2	21-11	22-8	23-4	24-0	24-7	25-2	25-8								
	16.0	18-6	19-3	19-11	20-7	21-2	21-9	22-4	22-10	23-4	23-10	24-3	24-8	25-2	25-7	25-11		
	19.2	17-5	18-1	18-9	19-5	19-11	20-6	21-0	21-6	21-11	22-5	22-10	23-3	23-8	24-0	24-5	24-9	25-2
	24.0	16-2	16-10	17-5	18-0	18-6	19-0	19-6	19-11	20-5	20-10	21-2	21-7	21-11	22-4	22-8	23-0	23-4
2 × 10	12.0	26-0																
	16.0	23-8	24-7	25-5														
	19.2	22-3	23-1	23-11	24-9	25-5												
	24.0	20-8	21-6	22-3	22-11	23-8	24-3	24-10	25-5	26-0								
F_b	12.0	711	769	825	880	932	983	1,033	1,082	1,129	1,176	1,221	1,266	1,310	1,354	1,396	1,438	1,480
	16.0	783	847	909	968	1,026	1,082	1,137	1,191	1,243	1,294	1,344	1,394	1,442	1,490	1,537	1,583	1,629
	19.2	832	900	965	1,029	1,090	1,150	1,208	1,265	1,321	1,375	1,429	1,481	1,533	1,583	1,633	1,682	1,731
	24.0	896	969	1,040	1,108	1,174	1,239	1,302	1,363	1,423	1,481	1,539	1,595	1,651	1,706	1,759	1,812	1,864

NOTE: The required bending design value, F_b, in pounds per square (× 0.00689 for N/mm²) is shown at the bottom of this table and is applicable to all lumber sizes shown. Spans are shown in feet-inches (1 foot = 304.8 mm, 1 inch = 25.4 mm) and are limited to 26 feet (7925 mm) and less.

Figure 28–3 This span table is suitable for determining ceiling joists with a live load of 40 psf and a dead load of 10 psf. *Reproduced from the Structural Engineering Design Provision of the* Uniform Building Code, *1994 edition. Courtesy International Conference of Building Officials.*

lightweight finishing material. The table in Figure 28–6 with a 20 psf load would be suitable for supporting most tiles. Vender catalogs should always be consulted to determine the required dead loads.

Once the live and dead loads have been determined, choose the appropriate table. Consider the size Hem-Fir rafter that should be used to support a roof over a room 18' wide covered with a gable roof. Assume the roof is supporting 235 lb composition shingles with a live load of 30 psf, a dead load of 10 psf, with 24" spacing. Since a gable roof is being framed, the actual horizontal span is only 9'-0". It can be seen in Figure 28–1 that a 2 × 6 has an Fb value of 1495. Since 1495 is so close to 1500, the 1500 column of Figure 28–4 can be used, producing a suitable span of 9'-9". The corresponding E value for 24" spacing is 1.19. Since this value is below the allowable safe limit of 1.6, a 2 × 6 Hem-Fir #2 rafter at 24" can be use to frame the roof.

If the same room were to be framed with DFL #2 members, the process to determine the rafter size would be similar. A 2 × 6 DFL #2 has a safe fiber bending working limit of 1235 with an E value of 1.6. Using the 1200 column of Figure 28–4, it can be seen that a 2 × 6 at 24" spacings cannot be used because the span is only 8'-8". Using a spacing of 19.2 will allow for a span of 9'-9" with an E value of 1.05.

In addition to listings based on the loads and deflection limits, some span tables list rafters based on usage and pitch. Common listings include rafters used with attics and rafters used for vaulted ceilings that will support either gypsum board or plaster. Tables are typically divided into rafters steeper than 3/12 or less than a 3/12 pitch. Figure 28–7 shows a table based on usage and roof pitch. Notice that the E value is listed directly below the span rather than at the bottom of the table.

RAFTERS WITH L/240 DEFLECTION LIMITATION

The allowable bending stress (F_b) and modulus of elasticity (E) used in this table shall be from Tables 23-I-X-1 and 23-I-X-2 only.

DESIGN CRITERIA:

Strength—Live load of 30 psf (1.44 kN/mm²) psf plus dead load of 10 psf (0.48 kN/m²) determines the required bending design value.

Deflection—For 30 psf (1.44 kN/m²) live load.

Limited to span in inches (mm) divided by 240.

Rafter Size (in) ×25.4 for mm	Spacing (in)	Bending Design Value, F_b (psi) ×0.00689 for N/mm²																					
		300	400	500	600	700	800	900	1000	1100	1200	1300	1400	1500	1600	1700	1800	1900	2000	2100	2200	2300	2400
2×6	12.0	6-2	7-1	7-11	8-8	9-5	10-0	10-8	11-3	11-9	12-4	12-10	13-3	13-9	14-2	14-8	15-1	15-6	15-11				
	16.0	5-4	6-2	6-10	7-6	8-2	8-8	9-3	9-9	10-2	10-8	11-1	11-6	11-11	12-4	12-8	13-1	13-5	13-9	14-1	14-5		
	19.2	4-10	5-7	6-3	6-10	7-5	7-11	8-5	8-11	9-4	9-9	10-1	10-6	10-10	11-3	11-7	11-11	12-3	12-7	12-10	13-2	13-6	
	24.0	4-4	5-0	5-7	6-2	6-8	7-1	7-6	7-11	8-4	8-8	9-1	9-5	9-9	10-0	10-4	10-8	10-11	11-3	11-6	11-9	12-0	12-4
2×8	12.0	8-1	9-4	10-6	11-6	12-5	13-3	14-0	14-10	15-6	16-3	16-10	17-6	18-1	18-9	19-4	19-10	20-5	20-11				
	16.0	7-0	8-1	9-1	9-11	10-9	11-6	12-2	12-10	13-5	14-0	14-7	15-2	15-8	16-3	16-9	17-2	17-8	18-1	18-7	19-0		
	19.2	6-5	7-5	8-3	9-0	9-9	10-6	11-1	11-8	12-3	12-10	13-4	13-10	14-4	14-10	15-3	15-8	16-2	16-7	16-11	17-4	17-9	
	24.0	5-9	6-7	7-5	8-1	8-9	9-4	9-11	10-6	11-0	11-6	11-11	12-5	12-10	13-3	13-8	14-0	14-5	14-10	15-2	15-6	15-10	16-3
2×10	12.0	10-4	11-11	13-4	14-8	15-10	16-11	17-11	18-11	19-10	20-8	21-6	22-4	23-1	23-11	24-7	25-4	26-0					
	16.0	8-11	10-4	11-7	12-8	13-8	14-8	15-6	16-4	17-2	17-11	18-8	19-4	20-0	20-8	21-4	21-11	22-6	23-1	23-8	24-3		
	19.2	8-2	9-5	10-7	11-7	12-6	13-4	14-2	14-11	15-8	16-4	17-0	17-8	18-3	18-11	19-6	20-0	20-7	21-1	21-8	22-2	22-8	
	24.0	7-4	8-5	9-5	10-4	11-2	11-11	12-8	13-4	14-0	14-8	15-3	15-10	16-4	16-11	17-5	17-11	18-5	18-11	19-4	19-10	20-3	20-8
2×12	12.0	12-7	14-6	16-3	17-9	19-3	20-6	21-9	23-0	24-1	25-2												
	16.0	10-11	12-7	14-1	15-5	16-8	17-9	18-10	19-11	20-10	21-9	22-8	23-6	24-4	25-2	25-11							
	19.2	9-11	11-6	12-10	14-1	15-2	16-3	17-3	18-2	19-0	19-11	20-8	21-6	22-3	23-0	23-8	24-4	25-0	25-8				
	24.0	8-11	10-3	11-6	12-7	13-7	14-6	15-5	16-3	17-0	17-9	18-6	19-3	19-11	20-6	21-2	21-9	22-5	23-0	23-6	24-1	24-8	25-2
E	12.0	0.15	0.23	0.32	0.43	0.54	0.66	0.78	0.82	1.06	1.21	1.36	1.52	1.69	1.86	2.04	2.22	2.41	2.60				
	16.0	0.13	0.20	0.28	0.37	0.47	0.57	0.68	0.80	0.92	1.05	1.18	1.32	1.46	1.61	1.76	1.92	2.08	2.25	2.42	2.60		
	19.2	0.12	0.18	0.26	0.34	0.43	0.52	0.62	0.73	0.84	0.95	1.08	1.20	1.33	1.47	1.61	1.75	1.90	2.05	2.21	2.37	2.53	
	24.0	0.11	0.16	0.23	0.30	0.38	0.46	0.55	0.65	0.75	0.85	0.96	1.08	1.19	1.31	1.44	1.57	1.70	1.84	1.98	2.12	2.27	2.41

NOTE: The required modulus of elasticity, E, in 1,000,000 pounds per square inch (psi) (×0.00689 for N/mm²) is shown at the bottom of this table, is limited to 2.6 million psi (17914 N/mm²) and less, and is applicable to all lumber sizes shown. Spans are shown in feet-inches (1 foot = 304.8 mm, 1 inch = 25.4 mm) and are limited to 26 feet (7925 mm) and less.

Figure 28–4 This span table is suitable for determining rafters with a live load of 30 psf and a dead load of 10 psf. *Reproduced from the Structural Engineering Design Provision of the* Uniform Building Code, *1994 edition. Courtesy International Conference of Building Officials.*

RAFTERS WITH L/240 DEFLECTION LIMITATION
The allowable bending stress (*F_b*) and modulus of elasticity (*E*) used in this table shall be from Tables 23-I-X-1 and 23-I-X-2 only.

DESIGN CRITERIA:
Strength—Live load of 30 psf (1.44 kN/mm²) plus dead load of 15 psf (0.72 kN/m²) determines the required bending design value.
Deflection—For 30 psf (1.44 kN/m²) live load.
Limited to span in inches (mm) divided by 240.

Bending Design Value, F_b (psi) — × 0.00689 for N/mm². Rafter Size (in) × 25.4 for mm.

Rafter Size (in)	Spacing (in)	300	400	500	600	700	800	900	1000	1100	1200	1300	1400	1500	1600	1700	1800	1900	2000	2100	2200	2300	2400	2500	2600	2700
2 × 6	12.0	5-10	6-8	7-6	8-2	8-10	9-6	10-0	10-7	11-1	11-7	12-1	12-6	13-0	13-5	13-10	14-2	14-7	15-0	15-4	15-8					
	16.0	5-0	5-10	6-6	7-1	7-8	8-2	8-8	9-2	9-7	10-0	10-5	10-10	11-3	11-7	11-11	12-4	12-8	13-0	13-3	13-7	13-11	14-2			
	19.2	4-7	5-4	5-11	6-6	7-0	7-6	7-11	8-4	8-9	9-2	9-6	9-11	10-3	10-7	10-11	11-3	11-6	11-10	12-2	12-5	12-8	13-0	13-3	13-6	
	24.0	4-1	4-9	5-4	5-10	6-3	6-8	7-1	7-6	7-10	8-2	8-6	8-10	9-2	9-6	9-9	10-0	10-4	10-7	10-10	11-1	11-4	11-7	11-10	12-1	12-4
2 × 8	12.0	7-8	8-10	8-10	10-10	11-8	12-6	13-3	13-11	14-8	15-3	15-11	16-6	17-1	17-8	18-2	18-9	19-3	19-9	20-3	20-8					
	16.0	6-7	7-8	8-7	9-4	10-1	10-10	11-6	12-1	12-8	13-3	13-9	14-4	14-10	15-3	15-9	16-3	16-8	17-1	17-6	17-11	18-4	18-9			
	19.2	6-0	7-0	7-10	8-7	9-3	9-10	10-6	11-0	11-7	12-1	12-7	13-1	13-6	13-11	14-5	14-10	15-2	15-7	16-0	16-4	16-9	17-1	17-5	17-9	
	24.0	5-5	6-3	7-0	7-8	8-3	8-10	9-4	9-10	10-4	10-10	11-3	11-8	12-1	12-6	12-10	13-3	13-7	13-11	14-4	14-8	15-0	15-3	15-7	15-11	16-3
2 × 10	12.0	9-9	11-3	12-7	13-9	14-11	15-11	16-11	17-10	18-8	19-6	20-4	21-1	21-10	22-6	23-3	23-11	24-6	25-2	25-10						
	16.0	8-5	9-9	10-11	11-11	12-11	13-9	14-8	15-5	16-2	16-11	17-7	18-3	18-11	19-6	20-1	20-8	21-3	21-10	22-4	22-10	23-5	23-11			
	19.2	7-8	8-11	9-11	10-11	11-9	12-7	13-4	14-1	14-9	15-5	16-1	16-8	17-3	17-10	18-4	18-11	19-5	19-11	20-5	20-10	21-4	21-10	22-3	22-8	
	24.0	6-11	8-0	8-11	9-9	10-6	11-3	11-11	12-7	13-2	13-9	14-4	14-11	15-5	15-11	16-5	16-11	17-4	17-10	18-3	18-8	19-1	19-6	19-11	20-4	20-8
2 × 12	12.0	11-10	13-8	15-4	16-9	18-1	19-4	20-6	21-8	22-8	23-9	24-8	25-7													
	16.0	10-3	11-10	13-3	14-6	15-8	16-9	17-9	18-9	19-8	20-6	21-5	22-2	23-0	23-9	24-5	25-2	25-10								
	19.2	9-4	10-10	12-1	13-3	14-4	15-4	16-3	17-1	17-11	18-9	19-6	20-3	21-0	21-8	22-4	23-0	23-7	24-2	24-10	25-5	25-11				
	24.0	8-5	9-8	10-10	11-10	12-10	13-8	14-6	15-4	16-1	16-9	17-5	18-1	18-9	19-4	20-0	20-6	21-1	21-8	22-2	22-8	23-3	23-9	24-2	24-8	25-2
E	12.0	0.13	0.19	0.27	0.36	0.45	0.55	0.66	0.77	0.89	1.01	1.14	1.28	1.41	1.56	1.71	1.86	2.02	2.18	2.34	2.51					
	16.0	0.11	0.17	0.24	0.31	0.39	0.48	0.57	0.67	0.77	0.88	0.99	1.10	1.22	1.35	1.48	1.61	1.75	1.89	2.03	2.18	2.33	2.48			
	18.2	0.10	0.15	0.22	0.28	0.36	0.44	0.52	0.61	0.70	0.80	0.90	1.01	1.12	1.23	1.35	1.47	1.59	1.72	1.85	1.99	2.12	2.26	2.41	2.55	
	24.0	0.09	0.14	0.19	0.25	0.32	0.39	0.46	0.54	0.63	0.72	0.81	0.90	1.00	1.10	1.21	1.31	1.43	1.54	1.66	1.78	1.90	2.02	2.15	2.28	2.41

NOTE: The required modulus of elasticity, *E*, in 1,000,000 pounds per square inch (psi) (× 0.00689 for N/mm²) is shown at the bottom of this table, is limited to 2.6 million psi (17914 N/mm²) and less, and is applicable to all lumber sizes shown. Spans are shown in feet-inches (1 foot = 304.8 mm, 1 inch = 25.4 mm) and are limited to 26 feet (7925 mm) and less.

Figure 28–5 This span table is suitable for determining rafters with a live load of 30 psf and a dead load of 15 psf. *Reproduced from the Structural Engineering Design Provision of the Uniform Building Code, 1994 edition. Courtesy International Conference of Building Officials.*

RAFTERS WITH L/240 DEFLECTION LIMITATION
The allowable bending stress (F_b) and modulus of elasticity (E) used in this table shall be from Tables 23-I-X-1 and 23-I-X-2 only.

DESIGN CRITERIA:
Strength—Live load of 30 psf (1.44 kN/mm^2) plus dead load of 20 psf (0.96 kN/m^2) determines the required bending design value.
Deflection—For 30 psf (1.44 kN/m^2) live load.
Limited to span in inches (mm) divided by 240.

Rafter Size (in) × 25.4 for mm	Spacing (in)	300	400	500	600	700	800	900	1000	1100	1200	1300	1400	1500	1600	1700	1800	1900	2000	2100	2200	2300	2400	2500	2600	2700
2 × 6	12.0	5-6	6-4	7-1	7-9	8-5	9-0	9-6	10-0	10-6	11-0	11-5	11-11	12-4	12-8	13-1	13-6	13-10	14-2	14-7	14-11	15-3	15-7	15-11		
	16.0	4-9	5-6	6-2	6-9	7-3	7-9	8-3	8-8	9-1	9-6	9-11	10-3	10-8	11-0	11-4	11-8	12-0	12-4	12-7	12-11	13-2	13-6	13-9	14-0	14-3
	19.2	4-4	5-0	5-7	6-2	6-8	7-1	7-6	7-11	8-4	8-8	9-1	9-5	9-9	10-0	10-4	10-8	10-11	11-3	11-6	11-9	12-0	12-4	12-7	12-10	13-1
	24.0	3-11	4-6	5-0	5-6	5-11	6-4	6-9	7-1	7-5	7-9	8-1	8-5	8-8	9-0	9-3	9-6	9-9	10-0	10-3	10-6	10-9	11-0	11-3	11-5	11-8
2 × 8	12.0	7-3	8-4	9-4	10-3	11-1	11-10	12-7	13-3	13-11	14-6	15-1	15-8	16-3	16-9	17-3	17-9	18-3	18-9	19-2	19-8	20-1	20-6	20-11		
	16.0	6-3	7-3	8-1	8-11	9-7	10-3	10-10	11-6	12-0	12-7	13-1	13-7	14-0	14-6	14-11	15-5	15-10	16-3	16-7	17-0	17-5	17-9	18-1	18-6	18-10
	19.2	5-9	6-7	7-5	8-1	8-9	9-4	9-11	10-6	11-0	11-6	11-11	12-5	12-10	13-3	13-8	14-0	14-5	14-10	15-2	15-6	15-10	16-3	16-7	16-10	17-2
	24.0	5-2	5-11	6-7	7-3	7-10	8-4	8-11	9-4	9-10	10-3	10-8	11-1	11-6	11-10	12-2	12-7	12-11	13-3	13-7	13-11	14-2	14-6	14-10	15-1	15-5
2 × 10	12.0	9-3	10-8	11-11	13-1	14-2	15-1	16-0	16-11	17-9	18-6	19-3	20-0	20-8	21-4	22-0	22-8	23-3	23-11	24-6	25-1	25-7				
	16.0	8-0	9-3	10-4	11-4	12-3	13-1	13-10	14-8	15-4	16-0	16-8	17-4	17-11	18-6	19-1	19-7	20-2	20-8	21-2	21-8	22-2	22-8	23-1	23-7	24-0
	19.2	7-4	8-5	9-5	10-4	11-2	11-11	12-8	13-4	14-0	14-8	15-3	15-10	16-4	16-11	17-5	17-11	18-5	18-11	19-4	19-10	20-3	20-8	21-1	21-6	21-11
	24.0	6-6	7-7	8-5	9-3	10-0	10-8	11-4	11-11	12-6	13-1	13-7	14-2	14-8	15-1	15-7	16-0	16-6	16-11	17-4	17-9	18-1	18-6	18-11	19-3	19-7
2 × 12	12.0	11-3	13-0	14-6	15-11	17-2	18-4	19-6	20-6	21-7	22-6	23-5	24-4	25-2	26-0											
	16.0	9-9	11-3	12-7	13-9	14-11	15-11	16-10	17-9	18-8	19-6	20-3	21-1	21-9	22-6	23-2	23-10	24-6	25-2	25-9						
	19.2	8-11	10-3	11-6	12-7	13-7	14-6	15-5	16-3	17-0	17-9	18-6	19-3	19-11	20-6	21-2	21-9	22-5	23-0	23-6	24-1	24-8	25-2	25-8		
	24.0	7-11	9-2	10-3	11-3	12-2	13-0	13-9	14-6	15-3	15-11	16-7	17-2	17-9	18-4	18-11	19-6	20-0	20-6	21-1	21-7	22-0	22-6	23-0	23-5	23-10
E	12.0	0.11	0.17	0.23	0.31	0.38	0.47	0.56	0.66	0.76	0.86	0.87	1.09	1.21	1.33	1.46	1.59	1.72	1.86	2.00	2.14	2.29	2.44	2.60		
	16.0	0.09	0.14	0.20	0.26	0.33	0.41	0.49	0.57	0.66	0.75	0.84	0.94	1.05	1.15	1.26	1.37	1.49	1.61	1.73	1.86	1.99	2.12	2.25	2.39	2.53
	19.2	0.09	0.13	0.18	0.24	0.30	0.37	0.44	0.52	0.60	0.68	0.77	0.88	0.95	1.05	1.15	1.25	1.36	1.47	1.58	1.70	1.81	1.93	2.05	2.18	2.31
	24.0	0.08	0.12	0.16	0.22	0.27	0.33	0.40	0.46	0.54	0.61	0.89	0.77	0.85	0.94	1.03	1.12	1.22	1.31	1.41	1.52	1.62	1.73	1.84	1.95	2.06

Bending Design Value, F_b (psi) × 0.00689 for N/mm^2

NOTE: The required modulus of elasticity, E, in 1,000,000 pounds per square inch (psi) (× 0.00689 for N/mm^2) is shown at the bottom of this table, is limited to 2.6 million psi (17914 N/mm^2) and less, and is applicable to all lumber sizes shown. Spans are shown in feet-inches (1 foot = 304.8 mm, 1 inch = 25.4 mm) and are limited to 26 feet (7925 mm) and less.

Figure 28–6 This span table is suitable for determining rafters with a live load of 30 psf and a dead load of 20 psf. *Reproduced from the Structural Engineering Design Provision of the* Uniform Building Code, *1994 edition. Courtesy International Conference of Building Officials.*

ALLOWABLE SPANS FOR HIGH-SLOPE RAFTERS, SLOPE OVER 3 IN 12 30 LBS. PER SQ. FT. LIVE LOAD (Light Roof Covering)

DESIGN CRITERIA: Strength—7 lbs. per sq. ft. dead load plus 30 lbs. per sq. ft. live load determines required fiber stress. Deflection—For 30 lbs. per sq. ft. live load. Limited to span in inches divided by 180. RAFTERS: Spans are measured along the horizontal projection and loads are considered as applied on the horizontal projection.

Rafter Size (IN)	Spacing (IN)	Allowable Extreme Fiber Stress in Bending F_b (psi).														
		500	600	700	800	900	1000	1100	1200	1300	1400	1500	1600	1700	1800	1900
2 × 4	12.0	5-3 / 0.27	5-9 / 0.36	6-3 / 0.45	6-8 / 0.55	7-1 / 0.66	7-5 / 0.77	7-9 / 0.89	8-2 / 1.02	8-6 / 1.15	8-9 / 1.28	9-1 / 1.42	9-5 / 1.57	9-8 / 1.72	10-0 / 1.87	10-3 / 2.03
	16.0	4-7 / 0.24	5-0 / 0.31	5-5 / 0.39	5-9 / 0.48	6-1 / 0.57	6-5 / 0.67	6-9 / 0.77	7-1 / 0.88	7-4 / 0.99	7-7 / 1.11	7-11 / 1.23	8-2 / 1.36	8-5 / 1.49	8-8 / 1.62	8-10 / 1.76
	24.0	3-9 / 0.19	4-1 / 0.25	4-5 / 0.32	4-8 / 0.39	5-0 / 0.47	5-3 / 0.55	5-6 / 0.63	5-9 / 0.72	6-0 / 0.81	6-3 / 0.91	6-5 / 1.01	6-8 / 1.11	6-10 / 1.21	7-1 / 1.32	7-3 / 1.43
2 × 6	12.0	8-3 / 0.27	9-1 / 0.36	9-9 / 0.45	10-5 / 0.55	11-1 / 0.66	11-8 / 0.77	12-3 / 0.89	12-9 / 1.02	13-4 / 1.15	13-10 / 1.28	14-4 / 1.42	14-9 / 1.57	15-3 / 1.72	15-8 / 1.87	16-1 / 2.03
	16.0	7-2 / 0.24	7-10 / 0.31	8-5 / 0.39	9-1 / 0.48	9-7 / 0.57	10-1 / 0.67	10-7 / 0.77	11-1 / 0.88	11-6 / 0.99	12-0 / 1.11	12-5 / 1.23	12-9 / 1.36	13-2 / 1.49	13-7 / 1.62	13-11 / 1.76
	24.0	5-10 / 0.19	6-5 / 0.25	6-11 / 0.32	7-5 / 0.39	7-10 / 0.47	8-3 / 0.55	8-8 / 0.63	9-1 / 0.72	9-5 / 0.81	9-9 / 0.91	10-1 / 1.01	10-5 / 1.11	10-9 / 1.21	11-1 / 1.32	11-5 / 1.43
2 × 8	12.0	10-11 / 0.27	11-11 / 0.36	12-10 / 0.45	13-9 / 0.55	14-7 / 0.66	15-5 / 0.77	16-2 / 0.89	16-10 / 1.02	17-7 / 1.15	18-2 / 1.28	18-10 / 1.42	19-6 / 1.57	20-1 / 1.72	20-8 / 1.87	21-3 / 2.03
	16.0	9-5 / 0.24	10-4 / 0.31	11-2 / 0.39	11-11 / 0.48	12-8 / 0.57	13-4 / 0.67	14-0 / 0.77	14-7 / 0.88	15-2 / 0.99	15-9 / 1.11	16-4 / 1.23	16-10 / 1.36	17-4 / 1.49	17-11 / 1.62	18-4 / 1.76
	24.0	7-8 / 0.19	8-5 / 0.25	9-1 / 0.32	9-9 / 0.39	10-4 / 0.47	10-11 / 0.55	11-5 / 0.63	11-11 / 0.72	12-5 / 0.81	12-10 / 0.91	13-4 / 1.01	13-9 / 1.11	14-2 / 1.21	14-7 / 1.32	15-0 / 1.43
2 × 10	12.0	13-11 / 0.27	15-2 / 0.36	16-5 / 0.45	17-7 / 0.55	18-7 / 0.66	19-8 / 0.77	20-7 / 0.89	21-6 / 1.02	22-5 / 1.15	23-3 / 1.28	24-1 / 1.42	24-10 / 1.57	25-7 / 1.72	26-4 / 1.87	27-1 / 2.03
	16.0	12-0 / 0.26	13-2 / 0.34	14-3 / 0.43	15-2 / 0.53	16-2 / 0.63	17-0 / 0.74	17-10 / 0.85	18-7 / 0.97	19-5 / 1.09	20-1 / 1.22	20-10 / 1-35	21-6 / 1.49	22-2 / 1.63	22-10 / 1.78	23-5 / 1.93
	24.0	9-10 / 0.19	10-9 / 0.25	11-7 / 0.32	12-5 / 0.39	13-2 / 0.47	13-11 / 0.55	14-7 / 0.63	15-2 / 0.72	15-10 / 0.81	16-5 / 0.91	17-0 / 1.01	17-7 / 1.11	18-1 / 1.21	18-7 / 1.32	19-2 / 1.43

NOTES: (1) The required modulus of elasticity (E) in 1,000,000 pounds per square inch is shown below each span. (2) Use single or repetitive member bending stress values (F_b) and modulus of elasticity values (E) from Tables Nos. 25-A-1 and 25-A-2 of the Uniform Building Code. For duration of load stress increases, see Section 2504 (c) 4 of the Uniform Building Code. (3) For more comprehensive tables covering a broader range of bending stress values (F_b) and modulus of elasticity values (E), other spacing of members and other conditions of loading, see U.B.C. Standard No. 25-21. (4) The spans in these tables are intended for use in covered structures or where moisture content in use does not exceed 19 percent.

Figure 28–7 A rafter span table suitable for sizing framing members to support lightweight roofing material at a pitch greater than 3/12. Notice that the values are listed in units of Fb rather than E. *Reproduced from the Dwelling Construction Under the* Uniform Building Code, 1991 edition. *Courtesy International Conference of Building Officials.*

CHAPTER

Sizing Joists and Rafters Using Span Tables Test

DIRECTIONS

Answer the questions with short, complete statements or drawings.

1. Letter your name and the date at the top of the sheet.
2. Letter the question number and provide the answer. You do not need to write out the question.

QUESTIONS

Question 28–1 List four common types of lumber used for framing throughout the country.

Question 28–2 How is fiber bending stress represented in engineering formulas?

Question 28–3 What does the term *modulus of elasticity* mean?

Question 28–4 How is modulus of elasticity represented in engineering formulas?

Question 28–5 Give the safe working stresses in fiber bending and modulus of elasticity for 2 × 6, 2 × 8, 2 × 10, and 2 × 12 for the #2 grade of the lumber used in your area.

Question 28–6 Using SPF #2, determine the size of floor joist that will be required to span 15′–0″ if spaced at 16″ o.c.

Question 28–7 You are considering the use of 2 × 8 Hem-Fir #2 floor joists to span 14′–0″. Will they work? Explain your answer. If they will fail, determine a suitable size and spacing for the span.

Question 28–8 Using 2 × 6 SP #2 at 16″ o.c. for ceiling joists, determine their maximum safe span.

Question 28–9 Using DFL #2 rafter/ceiling joist, determine the size of lumber required to span 16′–6″ with 16″ and 24″ spacings.

Question 28–10 Determine the floor joist size needed to span 13′–0″ to support a kitchen floor.

Question 28–11 Determine the smallest size of floor joist needed to span 13′–0″ to support a living room floor.

Question 28–12 What is the smallest size floor joist that can be used to span 15′–9″ beneath a kitchen?

Question 28–13 If Hem-Fir is used, what size floor joist would be needed to support a kitchen 12′–2″ wide?

Question 28–14 Use SPF to determine the smallest floor joist that could be used to span 12′–8″.

DIRECTIONS

Use UBC span tables to complete the following problems. Unless noted, use DFL #2 at standard spacing. Assume 30 lb live load for all rafters.

PROBLEMS

Problem 28–1 A living room is 17′–9″ wide. What is the smallest size SPF ceiling joist that could be used?

Problem 28–2 A contractor has bought a truckload of 2 × 6s 18′ long. Can they be used for ceiling joists spaced at 16″ o.c.?

Problem 28–3 What size southern pine rafter is needed for a home 28′ wide with a gable roof, 4/12 pitch with 235 composition shingles?

Problem 28–4 What is the maximum span allowed using 2 × 10 rafter ceiling joists for a 3/12 pitch using built-up roofing?

Problem 28–5 Determine the rafter size to be used to span 14′ and support tile roofing at a 6/12 pitch.

CHAPTER 29

Determining Simple Beams

As you advance in your architectural skills, the need to determine the size of structural members will occur frequently. There are several skills that you will need to develop in order to determine easily the size of structural members: the ability to distinguish loading patterns on the member, recognize standard engineering symbols used in beam formulas, recognize common causes of beam failure, and understand how to select beams to resist these tendencies.

▼ LOADING AND SUPPORT PATTERNS OF BEAMS

There are two common ways to load and support a beam. Loads can be uniformly distributed over the entire span of the beam or can be concentrated in one small area of a beam. In Chapter 27, uniformly distributed loads were discussed. A concentrated load is one that comes about as a result of a large load acting on only one area of a beam. Examples are a support post from an upper floor resting on a beam on a lower floor, the weight of a car being transferred through a wheel onto a floor system, or an air-conditioning unit resting on the roof members. Usually a beam is supported at each end. Common alternatives are to support a beam at the center of the span in addition to the ends or to provide support at one end and near the other end. The type of beam that extends past the support is called a *cantilevered beam*. Examples of each type of load and support system can be seen in Figure 29–1.

▼ DESIGN LOADS AND STRESSES

Design loads have been discussed in past chapters. See Figure 27–7 for a simplified list of design live (LL) and dead (DL) loads.

The allowable unit stresses for lumber have also been discussed in past chapters. Safe working loads for four common types of lumber were introduced in Figure 28–1. Those values are for 2 × material only. Figure 29–2 shows a partial listing of safe working stresses for beams. Stresses for beams are shown in a manner similar to stresses for joists. Figure 29–2 includes the now-familiar column of values for fiber bending stress (F_b) and modulus of elasticity (E) plus a column for horizontal shear (F_v).

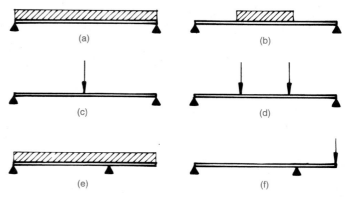

Figure 29–1   Common loading patterns on a beam include the following: (a) simple beam with a uniformly distributed load; (b) simple beam with a partially distributed load at the center; (c) simple beam with a concentrated load at the center; (d) simple beam with two equal concentrated loads placed symmetrically; (e) cantilevered beam with a uniform load; (f) cantilevered beam with a concentrated load at the free end.

Design Stress Values for Common Wood Beams			
	Fiber bending Stress, F_b	Horizontal shear, F_v	Modulus of elasticity, E
Douglas fir	1300	85	1,600,000
Eastern hemlock	1150	80	1,200,000
Western hemlock	1150	85	1,400,000
Hemlock-fir	1050	70	1,300,000
Southern pine	1200	85	1,500,000
Spruce-pine-fir	900	65	1,300,000

Figure 29–2   Partial listing of safe working values of common types of lumber used for beams. All values taken from Table 23–1–A–4 of the Uniform Building Code, 1994 edition. Courtesy The *International Conference of Building Officials*.

Loads typically affect structural members in five different ways. Each reaction can be seen in Figure 29–3. Bending or *deflection* (e) occurs as a result of gravity. As a load is placed on the beam, the beam will sag between its supports. As the span increases, the tendency to deflect increases. Deflection rarely causes a beam to break but greatly affects sheet rock, plaster, or glass which might be supported by the beam. As a beam settles, sheetrock, and plaster cracks, and doors start to stick. Two formulas will be used to determine deflection. The first is the legal limit of deflection which tells how much a building code will allow a specific beam to bend. Limits are typically set by the building code and are expressed as

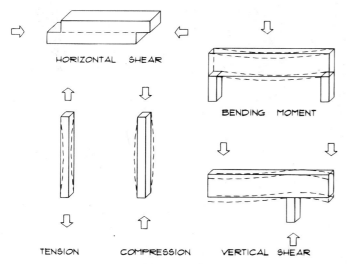

Figure 29–3 Forces acting on structural members.

a ratio of the length of the beam in inches over either 180, 240, or 360. These limits were introduced in Chapter 28. The second deflection formula determines how much a beam will sag under specific loading conditions.

Horizontal shear is a result of forces that affect the beam fibers parallel to the wood grain. Under severe bending pressure, adjacent wood fibers are pushed and pulled in opposite directions. The top portion of the beam nearest the load is under stress for compression. The edge of the beam away from the load is under stress from tension because the fibers tend to stretch as the beam deflects. The line where the compression and tension forces meet, called the *moment of inertia*, is the point at which the beam will fail. Horizontal shear has the greatest effect on beams with a relatively heavy load spread over a short span. Visualize a yardstick with supports at each end supporting a 2 lb load at the center. The yardstick will bow but probably not break. Move the supports toward the center of the beam so they are 1' apart. The beam will not bend but will be more prone to failure from breaking from the stress of horizontal shear.

Vertical shear is the tendency of a beam to fail perpendicular to the fibers of the beam from the two opposing forces. It is most typical where a beam or joist is cantilevered. The supported load causes the joist to bend about the support. The supporting post or column restricts the ability of the beam to bend.

Tension stresses attempt to lengthen a structural member. Tension is rarely a problem in residential and light commercial beam design. Tension is more likely to affect the intersection of beams than the beam itself. Beam details typically reflect a metal strap to join beams laid end to end, or a metal seat may be used to join two beams to a column.

Compression is the tendency to compress a structural member. Fibers of a beam tend to compress the portion of a beam resting on a column, but residential loads are

usually not sufficient to cause a structural problem. Posts are another area where compression can be seen. Wood is strongest along the grain. The loads in residential design typically are not large enough to cause problems in the design of residential columns. The length of the post is important in relation to the load to be supported, however. As the length of the post increases, the post tends to bend rather than compress. Structures with posts longer than 10' are often required by building departments to be approved by a structural engineer or architect. The forces of tension and compression will not be analyzed in this book.

▼ METHODS OF BEAM DESIGN

Four methods of beam design are used by professionals: a span computer, a computer program, wood design books, and, of course, the old-fashioned way of pencil, paper, and a few formulas.

A span computer is very similar to a slide rule. Figure 29–4 shows an illustration of a span computer made by the Western Wood Products Association. Beam sizes are determined by lining up values of the lumber to be used with a distance to be spanned.

Computer programs are available for many personal and business computers that will size beams. Each national wood distributor will supply programs that can determine spans and needed materials and then print out all appropriate stresses. These programs typically ask for loading information and then proceed to determine the span size within seconds. Using a computer program to solve beam spans is extremely easy, but you do have to be able to answer questions that require an understanding of the basics of beam design and loading.

Another practical method of sizing beams is through the use of books published by various wood associations.

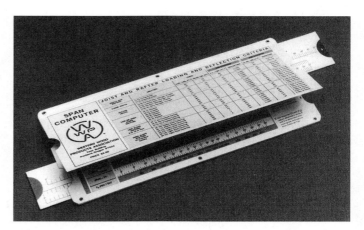

Figure 29–4 Span computer used to determine joist and beam sizes. *Courtesy Western Wood Products Association.*

Two of the most common span books used in architectural offices are *Wood Structural Design Data* from the National Forest Products Association and the *Western Woods Use Book* from the Western Wood Products Association. These books contain design information of wood members, standard formulas, and design tables that provide beam loads for a specific span. Figure 29–5 shows a partial listing for the 8′ span table. By following the instructions provided with the table, information on size, span, and loading patterns can be determined.

The final method of beam design is to use standard formulas to determine how the beam will be stressed. Figure 29–6 shows the formulas and the loading, shear, and moment diagrams for a simple beam. These are diagrams that can be drawn by the designer to determine where the maximum stress will occur. With a simple beam and a uniformly distributed load, the diagrams typically are not drawn because the results remain constant.

▼ NOTATIONS FOR FORMULAS

Standard symbols have been adopted by engineering and architectural communities to simplify the design of beams. Figure 29–7 gives a partial list of these notations. As a beginning designer, it is not important that you understand why the formulas work, but it is important that you know the notations to be used in the formulas.

▼ SIZING WOOD BEAMS

Wood beams can be determined by following these steps:

1. Determine the area to be supported by the beam.
2. Determine the weight supported by 1 linear foot of beam.
3. Determine the reactions.
4. Determine the pier sizes.
5. Determine the bending moment.
6. Determine the horizontal shear.
7. Determine the deflection.

Determining the Area to Be Supported

To determine the size of a beam, the weight the beam is to support must be known. To find the weight, find the area the beam is to support and multiply by the total loads. Figure 29–8 shows a sample floor plan with a beam of undetermined size. The beam is supporting an area 10′–0″ long (the span) multiplied by 10′–0″ wide (half of each floor joist supported). With an LL of 40 lb and a DL of 10 lb, this beam is holding a total of 5000 lb, which represents the total weight supported by the beam. In beam formulas, the total weight is represented by the letter W.

Determining Linear Weight

In some formulas only the weight per linear foot of beam is desired. This is represented by the letter w. In Figure 29–9, it can be seen that w is the product of the area to be supported, multiplied by the weight per sq ft. It can also be found by dividing the total weight (W) by the length of the beam. In our example, if the span (represented by L) is 10′, w = 5000/10 or 500 lb.

Determining Reactions

Even before the size of the beam is known, the supports for the beam can be determined. These supports for the beam are represented by the letter R, for reactions. The letter V is also sometimes used in place of the letter R. On a simple beam, half of the weight it is supporting will be dispersed to each end. In the example, W = 5000 lb and R = 2500 lb. At this point W, w, and R have been determined (W = 5000 lb, w = 500 lb, and R = 2500 lb.)

Determining Pier Sizes

The reaction from the beam will be transferred by a post to the floor and foundation system. Usually a concrete pier is used to support the loads at the foundation level. To determine the size of the pier, the working stress of the concrete and the bearing value of the soil must be taken into account.

Usually concrete with a working stresss of 2000 to 3500 psi is used in residential construction. The working stress of concrete specifies how much weight in pounds can be supported by each square inch (psi) of concrete surface. If each square inch of concrete can support 2500 lb, only 1 sq in. of concrete would be required to support a load of 2500 lb. The bearing value of the soil must also be considered.

See Figure 29–10 for safe soil-loading values. Most building departments use 2000 psf for the assumed safe working value of soil.

Don't skim over these numbers. The concrete is listed in pounds per square inch. Soil is listed in pounds per square foot. A pier supporting 2500 lb must be divided by 2000 lb (the soil-bearing value) to find the area of the pier needed. This will result in an area of 1.25 sq ft of concrete needed to support the load. See Figure 29–11 to determine the size of pier needed to obtain the proper soil support area.

Determining the Bending Moment

A *moment* is the tendency of a force to cause rotation about a certain point. In figuring a simple beam, W is the force and R is the point around which the force rotates. Determining the bending moment will calculate the size of the beam needed to resist the tendency for W to rotate around R.

To determine the size of the beam required to resist the force (W), use the following formula:

$$M = \frac{(w)(l)}{8}$$

WOOD BEAMS—SAFE LOAD TABLES

Symbols used in the tables are as follows:

F_b = Allowable unit stress in extreme fiber in bending, psi.

W = Total uniformly distributed load, pounds

w = Load per linear foot of beam, pounds

F_v = Horizontal shear stress, psi, induced by load W

E = Modulus of elasticity, 1000 psi, induced by load W for $l/360$ limit

Beam sizes are expressed as nominal sizes, inches, but calculations are based on net dimensions of S4S sizes.

SIZE OF BEAM		F_b									
		900	1000	1100	1200	1300	1400	1500	1600	1800	2000
8'- 0" SPAN											
2 x 14	W	3291	3657	4023	4389	4754	5120	5486	5852	6583	7315
	w	411	457	502	548	594	640	685	731	822	914
	F_v	124	138	151	165	179	193	207	220	248	276
	E	489	543	597	652	706	760	815	869	978	1086
6 x 8	W	3867	4296	4726	5156	5585	6015	6445	6875	7734	8593
	w	483	537	590	644	698	751	805	859	966	1074
	F_v	70	78	85	93	101	109	117	125	140	156
	E	864	960	1055	1152	1247	1343	1439	1535	1727	1919
4 x 10	W	3743	4159	4575	4991	5407	5823	6238	6654	7486	8318
	w	467	519	571	623	675	727	779	831	935	1039
	F_v	86	96	105	115	125	134	144	154	173	192
	E	700	778	856	934	1011	1089	1167	1245	1401	1556
3 x 12	W	3955	4394	4833	5273	5712	6152	6591	7031	7910	8789
	w	494	549	604	659	714	769	823	878	988	1098
	F_v	105	117	128	140	152	164	175	187	210	234
	E	576	640	704	768	832	896	959	1024	1151	1279
8 x 8	W	5273	5859	6445	7031	7617	8203	8789	9375	10546	11718
	w	659	732	805	878	952	1025	1098	1171	1318	1464
	F_v	70	78	85	93	101	109	117	125	140	156
	E	864	960	1055	1151	1247	1343	1439	1535	1727	1919
3 x 14	W	5486	6095	6705	7315	7924	8534	9143	9753	10972	12191
	w	685	761	838	914	990	1066	1142	1219	1371	1523
	F_v	124	138	151	165	179	193	207	220	248	276
	E	489	543	597	652	706	760	815	869	978	1086
4 x 12	W	5537	6152	6767	7382	7998	8613	9228	9843	11074	12304
	w	692	769	845	922	999	1076	1153	1230	1384	1538
	F_v	105	117	128	140	152	164	175	187	210	234
	E	576	640	703	768	832	896	960	1023	1151	1279
6 x 10	W	6204	6894	7583	8272	8962	9651	10341	11030	12409	13788
	w	775	861	947	1034	1120	1206	1292	1378	1551	1723
	F_v	89	98	108	118	128	138	148	158	178	197
	E	682	757	833	909	985	1061	1136	1212	1364	1515
3 x 16	W	7267	8075	8882	9690	10497	11305	12112	12920	14535	16150
	w	908	1009	1110	1211	1312	1413	1514	1615	1816	2018
	F_v	142	158	174	190	206	222	238	254	285	317
	E	424	472	519	566	613	660	708	755	849	944
4 x 14	W	7973	8859	9745	10631	11517	12403	13289	14175	15946	17718
	w	996	1107	1218	1328	1439	1550	1661	1771	1993	2214
	F_v	126	140	154	168	182	196	210	225	253	281
	E	480	533	586	640	693	746	800	853	960	1066

Figure 29–5 Partial listing of a typical span table. Once W, w, F_b, E, and F_v are known, spans can be determined for a simple beam. See Appendix A for complete listing. *From* Wood Structural Design Data. *Courtesy National Forest Products Association.*

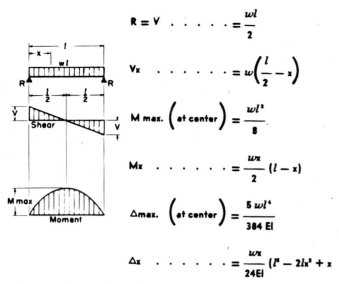

$$R = V \cdots \cdots = \frac{wl}{2}$$

$$V_x \cdots \cdots = w\left(\frac{l}{2} - x\right)$$

$$M \text{ max. }\left(\text{at center}\right) = \frac{wl^2}{8}$$

$$M_x \cdots \cdots = \frac{wx}{2}(l - x)$$

$$\Delta \text{max. }\left(\text{at center}\right) = \frac{5 wl^4}{384 \, EI}$$

$$\Delta_x \cdots \cdots = \frac{wx}{24EI}(l^3 - 2lx^2 + x$$

Figure 29–6 Loading, shear, and moment diagrams for simple beam with uniform loads showing where stress will affect a beam and the formulas for computing these stresses. *From Wood Structural Design Data. Courtesy National Forest Products Association.*

Beam Formula Notations		
b	=	breadth of beam in inches
d	=	depth of beam in inches
D	=	deflection due to load
E	=	modulus of elasticity
F_b	=	allowable unit stress in extreme fiber bending
F_v	=	unit stress in horizontal shear
I	=	moment of inertia of the section
l	=	span of beam in inches
L	=	span of beam in feet
M	=	bending or resisting moment
P	=	total concentrated load in pounds
S	=	section modulus
V	=	end reaction of beam
W	=	total uniformly distributed load in pounds
w	=	load per linear foot of beam in pounds

Figure 29–7 Common notations used in beam formulas.

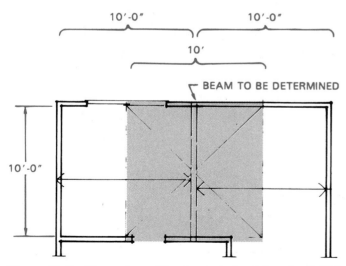

10'-0" 10'-0"

10'

BEAM TO BE DETERMINED

10'-0"

Figure 29–8 Floor plan with a beam to be determined. This beam is supporting an area of 100 sq ft, with an assumed weight of 50 lb per sq ft, W = 5000 lb.

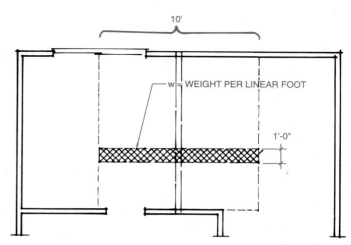

Figure 29–9 Determining linear weights. An area of 10 sq ft, with an assumed weight of 50 lb per sq ft, w = 500 lb.

TABLE NO. 29-B—ALLOWABLE FOUNDATION AND LATERAL PRESSURE

CLASS OF MATERIALS[2]	ALLOWABLE FOUNDATION PRESSURE LBS. SQ. FT.[3]	LATERAL BEARING LBS./SQ. FT. FT. OF DEPTH BELOW NATURAL GRADE[4]	LATERAL SLIDING[1]	
			COEF-FICIENT[5]	RESISTANCE LBS./SQ. FT.[6]
1. Massive Crystalline Bedrock	4000	1200	.79	
2. Sedimentary and Foliated Rock	2000	400	.35	
3. Sandy Gravel and/or Gravel (GW and GP)	2000	200	.35	
4. Sand, Silty Sand, Clayey Sand, Silty Gravel and Clayey Gravel (SW, SP, SM, SC, GM and GC)	1500	150	.25	
5. Clay, Sandy Clay, Silty Clay and Clayey Silt (CL, ML, MH and CH)	1000[7]	100		130

[1]Lateral bearing and lateral sliding resistance may be combined.
[2]For soil classifications OL, OH and PT (i.e., organic clays and peat), a foundation investigation shall be required.
[3]All values of allowable foundation pressure are for footings having a minimum width of 12

Figure 29–10 Safe soil-bearing values. *Reproduced from the Uniform Building Code, 1994 edition. Copyright 1994, with the permission of the publisher, International Conference of Building Officials.*

Pier Areas and Sizes	
ROUND PIERS	**SQUARE PIERS**
15" DIA. = 1.23 SQ FT	15" SQ = 1.56 SQ FT
18" DIA. = 1.77 SQ FT	18" SQ = 2.25 SQ FT
21" DIA. = 2.40 SQ FT	21" SQ = 3.06 SQ FT
24" DIA. = 3.14 SQ FT	24" SQ = 4.00 SQ FT
27" DIA. = 3.97 SQ FT	27" SQ = 5.06 SQ FT
30" DIA. = 4.90 SQ FT	30" SQ = 6.25 SQ FT
36" DIA. = 7.07 SQ FT	36" SQ = 9.00 SQ FT
42" DIA. = 9.60 SQ FT	42" SQ = 12.25 SQ FT

Figure 29–11 Common pier areas and sizes. By dividing the load to be supported by the soil-bearing pressure, the area of the concrete pier can be determined. Areas are shown for common pier sizes.

Once the moment is known, it can then be divided by the F_b of the wood to determine the size of beam required to resist the load. This whole process can be simplified by using the formula:

$$S = \frac{(3)(w)(L^2)}{(2)(F_b)}$$

Use the values for douglas fir from Figure 29–2, and apply this formula to the beam span from Figure 29–8:

$$S = \frac{(3)(500)(100)}{(2)(1300)} = \frac{150,000}{2600} = 57.69$$

This formula will determine the minimum section modulus required to support a load of 5000 lb. Look at the table that is Figure 29–12 to determine the size of lumber needed for the beam. A beam must be selected that has an S value larger than the S value found. By examining the S column of the table, you will find that a 4 × 12 has an S of 73.8 and 6 × 10 has an S value of 82.7. Each of these sizes of beams is suitable to resist the forces of bending moment.

Determining Horizontal Shear Values

To compute the size of a beam required to resist the forces of horizontal shear, use the following formula:

$$F_v = \frac{3(V)}{(2)(b)(d)} = \# < 85 \quad \left\| \begin{matrix} \text{safe design} \\ \text{value for DFL} \end{matrix} \right\|$$

For a simple beam, V = R. To determine b and d, use the actual size of the beam determined for bending. For this example, use the 4 × 12 value. The actual size of a 4 × 12 is 3.5″ × 11.25″. Figure 29–2 shows the product of (2bd) to be 78.75. These values can be inserted into the formula to find the shear on the beam.

$$F_v = \frac{(3)(V)}{(2)(b)(d)} = \frac{3 \times 2500}{78.75} = \frac{7500}{78.75} = 95.238 > 85 \quad \text{beam fails}$$

This answer is meaningless unless you know the safe shear value in horizontal shear for the type of wood in question. The safe limit for lumber in horizontal shear can be determined from the table in Figure 29–2. The value for DFL is 85. If the product of the formula is larger than the safe working stress, the beam fails. In the example above, 95.238 exceeds the limit for douglas fir. At this point, the designer must select a larger beam, and the formula must be reworked. The figures on the top of the formula will remain the same because these values are the result of the load, which has not changed. Insert the values for a 4 × 14 (2 · 3.5 · 13.5) beam. Now the formula should read:

$$F_v = \frac{7500}{94.5} = 79.37 < 85 \text{ beam safe}$$

Since this new value is less than 85, the beam works.

A 4 × 12 was found to have adequate stiffness to resist the load but not enough size to prevent horizontal shear from occurring. Either a 4 × 14 or a 6 × 10 beam will be needed. Even though the 6 × 10 has not been proven by the formula, you should be able to tell that it will work. The value in the (2bd) column of Figure 29–12 for a 6 × 10 beam is 104.5, which is larger than the value for the 4 × 14 beam. When the 104.5 value is inserted into the formula, it will produce a smaller F_v value, and so the 6 × 10 is a safer beam.

Another way to determine F_v is to multiply V times 3 and then divide the F_v value for the lumber being used.

$$F_v = \frac{3V}{85} = \# < \text{(2bd) value for selected bm from Figure 29–12}$$

$$F_v = \frac{3 \times 2500}{85} = \text{(minimum value needed. 2bd value must exceed the value.) } 88.23 = 4 \times 14 \text{ or } 6 \times 10$$

This will produce a minimum area required to resist the stress in horizontal shear. You have now determined

PROPERTIES OF STRUCTURAL LUMBER						
NOMINAL SIZE	(b)(d)	(2)(b)(d)	S	A	I	(384)(E)(I) *
2× 6	1.5 × 5.5	16.5	7.6	8.25	20.8	12,780
2× 8	1.5 × 7.25	21.75	13.1	10.875	47.6	29,245
2×10	1.5 × 9.25	27.75	21.4	13.875	98.9	60,764
2×12	1.5 × 11.25	33.75	31.6	16.875	177.9	109,302
2×14	1.5 × 13.25	39.75	43.9	19.875	290.8	178,668
4× 6	3.5 × 5.5	38.5	17.6	19.25	48.5	29,798
4× 8	3.5 × 7.25	50.75	30.7	25.375	111.0	68,198
4×10	3.5 × 9.25	64.75	49.9	32.375	230.8	141,804
4×12	3.5 × 11.25	78.75	73.8	39.375	415.3	255,160
4×14	3.5 × 13.5	94.5	106.3	47.250	717.6	440,893
6× 8	5.5 × 7.5	82.5	51.6	41.25	193.4	118,825
6×10	5.5 × 9.5	104.5	82.7	52.25	393.0	241,459
6×12	5.5 × 11.5	126.5	121.2	63.25	697.1	428,298
6×14	5.5 × 13.5	148.5	167.1	74.25	1127.7	692,859

*384 × EXI values are listed in units per million, with E value assumed to be 1.6 for DFL.

Figure 29–12 Structural properties of wood beams. The (2)(b)(d) and the (384)(E)(I) columns are part of formulas that are always needed. These two columns are set up for douglas fr. For different types of wood, use different values for these columns.

W, w, R, V, S, and F_v. The last value to be determined is the stress for modulus of elasticity.

Determining Deflection

Modulus of elasticity is the deflection, or sag, in a beam. Deflection limits are determined by building codes and are expressed as a fraction of an inch in relation to the span of the beam in inches. Limits set by the Uniform Building Code are: for floors and ceilings, 1/360; roofs under 3/12 pitch, 1/240; and roofs over 3/12 pitch, 1/180.

To determine deflection limits, the maximum allowable limit must first be known. Use the formula:

$$D_{max} = \frac{L \times 12}{360}$$

This formula requires the span in feet (L) to be multiplied by 12(12 inches per foot) and then divided by 360 (the safe limit for floors and ceiling).

In the example used in this chapter, L equals 10'. The maximum safe limit would be:

$$D_{max.} = \frac{L \times 12}{360} = \frac{10 \times 12}{360} = D = \frac{120}{360} = 0.33''$$

The maximum amount the beam is allowed to sag is 0.33". Now you need to determine how much the beam will actually sag under the load it is supporting.

The values for E and I need to be determined before the deflection can be determined. These values can be found in Figure 29–12. You will also need to know l^3 and W, but these values are not available in tables. W has been determined. For our example:

$$l = 10 \times 12 = 120''$$
$$l^3 = (l)(l)(l) = (120)(120)(120) = 1,728,000$$

Because the E value was reduced from 1,600,000 to 1.6, change the l^3 value from 1,728,000 to 1.728. Each of these numbers is much easier to use when calculating.

To determine how much a beam will sag, use the following formula:

$$D = \frac{5(W)(l^3)}{(384)(E)(I)} \text{ or } D = 22.5 \frac{(W)(l^3)}{(E)(I)}$$

The values determined thus far are W = 5000, w = 500, R = V = 2500, L = 10, l = 120, l^3 = 1.728 and I = 393 (see Figure 29–2, column I). Find also the value of (384)(E)(I) for a 6 × 10 beam from the table in Figure 29–2 (241,459). Now insert the values into the formula.

$$D = \frac{5(W)(L^3)}{384(E)(I)} = \frac{5 \times 5000 \times 1.728}{241,459} = \frac{43,200}{241,459} = 0.178''$$

Because 0.178" is less than the maximum allowable deflection of 0.33", a 6 × 10 beam can be used to support the loads.

▼ REVIEW

In what may have seemed like an endless string of formulas, tables, and values, you have determined the loads and stresses on a beam 10' long. Seven basic steps were required as follows:

1. Determine the values for W, w, L, l, l^3 and R.
   W = area to be supported × weight (LL + DL)
   w = W/L
   L = span in feet
   l = span in inches
   l^3 = (l)(l)(l)
   R = V = W/2
2. Determine S:

$$S = \frac{(3)(w)(L^2)}{(2)(F_b)}$$

3. Determine F_v:

$$F_v = \frac{(3)(V)}{(2)(b)(d)} = 85 \text{ or } \frac{(3)(V)}{85} = (2)(b)(d)$$

4. Determine the value for D_{max}

$$D = l/360 \text{ or } l/240 \text{ or } l/180.$$

5. Determine D:

$$D = \frac{(5)(W)(l^3)}{(384)(E)(I)}$$

6. Determine post supports: R = W/2
7. Determine piers size: R/soil bearing pressure

Figure 29–13 shows a floor plan with a ridge beam that needs to be determined. The beam will be SPF with no snow loads. Using the seven steps, determine the size of the beam.

Step 1. Determine the values.
         W = 10 × 12 × 40 = 4800 lb

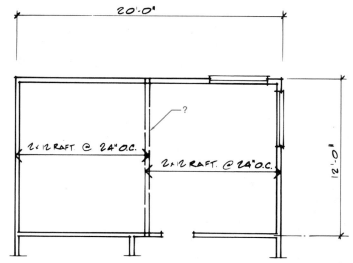

Figure 29–13 Sample floor plan with a beam of undetermined size.

$w = W/L = 4800/12 = 400$ lb

$R = V = 2400$

$L = 12'$

$L^2 = 144$

$l = 12' \times 12'' = 144$

$l^3 = 2.986$

$F_b = 900$

$F_v = 65$ max.

$E = 1.3$ max.

Step 2. Determine the section modulus.

$$S = \frac{(3)(2)(L^2)}{(2)(F_b)} = \frac{3 \times 400 \times 144}{2 \times 900} = \frac{172{,}800}{1800} = 96$$

Use a 4 × 14 or a 6 × 12 to determine F_v.

Step 3. Determine horizontal shear. The value from Table 29–2 is 65.

$$4 \times 14 \; F_v = \frac{(3)(V)}{(2)(b)(d)} = \frac{3 \times 2400}{2 \times 3.5 \times 13.5} = \frac{7200}{94.5} = 76.19 > 65$$

Since the computed value is larger than the table value, the 4 × 14 beam is inadequate. Try a 6 × 12.

$$6 \times 12 \; F_v = \frac{7200}{126.5} = 56.92$$

Finally, this beam works. Now solve for deflection.

Step 4. Determine D_{max}.

$$D_{max} = \frac{1}{180} = \frac{144}{180} = 0.8''$$

Step 5. Determine D:

$$D = \frac{(5)(W)(l^3)}{(384)(E)(I)} \quad \frac{5 \times 4800 \times 2.986}{384 \times 1.3 \times 697} = \frac{71{,}664}{347{,}942}$$

$$D = 0.20$$

Step 6. Determine reactions:

$$R = W/2 = 2400$$

Step 7. Determine piers. R/soil value (assume 2000 lb)

$$\frac{2400}{2000} = 1.2 \text{ sq ft}$$

According to the table (Figure 29–11) use either a 15″ diameter or 15″ square pier.

▼ SPAN TABLES

You have already been introduced to span tables. Now that you have labored sizing a beam by solving for S, F_v, and E, you can better understand why span tables should be used whenever possible. CABO now includes charts for sizing very simple beams for one- or two-story residences. The National Forest Products Association offers tables that are easy to use and accepted by most professionals as a design standard. Appendix A shows a partial listing of these tables for spans ranging from 6′–0″ through 16′–0″. Although shorter beams may be required in a residence, generally only the F_v value will need to be determined to find the beam size. Beams longer than 16′ are typically holding enough weight that a glu-lam beam should be used.

To use the information in Appendix A, "W" must be known. The type of wood must also be known so that the maximum values for F_b, F_v, and E can be used. In the beam solved in the last example, this value would be W = 4800#. If SPF is to be used, F_b = 900, F_v = 65, and E = 1.3. Because the beam is 12′ long, look in the 12′ span section of the table. Because the F_b is 900, your answer will come from this column. Proceed down this column until you find a W value that is equal to or larger than the W to be supported (4800 lb). 4 × 14 (5315) and 6 × 12 (6061) will work. Check the F_v values for each. This number must be less than 65. In our example, both seem to fail, but *this is not the case*. Consider what the 6 × 12 column has told you. If you were to have a 12′-long beam, and supported 6061 lbs, the beam would be under 71 units of stress in F_v. The beam only needs to resist 4800 lb, not 6061. In this case, you still have to figure F_v by using the formula $\frac{3v}{65}$ = 2bd. You determined this in the last example and know that a 6 × 12 will work. To determine if D is suitable, examine the D value for the 6 × 12. You will see the number 845. This represents .845 unit of stress. The maximum E value for SPF is 1.3. Since .845 is less than 1.3, the beam works.

Try solving the same beam for douglas fir. The values to be used are W = 4800, F_b = 1300, F_v = 85, and E = 1.6. Enter the 1300 F_b column until a weight is found equal to or greater than 4800 lb. You could use a 4 × 12. Notice the F_v is 101. Because 101 is greater than 85, the beam appears to fail. Instead of working the F_v formula, there is a short-cut. Because the beam only supports 4800 lb, go to the left of the W values until the W value is closer to 4800. As you move to the left in the W column, you'll find that at 4921 lb, the beam has F_v = 93. Since your beam holds only 4800 lb, you can continue to the left. At 4511 lb, the beam has an F_v of 85. 85 is the maximum F_v value, but at this point the 12′ beam cannot support the desired W of 4800.

Try a 4 × 14. Enter the F_b 1300 column. The maximum supported weight is 7678 lb. Using the W values for a 4 × 14, move to the left until you find a number close to 4800. The F_b 900 column is as close as you can get with a W value of 5315. A beam holding 5315 lb will have 84 units of F_v stress. A 4 × 14 beam is safe. Its E value is .720, less than 1.6, so E is safe also. Compare the values for a 6 × 10, and you will find that it, too, is safe. Enter the table in the 1300 F_b column and you will find the beam could support 5974 lb. Move to the left, and you will see the weight to be supported falls between 4596 and 5055 lb. Each has an F_v value less than tthe maximum of 85. Staying in the 5055 column (which is more than our

SPECIES AND COMMERCIAL GRADE	EXTREME FIBER BENDING F_b	HORIZONTAL SHEAR F_v	MODULUS OF ELASTICITY
DEL #1 22FV4 DF/DF	1300	85	1,600,000
	2200	165	1,700,000
Hem-Fir #1 22F E2 HF/HF	1050	70	1,300,000
	2200	155	1,400,000
SPE 22F-E-1 SP/SP	900	65	1,300,000
	2200	200	1,700,000

Figure 29–14 Comparative values of common framing lumber with laminated beams of equal materials. Values based on the 1991 *Uniform Building Code*.

PROPERTIES OF GLU-LAM BEAMS					
SIZE (b)(d)	S	A	(2)(b)(d)	I	(384)(E)(I) *
31/8 × 9.0	42.2	28.1	56.3	189.8	123,901
31/8 × 10.5	57.4	32.8	65.6	301.5	196,819
31/8 × 12.0	75.0	37.5	75.0	450.0	293,760
31/8 × 13.5	94.9	42.2	84.4	640.7	418,249
31/8 × 15.0	117.2	46.9	93.8	878.9	573,746
51/8 × 9.0	69.2	46.1	92.0	311.3	203,217
51/8 × 10.5	94.2	53.8	107.6	494.4	322,744
51/8 × 12.0	123.0	61.5	123.0	738.0	481,766
51/8 × 13.5	155.7	69.2	138.4	1,050.8	685,962
51/8 × 15.0	192.2	76.9	153.8	1,441.4	940,946
51/8 × 16.5	232.5	84.6	169.0	1,918.5	1,252,397
63/4 × 10.5	124.0	70.9	141.8	651.2	425,103
63/4 × 12.0	162.0	81.0	162.0	972.0	634,522
63/4 × 13.5	205.0	91.1	182.6	1,384.0	903,475
63/4 × 15.0	253.1	101.3	202.0	1,898.4	1,239,276
63/4 × 16.5	306.3	111.4	222.8	2,526.8	1,649,495

*All values for 384 × E × I are written in units per million. All E values are figured for Doug fir @ E = 1.7. Verify local conditions.

Figure 29–15 Structural properties of glu-lam beams. The (2)(b)(d) and the (384)(E)(I) columns are set up for douglas fir. If a different type of wood is to be used, you must use different values for these columns.

beam will have to support), the E value is 1.250, which is less than the maximum of 1.6 for douglas fir.

Laminated Beams

Two beam problems have now been solved using the tables that are found in this chapter. Both beams were chosen from standard dimensioned lumber. Often glulams are used because of their superior strength. Using a glu-lam can greatly reduce the depth of a beam as compared to conventional lumber. Glu-lam beams are determined in the same manner as standard lumber, but different values are used. See Figure 29–14. Figure 29–15 gives values for glu-lams made of douglas fir. You will also need to consult the building code for safe values for glu-lam beams. For beams constructed of douglas fir, the values are F_b = 2200, F_v = 165, and E = 1.7.

▼ SIMPLE BEAM WITH A LOAD CONCENTRATED AT THE CENTER

Often in light construction, load-bearing walls of an upper floor may not line up with the bearing walls of the floor below, as shown in Figure 29–16. This type of loading occurs when the function of the lower room dictates that no post be placed in the center. Since no post can be used on the lower floor, a beam will be required to span from wall to wall, and be centered below the upper level post. This beam will have no loads to support other than the weight from the post.

In order to size the beam needed to support the upper area, use the formulas presented in Figure 29–17. Notice that a new symbol, P, has been added to represent a point, or concentrated load. The point load is the sum of the reactions from each end of the beam.

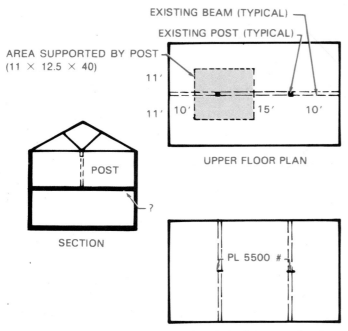

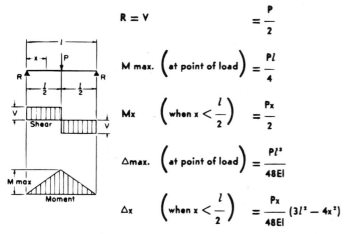

Figure 29–16 An engineer's sketch to help determine loads on a beam showing a load concentrated at the center.

Figure 29–17 Shear and moment diagrams for a simple beam with a load concentrated at the center. *Data reproduced courtesy Western Wood Products Association.*

22' BEAM @ LIVING RM.

$P = W = 5500$ # $V = 2750$ $L = 22'$ $l = 264$ $l^3 = 18,399744$

$M = \dfrac{(P)(l)}{4}$ • $M = \dfrac{(5500)(264)}{4}$ =

$\dfrac{1,452,000}{4} = 363,000 = M$

$S = \dfrac{M}{fb}$ • $\dfrac{363,000}{2200} = S = 165$ USE 5⅛ × 15 (S=192 or 6¾ × 13.5 (S=205

$fv = \dfrac{(3)(V)}{165} = 2bd = \dfrac{(3)(2750)}{165} = \dfrac{8250}{165} = 50$ BOTH BM. OK.

$D_{max.} = \dfrac{l}{360} = \dfrac{264}{360} .73$ $D = \dfrac{(P)(l^3)}{(48)(E)(I)}$ =

$D = \dfrac{(5500)(18.4)}{(48×1.7)(1384)} = \dfrac{101200}{112,934} = .89 > .73 = $ FAIL

? 5⅛ × 16½ = $I = 1441.4 = \dfrac{101200}{117,618} = .86 > .73 = $ FAIL

? 6¾ × 15 = $I = 1898.4 = \dfrac{101200}{154,909} = 65 < .73 = $ USE 6¾ × 15 f 2200 GLU-LAM BEAM

Figure 29–18 Worksheet for a beam with a load concentrated at the center. It is a good idea to include the location of the beam, a sketch of the loading, needed formulas, and the selected beam.

In Figure 29–18 three different beams have been selected that have the minimum required section modulus. Once the section modulus is known, the beam can be checked for horizontal shear and deflection. Because of the length of the beam, horizontal shear will not be a factor.

▼ SIMPLE BEAM WITH A LOAD CONCENTRATED AT ANY POINT

Figure 29–19 shows a sample floor plan that will result in concentrated off-center loads at the lower floor level. Figure 29–20 shows the diagrams and formulas that would be used to determine the beams of the lower floor plan. Figure 29–21 shows the worksheet for this beam.

▼ CANTILEVERED BEAM WITH A UNIFORM LOAD

This loading pattern typically occurs where floor joists of a deck cantilever past the supporting wall. One major difference of this type of loading is that the live loads of decks are almost double the normal floor live loads. Figure 29–22 shows the diagrams and formulas for a cantilevered beam or joist. Figure 29–23 shows the worksheet to determine the spacing of floor joist at a three ft deck cantilever.

Figure 29–18 shows a sample worksheet for this beam. Notice that the procedure used to solve for this beam is similar to the one used to determine a simple beam with uniform loads. With this procedure, you must determine the amount of point loads for the upper beams. The sketch shows the upper floor plan with each point load.

Once the point loads are known, determine M. Once M is known, divide it by the F_b value of the lumber to be used. This will provide the S value for the beam. Try to use a beam that has a depth equal to the depth of the floor joist.

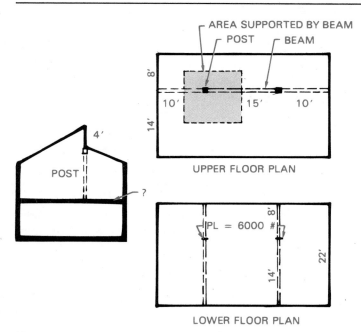

Figure 29–19 An engineer's sketch for a beam with a load concentrated at any point on the beam.

22' SPAN @ LOWER FLOOR

5500#

$a = 96$ $b = 168$

$l = 264$

R^1 R^2

$W = (4 + 7)(5 + 7.5)(40^\# = W = 5500^\#$

$l = 264''$

$R_1 = \dfrac{(P)(b)}{l} = \dfrac{(5500)(168)}{264} = \dfrac{924,000}{264} = 3500^\# = R_1$

$R_2 = P - R_1 = 5500 - 3500 = 2000 = R_2$

$M = \dfrac{(P)(a)(b)}{l} = \dfrac{(5500)(96)(168)}{264} = \dfrac{88,704,000}{264} = 336,000 = M$

$S = \dfrac{M}{Fb} = \dfrac{336,000}{2200} = 152.7$ use? $5\frac{1}{8} \times 13\frac{1}{2}$ or $6\frac{3}{4} \times 12$

$fv = \dfrac{(3)(V)}{165} = \dfrac{(3)(3500)}{165} = \dfrac{10,500}{165} = 63.6$ BOTH BEAMS OK.

$D = \dfrac{l}{360} = \dfrac{264}{360} = .73$ max. $E = \dfrac{(P \times a \times b \times a + 2b)\sqrt{3a(a+2b)}}{27(E)(l)(I)} =$

$D = \dfrac{(5500 \times 96 \times 168 \times 432)\sqrt{(288 \times 432)}}{(27 \times 1.7 \times 264)(I) = (12118)(I)} = \dfrac{13,516,525}{12,117.6\ (I)} =$

? $5\frac{1}{8} \times 13\frac{1}{2} = \dfrac{13,516,525}{12,733,174} = 1.06 > .75 = $ fail

? $5\frac{1}{8} \times 15 = \dfrac{13,516,525}{17,466,308} = .77 > .75 = $ fail

? $6\frac{3}{4} \times 15 = \dfrac{13,516,525}{23,004,051} = .58 < .750K$ | USE $6\frac{3}{4} \times 15$ f2200 GLU LAM BEAM |

Figure 29–21 Worksheet for a beam with a load concentrated off center.

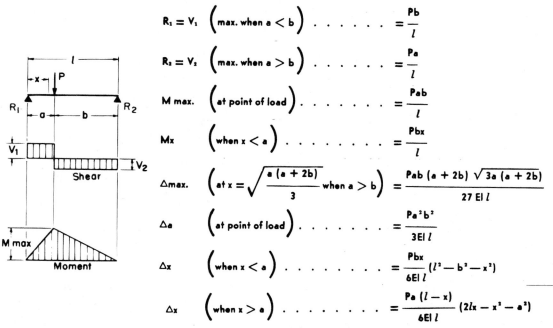

$R_1 = V_1$ (max. when $a < b$)	$= \dfrac{Pb}{l}$
$R_2 = V_2$ (max. when $a > b$)	$= \dfrac{Pa}{l}$
M max. (at point of load)	$= \dfrac{Pab}{l}$
M_x (when $x < a$)	$= \dfrac{Pbx}{l}$
Δmax. $\left(\text{at } x = \sqrt{\dfrac{a(a+2b)}{3}} \text{ when } a > b\right)$	$= \dfrac{Pab(a+2b)\sqrt{3a(a+2b)}}{27\,EI\,l}$
Δ_a (at point of load)	$= \dfrac{Pa^2b^2}{3EI\,l}$
Δ_x (when $x < a$)	$= \dfrac{Pbx}{6EI\,l}(l^2 - b^2 - x^2)$
Δ_x (when $x > a$)	$= \dfrac{Pa(l-x)}{6EI\,l}(2lx - x^2 - a^2)$

Figure 29–20 Shear and moment diagrams for a simple beam with a load concentrated at any point on the beam. *Courtesy Western Wood Products Association.*

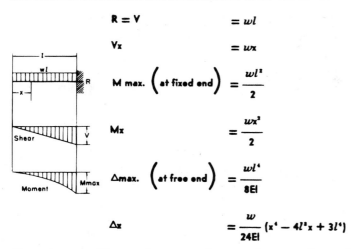

$$R = V = wl$$

$$V_x = wx$$

$$M \text{ max.} \left(\text{at fixed end} \right) = \frac{wl^2}{2}$$

$$M_x = \frac{wx^2}{2}$$

$$\Delta \text{max.} \left(\text{at free end} \right) = \frac{wl^4}{8EI}$$

$$\Delta_x = \frac{w}{24EI} (x^4 - 4l^3x + 3l^4)$$

Figure 29–22 Shear and moment diagrams for a cantilevered beam with a uniform load. *Data reproduced courtesy Western Wood Products Association.*

$$R = V = P$$

$$M \text{ max.} \left(\text{at fixed end} \right) = Pl$$

$$M_x = Px$$

$$\Delta \text{max.} \left(\text{at free end} \right) = \frac{Pl^3}{3EI}$$

$$\Delta_x = \frac{P}{6EI} (2l^3 - 3l^2x + x^3)$$

Figure 29–24 Shear and moment diagrams for a cantilevered beam with a concentrated load at the free end. *Data reproduced courtesy Western Wood Products Association.*

FLOOR JOIST, 36" CANTILEVER

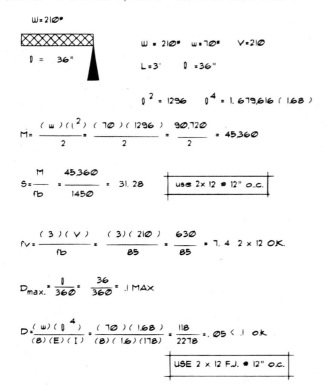

Figure 29–23 Worksheet for a cantilevered beam with a uniform load.

24" CANTILEVER @ WALL

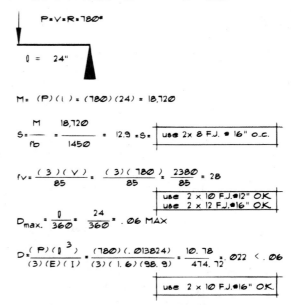

Figure 29–25 Worksheet for a cantilevered beam with a concentrated load at the free end.

▼ CANTILEVERED BEAM WITH A POINT LOAD AT THE FREE END

This type of beam results when a beam or joist is cantilevered and is supporting a point load. Floor joists that

are cantilevered and support a wall and roof can be sized by using the formulas in Figure 29–24. Figure 29–25 shows the worksheet for determining the size and spacing of floor joists to cantilever two ft and support a bearing wall.

Determining Simple Beams Test

DIRECTIONS

Answer the questions with short complete statements.

1. Letter your name, Determining Simple Beams with Nonuniform Loads Test, and the date at the top of the sheet.
2. Letter the question number and provide the answer. You do not need to write out the question.

QUESTIONS

Question 29–1   What is a simple beam?

Question 29–2   Explain the difference between a uniform and a concentrated load.

Question 29–3   List three common categories of stress that must be known before a beam size can be determined.

Question 29–4   Define the following notations: W, w, R, L, I, E, I, S, F_v, and F_b.

Question 29–5   List two methods of determining beams other than using formulas.

Question 29–6   List the values for W, w, and R for a floor beam that has an L of 10′ and is centered in a room 20′ wide.

PROBLEMS

Problem 29–1   A 4 × 6 DFL beam, with an L of 6′, and a W of 2800 lb, will be used to span an opening for a window. Will it fail in F_v?

Problem 29–2   A 4 × 14 DFL is being used as a ridge beam. L = 12′, W = 4650 lb. Will this beam have a safe E value? Assume D_{max} = 1/360.

Problem 29–3   If the soil bearing pressure is 1500 psf and the concrete has a strength of 2500 psi, what size footing is needed to support a load of 4600 lb?

Problem 29–4   What size DFL #2 girder will be required if w = 600 and L = 6′? Determine all necessary values to find the needed S, F_v and E stresses.

Problem 29–5   Determine the minimum size beam needed to span 14′ with a concentrated load of 3200 lb at the center. Assume the beam to be DFL. What size piers will be needed to resist the reactions if soil loads are 2000 psf?

Problem 29–6   Using southern pine, determine the size of beam needed to cantilever 3.5′ and support a concentrated load of 4000 lb at the free end.

Problem 29–7   Using SPF, determine what size beam will be needed to span 12′ with a concentrated load of 2500 lb placed 3′ from one end. What size piers will be required to support the loads if the soil pressure is 1500 psf?

Problem 29–8   Determine the minimum size beam needed to span 14′ with a concentrated load of 2200 lb in the center. Assume the beam to be DFL. What size piers will be needed to resist the reaction if soil loads are 2500 psf?

Problem 29–9   Using DFL #2, determine what size joist will be needed to cantilever 18″ at a deck if a spacing of 16″ is used.

Problem 29–10   Using DFL, determine what size beam will be needed to span 15′–6″ with a point load of 2750 lb at the midpoint. What size piers will be required to support the loads if soil pressure is 2000?

Problem 29–11   What size glu-lam beam will be required for a Ridge Beam with an L = 14.5′ and W = 5500 lb. Use fb 2200.

Problem 29–12   A residence has floor joists with a 24″ cantilever. A wall weighing 650 lb per linear foot will be supported at the end of the floor joist. What size and spacing should be used to support this load?

Problem 29–13   A 16′ long DFL laminated ridge beam will support 800 lb per linear foot. Determine what depth 5⅛ beam should be used.

Problem 29–14   The ridge beam in question 19 will be supported at one end by a post and at the other end by a beam 4′ long over a door. The point load from the ridge beam will be 12″ from one end of the 4′ header. What size door header should be used?

Problem 29–15   A two-level residence is 28′ wide with a gable roof with 24″ overhangs. The roof is cedar shakes at a 5/12 pitch, and local codes require a 30 lb live load. Wall weights are 8′ at the upper floor. There are several windows in the upper exterior wall that are 60″ wide. A load-bearing wall will be placed 13′ from the left exterior wall on the upper floor. The floor joists supporting the upper floor will cantilever 18″ past the lower wall on the right side of the residence so that the lower floor is only 26′–6″ wide. A bearing wall on the main floor will be 16′ from the left exterior wall. Part of the wall will be an opening. The walls of the lower floor will be 10 feet high. The lower floor will be supported by a girder directly below the bearing wall, with supports at 5′–0″ max. spacing. The owners of the house have two kids and a dog named Spot. Using span tables whenever possible, provide the size of all rafters, floor joists, beams, and piers for this residence. Use lumber common to your area for framing values.
Rafter =
Ceiling joists =
Upper floor joists (left) =
Upper floor joists (right) =
Lower floor joist =
5′–0″ headers @ upper floor =
8′–0″ header @ lower floor =
Girder =
The dog =

CHAPTER 30

Drawing Framing Plans

A *framing plan* is a drawing used to show the dimensions, framing members, and methods of resisting seismic and wind loads for a specific level of a structure. Office practice and the complexity of the structure determine the exact contents of the plan.

▼ FRAMING PLANS

When a simple one-level house using truss roof construction is drawn, architectural and structural information can often be combined on the floor plan, as seen in Figure 30–1. On a custom multilevel structure, a separate plan is usually developed to explain the roof framing as well as the architectural and structural requirements of each level. The floor plans are typically used to explain architectural information. See Figure 30–2. The framing plan is used to show the location of all walls and openings, header sizes, and structural connectors. Section markers to indicate where sections and details have been drawn are also placed on a framing plan. Figure 30–3 shows the framing plan that corresponds to the floor plan in Figure 30–2.

The framing plan is typically created from a sepia or Mylar copy of the floor plan showing the walls, doors, windows, and cabinets. When the floor plan is completed, it is with the thought that a cutting plane passed through the structure and removed the upper portion of the structure. The floor plan is drawn as if the viewer is looking down at the floor.

When the framing plan is drawn, it is with the premise that the viewer is lying on the floor, looking up. When the main framing plan is drawn for a two-level structure, it shows the layout of the walls for the main floor, and the beams and floor joists needed to frame the floor/ceiling above the main floor. The plan for the upper level shows the location of the walls on the upper floor and the members used to form the ceiling if western platform construction methods are used.

Rafters and other roof members are shown on the roof framing plan. If a room has a vaulted ceiling, the rafter/ceiling joist are typically shown on the framing plan. If the roof system is framed using trusses, these are often shown on the upper framing plan. If a complex roof shape is used, the trusses are shown on the roof framing plan. The floor system used to support the main floor is shown on the foundation plan. Section IX introduces the foundation and floor framing systems.

Generally as the framing plans are drawn, the structure is completed beginning with the highest level, and then moving down to the lowest. This method allows the loads to be accumulated so that lower-level support will be accurately sized. The residence that was drawn in Chapter 15 will be used later in this chapter to demonstrate this procedure.

Framing Members

A key element of the framing plan is to show and specify the location of headers, beams, posts, joists, and trusses used to frame the skeleton by the use of notes and symbols.

Headers are located over every opening in a wall. Headers over a door or window are typically not drawn and are referenced by a note. A beam is placed to control the span of a joist or rafter, or under a concentrated load, and is represented by parallel dashed lines. The distance between the lines is based on the thickness of the beam. If a built-up beam is to be provided, thin dashed lines representing the width of the individual members are drawn. Often the specification for the size and strength is placed parallel to the member.

Each beam and header will be supported by multiple studs, a post, or a steel column depending on the loads to be supported. When a post or column is hidden in a wall it may or may not be drawn. Posts or columns inside a room must be drawn using line quality to match that used to represent walls. Specifications for the post size or for connecting hardware are often specified on a 45° angle. Figure 30–4 shows common methods of specifying beams and posts.

Joists, rafters, and trusses are usually represented by a thin line with an arrow at each end. Many companies draw the line representing these members so that it extends from bearing point to bearing point. An alternative method is to show the joist symbol centered between the bearing points. The specification for the member is usually written parallel to the line and should include the size, type, and spacing of the member. If metal hangers are to be used to connect a joist to a beam, a note listing the connector is typically placed on a 45° angle and should include the size or model number and the manufacturer. If a vaulted ceiling is to be provided, the line dividing the vault from the flat ceiling must be drawn,

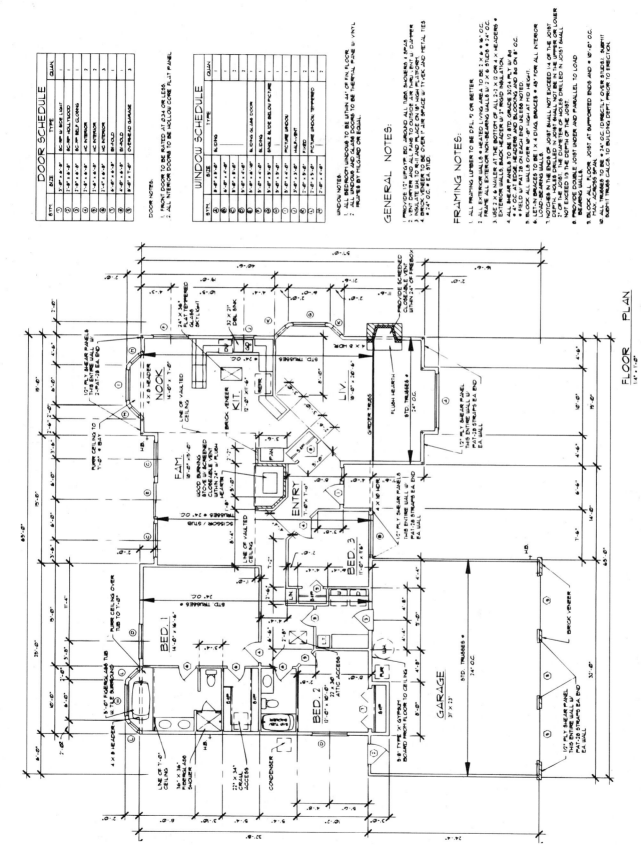

Figure 30–1 On a simple residence, the structural information can be shown on the floor plan. *Courtesy Steve Bloedel.*

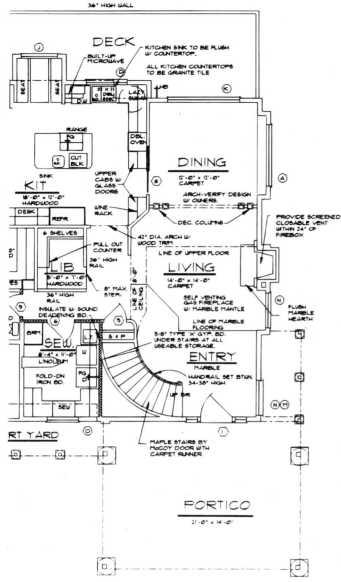

Figure 30–2 On a multilevel structure, typically only architectural information is placed on the floor plan.

located, and specified. Figure 30–5 shows common methods of representing joists, trusses, and vaulted ceilings.

Seismic and Wind Resistance

The severity of lateral loads will vary greatly depending on the area of the country where you work. Typically plywood shear panels, extra blocking, metal angles, and metal connectors to tie posts of the one level to members below them are common methods of resisting the forces caused by earthquakes or high winds. (See Chapter 25 for a review of these terms.) These members are determined by an engineer and placed on the framing plan by the drafter.

Plywood shear panels can be specified by a local note pointing to the area of the wall to be reinforced. If several

walls are to receive shear panels, a note to explain the construction of the panel can be placed as a general note, with a shortened local note used merely to locate their occurrence. Horizontal metal straps may be required to strengthen connections of the top plate or to tie the header into wall framing. These straps can be represented by a line placed on the side of the wall with a note to specify the manufacturer, the size, and the required blocking. Where walls can be reinforced with a let-in brace, a bold diagonal line is usually placed on the wall.

Metal straps used to tie trimmers, king studs, or posts of one level to another can be represented on each layer of the framing plan by a bold line where they occur. This vertical strap should be shown and specified on the framing plan for each level being connected. The specification should include the manufacturer, size, and required nailing.

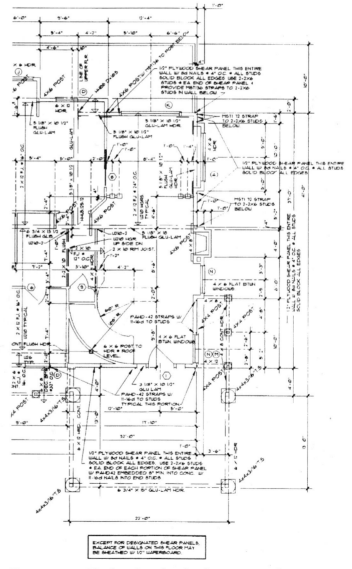

Figure 30–3 The framing plan for the structure shown in Figure 30–2 contains only the structural information.

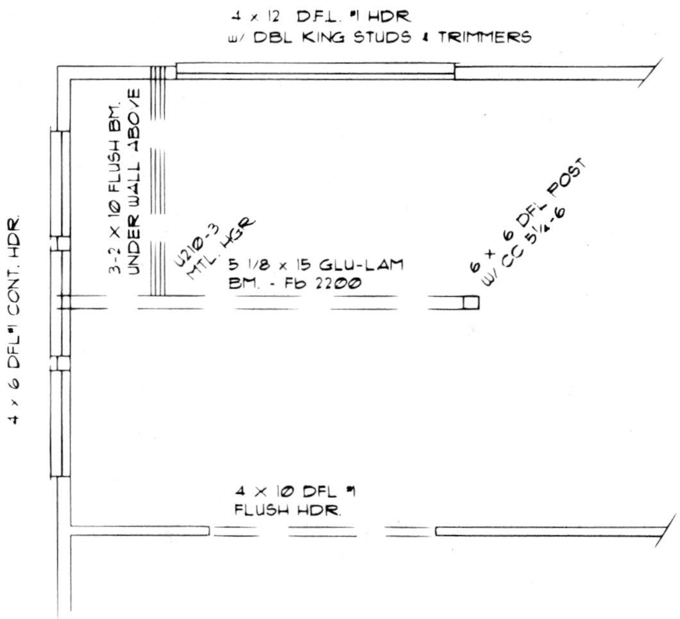

Figure 30–4 Headers over a window are typically specified by note but not drawn. Headers in a room must be drawn and specified.

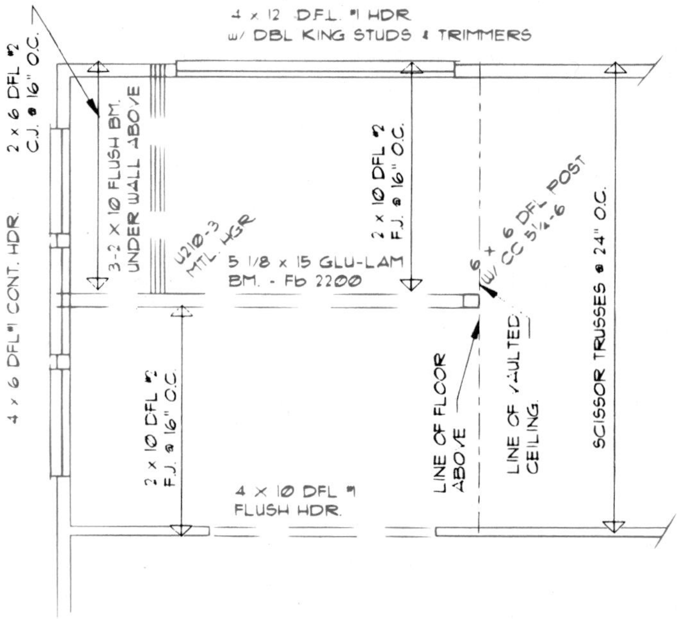

Figure 30–5 Joists, rafters, and trusses are each represented by an arrow that shows the direction of the span, with a note to specify the type, size, and spacing of the structural member.

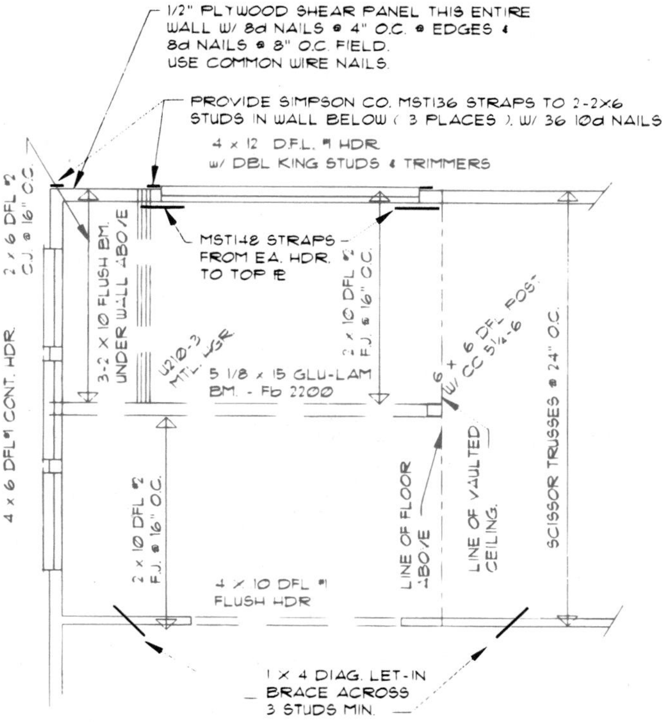

Figure 30–6 Material for resisting lateral loads is specified on the framing plan by note.

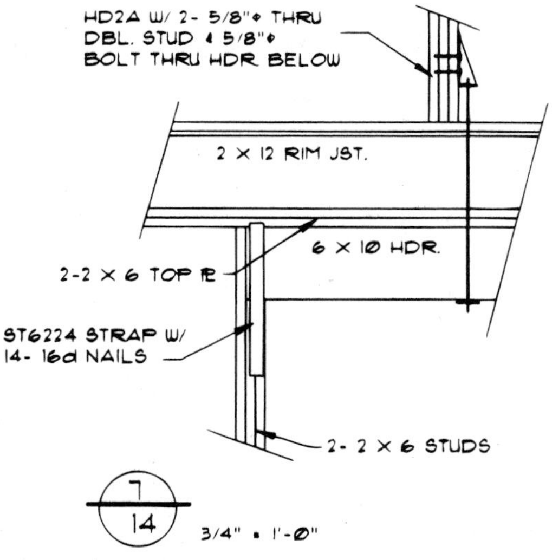

Figure 30–7 Metal strap and hold-down anchors, which are specified on the framing plan, are typically detailed to show exact placement.

Figure 30–6 shows how to represent shear panels, let-in braces, and metal straps. Straps are also shown in sections and details, as in Figure 30–7, and they can be shown on the exterior elevations. When the lowest floor level is to be attached to the foundation level, the metal connector is often represented by a cross. Figure 30–8 shows common methods of representing these ties on the framing plan.

Floors and roof members are often tied into the wall system to help resist lateral loads. In addition to the common nailing used to connect each member, metal angles are used to secure the truss or rafter to the top plate. Where lateral loads are severe, blocking is added between the trusses or rafters to keep these members from bending

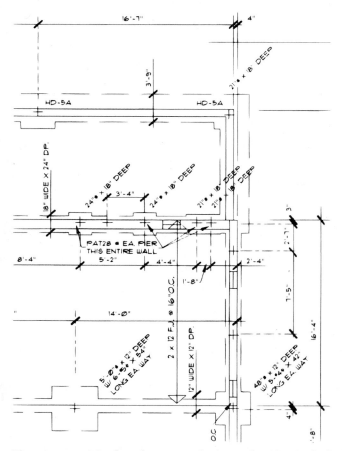

Figure 30–8 Metal anchors must be located and specified on the foundation plan.

under lateral stress. These areas where joist or rafters are stiffened to resist bending from lateral loads are often referred to as a *diaphragm*. The size, location, and framing method are specified by an engineer. The angles and blocking for the diaphragm must be clearly specified on the framing plan. The angles are typically represented in a section or detail but are not drawn on the framing plan. They should be specified by a note that contains the manufacturer, model number, spacing, and which members are to be connected. Areas that are to receive special blocking should be indicated on the framing plan. Figure 30–9 shows how a diaphragm can be represented on a framing plan.

Dimensions

Chapter 14 introduced the process of placing dimensions on the floor plan. The process for placing dimensions on the framing plan is exactly the same. If a separate framing plan is to be drawn, the floor plan is usually not dimensioned; instead, all dimensions are placed on the framing plans. In addition to dimensioning the walls, doors, and windows, the location of framing members often needs to be dimensioned. Beams that span between an opening in a wall are located by the dimensions, which locate the wall. Beams in the middle of a room must be located by dimensions.

Where joists extend past a wall, the length of the cantilever must be dimensioned. Joist size or type often varies when an upper level only partially covers another level. The limits of the placement of joists should also be

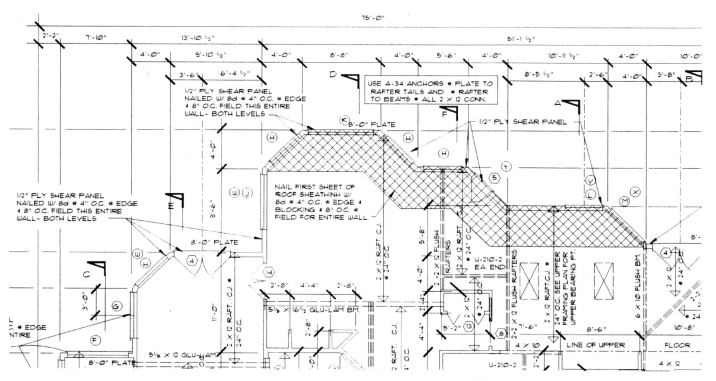

Figure 30–9 Special nailing to resist lateral forces can be shown on the framing plan.

dimensioned on a floor plan. Figure 30–10 shows how the location of structural members can be clarified for the framer.

Notes

The use of local notes to specify materials on the framing plan has been mentioned throughout this chapter. Many professionals also place general notes on the framing plan to ensure the compliance of the code that governs construction. These framing notes can be placed on the framing plan, with the sections and details, or on a separate specifications page that is included with the working drawings. Figure 30–11 shows common notes that might be included with the framing plan.

Section References

The framing plans are used as reference maps to show where cross-sections have been cut. Detail reference symbols are also placed on the framing plan to help the print reader understand material that is being displayed in the sections. (Chapter 34 will further explain section tags and their relationship to the sections and framing plans.) Figure 30–9 shows examples of section markers.

▼ COMPLETING A FRAMING PLAN

The order used to draw the framing plan depends on the method of construction and the level to be framed. A structure framed with trusses requires less detailing than the same one framed with rafters and ceiling joists. Interior beams will be greatly reduced if not entirely eliminated, when using trusses. Because a lower level is supporting more weight, beams tend to be shorter, requiring more and larger posts. Despite differences, framing plans also have similarities that can help in drafting the plan, no matter what level is to be drawn.

Once the walls have been located, dimensions can be placed to locate all walls and openings in them. Then individual framing members can be determined. Rafter

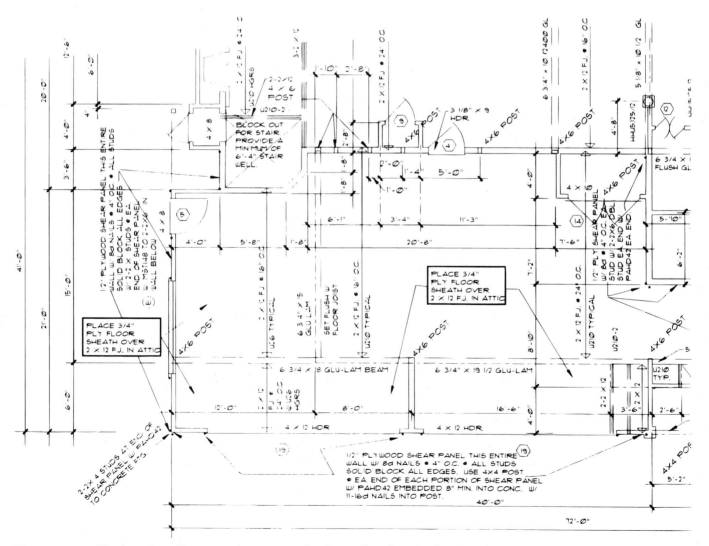

Figure 30–10 The location of beams and posts must be dimensioned on the framing plan.

FRAMING NOTES:

NOTES SHALL APPLY TO ALL LEVELS.
FRAMING STANDARDS ARE ACCORDING TO U.B.C.

1. ALL FRAMING LUMBER TO BE D.F.L. *2 MIN.
   ALL GLU-LAM BEAMS TO BE fb2200 V-4 DF/DF

2. FRAME ALL EXTERIOR WALLS W/ 2 x 6 STUDS @ 16" O.C. FRAME ALL
   EXTERIOR NON-BEARING WALLS W/ 2 x 6 STUDS @ 24" O.C. USE
   2 x 4 STUDS @ 16" O.C. FOR EXTERIOR UNHEATED WALLS AND FOR
   ALL INTERIOR WALLS UNLESS NOTED.

3. USE 2 x 6 NAILER AT THE BOTTOM OF ALL 2-2X12 OR 4 x HEADERS
   @ EXTERIOR WALLS. BACK HEADER W/ 2" RIGID INSULATION

4. ALL SHEAR PANELS TO BE 1/2" PLY. NAILED W/ 8 d'S @ 4" O.C.
   AT EDGE AND BLOCKING AND 6 d'S @ 8" O.C @ FIELD

5. PLYWOOD ROOF SHEATHING TO BE 1/2" STD. GRADE 32/16 PLY.
   LAID PERP. TO RAFTERS. NAIL W/ 8 d'S @ 6" O.C. @ EDGES AND
   12" O.C. AT FIELD

6. PROVIDE 3/4" STD. GRADE T. & G. PLY. FLOOR SHEATHING LAID
   PERP. TO FLOOR JOIST. NAIL W/ 10 d'S @ 6" O.C. @ EDGES AND
   BLOCKING & 12" O.C. @ FIELD. COVER W/ 3/8" HARDBOARD

7. BLOCK ALL WALLS OVER 10'-0" HIGH AT MID HEIGHT.

8. LET-IN BRACES TO BE 1 x 4 DIAG. BRACES @ 45° FOR ALL
   INTERIOR LOAD-BEARING WALLS.

9. NOTCHES IN THE ENDS OF JOIST SHALL NOT EXCEED 1/4 OF THE
   JOIST DEPTH. HOLES DRILLED IN JOIST SHALL NOT BE IN THE
   UPPER OR LOWER 2" OF THE JOIST. THE DIAMETER OF HOLES
   DRILLED IN JOIST SHALL NOT EXCEED 1/3 THE DEPTH OF THE
   JOIST.

10. PROVIDE DOUBLE JOIST UNDER AND PARALLEL TO LOAD
    BEARING WALLS.

11. BLOCK ALL FLOOR JOIST AT SUPPORTED ENDS AND @ 10'-0 O.C.
    MAX. ACROSS SPAN.

Figure 30–11 General notes can be used to ensure that minimum standards established by building codes will be maintained.

direction is determined by the shape of the roof, and its length is determined by the span for a specific size of the particular member. Rafter spans can be determined using the tables in Chapter 28. The process is similar if trusses are to be drawn, but the sizes of the trusses are determined by the manufacturer.

After the size and direction of the roof framing members are selected, loading bearing walls can be identified and headers selected for openings in the load-bearing walls. With the structural members specified, any dimensions required to locate structural members can be completed. Materials that can be specified with local notes can be added and the drawing completed by adding general notes.

Once the framing for the roof has been determined, framing for lower levels can be drawn. Bearing walls on the upper level need to have some method of supporting the loads as they are transferred downward. Support can be provided by stacking bearing walls or by transferring loads through floor joists to other walls. Figure 30–12 shows a simplified drawing of how loads can be transferred. Load bearing walls can be offset by the thickness of the floor joists and still be considered aligned. For example, if 2 × 10 floor joists are used, an upper wall could be offset 10" from a lower wall and still be considered to be aligned. When floor joists are cantilevered or used to transfer

vertical loads to lower walls that are not aligned, the tables in Chapter 28 cannot be used to determine joist sizes. These members need to be determined using the formulas found in Chapter 29. Once the bearing walls for the lowest floor level have been determined, the foundation can be completed.

▼ ROOF FRAMING PLANS

The roof plan was introduced in Chapter 20. The roof framing plan is similar, but in addition to showing the shape of the structure and the outline of the roof, it also shows the size and direction of the framing members used to frame the roof. The roof framing plan will vary depending on whether trusses or western platform construction methods are used. Depending on the simplicity of the structure, roof framing members may be drawn on the floor framing plans rather than having a separate roof framing plan.

Each method will be considered for gable, hip, and Dutch hip roofs for the residence drawn in Chapter 15 using both separate roof framing plans and individual framing plans. Each roof can be drawn by using the same steps to form the base drawing. If a roof plan has been drawn, it can be traced to form the base of the roof framing plan. To draw a roof framing plan use construction lines for the following steps. Use the line quality described throughout this chapter and Chapter 20 to complete this drawing.

Step 1. Draw the outline of the exterior walls.

Step 2. Draw any posts and headers required for covered porches.

Step 3. Draw the limits of the overhangs. Unless your instructor provides different instructions, use

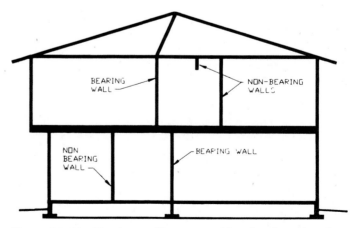

Figure 30–12 Bearing walls are considered to be aligned as long as the distance of the offset does not exceed the thickness of the floor joist. If the offset of the bearing walls is greater than the thickness of the floor joist, the size of the joist must be determined by using the formulas presented in Chapter 28.

24″ overhangs at eaves and 12″ overhangs at gable end walls.

Step 4. Locate all ridges.

Step 5. Locate all hip or valleys formed by roof intersections.

Step 6. Locate all roof openings, such as skylights and chimneys.

Step 7. Draw all items from steps 1 through 6 using finished line quality. Your drawing should now resemble Figure 30–13.

Gable Roof Framing Plan with Truss Framing

Step 1. Draw the boundaries of any areas to be vaulted. Assume the master bedroom will be vaulted. Provide dimensions to specify the limits.

Step 2. Draw and label the arrows to indicate the framing members. Assume standard/scissor trusses over the master bedroom with standard roof trusses at 24″ o.c. for the balance of the upper level and over the garage. Specify 4 × 6 beams at 32″ o.c. at the entry porch. Support the entry beams on 4 × 8 beams on 4 × 4 post with pc44 post caps by Simpson Company or equal.

Step 3. Provide local notes to specify any necessary straps or metal connectors. For this residence, provide an ST6224 tie strap from the 4 × 8 header to the garage top plate.

Step 4. Provide general notes to indicate any necessary framing material. Your drawing should now resemble Figure 30–14.

Gable Roof Framing Plan with Rafter Framing

Follow steps 1 through 7 for a truss roof framing plan to draw the base for this plan. The drawing should resemble Figure 30–13.

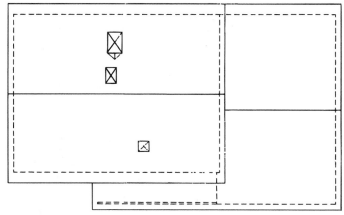

Figure 30–13 The roof plan serves as a base for the framing plan.

Step 1. Determine the location of interior walls, which are parallel to the ridge. Although the walls are not drawn on the framing plan, they can be used to support purlin braces.

Step 2. Draw the boundaries of any areas to be vaulted. Assume the master bedroom will be vaulted. Provide dimensions to specify the limits.

Step 3. Draw and label the arrows to indicate the framing members. Use the span tables from Chapter 28 to determine the required size. For this residence, DFL #2 will be used. Use 2 × 6 rafters at 24″ o.c. where possible. For this project the following framing will be used:

2 × 12 raft./c.j. over bedroom 1
4 × 14 DFL #1 exposed ridge beam
purlins at upper floor: 2 × 6
purlin over garage: 2 × 8
strongback over garage: 5⅛ × 13½ Fb 2200 glulam
beams at porch: 4 × 8
rafters at porch: 4 × 6 @ 32″ o.c.

Step 4. Provide purlins and strong backs at approximately the midpoint of rafter spans.

Step 5. Provide local notes to specify any necessary straps or metal connectors. For this residence, provide an ST6224 tie strap from the 4 × 8 header to the garage top plate. Support the entry beams on 4 × 4 post with pcc44 post caps by Simpson Company or equal.

Step 6. Provide general notes to indicate any necessary framing material. Your drawing should now resemble Figure 30–15.

Hip Roof Framing Plan with Truss Framing

Follow steps 1 through 6 for a truss roof framing plan to draw the base for this plan. The base drawing should resemble Figure 30–16.

Step 1. Determine the limit of the standard trusses based on the intersection of the hips with the ridge.

Step 2. Determine the limits of any required girder trusses. Because the roof of the structure has no perpendicular roofs, no girder trusses are required.

Step 3. Draw the boundaries, and dimension the limits of any areas to be vaulted. Because the framing members in the master bedroom will be spanning in two different directions, the room will not be vaulted for this roof plan.

Step 4. Determine the limits of the hip trusses. This distance is typically determined by the truss manufacturer and the pitch of the roof. For this plan assume 6′–0″ is required before hip trusses can be used.

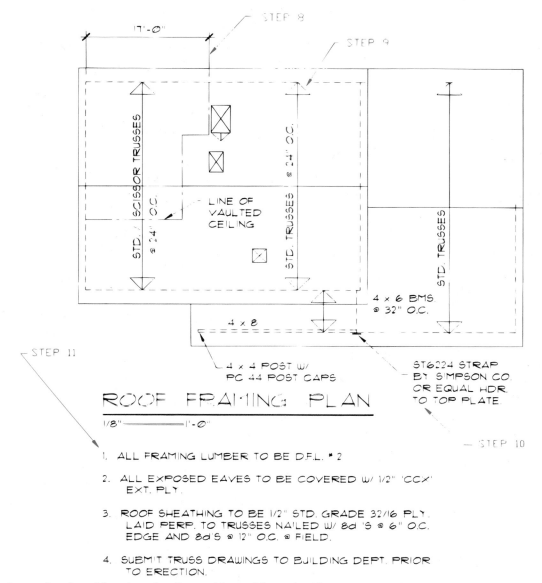

Figure 30–14 A completed roof framing plan for a gable roof framed with trusses.

Step 5. Draw and label the arrows to indicate the framing members. Assume standard trusses at 24″ o.c. over the entire upper floor and over the garage. Specify 4 × 6 beams at 32″ o.c. at the entry porch. Support the entry beams on 4 × 8 beams on 4 × 4 post with pc44 post caps by Simpson company or equal.

Step 6. Provide local notes to specify any necessary straps or metal connectors. For this residence, provide an ST6224 tie strap from the 4 × 8 header to the garage top plate.

Step 7. Provide general notes to indicate any necessary framing material. Your drawing should now resemble Figure 30–17.

Hip Roof Framing Plan with Rafter Framing

Follow steps 1 through 7 for a truss roof framing plan to draw the base for this plan. The drawing should resemble Figure 30–16. The layout of the roof will be similar to the layout of a stick framed gable roof.

Step 1. Determine the location of interior walls that are parallel to the ridge. These walls will support purlin braces.

Step 2. Draw the boundaries of any vaulted areas.

Step 3. Markers to show rafters can be drawn and labeled based on the span tables from Chapter 28.

Step 4. Locate purlins and strong backs where necessary based on rafter spans.

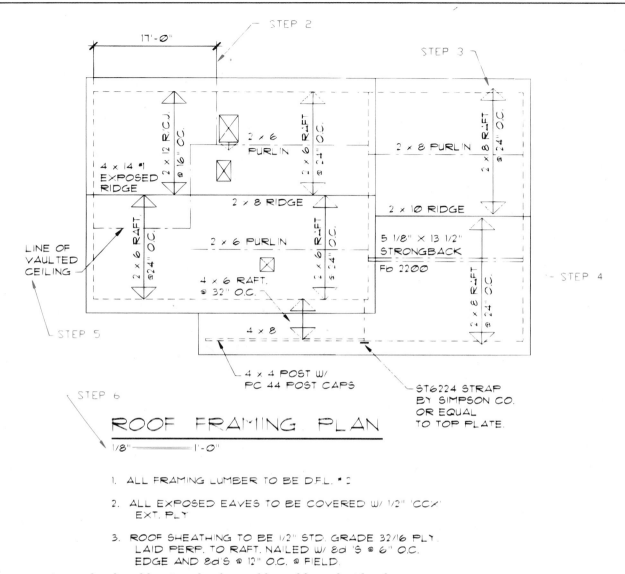

Figure 30–15 A completed roof framing plan for a gable roof framed with rafters.

Step 5. Provide local notes to specify any necessary straps or metal connectors.

Step 6. Provide general notes to indicate any necessary framing material. Your drawing should now resemble Figure 30–18.

Dutch Hip Roof Framing Plan with Truss Framing

Follow steps 1 through 6 for a truss roof framing plan to draw the base for this plan. The base drawing should resemble Figure 30–19.

Step 1. Determine the limit of the standard trusses based on the limits of the Dutch hip. For this plan, assume the gable end walls of the Dutch hip will be set at 6'–0" from the end walls.

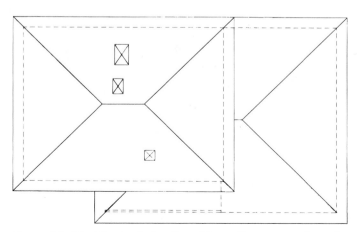

Figure 30–16 The roof plan for a hip roof serves as a base for the roof framing plan.

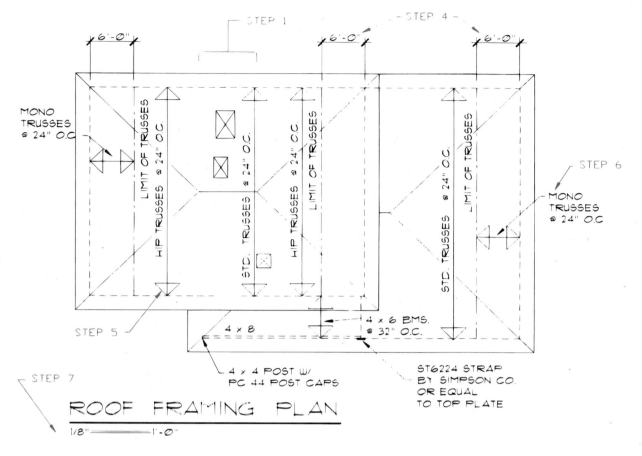

Figure 30–17 A completed roof framing plan for a hip roof framed with trusses.

Step 2. Determine the limits of any required girder trusses. Because the roof of the structure has no perpendicular roofs, no girder trusses will be required.

Step 3. Draw the boundaries, and dimension the limits of any areas to be vaulted. Because the framing members in the master bedroom will be spanning in two different directions, the room will not be vaulted for this roof plan.

Step 4. Draw and label the arrows to indicate the framing members.

Step 5. Provide local notes to specify any necessary straps or metal connectors.

Step 6. Provide general notes to indicate any necessary framing material. Your drawing should now resemble Figure 30–20.

Dutch Hip Roof Framing Plan with Rafter Framing

Follow steps 1 through 7 for a truss roof framing plan to draw the base for this plan. The drawing should resemble Figure 30–19.

Step 1. Lay out interior walls that are parallel to the ridge.

Step 2. Draw the boundaries, and dimension the limits of any areas to be vaulted. Because the framing members in the master bedroom will be spanning in two different directions, the room will not be vaulted for this roof plan.

Step 3. Draw and label the arrows to indicate the framing members. Use the span tables from Chapter 28 to determine the required size.

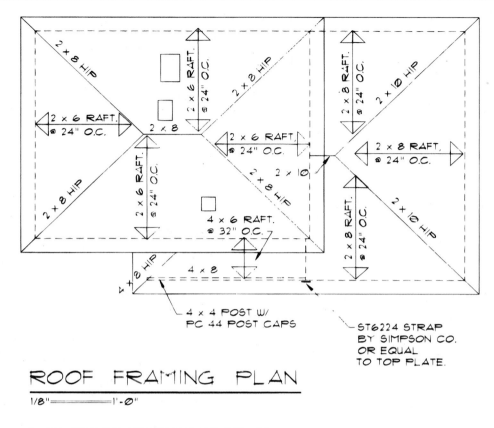

ROOF FRAMING PLAN

1/8"========1'-0"

1. ALL FRAMING LUMBER TO BE D.F.L. #2

2. ALL EXPOSED EAVES TO BE COVERED W/ 1/2" 'CCX' EXT. PLY.

3. ROOF SHEATHING TO BE 1/2" STD. GRADE 32/16 PLY. LAID PERP. TO RAFT. NAILED W/ 8d 'S @ 6" O.C EDGE AND 8d'S @ 12" O.C. @ FIELD.

4. MAXIMUM SPANS:

    2 × 8 RAFTERS @ 24" O.C. = 11'-0"
    " " @ 16" O.C. = 13'-5"
    " " @ 12" O.C. = 15'-6"

    2 × 6 RAFTERS @ 24" O.C. = 11'-6"
    " " @ 16" O.C. = 14'-0"
    " " @ 12" O.C. = 16'-3"

Figure 30–18 A completed roof framing plan for a hip roof framed with rafters.

Step 4. Provide purlins and strong backs where necessary based on rafter spans.

Step 5. Add dimensions to locate all openings, overhangs, and purlins.

Step 6. Provide local notes to specify any necessary straps or metal connectors.

Step 7. Provide general notes to indicate any necessary framing material. Your drawing should now resemble Figure 30–21.

▼ FLOOR FRAMING PLANS

The framing plan can be completed by using the base drawings used to draw the floor plans. Ideally a Mylar copy of the floor plan should be made once the walls, doors, windows, and cabinets are drawn. If a copy of the floor plan was not made prior to completing the floor plan, the layout can be traced. If a computer program is being used, layers containing the walls, windows, doors, and cabinets can be displayed, with material unrelated to the framing plan frozen. The base drawing for the upper framing plan should resemble Figure 30–22.

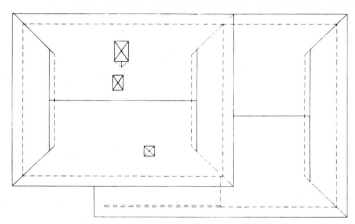

Figure 30–19 The roof plan for a Dutch hip roof serves as a base for the roof framing plan.

Step 1. If your floor plan was dimensioned, skip to step 9. If not, lay out all exterior dimension and extension lines using construction lines for each side of the structure. Establish lines for overall and major jogs from the exterior face of the wall.

Step 2. Lay out dimensions and extension lines to locate each interior wall that intersects with an exterior wall. Interior walls will be located from center to center, with center lines for extension lines.

Step 3. Lay out dimension and extension lines to locate exterior openings. Do not dimension doors in the garage on the framing plan.

Step 4. Lay out dimension and extension lines to locate all interior walls that do not intersect with an exterior wall.

Step 5. Place overall dimensions centered above each dimension line.

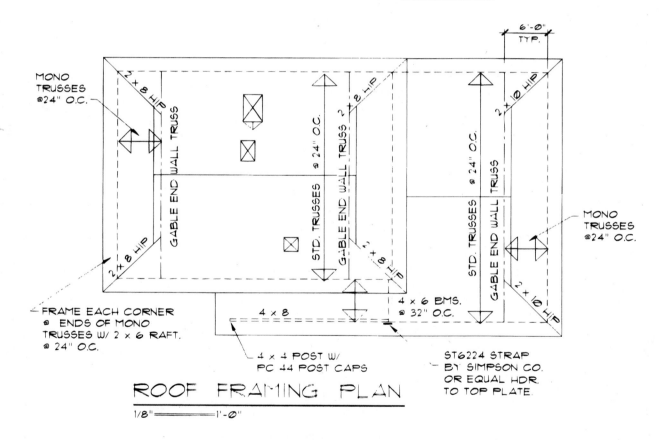

ROOF FRAMING PLAN

1/8" = 1'-0"

1. ALL FRAMING LUMBER TO BE D.F.L. #2

2. ALL EXPOSED EAVES TO BE COVERED W/ 1/2" 'CCX' EXT. PLY.

3. ROOF SHEATHING TO BE 1/2" STD. GRADE 32/16 PLY. LAID PERP. TO TRUSSES NAILED W/ 8d 'S @ 6" O.C. EDGE AND 8d's @ 12" O.C. @ FIELD.

4. SUBMIT TRUSS DRAWINGS TO BUILDING DEPT. PRIOR TO ERECTION.

Figure 30–20 A completed roof framing plan for a Dutch hip roof framed with trusses.

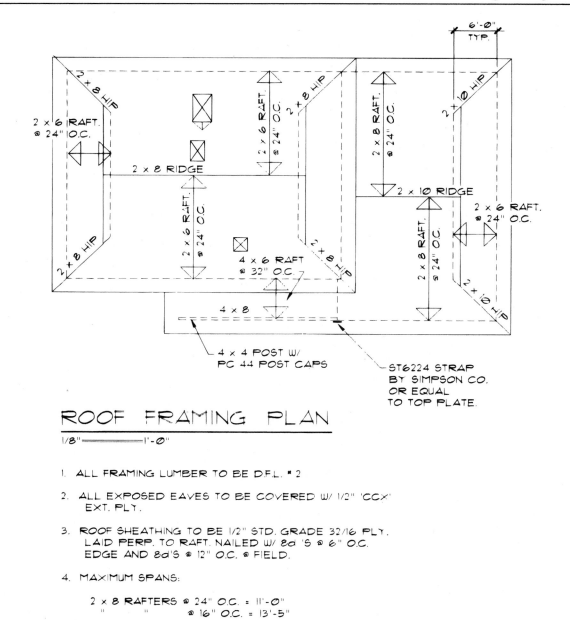

Figure 30–21 A completed roof framing plan for a Dutch hip roof framed with rafters.

Step 6. Specify dimensions that must be a specific size—for example, spaces for tubs, toilets, showers, and other manufactured items.

Step 7. Specify dimensions for areas that are required to be a specific size—for example, hallways, closets, and stairwells.

Step 8. Dimension all major jogs. Wherever possible, try to maintain modular sizes of the material being used. Although a distance from the edge of one wall to the center of the next wall may measure 12′–1″, 14′ joists would need to be pur-

chased. Reducing the distance to 12′–0″ or less would allow 12′ joists to be used.

Step 9. Dimension all remaining sizes, starting from the smallest to largest rooms.

Step 10. Coordinate all dimensions. The lines of dimensions closest to the structure should add up to the total seen on the next line of dimensions. Dimensions on the interior of the structure should add up to match any exterior dimensions. Your framing plan should now resemble Figure 30–23.

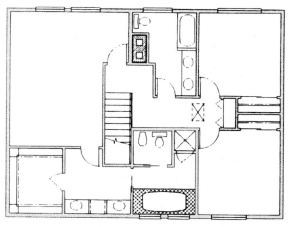

Figure 30–22 The upper floor plan can be used as the base to draw the upper floor framing plan. On a trussed framed roof, this plan can be used in place of the roof framing plan.

Step 11. Specify door and windows symbols, which were specified on the floor plan.

Step 12. Draw the boundaries of any areas to be vaulted. Assume the master bedroom will be vaulted. Provide dimensions to specify the limits.

Step 13. Draw and label the arrows to indicate the framing members. Assume standard/scissor trusses over the master bedroom with standard roof trusses at 24″ o.c. for the balance of the upper level. If a roof other than a truss-framed gable is to be shown on this plan, review the steps used throughout the layout roof framing plan.

Note: If the roof is to be stick framed, only the members used to show the support for the ceiling are shown on the framing plan. If a roof framing plan was drawn, *do not*

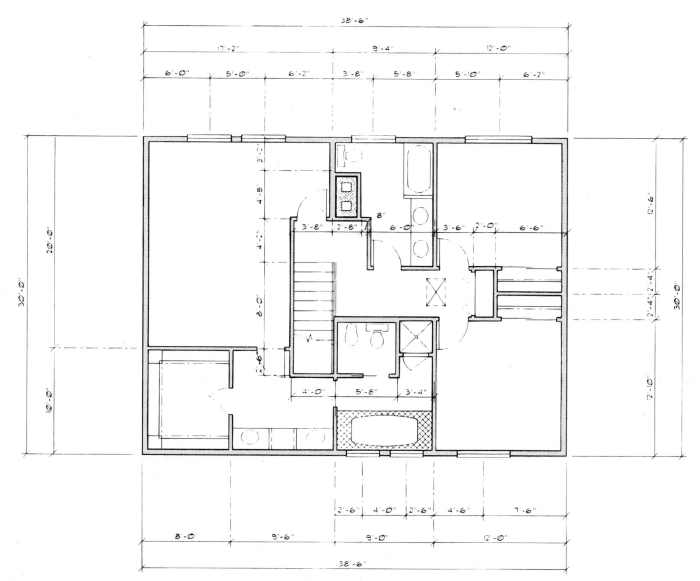

Figure 30–23 Rather than placing dimensions on a floor plan, many offices use the framing plan to display all dimensions.

repeat the material shown in steps 12 through 13 on the upper floor framing plan.

Step 14. Draw and specify all beams, headers, and posts.

Step 15. Provide local notes to specify any materials needed for resisting lateral loads caused by wind or seismic forces.

Note: Because requirements vary widely for each area of the country, verify with your instructor materials that are typical for your area.

Step 16. Provide local notes to specify any necessary straps or metal connectors.

Step 17. Provide general notes to indicate any necessary framing material. Your drawing should now resemble Figure 30–24. If the upper level was framed using western platform construction techniques, the plan would resemble Figure 30–25.

The framing plan for the main floor can be completed following the same procedure that was used to draw the upper floor. Because the upper floor of the residence drawn in Chapter 15 does not completely cover the lower floor, part of the framing plan will show floor joists to support the upper floor, and part of the plan will show roof framing members. If a separate roof framing plan was drawn, all roof framing material will be omitted from this plan.

Step 1. Dimension the framing plan following steps 1 through 9 of the upper framing plan.

Step 2. Specify door and window symbols specified on the floor plan.

Step 3. Draw the boundaries of any areas to be vaulted.

Step 4. Draw and label the arrows to indicate the framing members. Use the tables in Chapter 28 to determine the sizes of members for species of framing lumber common to your area.

Step 5. Using the tables and formulas of Chapter 29, draw and specify all beams, headers, and posts.

Step 6. Provide local notes to specify any materials needed for resisting lateral loads caused by wind or seismic forces. Verify with your instructor materials that are common to your area.

Step 7. Provide local notes to specify any necessary straps or metal connectors.

Step 8. Provide general notes to indicate any necessary framing material. Your drawing should now resemble Figure 30–26.

If the upper level is framed with joists, the lower floor will be affected. Figure 30–27 shows the changes required to the lower-floor framing plan.

The information for framing the floor will be placed on the foundation plan. Procedures for showing the support for the lower plan will be introduced in Chapter 33. Because this plan has a partial basement, the drafter will have the choice of where the floor framing over the basement can be shown. One option is to show a separate framing plan for just the basement, with the floor for the basement and the balance of the structure shown on the foundation plan. This plan would be drawn using the steps for the main framing plan and would resemble Figure 30–28. Because the plan is relatively simple, many professional offices would show the material for framing the main floor over the basement on the foundation plan.

▼ FRAMING PLAN LAYOUT PROBLEM

Use information from the floor, roof plan, and elevations to draw the framing plan for the residence that was started in Chapter 15. Use the span tables from Chapter 28 to determine all rafter sizes and the formulas in Chapter 29 to determine all beam sizes. Verify with your instructor the risk of wind and seismic damage that should be considered in the design.

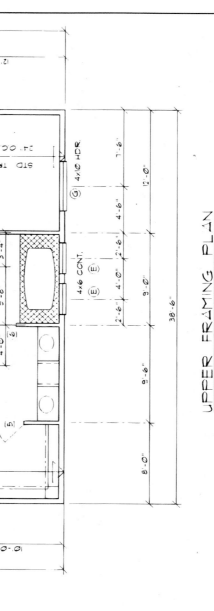

UPPER FRAMING PLAN

SCALE : ¼" = 1'-0"

FRAMING NOTES:

NOTES SHALL APPLY TO ALL LEVELS.
FRAMING STANDARDS ARE ACCORDING TO U.B.C.

1. ALL FRAMING LUMBER TO BE DFL. #2 MIN.
   ALL GLU-LAM BEAMS TO BE Ⅰb2200 V-4 DF/DF

2. FRAME ALL EXTERIOR WALLS W/ 2 × 6 STUDS @
   16" O.C. FRAME ALL EXTERIOR NON-BEARING
   WALLS W/ 2 × 6 STUDS @ 24" O.C. USE 2 × 4 STUDS
   @ 16" O.C. FOR EXTERIOR UNHEATED WALLS AND
   FOR ALL INTERIOR WALLS UNLESS NOTED.

3. USE 2 × 6 NAILER AT THE BOTTOM OF ALL 2-2 × 12
   OR 4 × HEADERS @ EXTERIOR WALLS. BACK
   HEADER W/ 2" RIGID INSULATION.

4. ALL SHEAR PANELS TO BE ½" PLY NAILED W/
   8d'S @ 4" O.C @ EDGE AND BLOCKING AND 6 d'S
   @ 8" O.C. @ FIELD.

5. PLYWOOD ROOF SHEATHING TO BE ½" STD GRADE
   32/16 PLY LAID PERP TO RAFTERS. NAIL W/ 8 d'S
   @ 6" O.C. EDGES AND 12" O.C. AT FIELD.

6. PROVIDE ¾" STD. GRADE T. & G. PLY. FLOOR
   SHEATH LAID PERP TO FLOOR JOIST. NAIL W/
   10d'S @ 6" O.C. @ EDGE & BLOCKING 12" O.C. @
   FIELD. COVER W/ ⅜" HARDBOARD.

7. BLOCK ALL WALLS OVER 10'-0" HIGH AT MID HEIGHT.

8. LET-IN BRACES TO BE 1 × 4 DIAG BRACES @ 45°
   FOR ALL INTERIOR LOAD-BEARING WALLS.

9. NOTCHES IN THE ENDS OF JOIST SHALL NOT
   EXCEED ¼ OF THE JOIST DEPTH. HOLES DRILLED
   IN JOIST SHALL NOT BE IN THE UPPER OR LOWER 2"
   OF THE JOIST. THE DIAMETER OF HOLES DRILLED IN
   JST. SHALL NOT EXCEED ⅓ THE DEPTH OF THE JST.

10. PROVIDE DOUBLE JOIST UNDER AND PARALLEL
    TO LOAD BEARING WALLS.

11. BLOCK ALL FLOOR JOIST AT SUPPORTED ENDS
    AND @ 10'-0" O.C. MAX. ACROSS SPAN.

Figure 30–24 The completed upper floor framing plan showing truss construction for the structure started in Chapter 15.

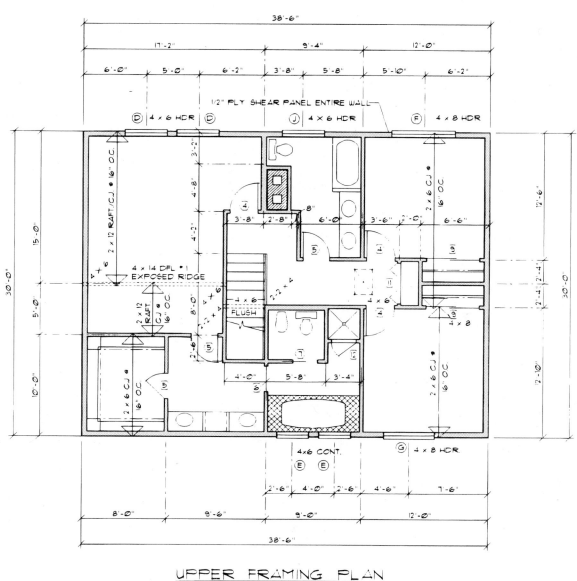

UPPER FRAMING PLAN

SCALE : ¼" = 1'-0"

Figure 30–25 The completed upper floor plan using western platform construction.

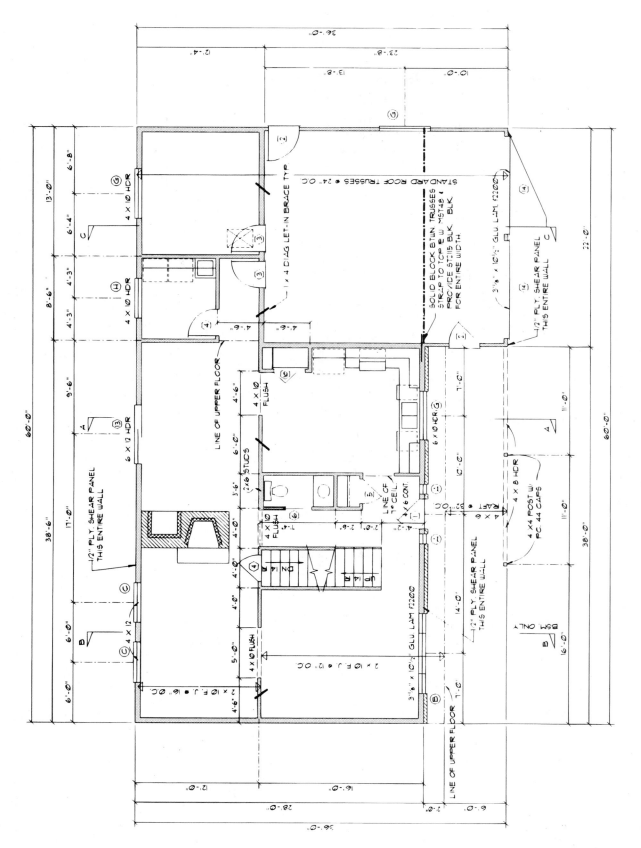

MAIN FRAMING PLAN

SCALE : ¼" = 1'-0"

Figure 30–26 The main floor framing plan for the residence started in Chapter 15. By separating the structural information from the architectural information, greater clarity is achieved.

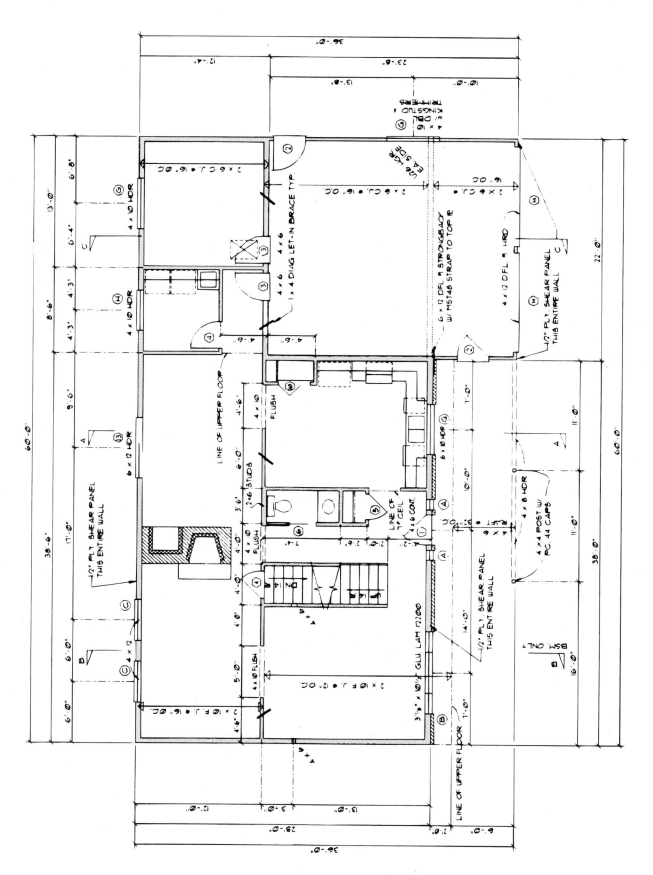

Figure 30–27 The completed main floor plan using western platform construction.

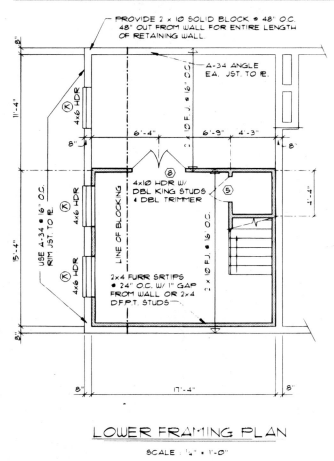

PROVIDE 2 x 10 SOLID BLOCK @ 48" O.C.
48" OUT FROM WALL FOR ENTIRE LENGTH
OF RETAINING WALL.

A-34 ANGLE
EA. JST. TO R.

Ⓚ 4x6 HDR

6'-4"   6'-9"   4'-3"

⑧

4x10 HDR W/
DBL KING STUDS
& DBL TRIMMER

⑤

LINE OF BLOCKING

Ⓚ 4x6 HDR

USE A-34 @ 16" O.C.
RIM JST. TO R.

2 x 10 F.J. @ 16" O.C.

Ⓚ 4x6 HDR

2x4 FURR SRTIPS
@ 24" O.C. W/ 1" GAP
FROM WALL OR 2x4
D.F.P.T. STUDS.

11'-4"   15'-4"

8"   17'-4"   8"

LOWER FRAMING PLAN

SCALE : ¼" = 1'-0"

Figure 30–28 The structural information for the basement can
be shown on a separate framing plan or on the foundation plan,
depending on the complexity of the structure.

SECTION IX

FOUNDATION PLANS

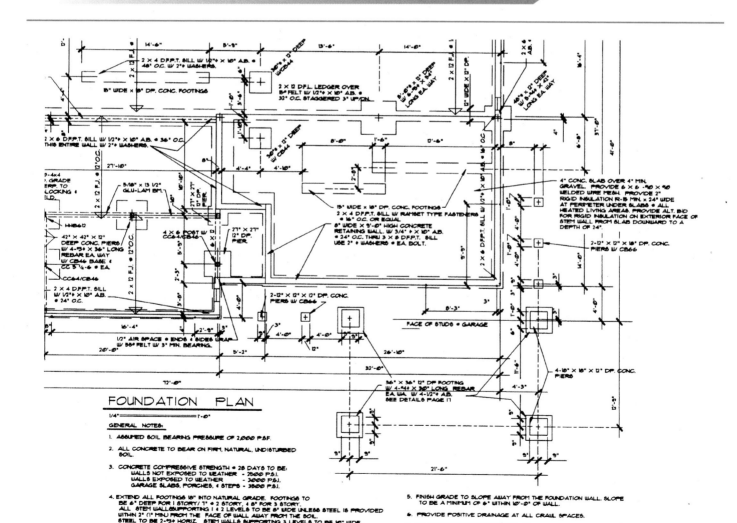

FOUNDATION PLAN
1/4" = 1'-0"

GENERAL NOTES:

1. ASSUMED SOIL BEARING PRESSURE OF 2,000 P.S.F.

2. ALL CONCRETE TO BEAR ON FIRM, NATURAL, UNDISTURBED SOIL.

3. CONCRETE COMPRESSIVE STRENGTH @ 28 DAYS TO BE:
   WALLS NOT EXPOSED TO WEATHER - 2,000 P.S.I.
   WALLS EXPOSED TO WEATHER - 3,000 P.S.I.
   GARAGE SLABS, PORCHES, & STEPS - 3,000 P.S.I.

4. EXTEND ALL FOOTINGS 18" INTO NATURAL GRADE. FOOTINGS TO BE 6" DEEP FOR 1 STORY / 7" @ 2 STORY, & 8" FOR 3 STORY. ALL STEM WALLS SUPPORTING 1 & 2 LEVELS TO BE 8" WIDE UNLESS STEEL IS PROVIDED WITHIN 2" (1" MIN.) FROM THE FACE OF WALL AWAY FROM THE SOIL. STEEL TO BE 2-#3 HORIZ. STEM WALLS SUPPORTING 3 LEVELS TO BE 10" WIDE.

5. FINISH GRADE TO SLOPE AWAY FROM THE FOUNDATION WALL. SLOPE TO BE A MINIMUM OF 6" WITHIN 10'-0" OF WALL.

6. PROVIDE POSITIVE DRAINAGE AT ALL CRAWL SPACES.

7. ALL FRAMING LUMBER TO BE D.F.L. #2.

Foundation Systems

All structures are required to have a foundation. The foundation provides a base to distribute the weight of the structure onto the soil. The weight, or loads, must be evenly distributed over enough soil to prevent them from compressing it. In addition to resisting the loads from gravity, the foundation must resist floods, winds, and earthquakes. Where flooding is a problem, the foundation system must be designed for the possibility that much of the supporting soil may be washed away. The foundation must also be designed to resist any debris that may be carried by flood waters.

The forces of wind on a structure can cause severe problems for a foundation. The walls of a structure act as a large sail. If the structure is not properly anchored to the foundation, the walls can be ripped away by the wind. Not only does wind try to push a structure sideways but upward as well. Because the structure is securely bonded together at each intersection, wind pressure will build under the roof overhangs and inside the structure and cause an upward tendency. Proper foundation design will resist this upward movement. Figure 31–1 shows an example of straps used to resist uplift.

Depending on the risk for seismic damage, special design considerations may be required of a foundation. Although earthquakes cause both vertical and horizontal movement, it is the horizontal movement that causes the most damage to structures. The foundation system must be designed so that it can move with the ground yet keep its basic shape. Steel reinforcing and welded wire mesh are often required to help resist or minimize damage due to the movement of the earth.

▼ SOIL CONSIDERATIONS

In addition to the forces of nature, the nature of the soil supporting the foundation must also be considered. The texture of the soil and the tendency of the soil to freeze will influence the design of the foundation system.

Soil Texture

The texture of the soil will affect its ability to resist the loads of the foundation. Before the foundation can be designed for a structure, the bearing capacity of the soil must be known. The soil bearing capacity is a design value specifying the amount of weight a square foot of soil can support. The bearing capacity of soil depends on its composition and the moisture content. The Uniform Building Code provides four basic soil classifications for soil:

1. Massive crystalline bedrock, assumed to support 4000 pounds per square foot (191.6 kPa).
2. Sedimentary, foliated rock, sandy gravel, and gravel, each assumed to be able to support 2000 psf (95.8 Kpa).
3. Sand and silty sand, each assumed to support 1500 psf (71.9 Kpa).
4. Clay and sandy clay, each assumed to support 1000 psf (47.9 Kpa).

Structures built on low-bearing capacity soils require footings that extend into stable soil or are spread over a wide area. Both options typically require the design to be approved by an engineer.

In residential construction, the type of soil can often be determined from the local building department. In commercial construction, a soils engineer is usually required

Figure 31–1 In addition to resisting the forces of gravity, a foundation must be able to resist the forces of uplift created by wind pressure acting on the structure. Anchor bolts and metal framing straps embedded into the concrete are two common methods of resisting the forces of nature. *Courtesy Tereasa Stark.*

to study the various types of soil that are at the job site and make recommendations for foundation design. The soil bearing values must be determined before a suitable material for the foundation can be selected. In addition to the texture, the tendency of freezing must also be considered.

The four common soil classifications are used to define natural, undisturbed soil; construction sites however, often include soil that has been brought to the site or moved on the site. Soil that is placed over the natural grade is called *fill material*. Fill material is often moved to a lot when an access road is placed in a sloping site, as seen in Figure 31–2. After a few years, vegetation covers the soil and gives it the appearance of natural grade. Footings resting on fill material will eventually settle under the weight of a structure. All discussion of foundation depths in this text refers to the footing depth into the natural grade.

Compaction

Fill material can be compacted to increase its bearing capacity. Compaction is typically accomplished by vibrating, tamping, rolling, or adding a temporary weight. There are three major ways to compact soil:

Static force: A heavy roller presses soil particles together.
Impact forces: A ramming shoe repeatedly strikes the ground at high speed.
Vibration: High-frequency vibration is applied to soil through a steel plate.

Large job sites are typically compacted by mobile equipment. Small areas in the construction site are generally compacted by hand-held mobile equipment. Granular soils are best compacted by vibrations, and soils containing large amounts of clay are best compacted by force. Each of these methods will reduce the amount of air voids contained between grains of soil. Proper compaction lessens the effect of settling and increases the stability of the soil, which increases the load-bearing capacity. The effects of frost damage are minimized in compacted soil because penetration of water into voids in the soil is minimized.

Moisture content is the most important factor of efficient soil compaction because it acts as a lubricant to help soil particles move closer together. Compaction should be completed under the supervision of a geotechnical engineer or other qualified expert who understands the measurements of soil moisture. Compaction is typically accomplished in lifts of from 6 through 12″ (150–300 mm). The soils or geotechnical engineer will specify the requirements for soils excavation and compaction. The drafter's job is to place the specifications of the engineer clearly on the plans.

Freezing

Don't confuse ground freezing with blizzards. Even in the warmer southern states, ground freezing can be a problem. Figure 31–3 shows the average frost penetration depths for the United States. A foundation should be built to a depth where the ground is not subject to freezing. Soil expands as it freezes and then contracts as it thaws. Expansion and shrinking of the soil will cause heaving in the foundation. As the concrete expands with the soil, the foundation can crack. As the soil thaws, water that cannot be absorbed by the soil can cause the soil to lose much of its bearing capacity, causing further cracking of the foundation. In addition to the geographic location, the soil type also affects the influences of freezing. Fine-grained soil is more susceptible to freezing because of its density than coarse-grained soil.

A foundation must rest on stable soil so that the foundation does not crack. The designer will have to verity the required depth of foundations with the local building department. Once the soil bearing capacity and the depth of freezing are known, the type of foundation system to be used can be determined.

Water Content

The amount of water the foundation will be exposed to, as well as the permeability of the soil, must also be considered in the design of the foundation. As the soil absorbs water, it expands, causing the foundation to heave. The amount of rain and the type of soil can also cause the foundation to heave. In areas of the country that receive extended periods of rainfall, little variation in the soil moisture content occurs, minimizing the risk of heaving caused by soil expansion. Greater danger results from shrinkage during periods of decreased rainfall. In areas of the country that receive only minimal rainfall followed by extended periods without moisture, soil shrinkage can cause severe foundation problems because of the greater moisture differential. To aid in the design of the foundation, each of the three model codes has included the Thornthwaite Moisture Index, which provides a listing of the amount of water that would be

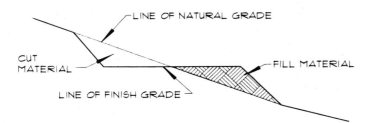

Figure 31–2 As access roads are created, soil is often cut away and pushed to the side of the roadway, creating areas of fill. Unless foundations extend into the natural grade, the structure will settle.

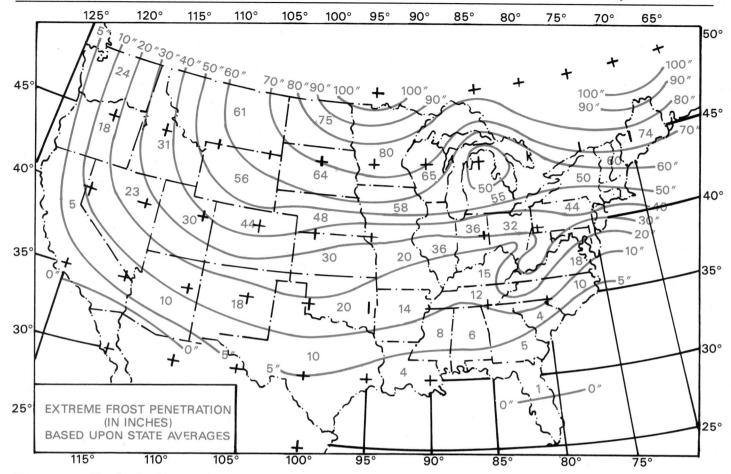

Figure 31–3 The depth of freezing can greatly affect the type of foundation used for a structure.

returned to the atmosphere by evaporation from the ground surface and transpiration if there was an unlimited supply of water to the plants and soil.

Concrete slab on grade is used primarily in dry areas from southern California to Texas, where the contrast between the dry soil under the slab and the damp soil beside the foundation provides a tendency for heaving to occur at the edge of the slab. If the soil beneath the slab experiences a change of moisture content after the slab is poured, the center of the slab can heave. The heaving can be resisted by reinforcing provided in the foundation and throughout the floor slab and by proper drainage. The effects of soil moisture will be investigated in Chapter 33, which considers concrete slab construction.

Surface and ground water must be properly diverted from the foundation so that the soil can support the building loads. Proper drainage also minimizes water leaking into the crawl space or basement, reducing mildew or rotting. Most codes require the finish grade to slope away from the foundation at a minimum slope of ½″ (13 mm) per 12″ (300 mm), for a minimum distance of 6′ (1800 mm). A 3 percent minimum slope is preferable for planted or grassy areas; a 1 percent slope is acceptable for paved areas.

Gravel or coarse-grained soils can be placed beside the foundation to increase percolation. Because of the void spaces between grains, coarse-grained soils are more permeable and drain better than fine-grained soils. A drain is often required to be placed beside the foundation at the base of a gravel bed in damp climates to facilitate drainage of water from the foundation. As the amount of water is reduced in the soil surrounding the foundation, the lateral loads imposed on the foundation are reduced. The weight of the soil becomes increasingly important as the height of the foundation wall is increased. Foundation walls enclosing basements should be waterproofed. Asphaltic emulsion is often used to prevent water penetration into the basement. Floor slabs placed below grade are required to be placed over a vapor barrier.

Radon

Structures built in areas of the country with high radon levels need to provide protection from this cancer-causing gas. The EPA (Environmental Protection Agency) has mapped the continental United States by county and identified high-risk areas to radon. The 1995

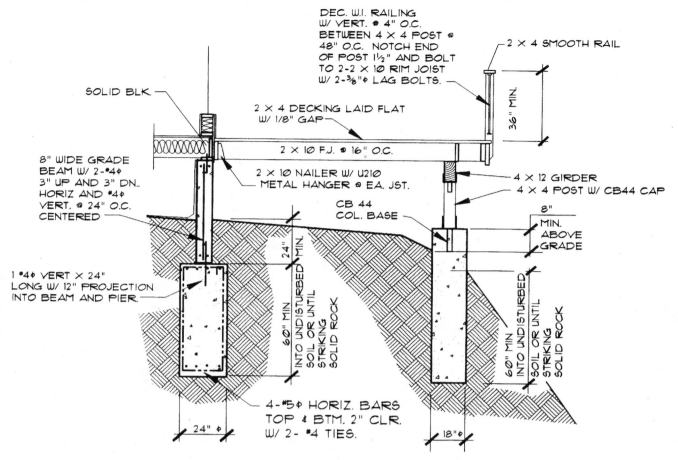

Figure 31–4 Shallow pilings are often made of concrete, which may or may not require steel reinforcement depending on the loads to be supported, the depth of the piling, and the design of the engineer. The piling on the left is being used to support a grade beam, which is a concrete beam placed below the finished grade to span between pilings.

revision of CABO adopted standards to protect structures in these areas from radon. In addition to the CABO guidelines, Florida, New Jersey, and Washington have their own codes for radon-resistant construction.

Common methods of reducing the buildup of radon can be achieved by making minor modifications to the gravel placed below basement slabs. A 4″ (100 mm) PVC vent can be placed in a minimum layer of 4″ (100 mm) gravel covered with 6 mil polyethylene sheathing. The plastic barrier should have a 12″ (300 mm) minimum lap at intersection. Any joints, cracks, or penetrations in the floor slab must be caulked. The vent must run under the slab until it can be routed up through the framing system to an exhaust point, which is a minimum of 10′ (3000 mm) from other openings in the structure and 12″ (300 mm) above the roof. The system should include rough-in electrical wiring for future installation of a fan located in the vent stack and a system-failure warning device. This can usually be accomplished by placing an electrical junction box in an attic space for future installation of a fan.

▼ TYPES OF FOUNDATIONS

The foundation is usually constructed of pilings, continuous footings, or grade beams.

Pilings

A *piling* is a type of foundation system that uses beams placed between columns to support loads of a structure. The columns may extend into the natural grade or be supported on other material that extends into stable soil. Piling foundations are typically used:

○ On steep hillside sites where it may not be feasible to use traditional excavating equipment.
○ Where the loads imposed from the structure exceed the bearing capacity of the soil.
○ On sites subject to flooding or other natural forces that might cause large amounts of soil to be removed.

Coastal property and sites near other bodies of water subject to flooding often use a piling foundation to keep the habitable space of the structure above the floodplain level. Typically a support beam is placed under or near each bearing wall. Beams are supported on a grid of vertical supports, which extend down to a more stable stratum of rock or dense soil. Beams can be steel, sawn or laminated wood, or prestressed concrete. Vertical supports may be concrete columns, steel tubes or beams, wood columns, or a combination of each material.

On shallow pilings a hole can be bored, and poured concrete with steel reinforcing can be used. Figure 31–4 shows an example of a detail for a poured concrete footing. If the vertical support is required to extend deeper than 10′ (3000 mm), a pressure-treated wood timber or steel column can be driven into the soil. Figure 31–5 shows a home supported on a piling foundation. Figure 31–6 shows an example of a steel piling and vertical supports used to support a wood floor system. Figure 31–7 shows the components of a piling plan for residence. In addition to the vertical columns and horizontal beams, diagonal steel cables are placed between the columns to resist lateral and rotational forces. An engineer must approve the design of the piling and the connection of the pilings to the superstructure.

In addition to resisting gravity loads, piling foundations must be able to resist forces from uplift, lateral force, and rotation. Figure 31–8 shows a detail of a concrete piling used to support steel columns, which in turn support the floor system of the residence, and a detail of a steel piling foundation, which is used to support steel columns above grade. Notice that the engineer has specified a system that uses braces approximately parallel to

Figure 31–6 When stable soil is deeper than 10′ (3048 mm), pilings are typically steel or pressure-treated timber. These pilings support the steel girders, which in turn will support the floor framing. *Courtesy Janice Jefferis.*

the ground to stabilize the top of the pilings from lateral loads. These braces would not be shown on the foundation plan but would be specified on the elevations, sections and details.

Continuous or Spread Foundations

The most typical type of foundation used in residential and light commercial construction is a continuous or spread foundation. This type of foundation consists of a footing and wall. A footing is the base of the foundation system and is used to displace the building loads over the soil. Figure 31–9 shows typical footings and how they are usually drawn on foundation plans. Footings are typically made of poured concrete and placed so that they extend below the freezing level. Figure 31–10 shows common footing sizes and depths as required by buildings codes. The strength of the concrete must also be specified, based on the location of the concrete in the structure and its chance of freezing. Figure 31–11 lists typical values.

When footings are placed over areas of soft soil or fill material, reinforcement steel is often placed in the footing. Concrete is extremely durable when it supports a load and is compressed but is very weak when in tension.

If the footing is resting on fill material, the bottom of the concrete footing will bend. As the footing bends, the concrete will be in tension. Steel is placed in the bottom of the footing to resist the forces of tension in concrete.

Figure 31–5 A piling foundation is often used when the job site is too steep for traditional foundation methods. *Courtesy American Plywood Association.*

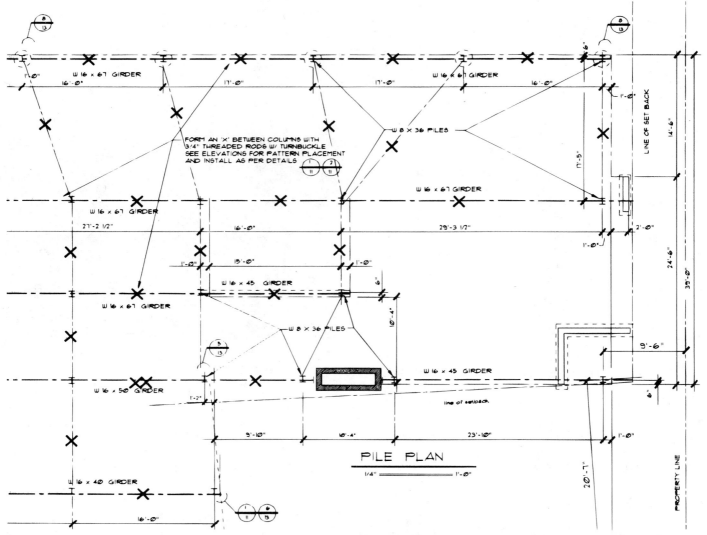

Figure 31–7 The foundation plan for a structure supported on pilings shows the location of all pilings, the beams that span between the pilings, and any diagonal bracing required to resist lateral loads. *Courtesy Residential Designs, AIBD.*

This reinforcing steel, or *rebar*, is usually not shown on the foundation plan but is specified in a note giving the size, quantity, and spacing of the steel. The size of rebar is represented by a number indicating the diameter of the bar in approximately eighths of an inch. Sizes typically range from ¼" to 2½" diameter. A ¼" diameter rebar is ⅜" or a #2 diameter; a ½" diameter bar is a #4. Many codes allow only rebar that is a #5 diameter or larger to be considered as reinforcing with the use of smaller steel considered as nonreinforced concrete. Steel sizes and numbers and their metric equivalent are:

Rebar Sizes		Metric Rebar	
#3	⁷⁄₁₆"	#10M	14 mm
#4	⁹⁄₁₆"	#15M	18 mm
#5	¹¹⁄₁₆"	#20M	22 mm
#6	⅞"	#25M	28 mm
#7	1"	#30M	34 mm
#8	1⅛"	#35M	40 mm
#9	1¼"	#45M	50 mm
#10	1⁷⁄₁₆"	#55M	62 mm
#11	1⅝"		
#14	1⅞"		
#18	2½"		

Steel rebar may be smooth, but most uses of steel require deformations to help the concrete bind to the steel. Typical steel deformation patterns can be seen in Figure 31–12. Placed between the deformations are the identification marks of the steel manufacturer. The first letter identifies the mill that produced the rebar, followed by the bar size and the type of steel. The strength of the steel used for rebar can also vary. Common grades associated

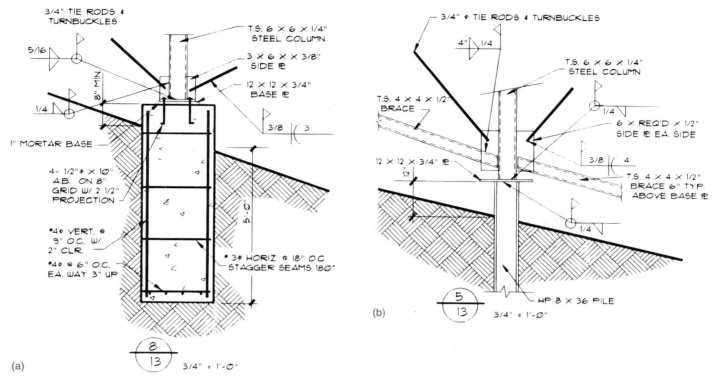

(a)

(b)

Figure 31–8 The drafter draws details of the pilings to supplement the foundation plan to ensure that all of the engineer's specifications can be clearly understood. (a) The construction of a concrete piling that supports steel columns. (b) The connection between a piling and a steel column that extends up to the superstructure of the residence. Because of the loads to be supported, the depth of the pilings, and a severe risk of seismic damage, the engineer has specified horizontal steel tube bracing to the top of the pilings. *Courtesy Residential Designs, AIBD.*

with residential and light commercial construction are grades 40, 50, 60, and 75.

The material used to construct the foundation wall and the area in which the building is to be located will affect how the wall and footing are tied together. If the wall and footing are made at different times, a *keyway* is placed in the footing. The keyway is formed by placing a 2 × 4 (50 × 100 mm) in the top of the concrete footing while the concrete is still wet. Once the concrete has set, the 2 × 4 (50 × 100 mm) is removed, leaving a keyway in the concrete. When the concrete for the wall is poured, it will form a key by filling in the keyway. If a stronger bond is desired, steel is often used to tie the footing to the foundation wall. Both methods of bonding the foundation wall to the footing can be seen in Figure 31–13. Footing steel is not drawn on the foundation plan, but is usually specified in a general note and shown in footing detail similar to that shown in Figure 31–14.

Grade Beams

To provide added support for a stem wall placed in unstable soil, a grade beam may be used in place of the foundation. A grade beam can be seen in Figure 31–4. The grade beam is similar to a wood beam that supports loads over a window. The grade beam is placed under the soil

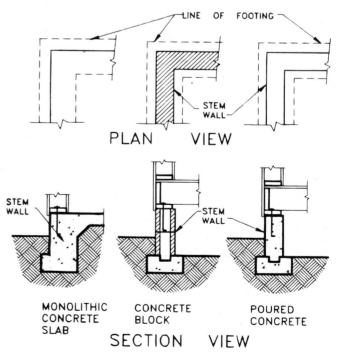

Figure 31–9 The footing is used to disperse building loads evenly into the soil. When drawn on the foundation plan, hidden lines are used.

Uniform Building Code Footing Requirements

Building Height	Footing Width (A)	Wall Width (B)	Footing Height (C)	Depth into Undisturbed Grade (D)
1 story	12″ (305 mm)	6″ (152 mm)	6″ (152 mm)	12″ (305 mm)
2 story	15″ (381 mm)	8″ (203 mm)	7″ (178 mm)	18″ (457 mm)
3 story	18″ (457 mm)	10″ (254 mm)	8″ (203 mm)	24″ (610 mm)

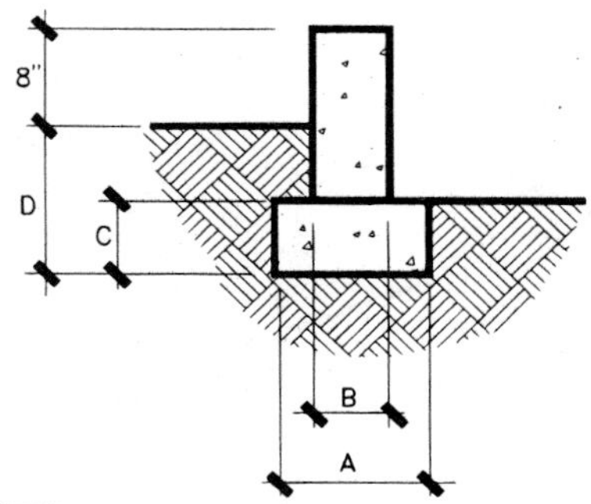

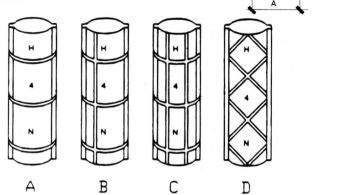

Figure 31–10 Footing design according to the Uniform Building Code and the Basic Natural Building Code. Verify exact depth with the local building department.

TYPE OR LOCATION OF CONCRETE	MINIMUM COMPREHENSIVE STRENGTH WEATHER POTENTIAL		
	NEGLIGIBLE	MODERATE	SEVERE
Basement walls and foundations not exposed to weather	2500	2500	2500
Basement slabs and interior slabs on grade, except garage floor slabs	2500	2500	2500
Basement walls, foundation walls, exterior walls, and other vertical concrete work exposed to weather	2500	3000	3000
Porches, concrete slabs and steps exposed to the weather, and garage floor slabs	2500	3000	3500

Figure 31–11 The compressive strength of concrete based on CABO.

below the stem wall and spans between stable supports. The support may be stable soil or pilings. The depth and reinforcing required for a grade beam are determined by the loads to be supported and are sized by an architect or engineer. A grade beam resembles a footing when drawn on a foundation plan. Steel reinforcing may be specified by notes and referenced to details rather than on the foundation plan.

Fireplace Footings

A masonry fireplace will need to be supported on a footing. Building codes usually require the footing to be a minimum of 12″ deep and extend 6″ past the face of the fireplace on each side. Figure 31–15 shows how fireplace footings are shown on the foundation plan.

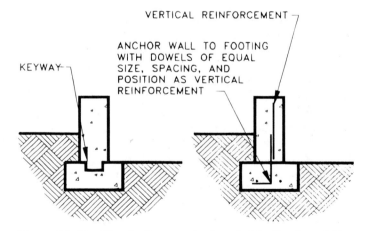

Figure 31–12 Concrete construction is often strengthened by adding steel reinforcing to the face of the concrete member that is in tension. To increase the bonding capacity of the concrete to the steel, ribs or deformations are added to the bar. Between the deformations, identification marks are imprinted to designate the mill that produced the steel, the bar size, and the type of steel.

Figure 31–13 The footing can be bonded to the foundation wall with a key or with steel. *Courtesy National Concrete Masonry Association.*

Veneer Footings

If masonry veneer is used, the footing must be wide enough to provide adequate support for the veneer. Depending on the type of veneer material to be used, the footing is typically widened by 4 to 6″. Figure 31–16 shows common methods of providing footing support for veneer.

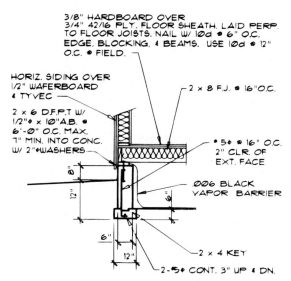

Figure 31–14 When steel is required in the foundation, it usually is specified in a general note or in a detail.

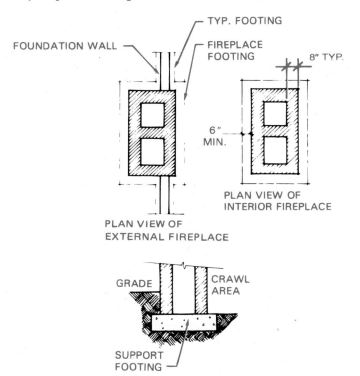

Figure 31–15 A masonry fireplace is required to have a 12″ deep footing that extends 6″ past the face of the fireplace.

Foundation Wall

The foundation wall is the vertical wall that extends from the top of the footing up to the first-floor level of the structure, as shown in Figure 31–9. The foundation wall is usually centered on the footing to help equally disperse the loads being supported. The height of the wall will extend either 6 or 8″ above the ground, depending on building code. This height reflects the minimum distance

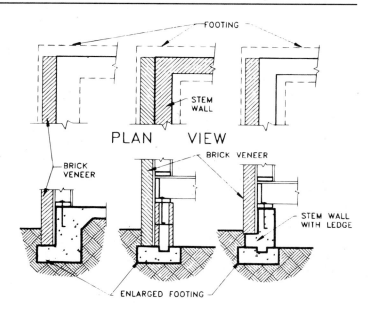

Figure 31–16 When a masonry veneer is added to a wall, additional footing width is needed to provide support.

that is required between wood-framing members and the grade. The required width of the wall varies depending on the code used. Figure 31–10 shows common wall dimensions. Some local codes require the stem wall for a one-level structure to be 8″ (203 mm) wide unless the wall is reinforced with vertical steel to resist problems created from freezing. Common methods of forming the footing and stem wall can be seen in Figure 31–17.

In addition to concrete block and poured concrete foundation walls, many companies are developing alternative methods of forming the stem wall. Blocks made of expanded polystyrene form (EPS) or other lightweight materials can be stacked into the desired position and fit together with interlocking teeth. EPS block forms can be assembled in a much shorter time than traditional form work and remain in place to become part of the finished wall. Reinforcing steel can be set inside the block forms in patterns similar to traditional block walls. Once the forms are assembled, concrete can be pumped into the forms using any of the common methods of pouring. The finished wall has an R-value between R-22 and R-35 depending on the manufacturer. Figure 31–18 shows a foundation being built using EPS forms.

The top of the foundation wall must be level. When the building lot is not level, the foundation wall is often stepped. This helps reduce the material needed to build the foundation wall. As the ground slopes downward, the height of the wall is increased, as shown in Figure 31–19. Foundation walls may not step more than 24″ (610 mm) in one step, and wood framing between the wall and any floor being supported may not be less than 14″ (356 mm)

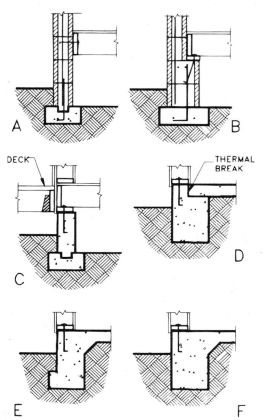

Figure 31–17 Common methods of forming stem walls. (A) Concrete masonry units, with a pressure-treated ledger to support wood floor joist. (B) Floor joist supported on a pressure-treated sill with concrete masonry units. (C) Floor joist supported by pressure-treated sill with joist supporting a wood deck hung from a ledger. (D) Concrete floor slab supported on foundation with an isolation joint between the stem wall and slab to help the slab maintain heat. (E) A stem wall with projected footing. (F) Stem wall and foundation of equal width. Although more concrete is used, it can be quickly formed saving time and material.

in height. The foundation wall also will change heights for a sunken floor. Figure 31–20 shows how a sunken floor can be represented on a foundation plan. Steel anchor bolts are placed in the top of the wall to secure the wood mudsill to the concrete. If concrete blocks are used, the cell in which the bolt is placed must be filled with grout. Figure 31–21 shows anchor bolts and lists minimum requirements of building codes. The mudsill is required to be pressure treated or be made of some other water-resistant wood so that it will not absorb moisture from the concrete. A 2″ (50 mm) round washer is placed over the bolt projecting through the mudsill before the nut is installed to increase the holding power of the bolts. The mudsill and anchor bolt are not drawn on the foundation plan but are typically specified with a note, as seen in Figure 31–20.

The mudsill must also be protected from termites in many parts of the country. Among the most common

Figure 31–18 Blocks made of expanded polystyrene foam provide both a permanent form for pouring the stem wall and insulation to protect the structure from heat loss. *Courtesy American Polysteel Forms.*

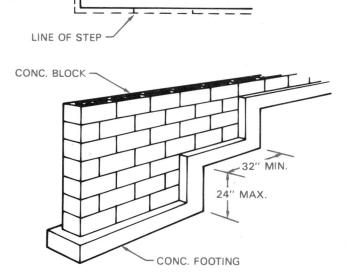

Figure 31–19 The footing and foundation wall are often stepped on sloping lots to save material.

methods of protection are the use of metal caps between the mudsill and the wall, chemical treatment of the soil around the foundation, and chemically treated wood near the ground. Figure 31–22 shows a metal termite shield in place. The metal shield is not drawn on the

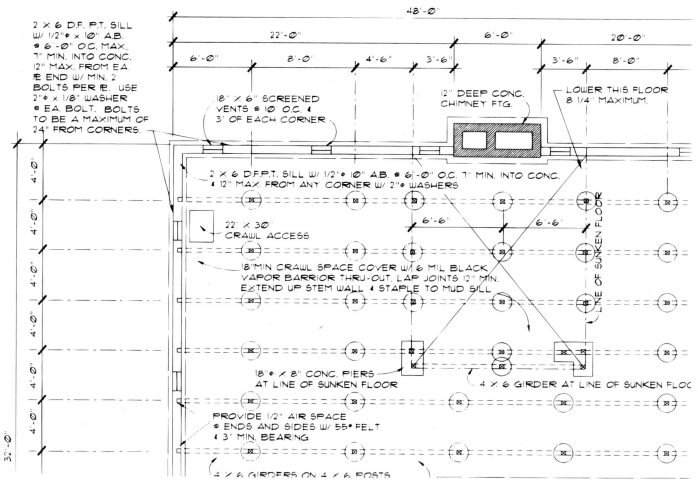

2 × 6 D.F. P.T. SILL W/ 1/2"⌀ × 10" A.B. ⌀ 6'-0" O.C. MAX, 7" MIN. INTO CONC. 12" MAX. FROM EA. ⌀ END W/ MIN. 2 BOLTS PER ⌀. USE 2" × 1/8" WASHER ⌀ EA. BOLT. BOLTS TO BE A MAXIMUM OF 24" FROM CORNERS.

18" × 6" SCREENED VENTS ⌀ 10' O.C. ⌀ 3' OF EACH CORNER

12" DEEP CONC. CHIMNEY FTG.

LOWER THIS FLOOR 8 1/4" MAXIMUM.

2 × 6 D.F.P.T. SILL W/ 1/2"⌀ 10" A.B. ⌀ 6'-0" O.C. 7" MIN. INTO CONC. ⌀ 12" MAX. FROM ANY CORNER W/ 2"⌀ WASHERS

22" × 30" CRAWL ACCESS

18"MIN CRAWL SPACE COVER W/ 6 MIL BLACK VAPOR BARRIOR THRU-OUT, LAP JOINTS 12" MIN. EXTEND UP STEM WALL ⌀ STAPLE TO MUD SILL

LINE OF SUNKEN FLOOR

18"⌀ × 8" CONC. PIERS AT LINE OF SUNKEN FLOOR

4 × 6 GIRDER AT LINE OF SUNKEN FLOOR

PROVIDE 1/2" AIR SPACE ⌀ ENDS AND SIDES W/ 55# FELT ⌀ 3" MIN. BEARING

4 × 6 GIRDERS ON 4 × 6 POSTS

Figure 31–20 Common components of a foundation plan are vents, a crawl access, fireplace footing, sunken floor, beam pockets, stem wall, footings, and piers. *Courtesy Michelle Cartwright.*

BUILDING CODE REQUIREMENTS FOR ANCHOR BOLTS		
	UBC/CABO	BOCA
Min. size	½⌀ (13 mm)	½⌀ (13 mm)
Depth into concrete	7" (178 mm)	8" (203 mm)
Depth into masonry	15" (381 mm)	15" (381 mm)
Max. spacing	6'-0" (1829 mm)	8'-0" (2438 mm)

Figure 31–21 Anchor bolts are set in concrete to hold down the wood framing members. *Photo courtesy John Black.*

foundation plan but is specified in a note near the foundation plan or in a foundation detail.

If the house is to have a wood flooring system, some method of securing support beams to the foundation must be provided. Typically a cavity or beam pocket is built into the foundation wall. The cavity provides both a method of supporting the beam and helps tie the floor and foundation system together. A 3" (75 mm) minimum amount of bearing surface must be provided for the beam. A ½" (13 mm) air space is provided around the beam in

the pocket for air circulation. Figure 31–23 shows common methods of beam support at the foundation wall. Air must also be allowed to circulate under the floor system. To provide ventilation under the floor, vents must be set into the foundation wall. Figure 31–24 shows how foundation vents are located in the foundation wall. If an access opening is not provided in the floor, an opening will need to be provided in the foundation wall for access to the crawl space under the floor. Figure 31–25 shows how the crawl space access can be shown on the foundation plan. Figure 31–20 shows how a crawl access, vents, and girder pockets are typically represented on a foundation plan.

If the foundation is for an energy-efficient structure, insulation is often added the wall. Two-inch (50 mm) rigid insulation is used to protect the wall from cold weather. This can be placed on either side of the wall. If the insulation is placed on the exterior side of the wall, the wall will retain heat from the building. This placement does cause problems in protecting the insulation from damage. Figure 31–26 shows exterior insulation placement.

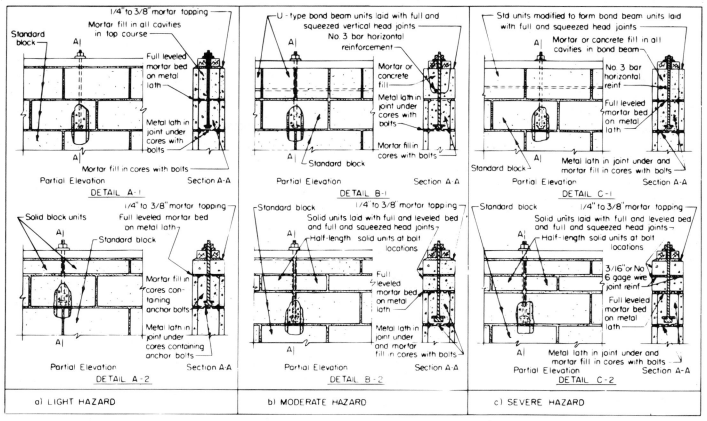

Figure 31–22 Recommended methods of capping concrete masonry foundations supporting wood frame construction in termite-infested areas. *Courtesy National Concrete Masonry Association.*

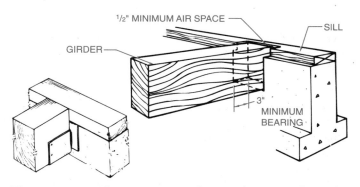

Figure 31–23 A beam seat, or pocket, can be recessed into the foundation wall or the beam may be supported by metal connectors. *Connectors courtesy Simpson Strong-Tie Company, Inc.*

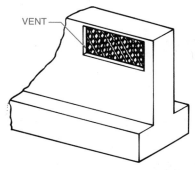

Figure 31–24 Foundation vents must be placed in the foundation wall to provide 1 sq ft ventilation for each 150 sq ft of crawl space. Vents must be placed within 3' of a corner.

In addition to supporting the loads of the structure, the foundation walls also must resist the lateral pressure of the soil pressing against the wall. When the wall is over 24″ (610 mm) in height, vertical steel is usually added to the wall to help reduce tension in the wall. Horizontal steel is also added to foundation walls in some seismic zones to help strengthen the wall. If required, the steel is typically specified in a note.

Retaining Walls

Retaining or basement walls are primarily made of concrete blocks or poured concrete. The material used will depend on labor trends in your area. The material used will affect the height of the wall. If concrete blocks are used, the wall is typically 12 blocks high from the top of the footing. If poured concrete is used, the wall will normally be 8′ (2438 mm) high from the top of the footing. This will allow 4 × 8 (1200 × 2440 mm) sheets of plywood to be used as forms for the sides of the wall.

BUILDING CODE REQUIREMENTS		
UBC	CABO	SBC
18 × 30″	18 × 24″	18 × 24″
(457 × 762 mm)	(457 × 610 mm)	(457 × 610 mm)

Figure 31–25 Minimum foundation access required if a crawl space is provided below the floor system.

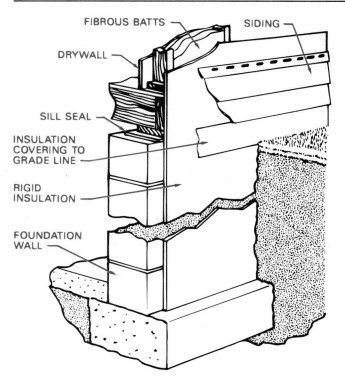

FIBROUS BATTS
SIDING
DRYWALL
SILL SEAL
INSULATION COVERING TO GRADE LINE
RIGID INSULATION
FOUNDATION WALL

Figure 31-26 Insulation is often placed on the foundation wall to help cut heat loss.

Regardless of which material is used, basement walls serve the same functions as the shorter foundation walls. Because of the added height, the lateral forces acting on the side of these walls are magnified. As seen in Figure 31–27, this lateral soil pressure tends to bend the wall inward, thus placing the soil side of the wall in compression and the interior face of the wall in tension. To resist this tensile stress, steel reinforcing may be required by the building department. The seismic zone will affect the size and placement of the steel. Figure 31–28 shows common patterns of steel placement. Figure 31–29 shows a typical foundation detail used to represent the steel placement of a retaining wall.

The footing for a retaining wall is usually 16″ (406 mm) wide and either 8″ or 12″ (200 or 300 mm) deep. The depth depends on the weight to be supported. Steel is typically extended from the footing into the wall. At the top of the wall, anchor bolts are placed in the wall in the same method as with a foundation wall. Anchor bolts for retaining walls are typically placed much closer together than for shorter walls. It is very important that the wall and the floor system be securely tied together. The floor system is used to help strengthen the wall and resist the soil pressure. Where seismic danger is great, metal angles are added to the anchor bolts to make the tie between the wall and the floor joist extremely rigid. Figure 31–30 shows common wall-to-floor connections for a retaining wall. These connections are shown on the foundation plan, as seen in Figure 31–31.

To reduce soil pressure next to the footing, a drain is installed. The drain is set at the base of the footing to collect and divert water from the face of the wall. Figure 31–32 shows how the drain is typically placed. Generally the area above the drain is filled with gravel so that subsurface water will percolate easily down to the drain and away from the wall. By reducing the water content of the soil, the lateral pressure on the wall is reduced. The drain

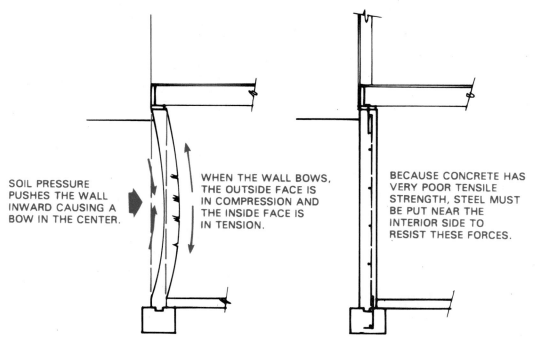

SOIL PRESSURE PUSHES THE WALL INWARD CAUSING A BOW IN THE CENTER.

WHEN THE WALL BOWS, THE OUTSIDE FACE IS IN COMPRESSION AND THE INSIDE FACE IS IN TENSION.

BECAUSE CONCRETE HAS VERY POOR TENSILE STRENGTH, STEEL MUST BE PUT NEAR THE INTERIOR SIDE TO RESIST THESE FORCES.

Figure 31-27 Stresses acting on a beam retaining wall. The wall serves as a beam spanning between each floor, and the soil is the supported load.

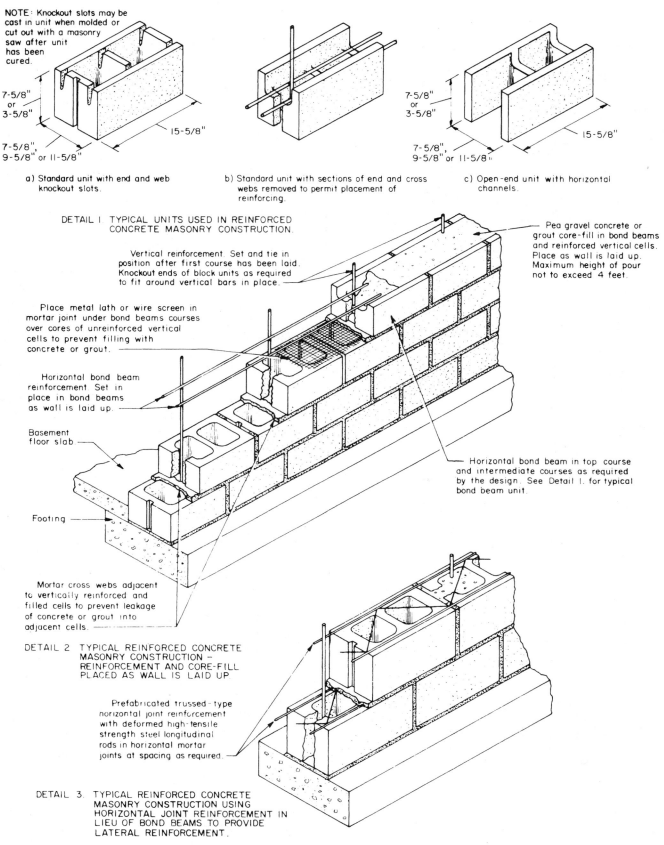

NOTE: Knockout slots may be cast in unit when molded or cut out with a masonry saw after unit has been cured.

7-5/8" or 3-5/8"

7-5/8", 9-5/8" or 11-5/8"

15-5/8"

a) Standard unit with end and web knockout slots.

b) Standard unit with sections of end and cross webs removed to permit placement of reinforcing.

7-5/8" or 3-5/8"

7-5/8", 9-5/8" or 11-5/8"

15-5/8"

c) Open-end unit with horizontal channels.

DETAIL 1. TYPICAL UNITS USED IN REINFORCED CONCRETE MASONRY CONSTRUCTION.

Vertical reinforcement. Set and tie in position after first course has been laid. Knockout ends of block units as required to fit around vertical bars in place.

Pea gravel concrete or grout core-fill in bond beams and reinforced vertical cells. Place as wall is laid up. Maximum height of pour not to exceed 4 feet.

Place metal lath or wire screen in mortar joint under bond beams courses over cores of unreinforced vertical cells to prevent filling with concrete or grout.

Horizontal bond beam reinforcement. Set in place in bond beams as wall is laid up.

Basement floor slab.

Footing

Horizontal bond beam in top course and intermediate courses as required by the design. See Detail 1. for typical bond beam unit.

Mortar cross webs adjacent to vertically reinforced and filled cells to prevent leakage of concrete or grout into adjacent cells.

DETAIL 2 TYPICAL REINFORCED CONCRETE MASONRY CONSTRUCTION – REINFORCEMENT AND CORE-FILL PLACED AS WALL IS LAID UP.

Prefabricated trussed-type horizontal joint reinforcement with deformed high-tensile strength steel longitudinal rods in horizontal mortar joints at spacing as required.

DETAIL 3. TYPICAL REINFORCED CONCRETE MASONRY CONSTRUCTION USING HORIZONTAL JOINT REINFORCEMENT IN LIEU OF BOND BEAMS TO PROVIDE LATERAL REINFORCEMENT.

Figure 31-28 Suggested construction details for reinforced concrete masonry foundation walls. *Courtesy National Concrete Masonry Association.*

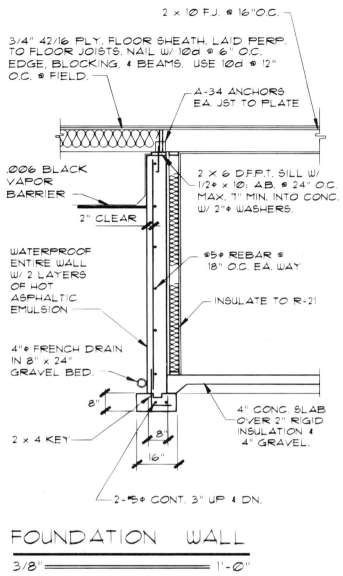

2 × 10 F.J. @ 16"O.C.

3/4" 42/16 PLY. FLOOR SHEATH. LAID PERP.
TO FLOOR JOISTS. NAIL W/ 10d @ 6" O.C.
EDGE, BLOCKING, & BEAMS. USE 10d @ 12"
O.C. @ FIELD.

A-34 ANCHORS
EA. JST TO PLATE

.006 BLACK
VAPOR
BARRIER

2" CLEAR

2 × 6 D.F.P.T. SILL W/
1/2⌀ × 10: A.B. @ 24" O.C.
MAX. 7" MIN. INTO CONC.
W/ 2"⌀ WASHERS.

WATERPROOF
ENTIRE WALL
W/ 2 LAYERS
OF HOT
ASPHALTIC
EMULSION

@5⌀ REBAR @
18" O.C. EA. WAY

INSULATE TO R-21

4"⌀ FRENCH DRAIN
IN 8" × 24"
GRAVEL BED.

8"

2 × 4 KEY

8"

16"

4" CONC. SLAB
OVER 2" RIGID
INSULATION &
4" GRAVEL.

2-#5⌀ CONT. 3" UP & DN.

FOUNDATION WALL

3/8" = 1'-∅"

Figure 31–29 The components of a concrete retaining wall are typically specified in a section or detail. Key components are the building material of the wall, the reinforcing material, floor attachment, waterproofing, and drainage method. *Courtesy Residential Designs, AIBD.*

is not drawn on the foundation plan but must be specified in a note on the foundation plan, sections, and foundation details.

No matter what the soil condition, the basement wall should be waterproofed to help reduce moisture from passing through the wall into the living area. Walls are typically covered with two coats of bituminous waterproofing material or fiber-reinforced asphaltic mastic.

Adding windows to the basement can help cut down the moisture content of the basement. This will sometimes require adding a window well to prevent the ground from being pushed in front of the window. Figure

31–31 shows how a window well, or areaway as it is sometimes called, can be drawn on the foundation plan.

Similar to a foundation wall, the basement wall also needs to be protected from termites. Figure 31–33 shows common ways that wood can be protected. When drawing the foundation plan, the metal shield will not be drawn, but it should be called out in a note.

Treated-Wood Basement Walls

Pressure-treated lumber can be used to frame both crawl space and basement walls. Treated-wood basement walls allow for easy installation of electrical wiring, insulation, and finishing materials.

Instead of a concrete foundation, a gravel bed is used to support the wall loads. Gravel is typically required to extend 6" (150 mm) on each side of the wall and be approximately 4" (100 mm) deep. The exact width and depth of the gravel bed are determined by an engineer based on the loads to be supported and the soil bearing capacity. A 2 × (50 mm) pressure-treated plate is laid on the gravel, and the wall is built of pressure-treated wood, as seen in Figure 31–34. Pressure-treated ½" (13 mm) thick C-D grade plywood with exterior glue is laid perpendicular to the studs and covered with 6 mil polyethylene and sealed with an adhesive.

Restraining Walls

As seen in Figure 31–35, when a structure is built on a sloping site, the masonry wall may not need to be full height. Although less soil is retained than with a full-height wall, more problems are encountered. Figure 31–36 shows the tendencies for bending for this type of wall. Because the wall is not supported at the top by the floor, the soil pressure must be resisted through the footing. This requires a larger footing than for a full-height wall. If the wall and footing connection is rigid enough to keep the wall in position, the soil pressure will tend to try to overturn the whole foundation. The extra footing width is to resist the tendency to overturn. Figure 31–37 shows an example of a detail that a drafter might be required to draw and how the wall would be represented on the foundation plan. Figure 31–38 shows how a restraining wall will be represented on a foundation plan. Depending on the slope of the ground being supported, a key may be required, as seen in Figure 31–39. A keyway on the top of the footing has been discussed. This key is added to the bottom of the footing to help keep it from sliding as a result of soil pressure against the wall. Generally the key is not shown on the foundation plan but is shown in a detail of the wall.

Interior Supports

Foundation walls and footings are typically part of the foundation system that supports the exterior shape of the

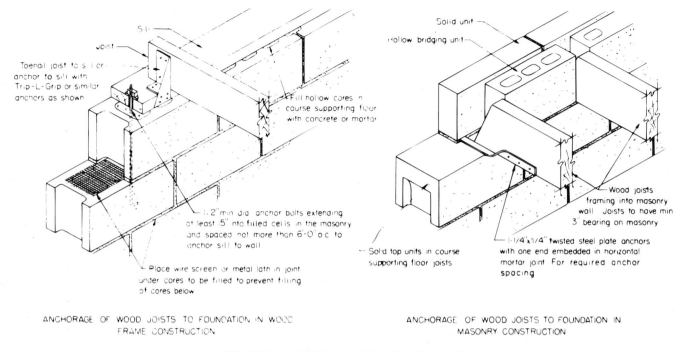

ANCHORAGE OF WOOD JOISTS TO FOUNDATION IN WOOD
FRAME CONSTRUCTION

ANCHORAGE OF WOOD JOISTS TO FOUNDATION IN
MASONRY CONSTRUCTION

SUGGESTED METHODS OF ANCHORING WOOD JOISTS
BEARING ON CONCRETE MASONRY FOUNDATION WALLS.

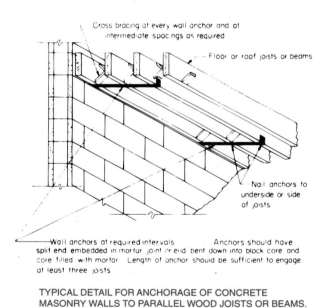

TYPICAL DETAIL FOR ANCHORAGE OF CONCRETE
MASONRY WALLS TO PARALLEL WOOD JOISTS OR BEAMS.

Figure 31–30 Metal angles and straps are typically used on basement walls to ensure a rigid connection between the floor and wall. *Courtesy National Concrete Masonry Association.*

structure. Interior loads are generally supported on spot footings, or piers, as seen in Figure 31–40. Piers are formed by using either preformed or framed forms, or by excavation. Pier depth is generally required to match that of the footings. Although Figure 31–40 shows the piers above the finished grade, many states require the mass of the footing to be placed below grade to prevent problems with freezing and shifting from seismic activity. The placement of piers will be determined by the type of floor system to be used and will be discussed in Chapter 32. The size of the pier will depend on the load being supported. Piers are usually drawn on the foundation plan with dotted lines, as shown in Figure 31–41.

Metal Connectors

Metal connectors are often used at the foundation level to resist the stresses from wind and seismic forces. Three

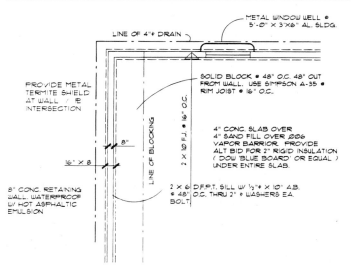

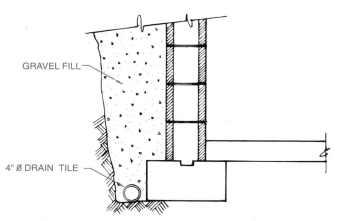

Figure 31–32 A drain is placed in a gravel bed to help disperse water away from the wall area.

Figure 31–31 Full-height basement walls require a window well to restrain the soil around windows. If bedrooms are located in a basement, windows must be within 44″ (1118 mm) of the finish floor. Many building codes require the floor to be blocked for extra rigidity to help the retaining wall resist lateral pressure.

used will determine how it is specified. Figure 31–43 shows how these connectors might be specified on the foundation plan.

▼ DIMENSIONING FOUNDATION COMPONENTS

of the most commonly used metal connectors are shown in Figure 31–42. Each is accompanied by a table of standard dimensions. The senior drafter and designer are required to determine the proper size connector and to specify it on the foundation plan. How the connector is

Throughout this chapter we have seen the typical components of a foundation system and how they are drawn on the foundation plan. The manner in which they are dimensioned is equally important to the construction crew. The line quality for the dimension and leader lines is the same as was used on the floor plan. Jogs in the

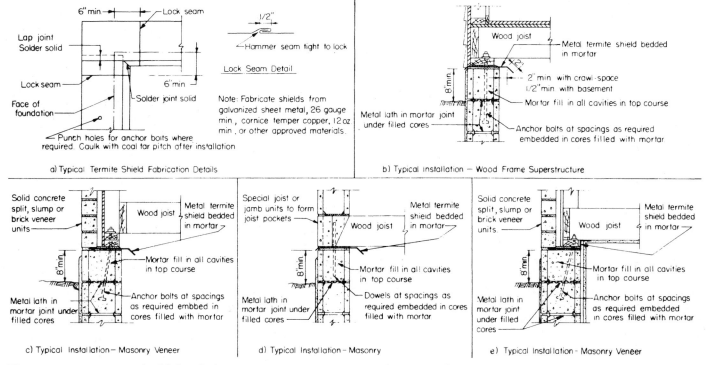

Figure 31–33 Termite shield details for retaining walls. *Courtesy National Concrete Masonry Association.*

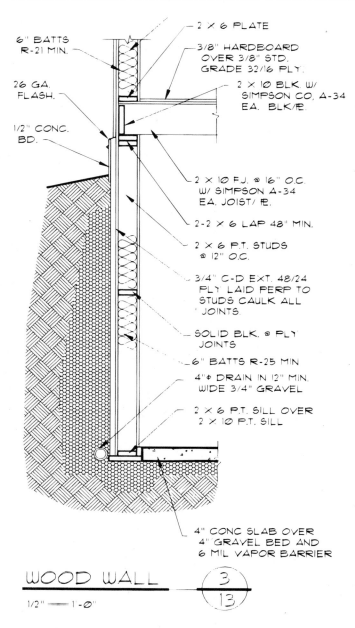

6" BATTS R-21 MIN.

2 × 6 PLATE

3/8" HARDBOARD OVER 3/8" STD. GRADE 32/16 PLY.

26 GA. FLASH.

2 × 10 BLK W/ SIMPSON CO. A-34 EA. BLK/℉.

1/2" CONC. BD.

2 × 10 F.J. @ 16" O.C. W/ SIMPSON A-34 EA. JOIST/ ℉.

2-2 × 6 LAP 48" MIN.

2 × 6 P.T. STUDS @ 12" O.C.

3/4" C-D EXT. 48/24 PLY LAID PERP TO STUDS CAULK ALL JOINTS.

SOLID BLK. @ PLY JOINTS

6" BATTS R-25 MIN

4"⌀ DRAIN IN 12" MIN. WIDE 3/4" GRAVEL

2 × 6 P.T. SILL OVER 2 × 10 P.T. SILL

4" CONC SLAB OVER 4" GRAVEL BED AND 6 MIL VAPOR BARRIER

WOOD WALL 3/13

1/2" = 1'-0"

GENERAL NOTES: ASSUMED SOIL PRESSURE 30° PER CUBIC FOOT.

WATER PROOF EXTERIOR SIDE OF WALL WITH 2 LAYER HOT ASPHALTIC EMULSION COVERED WITH 6 MIL. VAPOR BARRIER.

PROVIDE ALTERNATE BID FOR 2" RIGID INSULATION ON EXTERIOR FACE FOR FULL HEIGHT.

ALL MATERIAL BELOW UPPER TOP PLATE OF BASEMENT WALL TO BE PRESSURE PRESERVATIVELY TREATED IN ACCORDANCE WITH AWPA-C22 AND SO MARKED.

Figure 31–34 Stem walls and basement retaining walls can be framed using pressure-treated wood with no foundation. The size and spacing of members are determined by the loads being supported and the height of the wall.

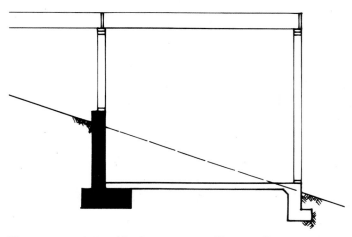

Figure 31–35 An 8′ high retaining wall is usually not required on sloping lots. A partial restraining wall with wood-framed walls above it is often used.

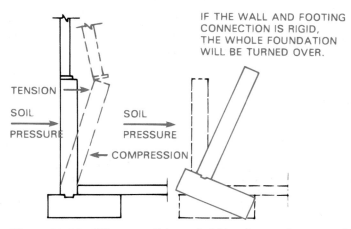

Figure 31–36 When a wall is not held in place at the top, soil pressure will tend to move the wall inward. The intersection of the wood and concrete walls is called a hinge point because of the tendency to move.

foundation wall are dimensioned using the same methods used on the floor plan. Most of the dimensions for major shapes will be exactly the same as the corresponding dimensions on the floor plan. When the foundation plan is drawn using a computer, the foundation plan is typically placed in the same drawing file as the floor plan. This will allow the overall dimensions for the floor plan to be displayed on the foundation plan if different layers are used to place the dimensions. Layers such as BASEDIM, FLOORDIM, and FNDDIMEN will help differentiate between dimensions for the floor and foundation.

A different method is used to dimension the interior walls of a foundation that is used on a floor plan. Foundation walls are dimensioned from face to face rather than face to center, as on a floor plan. Footing widths are usually dimensioned from center to center. Each type of dimension can be seen in Figure 31–39.

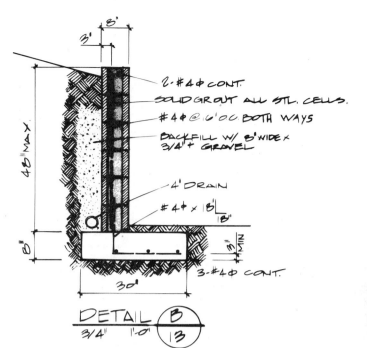

2-#4∅ CONT.

SOLID GROUT ALL STL. CELLS.

#4∅@16"OC BOTH WAYS

BACKFILL W/ 8'WIDE x 3/4"∅ GRAVEL

4"DRAIN

#4∅ x 18" @ 18"

3-#4∅ CONT.

48" MAX

8"

3" MIN

30'

DETAIL ⟨B/13⟩
3/4" = 1'-0"

Figure 31–37 A detail of a partial wall.

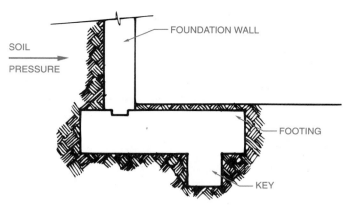

FOUNDATION WALL

SOIL PRESSURE

FOOTING

KEY

Figure 31–39 When soil pressure is too great for the wall to withstand, a key may be added to the bottom of the footing. The key provides added surface area to help resist sliding.

Figure 31–40 Concrete piers can be formed by using preformed or framed forms or by excavating. In the background, the footings steel is exposed to tie into the stem walls. Note the step in the footing to accommodate the slope of the site. *Courtesy Michelle Cartwright.*

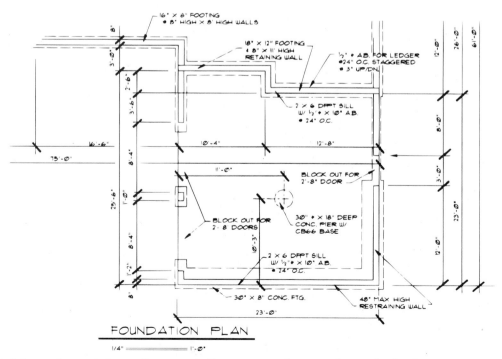

FOUNDATION PLAN
1/4" = 1'-0"

Figure 31–38 Retaining and restraining walls are typically represented in a similar method to footings and stem walls. Because of the added height, the width is typically specified on the foundation plan, and the building components are specified in a detail or section. *Courtesy Residential Designs, AIBD.*

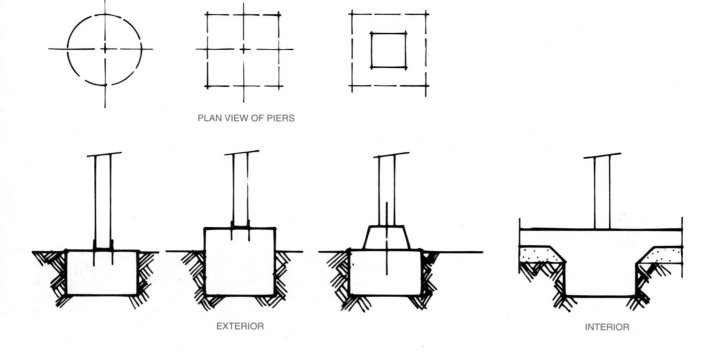

PLAN VIEW OF PIERS

EXTERIOR

INTERIOR

PIERS IN SIDE ELEVATION

Figure 31–41 Concrete piers are used to support interior loads. Most buildings codes require wood to be at least 8″ above the grade, which means that piers must also extend 8″ above the grade.

Model No.	DIMENSIONS		I.C.B.O. LOADS (12-16d NAILS)		U.B.C. Calc. (2)-½″ Dia.
	W	L	Vert. Up	Lateral	Vert. Up
PB44	3 9/16″	3 5/8″	1320	1320	—
PB46	5 1/2″	3 5/8″	1320	1320	—
PB66	5 1/2″	5 5/8″	1610	1610	3225
PB44R	4″	3 5/8″	1540	1540	—
PB46R	6″	3 5/8″	1320	1320	3225
PB66R	6″	5 5/8″	1610	1610	3225

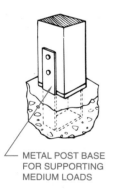

METAL POST BASE FOR SUPPORTING MEDIUM LOADS

Model No.	W	L	Material Stirrups	Bolts	Uplift Design Loads
CB44	3 9/16″	3 5/8″	3/16″x2″	(2) 5/8″	5030
CB46	3 9/16″	5 1/2″	3/16″x2″	(2) 5/8″	5030
CB48	3 9/16″	7 1/2″	3/16″x2″	(2) 5/8″	5030
CB5	5 1/4″	Specify	3/16″x3″	(2) 5/8″	5030
CB66	5 1/2″	5 1/2″	3/16″x3″	(2) 5/8″	5030
CB68	5 1/2″	7 1/2″	3/16″x3″	(2) 5/8″	5030
CB610	5 1/2″	9 1/2″	3/16″x3″	(2) 5/8″	5030
CB612	5 1/2″	11 1/2″	3/16″x3″	(2) 5/8″	5030
CB7	6 7/8″	Specify	1/4″x3″	(2) 3/4″	7230
CB88	7 1/2″	7 1/2″	1/4″x3″	(2) 3/4″	7230
CB810	7 1/2″	9 1/2″	1/4″x3″	(2) 3/4″	7230
CB812	7 1/2″	11 1/2″	1/4″x3″	(2) 3/4″	7230
CB9	8 7/8″	Specify	1/4″x3″	(2) 3/4″	7230
CB1010	9 1/2″	9 1/2″	1/4″x3″	(2) 3/4″	7230
CB1012	9 1/2″	11 1/2″	1/4″x3″	(2) 3/4″	7230
CB1212	11 1/2″	11 1/2″	1/4″x3″	(2) 3/4″	7230

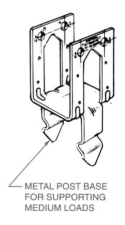

METAL POST BASE FOR SUPPORTING MEDIUM LOADS

Model No.	GA.	Dimensions					Bolt Attachment – Concrete		Stud	Average Test Ultimate	Design Load Value When Installed On Stud Thickness Of					Maximum Allowable Short Term Loading	Minimum Weld
		H	W	D	SO	CL	DIA.	Min. Embedment			1½″	2″	2½″	3″	3½″		
HD2	7	5 3/4	2 1/2	2 5/8	2 1/2	1 3/8	5/8	9″	(2)-5/8″MB	13,200	2065	2520	2520	2520	2520	3360	3/16″ × 2″
HD5	7	6 5/16	3 1/2	3 1/2	2 1/2		3/4	11″	(2)-3/4″MB	19,000	2485	2365	3610	3610	3610	4810	3/16″ × 8″
HD6	1/4	12 1/2	2 7/8	3 3/16	3″	1″		14″	(3)-3/4″MB	18,600	3730	4895	5410	5410	5410	6080	1/4″ × 2 1/2″
HD7	1/4	11 11/16	3 1/2	4″	2 7/8	2 1/8	1″	14″	(3)-7/8″MB	28,600	4355	5775	6500	6500	6500	8665	1/4″ × 9 1/2″
HD7	1/4	11 11/16	3 1/2	4″	2 7/8	2 1/8	1 1/8″	15″	(3)-7/8″MB	28,600	4355	5775	7195	7500	7500	9530	1/4″ × 9 1/2″
HD9	1/4	16 1/2	3 1/2	4 1/4	3 3/8	2 1/8	1 1/8″	27″	(3)-1″MB	—	—	—	—	—	11,210	14,940	1/4″ × 9 1/2″ Bottom, 3/16″ ea. face
HD12	1/4	20 1/2	3 1/2	4 1/4	3 3/8	2 1/8	1 1/8″	32″	(4)-1″MB	48,000	—	—	—	—	14,945	16,000	1/4″ × 9 1/2″ Bottom, 3/8″ ea. face
HD15	1/4	24 1/2	3 1/2	4 1/4	3 3/8	2 1/8	1 1/2″	38″	(5)-1″MB	55,300	—	—	—	—	18,336	18,330	1/4″ × 9 1/2″ Bottom, 3/8″ ea. face
HD2N	12	7 7/8	2 1/2	2″		1 1/8	5/8	9″	(2)-5/8″MB	8,800	2070	2520	2520	2520	2520	2520	1/8″ × 1″ (1/4″ face, 3/4″ top) ea. side
HD5N	10	9 3/8	2 5/8	2 1/4	2 3/16	1 1/8	3/4	11″	(2)-3/4″MB	11,600	2485	3265	3610	3610	3610	3610	1/8″ × 1″ (3/8″ face, 5/8″ top) ea. side
HD7N	7	12″	3 3/8	2 7/8	2 7/8	1 9/16	1″	14″	(2)-1″MB	20,300	3305	4410	5510	6490	6500	6500	3/16″ × 1 1/2″ (1/2″ face, 5/8″ top) ea. side

HOLD-DOWN ANCHORS FOR CONNECTING MEMBERS

Figure 31–42 Three of the most common types of metal connectors used on the foundation system. Drafters are often required to use vendor information when drawing foundation plans. *Courtesy Simpson Strong-Tie Company, Inc.*

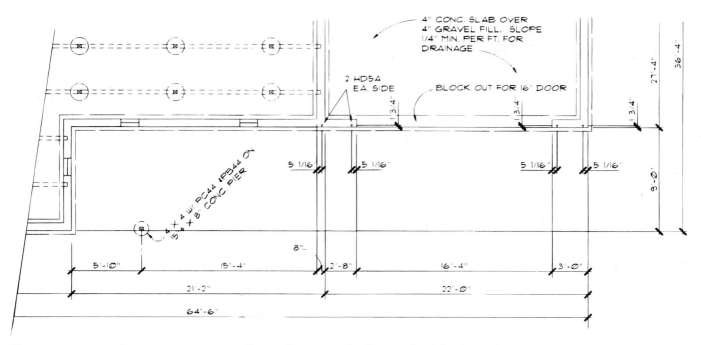

4" CONC. SLAB OVER
4" GRAVEL FILL. SLOPE
1/4" MIN. PER FT. FOR
DRAINAGE

2 HD5A
EA. SIDE

BLOCK OUT FOR 16' DOOR

Figure 31–43 Metal connectors are typically used to resist the forces of uplift, shear, flooding and seismic stresses. *Courtesy Michelle Cartwright.*

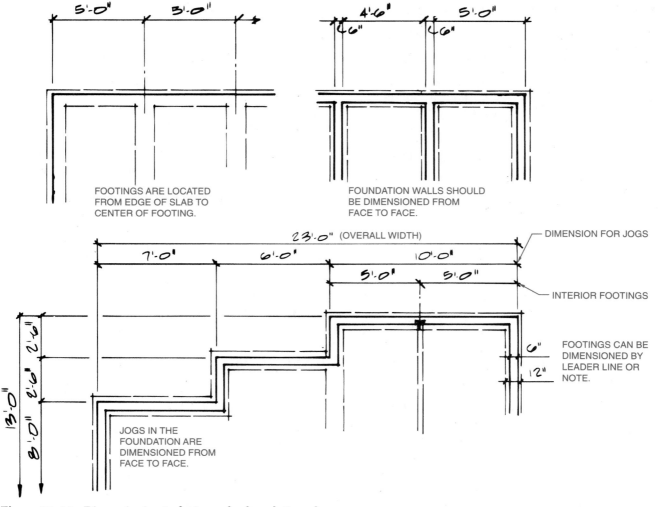

FOOTINGS ARE LOCATED
FROM EDGE OF SLAB TO
CENTER OF FOOTING.

FOUNDATION WALLS SHOULD
BE DIMENSIONED FROM
FACE TO FACE.

23'-0" (OVERALL WIDTH)

DIMENSION FOR JOGS

INTERIOR FOOTINGS

FOOTINGS CAN BE
DIMENSIONED BY
LEADER LINE OR
NOTE.

JOGS IN THE
FOUNDATION ARE
DIMENSIONED FROM
FACE TO FACE.

Figure 31–44 Dimensioning techniques for foundation plans.

CHAPTER 31

Foundation Systems Test

DIRECTIONS

Answer the questions with short, complete statements or drawings as needed.

1. Letter your name and the date at the top of the sheet.
2. Letter the question number and provide the answer. You do not need to write out the question.

QUESTIONS

Question 31–1   What are the major parts of the foundation system?

Question 31–2   What methods are used to determine the size of footings?

Question 31–3   List five forces that a foundation must withstand.

Question 31–4   How can the soil texture influence a foundation?

Question 31–5   List the major types of material used to build foundation walls.

Question 31–6   Why is steel placed in footings?

Question 31–7   Describe when a stepped footing might be used.

Question 31–8   What size footing should be used with a basement wall?

Question 31–9   What influences the size of piers?

Question 31–10   Describe the difference between a retaining and a restraining wall.

PROBLEMS

DIRECTIONS

Unless other instructions are given by your instructor, complete the following details for a one-level house. Use a scale of ¾″ = 1′–0″ with proper line weight and quality and provide notes required to label typical materials. Provide dimensions based on common practice in your area. Omit insulation unless specified.

Problem 31–1   Show a typical foundation with poured concrete with a 2 × 4 key, 2 × 6 f.j. at 16″ o.c., and 2 × 4 studs at 24″ o.c.

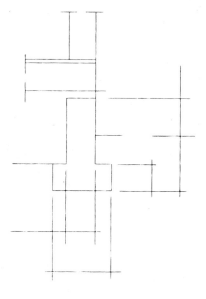

Problem 31–2   Show a typical post-and-beam foundation system. Show 4 × 6 girders @ 48″ o.c. parallel to the stem wall.

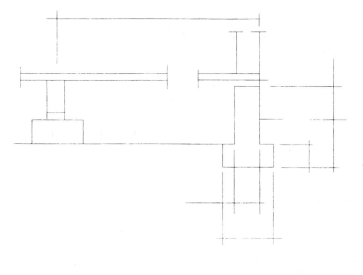

Problem 31–3 Draw a foundation showing a concrete floor system. Use #10 × 10-4″ × 4″ wwm 2″ dn. from top surface and 2 #4 bars centered in the footing 2″ up and down. Use 2 × 4 studs at 16″ o.c. for walls.

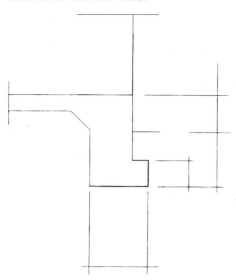

Problem 31–4 Draw a detail showing an interior footing for a concrete slab supporting a 2 × 4 stud-bearing wall. Anchor the wall with Ramset-type fasteners (or equal). Use the same slab reinforcing that was used in Problem 31–3.

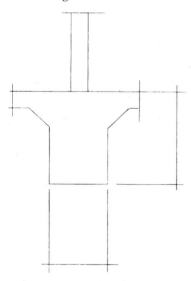

Problem 31–5 Design a retaining wall for an 8′ tall basement, using concrete blocks. Use 2 × 10 f.j. parallel to wall, w/wall, w/⅝″ dia. a.b. @ 32″ o.c. Show typical solid blocking and anchor rim joists with A35 anchors at 16″ o.c. Show #5 @ 18″ o.c. each way @ 2″ from tension side with 1-#5 in footing 2″ up. Use a scale of ½″ = 1′–0″.

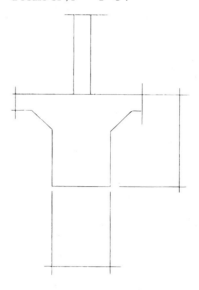

Problem 31–6 Show a detail of the intersection of a 48″ high concrete restraining wall. Use a 30″ wide footing with 2-#5 cont. 3″ up @ 12″ o.c. and #5 @ 24″ o.c. in wall 2″ from tension side. Use 4″ drain in 8 × 30″ gravel bed. Use 2 × 6 sill with ⅝″ a.b. @ 24″ o.c. Use an 8′ ceiling with 2 × 10 f.j. @ 16″ o.c. Cantilever 15″ past wall w/1″ exterior stucco. Use a scale of ½″ = 1′–0″.

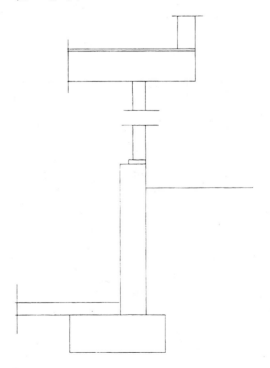

CHAPTER 32

Floor Systems and Foundation Support

The foundation plan not only shows the concrete footings and walls but also the members that are used to form the floor. Two common types of floor systems are typically used in residential construction: floor systems with a crawl space or basement below the floor system and a floor system built at grade level. Each has its own components and information that must be put on a foundation plan.

▼ ON-GRADE FOUNDATIONS

A concrete slab is often used for the floor system of residential or commercial structures. A concrete slab provides a firm floor system with little or no maintenance and generally requires less material and labor than conventional wood floor systems. The floor slab is usually poured as an extension of the foundation wall and footing, in what is referred to as monolithic construction. See Figure 32–1. Other common methods of pouring the foundation and floor system are shown in Figure 32–2.

A 3½″ (89 mm) concrete slab is the minimum thickness allowed by model building codes for residential floor slabs. Commercial slabs are often 5 or 6″ (127–152 mm) thick, depending on the floor loads to be supported. The slab is used only as a floor surface, not to support the weight of the walls or roof. If load-bearing walls must be supported, the floor slab must be thickened, as shown in Figure 32–3. If a load is concentrated in a small area, a pier may be placed under the slab to help disperse the weight, as seen in Figure 32–4.

Slab Joints

Concrete tends to shrink approximately 0.66″ per 100′ (16.7 mm/30480 mm) as the moisture in the mix hydrates and the concrete hardens. Concrete also continues to expand and shrink throughout its life based on the temperatures and the moisture in the supporting soil. This shrinkage can cause the floor slab to crack. To help control possible cracking, three types of joints can be placed in the slab: control, construction, and isolation joints. See Figure 32–5.

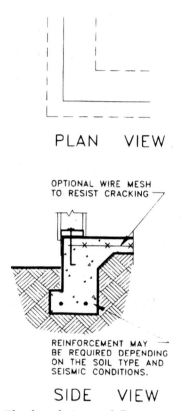

PLAN VIEW

OPTIONAL WIRE MESH TO RESIST CRACKING

REINFORCEMENT MAY BE REQUIRED DEPENDING ON THE SOIL TYPE AND SEISMIC CONDITIONS.

SIDE VIEW

Figure 32–1 The foundation and floor systems can often be constructed in one pour, saving time and money.

Control Joints. As the slab contracts during the initial drying process, tensile stress is created in the lower surface of the slab as it rubs against the soil. The friction between the slab and the soil causes cracking, which can be controlled by *control* or *contraction joints*. These joints do not prevent cracking, but they control where the cracks will develop in the slab. Control joints can be created by cutting the fresh concrete, placing a strip into the concrete during pouring, or sawing the concrete within 6 to 8 hours of placement. Joints are usually one-quarter of the slab depth. Because the slab has been weakened, any cracking due to stress will result along the joint. The American Concrete Institute (ACI) suggests that control joints be placed a distance in feet equal to

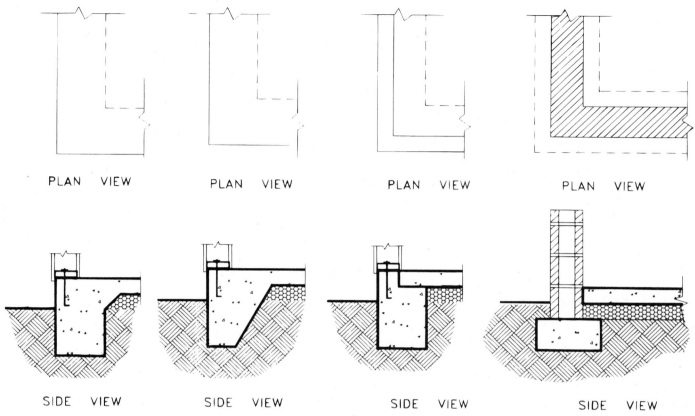

Figure 32–2   Common foundation and slab intersections.

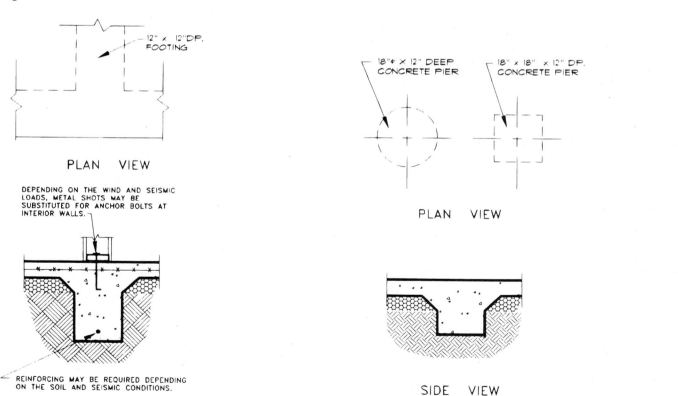

Figure 32–3   Footing and floor intersections at an interior load-bearing wall.

Figure 32–4   A concrete pier can be poured under loads concentrated in small areas. Piers are typically round or rectangular, with the size of the pier determined by the load to be supported.

CONTROL
JOINT

CONSTRUCTION
JOINT

ISOLATION
JOINT

Figure 32–5 Joints are placed in concrete to control cracking. A control joint is placed in the slab to weaken the slab and cause cracking to occur along the joint rather than throughout the slab. When construction must be interrupted, a construction joint is formed to increase bonding with the next day's pour. An isolation joint is provided to keep stress from one structural material from cracking another.

about two and a half times the slab depth in inches. For a 4″ (102 mm) slab, joints would be placed at approximately 10′ (3048 mm) intervals. The locations of control joints are usually specified in note form for residential slabs. The spacing and method or placement can be specified in the general notes place with the foundation plan.

Construction Joints. When concrete construction must be interrupted, a *construction joint* is used to provide a clean surface where work can be resumed. Because a vertical edge of one slab will have no bond to the next slab, a keyed joint is used to provide support between the two slabs. The key is typically formed by placing a beveled strip that is about one-fifth of the slab thickness and one-tenth of the slab thickness in width to the form used to mold the slab. The method used to form the joint can be specified in note form on the foundation notes, but the crew placing the concrete will determine the location.

Isolation Joints. *Isolation* or *expansion joints* are used to separate a slab from adjacent slabs, walls, columns, or other part of the structure. The joint prevents forces from an adjoining structural member from being transferred into the slab, causing cracking. These joints also allow for expansion of the slab caused by moisture or temperature. Isolation joints are typically between ¼ and ½″ (6–13 mm) wide. The location of isolation joints should be specified on the foundation plan. Because of the small size of residential foundations, isolation joints are not usually required. Chapter 46 discusses the drawing and specification of isolation joints in more detail.

Slab Placement

The slab may be placed above, below, or at grade level. Residential slabs are often placed above grade for hillside construction to provide a suitable floor for a garage. If the cost of filling between the slab and the natural grade is excessive, a platform can be used to support a lightweight concrete floor slab. Concrete is considered lightweight

depending on the amount of air that is pumped into the mixture during the manufacturing process. Above-grade residential slabs are typically supported by a wood-framed platform covered with plywood sheathing. Ribbed metal platforms are typically used for heavier floors found in commercial construction. The components of an above-ground concrete floor are typically noted on a framing plan but not drawn. The foundation plan shows the columns and footings used to support the increased weight of the floor. An example of a framing plan to support an above-ground concrete slab can be seen in Figure 32–6. The foundation plan for the same area can be seen in Figure 32–7. The construction process will also require details similar to those in Figure 32–8.

Slabs built below grade are most commonly used in basements. When used at grade, the slab is usually placed just above grade level. Most building codes require the top of the slab to be 8″ (203 mm) above the finish grade to keep structural wood away from the ground moisture.

Slab Preparation

When a slab is built at grade, approximately 8–12″ inches (200–300 mm) of topsoil is removed to provide a stable, level building site. Excavation usually extends about 5′ (1500 mm) beyond the building size to allow for the operation of excavating equipment needed to trench for the footings. Once forms for the footings have been set, fill material can be spread to support the slab. Most

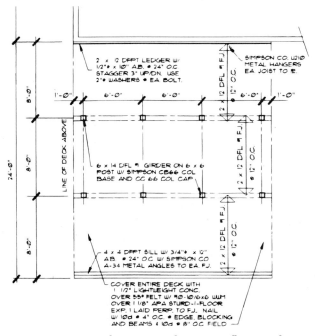

Figure 32–6 An above-ground concrete floor is often used for a garage floor and driveway for hillside residential construction. The framing plan shows the framing of the platform used to support the concrete slab.

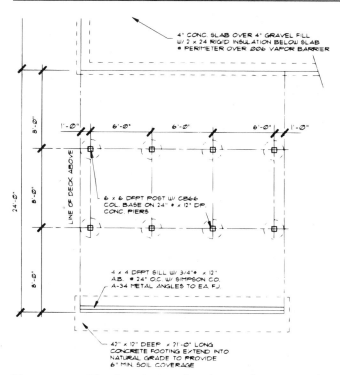

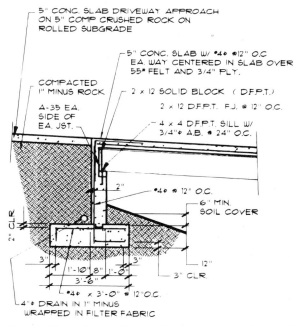

Figure 32–7 The foundation plan for an above-ground concerete slab typically shows the support piers for the framing platform.

Figure 32–8 In addition to the framing and foundation plans, details are used to clarify the concrete reinforcement.

codes require the slab to be placed on a 4″ (102 mm)-minimum base of compacted sand or gravel fill. The area for which the building is designed will dictate the type of fill material that is typically used. The fill material provides a level base for the concrete slab and helps eliminate cracking in the slab caused by settling of the ground

under the slab. The fill material is not shown on the foundation plan but is specified with a note.

Slab Reinforcement

When the slab is placed on more than 4″ (102 mm) of uncompacted fill, welded wire fabric should be specified to help the slab resist cracking. Spacings and sizes of wires of welded wire fabric are identified by style and designations. A typical designation specified on a foundation plan might be: 6 × 12—W16 × W8, where:

  6 = longitudinal wire spacing
  12 = transverse wire spacing
  16 = longitudinal wire size
  8 = transverse wire size

The letter *w* indicates smooth wire. A *D* can be used to represent deformed wire. Typically a steel mesh of number 10 wire in 6″ (150 mm) grids is used for residential floor slabs. Figure 32–9 shows an example of the mesh used in concrete slabs.

Steel reinforcing bars can be added to a floor slab to prevent bending of the slab due to expansive soil. While mesh is placed in a slab to limit cracking, steel reinforcement is placed in the concrete to prevent cracking due to bending. Steel reinforcing bars can be laid in a grid pattern near the surface of the concrete that will be in tension from bending. The placement of the reinforcement in the concrete is important to the effectiveness of the reinforcement. The amount of concrete placed around the steel is referred to as *coverage*. Proper coverage strengthens the bond between the steel and concrete and also protects the steel from corrosion if the concrete is exposed to chemicals, weather, or water. The proper coverage is also important to protect the steel from damage by fire. If steel is required to reinforce a residential concrete slab, an engineer will typically determine the size, spacing, coverage, and grade of bars to be used. As a general guideline to placing steel, the ACI recommends:

Figure 32–9 Welded wire mesh is often placed in concrete slabs to help reduce cracking.

○ For concrete cast against and permanently exposed to earth (footings): 3″ (75 mm) minimum coverage.

○ For concrete exposed to weather or earth such as basement walls: 2″ (50 mm) coverage for #6–18 bars and 1½″ (38 mm) coverage for #5 and W31/D31 wire or smaller.

○ For concrete not exposed to weather or in contact with ground: 1½″ (38) mm coverage for slabs, walls, and joist; ¾″ (19 mm) coverage for #11 bars and smaller; and 1½″ (38 mm) coverage for #14 and #18 bars.

Generally wire mesh and steel reinforcement is not shown on the foundation plan but is specified by a note.

Post Tensioned Concrete Reinforcement

The methods of reinforcement mentioned thus far assume that the slab will be poured over stable soil. Concrete slabs can often be poured over unstable soil by using a method of reinforcement known as *post tensioning*. This method of construction was originally developed in the late 1950s for reinforcement of slabs that were to be poured at ground level and then lifted into place for multilevel structures. Adopted for the use of residential slabs, post tensioning allows concrete slabs to be poured at grade over expansive soil. The technology has advanced sufficiently so that post tensioning is now widely used even on stable soils.

For design purposes, a concrete slab can be considered as a wide, shallow beam that is supported by a concrete foundation at the edge. While a beam may sag at the center due to loads and gravity, a concrete slab can either sag or bow depending on the soil conditions. Soil at the edge of a slab will be exposed to more moisture than soil near the center of the slab. This differential in moisture content can cause the edges of the slab to heave, resulting in tension being created in the bottom portion of the slab and compression in the upper portion. Center lift conditions can result as the soil beneath the interior of the slab becomes wetter and expands as the perimeter of the slab dries and shrinks, or as a combination of the two factors. As the center of the slab heaves, tension is created in the upper portion of the slab with compression in the lower portion. Because concrete is very poor at resisting stress from tension, the slab must be reinforced with a material such as steel that has a high tensile strength. Steel tendons with anchors can be extended through the slab as it is poured. Usually between 3 and 10 days after the concrete has been poured, these tendons can be stretched by hydraulic jacks, which place approximately 25,000 lb. of force on each tendon. The tendon force is transfered to the concrete slab through anchorage devices at the ends of the tendons. This process creates an internal compressive force throughout the slab, increasing the ability of the slab to resist cracking and heaving. Post tensioning usually allows for a thinner slab than normally would be

required to span over expansive conditions, elimination of other slab reinforcing, and elimination of most slab joints.

Two methods of post tensioning are typically used for residential slabs: flat slab and the ribbed slab method. The *flat slab* method uses steel tendons ranging in size from ⅜ to ½″ (10–13 mm) in diameter. Maximum spacings of tendons recommended by the Post Tensioning Institute (PTI) are:

⅜″ (9 mm) diameter: 5′–0″ (1524 mm) spacing.

⁷⁄₁₆″ (11 mm) diameter: 6′–10″ (1829 mm) spacing.

½″ (13 mm) diameter: 9′–0″ (2743 mm) spacing.

The exact spacing and size of tendons must be determined by an engineer based on the loads to be supported and the strength and conditions of the soil. When required, the tendons can be represented on the foundation plan, as seen in Figure 32–10. Details will also need to be provided to indicate how the tendons will be anchored, as well as showing the exact locations of the tendons. Figure 32–11 shows an example of a tendon detail. In addition to representing and specifying the steel throughout the floor system, the drafter will need to specify the engineer's requirements for the strength of the concrete at 28 days, the period when the concrete is to be stressed, as well as what strength the concrete should achieve before stressing.

A second method of post tensioning on on-grade floor slab is with the use of concrete ribs or beams placed below the slab. These beams reduce the span of the slab over the soil and provide increased support. The width, depth, and spacing are determined by the engineer based on the strength and condition of the soil and the size of the slab. Figure 32–12 shows an example of a beam detail that a drafter would be required to draw to show the reinforcing specified by the engineer. Figure 32–13

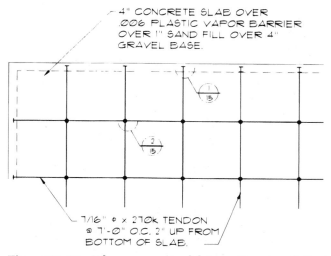

Figure 32–10 When a concrete slab is post tensioned, the tendons and anchors used to support the floor slab must be represented on the foundation plan.

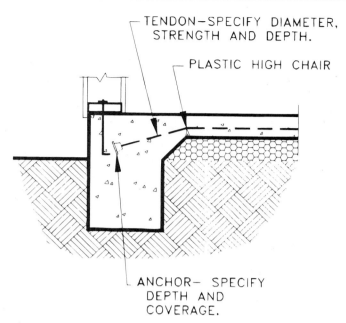

Figure 32–11 Tendon details should be drawn by the drafter to reflect the design of the engineer.

shows an example of how these beams could be shown on the foundation plan.

Moisture Protection

In most areas, the slab is required to be placed over 6 mil polyethylene sheet plastic to protect the floor from ground moisture. When a plastic vapor barrier is to be placed over gravel, a layer of sand should be specified to cover the gravel fill to avoid tearing the vapor barrier. An alternative is to use 55# rolled roofing in place of the plastic. The vapor barrier is not drawn on the foundation plan but is specified with a note on the foundation plan.

Slab Insulation

Depending on the risk of freezing, some municipalities require the concrete slab to be insulated to prevent heat loss. The insulation can be placed under the slab or on the outside of the stem wall. When placed under the slab,

a 2 × 24" (50 × 600 mm) minimum rigid insulation material should be used to insulate the slab. An isolation joint is typically provided between the stem wall and the slab to prevent heat loss through the stem wall. When placed on the exterior side of the stem wall, the insulation should extend past the bottom of the foundation. Care must be taken to protect exposed insulation on the exterior side of the wall. This can usually be done by placing a protective covering such as ½" (13 mm) concrete board over the insulation. Figure 32–14 shows common methods of insulating concrete floor slabs. Insulation is not shown on the foundation plan but is represented by a note, as seen in Figure 32–15, and specified in sections and footing details.

Plumbing and Heating Requirements

Plumbing and heating ducts must be placed under the slab before the concrete is poured. On residential plans, plumbing is usually not shown on the foundation plan. Generally the skills of the plumbing contractor are relied on for the placement of required utilities. Although piping runs are not shown, terminations such as floor drains are often shown and located on a concrete slab plan. Figure 32–15 shows how floor drains are typically represented. If heating ducts are to be run under the slab, they are usually drawn on the foundation plan as shown in Figure 32–16. Drawings for commercial construction usually include both a plumbing and mechanical plan.

Changes in Floor Elevation

The floor level is often required to step down to meet the design needs of the client. A stem wall is formed between the two floor levels and should match the required width for an exterior stem wall. The lower slab is typically thickened to a depth of 8" (200 mm) to support the change in elevation. Figure 32–15 shows how a lowered slab can be represented. The step often occurs at what will be the edge of a wall when the framing plan is complete. Great care must be taken to coordinate the dimensions of a floor plan with those of the foundation plan, so that walls match the foundation.

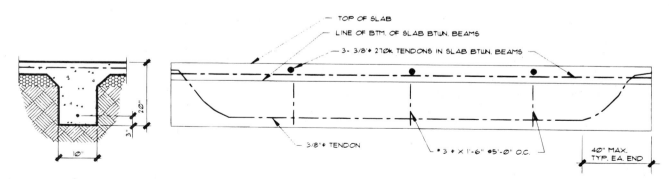

Figure 32–12 Concrete beams can be placed below the floor slab to increase the effects of the slab tensioning. Details must be drawn to indicate how the steel will be placed in the beam, as well as how the beam steel will interact with the slab steel.

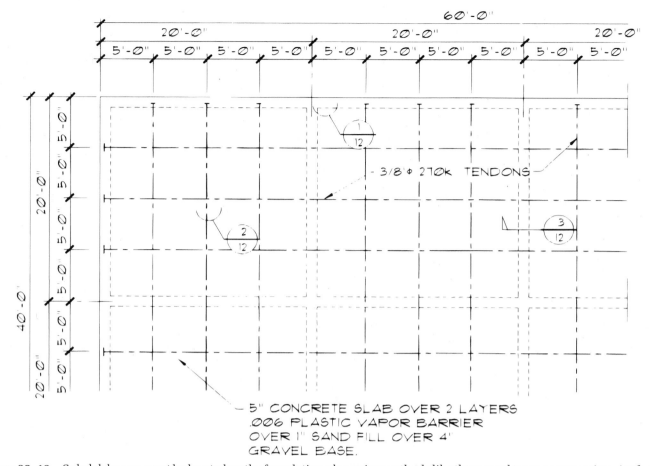

Figure 32–13   Subslab beams must be located on the foundation plan using methods like those used to represent an interior footing.

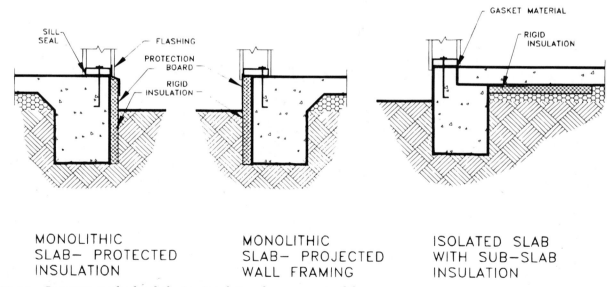

MONOLITHIC
SLAB— PROTECTED
INSULATION

MONOLITHIC
SLAB— PROJECTED
WALL FRAMING

ISOLATED SLAB
WITH SUB–SLAB
INSULATION

Figure 32–14   Common methods of placing insulation for a concrete slab.

Common Components Shown on a Slab Foundation

Refer to Figure 32–15. Dimensions are needed for each of the following items to provide location information.

○ Outline of slab
○ Interior footing locations
○ Changes in floor level
○ Floor drains

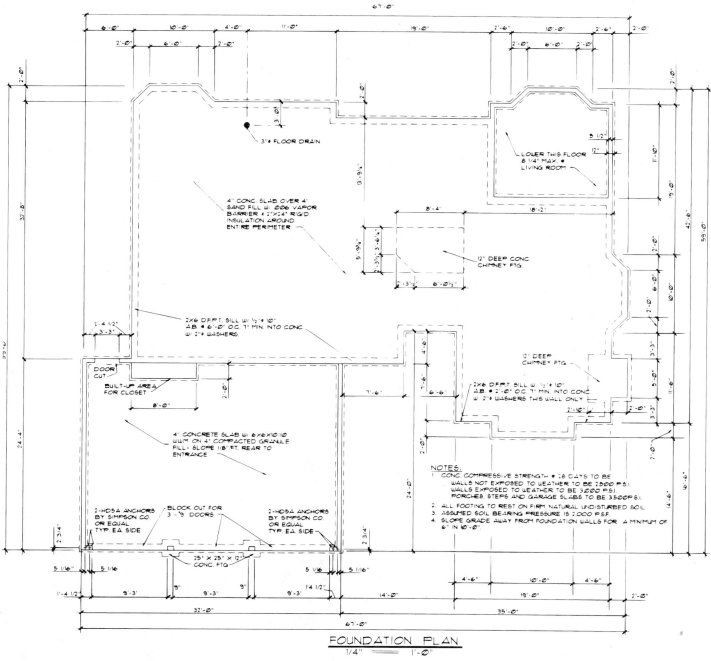

Figure 32–15 A foundation plan for a home with a concrete slab floor system.

○ Exterior footing locations
○ Ducts for mechanical
○ Metal anchors
○ Patio slabs

Common Components Specified by Note Only

See Figure 32–15.

○ Slab thickness and fill material

○ Wire mesh
○ Mudsill size
○ Vapor barriers
○ Pier sizes
○ Assumed soil strength
○ Reinforcing steel
○ Anchor bolt size and spacing
○ Insulation
○ Slab slopes
○ Concrete strength

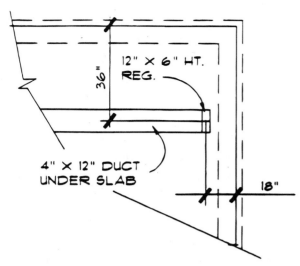

Figure 32–16 HVAC duct and registers that are to be located in the floor must be shown on the foundation plan.

▼ CRAWL SPACE FLOOR SYSTEMS

The crawl space is the area formed between the floor system and the ground. Building codes require a minimum of 18″ (457 mm) from the bottom of the floor to the ground and 12″ (305 mm) from the bottom of beams to the ground. Two common methods of providing a crawl space below the floor are the conventional method using floor joists and the post-and-beam system. An introduction to each system is needed to complete the foundation. Both floor systems were discussed in detail in relationship to the entire structure in Section VIII.

Joist Floor Framing

The most common method of framing a wood floor is with wood members called floor joists. Floor joists are used to span between the foundation walls and are shown in Figure 32–17. Sawn lumber ranging in size from 2 × 6 through 2 × 12 (50 × 150 to 50 × 300 mm) has traditionally been used to frame the floor platform. Contractors in many parts of the country are now framing with trusses or joists made from plywood or wood by-products known as engineered lumber. (These materials were discussed in Chapter 25.) The floor joists are usually placed at 16″ (400 mm) on center, but the spacing of joists may change depending on the span, the material used and the load to be supported.

Figure 32–18 shows a floor system framed with joists or stick framing methods. To construct this type of floor, a pressure-treated sill is bolted to the top of the foundation wall with the anchor bolts that were placed when the foundation was formed. The floor joists can then be nailed to the sill. With the floor joist in place, plywood floor sheathing is installed to provide a base for the finish floor. The size of the subfloor depends on the spacing of the floor joists and the floor loads that must be supported. Building codes differ

on the live loads that must be supported by the subfloor. The live load to be supported will also affect the thickness of the plywood to be used. The American Plywood Association (APA) has span tables for plywood from 7/16″ through 7/8″ (11.1–22.2 mm) to meet the various conditions that might be found in a residence. A subfloor can also be made from 1″ (25 mm) material such as 1 × 6 (25 × 150 mm) tongue-and-groove (T&G) lumber, although this method requires more labor. The floor sheathing is not represented on the foundation plan but is specified in a general note. Figure 32–19 shows common methods of drawing floor joists on the foundation plan.

When the distance between the foundation walls is too great for the floor joists to span, a girder is used to support the joists. A girder is a horizontal load-bearing member that spans between two or more supports at the foundation level. Depending on the load to be supported and the area you are in, either a wood laminated wood, engineered wood product, or steel member may be used for the girder and for the support post. The girder is usually supported in a beam pocket where it intersects the foundation wall. A concrete pier is put under the post to resist settling. Figure 32–20 shows methods of drawing the girders, posts, piers, and beam pocket on the foundation plan. Figure 32–21 shows a complete foundation plan using floor joists to support the floor.

Common Components Shown with a Joist Floor System

See Figure 32–21. Unless indicated with an asterisk, all components on the following pages require dimensions to provide location information.

Figure 32–17 A joist floor system typically places wood members at 16″ (400 mm) o.c., although spacings of 8, 12, and 24″ (200, 300, and 600 mm) are also used.

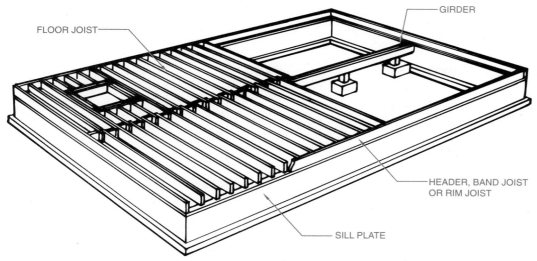

Figure 32–18 A floor framed with floor joist uses wood members ranging from 2 × 6 through 2 × 12 (50 × 150 to 50 × 300 mm) to span between supports.

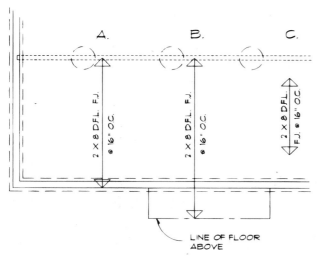

Figure 32–19 Common methods of representing floor joists in plan view. (a) An arrow is drawn from bearing point to bearing point. (b) When floor joists extend past the foundation wall, the length should be shown. (c) An arrow represents the direction of span and spacing.

○ Foundation wall
○ Door opening in foundation wall
○ Fireplace
○ Floor joist
○ Girders (bearing walls and floor support)
○ Girder pockets
○ Crawl access*
○ Exterior footings
○ Metal anchors
○ Fireplace footings
○ Outline of cantilevers
○ Interior piers
○ Changes in floor levels
○ Vents for crawl space*

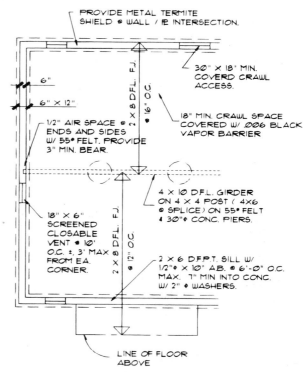

Figure 32–20 Common methods of showing joist floor components on a foundation plan.

Common Components Specified by Note Only

See Figure 32–21.

○ Floor joist size and spacing
○ Anchor bolt size and spacing
○ Insulation
○ Crawl height
○ Assumed soil strength
○ Subfloor material
○ Girder size

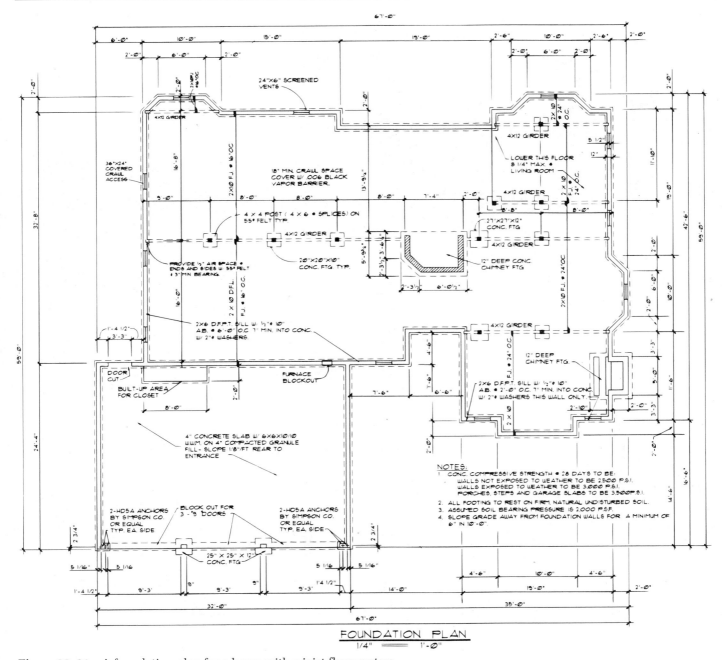

Figure 32–21 A foundation plan for a home with a joist floor system.

○ Mudsill size
○ Vapor barrier
○ Concrete strength
○ Wood type and grade

Post-and-Beam Floor Systems

A post-and-beam floor system is built using a standard foundation system. Rather than having floor joists span between the foundation walls, a series of beams are used to support the subfloor, as shown in Figure 32–22. Once the mudsill is bolted to the foundation wall, the beams are placed so that the top of each beam is flush with the top of the mudsill. The beams are usually placed at 48" (1200 mm) on center, but the spacing can vary depending on the size of the floor decking to be used. The area of the country

Figure 32–22 Girders are usually placed at 48" (1200 mm) o.c. with supports at 8'–0" (2400 mm) o.c. *Courtesy John Black.*

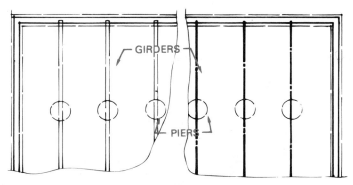

Figure 32–23 Common methods of drawing post-and-beam components in plan view.

where the structure is to be built affects how the subfloor is made. Generally 2″ (50 mm) thick material such as 2 × 6 (50 × 150) T & G boards laid perpendicular to the beams are used for a subfloor. Plywood with a thickness of 1⅛″ (28.5 mm) and an APA rating of STURD-I-FLOOR 2-4-1 with an exposure rating of EXP-1 can also be used to build a post-and-beam subfloor quickly and economically. When the subfloor is glued to the support beams the strength and quality of the floor is greatly increased by eliminating squeaks, bounce, and nail popping.

The beams are supported by wooden posts as they span between the foundation walls. Posts are usually placed at 8′ (2438 mm) on center but spacing can vary depending on the load to be supported. Each post is supported by a concrete pier. Beams, posts, and piers can be drawn on the foundation plan, as shown in Figure 32–23. Figure 32–24 shows a foundation plan with a post-and-beam floor system.

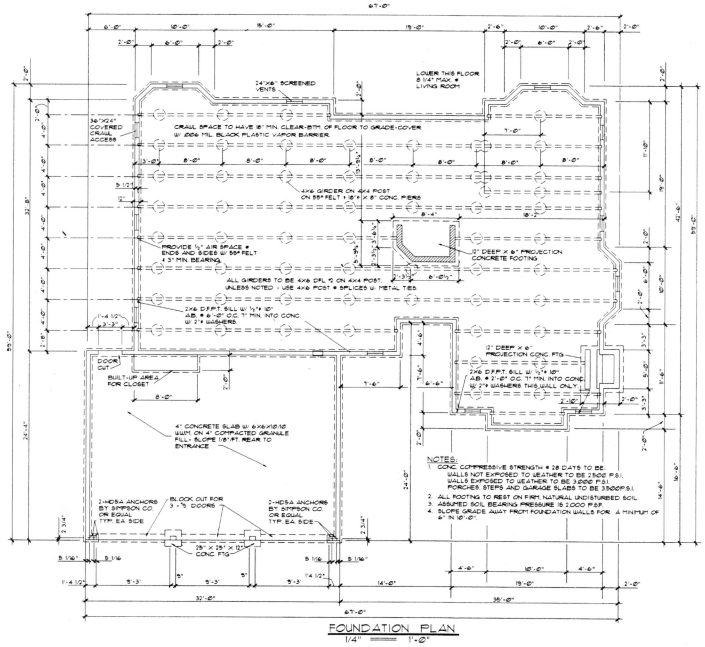

FOUNDATION PLAN
1/4″ = 1′-0″

Figure 32–24 A foundation plan for a home with a post-and-beam floor system.

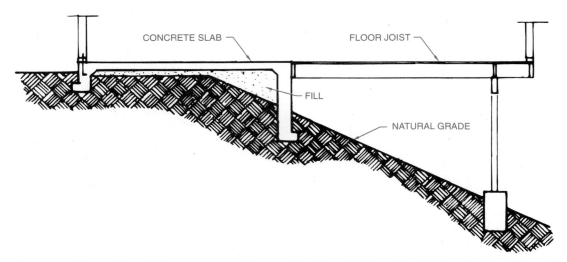

Figure 32–25 Concrete slab and floor joist systems are often combined on sloping sites to help minimize fill material.

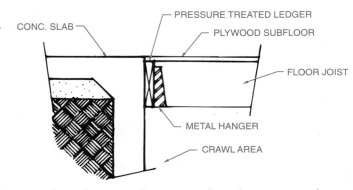

Figure 32–26 A ledger is used to provide anchorage to floor joists where they intersect the concrete slab. Metal joist hangers are used to join the joists to the ledger.

Common Components Shown for a Post-and-Beam System

See Figure 32–24. Unless indicated with an asterisk, all components require dimensions to provide location information.

- ○ Foundation wall
- ○ Openings in wall for doors
- ○ Fireplace
- ○ Girders (beams)
- ○ Girder pockets
- ○ Crawl access*
- ○ Exterior footings
- ○ Metal anchors
- ○ Fireplace footings
- ○ Piers
- ○ Changes in floor levels
- ○ Vents for crawl space*

Common Components Specified by Note only

See Figure 32–24.

- ○ Anchor bolt spacing

- ○ Mudsill size
- ○ Insulation
- ○ Minimum crawl height
- ○ Subfloor material
- ○ Girder size
- ○ Vapor barrier
- ○ Concrete strength
- ○ Wood type and grade
- ○ Assumed soil strength

Combined Floor Methods

Floor and foundation methods may be combined depending on the building site. This is typically done on partially sloping lots when part of a structure may be constructed with a slab and part of the structure with a joist floor system, as seen in Figure 32–25.

One component typically used when floor systems are combined is a ledger. A ledger is used to provide support for floor joists and subfloor when they intersect the concrete. Unless felt is placed between the concrete and the ledger, the ledger must be pressure-treated lumber. The ledger can be shown on the foundation plan, as shown in Figure 32–26.

CHAPTER 32 · Floor Systems and Foundation Support Test

QUESTIONS

Question 32–1 Why is a concrete slab called an on-grade floor system?

Question 32–2 What is the minimum thickness for a residential slab?

Question 32–3 Why are control joints placed in slabs?

Question 32–4 What is the minimum amount of fill placed under a slab?

Question 32–5 What thickness of vapor barrier is to be placed under a slab?

Question 32–6 What is the minimum height required in the crawl area?

Question 32–7 What is the purpose of a girder?

Question 32–8 How are floor joists attached to the foundation wall?

Question 32–9 How are girders supported at the foundation wall?

Question 32–10 What is a common spacing for beams in a post-and-beam floor?

CHAPTER 33

Foundation-Plan Layout

The foundation plan is typically drawn at the same scale as the floor plan that it will support. Although the floor plan can be traced to obtain overall sizes, this practice can lead to major errors in the foundation plan. If you trace a floor plan that is slightly out of scale, you will reproduce the same errors in the foundation plan. A better method is to draw the foundation plan using the dimensions that are found on a print of the floor plan. If the foundation cannot be drawn using the dimensions on the floor plan, your floor plan may be missing dimensions or contain errors. Great care needs to be taken with the foundation plan. If the foundation plan is not accurate, changes may be required that affect the entire structure.

This chapter includes guidelines for several types of foundation plans: joist construction, concrete slab, basement slab and joist, post and beam, and a piling foundation. Each is based on the floor plan that was used for examples in Section IV. Before attempting to draw a foundation plan, study the completed plan that precedes each example so you will know what the finished drawing should look like.

The foundation plan can be drawn by dividing the work into the six stages of foundation layout, drawing foundation members, drawing floor framing members, dimensioning, lettering, and evaluation. As you progress through the drawing, you will use several types of line quality. For layout steps, construction lines in nonreproducible blue with a 6H lead will be best. When drawing finished-quality lines, use the following:

A 5 mm lead, #0 pen, or sharp 3H lead for thin lines

A 7 mm lead, #2 pen, or sharp H lead for bold lines

A 9 mm lead, #3 pen, or H lead and draw two parallel lines very close for very bold lines

▼ CONCRETE SLAB FOUNDATION LAYOUT

The following steps can be used to draw a foundation with a concrete slab floor system. Not all steps will be required for every house. When the concrete slab foundation is complete, it should resemble the plan shown in Figure 33–5. Use construction lines for steps 1 through 9. Each step can be seen in Figure 33–1.

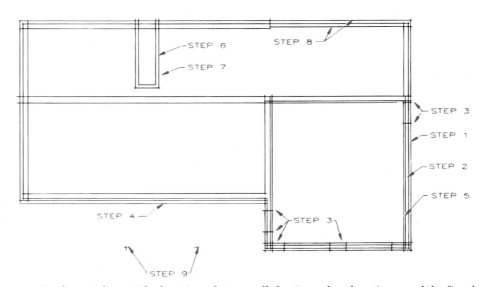

Figure 33–1   Use construction lines to lay out the location of stem wall, footings, door locations, and the fireplace for a concrete slab foundation plan.

Step 1. Using the dimensions on your floor plan, lay out the exterior edge of the slab. The edge of the slab should match the exterior side of the exterior walls on the floor plan.

Step 2. Draw the interior side of the stem wall around the slab at the garage area. See Figure 31–10 for a review of foundation dimensions.

Step 3. Block out doors in the stem walls. Allow for the door size plus 4″ (100 mm).

Step 4. Lay out a support ledge if brick veneer is to be used.

Step 5. Lay out the exterior footing width.

Step 6. Lay out the size of the fireplace based on measurements from the floor plan.

Step 7. Lay out the fireplace footing so that it extends 6″ (150 mm) minimum beyond the face of the fireplace.

Step 8. Lay out interior footings.

Step 9. Lay out any exterior piers that might be required for decks or porches.

Step 10. Darken all items that were drawn in steps 1–9. Use bold lines to draw steps 1–4 with finished-line quality. Use thin dashed lines to draw steps 4, 7, 8, and 9 with finished-line quality. Your drawing should now resemble the drawing in Figure 33–2.

See Figure 33–3 for steps 11 through 14. Because the following items are simple, they can be drawn without the use of construction lines. These items may or may not be required depending on your plan.

Step 11. Draw changes in the floor levels.

Step 12. Draw metal connectors.

Step 13. Draw floor drains.

Step 14. Draw heating registers.

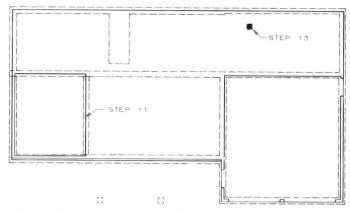

Figure 33–3 Add any finishing materials such as lower slabs and plumbing, electrical, and HVAC material that will be below the slab.

You now have all of the information drawn that is required to represent the floor and foundation systems. Follow steps 15 through 20 to place the required dimensions on the drawing. Use thin lines for all extension and dimension lines. Your drawing should resemble Figure 33–4 when complete.

Draw extension and dimension lines to locate:

Step 15. The overall size on each side of the foundation.

Step 16. Jogs in the foundation walls.

Step 17. Door openings in the stem wall.

Step 18. Interior footings.

Step 19. The fireplace.

Step 20. Heating and plumbing materials.

The final drawing procedure is to place dimensions and specify the materials that are to be used. Figure 33–5 is an example of the notes that are required on the foundation. Use the following steps to complete the foundation plan.

Step 21. Compute and neatly letter all dimensions in the appropriate location.

Step 22. Neatly letter all required general notes.

Step 23. Neatly letter all required local notes.

Step 24. Place a title and scale under drawing.

Step 25. Using a print of your plan, evaluate your drawing using the following checklist:

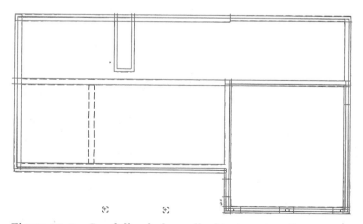

Figure 33–2 Carefully darken all objects using the proper finished-line quality. Construction lines should be so light that they do not need to be erased.

Slab Foundation-Plan Checklist

Correct Symbol and Location	Correct Structural Materials
Outline of slabs	Proper footing size
Walls	Proper footing location
Footings and piers	Proper pier size
Doors	Proper pier location
Ductwork	
Plumbing	
Floor slopes	

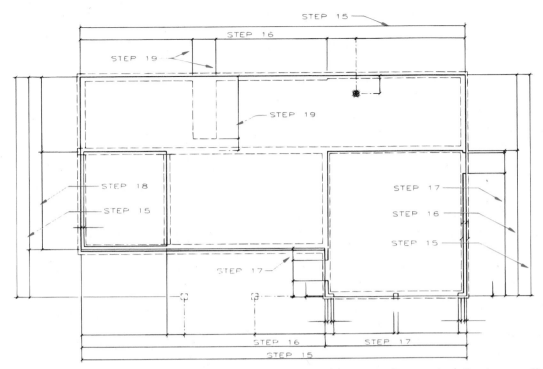

Figure 33–4 Prepare the plan to be dimensioned by adding dimension and extension lines to describe the overall, major jogs and each opening on each side of the foundation.

Slab Foundation Plan Checklist (cont.)

Required Local Notes	Required General Notes
Concrete slab thickness, fill, and reinforcement	Soil bearing pressure
Veneer ledges	Concrete strength
Door blockouts	Anchor bolt size and spacing
Fireplace footings	Vapor barriers
Pier sizes	Slab insulation
Footing sizes	Reinforcement
Floor drains	
Heating registers	

▼ FOUNDATION PLAN WITH JOIST CONSTRUCTION

The following steps (illustrated in Figure 33–6) can be used to draw a foundation plan showing continuous footings and floor joists. Your drawing should resemble Figure 33–5. Use construction lines for steps 1 through 9.

Step 1. Using the dimensions on your floor plan, lay out the exterior face of the foundation wall.

Step 2. Determine the foundation wall thickness and lay out the interior face of the foundation wall. See Figure 31–10 for a review of foundation footing and wall dimensions.

Step 3. Block out doors in the garage area. Typically the door size plus 4″ (100 mm) is provided as the foundation is formed. This allows for door framing.

Step 4. Block out a 30″ (760 mm) wide minimum crawl access.

Step 5. Lay out a support ledge if using masonry veneer.

Step 6. Lay out the size of the fireplace using the measurements on the floor plan.

Step 7. Lay out the footing width around the perimeter walls.

Step 8. Lay out the fireplace footing so it extends a minimum of 6″ (150 mm) past the face of the fireplace.

Step 9. Lay out any exterior piers required for porches or deck support.

Step 10. Lay out girder locations to support floor joists and load-bearing walls and changes in floor elevation.

Step 11. Locate the center of all support piers.

Step 12. Darken all items drawn in steps 1–9. Use bold lines to represent the materials drawn in steps 1–6. Use thin dashed lines for steps 7–9. When finished with this step, your drawing should resemble Figure 33–7.

See Figure 33–8 for steps 13 through 18. The items to be drawn in these steps are not drawn with construction lines. Because of their simplicity, these items can be drawn using finished lines and need not be traced.

Step 13. Using thin dashed lines, draw the girders.

Step 14. Draw arrows to represent the floor joist direction.

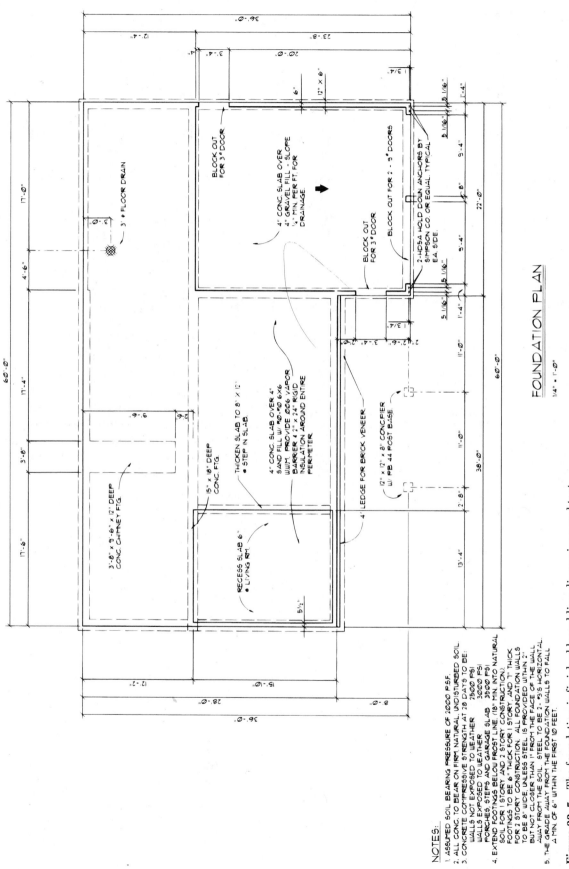

FOUNDATION PLAN
1/4" = 1'-0"

NOTES:
1. ASSUMED SOIL BEARING PRESSURE OF 2000 P.S.F.
2. ALL CONC. TO BEAR ON FIRM NATURAL UNDISTURBED SOIL.
3. CONCRETE COMPRESSIVE STRENGTH AT 28 DAYS TO BE:
   WALLS NOT EXPOSED TO WEATHER   2500 PSI
   WALLS EXPOSED TO WEATHER   3000 PSI
   PORCHES, STEPS AND GARAGE SLAB   3500 PSI
4. EXTEND FOOTINGS BELOW FROST LINE (18" MIN INTO NATURAL
   SOIL FOR 1 STORY AND 2 STORY CONSTRUCTION).
   FOOTINGS TO BE 6" THICK FOR 1 STORY, AND 7" THICK
   FOR 2 STORY CONSTRUCTION. ALL FOUNDATION WALLS
   TO BE 8" WIDE, UNLESS STEEL IS PROVIDED WITHIN 2"
   BUT NOT CLOSER THAN 1" FROM THE FACE OF THE WALL
   AWAY FROM THE SOIL. STEEL TO BE 2- #3S HORIZONTAL.
5. THE GRADE AWAY FROM THE FOUNDATION WALLS TO FALL
   A MIN OF 6" WITHIN THE FIRST 10 FEET.

Figure 33–5 The foundation is finished by adding dimensions and text.

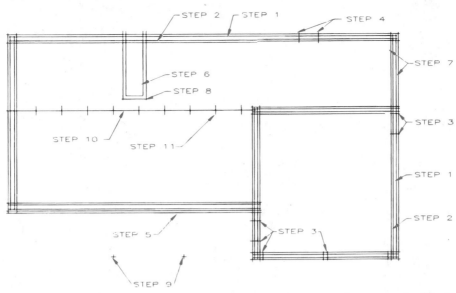

Figure 33–6 Use construction lines to lay out the location of stem wall, footings, door locations, the fireplace, and girder locations.

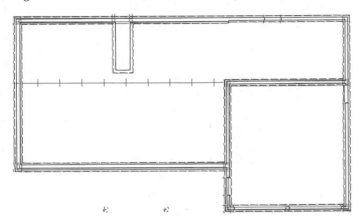

Figure 33–7 Carefully darken all objects using the proper finished-line quality. Construction lines should be so light that they do not need to be erased.

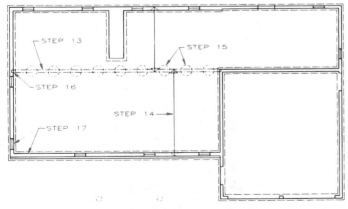

Figure 33–8 Draw the girders, joists, and vents.

Step 15. Draw the piers to support the girders.

Step 16. Draw beam pockets.

Step 17. Draw vents in the walls surrounding the crawl space. Use bold lines to represent the edges and thin lines to represent the vent.

Step 18. Crosshatch the masonry chimney if concrete blocks or brick is used.

You now have all of the information drawn that is required to represent the floor and foundation systems. Follow steps 19 through 24 to dimension these items. Use thin lines for all extension and dimension lines. When complete, your drawing should resemble Figure 33–9.

Draw extension and dimension lines to locate:

Step 19. The overall size of each side of the foundation.

Step 20. The jogs in the foundation wall.

Step 21. Door openings in the foundation wall.

Step 22. The fireplace.

Step 23. All girders and piers. Place extension lines on the outside of the foundation if possible.

Step 24. All metal connectors.

All materials have now been drawn and located. The final drawing procedure is to place dimensions and specify the material to be used. This is done with general and local notes in a method similar to that used on the floor plan. Figure 33–10 shows how the notes might appear in the foundation plan. Use the following steps as a guideline to completing the foundation plan.

Step 25. Compute all dimensions and place them in their appropriate locations.

Step 26. Neatly letter all required general notes.

Step 27. Neatly letter all required local notes.

Step 28. Place a title and scale under the drawing.

Step 29. Use a print of your plan to evaluate your drawing using the following checklist.

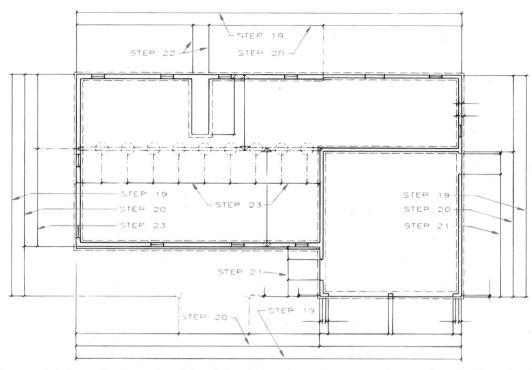

Figure 33–9 Prepare the plan to be dimensioned by adding dimension and extension lines to describe the overall, major jogs and each opening on each side of the foundation.

Foundation-Plan Checklist	
Correct Symbol and Location	**Correct Structural Materials**
Walls	Proper footing sizes
Footings	Proper wall size
Crawl access	Proper beam placement
Vents	Proper beam size
Doors	Proper joist size and direction
Girders	Proper pier locations and sizes
Joist	
Piers	
Required Local Notes	
Joist size and spacing	Veneer ledges
Beam sizes	Door blockouts
Beam pockets	Fireplace footing size
Metal connectors	Vent size and spacing
Anchor bolts and mudsill	
Garage slab thickness	
Fill and slope direction	
Required General Notes	**Required Dimensions**
Soil bearing information	Overall
Concrete strength	Jogs
Crawl space covering	Openings
Framing lumber grade and species	Girder locations
	Pier locations
	Metal connectors

▼ COMBINATION SLAB AND CRAWL SPACE PLANS

A structure may require a combined slab and floor joist system. The foundation plan will have similarities to both a foundation with joist and a slab foundation system. Figure 33–15 is an example of a foundation plan with a basement slab and a crawl space with joist construction. The following steps can be used to draw the foundation plan. Use construction lines for steps 1 through 12. Each step can be seen in Figure 33–11.

Step 1. Using the dimensions on your floor plan, lay out the exterior face of the foundation walls.

Step 2. Determine the wall thickness and lay out the interior face of the foundation walls.

Step 3. Block out door and window openings in the foundation walls.

Step 4. Block out a crawl access.

Step 5. Lay out a support ledge if masonry veneer is to be used.

Step 6. Lay out the size of the fireplace based on the size drawn on the floor plan.

Step 7. Lay out the footing under the fireplace.

Step 8. Lay out the footing width under all foundation walls.

Step 9. Lay out the footing width under all interior load-bearing walls.

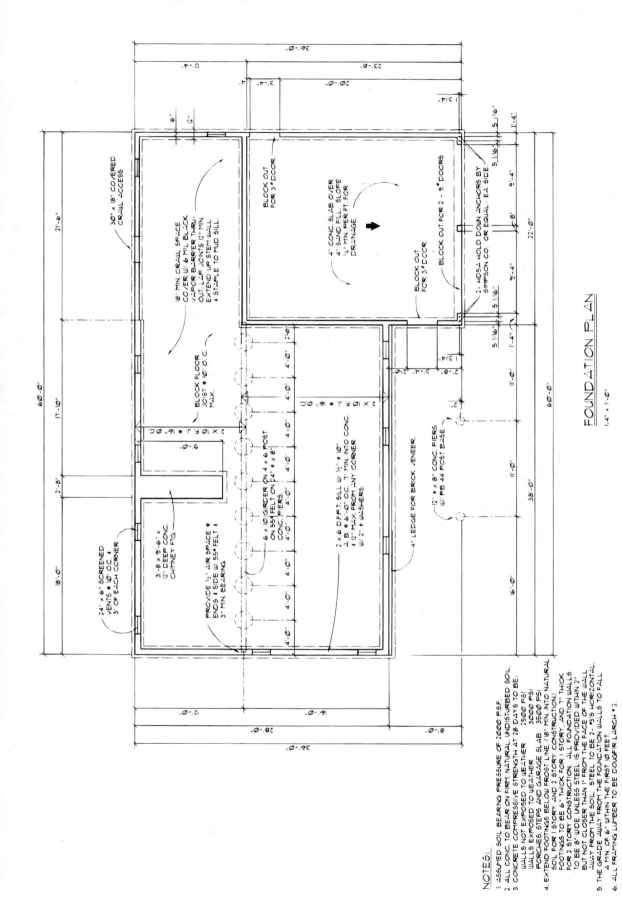

FOUNDATION PLAN
1/4" = 1'-0"

NOTES:
1. ASSUMED SOIL BEARING PRESSURE OF 2000 P.S.F.
2. ALL CONC. TO BEAR ON FIRM, NATURAL, UNDISTURBED SOIL.
3. CONCRETE COMPRESSIVE STRENGTH AT 28 DAYS TO BE:
   WALLS NOT EXPOSED TO WEATHER   2500 PSI
   WALLS EXPOSED TO WEATHER   3000 PSI
   PORCHES, STEPS AND GARAGE SLAB   3500 PSI
4. EXTEND FOOTINGS BELOW FROST LINE (18" MIN INTO NATURAL
   SOIL FOR 1 STORY AND 2 STORY CONSTRUCTION.
   FOOTINGS TO BE 6" THICK FOR 1 STORY, AND 7" THICK
   FOR 2 STORY CONSTRUCTION. ALL FOUNDATION WALLS
   TO BE 8" WIDE UNLESS STEEL IS PROVIDED WITHIN 2"
   BUT NOT CLOSER THAN 1" FROM THE FACE OF THE WALL
   AWAY FROM THE SOIL. STEEL TO BE 2 - #5'S HORIZONTAL.
5. THE GRADE AWAY FROM THE FOUNDATION WALLS TO FALL
   A MIN. OF 6" WITHIN THE FIRST 10 FEET.
6. ALL FRAMING LUMBER TO BE DOUGFIR LARCH #2.

Figure 33–10 The foundation is finished by adding dimensions and text.

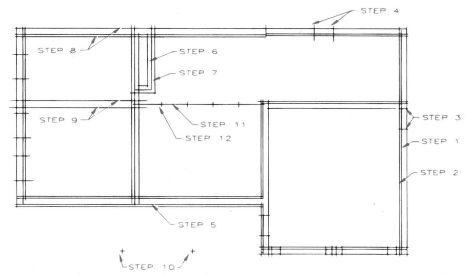

Figure 33–11 Use construction lines to lay out the location of stem wall, retaining walls, footings, door locations, and the fireplace for a foundation plan with a basement.

Step 10. Lay out any exterior piers required for porches and deck support.

Step 11. Lay out girder locations to support floor and load-bearing walls.

Step 12. Locate the center of all interior support piers.

Step 13. Darken all items drawn in steps 1–12. Use bold lines for steps 1–6. Use thin dashed lines for items 7–10 and step 12. When complete, your drawing should resemble Figure 33–12.

See Figure 33–13 for steps 14 through 21. These items have not been drawn with construction lines. Each can be drawn using finished-quality lines and need not be traced.

Step 14. Draw the girders using dashed lines.

Step 15. Draw piers with thin dashed lines.

Step 16. Draw beam pockets with thin lines.

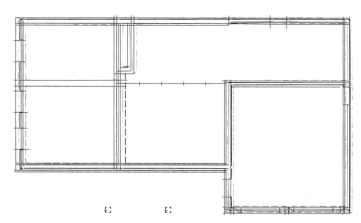

Figure 33–12 Carefully darken all objects using the proper finished-line quality. Construction lines should be so light that they do not need to be erased.

Step 17. Draw vents in the crawl area only.

Step 18. Draw windows with thin lines.

Step 19. Draw window wells using thin lines.

Step 20. If concrete blocks are used for the foundation walls, crosshatch the wall.

Step 21. Crosshatch the chimney.

You now have all of the information drawn that is needed to represent the floor and foundation systems. Follow steps 22 through 25 to dimension these items. Use thin lines for all extension and dimension lines. When complete, your drawing should resemble Figure 33–14.

Draw extension and dimension lines to locate:

Step 22. The overall size on each side of the foundation.

Step 23. Jogs in the foundation walls.

Step 24. All openings in the foundation walls except for vents.

Step 25. All girders and piers.

All materials have now been drawn and located. The final drawing procedure is to place dimensions and specify the materials that were used. Materials may be specified with general and local notes. Figure 33–15 is an example of the notes that can be found on the foundation plan. Use the following steps as guidelines to complete the foundation plan.

Step 26. Compute all dimensions and place them in the appropriate location.

Step 27. Neatly letter all required general notes.

Step 28. Neatly letter all required local notes.

Step 29. Place a title and scale under the drawing.

Step 30. Using a print of your plan, evaluate your drawing using the following checklist.

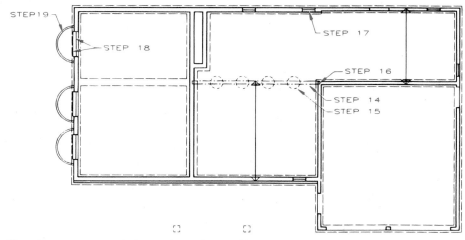

Figure 33–13   Draw all girders, piers, windows, and window wells.

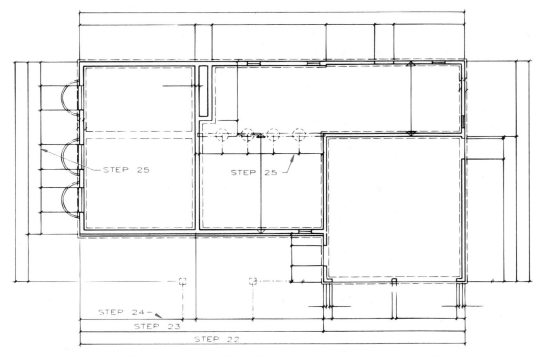

Figure 33–14   Prepare the plan to be dimensioned by adding dimension and extension lines to describe the overall, major jogs and each opening on each side of the foundation.

Checklist for Partial Slab and Floor Joist Foundation	
Correct Symbol and Location	**Correct Structural Materials**
Walls	Footing size
Footings	Wall size
Crawl access	Beam placement
Vents	Beam size
Doors	Joist size
Girders and joist	Pier location
Piers	Beam location
Slabs	Pier sizes

Required Local Notes	Required General Notes
Joist size and spacing	Soil bearing value
Beam size	Concrete strength
Beam pockets	Crawl space covering
Metal connectors	Framing lumber grade and
Anchor bolts	species
Mudsill	Slab insulation
Slab thickness	
Fill	
Slope direction of slab	

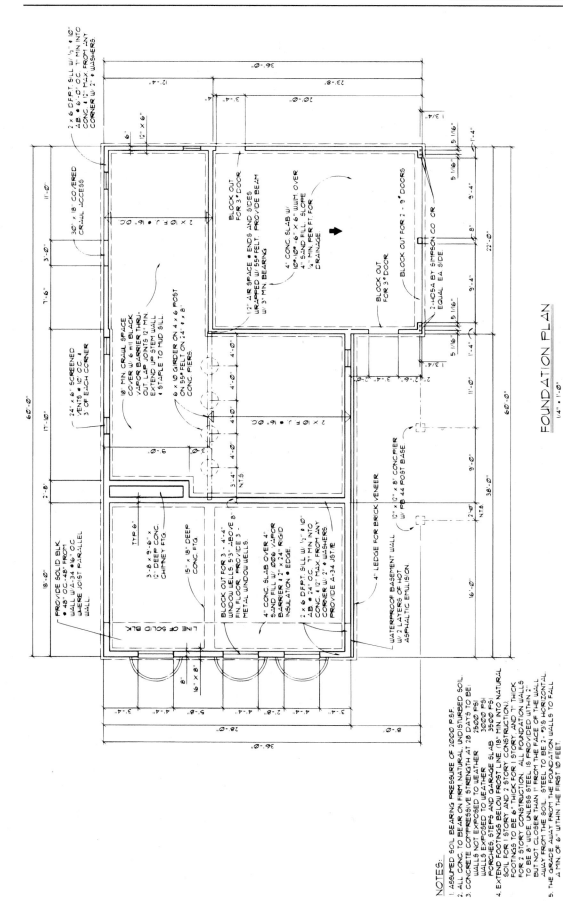

FOUNDATION PLAN
1/4" = 1'-0"

NOTES:
1. ASSUMED SOIL BEARING PRESSURE OF 2000 PSF.
2. ALL CONC. TO BEAR ON FIRM NATURAL UNDISTURBED SOIL.
3. CONCRETE COMPRESSIVE STRENGTH AT 28 DAYS TO BE:
   WALLS NOT EXPOSED TO WEATHER   1500 PSI
   WALLS EXPOSED TO WEATHER   3000 PSI
   PORCHES, STEPS AND GARAGE SLAB   3500 PSI
4. EXTEND FOOTINGS BELOW FROST LINE. (18" MIN. INTO NATURAL
   SOIL FOR 1 STORY AND 2 STORY CONSTRUCTION). ALL FOUNDATION WALLS
   FOOTINGS TO BE 6" THICK FOR 1 STORY, AND 7" THICK
   FOR 2 STORY CONSTRUCTION. ALL FOUNDATION WALLS
   TO BE 8" WIDE. UNLESS STEEL IS PROVIDED WITHIN 2"
   BUT NOT CLOSER THAN 1" FROM THE FACE OF THE WALL.
5. THE GRADE AWAY FROM THE SOIL. STEEL TO BE 2 - #5'S HORIZONTAL.
   AWAY FROM THE SOIL. STEEL TO BE 2 - #5'S HORIZONTAL
   A MIN. OF 6" WITHIN THE FIRST 10 FEET.

Figure 33–15 The foundation is finished by adding dimensions and text.

▼ STANDARD FOUNDATION WITH POST-AND-BEAM FLOOR SYSTEM

When your drawing is complete, it should resemble the plan in Figure 33–21. Use construction lines for drawing steps 1 through 12. Each step can be seen in Figure 33–16.

Step 1. Using the dimensions on your floor plan, lay out the exterior edge of the foundation wall.

Step 2. Lay out the interior face of the foundation wall.

Step 3. Block out doors in the garage area.

Step 4. Block out for a crawl access.

Step 5. Lay out the support ledge if masonry veneer is to be used.

Step 6. Lay out the size of the fireplace using measurements from the floor plan.

Step 7. Lay out the footing under the foundation walls.

Step 8. Lay out the footing under the fireplace.

Step 9. Lay out any exterior piers required to support porches and decks.

Step 10. Lay out the center of each load-bearing wall.

Step 11. Lay out all girders.

Step 12. Lay out the center for the piers to support the girders.

See Figure 33–17 for step 13.

Step 13. Darken all items in steps 1–12. Use bold lines to represent the material in steps 1–6. Use thin dashed lines to represent the material in steps 7–9, and 12.

Step 14. Use thin dashed lines or a very bold dashed line to represent the girders. When finished with this step, your drawing should resemble the drawing in Figure 33–18.

See Figure 33–19 for steps 15 through 18. The items to be drawn in these steps have not been drawn with con-struction lines. Because of their simplicity you can draw these items with finished lines.

Step 15. Draw beam pockets.

Step 16. Draw vents in the walls surrounding the crawl space.

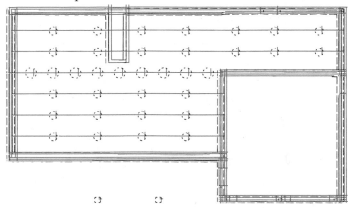

Figure 33–17 Carefully darken all objects using the proper finished-line quality. Construction lines should be so light that they do not need to be erased.

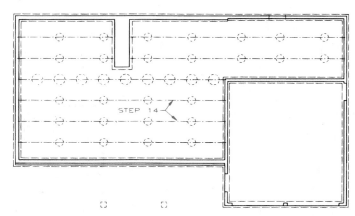

Figure 33–18 Draw the girders and the piers with finished-line quality.

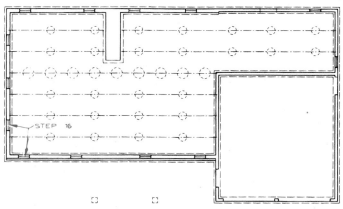

Figure 33–19 Add any finishing materials such as lowered slabs and vents.

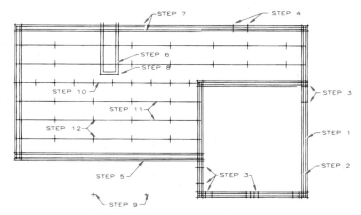

Figure 33–16 Use construction lines to lay out the location of stem wall, footings, door locations, the fireplace, and girder and pier locations for a post-and-beam foundation plan.

Step 17. Crosshatch walls formed using concrete blocks.

Step 18. Crosshatch the chimney.

You now have all of the information drawn to represent the floor and foundation systems. Follow steps 19 through 25 to dimension these items. Use thin lines for extension and dimension lines. When complete, your foundation plan should resemble Figure 33–20.

Draw extension and dimension lines to locate:

Step 19. The overall size of the foundation.

Step 20. The jogs in the foundation walls.

Step 21. All openings in the foundation walls except for vents.

Step 22. The fireplace.

Step 23. All girders.

Step 24. All piers.

Step 25. All metal connectors.

All materials have now been drawn to represent the floor and foundation systems. The final drafting procedure is to place dimensions and specify the material to be used. This is done by the use of general and local notes in a similar method that was used on the floor plan. Figure 33–21 shows how notes can be placed on the foundation plan. Complete the foundation plan using the following steps.

Step 26. Compute and neatly letter dimensions in the appropriate place.

Step 27. Neatly letter all required general notes.

Step 28. Place a title and scale below the drawing.

Step 29. Using a print of your plan, evaluate your drawing using the following checklist.

Post-and-Beam Checklist

Correct Symbol and Location	Correct Structural Materials
Walls	Proper footing sizes
Footings	Proper wall size
Crawl access	Proper beam placement
Vents	Proper pier placement
Doors	Veneer ledges
Girders and piers	
Line of slabs	
Fireplace	

Required Local Notes	Required General Notes
Beam size	Soil bearing values
Beam pockets	Concrete values
Metal connectors	Crawl space covering
Anchor bolts and mudsills	Crawl space insulation
Garage slab thickness, fill, slope direction	Framing lumber grade and species
Fireplace note	
Vent size and spacing	
Veneer ledges	
Door blockouts	

Required Dimensions	
Overall	Girder locations
Jogs	Pier locations
Wall openings	Metal locations
	Fireplace location

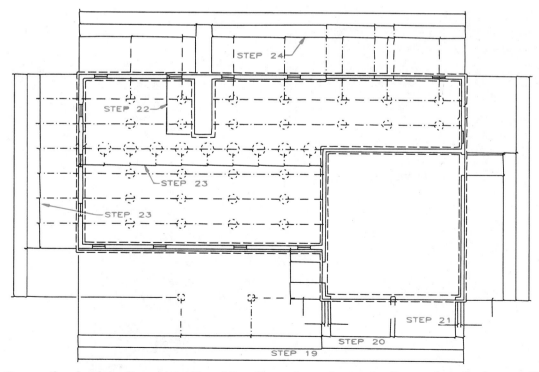

Figure 33–20 Prepare the plan to be dimensioned by adding dimension and extension lines to describe the overall, major jogs and each opening on each side of the foundation.

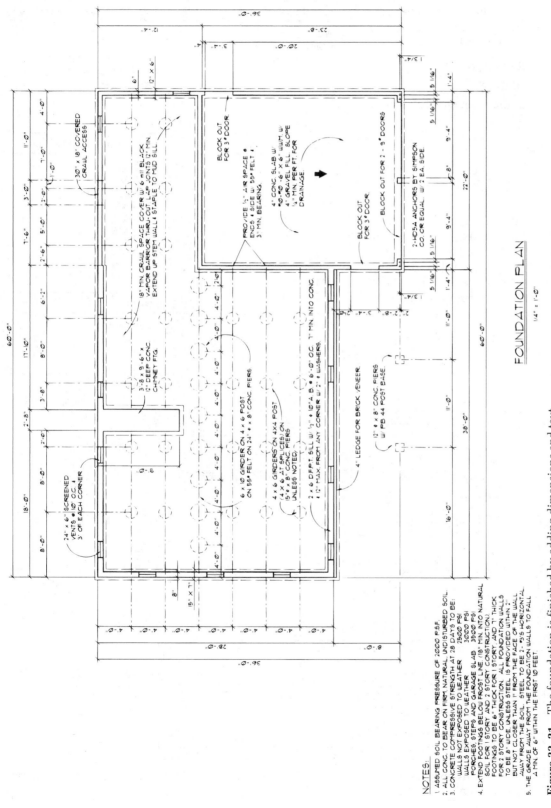

Figure 33–21 The foundation is finished by adding dimensions and text.

▼ PILING FOUNDATION AND JOIST FLOOR SYSTEM

When your drawing is complete, it should resemble the plan in Figure 33–26. Use construction lines for drawing steps 1 through 13. Each step can be seen in Figure 33–22.

Step 1. Outline the shape of the entire structure using the measurements from your floor plan.

Step 2. Lay out the shape of the slab.

Step 3. Lay out the foundation wall.

Step 4. Block out doors in the garage area.

Step 5. Lay out the support ledge if masonry veneer is to be used.

Step 6. If a masonry fireplace is to be used, lay out the size of the fireplace.

Step 7. Lay out the footing under the foundation walls.

Step 8. Lay out the footing under the fireplace.

Step 9. Lay out any piers or footings for interior walls or posts.

Step 10. Lay out the centerline of girders under load-bearing walls.

Step 11. Locate the center of pilings.

Step 12. Lay out the girder and pilings for any decks or porches.

Step 13. Lay out the centerline for any piers for decks and porches.

Step 14. Darken all items drawn with construction lines in steps 1–13. Use bold long-short-long dashed lines for step 1. Use bold lines for steps

2, 3, 4, 5, and 6. Use thin dashed lines to represent the material of steps 7, 8, 9, 11, 12, and 13. Use thin lines to represent the girders of steps 10 and 12. When this step is complete, your drawing should resemble Figure 33–23.

See Figure 33–24 for steps 15 through 19. These items have not been drawn using construction lines. Because of their simplicity, each may be drawn with finished-quality lines and need not be traced.

Step 15. Draw the ledger.

Step 16. Draw arrows to represent joist span and direction.

Step 17. Crosshatch the masonry for the chimney.

Step 18. Indicate crossbracing between wood or steel pilings.

Step 19. Place detail markers to indicate details to be drawn.

You now have all of the needed items drawn to represent the floor and foundation systems. Use steps 20 through 28 to locate these items. Use thin lines for the extension and dimension lines. When complete, your drawing should resemble Figure 33–25.

Draw extension and dimension lines to locate:

Step 20. The perimeter of the structure on each side.

Step 21. The limits of the slab.

Step 22. Jogs in the slab and foundation walls.

Step 23. Openings in the foundation walls.

Step 24. The fireplace.

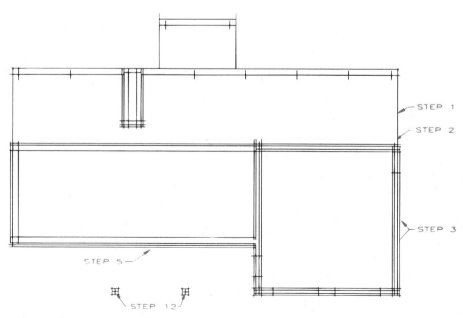

Figure 33–22 Use construction lines to lay out the location of stem wall, footings, door locations, slab outline, the fireplace, and girder and pier locations for a piling foundation plan.

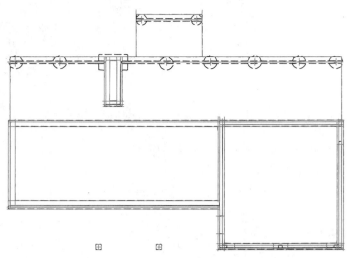

Figure 33–23 Carefully darken all objects using the proper finished-line quality. Construction lines should be so light that they do not need to be erased.

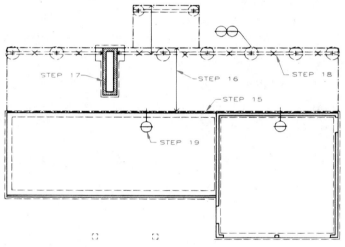

Figure 33–24 Add any finishing materials such as joists, column supports, vents, and detail markers.

Step 25. Girders.

Step 26. All pilings.

Step 27. All piers.

Step 28. Metal connectors.

All materials have now been drawn and located. The final drawing procedure is to place dimensions and specify the material to be used. This is done with the use of general and local notes. Figure 33–26 shows notes that are typically required. Use the following steps as a guideline to complete the foundation.

Step 29. Compute all dimensions and place them in the appropriate location.

Step 30. Neatly letter all required general notes.

Step 31. Neatly letter all required local notes.

Step 32. Place a title and scale below the drawing.

Step 33. Using a print of your plan, evaluate your drawing using the following checklist.

Piling Foundation and Joist Floor System Checklist

Correct Symbol and Location	Correct Structural Materials
Walls	Proper footing size
Footings	Proper wall size
Structure perimeter	Proper wall and footing location
Girders	Proper pier or piling location
Pilings or piers	Proper girder location
Metal connectors	Proper girder size
Openings in walls	Proper joist size
Fireplace	Proper joist direction

Required General Notes	Required Local Notes
Soil bearing pressure	Joist size and spacing
Concrete strength	Beam size
Crossbracing between	Post sizes
wood and steel	Metal connectors
pilings or posts	Pier or piling size
Framing lumber	Anchor bolt and mudsill size
grade and species	Veneer ledges
	Door blockouts in stemwall
	Fireplace footing size

Required Dimensions	
Overall	Girders
Jogs	Metal connectors
Pilings or piers	
Continuous footing	
centers	

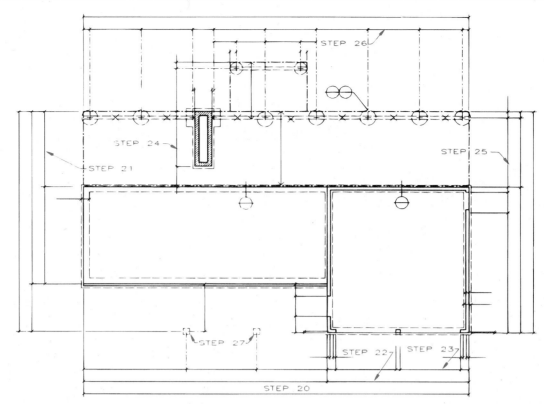

Figure 33–25 Prepare the plan to be dimensioned by adding dimension and extension lines to describe the overall, major jogs, upper floor lines, and each opening on each side of the foundation.

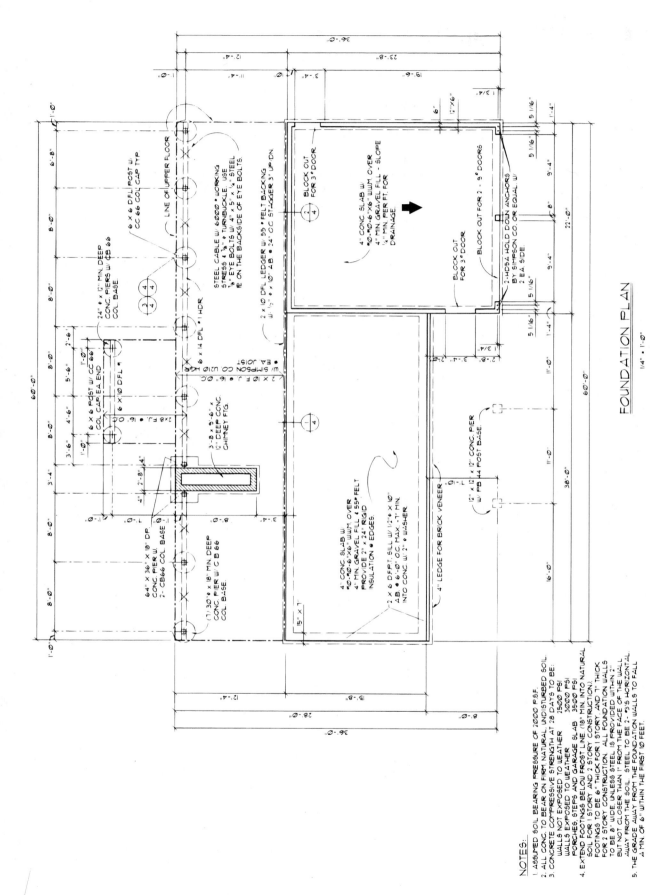

FOUNDATION PLAN
1/4" = 1'-0"

NOTES:
1. ASSUMED SOIL BEARING PRESSURE OF 2000 P.S.F.
2. ALL CONC. TO BEAR ON FIRM NATURAL UNDISTURBED SOIL.
3. CONCRETE COMPRESSIVE STRENGTH AT 28 DAYS TO BE:
   WALLS NOT EXPOSED TO WEATHER   1500 PSI
   WALLS EXPOSED TO WEATHER   3000 PSI
   PORCHES, STEPS AND GARAGE SLAB   3500 PSI
4. EXTEND FOOTINGS BELOW FROST LINE (18" MIN. INTO NATURAL
   SOIL FOR 1 STORY AND 2 STORY CONSTRUCTION).
   FOOTINGS TO BE 6" THICK FOR 1 STORY, AND 7" THICK
   FOR 2 STORY CONSTRUCTION. ALL FOUNDATION WALLS
   TO BE 8" WIDE UNLESS STEEL IS PROVIDED WITHIN 2"
   BUT NOT CLOSER THAN 1" FROM THE FACE OF THE WALL.
5. THE GRADE AWAY FROM THE SOIL, STEEL TO BE 2-#5 HORIZONTAL
   AWAY FROM THE FOUNDATION WALLS TO FALL
   A MIN. OF 6" WITHIN THE FIRST 10 FEET.

Figure 33–26   The foundation is finished by adding dimensions and text.

CADD Applications

USING CADD TO DRAW FOUNDATION PLANS

Because you do not have the same kind of accuracy problems when working with CADD as you have with manual drafting, the CADD floor plan may be used as an accurate basis for drawing the foundation plan. Display the floor plan on a layer, and then begin the foundation drawing directly over the floor plan on another layer. Use layers with prefixes such as FNDWALLS, FNDFOOT, FND-JOIST, FNDTEXT, and FNDDIMEN to keep the foundation plan separate from the floor-plan files. By using the OSNAP command, the line representing the outer side of the stem walls can be drawn, using the walls of the floor plan as a guide. The OFFSET command can be used to lay out the thickness of the stem walls and footings. Corners can be adjusted by using the FILLET or TRIM commands. The CHANGE command can be used to change the lines representing the footings from continuous to hidden. By following the step-by-step instructions for a particular floor type, the foundation plan can be completed.

Many of the dimensions used on the floor plan can also be used on the foundation plan. A layer such as BASEDIM can be used for placing dimensions required by the floor and foundation plans. General notes can be typed and stored as a WBLOCK and reused on future foundation plans. Many drafters also store lists of local notes required for a particular type of foundation as a WBLOCK and insert them into a drawing. Once inserted into the foundation plan, the notes can be moved to the desired position. When completed, the foundation can be stored separate from the floor plan to make plotting easier. Storing the foundation plan with the floor plan will save disk space, and proper use of layering can ease plotting. All foundation walls, bearing footings, and support beams will be in their correct locations. When you are finished drawing the foundation plan, turn OFF or FREEZE the floor plan layer to have a ready-to-plot foundation plan.

CHAPTER

33   Foundation-Plan Layout Test

DIRECTIONS

Answer the questions with short, complete statements.

1. Letter your name and the date at the top of the sheet.
2. Letter the question number and provide the answer. You do not need to write out the question.

QUESTIONS

Question 33–1   At what scale will the foundation be drawn?

Question 33–2   List five items that must be shown on a foundation plan for a concrete slab.

Question 33–3   What general categories of information must be dimensioned on a slab foundation?

Question 33–4   Show how floor joists are represented on a foundation plan.

Question 33–5   How large an opening should be provided in the stem wall for an 8'–0" wide garage door?

Question 33–7   How much space should be provided for a 3' entry door in a post-and-beam foundation? Explain your answer.

Question 33–8   How are the footings represented on a foundation plan?

Question 33–9   Show two methods of representing girders.

Question 33–10   What type of line quality is typically used to represent beam pockets?

Question 33–11   What are the disadvantages of tracing a print of the floor plan to lay out a foundation plan?

PROBLEMS

DIRECTIONS

1. Select from the following sketches the one that corresponds to the floor-plan problem from Chapter 16 that you have drawn. Draw the required foundation plan using the guidelines given in this chapter. If no sketch has been given for your foundation

plan, design a system that is suitable for the residence and your area of the country.

2. Draw the foundation plan using the same type and size of drawing material that you used for the floor plan.

3. Use the same scale that was used to draw the floor plan.

4. Refer to your floor plan to determine dimensions and position of load-bearing walls.

5. Use the sketch for reference only. Refer to the text of this chapter and class lecture notes for complete information.

6. When your drawing is complete, turn in a print to your instructor for evaluation.

Problem 33–1 Cape Cod. This floor plan requires a standard joist construction foundation and has a stick roof. A stick roof is constructed of ceiling joists and roof rafters that meet at a ridge board. The kind of roof is given so you can determine which partitions are load bearing.

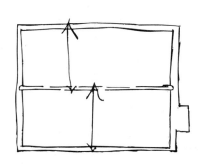

Problem 33–2 Cabin. This floor plan requires a piling foundation and has a stick roof. A stick roof is constructed of ceiling joists and roof rafters that meet at a ridge board. The kind of roof is given so you can determine which partitions are load bearing. Note the alternate foundation layout.

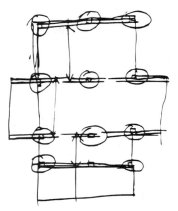
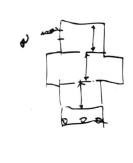

Problem 33–3 Ranch. This floor plan requires a post-and-beam foundation and has a truss roof. A truss roof requires no interior bearing walls. The kind of roof is given so you can determine which partitions are load bearing.

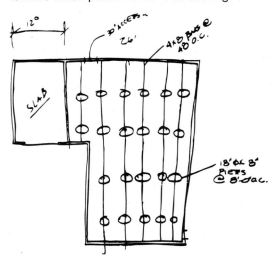

Problem 33–4 Spanish. This floor plan requires a concrete slab foundation and has a stick roof. A stick roof is constructed of ceiling joists and roof rafters that meet at a ridge board. The kind of roof is given so you can determine which partitions are load bearing.

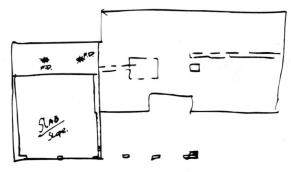

Problem 33–5 Early American. This floor plan will be drawn with a basement foundation system. The roof and floors will be framed using conventional framing methods. Provide support at the foundation level for all bearing walls. Determine the size of all footings and piers using the information in this section and Chapters 28 and 29.

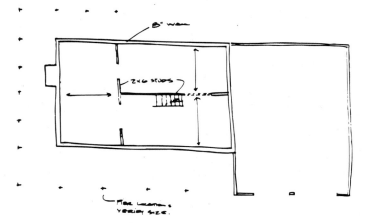

Problem 33–6   Contemporary. This floor plan will be drawn with a daylight basement. The roof and floors will be framed using conventional framing methods. Provide support at the foundation level for all bearing walls. Determine the size of all footings and piers using the information in this section and Chapters 28 and 29. Determine the size of windows and doors, and type of basement materials, based on common practices in your area or according to lectures by your instructor.

Problem 33–7 to 33–15   Duplex. Design a suitable foundation for the structure that you started in Chapter 15 using the information presented in this section and Chapters 41 and 42. Determine the type of roof framing system to be used, and design an appropriate foundation system. Sketch the foundation plan showing all required footings and piers and have it approved by your instructor prior to starting your drawing.

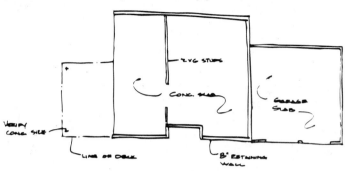

SECTION X

WALL SECTIONS AND DETAILS

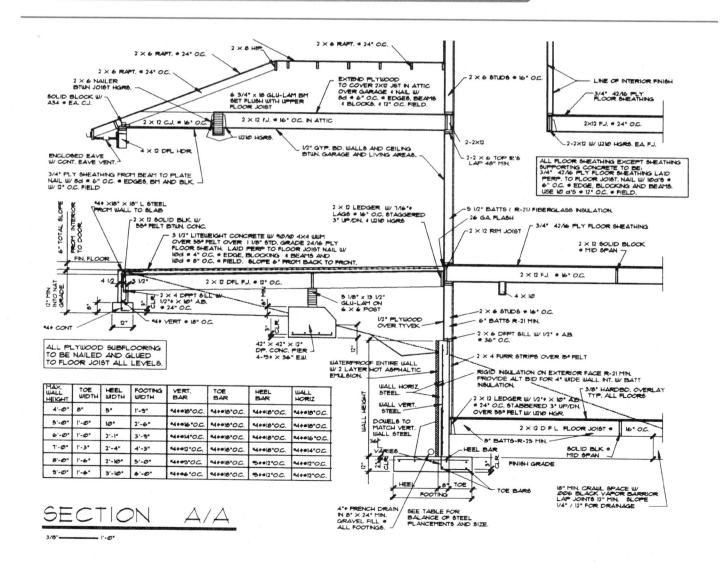

SECTION   A/A

3/8" ———— 1'-0"

MAX. WALL HEIGHT	TOE WIDTH	HEEL WIDTH	FOOTING WIDTH	VERT. BAR	TOE BAR	HEEL BAR	WALL HORIZ.
4'-0"	8"	8"	1'-9"	#4@18"O.C.	#4@18"O.C.	#4@18"O.C.	#4@18"O.C.
5'-0"	1'-0"	10"	2'-6"	#4@16"O.C.	#4@18"O.C.	#4@18"O.C.	#4@18"O.C.
6'-0"	1'-0"	2'-1"	3'-3"	#4@14"O.C.	#4@18"O.C.	#4@18"O.C.	#4@16"O.C.
7'-0"	1'-3"	2'-4"	4'-3"	#4@12"O.C.	#4@18"O.C.	#4@18"O.C.	#4@14"O.C.
8'-0"	1'-6"	2'-10"	5'-0"	#4@9"O.C.	#4@18"O.C.	5#@12"O.C.	#4@12"O.C.
9'-0"	1'-6"	3'-10"	6'-0"	#4@6"O.C.	#4@18"O.C.	5#@12"O.C.	#4@12"O.C.

CHAPTER 34

Sectioning Basics

Sections are drawn to show the vertical relationships of the structural materials called for on the floor, framing, and foundation plans. The sections show the methods of construction for the framing crew. Before drawing sections, it is important to understand the different types of sections, their common scales, and the relationship of the cutting plane to the section.

▼ TYPES OF SECTIONS

Three types of sections may be drawn for a set of plans: full sections, partial sections, and details.

For a simple home, only one section might be required to explain fully the types of construction to be used. Each major structural material must be drawn and specified as well as providing the vertical dimensions. A full section can be seen in Figure 34–1. Notice that the roof is framed with standard/scissor trusses, the exterior walls are framed with 2 × 6 (50 × 150 mm) studs, and the floor system is post and beam with an 8″ (200 mm) step. Vertical relationships for the roof pitch, wall height, window and door header heights, foundation, and crawl heights are also provided.

On a more complex structure, more than one section may be required to specify each major type of construction. Some offices use a combination of partial and full sections to explain the required construction procedures. A *partial section* will show only typical roof, wall, floor, and foundation information of one typical wall rather than the full structure. A partial section may be drawn at a scale of ⅜″ = 1′–0″, ½″ = 1′–0″, or ¾″ = 1′–0″, depending on office procedure. An example of a partial section can be seen in Figure 34–2. The partial section is typically supplemented with full sections drawn at a scale of ¼″ = 1′–0″ or ⅜″ = 1′–0″. Only material that has not been specified on the partial section is noted on the full sections. The use of partial sections is becoming increasingly popular in many professional offices using computers. Several different partial sections can be created to reflect major types of construction—for example, one- or two-level construction, truss or stick roof, post-and-beam or joist foundations, and various wall coverings. These partial sections can be stored on disk and inserted into each set of plans as needed.

Partial section can also be used on complex structures to serve as a reference for details of complicated areas.

Details are enlargements of specific areas of a structure and are typically drawn where several components intersect or where small members are required. Figure 34–3 shows an example of a partial section of a structure built on a piling foundation. Figure 34–4 shows examples of two of the details that relate to the partial section in Figure 34–3. Details are typically drawn at a scale of ½″ = 1′–0″ through 3″ = 1′–0″, depending on the complexity of the intersection.

Some offices have stock details. These are details of items such as footings that typically remain the same. By using a stock footing detail, the drafter saves the time of having to relabel all of this information each time this part is drawn in section. See Figure 34–5.

By combining the information on the framing plans and the sections, the contractor should be able to make accurate estimates of the amount of material required and the cost of completing the project. To help make the sections easier to read, sections have become somewhat standardized in several areas, including areas of scales and alignment.

▼ SCALE

Sections are typically drawn at a scale of ⅜″ = 1′–0″. Scales of ⅛ or ¼″ may be used for supplemental sections requiring little detail. A scale of ¾″ = 1′–0″ or larger may be required to draw some construction details. Several factors influence the choice of scale to use when drawing sections:

1. Size of the drawing sheet to be used.
2. Size of the project to be drawn.
3. Purpose of the section.
4. Placement of the section.

Factors 1 and 2 need little discussion. The floor plan determines the size of the project. Once the sheet size is selected for the floor plan, that size should be used throughout the entire project. The placement of the section as it relates to other drawings should have only a minor influence on the scale. It may be practical to put a partial section in a blank corner of a drawing, but don't let space dictate the scale.

The most important influence should be the use of the section. If the section is merely to show the shape of the project, a scale of ⅛″ = 1′–0″ would be fine. This type of

543

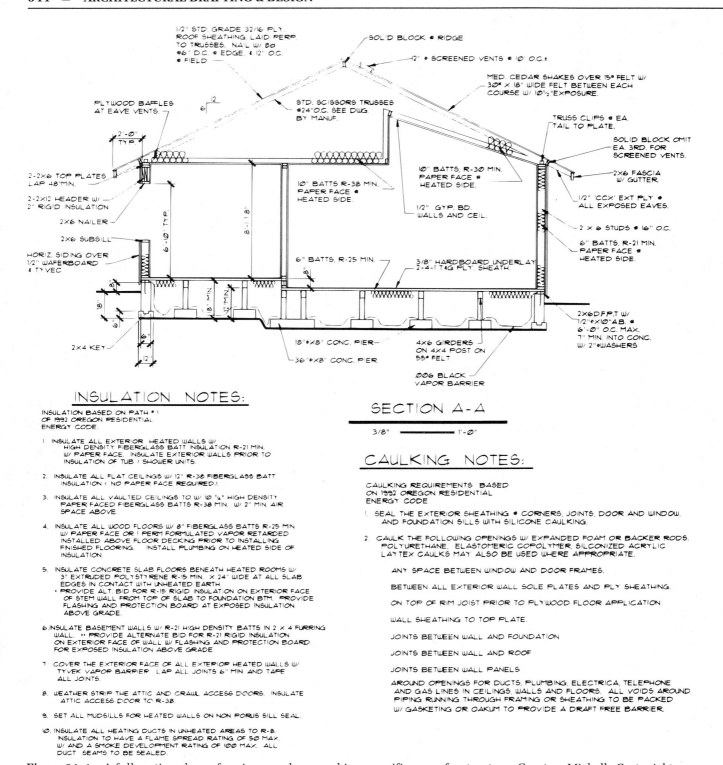

Figure 34–1 A full section shows framing members used in a specific area of a structure. *Courtesy Michelle Cartwright.*

section is rarely required in residential drawings but is often used when drawing apartments or office buildings. When used for residential projects, this type of section is used as a reference map to locate structural details. See Figure 34–6 for an example of a shape section.

The primary section is typically drawn at a scale of 3/8″ = 1′–0″. This scale provides two benefits, one to the print reader and one to the drafter. The main advantage of

using this scale is the ease of distinguishing each structural member. At a smaller scale, separate members such as the finished flooring and the rough flooring are difficult to draw and read. Without good clarity, problems could arise at the job site. At 3/8″ scale, the drafter will have a bigger drawing on which to place the notes and dimensions.

The scale of 1/2″ = 1′–0″ is not widely used in most offices. Using this scale does offer great clarity, but sections

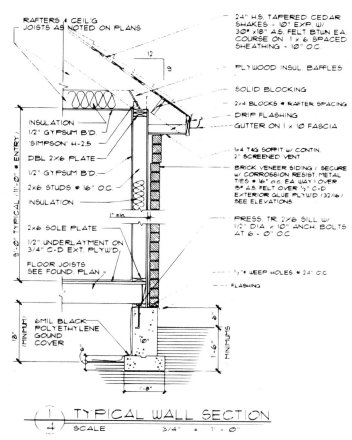

Figure 34–2 labels:

RAFTERS & CEIL'G JOISTS AS NOTED ON PLANS

24" H.S. TAPERED CEDAR SHAKES - 10" EXP. W/ 30° x18" A.S. FELT BTUN EA. COURSE ON 1 x 6 SPACED SHEATHING - 10" O.C.

PLYWOOD INSUL. BAFFLES

SOLID BLOCKING

2x4 BLOCKS @ RAFTER SPACING

DRIP FLASHING

GUTTER ON 1 x 10 FASCIA

INSULATION
1/2" GYPSUM B'D.
'SIMPSON' H-2.5
DBL 2x6 PLATE
1/2" GYPSUM B'D.
2x6 STUDS @ 16" O.C.
INSULATION

1x4 T&G SOFFIT W/ CONTIN. 2" SCREENED VENT

BRICK VENEER SIDING (SECURE W/ CORROSSION RESIST. METAL TIES @ 16" O.C. EA WAY) OVER 15# A.S. FELT OVER 1/2" C-D EXTERIOR GLUE PLYW'D (32/16) SEE ELEVATIONS

PRESS. TR. 2x6 SILL W/ 1/2" DIA. x 10" ANCH. BOLTS AT 6 - 0" OC

2x6 SOLE PLATE
1/2" UNDERLAYMENT ON 3/4" C-D EXT. PLYW'D
FLOOR JOISTS SEE FOUND. PLAN

1/2"Ø WEEP HOLES @ 24" O.C.

FLASHING

6 MIL BLACK POLYETHYLENE GOUND COVER

TYPICAL WALL SECTION
SCALE 3/4" = 1' - 0"

Figure 34–2 A partial section can be used to show typical roof, wall, floor, and foundation construction materials of a specific structure. *Courtesy Piercy Barclay Designers, Inc., CPBD.*

are so large that a great deal of paper is required to complete the project. Often, if more than one section must be drawn, the primary section is drawn at $3/8'' = 1'-0''$, and the other sections are drawn at $1/4'' = 1'-0''$. By combining drawings at these two scales, typical information can be placed on the $3/8''$ section, and the $1/4''$ sections are used to show variations with little detail.

▼ SECTION ALIGNMENT

In drawing sections, as with other parts of the plans, the drawing is read from the bottom or right side of the page. The cutting plane on the framing plan shows which way the section is being viewed. The arrows of the cutting plane should be pointing to the top or left side of the paper, depending on the area of the building being sectioned. See Figure 34–7. Where possible, the cutting plane should extend through the entire structure. The cutting plane can be broken for notes or dimensions to maintain clarity. On complex structures, the cutting plane can be jogged to show material clearly and avoid a second section from being drawn. Figure 34–8 shows an example of a framing plan with section markers.

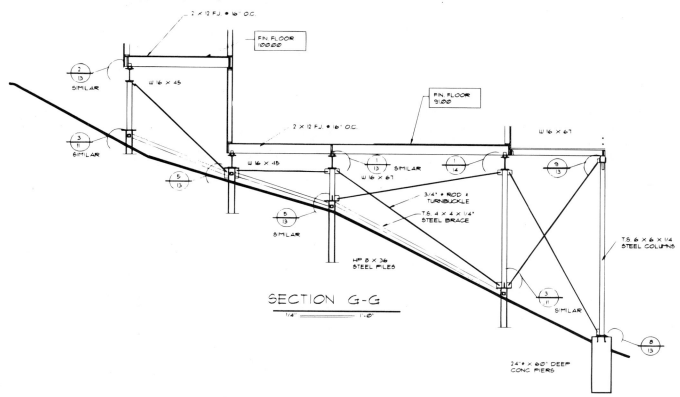

SECTION G-G
1/4" = 1' - 0"

Figure 34–3 A partial section can be used to provide supplemental information about an area of a structure. *Courtesy Residential Design, AIBD.*

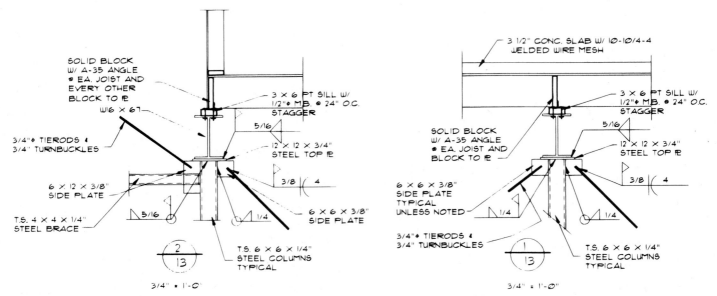

Figure 34–4 Details provide information about a small, complicated area of a structure, such as the intersection of various materials. These details are referenced to the partial section in Figure 34–3. *Courtesy Residential Designs, AIBD.*

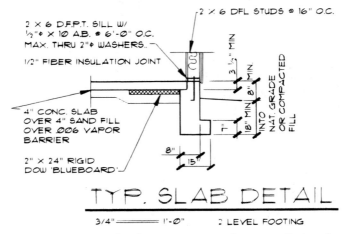

TYP. SLAB DETAIL

3/4" ══════ I'-Ø"   2 LEVEL FOOTING

Figure 34–5 Details of common areas of construction such as a footing are often referred to as a stock detail. A stock detail can be inserted into a drawing to save the time of drawing it for each application.

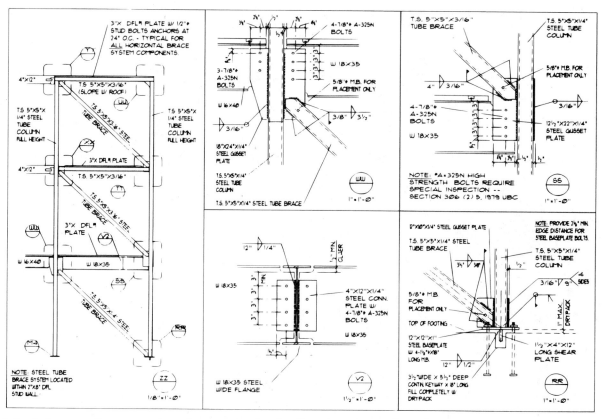

Figure 34–6 A shape section serves as a reference map for detail markers. Typically it shows no specific details of an intersection. *Courtesy Ken Smith AIA, StructureForm Masters, Inc.*

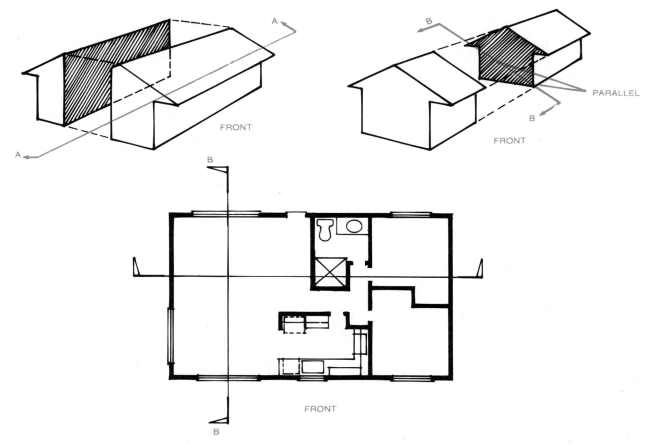

Figure 34–7 Cutting planes on the framing plan show the direction the section is to be viewed. Always try to keep the cutting plane arrows pointing to the left or to the top of the page.

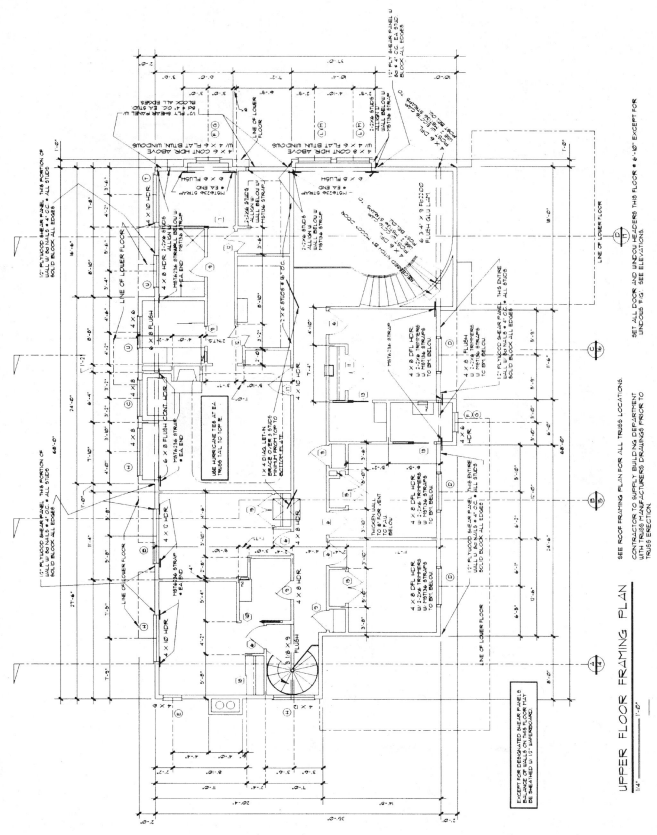

UPPER FLOOR FRAMING PLAN
1/4" = 1'-0"

Figure 34-8 Cutting plane markers may extend through a structure, be broken at text or dimensions to provide clarity, or be jogged to eliminate a second section. *Courtesy Residential Designs, AIBD.*

CHAPTER 34

Sectioning Basics Test

DIRECTIONS

Answer the questions with short, complete statements or drawings as needed.

1. Letter your name and the date at the top of the sheet.
2. Letter the question number and provide the answer. You do not need to write out the question.

QUESTIONS

Question 34–1 What is a full section?

Question 34–2 When could a partial section be used?

Question 34–3 What is a stock detail, and when would it be used?

Question 34–4 From which drawings does a drafter get the needed information to draw a section?

Question 34–5 Name the most common scale for drawing full sections.

Question 34–6 What factors influence the scale of a detail?

Question 34–7 What is a cutting plane, and how does it relate to a section?

Question 34–8 In which directions should the arrows on a cutting plane be pointing?

Question 34–9 What type of section might be drawn at a scale of $1/8'' = 1'-0''$?

Question 34–10 What factors influence the choice of scale for the section?

CHAPTER 35

Section Layout

Drawing sections will be accomplished in seven major stages. By following the step-by-step procedure for each stage, sections can be easily drawn and understood. Read through each stage, and carefully compare the step-by-step instructions with the corresponding illustrations.

▼ STAGE 1: EVALUATE NEEDS

Using a print of the floor and foundation plans, evaluate the major types of construction needed on the project. The differences in the construction required include the following:

○ The right side of the plan is one story, and the left side is two story.
○ The rear at right is a stick floor, and the front is a concrete slab.
○ The left side of the roof is vaulted and stick, while the right side is truss.

To provide the needed information to the framing crew, a minimum of two full sections is required. One section will cut through the family room and kitchen and show the stick floor, walls, the upper floor cantilever, and the truss roof. A full section will also be needed through the garage area in order to show the concrete and wooden floor con-struction. Partial sec-

tions should also be provided in order to show the stairs and the vaulted roof over bedroom 1.

▼ STAGE 2: LAY OUT THE SECTION

Use construction lines for this entire stage. Use a non-reproducible blue pencil or a 6H lead to draw all construction lines.

Having determined which sections need to be drawn, lay out the primary section at ⅜" = 1'–0". To determine sizes and locations of structural members, refer to the floor and foundation plans. For the layout of a post-and-beam, slab, or basement foundation, see Chapter 36. See Figure 35–1 for steps 1 through 5.

Stick Floor Layout

Step 1. Lay out the width of the building.
Step 2. Establish a baseline about 2" (50 mm) above the border.

Before laying out steps 3 through 8, it may be necessary to review the basics of foundation design in Section VIII. All sizes given in the following steps are based on the minimum standards of the Uniform Building Code (unless otherwise noted) and should be compared with local standards.

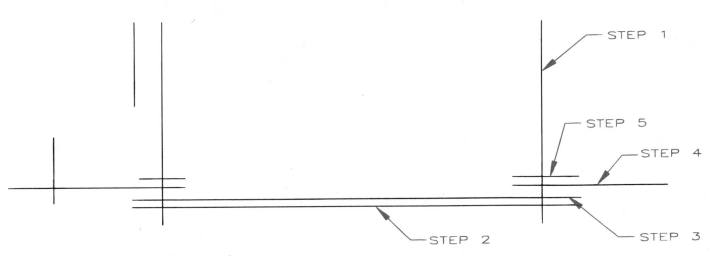

Figure 35–1 Initial steps of the section layout.

Uniform Building Code Footing Requirements

Building Height	Footing Width (A)	Wall Width (B)	Footing Height (C)	Depth into Undisturbed Grade (D)
1 story	12″ (305 mm)	6″ (152 mm)	6″ (152 mm)	12″ (305 mm)
2 story	15″ (381 mm)	8″ (203 mm)	7″ (178 mm)	18″ (457 mm)
3 story	18″ (457 mm)	10″ (254 mm)	8″ (203 mm)	24″ (610 mm)

Step 3. Measure up from the baseline the required thickness for a footing.

Step 4. Locate the finished grade.

Step 5. Locate the top of the stem wall. All wood is required to be 8″ minimum (200 mm) above finished grade.

See Figure 35–2 for steps 6 through 15.

Step 6. Block out the width of the stem wall.

Step 7. Block out 4″ (100 mm) wide by 8″ (200 mm) high ledge for brick veneer.

Step 8. Block out the required width for footings. Center the footing below the stem walls.

Step 9. Add 4″ (100 mm) in width to the footing to support the brick ledge.

Step 10. Locate the depth for the 2 × 6 (50 × 150 mm) mudsill. The top of the mudsill also becomes the bottom of the floor joists.

Step 11. Locate the depth of the floor joists.

Step 12. Lay out the plywood subfloor. Draw a line as close to the floor joist as possible to represent the plywood while still leaving a gap between the plywood and the joist.

Step 13. Lay out the girder.

Step 14. Lay out the post.

Step 15. Lay out the concrete piers centered under the post.

Wall Layout

With the foundation and floor system lightly drawn, proceed now to lay out the walls for the main level using construction lines. See Figure 35–3 for the layout of steps 16 through 22.

Step 16. Locate the top of the top plates. Measure up 8′–0″ (2400 mm) above the floor, and draw a line to represent the top of the top plate. In a one-level house, this would also represent the bottom of the ceiling.

Step 17. Locate the interior side of the exterior walls.

Step 18. Locate the interior walls.

After locating the interior walls, make sure that any interior bearing walls line up over a girder. If they do not line up, there is a mistake in your measurements or math calculations or a conflict between the dimensions of the floor and foundation plans.

Step 19. Lay out the tops of doors and windows. Headers for doors and windows are typically set at 6′–10″ (2080 mm) above the floor. Measure up from the plywood to locate the bottom of the header. The top will be drawn later.

Step 20. Lay out the subsills. To establish the subsill location, you must know the window size. This can be found on either the floor plan or on the window schedule. Measure down the required distance from the bottom of the header to establish the subsill location.

Step 21. Lay out the 7′–0″ (2130 mm) ceilings for soffits in areas such as kitchens, baths, or hallways.

Step 22. Lay out the patio post width and height.

Lay out the upper floor area following the same steps that were used for the lower level. See Figure 35–4 for steps 23 through 29.

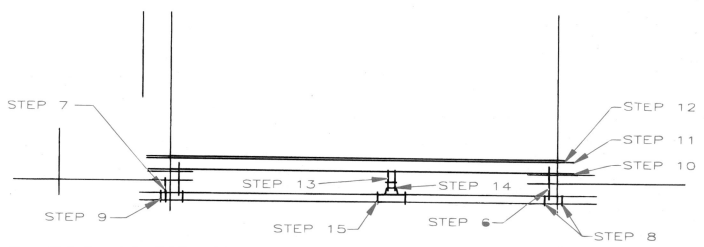

Figure 35–2 Layout of a stick floor system.

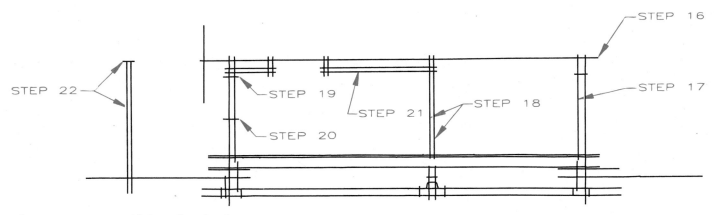

Figure 35–3 Layout of the ceiling level.

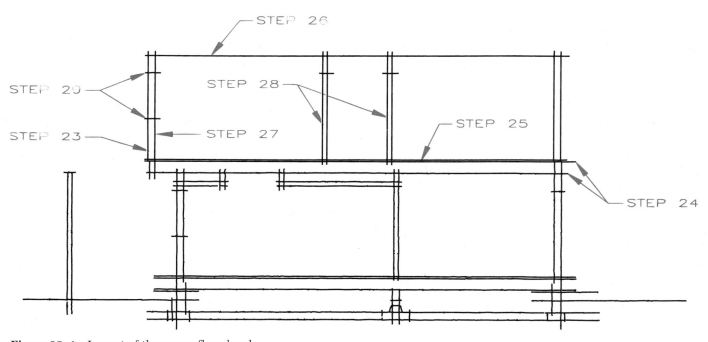

Figure 35–4 Layout of the upper floor level.

Step 23. Lay out exterior walls. Remember the 2′–0″ (600 mm) cantilever of the upper floor past the lower level walls.

Step 24. Locate the depth of the floor joists.

Step 25. Draw the plywood subfloor.

Step 26. Lay out the ceiling 8′–0″ (2400 mm) above the floor.

Step 27. Lay out the interior side of exterior walls.

Step 28. Lay out the interior walls.

Step 29. Lay out the windows.

Truss Roof Layout

Steps 30 through 37 can be seen in Figure 35–5. For the layout of other types of roofs, see Chapter 36.

Step 30. Locate the ridge.

Step 31. Locate the overhand on each side (see the elevations).

Step 32. Lay out the bottom side of the top chord. Starting at the intersection of the outer wall and the ceiling line, draw a line at a 5/12 pitch to match the pitch shown in the elevations. Do this for both sides. The lines should intersect at the ridge. A 5/12 pitch means that you measure up 5 units and over 12 units. A 5/12 pitch may also be drawn with a protractor or a drafting machine since it equals 22½°.

Step 33. Lay out the top side of the top chord. Assume chords to be 4″ (100 mm) deep.

Step 34. Lay out the plywood roof sheathing.

Step 35. Lay out the top side of the bottom chord.

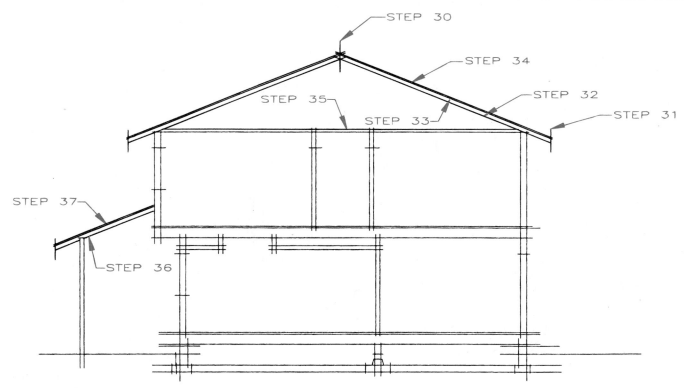

Figure 35–5 Truss roof layout.

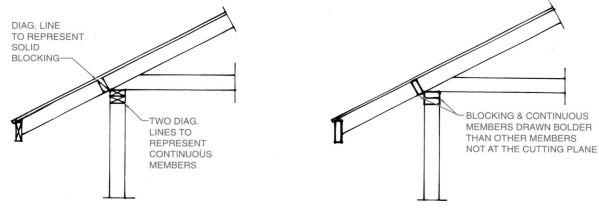

Figure 35–6 Finished-quality lines for structural members.

Step 36. Lay out the patio roof. Starting at the intersection of the inner wall and the ceiling, draw a line at a 5/12 pitch to represent the rafter.

Step 37. Lay out the 1″ (25 mm) tongue-and-groove roof sheathing.

The entire section has now been outlined.

▼ STAGE 3: FINISHED-QUALITY LINES— STRUCTURAL MEMBERS ONLY

When drawing with finished-quality lines, start at the roof and work down. This will help you keep the drawing clean. The steps used will be divided into truss roof, walls, upper floor, walls, lower floor, and foundation.

Continuous members shown in section are traditionally drawn with a diagonal cross (X) placed in the member. Blocking is drawn with one diagonal line (/) through the member. This method of representing sectioned members is time-consuming to draw and may be difficult to read on small-scale sections. Rather than drawing symbols, the members can be drawn much more easily with different line qualities, as shown in Figure 35–6.

Three types of lines are used to draw structural material for the sections:

Thin. Use a 5-mm lead, a #0 pen, or a sharp 3H lead.

Bold. Use a 9-mm lead, a #2 pen, or an H lead with a wedge tip.

Very bold. Use a 9-mm lead, and draw two parallel lines with no white space between them, or a #4 pen or an unpointed H lead.

As you read through these steps you will be asked to draw items that have not been laid out. These are items that can be drawn by eye and need not be traced. Just be careful to keep 2″ (50 mm) members in a uniform size so that they do not become 4″ (100 mm) members. When finished, your drawing should look like Figure 35–7.

Roof

To gain speed, draw all parallel lines on one side of the roof before drawing the opposite side. Roof steps 1 through 7 are shown in Figure 35–8.

Step 1. Use thin lines to draw the plywood sheathing.

Step 2. Draw the top chord with thin lines.

Step 3. Draw the eave blocking with bold lines.

Step 4. Repeat steps 1 through 3 on the opposite side of the roof.

Step 5. Draw the fascias with bold lines.

Step 6. Draw the bottom chord with thin lines.

Step 7. Draw the ridge blocking with a bold line.

Upper Walls and Floor

See Figure 35–9 for steps 8 through 18.

Step 8. Draw all double top plates with bold lines.

Step 9. Draw walls with thin lines.

Step 10. Draw all headers with bold lines.

Step 11. Draw all subsills with bold lines.

Step 12. Draw all bottom plates with bold lines.

Step 13. Draw exterior sheathing with thin lines.

Step 14. Draw the porch rafters and decking with thin lines.

Step 15. Draw fascia, block, header, and ledger for the porch roof with bold lines.

Step 16. Draw plywood subfloor with thin lines.

Step 17. Draw floor joists with thin lines.

Step 18. Draw floor blocks with bold lines.

Step 19. Repeat steps 8 through 13.

Lower Level Walls and Floor

See Figure 35–10 for steps 20 through 26.

Step 20. Draw furred ceiling with thin lines.

Step 21. Draw all blocks and ledgers for ceiling with bold lines.

Step 22. Repeat steps 16 through 18.

Step 23. Draw mudsills with bold lines.

Step 24. Draw girders with bold lines.

Step 25. Draw post with thin lines.

Step 26. Draw all exterior wall sheathing with thin lines.

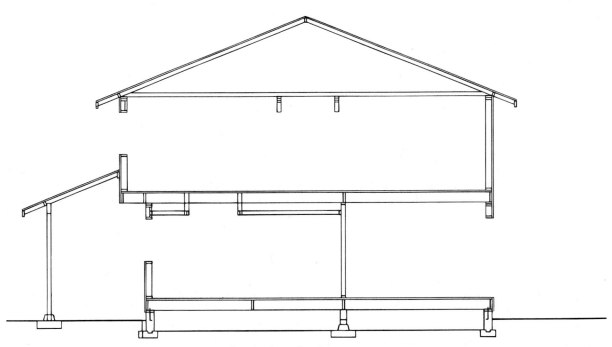

Figure 35–7 All structural members drawn with finished-quality lines.

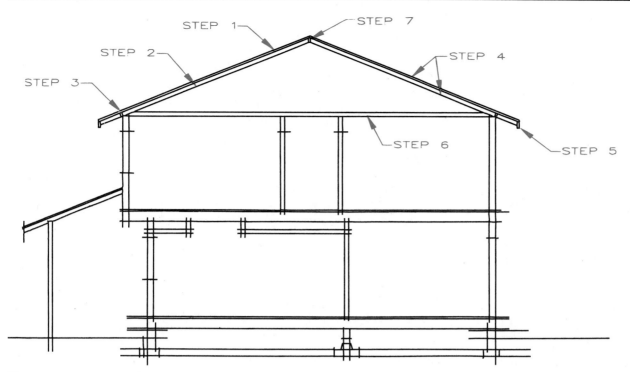

Figure 35–8 Truss roof construction drawn with finished-quality lines.

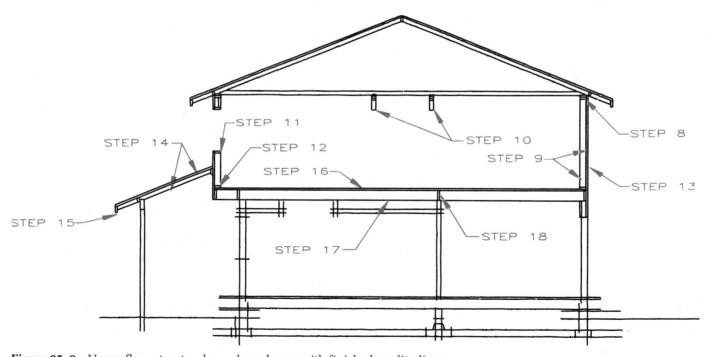

Figure 35–9 Upper floor structural members drawn with finished-quality lines.

Foundation

See Figure 35–11 for steps 27 through 30.

Step 27. Draw stem walls and footings with very bold lines.

Step 28. Draw the piers with very bold lines.

Step 29. Draw anchor bolts with bold lines.

Step 30. Use a very bold line to draw the finished grade.

All structural material is now drawn.

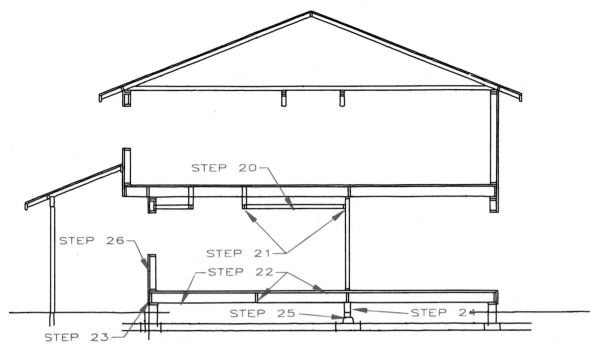

Figure 35–10 Structural members of the main floor level drawn with finished-quality lines.

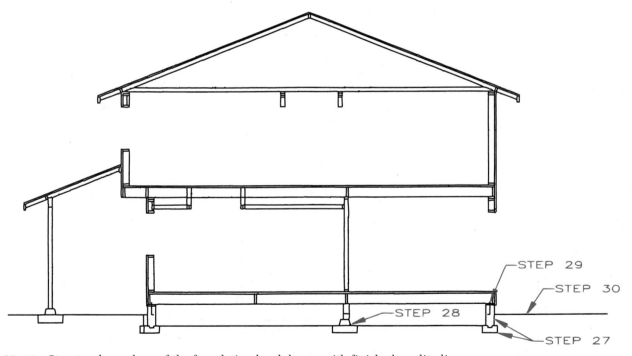

Figure 35–11 Structural members of the foundation level drawn with finished-quality lines.

▼ STAGE 4: DRAWING FINISHING MATERIALS

The material drawn in stage 3 forms the frame of the building. It seals the exterior from the weather and finishes the interior. Start at the roof and work down to the foundation to keep the drawing clean. Use thin lines for this entire stage unless otherwise noted. See Figure 35–12 for steps 1 through 14.

Roof

Step 1. Draw hurricane ties (approximately 3″ [75 mm] square).

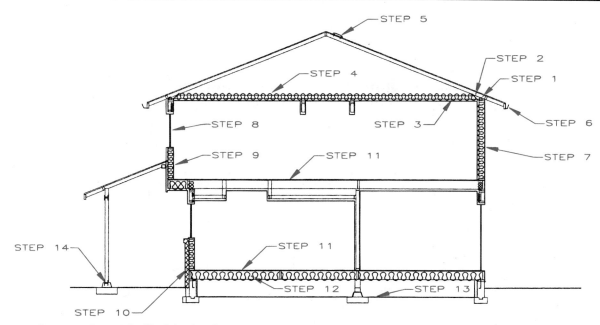

Figure 35–12 Cross section with all of the finishing materials for the roof, walls, floors, and foundations.

Step 2. Draw baffles with a bold line.

Step 3. Draw the finished ceiling.

Step 4. Draw insulation in the ceiling approximately 10″ (250 mm) deep.

Step 5. Draw ridge vents.

Step 6. Draw gutters.

Walls

Step 7. Draw the exterior siding.

Step 8. Draw all windows.

Step 9. Draw all interior finishes and insulation.

Step 10. Draw and crosshatch the brick veneer.

Floors and Crawl Spaces

Step 11. Draw the hardboard underlayment. Note that the plywood subfloor extends under all walls, but the hardboard underlayment does not go under any walls.

Step 12. Draw floor insulation.

Step 13. Draw the vapor barrier freehand, leaving a small space between the ground and the barrier.

Step 14. Draw the metal connectors and flashing for the porch.

▼ STAGE 5: DIMENSIONING

Now that the section is drawn, all structural members must be dimensioned so that the framing crew knows their vertical location. Figure 35–13 shows the needed dimensions. All leader lines should be thin, similar to those used on the floor and foundation plans. All lettering should be aligned.

Step 1. Floor to ceiling. Place a dimension line from the top of the floor sheathing to the bottom of the floor or ceiling joist.

Step 2. Determine the floor-to-floor dimensions. The walls were drawn 8′–0″ (2440 mm) high, but this is not their true height. The true height is the sum of the thickness of two top plates, the studs and the base plate. The plates are made from 2″ material, which is actually 1½″ thick. The studs are milled in 88⅝ or 92⅝″ lengths. Check with your instructor to see which size you should use. See Figure 35–14 to determine the true height of the walls.

Place the required leader and dimension lines to locate the following dimensions:

Step 3. Floor to bottom of bedroom windows (44″ (1120 mm) maximum).

Step 4. Finished grade to top of stem wall.

Step 5. Depth of footing into grade.

Step 6. Height of footing.

Step 7. Width of footing.

Step 8. Width of stem wall.

Step 9. Crawl space depth.

Step 10. Bottom of girders to grade.

Step 11. Eave overhangs.

Step 12. Cantilevers.

Step 13. Header heights from rough floor.

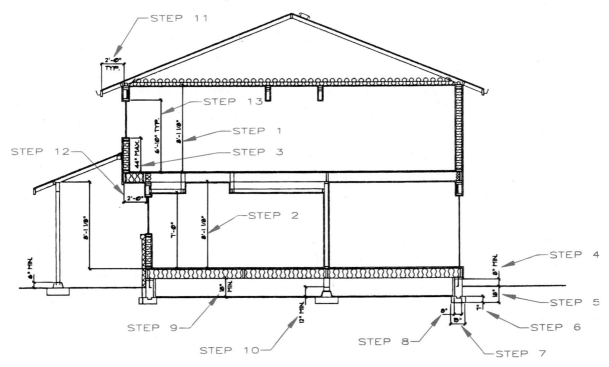

Figure 35–13 Cross section with all of the required dimensions.

Long Stud (in.)		Short Stud (in.)
3	2 Top plates	3
92⅝	Stud height	88⅝
+ 1½	Base plate	+ 1½
97⅛	Total height	93⅛
8′–1⅛″	Dimension to appear on section	7′–9⅛″

Figure 35–14 Determining floor to ceiling dimensions. Add the depth of the top plates, base plate, and the height of the studs. Notice that the heights for two common stud lengths are given.

Step 14. Height of brick veneer if necessary.

Step 15. Lean back and relax! You're almost done.

▼ STAGE 6: LETTERING NOTES

Everything that has been drawn and located must be explained. Place guidelines around the perimeter of the drawing and align the required notes on the guidelines, as seen in Figure 35–15. Doing so will help make your drawing neat and easy to read. Not all of the following notes will need to go on every section you draw. Typically, the primary section will be fully notated, and then other sections will have only construction notes. You will notice as you go through the list of general notes that some are marked (*). These marked items have several

options. Ask your instructor for help in determining which grade of material to use.

Remember that the following list is just a guideline. You, the drafter, will need to evaluate each section prior to placing the required notes. In an office setting your supervisor might give you a print with all of the notes that need to be placed on the original. As you gain confidence, you will probably be referred to a similar drawing for help in deciding which notes are needed. Eventually you will be able to draw and notate a section without any guidelines. The general notes that may appear on sections are as follows:

Roof

*1. RIDGE BLOCK, or 2 × ___ RIDGE BOARD.

2. SCREENED RIDGE VENTS @ 10′ O.C. ± .

*3. ½″ STD GRADE 32/16 PLY ROOF SHEATH. LAID PERP. TO TRUSSES. NAIL W/ 8d @ 6″ O.C. @ EDGE AND 8d @ 12″ O.C. FIELD.

or

1 × 4 SKIP SHEATHING W/ 3½″ SPACING.

*4. 235# (OR 300#) COMPO. ROOF SHINGLES OVER 15# FELT.

or

MED. CEDAR SHAKES OVER 15# FELT W/ 30# × 18″ WIDE FELT BETWEEN EACH COURSE W/ 10½″ EXPOSURE.

or

CONC. ROOF TILES BY (give manufacturer's name, and color and weight of tiles.) INSTALL AS PER MANUF. SPECS.

5. STD. ROOF TRUSSES @ 24″ O.C. SEE DRAW. BY MANUF.

or

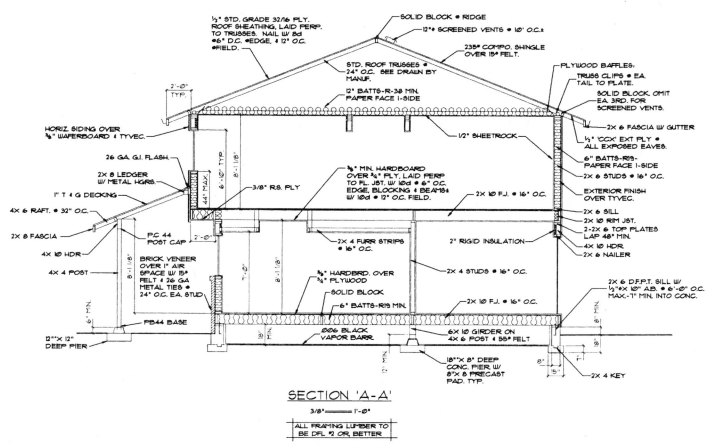

Figure 35–15 Cross section completed with all structural and finishing material drawn, dimensioned, and lettered.

2 × ___ RAFT. @ ___″ O.C.

2 × ___ C.J. @ ___″ O.C.

*6. ___ BATTS, R-___ PAPER FACE @ HEATED SIDE.

or

___ BLOWN-IN INSULATION R-___ MIN.

7. BAFFLES AT EAVE VENTS.

8. SOLID BLOCK—OMIT EA. 3rd FOR SCREENED VENTS.

9. TRUSS CLIPS @ EA. TAIL TO PLATE.

10. ___ × ___ FASCIA W/ GUTTER.

11. ½″ 'CCX' EXT. PLY @ ALL EXPOSED EAVES.

or

1 × 4 T&G DECKING @ ALL EXPOSED EAVES.

Walls

*1. 2—2 × ___ TOP PLATES, LAP 48″ MIN.

*2. 2 × ___ STUDS @ ___″ O.C.

*3. 2 × ___ SILL (Fill in the blank.)

*4. EXT. SIDING OVER ⅜″ PLY & 15 FELT

or

EXT. SIDING OVER ½″ WAFERBOARD & TYVEK.

*5. 3½ FIBERGLASS BATTS, R-11 MIN.—PAPER FACE ONE SIDE.

or

6″ FIBERGLASS BATTS, R-19, PAPER FACE ONE SIDE.

or

6″ FIBERGLASS BATTS R-21.

*6. ___ × ___ HEADER.

7. LINE OF INTERIOR FINISH.

8. SOLID BLOCK AT MID HEIGHT FOR ALL WALLS OVER 10′.

9. ⅝″ TYPE 'X' GYP. BD. FROM FLOOR TO BOTTOM OF ROOF SHEATH.

or

⅝″ TYPE 'X' GYP. BD. WALLS AND CEIL.

10. BRICK VENEER OVER 1″ AIR SPACE & 15# FELT W/ METAL TIES @ 24″ O.C. EA. STUD.

Floors and Foundation

1. ⅜″ MIN. HARDBOARD UNDERLAYMENT.

*2. ¾″ 42/16 PLY. FLOOR SHEATH. LAID PERP. TO FLOOR JOISTS. NAIL W/ 10d @ 6″ O.C. EDGE, BLOCKING, & BEAMS. USE 10d @ 12″ O.C. @ FIELD.

3. SOLID BLOCK @ 10′ O.C. MAX.

4. ___ BATTS—R-___ MIN. or (R-21).

*5. ___ × ___ GIRDERS (Fill in the blanks. See foundation plan to verify size.)

*6. 4 × 4 POST, 4 × 6 @ SPLICES, W/ METAL TIES ON 55# FELT ON ___ × ___ DEEP CONC. PIERS.

7. .006 BLACK VAPOR BARRIER or 55# ROLLED ROOFING.

*8. 2 × __ P.T. SILL W/ ½″ ⌀ A.B. @ 6′–0″ O.C. MAX.—7″ MIN. INTO CONC. W/ 2″ ⌀ WASHERS.

9. Give the section a title such as SECTION 'AA'.

10. Give the drawing a scale such as ⅜″ = 1′–0″.

11. List general notes near the title or near the title block.

    *ALL FRAMING LUMBER TO BE D.F.L. #2 OR BETTER. (Instead of D.F.L. you may need to substitute a different type of wood.)

    SEE SECTIONS 'BB' & 'CC' FOR BALANCE OF NOTES.

Give your eyes and fingers a well-deserved rest. You now have 99 percent of the section complete.

▼ STAGE 7: EVALUATE YOUR WORK

Don't assume because you did the work and it took a long time that the drawing is complete. Run a print, and evaluate your work for accuracy and quality. Don't just compare it to someone else's drawing. Use the checklist, and make sure your section matches the material and location that you have specified on the floor and foundation plans. Your best chance of finding your own mistakes is to get away from your drawing for an hour or two before checking it. Use the following checklist to evaluate your drawing before giving it to your instructor.

CADD Applications

DRAWING SECTIONS AND DETAILS WITH CADD

The steps for drawing sections presented in this chapter are used for both manual and computer-aided drafting. However, using the computer has some advantages: standard sections and details can be brought into the drawing from a menu library; placing notes on the section is easy with CADD. Also, some architectural CADD packages automatically draw a preliminary section from information you provide in relationship to the floor plan. This type of parametric design requires that an imaginary cutting plane line be placed through the floor plan in the desired section location. This is followed by computer prompts requesting information such as roof pitch and structural floor thicknesses. In programs of this type, the floor-plan walls are drawn with heights established, so these dimensions automatically convert to wall height information in the sectional view. All you do to complete the section is add material symbols, dimensions, and notes.

Some architectural CADD packages have template library menu overlays that contain typical section elements such as details, materials, and tags. After architects have used the CADD system awhile, they begin to save all typical or standard construction details as BLOCKS. These BLOCKS are commonly placed in a library manual for easy reference and can be called up and displayed at any time in any drawing. The standard details should be clearly labeled with an identification code for easy reference. Each CADD user should have a copy of the library reference manual, and every time a new detail is drawn, the reference manual should be updated with a drawing of the detail and the reference code. A typical CADD detail drawing is shown in Figure 35–16.

Many of the major window and door manufacturers, such as Pella and Anderson, have CADD software programs available that integrate with major CADD software packages. These handy additions are useful in providing floor plan layouts, complete specifications, and standard construction details, as shown in Figure 35–17.

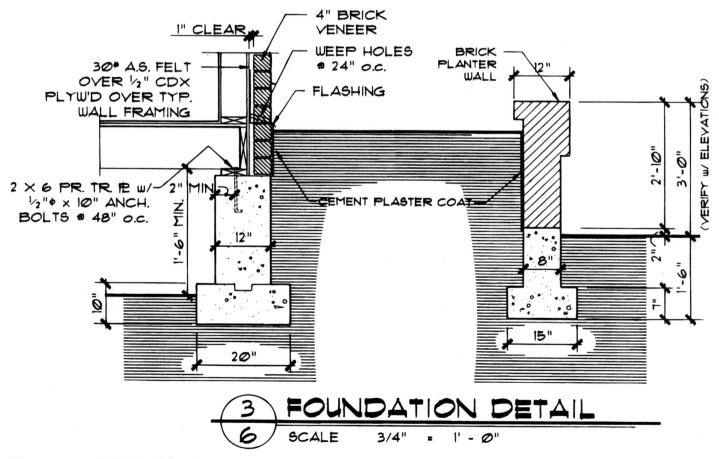

Figure 35–16 CADD detail drawing.

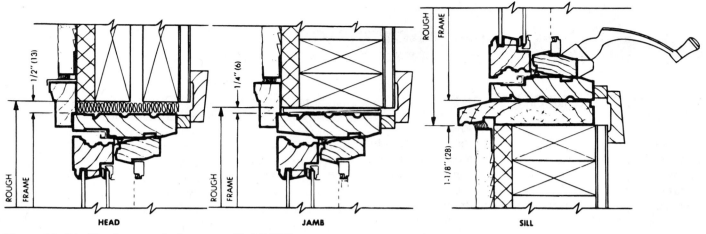

Figure 35–17 Use for manufacturer-supplied CADD program.

CHAPTER

35 Section Layout Test

DIRECTIONS

Answer the questions with short, complete statements or drawings as needed.

1. Letter your name and the date at the top of the sheet.
2. Letter the question number and provide the answer. You do not need to write out the question.

QUESTIONS

Question 35–1 Define the following terms.
a. Rafter
b. Truss
c. Ceiling joist
d. Collar tie
e. Jack stud
f. Rim joist
g. Chord
h. Sheathing

Question 35–2 On a blank sheet of paper, sketch a section view of a conventionally framed roof showing all interior supports.

Question 35–3 Give the typical sizes for the following materials.
a. Mudsill
b. Stud height
c. Roof sheathing
d. Wall sheathing
e. Floor decking
f. Underlayment

Question 35–4 List three common scales for drawing sections, and tell when each is best used.

Question 35–5 List the seven stages of drawing a section.

Question 35–6 List two different types of drawings for which stock drawings are typically used.

CHAPTER 36

Alternate Layout Techniques

Chapter 35 presented the basic methods for drawing a section with a crawl space, floor joists and truss roof. Those methods will not meet the need of all drafters because of the job site, the contractor's personal preference for framing, or the area of the country in which the house is to be built. This chapter provides you with other common construction methods to meet a variety of needs. Unless noted, all sizes given in the following steps are based on the minimum standards of the Uniform Building Code and should be compared with local standards.

▼ POST-AND-BEAM FOUNDATIONS

Section VIII presented an introduction to post-and-beam construction. This chapter will describe post-and-beam construction only as it applies to the foundation and lower floor.

Layout

Use construction lines for steps 1 through 14. See Figure 36–1.

Step 1. Lay out the width of the building.

Step 2. Lay out a baseline for the bottom of the footings.

Step 3. Lay out the footing thickness.

Step 4. Lay out the finished grade.

Step 5. Locate the top of the stem wall.

Step 6. Block out the width of the stem wall.

Step 7. Block out the width of the footings, centered under the stem wall.

Step 8. Block out a 4 × 8″ (100 × 200 mm) ledge for the brick veneer.

Step 9. Add 4″ (100 mm) width to the footing to support the brick ledge.

Step 10. Locate the 2 × 6″ (50 × 150 mm) mudsill.

See Figure 36–2 for steps 11 through 13.

Step 11. Lay out the depth of the 2″ (50 mm) floor decking.

Step 12. Lay out the girders and the support posts.

Step 13. Lay out the concrete support piers centered under the posts. Lay out the width first for each pier, and then draw one continuous guideline to represent the top of all piers.

Finished-Quality Lines—Structural Material

For drawing structural members with finished-quality lines, use line quality as described in Chapter 35 unless otherwise noted. See Figure 36–3 for steps 14 through 22.

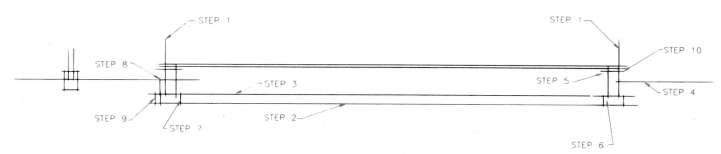

Figure 36–1 Layout for the post-and-beam foundation system using construction lines.

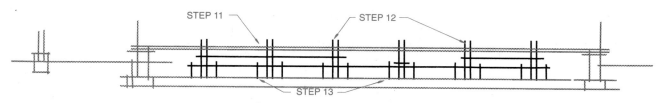

Figure 36–2 Layout of the interior supports for a post-and-beam foundation.

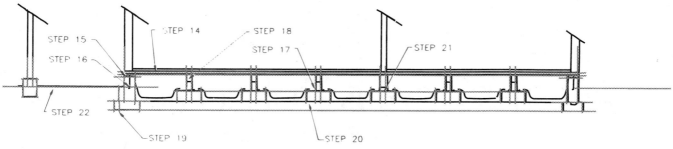

Figure 36–3 Finished-quality lines on the structural members of a post-and-beam foundation system.

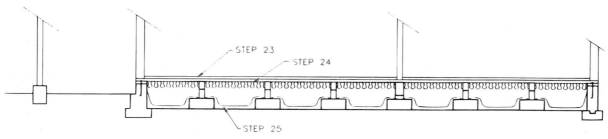

Figure 36–4 Finishing materials for a post-and-beam foundation system.

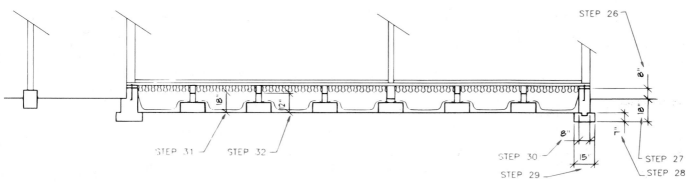

Figure 36–5 Required dimensions for a post-and-beam foundation system.

You will notice that some items will be drawn that have not been laid out. These are items that are simple enough that they can be drawn by estimation.

Step 14. With thin lines, draw the tongue and groove (T&G) decking. Do not show the actual T&G pattern. This is done only when details are shown at a larger scale such as ¾″ = 1′–0″.

Step 15. Draw the mudsills using bold lines.

Step 16. Draw anchor bolts with bold lines.

Step 17. Draw all posts with thin lines.

Step 18. Draw all girders with bold lines.

Step 19. Draw the stem walls and footings with very bold lines.

Step 20. Draw piers with very bold lines.

Step 21. Draw the post pads freehand with bold lines.

Step 22. Draw the finished grade with very bold lines.

Finished-Quality Lines—Finishing Material

The materials in this section will be used to seal the exterior walls from the weather and to finish the interior. Use thin lines for each step, unless otherwise noted. See Figure 36–4 for steps 23 through 25.

Step 23. Draw the hardboard subfloor from wall to wall.

Step 24. Draw the insulation.

Step 25. Draw the vapor barrier freehand.

Dimensioning

See Figure 36–5 for steps 26 through 32. Place the needed leader and dimension lines to locate the following dimensions:

Step 26. Finished grade to the top of the stem wall.

Step 27. Depth of the footing into grade.

Step 28. Height of the footing.

Step 29. Width of the footing.

Step 30. Width of the stem wall.

Step 31. Crawl space depth (18″ [457 mm] minimum).

Step 32. Bottom of girders to grade (12″ [305 mm] minimum).

Notes

Step 33. Letter the floor and foundation notes as shown in Figure 36–6. Typical notes should include the following.

    a. ⅜″ HARDBOARD OVERLAY

    b. 2 × 6 T&G DECKING

    c. 6″ BATTS. R-19 MIN.

    d. __ × __ GIRDERS

    e. 4 × 4 POST ON 55# FELT

    f. __″ DIA × __″ DEEP CONC. PIERS

    g. .006 BLACK VAPOR BARRIER or 55# ROLLED ROOFING

    h. 2 × 6 P.T. SILL W/ ½″ DIA × 10″ A.B. @ 6′–0″ O.C. MAX.—7″ MIN. INTO CONC. THRU 2″ DIA WASHERS.

Step 34. Complete the section as per Chapter 35.

▼ CONCRETE SLAB FOUNDATIONS

Structural Layout

Use light construction lines for steps 1 through 11. See Figure 36–7.

Step 1. Lay out the width of the building.

Step 2. Lay out a baseline for the bottom of the footings.

Step 3. Lay out the footing thickness.

Step 4. Lay out the finish grade.

Step 5. Locate the top of the slab. Building codes typically require all wood to be 8″ (200 mm) minimum above the finish grade.

Step 6. Lay out the bottom of the 4″ (100 mm) slab.

Step 7. Lay out 4″ (100 mm) of fill material under the slab.

Step 8. Lay out the width of the stem wall.

Step 9. Lay out the width of the footings.

Step 10. Lay out the 4 × 8 (100 × 200 mm) ledge for brick veneer.

Step 11. Block out the interior footings for load-bearing walls.

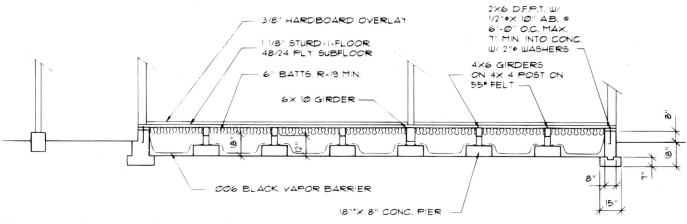

Figure 36–6 Typical notes for a post-and-beam foundation system.

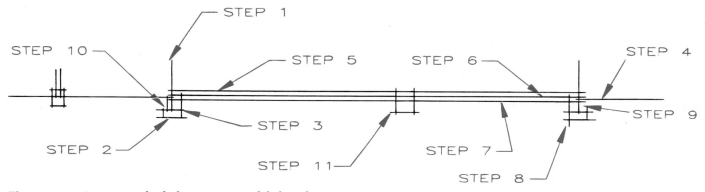

Figure 36–7 Layout methods for a concrete slab foundation.

Figure 36–8 Finished-quality lines for concrete slab structural materials.

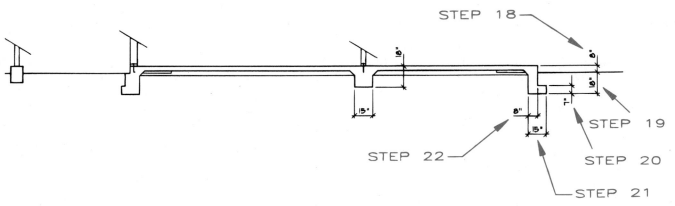

Figure 36–9 Required dimensions for a concrete slab foundation system.

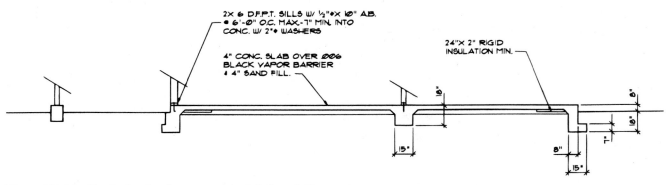

Figure 36–10 Typical notes for a concrete slab foundation system.

Finished Quality Lines

See Figure 36–8 for steps 12 through 17.

Step 12. Draw the outline of the concrete using very bold lines.

Step 13. Draw the 24 × 2″ (600 × 50 mm) insulation with thin lines.

Step 14. Draw the 4″ (100 mm) fill material with a thin line.

Step 15. Draw the mudsills with bold lines.

Step 16. Draw the anchor bolts with bold lines.

Step 17. Draw the finished grade with a very bold line.

Dimensions

See Figure 36–9 for steps 18 through 22. Place the required extension and dimension lines to locate the following dimensions:

Step 18. Finished grade to top of slab.

Step 19. Finished grade to bottom of footing.

Step 20. Height of footing.

Step 21. Width of footing.

Step 22. Width of stem wall.

Notes

Step 23. Letter the foundation notes as shown in Figure 36–10. Typical notes should include:

a. 2 × 6 PT. SILL W/ ½″ DIA × 10″ A.B. @ 6′–0″ O.C. MAX.—7″ MIN. INTO CONC. W/ 2″ DIA WASHERS.

b. 4″ CONC. SLAB OVER 4″ GRAVEL, 6 MIL VAPOR BARRIER, AND 2″ SAND FILL.

or

4″ CONC. SLB OVER 4″ GRAVEL FILL OVER 55# FELT.

c. 2″ × 24″ RIGID INSULATION.

Step 24. Complete the section as per Chapter 35.

▼ BASEMENT WITH CONCRETE SLAB

For the basement layout, use light construction lines. Depending on the seismic zone for which you are drawing, an engineer's drawing may be required for this kind of foundation. If so, you will need to follow engineer's design standards similar to those shown in Figure 36–11. Understanding engineer's design standards, or calculations (calcs), can be very frustrating. Calcs are generally divided into the areas of the item to be designed, the formulas to determine the size, and the solution.

The drafter does not need to understand the formulas that are used, but must be able to convert the solution into a drawing. The solution in Figure 36–11 is the notes that are listed under the heading USE. For an entry-level drafter, the engineer will usually provide a sketch similar

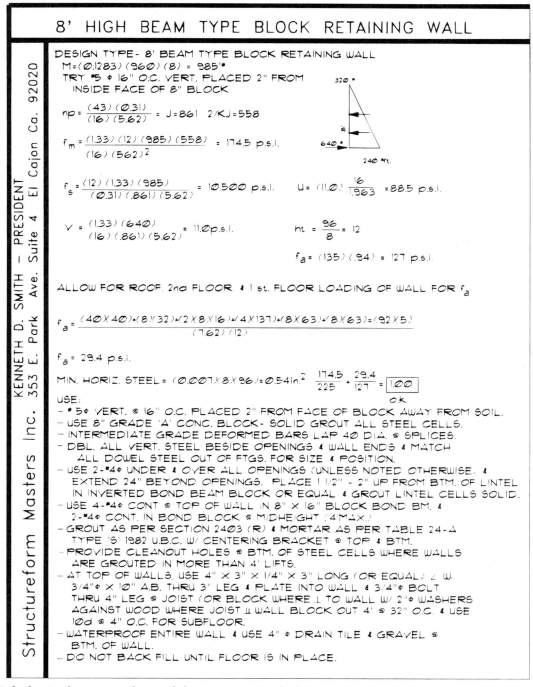

Figure 36–11 A drafter is often required to work from engineer's calculations when drawing retaining walls. Calculations usually show the math formulas to solve a problem and the written solution to the problem. It is the written solution to the problem that the drafter must make into a drawing. *Courtesy Ken Smith, Structureform Masters Inc.*

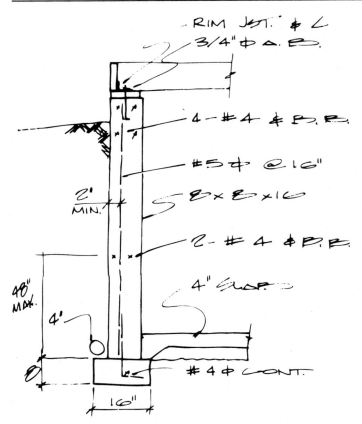

Figure 36–12 A sketch is often given to the drafter to help explain the written calculations. Many of the written calculations appear on the sketch but are not written in proper form.

to the drawing in Figure 36–12 to explain the calculations. By using the written calculations and the sketch, a drafter can make a drawing similar to Figure 36–18.

The basement can be drawn by following steps 1 through 13. See Figure 36–13.

Step 1. Lay out the width of the building.

Step 2. Lay out a baseline for the bottom of the footings.

Step 3. Measure up from the baseline for the footings. Footings thickness is typically 8″ (200 mm) deep for a one-story building and 12″ (300 mm) deep for two or more stories.

Step 4. Lay out the 4″ (100 mm) concrete slab.

Step 5. Lay out the 2″ (50 mm) mudsill. The height of the basement will vary.

Step 6. Lay out the wooden floor.

Step 7. Lay out the finished grade 8″ (200 mm) minimum below the mudsill.

Step 8. Block out the width of the retaining wall. Walls are usually made from 8″ (200 mm) thick material, but they can be made from 6″ (150 mm) thick material if the support steel is increased.

Step 9. Block out the width of the footing. Footings for this type of wall are usually 16″ (400 mm) wide, centered under the retaining wall.

Step 10. Lay out the 4″ (100 mm) ledge and add 4″ (100 mm) to the footing to support the brick veneer. An alternative to the ledge method is to form the wall 4″ (100 mm) out from the face of the exterior wall to support the brick. See Figure 36–14.

Step 11. Draw a line to show the 4″ (100 mm) fill material under the slab.

Step 12. Lay out the interior footings for bearing walls.

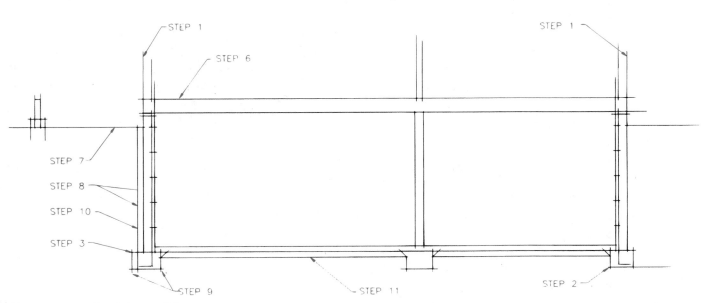

Figure 36–13 Retaining wall layout.

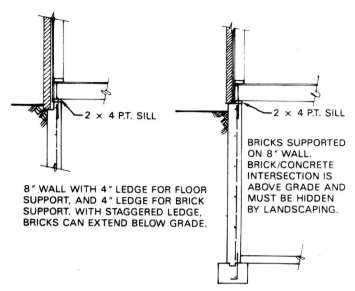

Figure 36–14 Common methods used for supporting brick veneer.

Step 13. Lay out the steel reinforcing for the wall. This will vary greatly depending on the seismic area you are in and how the wall is to be loaded. One typical placement pattern puts the vertical steel 2″ (51 mm) from the inside (tension) face, and the horizontal steel 18″ (457 mm) o.c. starting at the footing.

Finished-Quality Lines—Structural Material Only

See Figure 36–15 for steps 14 through 21.

Step 14. Draw the plywood flooring with thin lines.

Step 15. Draw the floor joists with thin lines. Draw the floor joists with bold lines when they are perpendicular to the wall.

Step 16. Draw the mudsills and anchor bolts with bold lines.

Step 17. Draw the walls with bold lines. If concrete blocks are used, draw the division lines of the blocks and crosshatch the blocks with thin lines.

Step 18. Draw the footings with bold lines.

Step 19. Draw the concrete slab with bold lines.

Step 20. Draw the wall steel with bold dashed lines.

Step 21. Complete the section as per Chapter 35.

Finishing Materials

See Figure 36–16 for steps 22 through 28.

Step 22. Draw the floor insulation with thin lines.

Step 23. Draw the finished grade with very bold lines.

Step 24. Draw the 4″ (100 mm) diameter drain with thin lines.

Step 25. Draw the 8 × 24″ (200 × 600 mm) gravel bed at the base of the wall.

Step 26. Draw the 4″ (100 mm) fill materials with thin lines.

Step 27. Draw the brick veneer and the crosshatching with thin lines.

Step 28. Draw the welded wire mesh with thin dashed lines.

Dimensions

See Figure 36–17 for steps 29 through 35. Place the needed leader and dimensions lines to locate the following dimensions:

Step 29. Floor to ceiling.

Step 30. Top of wall to finished grade.

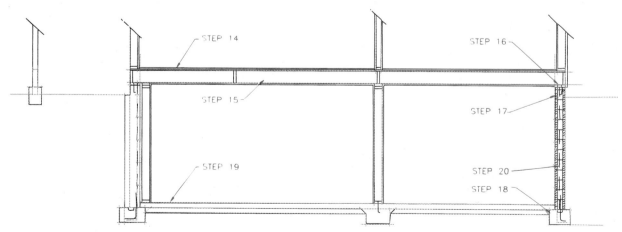

Figure 36–15 Finished-quality lines of the structural material for a basement foundation system. The left wall is drawn showing a poured concrete wall. The right wall is drawn showing 8 × 8 × 16 (200 × 200 × 400) concrete blocks. Generally the two wall systems are not used together.

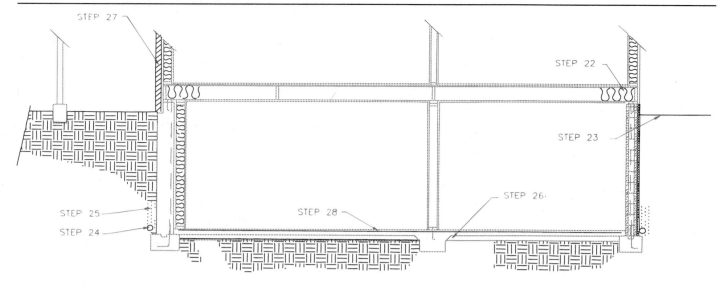

Figure 36–16 Finishing materials for a basement foundation system.

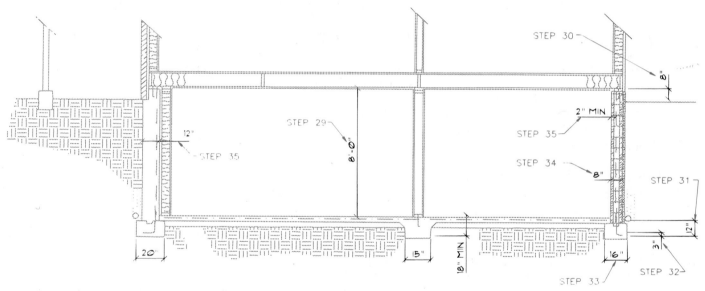

Figure 36–17 Required dimensions for a basement foundation system.

Step 31. Height of footing.

Step 32. Footing steel to bottom of footing.

Step 33. Footing width.

Step 34. Wall width.

Step 35. Edge of wall to vertical steel.

Notes

Step 36. Letter the section notes as shown in Figure 36–18. Remember that these notes are only guidelines and may vary slightly because of local standards. Typical notes should include the following:

a. 2 × FLOOR JOIST @ ___" O.C.

b. SOLID BLOCK @ 48" O.C.—48" OUT FROM WALL WHERE JOISTS ARE PARALLEL TO WALL.

c. 2 × ___ RIM JOIST W/ A-34 ANCHORS BY SIMPSON CO. OR EQUAL.

d. 2 × 6 P.T. SILL W/ ½" DIA × 10" A.B. @ 24" O.C. W/ 2" DIA WASHERS.

e. 8" CONC. WALL W/ #4 @ 18" O.C. EA. WAY. 8 × 8 × 16 GRADE 'A' CONC. BLOCKS W/ #4 @ 16" O.C. EA. WAY—SOLID GROUT ALL STEEL CELLS.

f. WATERPROOF THIS WALL W/ HOT ASPHALTIC EMULSION.

g. 4" DIA DRAIN IN 8 × 24 MIN. GRAVEL BED.

h. 2 × 4 KEY.

i. 4" CONC. SLAB OVER 4" GRAVEL FILL AND 55# ROLLED ROOFING.

or

4" CONC. SLAB OVER 4" SAND FILL AND 6 MIL VAPOR BARRIER, AND 2" SAND FILL.

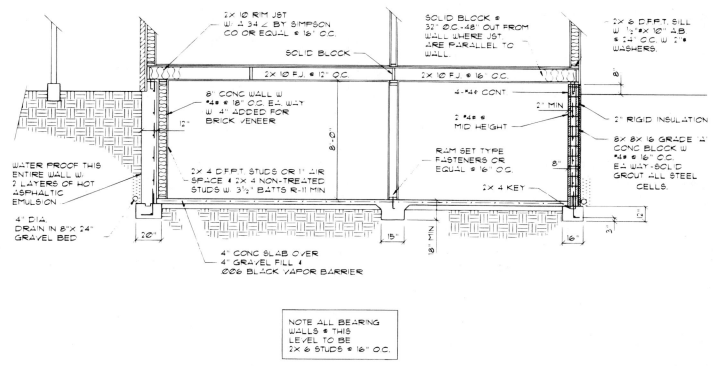

Figure 36–18 Typical notes for a basement foundation system.

Note: Wire mesh is often placed in the concrete slab to help control cracking. This mesh is typically called out on the sections and foundation plan as 6″ × 6″—#10 × #10 W.W.M.

▼ CONVENTIONAL (STICK) ROOF FRAMING

The two types of conventional roof framing presented are flat and vaulted ceilings. You will notice many similar features between the two styles.

Flat Ceiling

Use light construction lines to lay out the drawing. See Figure 36–19 for steps 1 through 7.

Step 1. Place a vertical line to represent the ridge.

Step 2. Locate each overhang.

Step 3. Lay out the bottom side of the rafter. Starting at the intersection of the ceiling and the interior side of the exterior wall, draw a line at a 5/12 pitch. Do this for both sides of the roof. Both lines should meet at the ridge line.

Step 4. Lay out the top side of the rafters.

Step 5. Lay out the top side of the ceiling joist.

Step 6. Lay out the ridge board.

Step 7. Lay out the fascias.

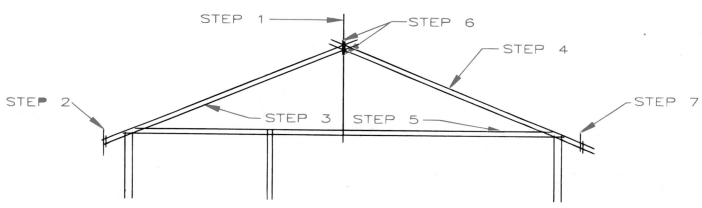

Figure 36–19 Layout of a conventionally framed roof.

Lay Out the Interior Supports

See Figure 36–20 for steps 8 through 12.

Step 8. Lay out the ridge brace and support.

Step 9. Lay out the purlin braces and supports.

Step 10. Lay out the strong backs.

Step 11. Lay out the purlins.

Step 12. Lay out the collar ties.

Finished-Quality Lines—Structural Members Only

See Figure 36–21 for steps 13 through 20.

Step 13. Draw the plywood sheathing with thin lines.

Step 14. Draw the rafters with thin lines.

Step 15. Draw the eave blocking with bold lines.

Step 16. Repeat steps 13 through 15 for the opposite side of the roof.

Step 17. Draw the fascias with bold lines.

Step 18. Draw the strong backs with bold lines.

Step 19. Draw the ridge board with bold lines.

Step 20. Draw blocking with bold lines.

Interior Support Finished-Quality Lines

See Figure 36–22 for steps 21 through 24.

Step 21. Draw all plates with bold lines.

Step 22. Draw all braces with thin lines.

Step 23. Draw all purlins with bold lines.

Step 24. Draw the collar tie with thin lines.

Dimensions

There are usually no dimensions required for the roof framing.

Finishing Materials

See Figure 36–23 for steps 25 through 29.

Step 25. Draw baffles with bold lines.

Step 26. Draw the finished ceiling with thin lines.

Step 27. Draw joist hangers with thin lines.

Step 28. Draw the 10″ thick insulation.

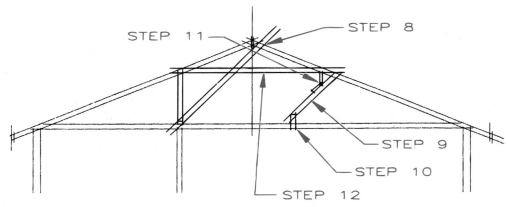

Figure 36–20 Layout of interior supports for a conventionally framed roof.

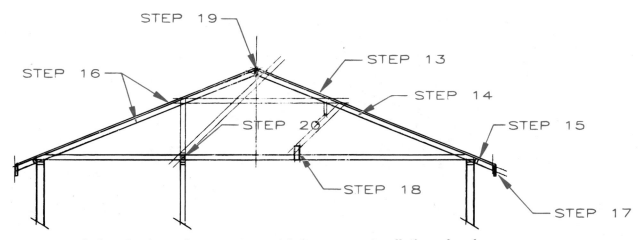

Figure 36–21 Finished-quality lines of structural materials for a conventionally framed roof.

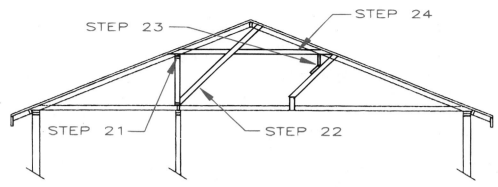

Figure 36–22 Finished-quality lines of interior supports.

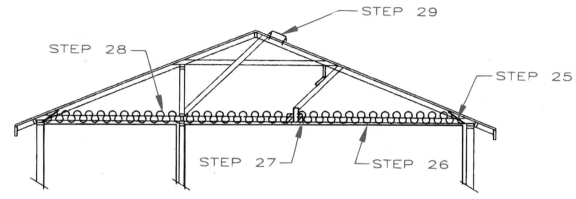

Figure 36–23 Finishing materials for a conventionally framed roof.

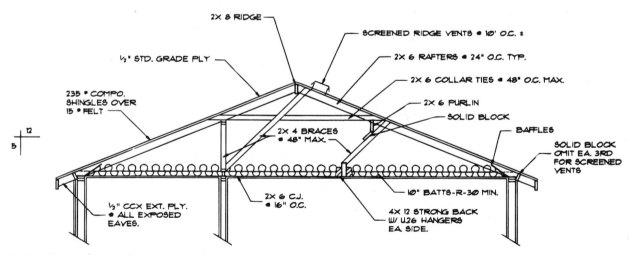

Figure 36–24 Typical notes for a conventionally framed roof.

Step 29. Draw the ridge vents.

Notes

See Chapter 35 for a complete listing of notes that apply to the roof. In addition to those notes, see Figure 36–24 for the notes needed to describe the interior supports. Roof notes should include the following:

a. 2 × ___ RIDGE BOARD.

b. 2 × ___ RAFTERS @ ___″ O.C.

c. 2 × ___ COLLAR TIE @ 48″ O.C. MAX.

d. 2 × ___ PURLIN.

e. 2 × 4 BRACE @ 48″ O.C. MAX.

▼ VAULTED CEILINGS

If the entire roof level is to have vaulted ceilings, the process of laying it out is very similar to the procedure for laying out a stick roof with flat ceilings. Usually, though, part of the roof has 2 × 6 rafters with flat ceilings and a vaulted ceiling using 2 × 12 rafter/ceiling joists. In order to make the two roofs align on the outside, the plate must be lowered in the area having the vaulted ceilings. See Figure 36–25 for steps 1 through 7 of laying out the vaulted ceiling. Use construction lines for all steps.

Step 1. Lay out the overhangs.

Step 2. Lay out the bottom side of the 2 × 6 rafter.

Step 3. Lay out the top side of the 2 × 6 rafter. For this example, this size rafter is used throughout the balance of the house, not the 2 × 12 rafter/ceiling joist.

Step 4. Measure down from the top side of the rafter drawn in step 3 to establish the bottom of the 2 × 12 rafter/ceiling joist. Notice that the 2 × 12 extends below the top plate of the wall.

Step 5. Repeat steps 2 through 4 for the opposite side of the roof.

Step 6. Lay out the ridge beam.

Step 7. Lay out the fascias and the rafter tail cut for the 2 × 12 rafter/ceiling joists.

Finished-Quality Lines—Structural Members Only

See Figure 36–26 for steps 8 through 12.

Step 8. Draw the plywood sheathing with thin lines.

Step 9. Draw the rafter/ceiling joist with thin lines.

Step 10. Draw all blocks, plates, and fascias with bold lines.

Step 11. Draw the ridge beam with bold lines.

Step 12. Repeat steps 8 through 10 for the opposite side of the roof.

Finishing Materials

See Figure 36–27 for steps 13 through 17.

Step 13. Draw the baffles with bold lines.

Step 14. Draw the insulation with thin lines.

Step 15. Draw the ceiling interior finish with thin lines.

Step 16. Draw the ridge vents and notches with thin lines.

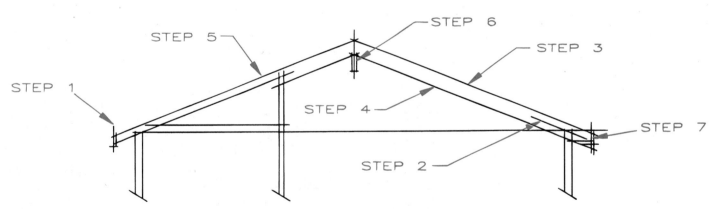

Figure 36–25   Layout of a vaulted ceiling.

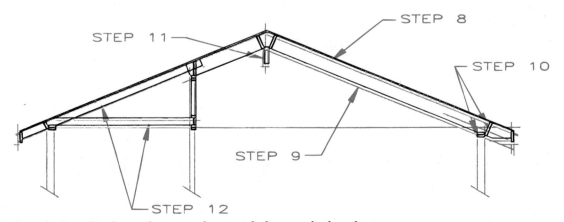

Figure 36–26   Finished-quality lines of structural materials for a vaulted roof.

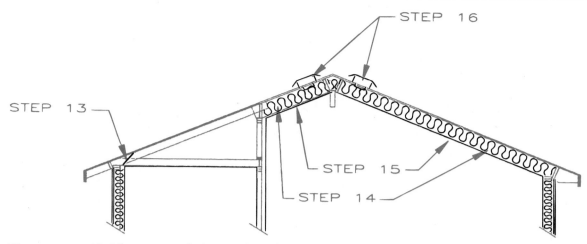

Figure 36–27 Finishing materials for a vaulted roof.

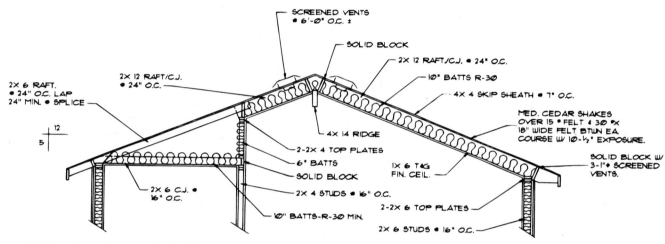

Figure 36–28 Notes for a vaulted ceiling.

Step 17. Draw the metal hangers and crosshatch with thin lines. Depending on the ridge beam option that you drew, joist hangers may not be required.

Dimensions

No dimensions are needed to describe the roof. You will need to give the floor-to-plate dimension. List this dimension as 7′–6″ ± VERIFY AT JOB SITE. By telling the framer to verify the height at the job site, you are alerting the framer to a problem. This is a problem that can be solved much more easily at the job site than at the drafting table.

Notes

See Chapter 35 for a complete listing of notes that apply to the roof. In addition to those notes, see Figure 36–28 for the notes needed to describe the roof completely. The notes should include the following:

a. __ × __ RIDGE BEAM.
b. 2 × __ RAFT./C.J. @ __″ O.C.
c. U-210 JST. HGR. BY SIMPSON CO. OR EQUAL.
d. SCREENED RIDGE VENTS @ EA. 3RD. SPACE. NOTCH RAFT. FOR AIR FLOW.
e. SOLID BLOCK W/ 3–1″ DIA SCREENED VENTS @ EA. SPACE.
f. NOTCH RAFTER TAILS AS REQD.
g. LINE OF INTERIOR FINISH.
h. __ BATTS—R-__ MIN.

▼ GARAGE/RESIDENCE SECTION

Drawing this section is similar to drawing the sections just described since it is a combination of several types of sections. The steps needed to draw this section can be seen in Figures 36–29 through 36–33.

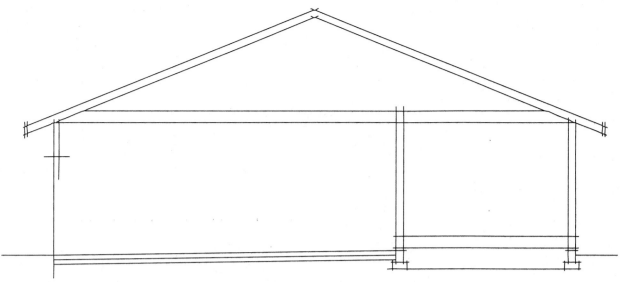

Figure 36–29 Layout of a section drawn through the garage and living area.

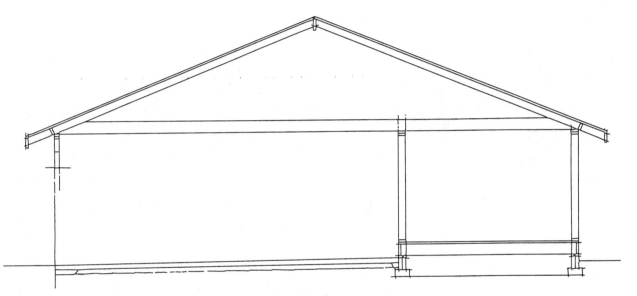

Figure 36–30 Finished-quality lines for structural materials.

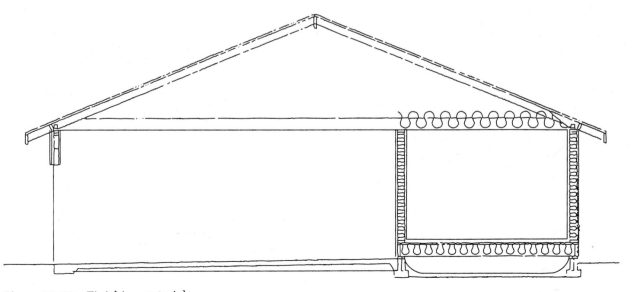

Figure 36–31 Finishing materials.

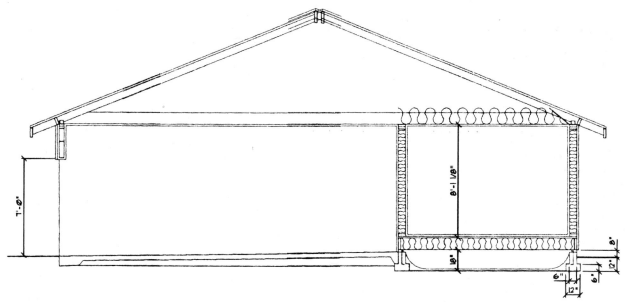

Figure 36–32 Typical dimensions.

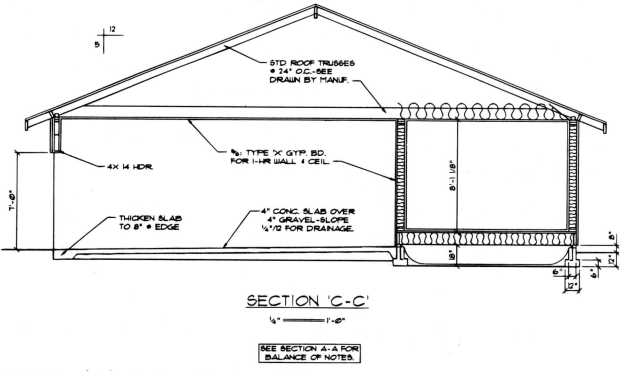

SECTION 'C-C'

¼" ========= 1'-0"

SEE SECTION A-A FOR
BALANCE OF NOTES.

STD ROOF TRUSSES
@ 24" O.C.-SEE
DRAWN BY MANUF.

⅝: TYPE 'X' GYP. BD.
FOR 1-HR WALL & CEIL.

4X 14 HDR

THICKEN SLAB
TO 8" @ EDGE

4" CONC. SLAB OVER
4" GRAVEL-SLOPE
¼"/12 FOR DRAINAGE.

Figure 36–33 A completed section through the garage and living area.

CHAPTER
36 Alternate Layout Techniques Test

DIRECTIONS

Answer the following questions with short, complete sentences.

1. Letter your name and the date at the top of the sheet.
2. Letter the part number and provide the name of that part.

QUESTIONS

Question 36–1 Sketch and dimension an exterior footing for a concrete slab supporting two floors.

Question 36–2 What thickness of floor decking is typically used for floors using the post-and-beam method?

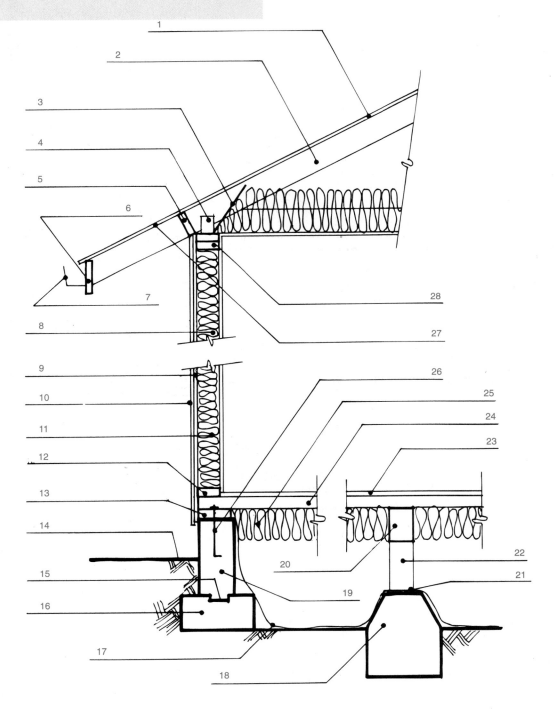

Question 36–3 What type of line quality is used to represent the mudsill?

Question 36–4 How are stem walls represented on the finished drawing?

Question 36–5 What is the thickness of the vapor barrier used under a crawl space?

Question 36–6 The top of the concrete slab must be __" above the finished grade.

Question 36–7 Basement walls are typically __" wide.

Question 36–8 What determines the amount of structural steel placed in a retaining wall?

Question 36–9 Describe a common blocking pattern that is typically used to support the top of a concrete retaining wall.

Question 36–10 Sketch and label the framing for a stick framed roof showing rafters, ceiling joists, ridge, purlins, purlin blocks, and braces.

Question 36–11 On a separate sheet of paper, identify each part of the drawing for this question.

Question 36–12 On a separate sheet of paper, identify each part of the drawing for this question.

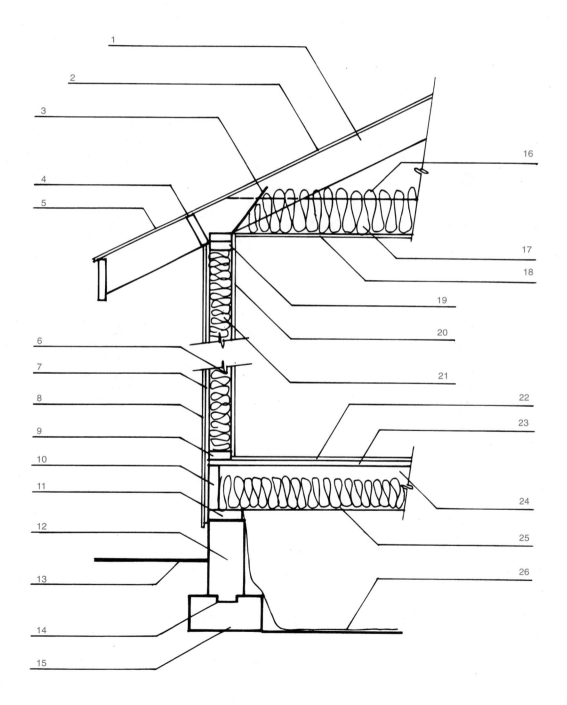

Question 12

PROBLEMS

Problem 36–1 Cape Cod. Using a print of your floor and foundation plans, evaluate the construction methods used and determine which sections and partial sections need to be drawn. Lay out a freehand sketch of the sections you will need showing all wall locations, floor supports, and roof construction. Include sections showing stair and masonry fireplace construction if required. Use a C-size drawing sheet to lay out and draw all needed sections. Choose a scale based on the reading material in this section and use the examples given in each chapter to help you complete your drawings.

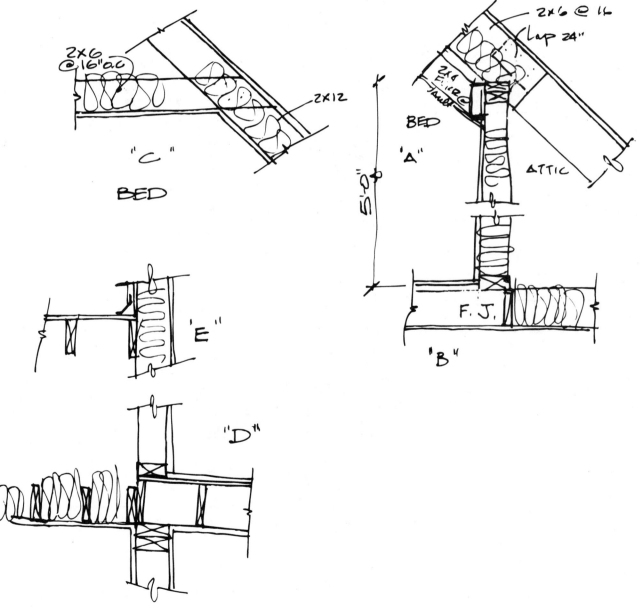

Problem 36–1

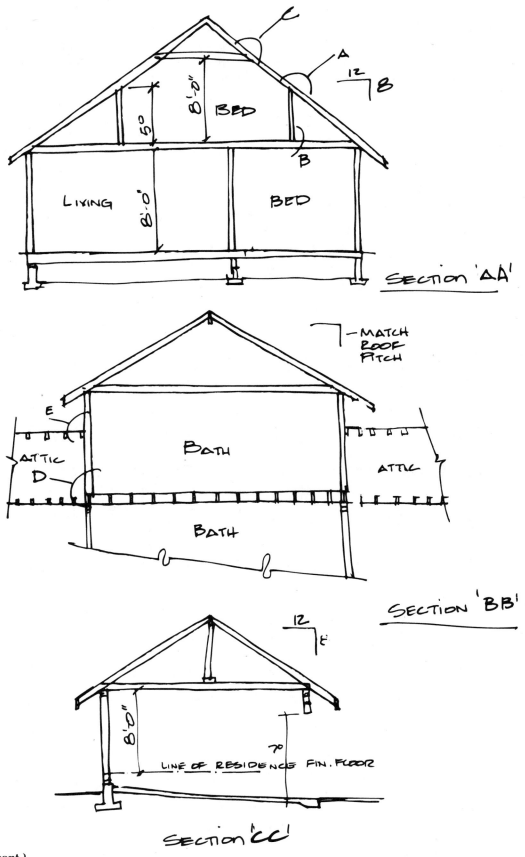

Section 'AA'

—MATCH ROOF PITCH

Section 'BB'

Section 'CC'

LINE OF RESIDENCE FIN. FLOOR

Problem 36–1 (cont.)

Problem 36–2 Cabin. Using a print of your floor and foundation plans, evaluate the construction methods used and determine which sections and partial sections need to be drawn. Lay out a freehand sketch of the sections you will need showing all wall locations, floor supports, and roof construction. Include sections showing cantilevers and stair construction. Use a C-size drawing sheet to lay out and draw all needed sections.

Choose a scale based on the reading material in this section and use the examples given in each chapter to help you complete your drawings. Assume:

2 × 12 rafters @ 16″ o.c. w/10″ batts.
2 × 10 floor joists at 16″ o.c. at the upper floor level.
2 × 6 exterior studs at 16″ o.c.

*NOTE: Lightly lay out stair run to see if stair headroom will affect desired pitches.

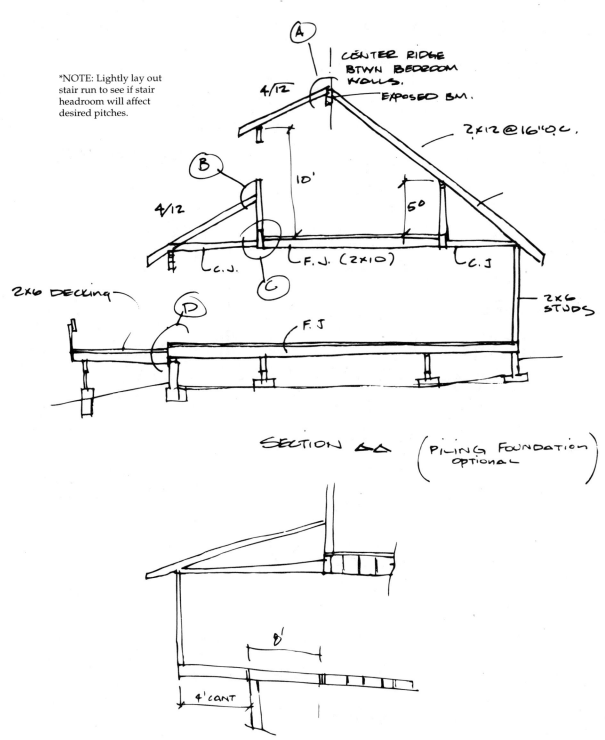

Problem 36–2

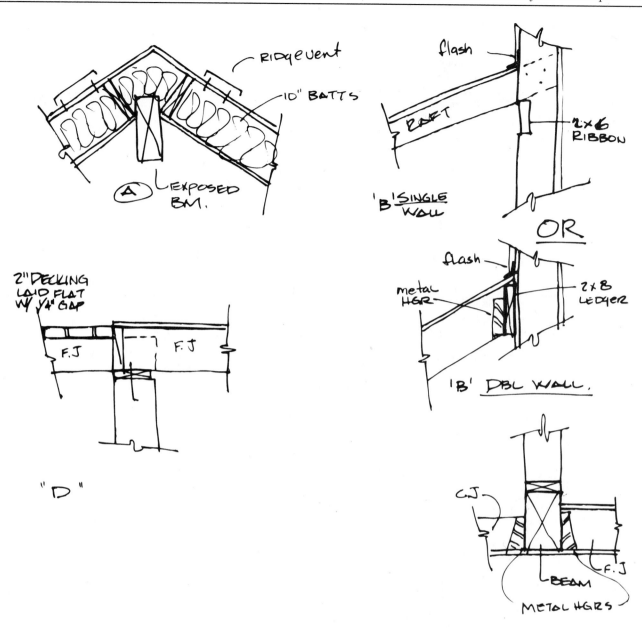

RIDGE VENT

10" BATTS

(A) EXPOSED BM.

flash

RAFT

'B' SINGLE WALL

2×6 RIBBON

OR

flash

metal HGR

2×8 LEDGER

'B' DBL WALL.

2" DECKING LAID FLAT W/ ¼" GAP

F.J   F.J

"D"

C.J

F.J

BEAM

METAL HGRS

'C'

Problem 36–2 (cont.)

Problem 36–3 Ranch. Using a print of your floor and foundation plans, evaluate the construction methods used and determine which sections and partial sections need to be drawn. Lay out a freehand sketch of the sections you will need showing all wall locations, floor supports, and roof construction. Use a C-size drawing sheet to lay out and draw all needed sections. Choose a scale based on the reading material in this section and use the examples given in each chapter to help you complete your drawings. Ask your instructor what type of floor system to use. Assume:

Truss roof @ 24" o.c.
Scissor trusses over living room. Interior pitch @ 3/12.

Use 2 × 6 rafters in areas where trusses cannot be used. Determine spacing.

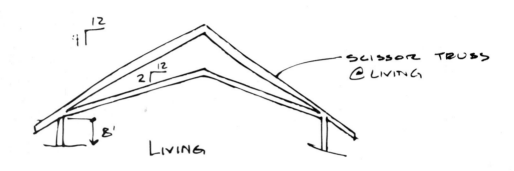

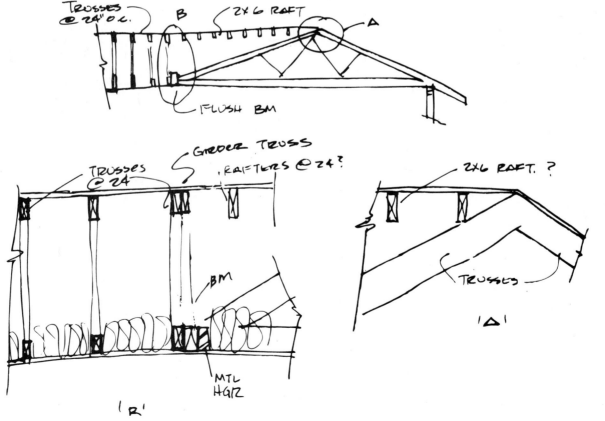

Problem 36–3

Problem 36–4   Spanish. Using a print of your floor and foundation plans, evaluate the construction methods used and determine which sections and partial sections need to be drawn. Lay out a freehand sketch of the sections you will need showing all wall locations, floor supports, and roof construction. Include a section showing masonry fireplace construction, if required. Use a C-size drawing sheet to lay out and draw all needed sections. Choose a scale based on the reading material in this section and use the examples given in each chapter to help you complete your drawings. Ask your instructor what type of floor system to use. Assume:

Rafter/C.J. @ 16″ o.c.
¼″ per 12″ min. roof pitch.

Use 2 × 6 rafters in areas where trusses can't be used. Determine spacing.

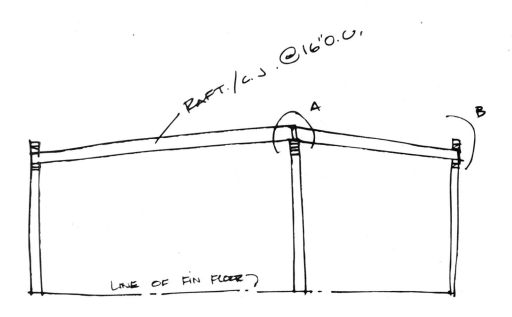

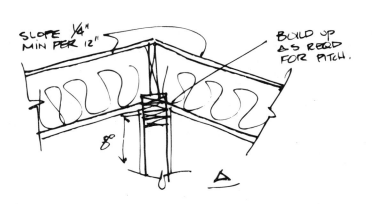

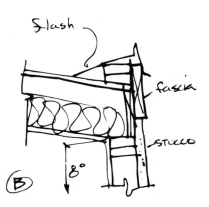

Problem 36–4

Problem 36–5 Early American. Using a print of your floor and foundation plans, evaluate the construction methods used and determine which sections and partial sections need to be drawn. Lay out a freehand sketch of the sections you will need showing all wall locations, floor supports, and roof construction. Include sections showing stair and masonry fireplace construction, if required. Use a C-size drawing sheet to lay out and draw all needed sections. Choose a scale based on the reading material in this section and use the examples given in each chapter to help you complete your drawings. Assume:

Rafter/C.J. @ 16″ o.c. over master bedroom.

Rafters @ 24″ o.c. except at vaulted ceilings.

Determine proper sizes for floor joists.

Unless your instructor tells you otherwise, use concrete blocks for the basement walls.

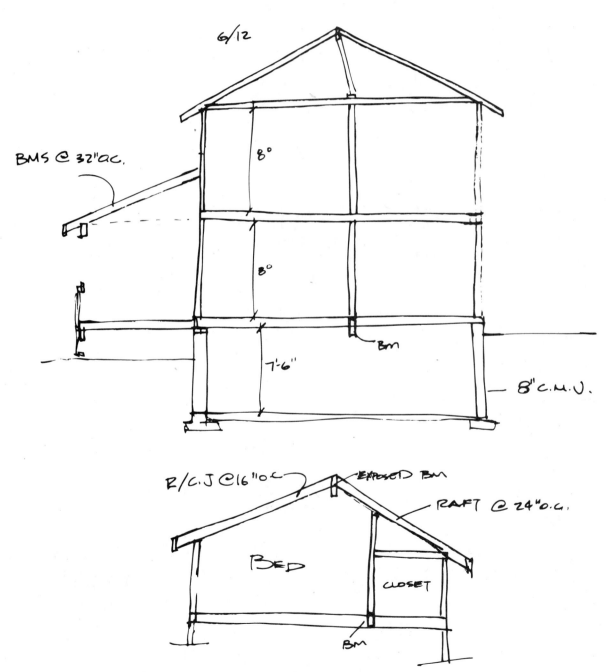

Problem 36–5

Problem 36–6   Contemporary. Using a print of your floor and foundation plans, evaluate the construction methods used and determine which sections and partial sections need to be drawn. Lay out a freehand sketch of the sections you will need showing all wall locations, floor supports, and roof construction. Include sections showing stair and masonry fireplace construction. Use a C-size drawing sheet to lay out and draw all needed sections. Choose a scale based on the reading material in this section and use the examples given in each chapter to help you complete your drawings. Determine proper sizes for all framing members.

Problem 36–7 to 36–15   Using a print of your floor and foundation plans, determine which sections for the project started in Chapter 15 should be drawn. Use energy-efficient construction methods suitable for your area. Check with your local building department to determine if special treatment is required for the wall between the two units for the duplex. Use D-size vellum to lay out and complete all required sections. Choose a scale that will allow construction information to be clearly shown. Determine proper sizes for all framing members.

*NOTE: Lightly lay out stair run to determine stair headroom, and roof height and pitch.

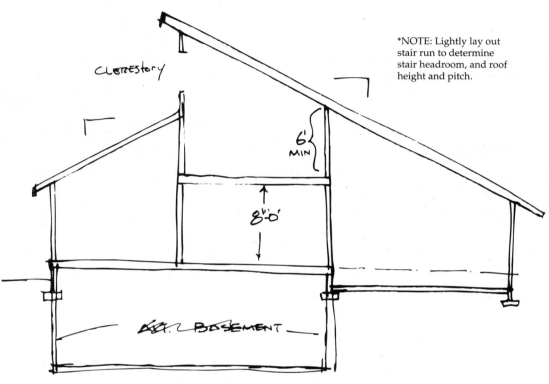

Problem 36–6

Stair Construction and Layout

Stairs were introduced with floor plans in Section IV. Minimal information was provided in that section so that you could draw stairs on floor plans. This chapter will show you how to draw stairs in section. Step-by-step instructions will be given for the layout and drawing of straight run, open, and U-shaped stair layouts.

▼ STAIR TERMINOLOGY

There are several basic terms you will need to be familiar with when working with stairs. Each can be seen in Figure 37–1:

○ *Run:* The horizontal distance from end to end of the stairs.

○ *Rise:* The vertical distance from top to bottom of the stairs.

○ *Tread:* The horizontal step of the stairs. It is usually made from 1″ (25 mm) material on enclosed stairs and 2″ (50 mm) material for open stairs. Tread width is the measurement from the face of the riser to the nosing. The nosing is the portion of the tread that extends past the riser.

○ *Riser:* The vertical backing between the treads. It is usually made from 1″ (25 mm) material for enclosed stairs and is not used on open stairs.

○ *Stringer or stair jack:* The support for the treads. A 2 × 12 (50 × 300) notched stringer is typically used for enclosed stairs. For an open stair a 4 × 14 (100 × 350) is common, but sizes vary greatly. Figure 37–2 shows the stringers, risers, and treads.

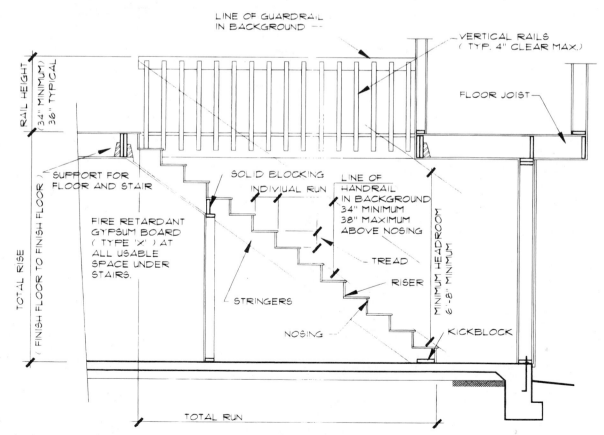

Figure 37–1 Common stair terms.

○ *Kick block or kicker:* Used to keep the bottom of the stringer from sliding on the floor when downward pressure is applied to the stringer.

○ *Headroom:* The vertical distance measured from the tread nosing to a wall or floor above the stairs. Building codes will specify a minimum size.

○ *Handrail:* The railing that you slide your hand along as you walk down the stairs.

○ *Guardrail:* The railing placed around an opening for the stairs.

Type X gypsum (GYP.) board ⅝″ (16 mm) thick is required by the UBC for enclosing all usable storage space under the stairs. This is a gypsum board that has a 1 hr. fire rating. Figure 37–3 shows common stair dimensions from three codes.

▼ DETERMINING RISE AND RUN

Building codes dictate the maximum rise of the stairs. To determine the actual rise, the total height from floor to floor must be known. Review Chapter 35. By adding the floor-to-ceiling height, the depth of the floor joist, and the depth of the floor covering, the total rise can be found. The total rise can then be divided by the maximum allowable rise to determine the number of steps required, as shown in Figure 37–4.

Once the required rise is determined, this information should be stored in your memory for future reference. Of the residential stairs you will lay out in your career as a drafter, probably 99 percent will have the same rise. So with a standard 8′–0″ (2430 mm) ceiling, you will always need 14 risers.

Figure 37–2 Stairs with stringers in place, ready for the treads and risers. *Courtesy Angie Renner.*

Once the rise is known, the required number of treads can be found easily since there will always be one fewer tread than the number of risers. Thus, a typical stair for a house with 8′–0″ (2430 mm) ceilings will have 14 risers and 13 treads. If each tread is 10½″ (267 mm) wide, the total run can be found by multiplying 10½″ (270 mm) (the width) by 13 (the number of treads required). With this basic information, you are now ready to lay out the stairs. The layout for a straight stairway will be described first.

▼ STRAIGHT STAIR LAYOUT

The straight-run stair is a common type of stair that will need to be drawn. It is a stair that goes from one floor to another in one straight run. An example of a straight-run stair can be seen in Figure 37–5. See Figure 37–6 for steps 1 through 4.

Step 1. Lay out walls that may be near the stairs.

Step 2. Lay out each floor level.

Step 3. Lay out one end of the stairs. If no dimensions are available on the floor plans, scale your drawing.

Step 4. Figure and lay out the total run of the stairs.

Step 5. Lay out the required risers. See Figure 37–7.

Step 6. Lay out the required treads. See Figure 37–7.

Step 7. Lay out the stringer, and outline the treads and risers, as shown in Figure 37–8.

Drawing the Stairs with Finished-Quality Lines

Use a thin line to draw the stairs unless otherwise noted. See Figure 37–9 for steps 8 through 11.

Step 8. Draw the treads and risers.

Step 9. Draw the bottom side of the stringer.

Step 10. Draw the upper stringer support or support wall where the stringer intersects the floor.

Step 11. Draw metal hangers and crosshatch them if there is no support wall.

See Figure 37–10 for steps 12 through 17.

Step 12. Draw the kick block with bold lines.

Step 13. Draw any intermediate support walls.

Step 14. Draw solid blocking with bold lines.

Step 15. Draw the ⅝″ (16 mm) type X gypsum board.

Step 16. Draw any floors or walls that are over the stairs.

Step 17. Draw the handrail.

Dimensions and Notes

See Figure 37–11 for steps 18 through 23. Place the required leader and dimension lines to locate the following needed dimensions:

Type	UBC	CABO/BOCA*	SBC
Straight stairs			
Rise	8″ (203 mm) max.	8¼″ (210 mm) max.	7¾″ (197 mm) max.
Run	9″ (229 mm) max.	9″ (229 mm) min.†	>8<9″ (229 mm) min. excluding nosing 10″ (254 mm) including nosing
Headroom	6′-6″ (1981 mm) min.	6′-6″ (1981 mm) min. 6′-8″ (2032 mm) min.	6′-6″ (1981 mm) min.
Tread width	30″ (762 mm) min.	36″ (914 mm) min.	36″ (914 mm) min.
Handrail	30″ (762 mm) min. 34″ (864 mm) max.	30″ (762 mm) min.	30″ (762 mm) min. 34″ (864 mm) max.
Guardrail	36″ (914 mm) min.	36″ (914 mm) min.	36″ (914 mm) min.
Winders			
Tread depth	9″ (129 mm) @ 12″ (305 mm) for each code 6″ (152 mm) min.	6″ (152 mm) min.	6″ (152 mm) min.
Spiral			
Tread depth	7½″ (191 mm) @ 12″ (305 mm) for each code		
Rise	9½″ (241 mm) max.	9½″ (241 mm) max.	9½″ (241 mm) max.
Radius	26″ (660 mm) min.	26″ (660 mm) min.	26″ (660 mm) min.

*CABO/BOCA requirements are the same except as noted.

†Treads and risers of required stairs shall be proportioned so that the sum of two risers and one tread, excluding nosing, is not less than 24″ (610 mm) or more than 25″ (635 mm).

Figure 37–3 Basic stair dimensions according to the four national codes.

Step 1. Determine the total rise in inches.

19	¾	plywood
235	9¼	floor joist
76	3	top plates
2353	92⅝	studs
38	1½	bottom plate
2721 mm	107.125″	total rise

Step 2. From the number of risers required. Divide rise by 8 (8″ [203 mm] is the maximum rise of UBC).

$$203\overline{)2721}^{\,13.4} \qquad 8\overline{)107.125}^{\,13.4}$$

Since you cannot have .4 risers, the number will be rounded up to 14 risers.

Step 3. Find the number of treads required.
Number of treads equal Rise − run.
14 − 1 = 13 treads required.

Step 4. Multiply the length of each tread by the number of treads to find the total run.

Figure 37–4 Determining the rise and run needed for a flight of stairs.

Step 18. Total rise

Step 19. Total run

Step 20. Rise

Step 21. Run

Step 22. Headroom

Step 23. Handrail

See Figure 37–12 for typical notes that are placed on stair sections. Have your instructor specify local variations.

▼ OPEN STAIRWAY LAYOUT

An open stairway is similar to the straight, enclosed stairway. It goes from one level to the next in a straight run. The major difference is that with the open stair, there are no risers between the treads. This allows for viewing from one floor to the next, creating an open feeling. See Figure 37–13 for an open stairway.

Step 1. Lay out the stairs following steps 1 through 4 of the enclosed stair layout.

Step 2. Lay out the 3 × 12 (75 × 300 mm) treads. See Figure 37–14.

Step 3. Lay out the 14″ (360 mm) deep stringer centered on the treads.

Finished-Quality Lines

Use thin lines for each step unless otherwise noted. See Figure 37–15 for steps 4 through 9.

Step 4. Draw the treads with a bold line.

Step 5. Draw the stringer.

Step 6. Draw the upper stringer supports with bold lines.

Step 7. Draw the metal hangers for the floor and stringer.

Step 8. Draw any floors or walls that are near the stairs.

Step 9. Draw the handrail.

Step 10. Place the required leader and dimension lines to provide the needed dimensions. See steps 18 through 24 of the enclosed stair layout for a guide to the needed dimensions. See Figure 37–16.

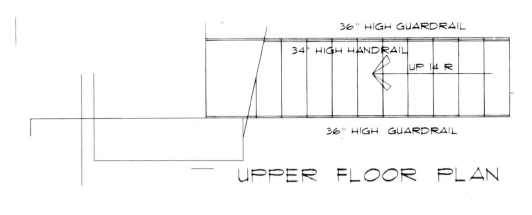

36" HIGH GUARDRAIL

34" HIGH HANDRAIL

UP 14 R

36" HIGH GUARDRAIL

UPPER FLOOR PLAN

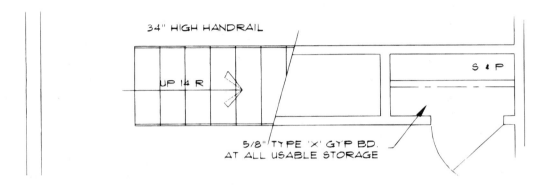

34" HIGH HANDRAIL

UP 14 R

S & P

5/8" TYPE 'X' GYP BD.
AT ALL USABLE STORAGE

LOWER FLOOR PLAN

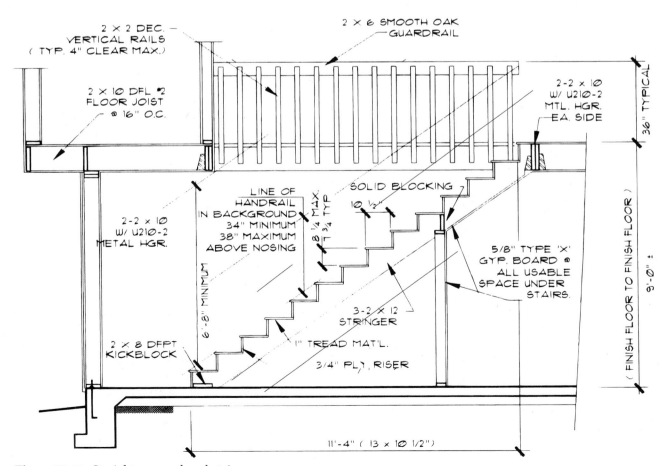

2 × 2 DEC.
VERTICAL RAILS
(TYP. 4" CLEAR MAX.)

2 × 6 SMOOTH OAK
GUARDRAIL

2 × 10 DFL #2
FLOOR JOIST
@ 16" O.C.

2-2 × 10
W/ U210-2
MTL. HGR.
EA. SIDE

36" TYPICAL

2-2 × 10
W/ U210-2
METAL HGR.

SOLID BLOCKING

10 1/2"

LINE OF
HANDRAIL
IN BACKGROUND
34" MINIMUM
38" MAXIMUM
ABOVE NOSING

8 1/4 MAX.
1 3/4 TYP

6'-8" MINIMUM

5/8" TYPE 'X'
GYP. BOARD @
ALL USABLE
SPACE UNDER
STAIRS.

(FINISH FLOOR TO FINISH FLOOR)

9'-0" ±

3-2 × 12
STRINGER

1" TREAD MAT'L.

3/4" PLY. RISER

2 × 8 DFPT
KICKBLOCK

11'-4" (13 × 10 1/2")

Figure 37–5   Straight-run enclosed stairs.

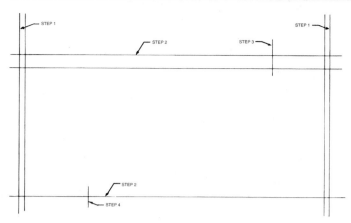

Figure 37-6 Lay out walls, floor, and each end of the stairs.

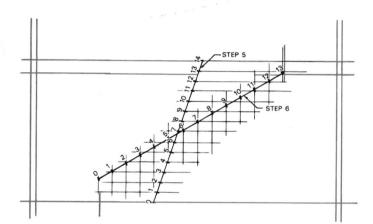

Figure 37-7 Layout of the risers and treads.

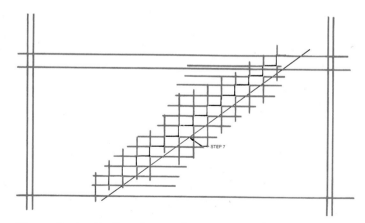

Figure 37-8 Outline the treads, stringer, and risers.

Step 11. Place the required notes on the section. Use Figure 37-16 as a guide. Have your instructor specify local variations.

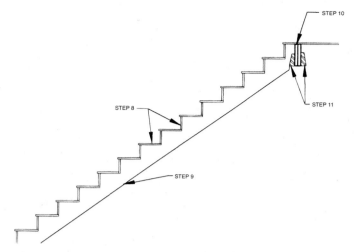

Figure 37-9 Finished-quality lines for treads, risers, and the stringer.

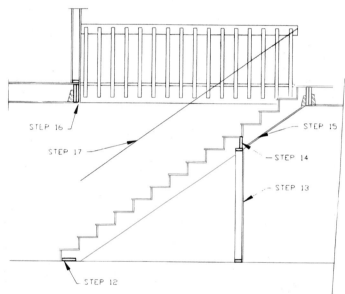

Figure 37-10 Finishing materials.

▼ U-SHAPED STAIRS

The U-shaped stair is often used in residential design. Rather than going up a whole flight of steps in a straight run, this stair layout introduces a landing. The landing is usually located at the midpoint of the run, but it can be offset depending on the amount of room allowed for stairs on the floor plan. Figure 37-17 shows what a U-shaped stair looks like on the floor plan and in section.

The stairs may be either open or enclosed, depending on the location. The layout of the stair is similar to the layout of the straight-run stair. It requires a little more planning in the layout stage because of the landing. Lay out the distance from the start of the stairs to the landing, based on the floor-plan measurements. Then proceed using a method similar to that used to draw the straight-run stair. See Figure 37-18 for help in laying out the section.

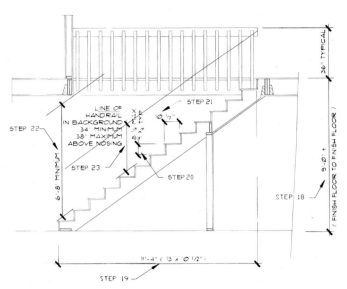

Figure 37–11 Stair dimensions.

▼ EXTERIOR STAIRS

It is quite common to need to draw sections of exterior stairs on multilevel homes. Figure 37–19 shows two different types of exterior stairs. Although there are many variations, these two options are common. Both can be laid out by following the procedure for the straight run stairs.

There are some major differences in the finishing materials. Notice there is no riser on the wood stairs, and the tread is thicker than the tread of an interior step. Usually the same material that is used on the deck is used for tread material. In many parts of the country, a nonskid material should also be called for to cover the treads.

The concrete stair can also be laid out by following the straight run stairs. Once the risers and run have been marked off, the riser can be drawn. Notice that the riser is drawn on a slight angle. It can be drawn at about 10° and not labeled. This is something the flatwork crew will determine at the job site, depending on their forms.

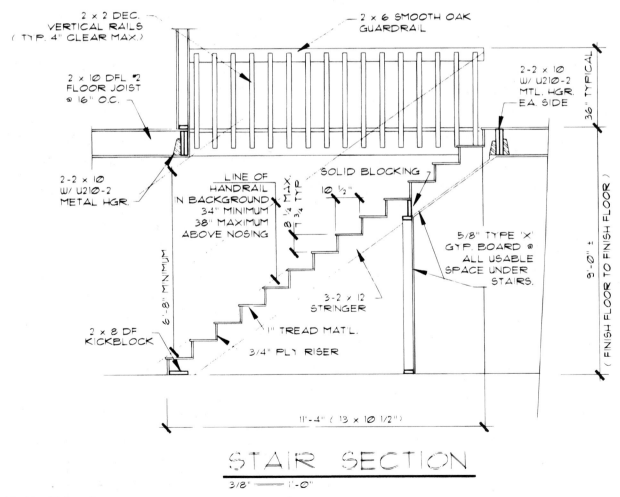

STAIR SECTION

3/8" = 1'-0"

Figure 37–12 Stair notes.

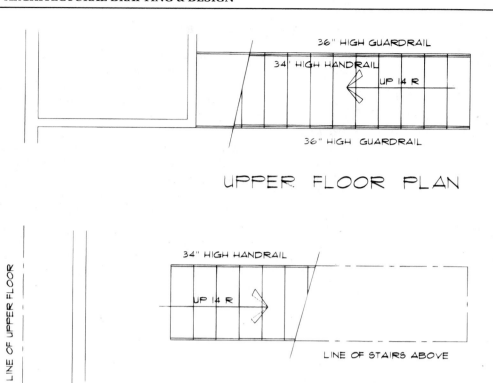

36" HIGH GUARDRAIL

34" HIGH HANDRAIL

UP 14 R

36" HIGH GUARDRAIL

UPPER FLOOR PLAN

LINE OF UPPER FLOOR

34" HIGH HANDRAIL

UP 14 R

LINE OF STAIRS ABOVE

LOWER FLOOR PLAN

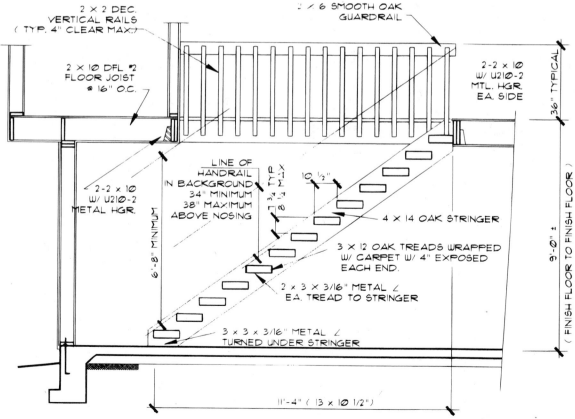

2 × 2 DEC. VERTICAL RAILS (TYP. 4" CLEAR MAX.)

2 × 6 SMOOTH OAK GUARDRAIL

2 × 10 DFL #2 FLOOR JOIST @ 16" O.C.

2-2 × 10 W/ U210-2 MTL. HGR. EA. SIDE

36" TYPICAL

2-2 × 10 W/ U210-2 METAL HGR.

LINE OF HANDRAIL IN BACKGROUND 34" MINIMUM 38" MAXIMUM ABOVE NOSING

6'-8" MINIMUM

1 3/4" TYP 8 1/4"

10 1/2"

4 × 14 OAK STRINGER

3 × 12 OAK TREADS WRAPPED W/ CARPET W/ 4" EXPOSED EACH END.

2 × 3 × 3/16" METAL ∠ EA. TREAD TO STRINGER

3 × 3 × 3/16" METAL ∠ TURNED UNDER STRINGER

9'-0" ± (FINISH FLOOR TO FINISH FLOOR)

11'-4" (13 × 10 1/2")

Figure 37–13   Open stairway layout.

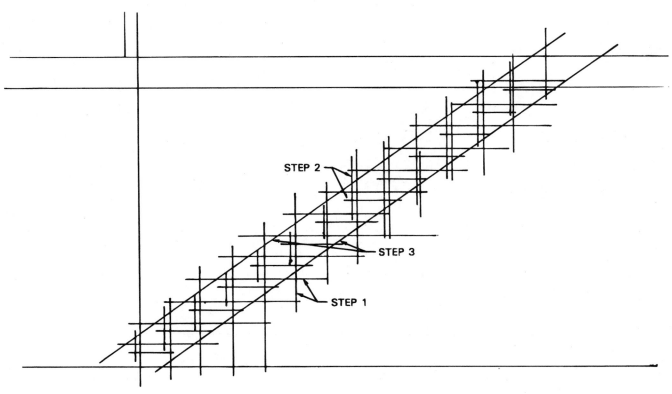

Figure 37–14 Lay out the open tread stairs.

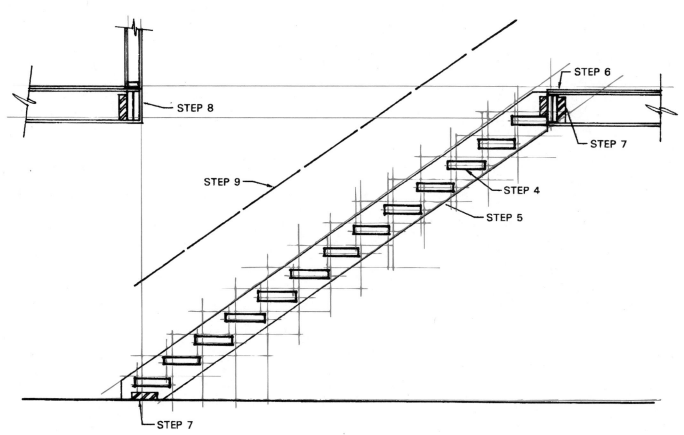

Figure 37–15 Finished-quality lines for structural material on open stairs.

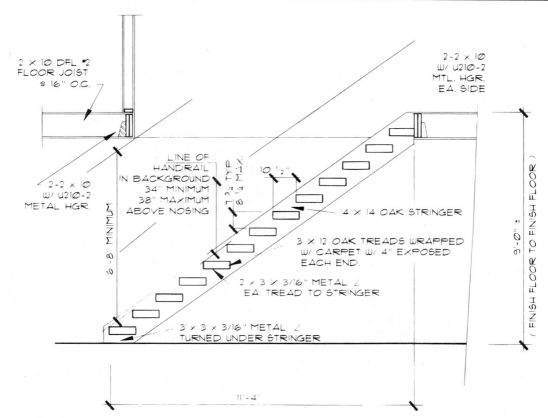

Figure 37–16 The completed open stair with dimensions and notes added.

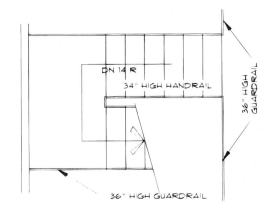

UPPER FLOOR PLAN

DN 14 R

34" HIGH HANDRAIL

36" HIGH GUARDRAIL

36" HIGH GUARDRAIL

LOWER FLOOR PLAN

5/8" TYPE 'X' GYP BD.
AT ALL USABLE STORAGE

5 R

34" HIGH HANDRAIL

UP 14 R

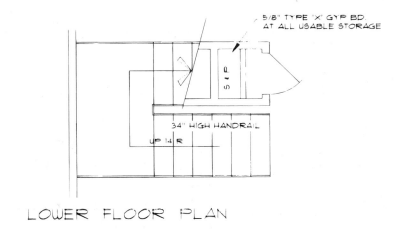

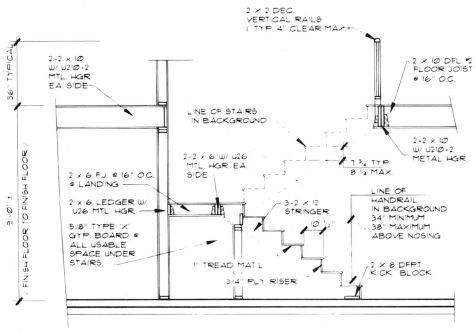

2 X 2 DEC.
VERTICAL RAILS
(TYP. 4" CLEAR MAX.)

2 X 10 DFL #2
FLOOR JOIST
@ 16" O.C.

36" TYPICAL

2-2 x 10
W/ U210-2
MTL. HGR.
EA. SIDE

LINE OF STAIRS
IN BACKGROUND

2-2 x 10
W/ U210-2
METAL HGR.

7 3/4 TYP
8 1/4 MAX.

2-2 x 6 W/ U26
MTL. HGR. EA.
SIDE

2 x 6 F.J. @ 16" O.C.
@ LANDING

2 x 6 LEDGER W/
U26 MTL. HGR.

3-2 x 12
STRINGER

LINE OF
HANDRAIL
IN BACKGROUND
34" MINIMUM
38" MAXIMUM
ABOVE NOSING

9'-0" ±

(FINISH FLOOR TO FINISH FLOOR)

5/8" TYPE 'X'
GYP. BOARD @
ALL USABLE
SPACE UNDER
STAIRS.

10 1/2"

1" TREAD MAT'L

3/4" PLY RISER

2 X 8 DFPT
KICK BLOCK

Figure 37–17 The U-shaped stair.

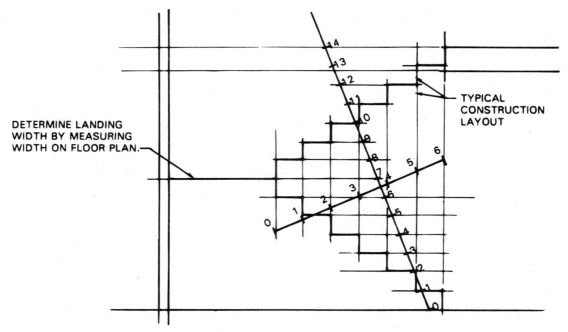

DETERMINE LANDING
WIDTH BY MEASURING
WIDTH ON FLOOR PLAN.

TYPICAL
CONSTRUCTION
LAYOUT

Figure 37–18 The layout of U-shaped stair runs.

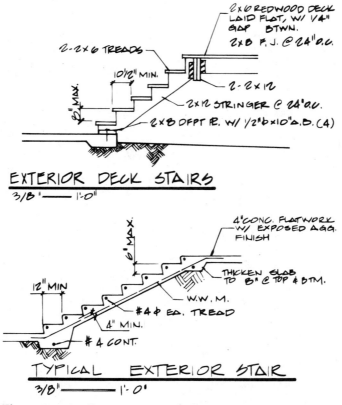

2-2×6 TREADS

10½" MIN.

2" MAX.

2×6 REDWOOD DECK
LAID FLAT, W/ ¼"
GAP BTWN.
2×8 F. J. @ 24" O.C.

2-2×12

2×12 STRINGER @ 24" O.C.

2×8 OFPT R. W/ ½"b×10"A.B. (4)

EXTERIOR DECK STAIRS
3/8" ———— 1'-0"

6" MAX.

12" MIN

4" MIN.

4" CONC. FLATWORK
W/ EXPOSED AGG.
FINISH

THICKEN SLAB
TO B" @ TOP & BTM.

W.W.M.

#4 φ EA. TREAD

#4 CONT.

TYPICAL   EXTERIOR STAIR
3/8" ———— 1'-0'

Figure 37–19 Common types of exterior stairs.

CADD Applications

DRAWING STAIRS WITH CADD

A stair section can be drawn using a CAD system by following the manual layout procedure. Commands such as ARRAY, OFFSET, TRIM, and FILLET will quickly reproduce repetitive elements of the stair. The section is even easier to draw with a parametric CADD system. All you have to do is pick the starting point of the stairs, specify the number of steps required, give the stair width and direction, and give the total rise; the program then automatically calculates the rise of each step. The program asks you to provide handrails with or without ballisters and provides you with several op-tions of handrail ends. After you have given the re-quired information, the stair is automatically drawn.

There are also CADD stair detailing systems that re-duce detailing from hours to minutes. The CADD pro-gram uses your specifications for type of stair construc-tion, total rise, and total run. It then automatically calculates the individual rise and run. You specify the tread, riser type, thickness, stringer dimensions, and railing specifications. After all design variables are in-put, the computer automatically draws, completely di-mensions and labels the stair section.

CHAPTER 37   Stair Construction and Layout Test

DIRECTIONS

Answer the questions with short, complete statements or drawings as needed.

1. Letter your name and the date at the top of the sheet.
2. Letter the question number and provide the an-swer. You do not need to write out the question.

QUESTIONS

Question 37–1   What is a tread?

Question 37–2   What is the minimum headroom required for a residential stair?

Question 37–3   What is the maximum individual rise of a step?

Question 37–4   What member is used to support the stairs?

Question 37–5   What spacing is required between the verti-cals of a railing?

Question 37–6   Describe the difference between a handrail and a guardrail.

Question 37–7   How many risers are required if the height be-tween floors is 10′ (3048 mm).

Question 37–8   Sketch three common stair types.

Question 37–9   If a run of 10″ (255 mm) is to be used, what will be the total run when the distance between floors is 9′ (2700 mm)?

Question 37–10   What is a common size for treads on an open-tread layout?

CHAPTER 38

Fireplace Construction and Layout

Thought was given to the type and location of the fireplace when the floor plan was drawn. As the sections are being drawn, consideration must be given to the construction of the fireplace and chimney. The most common construction materials for a fireplace and chimney are masonry or metal. The metal, or zero clearance, fireplace is manufactured and does not require a section to explain its construction.

▼ FIREPLACE TERMS

Fireplace construction has its own set of vocabulary that a drafter should be familiar with to draw a section. Figure 38–1 shows each of the major components of a fireplace and chimney.

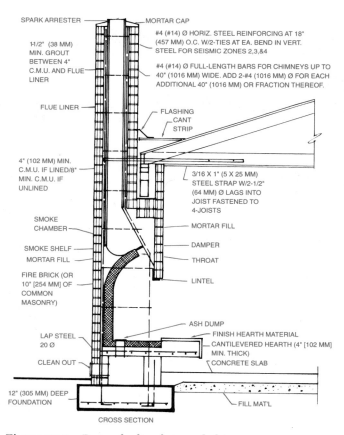

SPARK ARRESTER

MORTAR CAP

1-1/2" (38 MM) MIN. GROUT BETWEEN 4" C.M.U. AND FLUE LINER

#4 (#14) Ø HORIZ. STEEL REINFORCING AT 18" (457 MM) O.C. W/2-TIES AT EA. BEND IN VERT. STEEL FOR SEISMIC ZONES 2,3,&4

#4 (#14) Ø FULL-LENGTH BARS FOR CHIMNEYS UP TO 40" (1016 MM) WIDE. ADD 2-#4 (1016 MM) Ø FOR EACH ADDITIONAL 40" (1016 MM) OR FRACTION THEREOF.

FLUE LINER

FLASHING
CANT
STRIP

4" (102 MM) MIN. C.M.U. IF LINED/8" MIN. C.M.U. IF UNLINED

3/16 X 1" (5 X 25 MM) STEEL STRAP W/2-1/2" (64 MM) Ø LAGS INTO JOIST FASTENED TO 4-JOISTS

MORTAR FILL

SMOKE CHAMBER

DAMPER

SMOKE SHELF
MORTAR FILL

THROAT

FIRE BRICK (OR 10" [254 MM] OF COMMON MASONRY)

LINTEL

LAP STEEL 20 Ø

ASH DUMP

FINISH HEARTH MATERIAL

CANTILEVERED HEARTH (4" [102 MM] MIN. THICK)

CLEAN OUT

CONCRETE SLAB

12" (305 MM) DEEP FOUNDATION

FILL MAT'L

CROSS SECTION

Figure 38–1   Parts of a fireplace and chimney.

The *fireplace opening* is the area between the side and top faces of the fireplace. The size of the fireplace opening should be given much consideration since it is important for the appearance and operation of the fireplace. If the opening is too small, the fireplace will not produce a sufficient amount of heat. If the opening is too large, the fire could make a room too hot. The Masonry Institute of America suggests that the opening be approximately one-thirtieth of the room area for small rooms, and one-sixty-fifth of the room area for large rooms. Figure 38–2 shows suggested fireplace opening sizes compared to room size. The ideal dimensions for a single-face fireplace have been determined to be 36″ (900 mm) wide and 26″ (660 mm) high. These dimensions may be varied slightly to meet the size of the brick or to fit other special dimensions of the room.

The *hearth* is the floor of the fireplace and consists of an inner and outer hearth. The inner hearth is the floor of the firebox. The hearth is made of fire-resistant brick and holds the burning fuel. An ash dump is usually located in the inner hearth. The *ash dump* is an opening in the hearth that the ashes can be dumped into. The ash dump normally is covered with a small metal plate, which can be removed to provide access to the ash pit. The *ash pit* is the space below the fireplace where the ashes can be stored.

The outer hearth may be made of any incombustible material. The material is usually selected to blend with other interior design features and may be brick, tile, marble, or stone. The outer hearth protects the combustible floor around the fireplace. Figure 38–3 shows the minimum sizes for the outer hearth.

The fireplace opening is the front of the firebox. The *firebox* is where the combustion of the fuel occurs. The side should be slanted slightly to radiate heat into the room. The rear wall should be curved to provide an upward draft into the upper part of the fireplace and chimney. The firebox is usually constructed of fire-resistant brick set in fire-resistant mortar. Figure 38–3 shows minimum wall thickness for the firebox.

The firebox depth should be proportional to the size of the fireplace opening. By providing a proper depth, smoke will not discolor the front face (breast) of the fireplace. With an opening of 36″ × 26″ (900 × 660 mm), a depth of 20″ (500 mm) should be provided for a single-face fireplace. Figure 38–4 lists recommended fireplace opening-to-depth proportions.

Suggested Width of Fireplace Openings Appropriate to Room Size

Size of Room		If in Short Wall		If in Long Wall	
(Ft)	(mm)	(in.)	(mm)	(in.)	(mm)
10 × 14	(3000 × 4300)	24	(600)	24–32	(600–800)
12 × 16	(3600 × 4900)	28–36	(700–900)	32–36	(800–900)
12 × 20	(3600 × 6100)	32–36	(800–900)	36–40	(900–1000)
12 × 24	(3600 × 7300)	32–36	(800–900)	36–48	(900–1200)
14 × 28	(4300 × 8600)	32–40	(800–1000)	40–48	(1000–1200)
16 × 30	(4900 × 9100)	36–40	(900–1000)	48–60	(1200–1500)
20 × 36	(6100 × 11000)	40–48	(1000–1200)	48–72	(1200–1800)

Figure 38–2　The size of the fireplace should be proportioned to the size of the room. This will give both a pleasing appearance and a fireplace that will not overheat the room. *Reprinted from* Residential Fireplace and Chimney Construction Details and Specifications, *with permission from the Masonry Institute of America, Los Angeles, CA.*

Above the fireplace opening is a lintel. The *lintel* is a steel angle that supports the fireplace face. The throat of a fireplace is the opening at the top of the firebox that opens into the chimney. The throat should be able to be closed when the fireplace is not in use. This is done by installing a damper. The *damper* extends the full width of the throat and is used to prevent heat from escaping up the chimney when the fireplace is not in use. When fuel is being burned in the firebox, the damper can be opened to allow smoke from the firebox into the smoke chamber of the chimney. The *smoke chamber* acts as a funnel between the firebox and the chimney. The shape of the smoke chamber should be symmetrical so that the chimney draft pulls evenly and creates an even fire in the firebox. The smoke chamber should be centered under the flue in the chimney and directly above the firebox. A *smoke shelf* is located at the bottom of the smoke chamber behind the damper. The smoke shelf prevents down drafts from the chimney from entering the firebox.

The *chimney* is the upper extension of the fireplace and is built to carry off the smoke from the fire. The main components of the chimney are the flue, lining, anchors, cap, and spark arrester. The wall thickness of the chimney will be determined by the type of flue construction. Figure 38–3 shows the minimum wall thickness for chimneys.

The *flue* is the opening inside the chimney that allows the smoke and combustion gases to pass from the firebox away from the structure. A flue may be constructed of normal masonry products or may be covered with a flue liner. The size of the flue must be proportional to the size of the firebox opening and the number of open faces of the fireplace. A flue that is too small will not allow the fire to burn well and will cause smoke to exit through the front of the firebox. A flue that is too large for the firebox will cause too great a draft through the house as the fire draws its combustion air. Flue sizes are generally required to equal either one-eighth or one-tenth of the fireplace opening. Figure 38–5 shows recommended areas for residential fireplaces.

A *chimney liner* is usually built of fire clay or terra cotta. The liner is built into the chimney to provide a smooth surface to the flue wall and to reduce the width of the chimney wall. The smooth surface of the liner will help to reduce the buildup of soot, which could cause a chimney fire. The *chimney cap* is the sloping surface on the top of the chimney. The slope prevents rain from collecting on the top of the chimney. The flue normally projects 2 to 3″ (50–75 mm) above the cap so that water will not run down the flue. The *chimney hood* is a covering that may be placed over the flue for protection from the elements. The hood can be made of either masonry or metal. The masonry cap is built so that openings allow for the prevailing wind to blow through the hood and create a draft in the flue. The metal hood can usually be rotated by wind pressure to keep the opening of the hood downwind and thus prevent rain or snow from entering the flue. A *spark arrester* is a screen placed at the top of the flue inside the hood to prevent combustibles from leaving the flue.

Chimney Reinforcement

A minimum of four ½″ (13 mm) diameter (#4) vertical reinforcing bars should be used in the chimney extending from the foundation up to the top of the chimney. Typically these vertical bars are supported at 18″ (450 mm) intervals with ¼″ (6 mm) horizontal rebar. Rebar is also installed when the vertical steel is bent for a change in chimney width. In addition to the reinforcement in the chimney, the chimney must be attached to the structure. This is typically done with *steel anchors* that connect the fireplace to the framing members at each floor and ceiling level. Steel straps approximately 3/16 × 1″ (5 × 25 mm) are embedded into the grout of the chimney or wrapped around the reinforcing steel and then bolted to the framing members.

▼ DRAWING THE FIREPLACE SECTION

The fireplace section can be a valuable part of the working drawings. Although a fireplace drawing is required by most building departments when a home has a masonry fireplace and chimney, a drafter may not be required to

General Code Requirements

ITEM	Letter	Uniform Building Code* 1994	FHA &VA	Local or Special Requiremens
Hearth Slab Thickness	A	4″	4″	
Hearth Slab Width (Each side of opening)	B	8″ Fireplace opg. <6 sq. ft. 12″ Fireplace opg. ≥6 sq. ft.	8″	
Hearth Slab Length (Front of opening)	C	16″ Firepl. opg. <6 sq. ft. 20″ Firepl. opg. ≥6 sq. ft.	16″	
Hearth Slab Reinforcing	D	Reinforced to carry its own weight and all imposed loads.	Required if cantilevered in connection with raised wood floor construction.	
Thickness of Wall of Firebox	E	10″ common brick or 8″ where a fireback lining is used. Jts. in fireback ¼″ max.	8″ including minimum 2″ fireback lining—12″ when no lining is provided.	
Distance from Top of Opening to Throat	F	6″	6″ min.; 8″ recommended.	
Smoke Chamber Edge of Shelf	G		½″ offset. 6″ plus paraging may be omitted if wall thickness is 8″ or more of solid masonry. Form damper is required.	
Rear Wall—Thickness		6″		
Front & Side Wall—Thickness		8″		
Chimney Vertical Reinforcing	**H	Four #4 full length bars for chimney up to 40″ wide. Add two #4 bars for each additional 40″ or fraction of width or each additional flue.	Four #4 bars full length, no splice unless welded.	
Horizontal Reinforcing	J	¼″ ties at 18″ and two ties at each bend in vertical steel.	¼″ bars at 24″	
Bond Beams	K	No specified requirements. L.A. City requirements are good practice	Two ¼″ bars at top bond beam 4″ high. Two ¼″ bars at anchorage bond beam 5″ high.	
Fireplace Lintel	L	Incombustible material	2½″ × 3″ × 3⁄16″ angle with 3″ end bearing	
Walls with Flue Lining	M	Brick with grout around lining. 4″ min. from flue lining to outside face of chimney.	Brick with grout around lining. 4″ min. from flue lining to outside face of chimney.	
Walls with Unlined Flue	N	8″ Solid masonry	8″ Solid masonry	
Distance Between Adjacent Flues	O	4″ including flue liner	4″ wythe for brick	
Effective Flue Area (Based on Area of Fireplace Opening)	P	Round lining–1/12 or 50 sq.in.min. Rectangular lining 1/10 or 64 sq. in. min. Unlined or lined with firebrick—1/8 or 100 sq. in. min.	1/10 for chimneys over 15′ high and over 1/8 for chimneys less than 15″ high	
Clearances Wood Frame	R	1″ when outside of wall or ½″ gypsum board 1″ when entirely within structure	¾″ from subfloor or floor or roof sheathing. 2″ from framing members	
Combustible Material		6″ min. to fireplace opening. 12″ from opening when material projecting more than1/8 for ea. 1″	3½″ to edge of fireplace 12″ from opening when projecting more than 1½″	
Above Roof		2′ at 10′	2′ at 10′	
Anchorage Strap	S	3⁄16″ × 1″	¼″ × 1″	
Number		2	2	
Embedment into chimney		12″ hooked around outer bar w/6″ ext.	18″ hooked around outer bar	
Fasten to		4 Joists	3 joists	
Bolts		Two ½″ Dia.	Two ½″ Dia.	
Footing Thickness	T	12″ mn.	8″ min. for 1 story chimney 12″ min. for 2 story chimney	
Width		6″ each side of fireplace wall	6″ each side of fireplace wall.	
Outside Air Intake	U	Optional	Required	6 sq. in. minimum area (California Energy Commission Requirement)
Glass Screen Door		Optional	Required but shall not interfere with energy conservation improving devices	Required but shall not interfere with energy conservation improving devices

*Applies to Los Angeles County and Los Angeles City requirements.
**H EXCEPTION. Chimneys constructed of hollow masonry units may have vertical reinforcing bars spliced to footing dowels, provided that the splice is inspected prior to grouting of the wall.

Figure 38–3 General code requirements for fireplace and chimney construction. The letters in the second column will be helpful in locating a specific item in Figures 38–6 through 38–9. *Reprinted from* Residential Fireplace and Chimney Construction Details and Specifications, *with permission from the Masonry Institute of America, Los Angeles, CA.*

Fireplace Type	Width of Opening (w)		Height of Opening (h)		Depth of Opening (d)	
	(in.)	(mm)	(in.)	(mm)	(in.)	mm
Single Face	28	700	24	600	20	500
	30	760	24	600	20	500
	30	760	26	660	20	500
	36	900	26	660	20	500
	36	900	28	700	22	560
	40	1000	28	700	22	560
	48	1200	32	800	25	635
Two faces	34	865	27	685	23	585
adjacent "L"	39	990	27	685	23	585
or corner type.	46	1170	27	685	23	585
	52	1320	30	760	27	635
Two faces*	32	800	21	530	30	760
opposite	35	890	21	530	30	760
Look through	42	1070	21	530	30	760
	48	1200	21	530	34	865
Three face*	39	990	21	530	30	760
2 long, 1 short	46	1170	21	530	30	760
3-way opening	52	1320	21	530	34	865
Three face*	43	1090	27	685	23	585
1 long, 2 short	50	1270	27	685	23	585
3-way opening	56	1420	30	760	27	686

*Fireplaces open on more than front and one end are **not** recommended.

Figure 38–4 Fireplace opening-to-depth proportion guide. *Reprinted from* Residential Fireplace and Chimney Construction Details and Specifications, *with permission from the Masonry Institute of America, Los Angeles, CA.*

draw a fireplace section. Because of the similarities of fireplace design, a fireplace drawing is often kept on file as a stock detail at many offices. When a house that has a fireplace is being drawn, the drafter needs only to get the stock detail from a file, make the needed copies, and attach them to the prints of the plans. Figure 38–6 shows a stock detail.

In some areas, rather than the drafters providing the drawing, a standard fireplace drawing by the Masons Association may be attached to the plans, as shown in Figure 38–7. Not all building departments will accept these drawings, so check with your building departments before you submit plans for a permit.

Fireplace Guidelines

If you are required to draw a fireplace section, use the drawings by the Masonry Institute of America as a guide. Fireplace drawings are also included in some codes such as *Dwelling Construction Under the Uniform Building Code* and would be helpful for the beginning drafter to use as a guide.

A print of the floor plans will be needed to help determine wall locations and the size of the fireplace. As the drawing for the fireplace is started, use construction lines. See Figure 38–8 for steps 1 through 4.

Layout of the Fireplace

Step 1. Lay out the width of the fireplace (20″ or 500 mm) for the firebox and 8″ (200 mm) for the rear wall.

Step 2. Lay out all walls, floors, and ceilings that are near the fireplace. Be sure to maintain the required minimum distance from the masonry to the wood as determined by the code in your area.

Step 3. Lay out the foundation for the fireplace. The footing is typically 6″ (150 mm) wider on each side than the fireplace and 12″ (300 mm) deep.

Step 4. Lay out the hearth in front of the fireplace. The hearth will vary in thickness depending on the type of finishing material used. If the fireplace is being drawn for a house with a wood floor, the hearth will require a 4″ (100 mm) minimum concrete slab projecting from the fireplace base to support the finished hearth.

See Figure 38–9 for steps 5 through 8.

Step 5. Lay out the firebox. Assume that a 36 × 26 × 20″ (900 × 660 × 500 mm) firebox will be used. See Figure 38–10 for guidelines for laying out the firebox.

Step 6. Determine the size of flue required. A 36 × 26″ opening has an area of 936 sq in. By examining Figure 38–11 you can see that a flue with an area of 91 sq in. is required. The flue walls can be constructed using either 8″ (200 mm) of masonry or 4″ (100 mm) of masonry and a clay flue liner. If a liner is used, a 13″ (330 mm) round liner is the minimum size required.

Step 7. Lay out the flue with a liner. Draw the interior face of the liner. The thickness will be shown later.

Step 8. Determine the height of the chimney. Most building codes require the chimney to project 24″ (600 mm) minimum above any construction within 10′ (3000 mm) of the chimney. See Figure 38–12.

Drawing with Finished-Quality Lines

Step 9. Use bold lines to outline all framing materials as shown in Figure 38–13.

See Figure 38–14 for steps 10 through 12.

Step 10. Use bold lines to outline the masonry, flue liner, and hearth.

Step 11. Use thin lines at a 45° angle to crosshatch the masonry.

Step 12. Crosshatch the firebrick with lines at a 45° angle so that a grid is created.

Type of Fireplace	Width of Opening w in.	Height of Opening h in.	Depth of Opening d in.	Area of Fireplace opening for flue determination sq. in.	Flue Size Required at 1/10 Area of fireplace opening	Flue Size Required at 1/8 Area of Fireplace opening (FHA Requirement)*
	28	24	20	672	8½ × 13	8½ × 17
	30	24	20	720	8½ × 17	13″ round
	30	26	20	780	8½ × 17	10 × 18
	36	26	20	936	10 × 18	13 × 17
	36	28	22	1008	10 × 18	10 × 21
	40	28	22	1120	10 × 18	10 × 21
	48	32	25	1536	13 × 21	10 × 21
	60	32	25	1920	17 × 21	21 × 21
	34	27	23	1107	10 × 18	10 × 21
	39	27	23	1223	10 × 21	13 × 21
	46	27	23	1388	10 × 21	13 × 21
	52	30	27	1884	13 × 21	17 × 21
	64	30	27	2085	17 × 21	21 × 21
	32	21	30	1344	13 × 17 or 10 × 21	17 × 17 or 13 × 21
	35	21	30	1470	17 × 17 or 13 × 21	17 × 21
	42	21	30	1764	17 × 21	17 × 21
	48	21	34	2016	17 × 21	21 × 21
	39	21	30	1638	13 × 21 or 17 × 17	17 × 21
	46	21	30	1932	17 × 21	21 × 21
	52	21	34	2184	17 × 21	21 × 21
	43	27	23	1782	13 × 21 or 17 × 17	17 × 21
	50	27	23	1971	17 × 21	17 × 21
	56	30	27	2490	21 × 21	21 × 21 or
	68	30	27	2850	13 × 21 & ▲ 10 × 21	2 — 10 × 21 ▲ 2 — 13 × 21 or 2 — 17 × 17 or ▲ 10 × 18 & 17 × 21

*FHA Requirement if chimney is less than 15′ high. Use 1/10 ratio if chimney is 15′ or more in height.
▲ Rather than using 2 flue liners in chimney, many times the flue is left unlined with 8″ masonry walls. Unlined flues must have a minimum area of 1/8 fireplace opening.

Figure 38–5 Recommended flue areas for residential fireplaces. *Reprinted from* Residential Fireplace and Chimney Construction Details and Specifications, *with permission from* the *Masonry Institute of America, Los Angeles, CA.*

See Figure 38–15 for steps 13 through 15.

Step 13. Draw all reinforcing steel with bold lines. Verify with local building codes to determine what steel will be required.

Step 14. Draw the lintel with a very bold line.

Step 15. Draw the damper with a very bold line.

See Figure 38–16 for steps 16 and 17.

Step 16. Dimension the drawing. Items that must be dimensionsed include the following:
   a. Height above roof
   b. Width of hearth
   c. Width and height of firebox
   d. Width and depth of footings
   e. Wood to masonry clearance

Step 17. Place notes on the drawing using Figure 36–16 as a guide. Notes will vary depending on the code that you follow and the structural material that has been specified on the floor and foundation plans.

▼ FIREPLACE ELEVATIONS

The fireplace elevation is the drawing of the fireplace as viewed from inside the home. Fireplace elevations typically show the size of the firebox and the material that will be used to decorate the face of the fireplace. Fireplace elevations can be drawn by using the following steps. See Figure 38–17 for steps 1 through 4.

Elevation Layout

Step 1. Lay out the size of the wall that will contain the fireplace.

Step 2. Lay out the width of the chimney.

Step 3. Lay out the height and width of the hearth if a raised hearth is to be used.

Step 4. Lay out the fireplace opening.

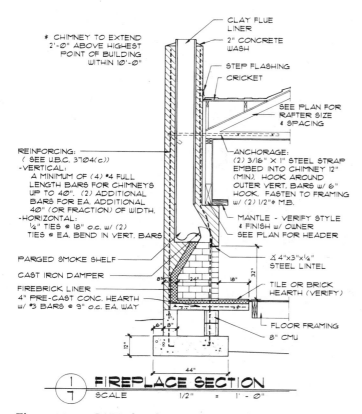

Figure 38–6 CADD fireplace section.

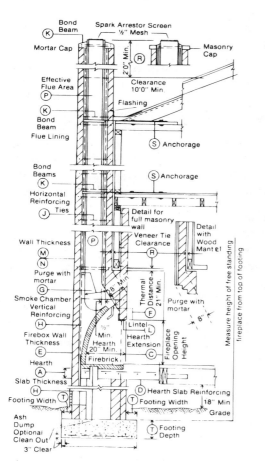

Figure 38–7 A brick fireplace and chimney with a wood-framed floor. The circled letters refer to item references listed in Figure 38–3. *Reprinted from* Residential Fireplace and Chimney Construction Details and Specifications, *with permission from the Masonry Institute of America, Los Angeles, CA.*

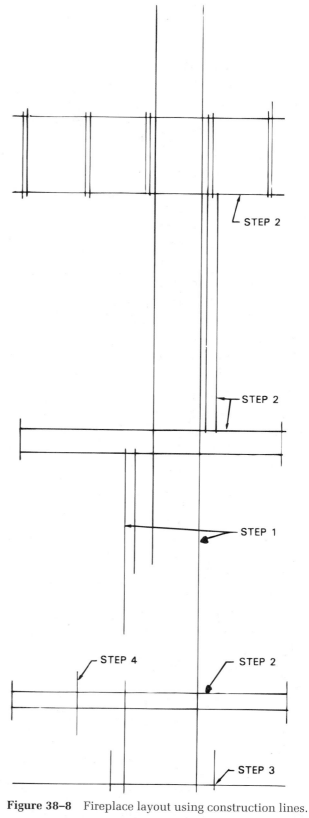

Figure 38–8 Fireplace layout using construction lines.

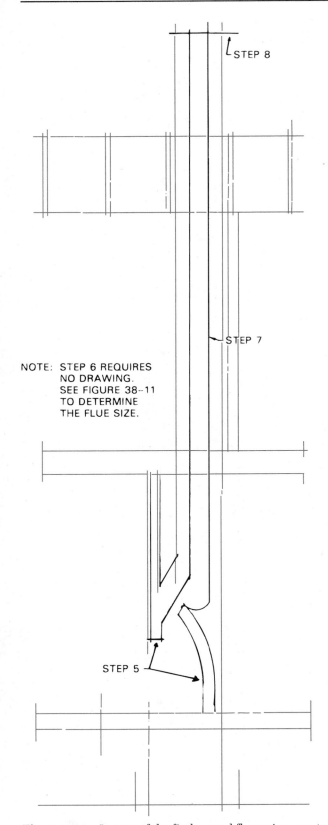

STEP 8

STEP 7

NOTE: STEP 6 REQUIRES
NO DRAWING.
SEE FIGURE 38-11
TO DETERMINE
THE FLUE SIZE.

STEP 5

Figure 38–9 Layout of the firebox and flue using construction lines.

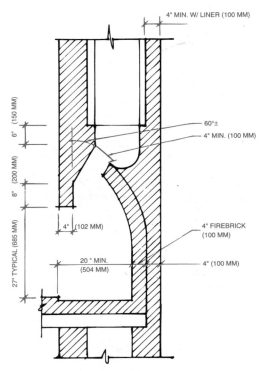

4" MIN. W/ LINER (100 MM)

6" (150 MM)

8" (200 MM)

27" TYPICAL (685 MM)

60°±

4" MIN. (100 MM)

4" (102 MM)

4" FIREBRICK (100 MM)

20 " MIN. (504 MM)

4" (100 MM)

Figure 38–10 Firebox layout.

Finished-Quality Lines

Usually the designer will provide the drafter with a rough sketch similar to Figure 38–18 to describe the finishing materials on the face of the fireplace. By using this sketch and the methods described in Chapter 21, the fireplace elevation can be drawn. See Figure 38–19. Once the finishing materials have been drawn, the elevation can be lettered and dimensioned as seen in Figure 38–20. Notes will need to be placed on the drawing to describe all materials. Dimensions will be required to describe:

1. Room height
2. Hearth height
3. Fireplace opening height and width
4. Mantel height

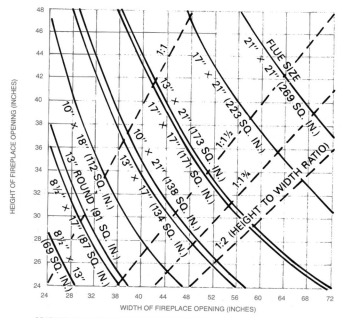

GRAPH TO DETERMINE THE PROPER FLUE SIZE FOR A SINGLE-FACE FIREPLACE, BY
MATCHING THE WIDTH AND HEIGHT OF THE FIREPLACE OPENING, THE MINIMUM FLUE SIZE
CAN BE DETERMINED. ONCE THE MINIMUM FLUE SIZE HAS BEEN DETERMINED, THE
CHIMNEY SIZE CAN BE DETERMINED.

Nominal Dimension of Flue Lining	Actual Outside Dimensions of Flue Lining	Effective Flue Area	Max. Area of Fireplace Opening	Minimum Outside Dimension of Chimney
8½″ Round	8½″ Round	39 sq. in.	390 sq. in.	17″ × 17″
8½″ × 13″ oval	8½″ × 12¾″	69 sq. in.	690 sq. in.	17″ × 21″
8½″ × 17″ oval	8½″ × 16¾″	87 sq. in.	870 sq. in.	17″ × 25″
13″ Round	12¾″ Round	91 sq. in.	1092 sq. in.	21″ × 21″
10″ × 18″ oval	10 × 17¾″	112 sq. in.	1120 sq. in.	19″ × 26″
10″ × 21″ oval	10″ × 21″	138 sq. in.	1380 sq. in.	19″ × 30″
13″ × 17″ oval	12¾″ × 16¾″	134 sq. in.	1340 sq. in.	21″ × 25″
13″ × 21″ oval	12¾″ × 21″	173 sq. in.	1730 sq. in.	21″ × 30″
17″ × 17″ oval	16¾″ × 16¾″	171 sq. in.	1710 sq. in.	25″ × 25″
17″ × 21″ oval	16¾″ × 21″	223 sq. in.	2230 sq. in.	25″ × 30″
21″ × 21″ oval	21″ × 21″	269 sq. in.	2690 sq. in.	30″ × 30″

Figure 38–11   Flue and chimney sizes based on the area of the fireplace opening. *Reprinted from* Residential Fireplace and Chimney Construction Details and Specifications, *with permission from the Masonry Institute of America, Los Angeles, CA.*

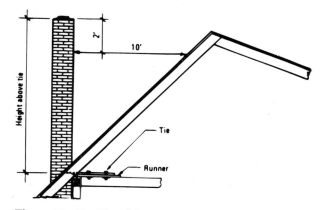

Figure 38–12   The chimney is required to extend 2′ (610 mm) above any part of the structure that is within 10′. *Reprinted from* Residential Fireplace and Chimney Construction Details and Specifications, *with permission from the Masonry Institute of America, Los Angeles, CA.*

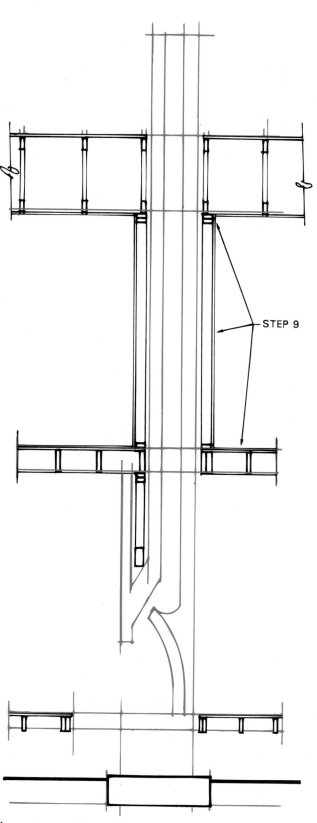

Figure 38–13   Draw all framing members with finished-quality lines.

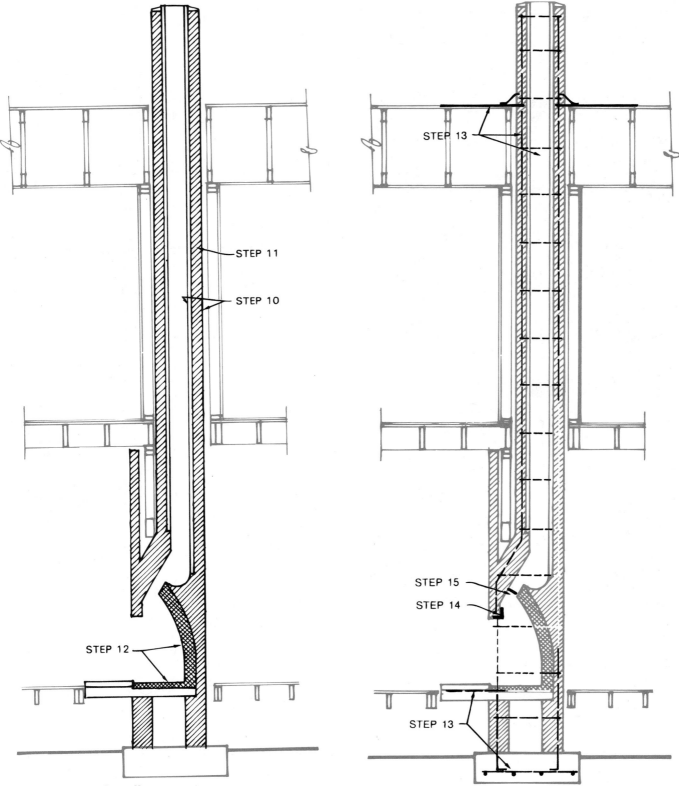

STEP 11

STEP 10

STEP 12

Figure 38–14 Darken all masonry.

STEP 13

STEP 15

STEP 14

STEP 13

Figure 38–15 Draw all reinforcing steel.

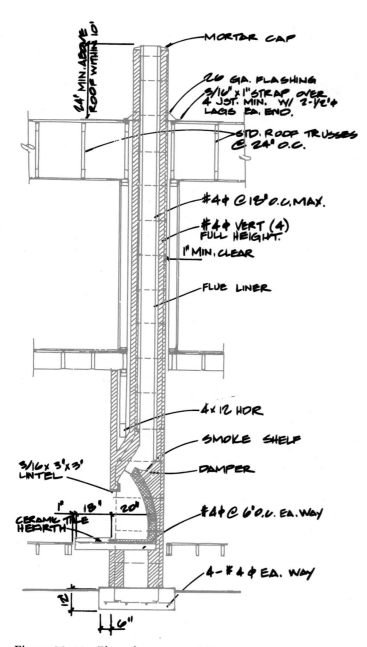

MORTAR CAP

24' MIN. ABOVE ROOF WITHIN 10'

26 GA. FLASHING
3/16" x 1" STRAP OVER.
4 JST. MIN. W/ 2-1/2"∅
LAGS EA. END.

STD. ROOF TRUSSES
@ 24" O.C.

#4 @ 18" O.C. MAX.

#4 VERT (4)
FULL HEIGHT.

1" MIN. CLEAR

FLUE LINER

4 x 12 HDR

SMOKE SHELF

DAMPER

3/16 x 3' x 3'
LINTEL

1"   18'   20'

CERAMIC TILE
HEARTH

#4 @ 6' O.C. EA. WAY

4 - #4 ∅ EA. WAY

12"

6"

Figure 38–16   Place the notes and dimensions.

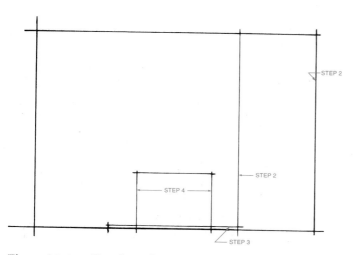

STEP 2

STEP 2

STEP 4

STEP 3

Figure 38–17   Fireplace elevation layout.

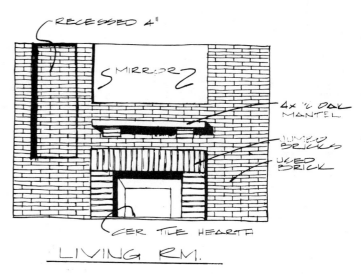

RECESSED 4"

MIRROR

4 x 12 OAK MANTEL

JUMBO BRICKS

USED BRICK

CER. TILE HEARTH

LIVING RM.

Figure 38–18   The drafter can often use the designer's preliminary sketch as a guide when drawing the finishing materials for a fireplace elevation.

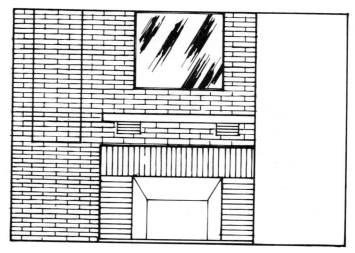

Figure 38–19   Drawing the finished elevation.

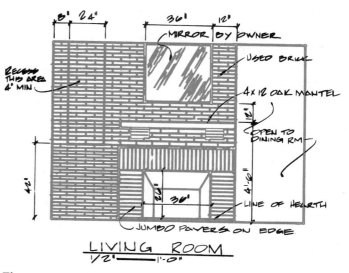

8'   24"   36'   12'

MIRROR BY OWNER

USED BRICK

RECESS THIS AREA 4" MIN

4 x 12 OAK MANTEL

OPEN TO DINING RM.

42'

12"

36'

LINE OF HEARTH

JUMBO PAVERS ON EDGE.

LIVING   ROOM
1/2" ——— 1'-0"

Figure 38–20   Lettering and dimensions for a fireplace elevation.

▼ CADD Applications

DRAWING FIREPLACE SECTIONS AND ELEVATIONS WITH CADD

In many cases the architectural firm draws a number of standard fireplace sections and details. These typical sections are then saved in the CADD symbols library and called up when necessary to insert on a drawing. The standard sections may be completely dimensioned and noted, or the dimensions and specific notes may be added after the section is brought into the drawing.

Drawing fireplace elevations with CADD is easy because a fireplace elevation is normally either a fireplace box and a surrounding mantel, or a fireplace box and a masonry wall. In either case the job is simple. If you are

drawing fireplace elevations with a surrounding mantel, draw a few standard applications and add one of them to the drawing at any time. Figure 38–21 shows a standard mantel in a fireplace elevation. With most CADD programs, these symbols are dragged onto the drawing where the computer gives you a chance to scale the display up or down. This provides maximum flexibility while inserting the drawing. Once the standard mantel has been added to the drawing, it can be increased or decreased in size with commands such as SCALE, TRIM, or STRETCH. If the fireplace elevation is a masonry wall, all you do is define the area with lines and add elevation symbols such as stone or brick, as shown in Figure 38–22.

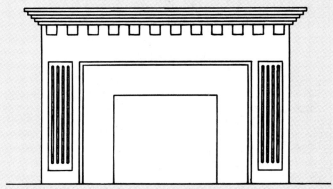

Figure 38–21 CADD mantel.

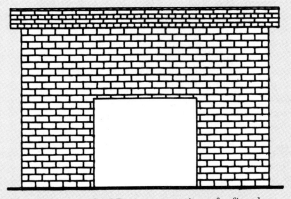

Figure 38–22 CADD representation of a fireplace masonry wall.

CHAPTER

38 Fireplace Construction and Layout Test

DIRECTIONS

Answer the questions with short, complete statements or drawings as needed.

1. Letter your name and the date at the top of the sheet.
2. Letter the question number and provide the answer. You do not need to write out the question.

QUESTIONS

Question 38–1 What purpose does a damper serve?

Question 38–2 What parts does the throat connect?

Question 38–3 What is the most common size of fireplace opening?

Question 38–4 What is the required flue area for a fireplace opening of 44 × 26″ (1118 × 660 mm)?

Question 38–5 What flues could be used for a fireplace opening of 1340 sq in. (864514 mm^2)?

Question 38–6 How is masonry shown in cross section?

Question 38–7 Why is a fireplace elevation drawn?

Question 38–8 Why is fireplace information often placed in stock details?

Question 38–9 Where should chimney anchors be placed?

Question 38–10 How far should the hearth extend in front of the fireplace with an opening of 7 sq ft (.65 m^2)?

SECTION XI

ARCHITECTURAL RENDERING

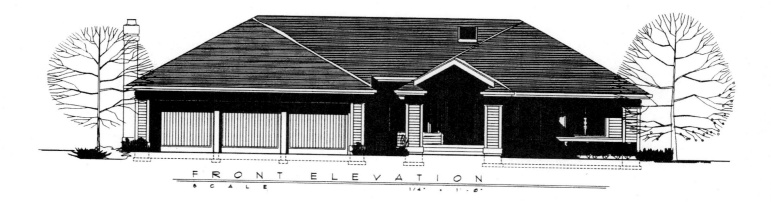

FRONT ELEVATION
SCALE   1/4" = 1'-0"

CHAPTER 39

Presentation Drawings

The term *presentation drawings* was introduced during the discussion of the design sequence in Chapter 1. As the residence was being designed, the floor plan and a rendering were drawn to help the owners understand the designer's ideas. Presentation drawings are the drawings that are used to convey basic design concepts from the design team to the owner or other interested persons. Presentation drawings are a very important part of public hearings and design reviews as a structure is studied by government and private agencies to determine its impact on the community. In residential architecture, presentation drawings are frequently used to help advertise existing stock plans.

Your artistic ability, the type of drawing to be done, and the needs of the client will affect how the presentation drawing will be done and who will produce the drawings. Some offices have an architectural illustrator do all presentation drawings. An illustrator combines the skills of an artist with the techniques of drafting. Other offices allow design drafters to make the presentation drawings. A drafter may be able to match the quality of an illustrator but not the speed. Many of the presentation drawings that are contained throughout this book would take an illustrator only a few hours to draw.

▼ TYPES OF PRESENTATION DRAWINGS

The same types of drawings that are required for the working drawings can also be used to help present the design ideas. Because the owner, user, or general public may not be able to understand the working drawings fully, each can be drawn as a presentation drawing to present basic information. The most common types of presentation drawings are renderings, elevations, floor plans, plot plans, and sections.

Renderings

Renderings are the best type of presentation drawing for showing the shape or style of a structure. The term *rendering* can be used to describe an artistic process applied to a drawing. Each of the presentation drawings can be rendered using one of the artistic styles soon to be discussed. Rendering can also be used to refer to a drawing created using the perspective layout method, which

will be presented in Chapter 40. Although rendering is the artistic process used on a perspective drawing, the term is also often applied to the drawings.

Figure 39–1 shows an example of a rendering. A rendering is used to present the structure as it will appear in its natural setting. Exterior renderings are typically drawn using a two-point perspective method. A rendering can also be very useful for showing the interior shape and layout of a room, as seen in Figure 39–2. Interior renderings are usually drawn using a one-point perspective method.

Elevations

A rendered elevation is often used as a presentation drawing to help show the shape of the structure. An example of a presentation elevation can be seen in Figure 39–3. This type of presentation drawing gives the viewer an accurate idea of the finished product, while the working drawing is aimed at providing the construction crew information about the materials that they will be installing. The rendered elevation helps show the various changes in surface much better than the working elevation. Although the rendered elevation does not show the depth as well as a rendering does, it allows the viewer a clearer understanding of the project without having to invest the time or money required to draw a rendering. A rendered elevation usually includes all of the material shown on a working elevation with the addition of shades and plants. Depending on the artistic level of the drafter, people and automobiles are often shown also.

Floor Plans

Floor plans are often used as presentation drawings to convey the layout of interior space. Similar to the preliminary floor plan in the design process, a presentation floor plan is used to show room relationships, openings such as windows and doors, and basic room sizes. Furniture and traffic patterns are also usually shown. Figure 39–4 provides an example of a presentation floor plan.

Plot Plans

A rendered plot plan is used to show how the structure will relate to the job site and to the surrounding area. The placement of the building on the site and the north arrow are the major items shown. As seen in Figure 39–5,

Figure 39–1 The most common type of presentation drawing is a rendering, or perspective, drawing. The rendering presents an image of how the structure will look when complete. *Courtesy Paul Franks.*

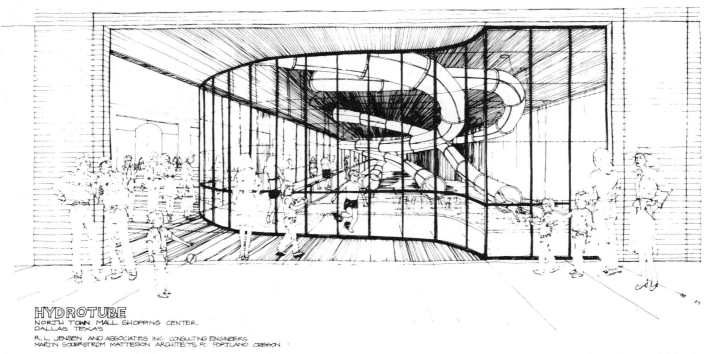

Figure 39–2 Renderings are often helpful for showing the relationships of interior spaces. *Courtesy Paul Franks, and Martin Soderstrom, Matteson Architects.*

streets, driveways, walkways, and plantings are usually shown. Although most of these items are shown on the working plot plan, the presentation plot plan presents this material in a more artistic fashion than shown on the plot plan in the set of working drawings.

Sections

Sections are often drawn as part of the presentation drawings to show the vertical relationships within the structure. As seen in Figure 39–6, a section can be used to

Figure 39–3 A presentation elevation helps show the shape of the structure without requiring the amount of time that is required to draw a rendering. *Courtesy Jeff's Residential Designs.*

show the changes of floor or ceiling levels, vertical relationships, and sun angles. Working sections show these same items with emphasis on structural materials. The presentation sections may show some structural material, but the emphasis is on spatial relationships.

▼ METHODS OF PRESENTATION

No matter the type of drawing to be made, each can be drawn on any one of several different media. The drawings are made using different materials and line techniques.

Common Media

Common media for presentation drawings include sketch paper, vellum, Mylar, and illustration board. Sketch paper is often used for the initial layout stages of presentation drawings, and some drawings are even done in their finished state on sketch paper. Most drawing materials will adhere to sketch paper, but it does not provide good durability. Drafting vellum can be a very durable drawing medium but will wrinkle when used with some materials. Vellum is best used with graphite or ink. Polyester film is often used in presentation drawings because of the ease of correcting errors. Polyester film provides both durability and a quality surface for making photographic reproductions. Ink is the most common element used on polyester film, but some types of watercolors and polyester leads can also be used.

Illustration board is extremely durable and provides a suitable surface for all types of drawing materials. When the materials are to be photographed, a white board is usually used. Beige, gray, and light blue board also can give a pleasing appearance for presentation drawings. Because of the difficulty of correcting mistakes, illustration board is not usually used by beginners. Until experience is gained, presentation drawings can be drawn on vellum and then mounted on illustration board.

When drawing on illustration board, the drafter must plan the arrangement of the entire layout carefully prior to placing the first lines. Drawings are usually drawn on sketch paper first and then transferred to the illustration

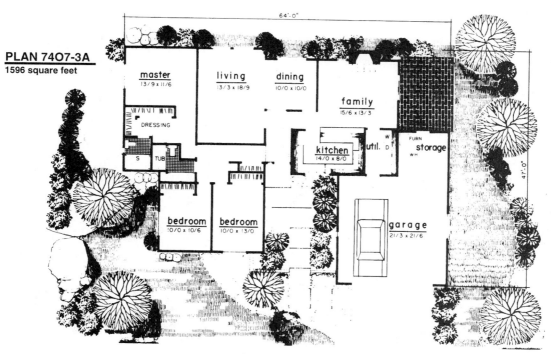

Figure 39–4 A floor plan is often part of the presentation drawings and is used to show room relationships. *Courtesy Piercy & Barclay Designers, Inc.*

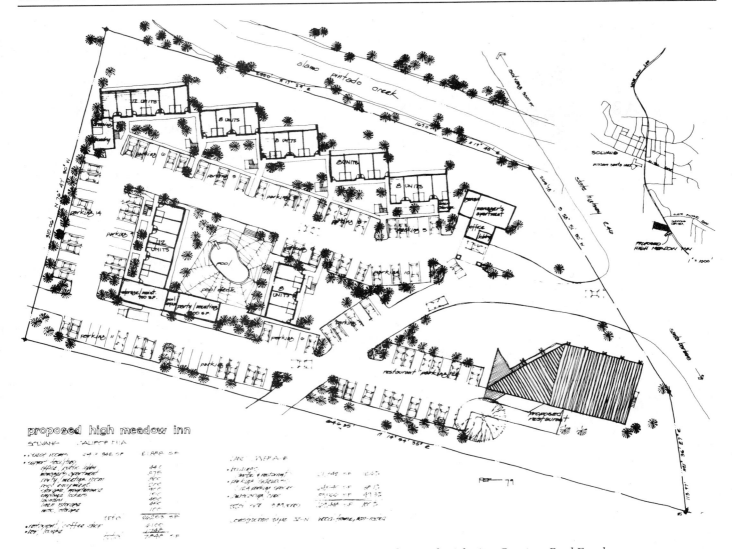

Figure 39–5 A rendered plot plan is used to show how structures relate to the job site. *Courtesy Paul Franks.*

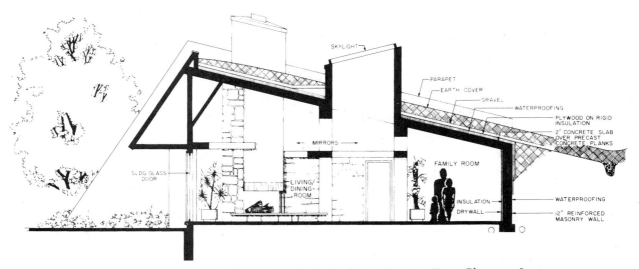

Figure 39–6 Presentation sections are used to show vertical relationships. *Courtesy Home Planners, Inc.*

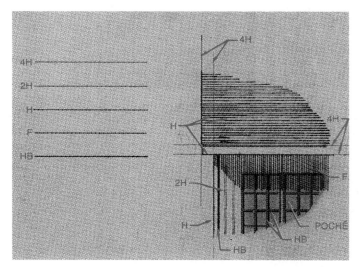

Figure 39–7 Types of graphite lines that are typically used on renderings.

which pen points will be used. Common points used for presentation work are numbers 0, 1, and 2. Many drafters and illustrators also add the use of 000, 00, and number 4 points to get a greater variation in line contrast. Figure 39–8 shows examples of various line weights available for ink work.

Color is often added to a presentation drawing by the use of colored pencils, markers, or pastels. The skill of the drafter and the use of the drawing will affect where the color is to be placed. Color is usually added to the original drawing when the illustration is to be reproduced. When the drawing is to be displayed, color is often placed on the print rather than to an original drawing. Best results are achieved when color is used to highlight a drawing rather than coloring every item in the drawing. The use of water colors is common among professional illustrators because of the lifelike presentations that can be achieved. See Figure 39–9.

board. To transfer a drawing onto the board, lightly shade the back side of the sketch with a soft graphite lead, which will act as carbon paper when the original lines are traced to transfer them onto illustration board. To avoid smearing, graphite should be placed only over each line to be transferred and not over the entire backside of the drawing.

A pleasing effect can be achieved by combining media. A rendering can be drawn using overlay principles and combining vellum or polyester film, and illustration board. For example, the structure may be drawn on illustration board and the landscaping drawn in color on polyester film. Because of the air space between the polyester film and the illustration board, an illustration of depth is created.

Drawing Materials

Common materials used to draw presentation drawings are graphite, ink, colored pencil, felt tip pens, and watercolor.

Because of the ease of correcting errors, graphite is a good material for the beginning drafter to experiment with. Common leads that are used for presentation work are 4H or 5H for layout work and 2H, H, F, HB, and 4B for drawing object lines. The choice of lead depends on the object to be drawn and the surface of the media being drawn upon. Figure 39–7 shows the different qualities that can be achieved with graphite on vellum.

Because of its reproductive qualities, ink is commonly used for presentation drawings. Ink lines have a uniform density that can be easily reproduced photographically. Careful planning is required when working with ink to ensure that the drawing is properly laid out prior to inking and to determine the order in which the lines will be drawn. With proper planning, you can draw in one area while the ink dries in another area of the drawing.

The complexity of the drawing affects the size of the pen points to be used. The technique to be used dictates

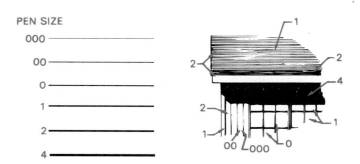

Figure 39–8 Common pen point sizes used for presentation work. The size of the drawing and the amount of detail to be shown affects the size of point to be used.

Figure 39–9 Renderings drawn with opaque water color can be reproduced easily in color or black and white. *Courtesy Home Building Plan Service, Inc.*

Line Techniques

Two techniques commonly used for drawing lines in presentation drawing are mechanical and freehand methods. Each uses the same methods for layout but varies in the method of achieving finished line quality. Figure 39–10 shows an example of a rendering drawn mechanically using a straightedge to produce lines. If a more casual effect is desired, initial layout lines can be traced without using tools. Figure 39–11 shows an example of a freehand rendering.

Figure 39–10 Renderings drawn with tools to provide straight lines. *Courtesy Piercy & Barclay Designers, Inc.*

In addition to line methods, many styles of lettering are also used on presentation drawings. Lettering may be placed with mechanical methods such as with a lettering guide or rub-ons. Many illustrations are lettered with freehand lettering similar to Figure 39–12.

▼ RENDERING PROCEDURE FOR DRAWINGS

Because of the number of steps required to make a perspective drawing, one- and two-point perspectives will be covered in the following chapters. Other drawings can be rendered as follows.

Elevations

Presentation elevations can be drawn by following the initial layout steps described in Chapter 22. The elevations are usually drawn at the same scale as the floor plan. Once the elevation is drawn with construction lines, the drafter should plan the elements to be included in the presentation. For our example, this will be limited to plants and shading. Care must be taken to ensure that the surroundings in the drawing do not overshadow the main structure.

Plants. Your artistic ability will determine how plants are to be drawn. Rub-on plants are great time savers, but

Figure 39–11 Presentation drawings are often drawn using freehand methods. *Courtesy Paul Franks.*

ROOF PLAN

FOUNDATION PLAN

PLOT PLAN

Figure 39–12 Lettering can be drawn by using mechanical methods, such as a LeRoy Lettering Guide, rub-ons, or with freehand techniques.

they should not be your only means of representing plants because of the expense and limited selection of rub-on plants.

The method of drawing plantings should match the method used to draw the structure. No matter the method, plants are typically kept very simple. The area of the country where the structure will be built determines the types of plants that are shown on the elevation. Figure 39–13 shows common types of trees that can be placed on the elevations. Be sure to use plants that are typically seen in your area.

Start the presentation elevations using the same methods used on the working drawings. Sketch the area where the plants will be placed, as seen in Figure 39–14. Plants that will be in the foreground should be drawn prior to drawing the structure with finished quality lines. Figure 39–15 shows the foreground planting in place. Once the foreground trees have been drawn, the material on the elevations can be drawn by following the same steps that were used for the working drawings. The elevation should now resemble the drawing in Figure 39–16.

Shading. Shadows are often used on presentation drawings to show surface changes and help create a sense of realism. Before shadows can be drawn, a light source must be established. A light source should be selected that will give the best presentation. The source of light will be influenced by the shape of the building. This can

Figure 39–13 Common types of trees typically placed on presentation elevations. Trees should be simple and not distract from the structure. *Courtesy Home Building Plan Service, Inc.; Home Planners, Inc.; and Piercy & Barclay Designers, Inc.*

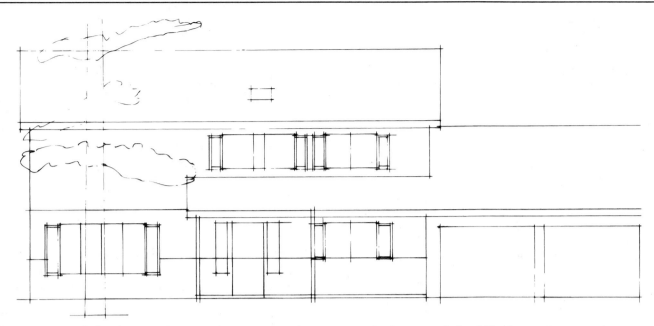

Figure 39–14 With the elevation drawn with construction lines, trees in the foreground should be drawn. Be sure to draw trees that grow in the vicinity of where the structure will be built.

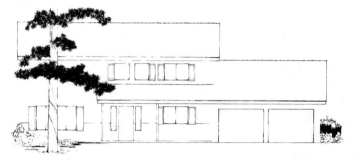

Figure 39–15 Plantings that will be in the foreground should be drawn before drawing the elevation.

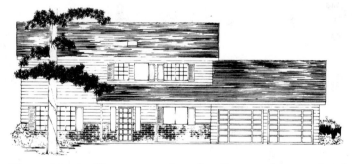

Figure 39–16 The presentation elevation can be completed using similar steps to those used to draw elevations in Chapter 22. Material that is in the background such as the chimney is often omitted for clarity.

be seen in Figure 39–17. Even though the sun may strike the structure as shown on the left, so little of the structure is left unshaded that it is hard to determine the appearance of the building. By moving the light source to the other side of the drawing, as seen at the right, more of the features of the building surface are identified.

Once the light source has been selected, the amount of shadow to be seen should be determined. Figure 39–18 shows the effect of moving the light source away from the horizon. Determine where shadows will be created and the amount of shade that will be created by projecting a line from the light source across the surface to be shaded, as seen in Figure 39–19. If you are not happy with the amount of shade, raise or lower the light source. Once the first shadow is projected, the proportions of the other shadows must be maintained. Be sure to use a very light line as you are lining out the location of shadows.

Shade can be drawn using several different methods. Figure 39–20 shows common methods of using graphite or ink. Determine the method of shading you will be using, and then lay out the area where the shadows will be placed. The final drawing procedure is to draw the small shadows created by changes in surface material. Figure 39–21 shows common areas that should receive shading.

With all material now drawn and rendered, basic materials should be specified. Complete specifications do not need to be given for the products to be used, but general specifications should be included. Figure 39–22 shows examples of notes that are often placed on the elevations.

Floor Plan

The use of the presentation floor plan will determine the scale at which it will be drawn. If the plan is to be mounted and displayed, the layout space will determine the scale to be used. When the plan will be reproduced, plans are typically drawn at a scale of either $\frac{1}{4}'' = 1'-0''$

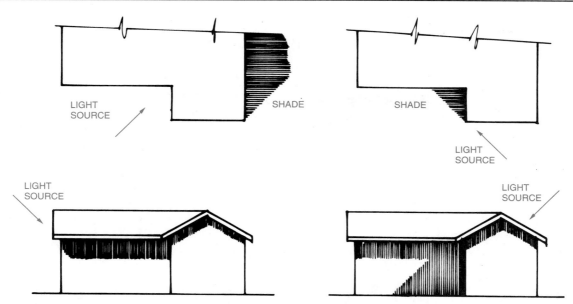

Figure 39–17 The shape of the structure influences the selection of the light source. Select a source that will accent depth but will not dominate a surface.

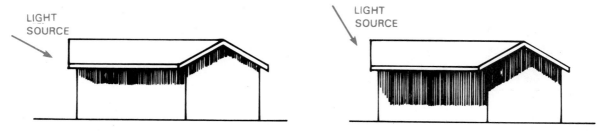

Figure 39–18 As the light source is moved above the horizon line, the depth of the shadow will be increased.

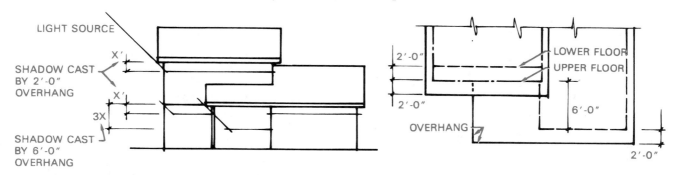

Figure 39–19 Shadows are placed by determining a light source. The depth of the shadow can be determined by the illustrator. Select a depth that does not cover important information. Once the first shadow has been drawn, keep other shadows in proportion. Keep shadow outlines very light.

or ⅛″ = 1′–0″ and then photographically reduced. The presentation floor plan can be drawn using many of the same steps that were described in Chapter 15.

Once the walls are drawn, the symbols for plumbing and electrical appliances can be drawn. Closets and storage areas are usually represented on the floor plan. These symbols can be seen in Figure 39–23. Furniture is often shown on the floor plan to show possible living arrangements. Furniture can be sketched, drawn with templates, or put in place with rub-ons. When brick, stone, or tile is used for a floor covering, it is often shown on the floor plan. A directional arrow may be shown on a floor plan to help orient the viewer to the position of the sun or to surrounding landmarks. The symbols for brick and tile floor covering have been added to the floor plan of Figure 39–24.

Written specifications on the floor plan are usually limited to room types and sizes, appliances, and general titles. Figure 39–25 shows the completed presentation floor plan for the plan that was presented in Chapter 15.

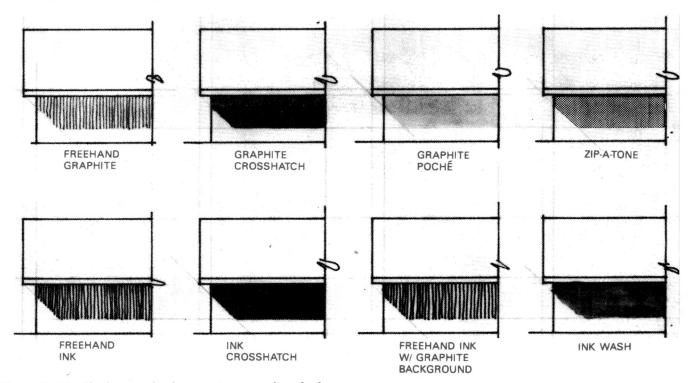

Figure 39–20 Shading can be drawn using several methods.

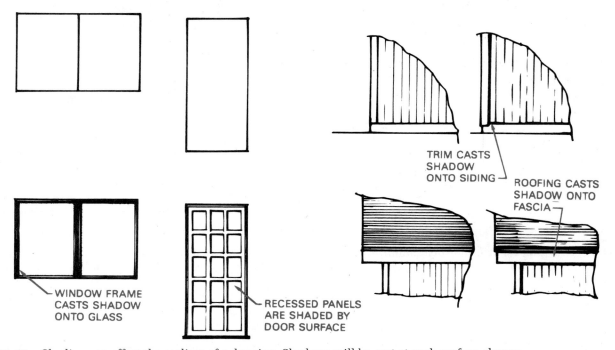

Figure 39–21 Shading can affect the realism of a drawing. Shadows will be cast at each surface change.

Plot Plan

The plot plan can be drawn by following many of the steps of Chapter 12. Once the lot and structure have been outlined, walks, driveways, decks, and pools should be drawn.

Once the structure is located, plantings and walkways are usually indicated on the plan, as shown in Figure 39–26. As with the other types of presentation drawings, basic sizes are specified on the plot plan as shown on Figure 39–27.

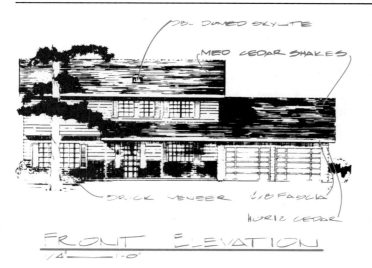

FRONT ELEVATION
1/4" = 1'-0"

Figure 39–22 Notes are added to the presentation elevation to explain major surface materials but are not as specific as the notes on the working elevation.

Figure 39–23 With the walls, doors, and windows drawn, the cabinets, stairs, and other interior features can be added.

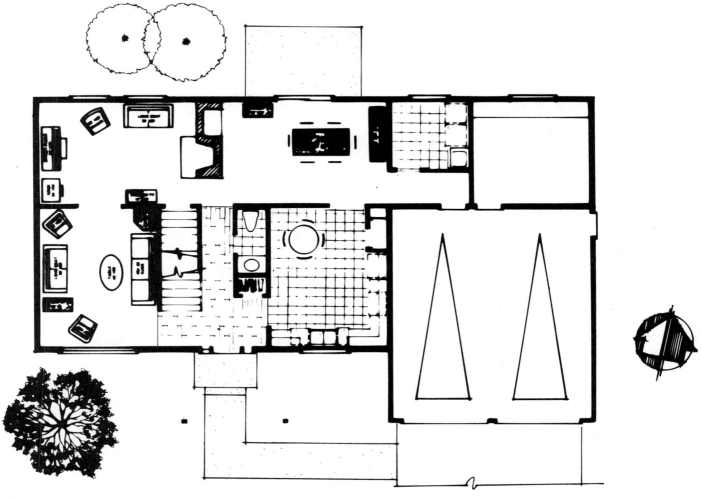

Figure 39–24 Floor materials are drawn to help clarify the design.

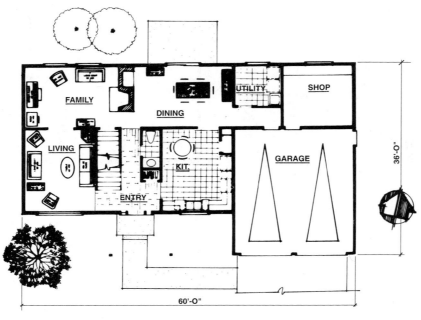

MAIN FLOOR PLAN

Figure 39–25 The completed floor plan is a useful tool for presenting the arrangement of interior space.

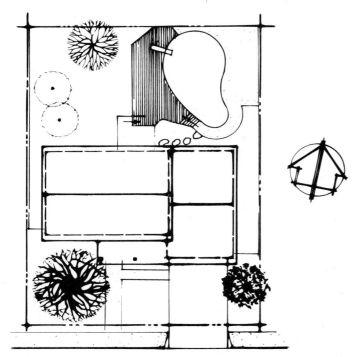

Figure 39–26 Landscaping, decks, and pools should be shown on the presentation plot plan to show how the structure will blend with its surroundings.

Sections

Section presentation drawings are made by using the same methods described in Chapters 35 and 36. Sections are usually drawn at a scale similar to the presentation floor plan. The initial section layout for the house of

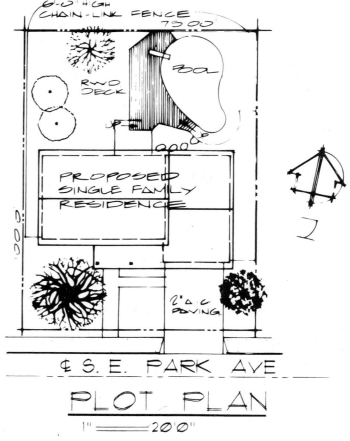

PLOT PLAN
1" = 20'0"

Figure 39–27 Major items should be specified on the plot plan to help the viewer better understand the project.

Chapter 15 can be seen in Figure 39–28. Lines of sun angles at various times of the year and lines of sight are typically indicated on the section. Room names and ceiling heights are also specified on the section. Methods of presenting this type of information can be seen in Figure 39–29.

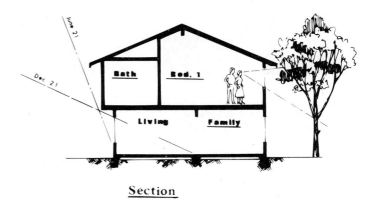

Section

Figure 39–29 The presentation section is often used to show sun angles and views. Label rooms so that the viewer will understand which areas are being shown.

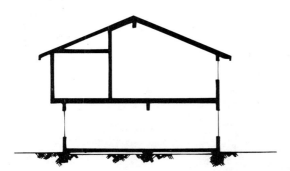

Figure 39–28 The initial layout of the section can be drawn using procedures similar to those that were used for the layout of the working section in Chapters 35 and 36.

CADD Applications

USING CADD TO MAKE PRESENTATION DRAWINGS

When CADD is used to draw elevations (Chapters 21 and 22) the material representations look so realistic that they may also be used as presentation drawings. Most architectural drafters even add such things as trees and cars to the drawings, because with CADD these items can be placed on the drawing in seconds. You simply pick a symbol from the tablet menu library and add it to the elevation in the desired location. This process is so easy that elevation drawings often contain such presentation features as landscaping and shading.

Preparing three dimensional (3-D) drawings is easy with CADD. Some of the powerful CADD packages, such as AutoCAD's AEC Architectural, allow you to generate a 3-D drawing from the floor plan layout automatically. Wall and header heights are established as you draw the floor plan. When it is complete, you can view the drawing in 3-D from any selected point in space. You can change the viewpoint until you find the view that displays the building best. Some architects have found that a combination of 3-D CADD presentation of the building and the artistic addition of landscaping features creates beautiful presentation drawings in a shorter period of time.

There are also CADD rendering programs available that let you turn 3-D line drawings into realistically shaded pictures. AutoSHADE by Autodesk is one such program. You can create a drawing that looks like a photographer's image by adjusting camera and lighting locations. You can produce both color and black-and-white drawings with this process. Figure 39–30 shows a drawing rendered with software by Big D.

Figure 39–30 Renderings using Big D software create a realistic drawing. *Courtesy Graphics Software, Inc.*

 Presentation Drawings Test

DIRECTIONS

Answer the questions with short complete statements or drawings as needed.

1. Letter your name and the date at the top of the sheet.
2. Letter the question number and provide the answer. You do not need to write out the question.

QUESTIONS

Question 39–1 What is the use of presentation drawings?

Question 39–2 What is the major difference between the information provided on a presentation drawing and a working drawing?

Question 39–3 What are some of the factors that might influence the selection of the person who will draw the presentation drawings?

Question 39–4 List and describe the major information found on each type of presentation plan.

Question 39–5 What type of drawing is best used to show the shape and style of a structure?

Question 39–6 List four different media that are often used for presentation drawings.

Question 39–7 Which drawing material is most often used for presentation drawings?

Question 39–8 What common methods are used to add color to a drawing?

Question 39–9 At what scale are sections usually drawn?

Question 39–10 What scale is typically used when a floor plan will be reproduced photographically?

CHAPTER 40

Perspective Drawing Techniques

All of the drawings that you have done so far have been orthographic projections, which have allowed the size and shape of a structure to be accurately presented in a drawing. To develop a drawing similar to what is seen when looking at the project requires the use of perspective drawing methods. Perspective drawing methods present structures very much as they appear in their natural setting. Figure 40–1 presents two elevations for a residence. This same residence can be seen in a perspective drawing in Figure 40–2, and in a photograph in Figure 40–3. As you can see, the perspective drawing resembles the photo much more than does the working drawing.

Figure 40–2   A two-point perspective is a drawing method that closely resembles a photograph. *Courtesy Jeff's Residential Designs.*

Figure 40–3   Many people who are unable to interpret orthographic drawings are able to visualize a project when they see a photograph. *Courtesy Jeff's Residential Designs.*

▼ TYPES OF PERSPECTIVE DRAWINGS

Three methods are used to draw perspective drawings: the three-, two-, and one-point methods. The three-point method is primarily used to draw very tall multilevel structures as seen in Figure 40–4 and will not be covered in this text.

Figure 40–2 is an example of a two-point perspective. Notice that all horizontal lines merge into the horizon. With this method, at least two surfaces of the structure will be seen. Two-point perspective is generally used to present the exterior views of a structure, but the technique can also be used to present interior shapes, as seen in Figure 40–5.

The one-point perspective method is used primarily for presenting interior space layouts. Figure 40–6 shows an example of an interior space drawn using the one-point

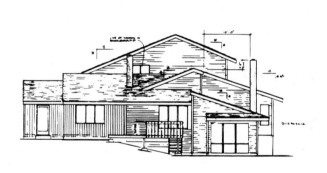

⌂ WEST ELEVATION

SOUTH ELEVATION ⌂

Figure 40–1   Orthographic drawings are used to present information regarding the construction of a structure. Many people outside the construction field have a difficult time interpreting those drawings. *Courtesy Jeff's Residential Designs.*

Figure 40–4 A three-point perspective drawing method is used to present tall structures. *Courtesy Olympia & York Properties, Inc.*

Figure 40–5 Although the two-point perspective is primarily used to present the exterior of a structure, it can also be used to present large interior areas. *Courtesy Paul Franks.*

perspective method. Notice that all lines merge into one central point. The one-point method can also be used to present exterior views such as courtyards.

Figure 40–6 The one-point perspective method is used to present interior views of a structure. *Courtesy Paul Franks.*

▼ PERSPECTIVE TERMS

In the drawing of perspective views, six terms are used frequently: ground line, station point, horizon line, vanishing points, picture plane, and the true-height line. The relationship of these lines, points, and planes can be seen in Figure 40–7.

The Ground Line (G.L.)

The *ground line* represents the horizontal surface at the base of the perspective drawing. It is this surface that gives the viewer a base from which to judge heights. The ground line is used as a base when making vertical measurements.

The Station Point (S.P.)

The *station point* represents the position of the observer's eye. Figure 40–8 shows the theoretical positioning of the station point. The station point will be used in a manner similar to a vanishing point. Since the S.P. represents the location of the observer, all measurements of width will converge at the station point.

Figure 40–9 shows the method used to find the horizontal location of a surface. A straight line is projected from each corner of the surface in the plan view to the S.P. From the intersection of each line with the picture plane, another line is projected at 90° to the picture plane down to the area where the perspective will be drawn. When a surface extends past the picture plane (P.P.), a line is projected from the S.P. to the surface and then up to the P.P., as shown in Figure 40–10.

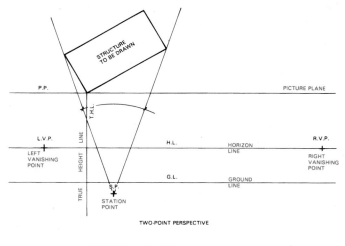

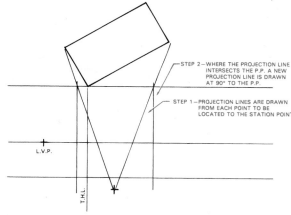

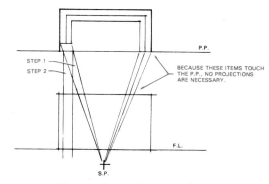

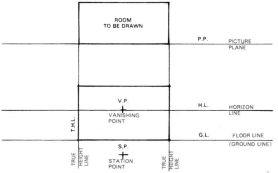

Figure 40–7 When drawing perspectives, six terms are used repeatedly: ground line, station point, horizon line, vanishing points, picture plane, and true-height line.

Figure 40–9 The station point is used to project the width of the object to be drawn.

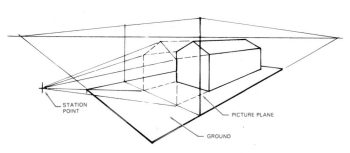

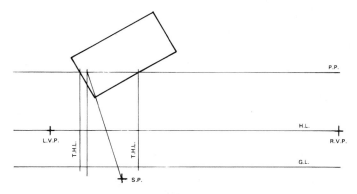

Figure 40–8 The station point represents the location of the viewer. All points on the floor plan will converge at the station point (S.P.).

Figure 40–10 When a structure extends past the picture plane, a line is projected from the station point through the point up to the picture plane. From this point on the picture plane, a line, perpendicular to the picture plane, is projected to the drawing area.

The location of the station point affects the total size of the finished drawing. Figure 40–11 shows the effect of moving the station point in relationship to the picture plane. As the distance between the S.P. and the P.P. is decreased, the width of the drawing will be decreased. Likewise, as the distance is increased between the S.P. and the P.P., the perspective will become larger.

The station point can also affect the view of each surface to be presented in the perspective drawing. The station point can be placed anywhere on the drawing but is usually placed so that the structure will fit within a cone

that is 30° wide. This can be seen in Figure 40–7. As the S.P. is moved in a horizontal direction, the width of each surface to be projected will be affected. Figure 40–12 shows the effect of moving the S.P.

The Horizon Line (H.L.)

The *horizon line* is drawn parallel to the ground line and represents the intersection of ground and sky. As the

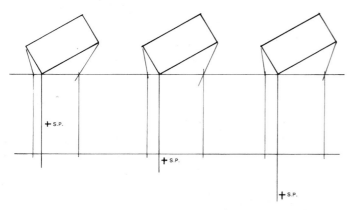

Figure 40–11 As the distance between the picture plane and the station point is increased, the width of the drawing is also increased.

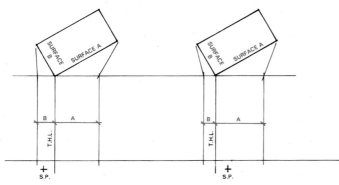

Figure 40–12 As the station point is shifted to the left or right of the true-height line, the width of each surface is affected.

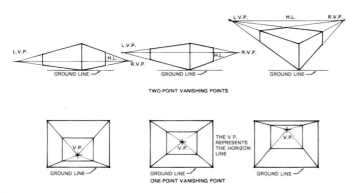

Figure 40–13 Changing the distance from the horizon line to the ground line affects the view of the structure.

distance between the horizon line and ground line is varied, the view of the structure will be greatly affected. Figure 40–13 shows the differences that can be created as the horizon line is varied. The H.L. is usually placed at an eye level of between 5 to 6′ (1500–1800 mm). It can be placed anywhere on the drawing, depending on the view desired. When placed above the highest point in the elevation, the viewer will be able to look down on the structure. Moving the H.L. above eye level can be very helpful if a roof with an intricate shape needs to be displayed.

The H.L. can also be placed below the normal line of vision. This placement allows the viewer to see items that are normally hidden by eave overhangs. Avoid placing the horizon line at the top or bottom of the object to be drawn.

Vanishing Points (V.P.)

When drawing one- or two-point perspectives, the *vanishing point* or points will always be on the horizon line. As the structure is drawn, all horizontal lines will converge at the horizon line, as seen in Figure 40–14. On a one-point perspective the vanishing point may be placed anywhere on the horizon line. Figure 40–15 shows the effect of changing the location of the vanishing point.

The vanishing points will be on the horizon line in a two-point perspective, but their location will be determined by the location of the station point and the angle of the floor plan to the picture plane. Figure 40–16 shows how the vanishing points are established. With the floor plan and S.P. established, lines parallel to the floor plan can be projected from the S.P. These lines are extended from the S.P. until they intersect the P.P. A line at 90° to the P.P. is extended from this intersection down to the horizon line. The point where this line intersects the horizon line forms the vanishing point. Be sure that you project the line from the S.P. all the way to the P.P. A common mistake in the layout is to project the angle directly onto the horizon line. This will result in a perspective that is greatly reduced in size.

Picture Plane (P.P.)

The *picture plane* represents the plane that the view of the object is projected onto. Figure 40–18 shows how the theory of the picture plane affects a perspective drawing. The picture plane is represented by a horizontal line when drawing perspectives. It is a reference line on

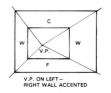

Figure 40–14 In a two-point perspective, all horizontal lines will converge at the vanishing points.

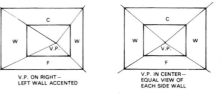

Figure 40–15 In a one-point perspective, the vanishing point may be placed anywhere. As the vanishing point (V.P.) is shifted from side to side, the shape of the walls is changed. The station point (S.P.) should always be kept below the V.P.

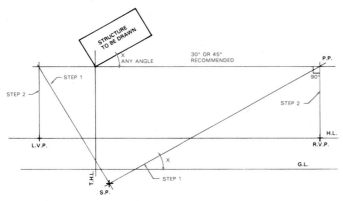

Figure 40–16 The vanishing points on a two-point perspective can be located with two easy steps. Start by projecting lines parallel to the structure from the station point (S.P.) to the picture plane (P.P.). Where these lines intersect the P.P., project a second line down to the horizon line at a 90° angle. Each vanishing point (V.P.) is located where the second projection line intersects the horizon line.

which the floor plan of the structure to be drawn is placed. Any point of the structure that is on the picture plane will be drawn in its true size in the perspective drawing. Parts of the structure that are above the picture plane will appear smaller in the perspective drawing, and parts of the structure that lie below the picture plane line will become larger. The relationship of the structure to the picture plane can be seen in Figure 40–17.

The relationship of the floor plan to the picture plane not only controls the height of the structure but also the amount of surface area that will be exposed. As a surface of the floor plan is rotated away from the picture plane, its length will be shortened in the perspective drawing. Figure 40–18 shows the effect of rotating the floor plan away from the picture plane.

The structure to be drawn can be placed at any angle to the picture plane. Using an angle of 45°, 30°, or 15° will be helpful in some of the projecting steps if you use a drafting machine. The key factor in choosing the angle should

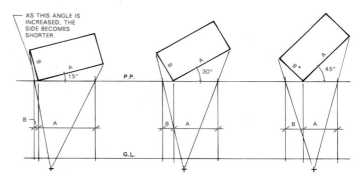

Figure 40–18 As a surface is rotated away from the picture plane (P.P.), it will become foreshortened in the perspective view.

be the desired effect of the presentation drawing. If one side of the structure is very attractive and the other side to be shown is plain, choose an angle that will shorten the plain side.

True-height Line (T.H.L.)

In the discussion of the picture plane, you learned that where the object to be drawn touches the picture plane is the only place where true height can be determined. Often in a two-point perspective, only one corner of the structure touches the picture plane line. The line projected from that point is called the *true-height line*. All other surfaces must have their height projected to this line and then projected to the vanishing point. To save time, an elevation is often used to project the height of the structure to the true-height line. This method can be seen in Figure 40–19.

Projecting true heights tends to be the most difficult aspect of drawing the perspective drawing for the beginning illustrator. *All* heights can be projected from the elevation to the T.H.L. Once projected to the T.H.L., the height must be taken to the vanishing point, as shown in Figure 40–20. If the height is for a surface that is not on the picture plane, an additional step is required, as seen in Figure 40–21. This procedure requires the height to be projected to the T.H.L., back toward the V.P., and then back to the true location.

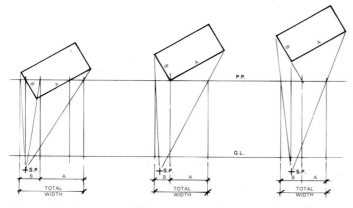

Figure 40–17 Areas of the structure that extend past the picture plane will appear enlarged. When the structure is behind the picture plane (P.P.), its size will be reduced.

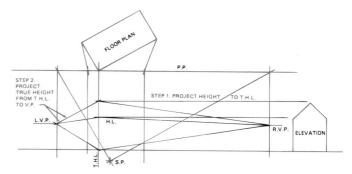

Figure 40–19 Heights from an elevation are projected to the true-height line (T.H.L.) and then to the vanishing point (V.P.).

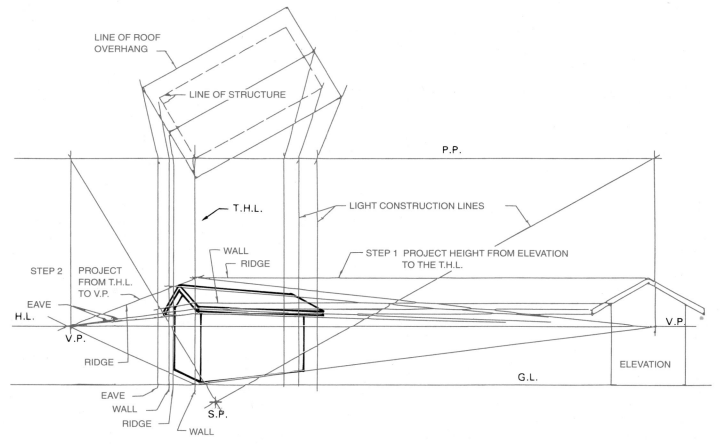

Figure 40–20 All heights are initially projected to the true-height line (T.H.L.) and from the T.H.L. to a vanishing point (V.P.). The actual height is established where the height projection line intersects the location line from the plan view.

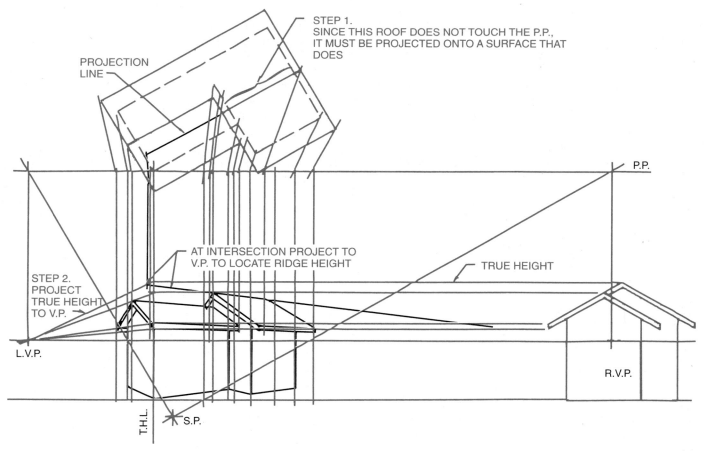

Figure 40–21 When a surface does not touch the picture plane, it must be projected to a surface that does and then to the station point.

Combining the Effects

It takes plenty of experience to help get over the initial confusion caused by the relationship of these points, lines, and planes. On your initial drawings, keep one corner of the floor plan on the P.P. Experiment with changing the drawing size once you feel comfortable with the process of projecting heights. A typical layout for your first few drawings would be to keep the plan view at either a 30° or 45° angle to the picture plane. Using an angle of less than 30° will usually place one of the vanishing points off your layout area. Keep the distance from the S.P. to the P.P. about twice the height of the object to be drawn and near the true-height line to help avoid distortion of a surface. As you begin to feel confidence with the perspective method, start by changing one variable, then another, and then change others as needed.

▼ TWO-POINT PERSPECTIVE DRAWING METHOD

There are numerous ways to draw two-point perspectives, each with advantages and disadvantages. As you continue with your architectural training, you will be exposed to many other methods. Don't be afraid to experiment with a method. Keep in mind that this chapter is intended to be only an introduction to perspective methods.

To begin your drawing, you will need a print of your floor plan and an elevation. You will find it helpful to draw the outline of the roof on the print of the floor plan. You will also need a sheet of paper about 4' (1200 mm) long and 2' (600 mm) high on which to do the initial layout work. Butcher paper or the back of a sheet of blueprint paper provides a durable surface for drawing. The use of colored pencils can also help you keep track of various lines. Some students like to draw all roof lines in one color, all wall lines in a second color, and all window and door lines in a third color. The residence that was drawn in Chapter 15 will be drawn for our example.

Setup of Basic Elements

See Figure 40–22 for steps 1 through 10.

Step 1. Draw a line at the top of your paper to represent the P.P.

Step 2. Tape a print of your floor plan on the paper so that a corner of the building touches the P.P. For this example, a 30° angle between the long side of the floor plan and the P.P. was used.

Step 3. Draw the line of any cantilevers for the upper floor.

Step 4. Draw the outline of the roof on the floor plan.

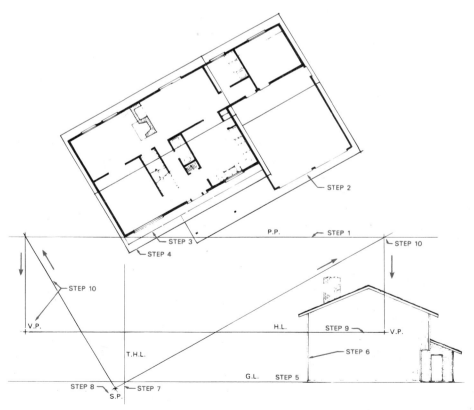

Figure 40–22 Initial layout for a two-point perspective. Establish the picture plane (P.P.), horizon line (H.L.), ground line (G.L.), true-height line (T.H.L.), station point (S.P.), and the vanishing points (V.P.s).

Step 5. Establish a ground line. The distance from the G.L. to the P.P. should be greater than the height of the structure to be drawn to avoid overlapping layout lines.

Step 6. Tape a print of the elevation to the side of the drawing area with the ground of the elevation touching the G.L.

Step 7. Establish the true-height line by projecting the intersection of the floor plan and the P.P. down at 90° from the P.P.

Step 8. Establish the station point. Keep the S.P. within an inch of the T.H.L., and a distance of about twice the height of the object to be drawn.

Step 9. Establish the horizon line at about 6 scale ft. above the ground.

Step 10. Establish the vanishing points. Start at the S.P. and project lines parallel to the floor plan up to the P.P. From the point where these lines intersect the P.P., project a line down at 90° to the H.L. The V.P.s are established where this line intersects the H.L.

You are now ready to project individual points from the floor plan down to the area where the perspective drawing will be drawn. Start by determining the maximum size of the house.

Step 11. Project lines from each corner of the roof down to the S.P. Even though the lines converge on the S.P., stop the lines just after they pass the P.P. See Figure 40–23.

See Figure 40–24 for steps 12 through 14.

Step 12. Project a line down from the P.P. to the drawing area to represent each corner of the roof.

Step 13. Project the height for the roof onto the T.H.L.

Step 14. Project the heights of the roof on the T.H.L. back to each V.P. As width and height lines intersect, you can begin to lay out the basic roof shape.

Step 15. Continue to project height and width lines for each roof surface. Work from front to back and top to bottom. When all roof surfaces are projected, your drawing will resemble Figure 40–25.

You now have the roof shape blocked out plus a multitude of lines that may have you getting tense. Don't panic. To help you keep track of the lines, use colored pencils for the walls' projection lines. See Figure 40–26 for steps 16 and 17.

Step 16. Project a line from the corner of each wall to the S.P., stopping at the P.P.

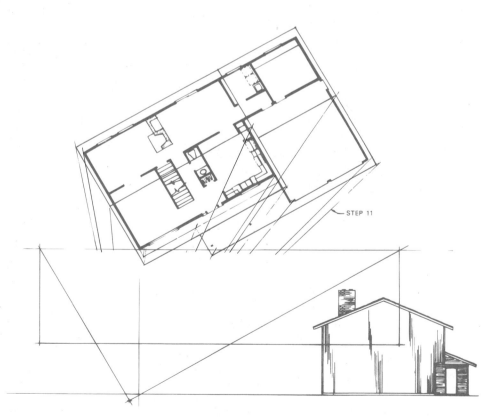

Figure 40–23 Layout of the roof shape. Project corner locations from the plan to the station point (S.P.).

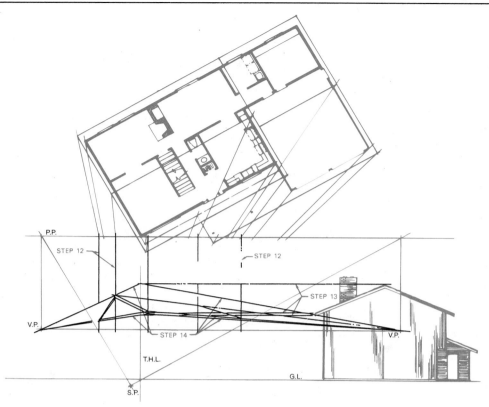

Figure 40–24 The roof shape can be drawn by projecting the width and height lines. The roof plan must be drawn on the floor plan before it can be projected to the drawing area.

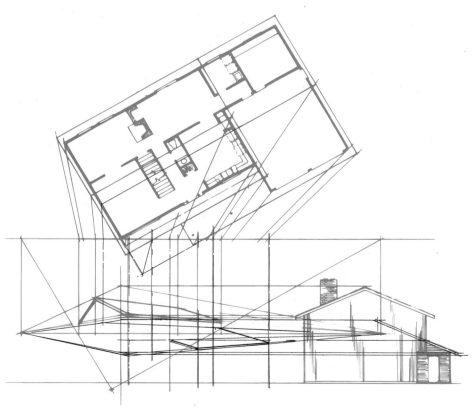

Figure 40–25 Lay out the upper roof by projecting heights from the elevation to the true-height line (T.H.L.) and then back to the vanishing points (V.P.s). Corresponding points from the picture plane (P.P.) are projected down to the drawing area.

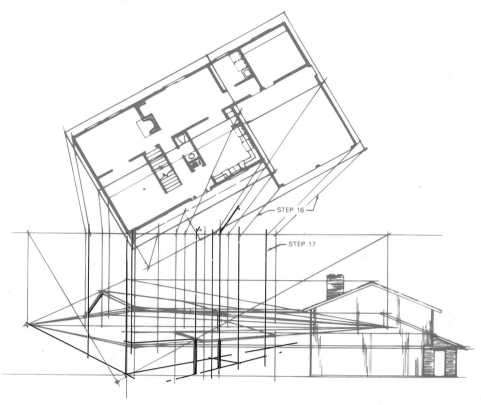

Figure 40–26   The shape of the two-point perspective can be finished by connecting the height and width lines. Work from top to bottom and from front to back.

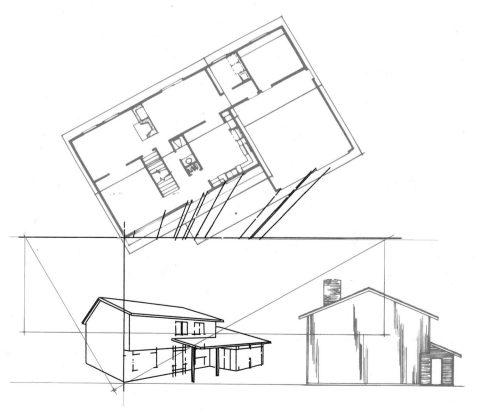

Figure 40–27   The perspective drawing is now complete. Proceed to Chapter 41 for an explanation on how to render the drawing with finished quality lines.

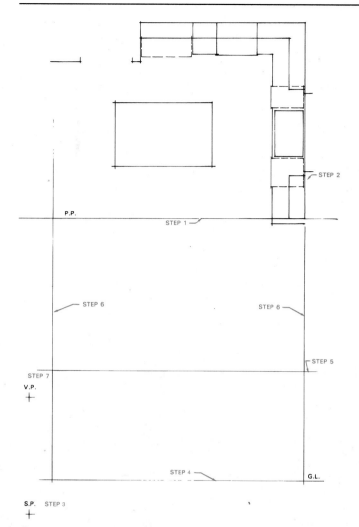

Figure 40–28 The initial layout for a one-point perspective. Start by taping a print of the floor plan to the picture plane.

Step 17. Project a line from the P.P. to the drawing area to represent each wall.

Step 18. Using a different color than was used for roof and wall projections, plot the outline for all windows and doors. Project the height from the elevations and the locations from the floor plan.

Your drawing should now resemble Figure 40–27, and is now ready to be transformed into a rendered presentation drawing. The steps for completing this process will be discussed in the following chapter.

▼ ONE-POINT PERSPECTIVE DRAWING METHOD

Much of the process for drawing the one-point perspective is similar to drawing the two-point perspective. Differences in procedure will be explained in a step-by-step process. To begin your drawing, you will need a sheet of butcher paper and a print of your floor plan. Depending

on the complexity of the area to be drawn, you may need to draw an elevation. If you are working on a simple interior, you may be able to mark the heights on the T.H.L. rather than drawing an elevation. Depending on the size of the area you will be drawing, the floor plan may need to be enlarged. Often your drawing will seem very small if you start with a ¼″ = 1′–0″ floor plan. Because kitchens are such a common item to be drawn in presentation drawings, the kitchen from the residence in Chapter 15 will be drawn for our example. See Figure 40–28 for steps 1 through 7.

Step 1. Establish the picture plane.

Step 2. Tape a print of the floor plan to the P.P. Remember to redraw the floor plan at a larger scale.

Step 3. Locate the S.P. using the same consideration as with the two-point method.

Step 4. Establish the ground line. In a one-point perspective, this line represents the floor.

Step 5. Measure up from the ground line, and establish the ceiling line. Notice that both lines will be parallel. All lines that are parallel to the P.P. will remain parallel in the drawing.

Step 6. Establish the width of the drawing by projecting the points where the walls intersect the P.P. down to the drawing area.

Step 7. Establish a vanishing point. The V.P. can go at any height but should be kept above the S.P.

See Figure 40–29 for steps 8 through 10.

Step 8. Project lines from each of the room corners to the V.P. to begin establishing the side walls, floor, and ceiling.

Step 9. Project lines from each room corner of the plan to the S.P. Extend the lines to just past the P.P. to avoid construction lines in the area where the perspective will be drawn.

Step 10. Lay out the floor, walls, and ceiling. These can be determined where the wall line intersects the lines that were projected from the floor and ceiling to the V.P.

See Figure 40–30 for steps 11 through 14.

Step 11. Measure the height of the cabinets on the cabinet elevations of Chapter 23, and project these heights on either of the T.H.L. In a one-point perspective, each side of the drawing is a T.H.L.

Step 12. Lay out the shape of the base cabinet. This can be done by projecting the width from the floor plan and measuring the height on the T.H.L. Project the height back to the V.P.

Step 13. Lay out the cabinet on the back wall by projecting the intersection point on the floor plan down onto the perspective.

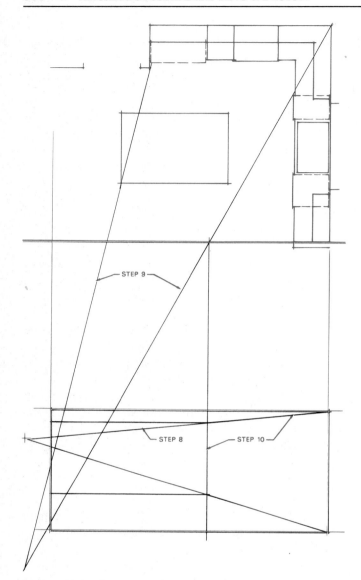

Figure 40–29 Establishing the walls, floor, and ceiling.

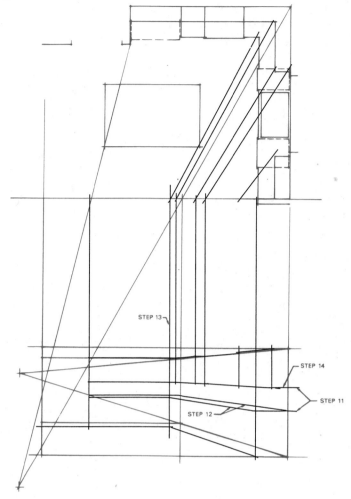

Figure 40–30 Lay out the basic cabinet shapes by projecting each location from the floor plan.

Step 14. Lay out the upper cabinet following the same procedure used to lay out the base cabinet.

See Figure 40–31 for steps 15 through 17. Similar to the two-point perspective, your drawing is probably starting to get cluttered with projection lines. Don't hesitate to use multiple colors for projection lines.

Step 15. Lay out the widths for all cabinets and appliances by projecting their widths to the S.P. and then down to the drawing. To determine heights, mark the true height on the T.H.L. and then project the height back to the V.P.

Step 16. Block out any windows.

Step 17. Block out soffits, skylights, or other items in the ceiling.

You now have the basic materials drawn for the one-point perspective. Your drawing should resemble the drawing in Figure 40–32. Chapter 41 will introduce you to the basics of rendering the perspective drawing.

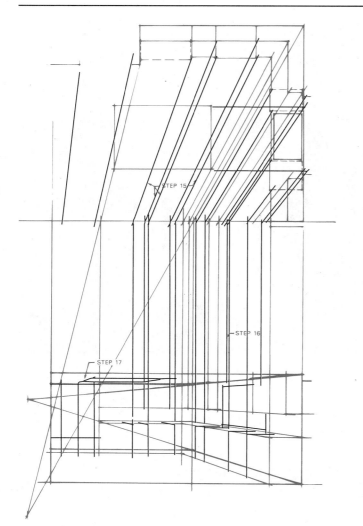

Figure 40–31 Block out individual cabinets. Use drawings of the cabinet elevations to establish each location.

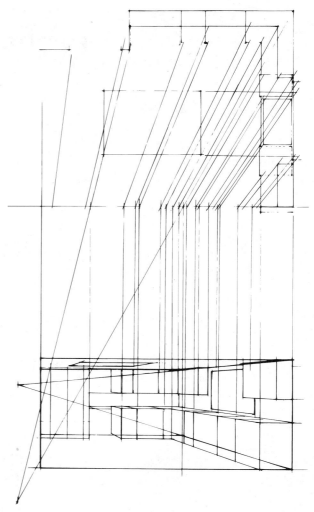

Figure 40–32 The completed one-point perspective ready to be rendered. See Chapter 41 for an explanation of how to render the drawing with finished-quality lines.

▼ CADD Applications

USING CADD TO MAKE PERSPECTIVES

Preparing perspective drawings is easy with CADD. Some of the powerful CADD packages allow you to automatically generate a perspective drawing from the floor plan layout. Wall and header heights are established while you draw the floor plan. When the floor plan is complete you can view the drawing in perspective from any selected point in space. You can change the view point until you find the perspective view that displays the building best. When you move the view point, the CADD system automatically adjusts the picture plane, station point, and vanishing points, but the ground plane always remains the same. Interior perspectives may be drawn in the same manner, or you can create the drawing directly in 3-D. Some systems have a split screen that allows you to draw in plan view on one part of the screen while generating a 3-D view on the other. The perspective of a residence drawn using CADD is shown in Figure 40–33.

Figure 40–33 CADD perspective drawing of a multifamily housing project. *Courtesy Soderstrom Architects PC.*

CHAPTER

40 Perspective Drawing Techniques Test

DIRECTIONS

Answer the questions with short, complete statements or drawings as needed.

1. Letter your name and the date at the top of the sheet.
2. Letter the question number and provide the answer. You do not need to write out the question.

QUESTIONS

Question 40–1 What are the major types of perspective drawings?

Question 40–2 On a two-point perspective, where will all horizontal lines meet?

Question 40–3 Explain the effect of moving the station point.

Question 40–4 What are the most common uses for a one-point perspective?

Question 40–5 Sketch the method of locating the vanishing points for a two-point perspective and describe each step that is used.

Question 40–6 How does the placement of the floor plan affect the layout of a perspective?

Question 40–7 Where is the actual height of the structure shown in the perspective?

Question 40–8 How are three-point perspectives typically used?

Question 40–9 What drawings are required to draw a two-point perspective?

Question 40–10 What represents the horizontal surface at the base of the perspective?

DIRECTIONS

1. Lay out the following problems on butcher paper.
2. When complete, trace the object onto vellum using either ink or graphite.

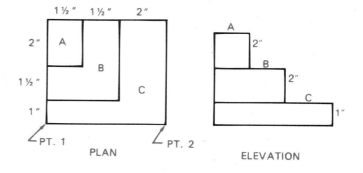

PROBLEMS

Problem 40–1 Draw the following object as a perspective drawing.
1. Enlarge 3×.
2. Place the G.L. 7″ away from the P.P.
3. Set the H.L. 5½″ above the G.L.
4. Set the S.P. ⅜″ below the G.L. and ¼″ to the left of the T.H.L.
5. Set point 2 on the P.P. with the line 1–2 at a 30° angle to the P.P.
6. Lay out the perspective drawing. Leave all construction lines.
7. Darken all object lines using graphite.
8. Repeat steps 1 through 7 using an H.L. that is 3″ above the G.L.
9. Repeat steps 1 through 7 using an S.P. that is ⅜″ below the ground and ¼″ to the right of the T.H.L.

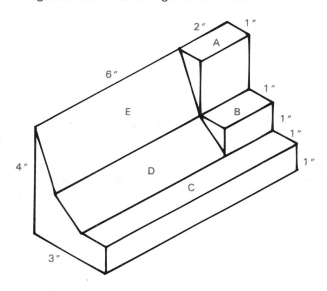

Problem 40–2 Make a perspective drawing of the following object. Place the plan at a 45° angle to the P.P. Select your own V.P. and S.P. Leave all construction lines. Darken all object lines. Draw the object a second time at 15° to the P.P.

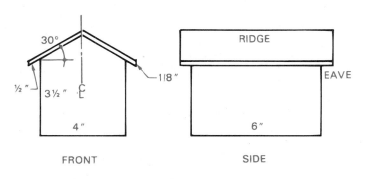

Problem 40–3 Draw a perspective drawing of the following object. Select a horizon line that will not accent the height. Select an angle for the floor plan placement that will equally accent each surface. Set point 1 on the P.P. Trace this drawing onto a sheet of vellum using ink.

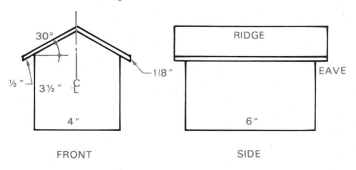

Problem 40–4 Make two perspective drawings of the following object.
1. Set point 1 on the P.P.
2. Draw perspective A with the H.L. above the ridge.
3. Draw perspective B with the H.L. below the eave.
4. Use the same S.P. for each drawing.
5. Trace your drawings onto a sheet of vellum using ink.

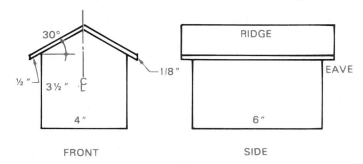

Problem 40–5 Draw a one-point perspective of the following object at a scale of ½″ = 1′–0″. Draw a floor plan on the P.P. Set the G.L. 8″ below the P.P. Set the V.P. in the center of the drawing at a point above B. Set the S.P. 4″ directly below the V.P. Darken all object lines using graphite. Leave all construction lines.

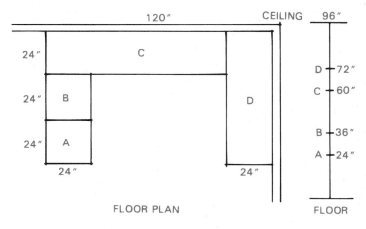

Problem 40–6 Draw a one-point perspective of the following object at a scale of ½″ = 1′–0″. Draw the floor plan on the P.P. Establish your own G.L. and S.P. Set the V.P. in the left third of the drawing, at a point above D. Darken all object lines using graphite. Leave all construction lines. Redraw this object using steps 1 through 3. Set the V.P. at any point above the G.L. in the right third of the drawing. Darken all object lines using graphite. Leave all construction lines.

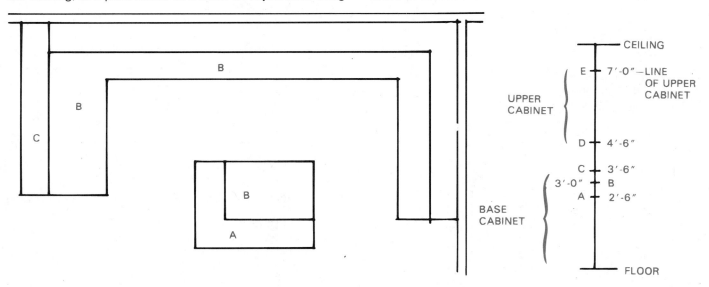

Problem 40–7 Using a print of your floor plan and elevation, draw a two-point perspective of the residence that you have drawn for your project. Select a view that will not accent one surface greatly over the other. Leave all lines as construction lines. This drawing will be used in the next chapter for your rendering project.

Problem 40–8 Using a print of your floor plan and cabinet elevations, draw a one-point perspective of one of the rooms. Do all work on butcher paper. Leave all lines as construction lines. This drawing will be used in the next chapter for your rendering project.

CHAPTER 41

Rendering Methods for Perspective Drawings

Figure 40–27 showed an example of a perspective drawing. It presents a better view of the structure than an orthographic drawing, but it still does not appear very realistic. In order to transform the perspective drawing into a realistic presentation of the structure, the drawing must be rendered. You have already been exposed to the process of rendering other types of presentation drawings. Similar materials and media are used when rendering perspective drawings. A rendered perspective drawing typically shows depth, shading, reflections, texture, and entourage, or surroundings.

More than any other type of presentation drawing, a rendered perspective can be an intimidating project for the beginning student. On other forms of presentation drawings, you will be able to use standard drafting tools to form the lines. When rendering perspective drawings, you will need to rely on your artistic ability. Don't panic if you feel you will never be able to draw a rendering. The techniques needed require lots of practice, but they can be mastered. Start by developing a scrapbook of renderings done by professional illustrators in various materials. Also collect photos of people, trees, and cars. Until you gain confidence in your artistic ability, you can use one of the photos or renderings from your scrapbook as a guide, or, if necessary, you could even trace it.

Your renderings will also be enhanced if you will take the time to notice and sketch the things around you. Pay special attention to trees and shadows that are cast by and onto buildings. You will notice that, depending on the location and intensity of sunlight, surfaces take on different appearances. Try to reproduce these effects in your sketches.

▼ PRESENTING DEPTH

Figure 41–1 provides an example of how depth and surface texture can be shown. The perspective drawing itself shows the illusion of depth by presenting three surface areas in one view. On a two-point perspective, the feeling of depth is often shown at such areas as windows, doors, fascias, surface intersections, and trim boards. The feeling of depth can be created by using offsetting lines and

by using different line weights. Figure 41–2 shows examples of how depth can be created.

The same guidelines for showing depth should be used on the one-point perspective. Cabinet doors, drawers, and appliances typically project past the face of a cabinet and should be shown on the perspective drawing. Figure 41–3 shows an example of how depth can be shown on a one-point perspective.

Figure 41–1 A rendered perspective drawing can show depth and texture, creating a realistic look to the drawing. *Courtesy Home Building Plan Service.*

Figure 41–2 Depth can be shown at surface changes by using thicker lines to represent shadows.

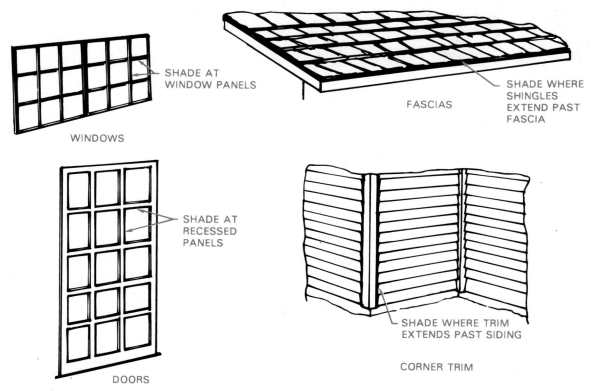

Figure 41–3 Depth can be seen in this one-point rendering. *Courtesy Shakertown Corp.*

▼ SHADOWS

Shadows provide an excellent method of presenting depth. Shadows are cast anytime a surface blocks light from striking another surface. Figure 41–4 shows an example of the effect of shadows on a structure. The principles for projecting shadows in perspective are similar to the methods used in orthographic drawings. Before any shadows can be projected, a light source must be established using the same guidelines that were used for presentation elevations. Keep the light source in a position so that major features of the structure will not be hidden.

Shadows are usually projected at an angle of 30°, 45°, or 50°. To save time in projecting, use a standard triangle or drafting machine setting. Figure 41–5 shows the effects of changing the angle of the light source.

Figure 41–4 Shades and shadows are used to show depth. *Courtesy Piercy & Barclay Designers, Inc. C.P.B.D.*

Two principles affect the layout of shadows if the light source is drawn parallel to the picture plane.

1. A vertical wall or surface casts a shadow on the adjacent ground or horizontal surface in the *direction in which the light rays are traveling.*
2. The shadow on a plane, caused by a line parallel to the plane, forms a shadow line *parallel to the line.* Both converge at the same vanishing point.

Both principles can be seen in Figure 41–6. The vertical lines cast their shadows according to the first rule, and the horizontal lines cast a shadow according to the second rule. These same two principles can be applied as the shadows are cast onto other parts of the structure, as seen in Figure 41–7. When a structure has an overhang, the same principles can be applied to project the shaded areas.

You will often be required to project the height of a horizontal surface onto an inclined surface. This can be done by following the procedure shown in Figure 41–8. A similar problem is created when trying to project the shadow of an inclined surface onto a flat surface. Figure 41–9 shows how the shadow created by a roof can be projected onto the ground.

One of the problems of casting shadows parallel to the picture plane is that this method requires one of the two wall surfaces to be in the shade. Material can still be seen when it is in the shade, but it will not be as prominent as

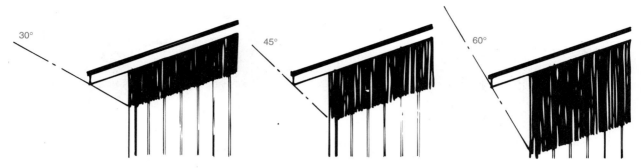

Figure 41–5 As the angle of light becomes steeper, the length of shadow becomes longer.

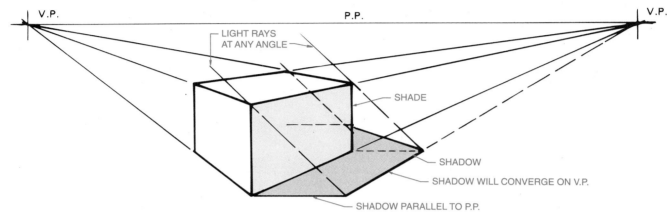

Figure 41–6 A vertical surface casts a shadow on a horizontal surface in the direction that the light rays are traveling. The shadow that is created is parallel to the surface that created it and converges at the same vanishing point.

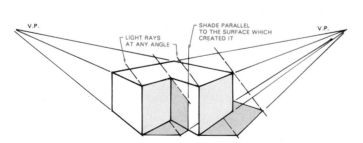

Figure 41–7 Shadows cast onto a vertical wall are projected using the same methods used to project shadows onto the ground.

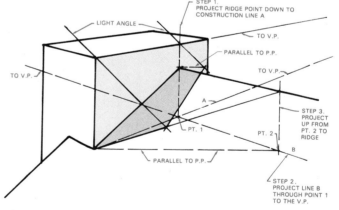

Figure 41–8 A method of projecting shadows onto an angled surface.

the same material in direct sunlight. Shading film, ink wash, and graphite shading are the most common methods of showing shaded areas. These same materials are used to show shadows, but they must be applied using a denser pattern. Methods similar to those for drawing shadows can be used on perspective drawings that were used on the rendered elevation.

If you would like to have both surfaces of a structure in direct sunlight, the light source will need to be moved so that it is no longer parallel to the picture plane. Figure 41–10 shows a rendering drawn with the light source behind the viewer so that both sides are in sunlight.

▼ REFLECTIONS

Reflections in perspective drawings are typically caused by water or glass. When casting a reflection caused by water, the point to be projected will fall directly below that point and the depth of the reflection will be equal to the distance that the point is above the water's surface. Figure 41–11 shows how these two principles can be used to project a reflection. Figure 41–12 shows another method for rendering water.

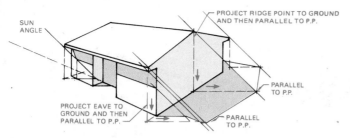

Figure 41–9 Projecting shadows from an inclined roof onto a horizontal surface.

Figure 41–10 The sun is often placed behind the viewer's back so that shadows will be cast while each surface remains in the sunlight. *Courtesy Piercy & Barclay Designers, Inc.*

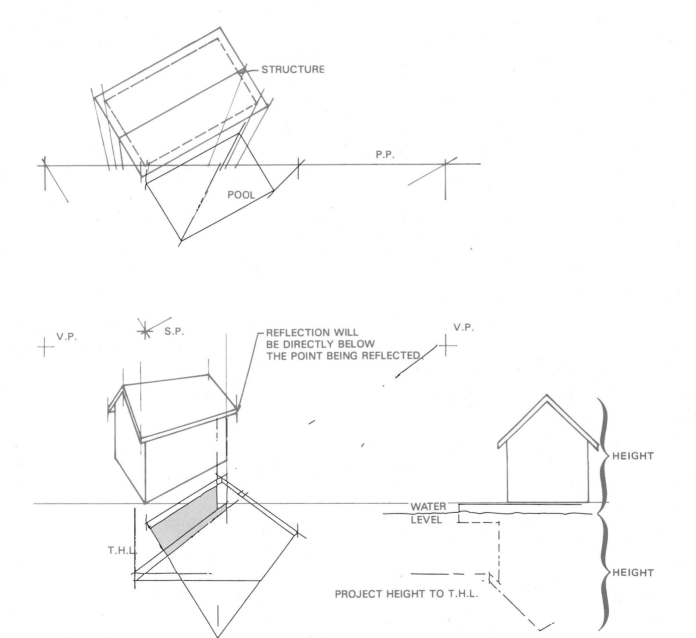

Figure 41–11 Projecting reflections on water. The reflected image will fall directly below the point to be projected. The depth of the reflection will be equal to the distance that the point is above the reflecting surface.

Figure 41–12 Reflections on water. *Courtesy Home Planners, Inc.*

Large areas of glass in a perspective drawing can often appear very dull unless rendered. Figure 41–13 shows common methods of rendering glass. Glass is often rendered solid black with small areas of white to indicate reflections. Glass can also be left unrendered, and all material in the background can be shown. A third alternative is to use an ink wash, a combination of water and ink applied with a small paintbrush that gives a gray tone rather than the solid black color of ink. Small areas of reflection are often left white or shaded with a soft graphite.

▼ TEXTURE

An important part of the rendered perspective is to show the texture of the materials used to build the structure. Most materials can be drawn on the perspective drawing in styles similar to those used to draw an elevation. Al-

though the material will appear similar, the actual lines used to represent the material should be different on a perspective drawing.

Roofing Materials

Shingles, tiles, and metal panels are the major types of materials that will be rendered on the roof. Horizontal construction lines that represent each course of shingles or tiles are usually drawn using tools and then traced using freehand techniques, as shown in Figure 41–14. Bar tiles are represented using methods similar to shingles. Curved tiles can be drawn as shown in Figure 41–15.

Siding or Paneling

Siding will be drawn using either horizontal or vertical lines. If horizontal siding is used, the lines representing the siding will merge at the vanishing points. When horizontal lap siding intersects another surface, a sawtooth shade pattern will result.

Vertical siding can be drawn using the same methods used on the rendered elevation. On some types of siding, both thickness and texture must be drawn. Figure 41–16

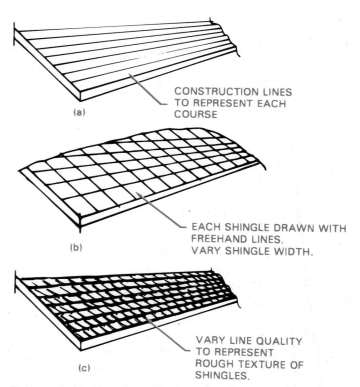

Figure 41–14 Drawing shingles. Shingles can be drawn by lightly drawing lines to the vanishing point (V.P.) to represent each course. Lines can then be drawn to represent each shingle. With each shingle drawn, the original horizontal lines should then be redrawn freehand to represent the rough texture of shingles.

Figure 41–13 Glass can be rendered using several techniques. *Courtesy Home Building Plan Service.*

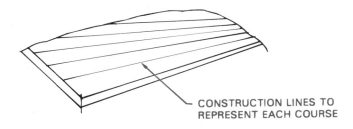

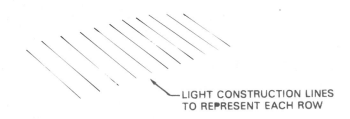

CONSTRUCTION LINES TO
REPRESENT EACH COURSE

LIGHT CONSTRUCTION LINES
TO REPRESENT EACH ROW

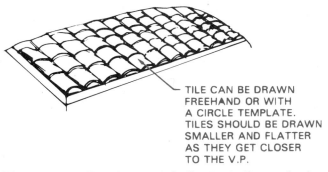

TILE CAN BE DRAWN
FREEHAND OR WITH
A CIRCLE TEMPLATE.
TILES SHOULD BE DRAWN
SMALLER AND FLATTER
AS THEY GET CLOSER
TO THE V.P.

Figure 41–15 Drawing curved tiles is similar to the layout procedure used when drawing shingles. Tiles should be drawn smaller as they get closer to the vanishing point (V.P.).

shows common methods of drawing siding. Paneling is usually drawn using a combination of vertical straightedge and freehand lines to represent the wood grain.

▼ ENTOURAGE

Entourage, the term for the surroundings of a building, consists of ground cover, trees, people, and cars. The entourage helps to create an attractive, realistic drawing. In addition it helps define the scale of the drawing. The illustrator must be very careful that the entourage is not allowed to compete with the structure. The surroundings should accent rather than draw attention away from the building. Figure 41–17 shows an example of entourage that blends well with the structure. Notice that rocks, plants, and trees help present the residence as it might appear but do not draw attention away from the actual structure.

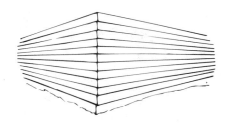

HORIZONTAL SIDING
EXTENDS TO EACH V.P.

VERTICAL SIDING IS
REPRESENTED BY
VERTICAL LINES WITH
RANDOM WEIGHT.
SPACING BETWEEN THE
LINES DIMINISHES AS
THE LINES GET CLOSER
TO THE V.P.

Figure 41–16 Drawing horizontal or vertical siding.

Figure 41–17 Entourage should be used to help present the structure in a lifelike setting. *Courtesy Piercy & Barclay Designers, Inc. C.P.B.D.*

Plants

Plants can be drawn by hand or applied as rub-ons. A wide variety of rub-on trees and plants are available for use on perspectives. If rub-ons are to be used, they should be placed on the drawing after the line work has been completed. This will help keep the plants from being peeled off the illustration as you draw. Care must be taken to outline the shape of the plant so that the lines of the structure are not placed where the plants will go. If a diazo copy will be made, entourage can be placed on the back of the drawing. If the drawing will be photographically reproduced, the rub-ons must be placed on the front side of the drawing material.

Although rub-on plants may be convenient, every drafter should know how to draw entourage. Freehand entourage can best be drawn by spending time observing and sketching various kinds of trees and plants. The method used to draw the entourage should be consistent with the method used to draw the structure. A wide variety of plants can be seen by studying the rendered perspective drawings presented throughout the text. Practice drawing trees and plants that typically grow in your area until you gain confidence in your work. Keep trees in proportion to the structure. Heights for plants can be projected using methods similar to those used for the structure.

Another common element that should be included when drawing plantings is a method of drawing a ground cover such as grass. Broad strokes of a soft lead are often used to represent grass and the shape of the ground as shown in Figure 41–18. Ground cover provides a method to fill in the white space around a structure without distracting the viewer from it.

People

Many illustrators include people as part of their entourage. If properly drawn, people can help set the scale of the structure and give depth to the drawing. The heights of the people in your drawing should be projected from the vanishing point to the area where they will be placed, so that they will be consistent with the height of the structure. Figure 41–19 shows how people can be drawn in a rendering. As you study the people shown in the renderings throughout this text, you will notice that some of the figures have been outlined, while others have been finely detailed.

The method used to draw people should be consistent with the scale of the drawing, and their location should not detract from the main focus of the drawing. Depending on your artistic ability, you may want to collect from magazines and clothes catalogs illustrations of people of various sizes posed in different positions. Such pictures could then be traced onto your rendering.

Furniture

For an interior rendering, furniture and people together help to provide a reference for scale. Furniture should be kept simple and consistent with the line quality of the drawing. Figure 41–20 shows how the use of furniture, people, and interior plants can help draw the viewer into the drawing to create a sense of realism rather than showing just a drawing of a room.

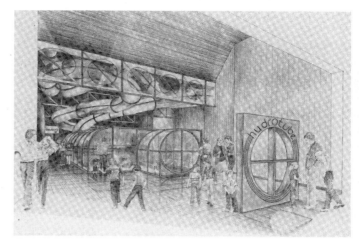

Figure 41–19 People are often placed within a drawing to give the viewer a sense of belonging. *Courtesy Paul Franks and Gary M. Larsen, Architects.*

Figure 41–20 Furniture helps show how a space can be used. *Courtesy Paul Franks.*

Cars

Cars can be a very attractive addition to the rendering but can easily overshadow the structure. Be careful not to detail the car so completely that the viewer is thinking about the car rather than the structure. Care must be taken in the placement of the car so that it blends with the surroundings. Drawing people near cars helps tie the scale of the entourage.

Content

You've been exposed to many of the elements that are typically placed in renderings, but you still must consider which elements should be used on your rendering and where these elements should be placed. The entourage should be consistent with the purpose of the drawing and not obscure the structure. The following three guidelines

Figure 41–18 Ground cover will often show individual plants. *Courtesy Home Building Plan Service.*

are helpful in planning your drawing in order to maintain a proper relation between the structure and the entourage:

1. Use only entourage that is necessary to show the location, scale, and usage of the structure.
2. Draw the entourage with only enough detail to illustrate the desired material.
3. Never obscure structural elements with entourage.

▼ PLANNING THE DRAWING

Before rendering a presentation drawing, do several sketches to determine the effectiveness of a layout. Consideration should be given to the placement of the building, people, landscaping, and contrasting values that will be created by the layout. Figure 41–21 is an example of a rendering that has made good use of contrast to help present the structure. Entourage has been used to create black and white surfaces against which the structure can be contrasted.

The type of drawing that you are rendering should affect the type of balance that is used in your presentation. When making presentation drawings of residential projects, an informal method of balance will typically be most appropriate. Figure 41–22 shows an example of a presentation drawing using formal balance. Notice that the entourage is used as a frame to present the structure.

Figure 41–23 shows an example of informal balance with the mass of the structure and the entourage offsetting each other.

Figure 41–22 Formal balance in presentation can be seen in the location of trees used to frame the residence. *Courtesy Home Building Plan Service.*

Figure 41–23 Informal balance can be seen in this drawing as the trees are used to offset the mass of the structure. *Courtesy Home Building Plan Service.*

▼ TWO-POINT PERSPECTIVE RENDERING

See Figure 41–24 for steps 1 through 3.

Step 1. Tape your perspective layout to your drawing surface. If you are using graphite and vellum, cover your layout with a thin sheet of sketch paper. This will keep the layout from being traced onto the back of your vellum as you trace the layout.

Figure 41–21 Entourage should be used to accent the structure. *Courtesy Home Building Plan Service.*

Step 2. Lay out areas in the foreground where entourage will be placed.

Step 3. Lay out areas that will be shaded.

See Figure 41–25 for steps 4 through 6.

Step 4. Draw trees that will be placed in the foreground.

Step 5. Draw people and cars that will be placed on the drawing. On residential renderings, people and cars are not usually drawn.

Step 6. Draw entourage to help tie the structure to the ground.

Step 7. Render the structure by first drawing material in the foreground and then work into the background. See Figure 41–26.

See Figure 41–27 for steps 8 through 10.

Step 8. Draw major areas of shades and shadows.

Step 9. Highlight areas in materials that will cast shadows.

Step 10. Letter the job title or owner's name in an appropriate place.

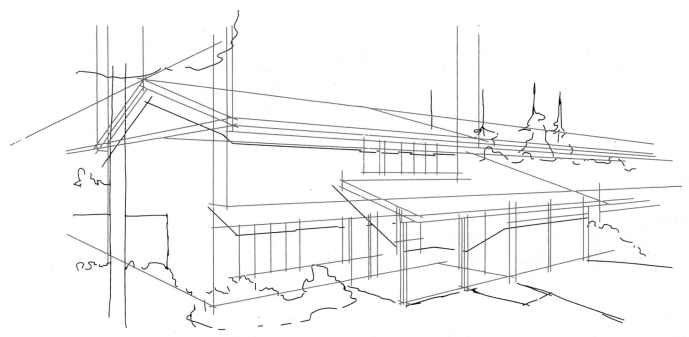

Figure 41–24 With the perspective drawn with construction lines, entourage should be sketched in. Notice that features that are farthest from the viewer are often omitted to enhance the presentation.

Figure 41–25 Draw the foreground entourage prior to drawing the structure. Background and ground cover can also be added.

Figure 41–26 The structure can be rendered once the foreground entourage is drawn. Work from the top to bottom, and from front to back.

Figure 41–27 Highlights are added to surface changes.

▼ ONE-POINT PERSPECTIVE RENDERING

The one-point perspective can be drawn by following the same procedure that was used for a two-point perspective. The entourage that will be shown for an interior view will be different from an exterior illustration; however, the interior rendering must be drawn in the same sequence as an exterior view. That is, draw the foreground entourage before drawing the cabinets or furniture in the background. See Figure 41–28.

Figure 41–28   The perspective shown in Figure 40–32 is rendered to show depth and texture.

CHAPTER
41   Rendering Methods for Perspective Drawings Test

DIRECTIONS

Answer the questions with short, complete statements or drawings as needed.

1. Letter your name and the date at the top of the sheet.
2. Letter the question number and provide the answer. You do not need to write out the question.

QUESTIONS

Question 41–1   What two principles affect the layout of shade and shadows?

Question 41–2   Sketch an example of how shakes appear in a rendering.

Question 41–3   List three steps that could help improve a drafter's rendering technique.

Question 41–4   At what angles are shadows usually projected? Explain.

Question 41–5   If you would like to have both surfaces in a two-point perspective in direct sunlight, where would you place the light source?

Question 41–6   What is entourage?

Question 41–7   List three guidelines for planning entourage.

Question 41–8   Sketch four examples of how glass can be rendered.

Question 41–9   Sketch examples of how depth can be shown at a window.

Question 41–10   Describe the elements to be considered in planning a rendering.

PROBLEMS

Problem 41–1   Using your two-point perspective drawing from Problem 40–7, render the residence using either graphite or ink. Draw entourage by hand; do not use rub-ons. Use the examples found in this text as guidelines for drawing your entourage.

Problem 41–2   Using your one-point perspective drawing from Problem 40–8, render the interior illustration using either graphite or ink. Draw entourage freehand; do not use rub-ons. Use the examples found in this text as guidelines for drawing your entourage.

GENERAL CONSTRUCTION SPECIFICATIONS AND SUPERVISION

WINDOW NOTES:

1. ALL WINDOWS TO BE U-.40 MIN.   PROVIDE ALT. BIDD FOR U=.54 IF FLAT CEILINGS ARE INSULATED TO R=49

2. ALL WINDOWS TO BE ANDERSON OR EQUAL.

3. ALL BEDROOM WINDOWS TO BE WITH IN 44" OF FIN. FLOOR

4. ALL WINDOW HEADERS TO BE SET AT 6'-10" FOR UPPER FLOOR. MAIN FLOOR HEADERS TO BE SET @ 6'-10" EXCEPT FOR FAMILY, FAM/DINING, KITCHEN, AND DINING ROOM HEADERS TO BE SET AT 7'-4" ABOVE FIN. FLOOR. LOWER FLOOR TO BE SET AT 8'-0" UNLESS NOTED.

5. UNLESS NOTED ALL HEADERS OVER EXTERIOR DOORS AND WINDOWS TO BE 4" WIDE WITH 2" RIGID INSULATION BACKER. SEE TYPICAL SECTION.

FRAMING NOTES:

NOTES SHALL APPLY TO ALL LEVELS.
FRAMING STANDARDS ARE ACCORDING TO U.B.C. OR C.A.B.O.
WITH THE STRICTEST ALTERNATIVE LISTED.

1. ALL FRAMING LUMBER TO BE D.F.L. #2 MIN. ALL GLU-LAM BEAMS TO BE fb2200, V-4, DF/DF

2. FRAME ALL EXTERIOR WALLS W/ 2 X 6 STUDS @ 16" O.C. FRAME ALL EXTERIOR NON-BEARING WALLS W/ 2 X 6 STUDS @ 24" O.C.

3. USE 2 X 6 NAILER AT THE BOTTOM OF ALL 2-2X12 OR 4 X HEADERS @ EXTERIOR WALLS. BACK HEADER W/ 2" RIGID INSULATION.

4. ALL SHEAR PANELS TO BE 1/2" PLY NAILED W/ 8 d'S @ 4" O.C. AT EDGE AND BLOCKING AND 8d'S @ 8" O.C. @ FIELD.

5. PLYWOOD ROOF SHEATHING TO BE 1/2" STD. GRADE 32/16 PLY. LAID PERP TO RAFTERS. NAIL W/ 8d'S @ 6" O.C. @ EDGES AND 12" O.C. AT FIELD

6. PROVIDE 3/4" STD. GRADE T. & G. PLY. FLOOR SHEATHING LAID PERP. TO FLOOR JOIST. NAIL W/ 10 d'S @ 6" O.C. @ EDGES AND BLOCKING AND 12" O.C. @ FIELD. COVER W/ 3/8" HARDBOARD.

7. BLOCK ALL WALLS OVER 10'-0" HIGH AT MID HEIGHT.

8. LET-IN BRACES TO BE 1 X 4 DIAG. BRACES @ 45° FOR ALL INTERIOR LOAD-BEARING WALLS.

9. NOTCHES IN THE ENDS OF JOIST SHALL NOT EXCEED 1/4 OF THE JOIST DEPTH. HOLES DRILLED IN JOIST SHALL NOT BE IN THE UPPER OR LOWER 2" OF THE JOIST. THE DIAMETER OF HOLES DRILLED IN JOIST SHALL NOT EXCEED 1/3 THE DEPTH OF THE JOIST.

10. PROVIDE DOUBLE JOIST UNDER AND PARALLEL TO LOAD BEARING WALLS.

11. BLOCK ALL FLOOR JOIST AT SUPPORTED ENDS AND @ 10'-0 O.C. MAX. ACROSS SPAN.

FASTENER SCHEDULE

DESCRIPTION OF BUILDING MATERIAL	NUMBER & TYPE OF FASTENERS (1,2,3,5,)	SPACING OF FASTENERS
JOIST TO SILL OR GIRDER, TOE NAIL	3- 8d	
1 X 6 SUBFLOOR OR LESS TO EACH JOIST, FACE NAIL	2-8d / 2 STAPLES, 1 3/4"	
WIDER THAN 1 X 6 SUBFLOOR TO EA. JOIST, FACE NAIL	3-8d / 4 STAPLES, 1 3/4"	
2" SUBFLOOR TO JOIST OR GIRDER BLIND AND FACE NAIL	2-16d	
SOLE PLATE TO JOIST OR BLOCKING FACE NAIL	16d	16" O.C.
TOP OR SOLE PLATE TO STUD, END NAIL	2-16d	
STUD TO SOLE PLATE, TOE NAIL	3-8d OR 2-16d	
DOUBLE STUDS, FACE NAIL	16d	24" O.C.
DOUBLE TOP PLATE, FACE NAIL	16d	24" O.C.
TOP PLATES, TAPS & INTERSECTIONS, FACE NAIL	2-16d	
CONTINUED HEADER, TWO PIECES	16d	16" O.C. ALONG EA EDGE
CEILING JOIST TO PLATE, TOE NAIL	3-8d	
CONTINUOUS HEADER TO STUD, TOE NAIL	4-8d	
CEILING JOIST, TAPS OVER PARTITIONS, FACE NAIL	3-16d	
CEILING JOIST TO PARALLEL RAFTERS, FACE NAIL	3-16d	
RAFTERS TO PLATE, TOE NAIL	2-16d	
1" BRACE TO EA STUD & PLATE FACE NAIL	2-8d / 2 STAPLES 1 ¾"	
1 X 6 SHEATHING TO EA. BEARING FACE NAIL	2-8d / 2 STAPLES 1¼"	
1 X 8 SHEATHING TO EA. BEARING FACE NAIL	2-8D / 3 STAPLES, 1½"	
WIDER THAN 1 X 8 SHEATHING TO EACH BEARING, FACE NAIL	3-8d / 4 STAPLES 1¼"	
BUILT-UP CORNER STUDS	16D	
BUILT-UP GIRDER AND BEAMS	16d	32" O.C. @ TOP & BTM. AND STAGGERED 2-20d @ EA. END & @ EA. SPLICE
2" PLANKS	2-16d	AT EACH BEARING
ROOF RAFTERS TO RIDGE, VALLEY OR HIP RAFTERS: TOENAIL, FACE NAIL	4-16d / 3-16d	
RAFTER TIES TO RAFTER, FACE NAIL	3-8d	
PLYWOOD & PARTICLEBOARD, ROOF & WALL SHEATHING TO FRAME		

CHAPTER 42

General Construction Specifications

▼ CONSTRUCTION SPECIFICATIONS

Building plans, including all of the elements that make up a complete set of residential or commercial drawings, contain most general and some specific information about the construction of the structure. However, it is very difficult to provide all of the required information on a set of plans. Schedules, as discussed in detail in Section IV, provide a certain amount of subordinate information. Information that cannot be clearly or completely provided on the drawing or in schedules are provided in construction specifications. Specifications are an integral part of any set of plans.

Most lenders have a format for giving residential construction specifications. The Federal Housing Administration (FHA) or the Federal Home Loan Mortgage Corporation (FHLMC) has a specification format entitled Description of Materials. This specifications form is used widely in a revised or identical manner by most residential construction lenders. The same form is used by Farm Home Administration (FmHA) and by the Veterans Administration (VA). Figure 42–1 shows a completed FHA Description of Materials form for a typical structure. These plans, construction specifications, and building contract together become the legal documents for the construction project. These documents should be prepared very carefully in cooperation with the architect, client, and contractor. Any deviation from these documents should be approved by all three parties. When brand names are used, a clause specifying "or equivalent" may be added. This means that another brand of equivalent value to the one specified may be substituted with the construction supervisor's approval.

▼ TYPICAL MINIMUM CONSTRUCTION SPECIFICATIONS

Minimum construction specifications as established by local building officials vary from one location to the next, and their contents are dependent on specific local requirements, climate, codes used, and the extent of coverage. Verify the local requirements for a construction project as they may differ from those given here. The following are some general classifications of construction specifications.

Room Dimensions

- ○ Minimum room size is to be 70 sq ft.
- ○ Ceiling height minimum is to be 7'–6" in 50 percent of area except 7'–0" may be used for bathrooms and hallways.

Light and Ventilation

- ○ Minimum window area is to be $\frac{1}{10}$ floor area with not less than 10 sq ft for habitable rooms and 3 sq ft for bathrooms and laundry rooms. Not less than one-half of this required window area is to be openable. Every sleeping room is required to have a window or door for emergency exit. Windows with an openable area of not less than 5 sq ft with no dimension less than 22" meet this requirement, and the sill height is to be not more than 44" above the floor.
- ○ Glass subject to human impact is to be tempered glass.
- ○ Glass doors in shower and tub enclosures are to be tempered glass or fracture-resistant plastic.
- ○ Attic ventilation is to be a minimum of $\frac{1}{300}$ of the attic area, one-half in the soffit and one-half in the upper area.
- ○ Bathroom and kitchen fans and dryer are to vent directly outside.

Foundation

- ○ Concrete mix is to have a minimum ultimate compressive strength of 2500 psi at 28 days and shall be composed of 1 part cement, 3 parts sand, 4 parts of 1" maximum size rock, and not more than 7½ gallons of water per sack of cement.
- ○ Foundation mud sills, plates, and sleepers are to be pressure treated or of foundation-grade redwood. All footing sills shall have full bearing on the footing wall or slab and shall be bolted to the foundation with ½ × 10" bolts embedded at least 7" into the concrete or reinforced masonry, or 15" into unreinforced grouted masonry. Bolts shall be spaced not to exceed 6' on center with bolts not over 12" from cut end of sills.
- ○ Crawl space shall be ventilated by an approved mechanical means or by openings with a net area not less than 1½ sq ft for each 25 linear ft of exterior wall.

FHA Form 2005
VA Form 26-1852
Form FmHA 424-2
Rev. 4/77

U. S. DEPARTMENT OF HOUSING AND URBAN DEVELOPMENT
FEDERAL HOUSING ADMINISTRATION
For accurate register of carbon copies, form
may be separated along above fold. Staple
completed sheets together in original order.

Form Approved
OMB No. 63-R0055

☐ Proposed Construction

DESCRIPTION OF MATERIALS

No. _____
(To be inserted by FHA, VA or FmHA)

☐ Under Construction

Property address _____ LAKE HOUSE _____ City _____ State _____

Mortgagor or Sponsor _____
(Name) *(Address)*

Contractor or Builder _____
(Name) *(Address)*

INSTRUCTIONS

1. For additional information on how this form is to be submitted, number of copies, etc., see the instructions applicable to the FHA Application for Mortgage Insurance, VA Request for Determination of Reasonable Value, or FmHA Property Information and Appraisal Report, as the case may be.
2. Describe all materials and equipment to be used, whether or not shown on the drawings, by marking an X in each appropriate check-box and entering the information called for in each space. If space is inadequate, enter "See misc." and describe under item 27 or on an attached sheet. THE USE OF PAINT CONTAINING MORE THAN THE PERCENTAGE OF LEAD BY WEIGHT PERMITTED BY LAW IS PROHIBITED.
3. Work not specifically described or shown will not be considered unless

required, then the minimum acceptable will be assumed. Work exceeding minimum requirements cannot be considered unless specifically described.
4. Include no alternates, "or equal" phrases, or contradictory items. (Consideration of a request for acceptance of substitute materials or equipment is not thereby precluded.)
5. Include signatures required at the end of this form.
6. The construction shall be completed in compliance with the related drawings and specifications, as amended during processing. The specifications include this Description of Materials and the applicable Minimum Property Standards.

1. EXCAVATION:
Bearing soil, type ___ SANDY LOAM _____

2. FOUNDATIONS:
Footings: concrete mix _____; strength psi _2500 @ 28 D._ Reinforcing _2 - #4_
Foundation wall: material __CONCRETE BLOCK_____ Reinforcing _HORIZONTAL @ 24"_
Interior foundation wall: material _CONCRETE BLOCK___ Party foundation wall _____
Columns: material and sizes _TS3 1/2 X 3 1/2 X .250_ Piers: material and reinforcing _____
Girders: material and sizes _SEE DRAWINGS_____ Sills: material _2 X 10/CELLOTEX SILL SEAL_
Basement entrance areaway _____ Window areaways _____
Waterproofing _2 COATS ASPHALT BELOW GRADE__ Footing drains _____
Termite protection _____
Basementless space: ground cover _____; insulation _R-7.5 FOAM BD._; foundation vents ____
Special foundations _____
Additional information: _3/8"ø X 12" ANCHOR BOLTS @ 3'-0"_

3. CHIMNEYS:
Material _TRIPLE-WALL STEEL_____ Prefabricated *(make and size)* _MAJESTIC 9"_
Flue lining: material _STAINLESS STEEL_ Heater flue size _____ Fireplace flue size _____
Vents *(material and size):* gas or oil heater _____; water heater _____
Additional information: _INSTALLED IN ACCORDANCE W/MAJESTIC INSTRUCTIONS_

4. FIREPLACES:
Type: ☒ solid fuel; ☐ gas-burning; ☐ circulator *(make and size)* _MAJESTIC aM28_ Ash dump and clean-out _____
Fireplace: facing _FACE BRICK_; lining _____; hearth _FACE BRICK_; mantel _____
Additional information: _____

5. EXTERIOR WALLS:
Wood frame: wood grade, and species _HEM-FIR #2_ ☒ Corner bracing. Building paper or felt _3/4"CDX PLY. WD._
Sheathing _R-5.4 FOAM BD._; thickness _3/4"_; width _____; ☐ solid; ☐ spaced ___ " o. c.; ☐ diagonal; _____
Siding _PLYWOOD_; grade _EXT._; type _T1-11_; size _1/2"_; exposure ___"; fastening _AL. NAILS_
Shingles _____; grade _____; type _____; size _____; exposure _____; fastening _____
Stucco _____; thickness ___"; Lath _____; weight ____ lb.
Masonry veneer _____ Sills _____ Lintels _____ Base flashing _____
Masonry: ☐ solid ☐ faced ☐ stuccoed; total wall thickness ___"; facing thickness ___"; facing material _____
Backup material _____; thickness ___"; bonding _____
Door sills _OAK_ Window sills _____ Lintels _____ Base flashing _____
Interior surfaces: dampproofing, ___ coats of _____; furring _1 X 3 @ 16" O.C._
Additional information: _____
Exterior painting: material _STAIN_; number of coats _1_
Gable wall construction: ☒ same as main walls; ☐ other construction _____

6. FLOOR FRAMING:
Joists: wood, grade, and species _DOUG. FIR. #2_; other _____; bridging _____; anchors _____
Concrete slab: ☒ basement floor; ☐ first floor; ☒ ground supported; ☐ self-supporting; mix _2500 PSI @ 28 DAYS_; thickness _4_";
reinforcing _6 X 6 - 10/10 WWM_; insulation _____; membrane _4-MIL POLYETHYLENE_
Fill under slab: material _R.O.B. GRAVEL_; thickness _6_". Additional information: _R-14.6 FOAM BD._
INSULATION UNDER HEAT SINK

7. SUBFLOORING: *(Describe underflooring for special floors under item 21.)*
Material: grade and species _CD GRADE PLYWOOD_; size _1/2"_; type _____
Laid: ☐ first floor; ☐ second floor; ☐ attic _____ sq. ft.; ☐ diagonal; ☐ right angles. Additional information: _ALL FRAMED_
FLOOR NOT DESCRIBED UNDER ITEM 21 TO HAVE 3/8" PLY. WD. UNDERLAY AT RIGHT ANGLES.

8. FINISH FLOORING: *(Wood only. Describe other finish flooring under item 21.)*

LOCATION	ROOMS	GRADE	SPECIES	THICKNESS	WIDTH	BLDG. PAPER	FINISH
First floor	LOFT	SELECT	R. OAK	25/32	2 1/2	YES	FILL & 2 COATS POLYURETHANE
Second floor							
Attic floor	sq. ft						

Additional information: _____

FHA Form 2005
VA Form 26-1852
Form FmHA 424-2

DESCRIPTION OF MATERIALS

Figure 42-1 Completed FHA Description of Materials for a structure. *From Huth,* Understanding Construction Drawings, *Delmar Publishers Inc.*

DESCRIPTION OF MATERIALS

9. PARTITION FRAMING:
Studs: wood, grade, and species ___HEM-FIR #2___ size and spacing ___2 X 4 / 16" O.C.___ Other _____
Additional information: _____

10. CEILING FRAMING:
Joists: wood, grade, and species ___HEM-FIR #2___ Other _____ Bridging _____
Additional information: _____

11. ROOF FRAMING:
Rafters: wood, grade, and species ___HEM-FIR #2___ Roof trusses (see detail): grade and species _____
Additional information: _____

12. ROOFING:
Sheathing: wood, grade, and species ___1/2" CDX PLY. WD.___ ; ☐ solid; ☐ spaced _____" o.c.
Roofing ___COMP. SHINGLES___ ; grade ___#235___ ; size _____ ; type _____
Underlay ___ASPHALT SATURATED FELT___ ; weight or thickness ___#15___ ; size _____ ; fastening _____
Built-up roofing _____ ; number of plies _____ ; surfacing material _____
Flashing: material ___ALUMINUM___ ; gage or weight ___28 GA.___ ; ☐ gravel stops; ☐ snow guards
Additional information: ___ALUMINUM DRIP EDGE___

13. GUTTERS AND DOWNSPOUTS:
Gutters: material _____ ; gage or weight _____ ; size _____ ; shape _____
Downspouts: material _____ ; gage or weight _____ ; size _____ ; shape _____ ; number _____
Downspouts connected to: ☐ Storm sewer; ☐ sanitary sewer; ☐ dry-well. ☐ Splash blocks: material and size _____
Additional information: _____

14. LATH AND PLASTER
Lath ☐ walls, ☐ ceilings: material _____ ; weight or thickness _____ Plaster: coats _____ ; finish _____
Dry-wall ☒ walls, ☒ ceilings: material ___GYPSUM___ ; thickness ___1/2"___ ; finish ___PAINTED___
Joint treatment ___TAPE & COMPOUND WITH METAL CORNER BEADS___

15. DECORATING: *(Paint, wallpaper, etc.)*

ROOMS	WALL FINISH MATERIAL AND APPLICATION	CEILING FINISH MATERIAL AND APPLICATION
Kitchen _____	LOW LUSTER ENAMEL - - 2 COATS	ALL CEILINGS - - ONE COAT
Bath _____	VINYL WALLCOVERING	FLAT LATEX/ONE COAT
Other _____	FLAT LATEX - - 2 COATS	STIPPLED LATEX

Additional information: _____

16. INTERIOR DOORS AND TRIM:
Doors: type _____ ; material ___WOOD DOORS BIRCH OR PINE___ ; thickness _____
Door trim: type ___RANCH___ ; material ___PINE___ Base: type ___RANCH___ ; material ___PINE___ ; size ___3 1/2"___
Finish: doors ___SEMIGLOSS LATEX ENAMEL - 2 CTS___ ; trim ___SEMIGLOSS LATEX ENAMEL - - 2 COATS___
Other trim *(item, type and location)* ___BEAM CASINGS - - PINE W/2 COATS SEMIGLOSS; RAILING & LADDER - - WOOD___
Additional information: ___PARTS, SANDED & 2 COATS POLYURETHANE - - METAL PARTS, 2 COATS RUST-RESISTANT SEMIGLOSS ENAMEL___

17. WINDOWS:
Windows: type ___SLIDING___ ; make ___CAPITOL___ ; material ___THERM-BRK. AL.___ ; sash thickness _____
Glass: grade ___INSULATING___ ; ☐ sash weights; ☐ balances. type _____ ; head flashing ___INTEGRAL FLANGE___
Trim: type ___GYP. BD. RETURN___ , material _____ Paint ___AS WALL___ ; number coats _____
Weatherstripping: type ___FACTORY INSTALLED___ ; material _____ Storm sash, number _____
Screens: ☒ full, ☐ half. type _____ ; number _____ ; screen cloth material ___ALUMINUM___
Basement windows: type _____ ; material _____ ; screens, number _____ ; Storm sash, number _____
Special windows ___SKYLIGHT - - SKYVUE #DV 2852___
Additional information: ___WOOD DRIP CAPS___

18. ENTRANCES AND EXTERIOR DETAIL:
Main entrance door: material ___HOLLOW CORE STEEL___ ; width ___3'-0"___ ; thickness ___1 3/4"___. Frame: material ___WOOD___ , thickness _____"
Other entrance doors: material ___SAME___ ; width _____ ; thickness _____". Frame: material _____ ; thickness _____"
Head flashing _____ Weatherstripping: type ___MAGNETIC___ saddles ___OAK___
Screen doors: thickness _____"; number _____ ; screen cloth material _____ Storm doors: thickness _____"; number _____
Combination storm and screen doors: thickness _____"; number _____ ; screen cloth material _____
Shutters: ☐ hinged; ☐ fixed. Railings _____ , Attic louvers _____
Exterior millwork: grade and species ___#1 PINE___ Paint ___STAIN___ ; number coats ___1___
Additional information: ___ALL DECK LUMBER TO BE TREATED W/COPPER-BASE PRESERVATIVE___

19. CABINETS AND INTERIOR DETAIL:
Kitchen cabinets, wall units: material ___SCHEIRCH, GARDEN GROVE___ ; lineal feet of shelves _____ ; shelf width _____
Base units: material ___GARDEN GROVE___ ; counter top ___PLASTIC LAMINATE___ ; edging _____
Back and end splash ___MOLDED___ Finish of cabinets ___FACTORY FINISHED___ ; number coats _____
Medicine cabinets: make ___MIAMI CAREY___ ; model ___UP-CRP3418/DWN-CRP-306-AL___
Other cabinets and built-in furniture ___BENCH IN L.R. - - AC PLYWOOD, SANDED, 1 COAT FIRZITE___
Additional information: ___2 COATS SEMIGLOSS ENAMEL___

20. STAIRS:

STAIR	TREADS		RISERS		STRINGS		HANDRAIL		BALUSTERS	
	Material	Thickness	Material	Thickness	Material	Size	Material	Size	Material	Size
Basement _____	OAK	5/4	PINE	1" NOM.	PINE	5/4				
Main _____	OAK	5/4	PINE	1" NOM.	PINE	5/4	OAK	5/4" X 3"	STEEL	1"
Attic _____										

Disappearing: make and model number _____
Additional information: _____

Figure 42–1 Continued.

Openings shall be covered with not less than ¼" or more than ½" of corrosion-resistant wire mesh. If the crawl space is to be heated, closeable covers for vent openings shall be provided. Water drainage and 6 mil black ground cover shall be provided in the crawl space.

21. SPECIAL FLOORS AND WAINSCOT: *(Describe Carpet as listed in Certified Products Directory)*

	LOCATION	MATERIAL, COLOR, BORDER, SIZES, GAGE, ETC.	THRESHOLD MATERIAL	WALL BASE MATERIAL	UNDERFLOOR MATERIAL
FLOORS	Kitchen	QUARRY TILE (N.I.C.)	OAK	PINE	1" CONC.
	Bath	UNGLAZED CERAMIC TILE	MARBLE	TILE	1" CONC.
	L.R. & D.R.	QUARRY TILE (N.I.C.)		PINE	CONC.

	LOCATION	MATERIAL, COLOR, BORDER, CAP. SIZES, GAGE, ETC.	HEIGHT	HEIGHT OVER TUB	HEIGHT IN SHOWERS (FROM FLOOR)
WAINSCOT	Bath	CERAMIC TILE	4'-0"	4'-0" FROM TUB	TO CLG.

Bathroom accessories: ☒ Recessed; material ____CHINA____; number ___5___; ☒ Attached; material ___CHINA___; number ___7___
Additional information: _____

22. PLUMBING:

FIXTURE	NUMBER	LOCATION	MAKE	MFR'S FIXTURE IDENTIFICATION NO.	SIZE	COLOR
Sink	1	KITCHEN	MOEN	S3322-4		S.S.
Lavatory	3		UNIVERSAL-RUNDLE	163330	PER DWGS.	BY OWNER
Water closet	3		UNIVERSAL-RUNDLE	4055-15		BY OWNER
Bathtub	1		UNIVERSAL-RUNDLE	118-2	5'-0"	BY OWNER
Shower over tub △	1		MOEN			CROME
Stall shower △	1		KINKEAD INDUSTRIES	MARBLEMOLD	3'-0" X 3'-0"	
Laundry trays						

△☒ Curtain rod   △☒ Door   ☐ Shower pan: material ____SYNTHETIC HARD RUBBER____
Water supply: ☐ public; ☐ community system; ☒ individual (private) system. ★
Sewage disposal: ☐ public; ☐ community system; ☐ individual (private) system. ★
★ Show and describe individual system in complete detail in separate drawings and specifications according to requirements.
House drain (inside): ☐ cast iron; ☐ tile; ☒ other ___PVC___   House sewer (outside): ☐ cast iron; ☐ tile; ☒ other ___PVC___
Water piping: ☐ galvanized steel; ☒ copper tubing; ☐ other _____ Sill cocks, number ___2___
Domestic water heater: type ___ELECTRIC___; make and model ___STATE-CENSIBLE___; heating capacity _____
_____ gph. 100° rise. Storage tank: material ___GLASS___; capacity ___40___ gallons.
Gas service: ☐ utility company; ☐ liq. pet. gas; ☐ other _____ Gas piping: ☐ cooking; ☐ house heating.
Footing drains connected to: ☐ storm sewer; ☐ sanitary sewer; ☐ dry well. Sump pump; make and model _____
_____; capacity _____; discharges into _____

23. HEATING:
☐ Hot water. ☐ Steam. ☐ Vapor. ☐ One-pipe system. ☐ Two-pipe system.
☐ Radiators. ☐ Convectors. ☐ Baseboard radiation. Make and model _____
Radiant panel: ☐ floor; ☐ wall; ☐ ceiling. Panel coil: material _____
☐ Circulator. ☐ Return pump. Make and model _____; capacity _____ gpm.
Boiler: make and model _____ Output _____ Btuh.; net rating _____ Btuh.
Additional information: _____
Warm air: ☐ Gravity. ☐ Forced. Type of system _____
Duct material: supply _____; return _____ Insulation _____, thickness _____ ☐ Outside air intake.
Furnace: make and model _____ Input _____ Btuh.; output _____ Btuh.
Additional information: _____
☐ Space heater; ☐ floor furnace; ☐ wall heater. Input _____ Btuh.; output _____ Btuh.; number units _____
Make, model _____ Additional information: _____
Controls: make and types _____
Additional information: _____
Fuel: ☐ Coal; ☐ oil; ☐ gas; ☐ liq. pet. gas; ☐ electric; ☐ other _____; storage capacity _____
Additional information: _____
Firing equipment furnished separately: ☐ Gas burner, conversion type. ☐ Stoker: hopper feed ☐; bin feed ☐
Oil burner: ☐ pressure atomizing; ☐ vaporizing _____
Make and model _____ Control _____
Additional information: _____
Electric heating system: type ___BASEBOARD RESISTANCE___ Input _____ watts; @ _____ volts; output _____ Btuh.
Additional information: ___SEE SCHEDULE___
Ventilating equipment: attic fan, make and model _____; capacity _____ cfm.
kitchen exhaust fan, make and model ___GENERAL ELECTRIC #JV330___
Other heating, ventilating, or cooling equipment _____

24. ELECTRIC WIRING:
Service: ☐ overhead; ☒ underground. Panel: ☐ fuse box; ☒ circuit-breaker; make___SQUARE D___ AMP's ___200___ No. circuits ___40___
Wiring: ☐ conduit; ☐ armored cable; ☒ nonmetallic cable; ☐ knob and tube; ☐ other _____
Special outlets: ☒ range; ☒ water heater; ☐ other ___CLOTHES DRYER___
☐ Doorbell. ☒ Chimes. Push-button locations ___KITCHEN ENTRANCE___ Additional information: ___ONE TELEPHONE___
___JACK IN EACH LIVING SPACE TO BE COMPATIBLE W/LOCAL TELEPHONE UTILITY___

25. LIGHTING FIXTURES:
Total number of fixtures ___22___ Total allowance for fixtures, typical installation, $ ___$450___
Nontypical installation _____
Additional information: _____

3   DESCRIPTION OF MATERIALS

Figure 42–1 Continued.

DESCRIPTION OF MATERIALS

26. INSULATION:

Location	Thickness	Material, Type, and Method of Installation	Vapor Barrier
Roof	9"	R-30 FIBERGLASS	4 MIL POLYETHYLENE
Ceiling			
Wall	6"	R-19 FIBERGLASS	4 MIL POLYETHYLENE
Floor			
MASONARY WALLS - - 1", R-7.2 FOAM BD BETWEEN FURRING			

27. MISCELLANEOUS: *(Describe any main dwelling materials, equipment, or construction items not shown elsewhere; or use to provide additional information where the space provided was inadequate.   Always reference by item number to correspond to numbering used on this form.)*

2. HEATSINK – 12" R.Q.B. GRAVEL MINIMUM OVER UNEXCAVATED EARTH. BLOCK UNDER CONCRETE AT HEAT SINK TO BE LAID FLAT W/3/8" JOINTS & NO MORTAR.

19. VANITIES - - SCHEIRICH GARDEN GROVE

HARDWARE: *(make, material, and finish.)* KWIK-SET: ENTRANCES - - BRONZE
BATHS - - CHROME
OTHERS - - BRASS

SPECIAL EQUIPMENT: *(State material or make, model and quantity.   Include only equipment and appliances which are acceptable by local law, custom and applicable FHA standards.   Do not include items which, by established custom, are supplied by occupant and removed when he vacates premises or chattles prohibited by law from becoming realty.)*
APPLIANCES NOT INCLUDED ABOVE - - NOT IN CONTRACT

PORCHES: DECKS ACCORDING TO DRAWINGS - - ALL LUMBER TO BE CCA TREATED OR EQUAL. FASTENERS TO BE GALVANIZED STEEL.

TERRACES:

GARAGES: ACCORDING TO PLAN, NO INSULATION, OVERHEAD DOOR - - WOOD FRAME W/HARDBOARD PANELS, ONE SECTION GLAZED.

WALKS AND DRIVEWAYS:
Driveway: width __SEE DWG__ ; base material __GRADE__ ; thickness _____ "; surfacing material __CLEAN GRAVEL__ ; thickness _6_ "
Front walk: width _3'-0"_ ; material __STONE__ ; thickness _4_ ". Service walk: width _____ ; material _____ ; thickness _____ "
Steps: material _____ ; treads _____ "; risers _____ ". Cheek walls _____

OTHER ONSITE IMPROVEMENTS:
(Specify all exterior onsite improvements not described elsewhere, including items such as unusual grading, drainage structures, retaining walls, fence, railings, and accessory structures.)

LANDSCAPING, PLANTING, AND FINISH GRADING:
Topsoil __2__ " thick: ☒ front yard; ☒ side yards; ☒ rear yard to __PROPERTY LINE__ feet behind main building.
Lawns *(seeded, sodded, or sprigged)*: ☒ front yard __SEED__ ; ☒ side yards __SEED__ ; ☒ rear yard __SEED__
Planting: ☐ as specified and shown on drawings; ☐ as follows:
_____ Shade trees, deciduous, _____ " caliper.   _____ Evergreen trees. _____ ' to _____ ', B & B.
_____ Low flowering trees, deciduous, _____ ' to _____ '   _____ Evergreen shrubs. _____ ' to _____ ', B & B.
_____ High-growing shrubs, deciduous, _____ ' to _____ '   _____ Vines, 2-year _____
_____ Medium-growing shrubs, deciduous, _____ ' to _____ '
_____ Low-growing shrubs, deciduous, _____ ' to _____ '

IDENTIFICATION.—This exhibit shall be identified by the signature of the builder, or sponsor, and/or the proposed mortgagor if the latter is known at the time of application.

Date _____   Signature _____

Signature _____

Figure 42–1 Continued.

○ Access to crawl space is to be a minimum of 18 × 24″.

○ Basement foundation walls with a height of 8′ or less supporting a well-drained porous fill of 7′ or less, with soil pressure not more than 30 lb per sq ft equivalent fluid pressure, and with the bottom of the wall supported from inward movement by structural floor systems may be of plain concrete with 8″ minimum thickness and minimum ultimate compressive strength of 2500 psi at 28 days. Basement walls supporting backfill and not meeting these criteria shall be designed in accordance with accepted engineering practices.

○ Concrete forms for footings shall conform to the shape, lines, and dimensions of the members as called for on the plans and shall be substantial and sufficiently tight to prevent leakage of mortar and slumping out of concrete in the ground contact area.

Framing

○ Lumber. All joists, rafters, beams, and posts 2 to 4″ thick shall be No. 2 Grade Douglas fir-larch or better. All posts and beams 5″ and thicker shall be No. 1 Grade Douglas fir-larch or better.

○ Beams (untreated) bearing in concrete or masonry wall pockets shall have air space on sides and ends. Beams are to have not less than 4″ of bearing on masonry or concrete.

○ Wall bracing. Every exterior wood stud wall and main cross partition shall be braced at each end and at least every 25′ of length with 1 × 4 diagonal let-in braces or equivalent.

○ Joists are to have not less than 1½″ of bearing on wood or metal nor less than 3″ on masonry.

○ Floor joists are to have solid blocking at each support and at the ends except when the end is nailed to a rim joist or adjoining studs. Joists 2 × 14 or larger are to have bridging at maximum intervals of 8′.

○ Two in. clearance is required between combustible material and the walls of an interior fireplace or chimney. One in. clearance is required when the chimney is on an outside wall. (½″ moisture-resistant gypsum board may be used in lieu of the 1″ clearance requirement).

○ Rafter purlin braces are to be not less than 45° to the horizontal.

○ Rafters, when not parallel to ceiling joists, are to have ties that are 1 × 4 minimum spaced not more than 4′ on center.

○ Provide a double top plate with a minimum 48″ lap splice.

○ Metal truss tie-downs are to be required for manufactured trusses.

○ Plant manufactured trusses (if used) shall be of an approved design with an engineered drawing.

○ Fire blocking shall be provided for walls over 10′–0″ in height, also for horizontal shafts 10′–0″ on center, and for any concealed draft opening.

○ Garage walls and ceiling adjacent to or under dwelling require one-hour fire-resistant construction on the garage side. A self-closing door between the garage and dwelling is to be a minimum 1⅜″ solid core construction.

○ Ceramic tile, or approved material, is to be used in a water-splash area.

○ Building paper, or other approved material, is to be used under siding.

○ Framing in the water-splash area is to be protected by waterproof paper, waterproof gypsum, or other approved substitute.

○ Post-and-beam connections. A positive connection shall be provided between beam, post, and footing to ensure against uplift and lateral displacement. Untreated posts shall be separated from concrete or masonry by a rust-resistant metal plate or impervious membrane and be at least 6″ from any earth.

Stairways

○ Maximum rise is to be 8″, minimum run 9″, minimum headroom to be 6′–6″, and minimum width to be 30″.

○ Winding and curved stairways are to have a minimum inside tread width of 6″.

○ Enclosed usable space under stairway is to be protected by one-hour fire-resistant construction (⅝″ type X gypsum board).

○ Handrails are to be from 30 to 34″ above tread nosing, and intermediate rails are to be such that no object 5″ in diameter can pass through.

○ Generally, for commercial or public structures, all unenclosed floor and roof openings, balconies, decks, and porches more than 30″ above grade shall be protected by a guardrail not less than 42″ in height with intermediate rails or dividers such that no object 9″ in diameter can pass through. Generally, guardrails for residential occupancies may be not less than 36″ in height. Specific applications are subject to local or national building codes.

Roof

○ Composition shingles on roof slopes between 4 to 12 and 7 to 12 shall have an underlayment of not less than 15 lb felt. For slopes from 2 to 12 to less than 4 to 12. Building Department approval of roofing manufacturers' low-slope instructions is required.

○ Shake roofs require solid roof sheathing (in lieu of solid sheathing, spaced sheathing may be used but shall not be less than 1 × 4 with not more than 3″ clearance between) with an underlayment of not less

than 15 lb felt with an interlace of not less than 30 lb felt. For slopes less than 4 to 12, special approval is required.
○ Attic scuttle is to have a minimum of 22 by 30″ of headroom above.

Chimney and Fireplace

○ Reinforcing. Masonry constructed chimneys extending more than 7′ above the last anchorage point (example: roof line) must have not less than four number 4 steel reinforcing bars placed vertically for the full height of the chimney with horizontal ties not less than ¼″ diameter spaced at not over 18″ intervals. If the width of the chimney exceeds 40″, two additional number 4 vertical bars shall be provided for each additional flue or for each additional 40″ in width or fraction thereof.
○ Anchorage. All masonry chimneys over 18′ high shall be anchored at each floor and/or ceiling line more than 6′ above grade, except when constructed completely within the exterior walls of the building.

Thermal Insulation and Heating

○ Thermal designs employing the R factor must meet minimum R factors as follows:
  a. Ceiling or roof: flat R–38, vaulted R–30.
  b. Walls: R–21, vapor barrier required. Minimum one permeability rating.
  c. Floors over unheated crawl space or basements: R–25 including reflective foil.
  d. Basement walls: R–21.
  e. Slab-on-Grade: R–15 around perimeter a minimum of 18″ horizontally or vertically.
○ Thermal glazing. Heated portions of buildings located in the 5000 or less degree-day zone do not require thermal glazing on that portion of the glazing that is less than 20 percent of the total area of exterior walls including doors and windows. Heated portions of buildings located in zones over 5000 degree days shall be provided with special thermal glazing in all exterior wall areas.
○ Duct insulation. Supply and return air ducts used for heating and/or cooling located in unheated attics, garages, crawl spaces, or other unheated spaces other than between floors or interior walls shall be insulated with an R–8 minimum.
○ Heating. Every dwelling unit and guest room shall be provided with heating facilities capable of maintaining a room temperature of 70° F at a point 3′ above the floor.

Fire Warning System

○ Every dwelling shall be provided with approved detectors of products of combustion mounted on the ceiling or a wall within 12″ of the ceiling at a point centrally located in the corridor or area giving access to and not over 12′ from rooms used for sleeping. Where sleeping rooms are on an upper level, the detector shall be placed at the high area of the ceiling near the top of the stairway.

▼ SPECIFICATIONS FOR COMMERCIAL CONSTRUCTION

Specifications for commercial construction projects are often more complex and comprehensive than the documents for residential construction. Commercial project specifications may provide very detailed instructions for each phase of construction. Specifications may establish time schedules for the completion of the project. Also, in certain situations, the specifications include inspections that are in conjunction with or in addition to required local jurisdiction inspections.

Construction specifications often follow the guidelines of the individual architect or engineering firm although a common format has been established by the Construction Specification Institute (CSI). The CSI format is made up of 17 major divisions. Within each major division are several subsections. The major categories are numbered in order beginning at Division 0 through Division 16. The subdivisions of each general category are numbered with five-digit numbers. For example, one of the division 15 subsections is numbered 15050. When specific divisions or subdivisions are not required as elements of the project, they are excluded. The numbering system continues as normal with these unnecessary elements left. Figure 42–2 shows the categorical breakdown of the CSI system.

The CSI format is an effective system for indexing large project specifications. Much of the CSI format cannot be effectively used in some light commercial or residential construction. When comprehensive specifications are required for residential construction, an abridged format of the CSI system may be used. This short form may include the major divisions and exclude the subdivisions.

DIVISION 0 - BIDDING AND CONTRACT REQUIREMENTS

00010	PRE-BID INFORMATION
00100	INSTRUCTIONS TO BIDDERS
00200	INFORMATION AVAILABLE TO BIDDERS
00300	BID/TENDER FORMS
00400	SUPPLEMENTS TO BID/TENDER FORMS
00500	AGREEMENT FORMS
00600	BONDS AND CERTIFICATES
00700	GENERAL CONDITIONS OF THE CONTRACT
00800	SUPPLEMENTARY CONDITIONS
00950	DRAWINGS INDEX
00900	ADDENDA AND MODIFICATIONS

SPECIFICATIONS—DIVISIONS 1-16

DIVISION 1 - GENERAL REQUIREMENTS

01010	SUMMARY OF WORK
01020	ALLOWANCES
01030	SPECIAL PROJECT PROCEDURES
01040	COORDINATION
01050	FIELD ENGINEERING
01060	REGULATORY REQUIREMENTS
01070	ABBREVIATIONS AND SYMBOLS
01080	IDENTIFICATION SYSTEMS
01100	ALTERNATES/ALTERNATIVES
01150	MEASUREMENT AND PAYMENT
01200	PROJECT MEETINGS
01300	SUBMITTALS
01400	QUALITY CONTROL
01500	CONSTRUCTION FACILITIES AND TEMPORARY CONTROLS
01600	MATERIAL AND EQUIPMENT
01650	STARTING OF SYSTEMS
01660	TESTING, ADJUSTING, AND BALANCING OF SYSTEMS
01700	CONTRACT CLOSEOUT
01800	MAINTENANCE MATERIALS

DIVISION 2 - SITE WORK

02010	SUBSURFACE INVESTIGATION
02050	DEMOLITION
02100	SITE PREPARATION
02150	UNDERPINNING
02200	EARTHWORK
02300	TUNNELLING
02350	PILES, CAISSONS AND COFFERDAMS
02400	DRAINAGE
02440	SITE IMPROVEMENTS
02480	LANDSCAPING
02500	PAVING AND SURFACING
02580	BRIDGES
02590	PONDS AND RESERVOIRS
02600	PIPED UTILITY MATERIALS AND METHODS
02700	PIPED UTILITIES
02800	POWER AND COMMUNICATION UTILITIES
02850	RAILROAD WORK
02880	MARINE WORK

DIVISION 3 - CONCRETE

03010	CONCRETE MATERIALS
03050	CONCRETING PROCEDURES
03100	CONCRETE FORMWORK
03150	FORMS
03180	FORM TIES AND ACCESSORIES
03200	CONCRETE REINFORCEMENT
03250	CONCRETE ACCESSORIES
03300	CAST-IN-PLACE CONCRETE
03350	SPECIAL CONCRETE FINISHES
03360	SPECIALLY PLACED CONCRETE
03370	CONCRETE CURING
03400	PRECAST CONCRETE
03500	CEMENTITIOUS DECKS
03600	GROUT
03700	CONCRETE RESTORATION AND CLEANING

DIVISION 4 - MASONRY

04050	MASONRY PROCEDURES
04100	MORTAR
04150	MASONRY ACCESSORIES
04200	UNIT MASONRY
04400	STONE
04500	MASONRY RESTORATION AND CLEANING
04550	REFRACTORIES
04600	CORROSION RESISTANT MASONRY

DIVISION 5 - METALS

05010	METAL MATERIALS AND METHODS
05050	METAL FASTENING
05100	STRUCTURAL METAL FRAMING
05200	METAL JOISTS
05300	METAL DECKING
05400	COLD-FORMED METAL FRAMING
05500	METAL FABRICATIONS
05700	ORNAMENTAL METAL
05800	EXPANSION CONTROL
05900	METAL FINISHES

DIVISION 6 - WOOD AND PLASTICS

06050	FASTENERS AND SUPPORTS
06100	ROUGH CARPENTRY
06130	HEAVY TIMBER CONSTRUCTION
06150	WOOD-METAL SYSTEMS
06170	PREFABRICATED STRUCTURAL WOOD
06200	FINISH CARPENTRY
06300	WOOD TREATMENT
06400	ARCHITECTURAL WOODWORK
06500	PREFABRICATED STRUCTURAL PLASTICS
06600	PLASTIC FABRICATIONS

DIVISION 7 - THERMAL AND MOISTURE PROTECTION

07100	WATERPROOFING
07150	DAMPPROOFING
07200	INSULATION
07250	FIREPROOFING
07300	SHINGLES AND ROOFING TILES
07400	PREFORMED ROOFING AND SIDING
07500	MEMBRANE ROOFING
07570	TRAFFIC TOPPING
07600	FLASHING AND SHEET METAL
07800	ROOF ACCESSORIES
07900	SEALANTS

DIVISION 8 - DOORS AND WINDOWS

08100	METAL DOORS AND FRAMES
08200	WOOD AND PLASTIC DOORS
08250	DOOR OPENING ASSEMBLIES
08300	SPECIAL DOORS
08400	ENTRANCES AND STOREFRONTS
08500	METAL WINDOWS
08600	WOOD AND PLASTIC WINDOWS
08650	SPECIAL WINDOWS
08700	HARDWARE
08800	GLAZING
08900	GLAZED CURTAIN WALLS

DIVISION 9 - FINISHES

09100	METAL SUPPORT SYSTEMS
09200	LATH AND PLASTER
09230	AGGREGATE COATINGS
09250	GYPSUM WALLBOARD
09300	TILE
09400	TERRAZZO
09500	ACOUSTICAL TREATMENT
09550	WOOD FLOORING
09600	STONE AND BRICK FLOORING
09650	RESILIENT FLOORING
09680	CARPETING
09700	SPECIAL FLOORING
09760	FLOOR TREATMENT
09800	SPECIAL COATINGS
09900	PAINTING
09950	WALL COVERING

Figure 42–2 Construction Specifications Institute (CSI) format for specifications. *Reproduced from MASTERFORMAT, courtesy Construction Specifications Institute.*

DIVISION 10 - SPECIALITIES

10100	CHALKBOARDS AND TACKBOARDS
10150	COMPARTMENTS AND CUBICLES
10200	LOUVERS AND VENTS
10240	GRILLES AND SCREENS
10250	SERVICE WALL SYSTEMS
10260	WALL AND CORNER GUARDS
10270	ACCESS FLOORING
10280	SPECIALTY MODULES
10290	PEST CONTROL
10300	FIREPLACES AND STOVES
10340	PREFABRICATED STEEPLES, SPIRES, AND CUPOLAS
10350	FLAGPOLES
10400	IDENTIFYING DEVICES
10450	PEDESTRIAN CONTROL DEVICES
10500	LOCKERS
10520	FIRE EXTINGUISHERS, CABINETS, AND ACCESSORIES
10530	PROTECTIVE COVERS
10550	POSTAL SPECIALTIES
10600	PARTITIONS
10650	SCALES
10670	STORAGE SHELVING
10700	EXTERIOR SUN CONTROL DEVICES
10750	TELEPHONE ENCLOSURES
10800	TOILET AND BATH ACCESSORIES
10900	WARDROBE SPECIALTIES

DIVISION 11 - EQUIPMENT

11010	MAINTENANCE EQUIPMENT
11020	SECURITY AND VAULT EQUIPMENT
11030	CHECKROOM EQUIPMENT
11040	ECCLESIASTICAL EQUIPMENT
11050	LIBRARY EQUIPMENT
11060	THEATER AND STAGE EQUIPMENT
11070	MUSICAL EQUIPMENT
11080	REGISTRATION EQUIPMENT
11100	MERCANTILE EQUIPMENT
11110	COMMERCIAL LAUNDRY AND DRY CLEANING EQUIPMENT
11120	VENDING EQUIPMENT
11130	AUDIO-VISUAL EQUIPMENT
11140	SERVICE STATION EQUIPMENT
11150	PARKING EQUIPMENT
11160	LOADING DOCK EQUIPMENT
11170	WASTE HANDLING EQUIPMENT
11190	DETENTION EQUIPMENT
11200	WATER SUPPLY AND TREATMENT EQUIPMENT
11300	FLUID WASTE DISPOSAL AND TREATMENT EQUIPMENT
11400	FOOD SERVICE EQUIPMENT
11450	RESIDENTIAL EQUIPMENT
11460	UNIT KITCHENS
11470	DARKROOM EQUIPMENT
11480	ATHLETIC, RECREATIONAL, AND THERAPEUTIC EQUIPMENT
11500	INDUSTRIAL AND PROCESS EQUIPMENT
11600	LABORATORY EQUIPMENT
11650	PLANETARIUM AND OBSERVATORY EQUIPMENT
11700	MEDICAL EQUIPMENT
11780	MORTUARY EQUIPMENT
11800	TELECOMMUNICATION EQUIPMENT
11850	NAVIGATION EQUIPMENT

DIVISION 12 - FURNISHINGS

12100	ARTWORK
12300	MANUFACTURED CABINETS AND CASEWORK
12500	WINDOW TREATMENT
12550	FABRICS
12600	FURNITURE AND ACCESSORIES
12670	RUGS AND MATS
12700	MULTIPLE SEATING
12800	INTERIOR PLANTS AND PLANTINGS

DIVISION 13 - SPECIAL CONSTRUCTION

13010	AIR SUPPORTED STRUCTURES
13020	INTEGRATED ASSEMBLES
13030	AUDIOMETRIC ROOMS
13040	CLEAN ROOMS
13050	HYPERBARIC ROOMS
13060	INSULATED ROOMS
13070	INTEGRATED CEILINGS
13080	SOUND, VIBRATION, AND SEISMIC CONTROL
13090	RADIATION PROTECTION
13100	NUCLEAR REACTORS
13110	OBSERVATORIES
13120	PRE-ENGINEERED STRUCTURES
13130	SPECIAL PURPOSE ROOMS AND BUILDINGS
13140	VAULTS
13150	POOLS
13160	ICE RINKS
13170	KENNELS AND ANIMAL SHELTERS
13200	SEISMOGRAPHIC INSTRUMENTATION
13210	STRESS RECORDING INSTRUMENTATION
13220	SOLAR AND WIND INSTRUMENTATION
13410	LIQUID AND GAS STORAGE TANKS
13510	RESTORATION OF UNDERGROUND PIPELINES
13520	FILTER UNDERDRAINS AND MEDIA
13530	DIGESTION TANK COVERS AND APPURTENANCES
13540	OXYGENATION SYSTEMS
13550	THERMAL SLUDGE CONDITIONING SYSTEMS
13560	SITE CONSTRUCTED INCINERATORS
13600	UTILITY CONTROL SYSTEMS
13700	INDUSTRIAL AND PROCESS CONTROL SYSTEMS
13800	OIL AND GAS REFINING INSTALLATIONS AND CONTROL SYSTEMS
13900	TRANSPORTATION INSTRUMENTATION
13940	BUILDING AUTOMATION SYSTEMS
13970	FIRE SUPPRESSION AND SUPERVISORY SYSTEMS
13980	SOLAR ENERGY SYSTEMS
13990	WIND ENERGY SYSTEMS

DIVISION 14 - CONVEYING SYSTEMS

14100	DUMBWATERS
14200	ELEVATORS
14300	HOISTS AND CRANES
14400	LIFTS
14500	MATERIAL HANDLING SYSTEMS
14600	TURNTABLES
14700	MOVING STAIRS AND WALKS
14800	POWERED SCAFFOLDING
14900	TRANSPORTATION SYSTEMS

DIVISION 15 - MECHANICAL

15050	BASIC MATERIALS AND METHODS
15200	NOISE, VIBRATION, AND SEISMIC CONTROL
15250	INSULATION
15300	SPECIAL PIPING SYSTEMS
15400	PLUMBING SYSTEMS
15450	PLUMBING FIXTURES AND TRIM
15500	FIRE PROTECTION
15600	POWER OR HEAT GENERATION
15650	REFRIGERATION
15700	LIQUID HEAT TRANSFER
15800	AIR DISTRIBUTION
15900	CONTROLS AND INSTRUMENTATION

DIVISION 16 - ELECTRICAL

16050	BASIC MATERIALS AND METHODS
16200	POWER GENERATION
16300	POWER TRANSMISSION
16400	SERVICE AND DISTRIBUTION
16500	LIGHTING
16600	SPECIAL SYSTEMS
16700	COMMUNICATIONS
16850	HEATING AND COOLING
16900	CONTROLS AND INSTRUMENTATION

Figure 42–2 Continued.

CHAPTER

 General Construction Specifications Test

DIRECTIONS

Answer the questions with short, complete statements or drawings as needed.

1. Letter your name and the date at the top of the sheet.
2. Letter the question number and provide the answer. You do not need to write out the question.

QUESTIONS

Question 42–1 Give a general definition of construction specifications.

Question 42–2 Identify the basic differences between residential and commercial specifications.

Question 42–3 List four factors that influence the specific requirements of minimum construction specifications established by local building officials.

Question 42–4 Using general terms, list the typical minimum construction requirements for the following categories:
a. Room dimensions
b. Light and ventilation
c. Foundation
d. Framing
e. Stairways
f. Roof
g. Chimney and fireplace
h. Thermal insulation and heating
i. Fire warning system
  A partial example would be as follows:
  4. d. Framing
     1. lumber grades
     2. beams bearing area

Question 42–5 Describe the format established by the Construction Specifications Institute (CSI) for construction specifications.

PROBLEM

Problem 42–1 Obtain a blank copy of the FHA Description of Materials form from your instructor or local FHA office. There is a blank form found in the Instructor's Guide for this text, which may be copied as needed for this problem. Using your set of architectural plans for the residence you have been drawing as a continuing problem, complete the FHA form. If you have more than one set of plans, select only one set or have your instructor select the set of plans to use.

CHAPTER 43

Construction Supervision Procedures

The architect or designer may often be involved with several phases of preparation before construction of the project begins. The level of commitment may go beyond the preparation of plans to the complete supervision of the entire construction project. There are several items that any client may need before construction starts, depending on the complexity or unique requirements of the project. Architects are often involved with zone changes when necessary, specification preparation, building permit applications, bonding requirements, client financial statement, lender approval, and building contractor estimates and bid procurement.

▼ LOAN APPLICATIONS

Loan applications vary depending on the requirements of the lender. Most applications for construction financing contain a variety of similar information. The FHA-proposed construction appraisal requirements are, in part, as follows:

1. Plot plan (three copies per lot).
2. Prints (three copies per plan). Prints should include the following:
   Four elevations: front, rear, right side, and left side.
   Floor plan(s).
   Foundation plan.
   Wall section(s).
   Roof plan.
   Cross sections of exterior walls, stairs, etc.
   Cabinet detail including a cross section.
   Fireplace detail (manufacturer's detail if fireplace is prefabricated).
   Truss detail and engineering (include name and address of supplier).
   Heating plan. If forced air, show size and location of ducts and registers plus cubic feet per minute (CFM) at each register, and furnace Btus.
   Location of wall units with watts and CFM at register.
3. Specifications (one copy per print). Areas commonly missed on specifications are the following:
   Lists all appliances with make and model number.
   List smoke detector with make and model number and "Direct Wire."
   If landscaped, supply a typical detail for each plan type.

List plumbing fixtures with make, model number, size, and color.
List carpet by brand, style, and directory number.
List carpet pad by brand, style, thickness, and directory number.
Include name and address of manufacturer of trusses.
Include name of manufacturer of fireplace and model number if applicable.

4. Public Utility District (PUD) or heating contractor heat loss calculation (one per specification sheet). A room-by-room Btu heat loss count showing the total Btu heat loss for the dwelling as well as the total output of the heating system.
5. Proposed sale price.
6. Copy of earnest money agreement in blank or completed if a presale.
7. If a presale, the buyer must sign the specification sheet as well as the builder.

▼ INDIVIDUAL APPRAISAL REQUIREMENTS

The subdivision should be FHA approved, or a letter must accompany the submission with the following information:

1. Total number of lots in the subdivision.
2. Evidence of acceptance of streets and utilities by local authorities for continued maintenance.
3. One copy of the recorded plat and covenants. Covenants are legal agreements setting the conditions and restrictions for the property.

▼ MASTER APPRAISAL REQUIREMENTS

A minimum of five lots for each plan type is required for a master. The subdivision must be FHA or VA approved and contain the following information:

1. Location map and a copy of the recorded plat and covenants.
2. A letter from the builder that includes the following information:
   a. The number of lots in the area owned by the builder and the number of lots proposed to be built upon.

b. The number of homes presently under construction and the number completed and unsold.

If the builder has not worked with the FHA previously, an equal employment opportunity certificate and an affirmative fair housing marketing plan must be submitted for either a master or individual appraisal request.

▼ CHANGE ORDERS

Any physical change in the plans or specifications should be submitted to the FHA on FHA Form 2577. It should be noted whether the described change will be an increase or decrease in value and by what dollar amount. It must be signed by the lender, the builder, and, if sold, the purchaser, who must also sign prior to submission to the FHA.

Most lending institutions' applications are not as comprehensive as the FHA's, while other may require additional information. The best practice is to research the lender to identify the needed information. A well-prepared set of documents and drawings usually stands a better chance of funding.

▼ BUILDING PERMITS

The responsibility of completing the building permit application may fall to the architect, designer, or builder. The architect should contact the local building official to determine the process to be followed. Generally the building permit application is a basic form that identifies the major characteristics of the structure to be built, the legal description and location of the property, and information about the applicant. The application is usually accompanied by two sets of plans and up to five sets of plot plans. The fee for a building permit usually depends on the estimated cost of construction. The local building official determines the amount based on a standard schedule at a given cost per square foot. The fees are often divided into two parts: a plan-check fee paid upon application and a building-permit fee paid when the permit is received. There are other permits and fees that may or may not be paid at this time. In some cases the mechanical, sewer, plumbing, electrical, and water permits are obtained by the general or subcontractors. Water and sewer permits may be expensive depending on the local assessments for these utilities.

▼ CONTRACTS

Building contracts may be very complex documents for large commercial construction or short forms for residential projects. The main concern in the preparation of the contract is that all parties understand what will be specifically done, in what period of time, and for what reimbursement. The contract becomes an agreement between the client, general contractor, and architect. Figure 43–1 shows an example of a typical building contract.

It is customary to specify the date by which the project is to be completed. On some large projects, dates for completion of the various stages of construction are specified. Usually the contractor receives a percentage of the contract price for the completion of each stage of construction. Payments are typically made three or four times during construction. Another method is for the contractor to receive partial payment for the work done each month. Verify the method used with the lending agency.

The owner is usually responsible for having the property surveyed. The architect may be responsible for administering the contract. The contractor is responsible for the construction and security of the site during the construction period.

Certain kinds of insurance are required during construction. The contractor is required to have liability insurance. This protects the contractor against being sued for accidents occurring on the site. The owner is required to have property insurance, including fire, theft, and vandalism. Workman's compensation is another form of insurance that provides income for contractors' employees if they are injured at work. The contractor must be licensed, bonded, and insured under the requirements of the state where the construction occurs.

The contract describes conditions under which the contract may be ended. Contracts may be terminated if one party fails to comply with the contract, when one of the parties is disabled or dies, and for several other reasons.

There are two kinds of contracts in use for most construction—the fixed-sum and the cost-plus contract, each offering certain advantages and disadvantages.

Fixed-sum (sometimes called *lump-sum*) contracts are used most often. With a fixed-sum contract, the contractor agrees to complete the project for a certain amount of money. The greatest advantage of this kind of contract is that the owner knows in advance exactly what the cost will be. However, the contractor does not know what hidden problems may be encountered, and so the contractor's price must be high enough to cover unforseen circumstances, such as excessive rock in the excavation of sudden increases in the cost of materials.

A *cost-plus* contract is one in which the contractor agrees to complete the work for the actual cost, plus a percentage for overhead and profit. The advantage of this type of contract is that the contractor does not have to allow for unforseen problems. A cost-plus contract is also useful when changes are to be made during the course of construction. The main disadvantage of this kind of contract is that the owner does not know exactly what the cost will be until the project is completed.

▼ COMPLETION NOTICE

The *completion notice* is a document that should be posted in a conspicuous place on or adjacent to the structure. This legal document notifies all parties involved in the project that work has been substantially completed.

FORM No. 144—BUILDING CONTRACT (Fixed Price—No Service Charge).

TN

THIS AGREEMENT, Made the ....day of ...., 19..........., by and between ...., hereinafter called the Contractor, and ...., hereinafter called the Owner, WITNESSETH:

The parties hereto, each in consideration of the promises of the other, agree as follows:

ARTICLE I: The contractor shall and will perform all the work for the

as shown on the drawings and described in the specifications therefor prepared by ....
....;
said drawings, specifications and this contract hereinafter, for brevity, are called "contract documents"; they are identified by the signatures of the parties hereto and hereby are made a part hereof. All said work is to be done under the direction of ....
....who, for brevity hereinafter is designated as "supervisor." (Publisher's note: If the owner himself is to supervise said work, simply insert the word "owner" in the blank space immediately preceding.) The supervisor's decision as to the true construction and meaning of the drawings and specifications shall be final and binding upon both parties. All of said drawings and specifications including those hereinafter mentioned have been and will be prepared by the owner at his expense and are to remain his property; said drawings and specifications are loaned to the contractor for the purposes of this contract and at the completion of the work are to be returned to the owner; none of said contract documents shall be used by, submitted or shown to third parties without owner's written consent.

ARTICLE II: The contractor shall commence work within .... days from the date hereof and substantially complete the same on or before ...., 19......... . At all times the supervisor shall have access to said work for the purpose of inspecting the same and the progress thereof. Should completion be delayed by reason of the fault of the owner or of any other contractor employed by him or by fire, casualty, strikes, delays in obtaining materials or other reasons beyond the contractor's control, then the completion date shall be extended for a period equivalent to the time lost for such reasons. Should the parties be unable to agree as to the period of such extension, the question shall be referred to arbitration as hereunder provided. However, the contractor shall take special precautions to protect his work during freezing weather and shall be fully responsible for the effect of such weather upon said work.

ARTICLE III: Subject to the provisions for adjustment set forth in ARTICLE V hereof, the owner shall pay to the contractor for the performance of this contract, in current funds, the sum of $ ...., payable at the following times:

NOTE—This form not suitable for use as a retail installment contract where a finance charge is being made.

Figure 43–1 Standard building contract. *Courtesy Stevens-Ness Law Publishing Co.*

Sales tax, if any, shall be paid by the owner in addition to the fixed price mentioned above. Should any progress payments be provided for above, the same shall not include or be based upon any salary, allowance or compensation to the contractor, if an individual, or any officer of the contractor, if a corporation, nor shall it include any of the contractor's overhead or general expenses of any kind; before any such progress payment is made, the contractor shall deliver to the supervisor receipts, vouchers or other evidence satisfactory to the supervisor showing contractor's payment for materials, labor and other items for which the contractor seeks payment, including payments to subcontractors, if any. After three days' written notice to the contractor, bills for labor or materials not paid by the contractor when due, may be paid by the owner and deducted from any payment due or to become due to the contractor. After similar notice, liens, if any are filed, including attorney's fees and costs claimed therein, may be paid, settled or compromised by the owner and amounts paid therefor shall likewise be deducted. However, the contractor shall have the right to contest any such bills, claims or liens.

Final payment shall be made within .... days after the completion of said work as certified in writing by the supervisor; however, before the latter shall so certify, the contractor shall submit evidence satisfactory to the supervisor that all payrolls, material bills and other indebtedness connected with the work have been fully paid, including those incurred by each and all of contractor's subcontractors. Provided always, that no payment made to the contractor pursuant to the terms hereof shall be construed as an acceptance of any work or materials not in accordance with the contract documents.

ARTICLE IV: In his performance of said work, contractor shall obtain at his own expense all necessary permits and comply with all applicable laws, ordinances, building codes and regulations of any public authority and be responsible for any infraction or violation thereof and any expense or damages resulting from any such infraction or violation. If the parties are unable to agree upon the dollar amount of contractor's responsibility under this paragraph, the matter shall be referred to arbitration as hereinafter provided. Any work claimed by the supervisor to be defective shall be uncovered by the contractor so that a complete inspection may be made; the contractor further agrees promptly (1) to remove from the job site all materials, whether or not incorporated in the work, condemned by any public authority, (2) to take down and remove all portions of the work likewise condemned or deemed by the supervisor as failing in any way to conform to any of said contract documents and (3) to replace all faulty work and materials.

ARTICLE V: No eliminations or alterations shall be made in the work except upon written order of the supervisor. Should any such eliminations or alterations require new plans or specifications, the owner shall supply the same at his expense. Should any of said eliminations or alterations require an adjustment of the agreed price (upward or downward) such adjustment shall be evidenced by the written agreement of the parties. Should they not be able so to agree, the work shall go on nevertheless under the order mentioned above and the determination of the proper adjustment shall be referred to arbitration as hereinafter provided.

ARTICLE VI: The owner reserves the right to let other contracts in connection with the improvement of which the work herein undertaken by the contractor is a part. In such event, due written notice of such other contracts shall be given promptly to the contractor and the latter shall afford said other contractors a reasonable opportunity for the storage of their materials and the execution of their contracts and shall properly coordinate his work within theirs. In this connection, should the contractor suffer loss by reason of any delay brought about by said other contractors, the owner agrees to reimburse the contractor for such loss; on the other hand, the contractor agrees that if he shall delay the work of said other contractors so as to cause loss for which the owner shall become liable, then he shall reimburse the owner for any such loss. If the parties are unable to agree as to the amounts so to be reimbursed, all questions relative thereto shall be submitted to arbitration as hereinafter provided.

ARTICLE VII: The contractor may subcontract any part of said work but not the whole thereof. Within seven days after entering into any such subcontract, the contractor shall notify the supervisor in writing of the names of said subcontractors and the work to be undertaken by each of them. In this connection, the contractor shall be fully responsible to the owner for the acts and omissions of any of said subcontractors or of persons either directly or indirectly employed by them. Nothing contained herein shall create any contractual relation between any such subcontractor and the owner.

ARTICLE VIII: At no time shall the contractor or any of his subcontractors employ on the work any unfit person or anyone not skilled in the work assigned to him. Any employee adjudged by the supervisor to be incompetent or unfit immediately shall be discharged and shall not again be employed upon the work. Should the contractor at any time be adjudged a bankrupt or should a receiver be appointed for his affairs or should he neglect to supply sufficient properly skilled workmen or supply materials of the proper quality or fail in any respect to prosecute the work with promptness and diligence (except because of matters for which an extension of the completion date is above provided for) or comply with said contract documents or any thereof, then in any of such events, after seven days' written notice to the contractor, the owner may, if the contractor is still in default, terminate the contractor's right to continue said work and may take exclusive possession of the premises and of all materials, tools and appliances thereon and finish the work by whatever method he may deem expedient. In such case the contractor shall not be entitled to receive any further payment until the work is finished. If the unpaid balance of the contract price shall exceed the expense of finishing the work, including compensation to the owner for additional managerial and administrative expenses, such excess shall be paid to the contractor; however, if such expense shall exceed such unpaid balance, the contractor shall pay the difference to the owner. If the parties are unable to agree upon the amounts so to be paid, the question shall be submitted to arbitration as hereinafter provided.

ARTICLE IX: All materials incorporated in any structure in connection with said work by the contractor shall, as soon as incorporated, become the property of the owner. At all times the owner, at his expense, shall effect and maintain fire insurance, with extended coverage, upon the entire structure on which the work under this contract is to be done, in an amount equal to the full insurable value thereof; said insurance shall cover materials on the work site intended by the contractor to be incorporated into said structure but not yet incorporated as well as contractor's temporary buildings incident to the said work. The insured in such policy or policies shall include the owner, the contractor and such other persons as either of them may designate. Loss, if any, shall be made payable to said insured as their respective interests may appear. Certificates showing the existence of such insurance shall be delivered to the contractor if he so requests. The owner shall have power, in his sole discretion to adjust and settle any loss with the insurer which he may deem reasonable. If loss should occur and the parties hereto are unable to agree as to the division of the proceeds thereof, the question as to the amount as to which each insured shall be entitled shall be referred to arbitration as hereinafter provided.

ARTICLE X: At all times the contractor shall take all necessary precautions for the safety of persons on the work by whomsoever employed; he shall comply with all workers' compensation and similar legislation and further shall maintain at his expense public liability insurance against claims for damages because of bodily injury, including death and property damage, which may arise during his operations and those of all subcontractors under him. The insured in all such liability policies shall be the parties hereto and any others which they, or either of them, shall designate. The said insurance shall be written for not less than $ .... for injuries, including death, to any one person in any one accident; not less than $ .... for bodily injury, including death, to more than one person in any one accident, and $ .... property damage. The contractor shall deliver to the owner within ten days after the date hereof, one or more certificates from a responsible insurance company or companies satisfactory to the owner, showing the existence of such insurance. No such insurance shall be cancelled without ten days' prior written notice to the owner.

ARTICLE XI: All disputes, claims or questions subject to arbitration under this contract shall be submitted to three arbitrators, one to be designated by the owner, one by the contractor and the two thus selected to choose the third arbitrator; each party hereto shall have the right to appear before said arbitrators either in person, by attorney or other representative and to present witnesses or evidence, if desired; the decision of the majority of said arbitrators shall be final, binding and conclusive upon all parties hereto; the parties further agree that the decision of the arbitrators shall be a condition precedent to any right of legal action which either party hereto may have against the other. The work herein contracted for shall not be delayed during any arbitration proceedings except by mutual written agreement of the parties. The expense of such arbitration shall be shared equally by the parties hereto.

Figure 43–1 Continued.

ARTICLE XII: The contractor shall keep the premises (especially that part thereof under the floors thereof) free from accumulation of waste materials or rubbish and at the completion of the work shall remove all of his tools, scaffoldings and supplies and leave the premises broom-clean, or its equivalent.

ARTICLE XIII: If the owner should require a completion bond from the contractor, the premium therefor shall be added to the contract price and paid by the owner on delivery of said bond to him.

ARTICLE XIV: If the contractor employs a foreman or superintendent on said work, all directions and instructions given to the latter shall be as binding as if given to the contractor.

ARTICLE XV: The contractor agrees at all times to keep said work and the real estate on which the same is to be constructed free and clear of all construction and materialmen's liens, including liens on behalf of any subcontractor or person claiming under any such subcontractor and to defend and save the owner harmless therefrom.

ARTICLE XVI: In all respects the contractor shall be deemed to be an independent contractor.

ARTICLE XVII: In the event of any suit or action arising out of this contract, the losing party therein agrees to pay to the prevailing party therein the latter's costs and reasonable attorney's fees to be fixed by the trial court and in the event of an appeal, the prevailing party's costs and reasonable attorney's fees in the appellate court to be fixed by the appellate court.

ARTICLE XVIII: Any notice given by one party hereto to the other shall be sufficient if in writing, contained in a sealed envelope with postage thereon fully prepaid and deposited in the U. S. Registered Mails; any such notice conclusively shall be deemed received by the addressee thereof on the day of such deposit. If such notice is intended for the owner, the envelope containing the same shall be addressed to the owner at the following address: ....
....

and if intended for the contractor, if addressed to .... ,
....

ARTICLE XIX: In construing this contract and where the context so requires, the singular shall be deemed to include the plural, the masculine shall include the feminine and the neuter and all grammatical changes shall be made and implied so that this contract shall apply equally to individuals and to corporations; further, the word "work" shall mean and include the entire job undertaken to be performed by the contractor as described in the contract documents, and each thereof, together with all services, labor and materials necessary to be used and furnished to complete the same, except for the preparation of the said plans and specifications and further except the compensation of the said supervisor.

ARTICLE XX: The parties hereto further agree

IN WITNESS WHEREOF, the parties have hereunto set their hands in duplicate.

....
CONTRACTOR   OWNER

....

Figure 43–1 Continued.

There may be a very small part of work or cleanup to be done, but the project, for all practical purposes, is complete. The completion notice must be recorded in the local jurisdiction. Completion notices served several functions. Subcontractors and suppliers have a certain given period of time to file a claim or lien against the contractor or client to obtain reimbursement for labor or materials that have not been paid. Lending institutions often hold a percentage of funds for a given period of time after the completion notice has been posted. It is important for the contractor to have this document posted so the balance of payment can be obtained. The completion notice is often posted in conjunction with a final inspection that includes the local building officials and the client. Some lenders require that all building inspection reports be submitted before payment is given and may also require a private inspection by an agent of the lender. Figure 43–2 shows a sample completion notice.

▼ BIDS

Construction bids are often obtained by the architect for the client. The purpose is to get the best price for the best work. Some projects require that work be given to the lowest bidder, while other projects do not necessarily go to the lowest bid. In some situations, especially in the private sector, other factors are considered, such as an evaluation of the builder's history based on quality, ability to meet schedules, cooperation with all parties, financial stability, and license, bond, and insurance.

The bid becomes part of the legal documents for completion of the project. The legal documents may include plans, specifications, contracts, and bids. The following items are part of a total analysis of costs for residential building construction. The architect and client should clearly know what the bid includes.

1. Plans
2. Permits, fees, specifications
3. Roads and road clearing
4. Excavation
5. Water connection (well and pump)
6. Sever connection (septic)
7. Foundation, waterproofing
8. Framing, including materials, trusses, and labor
9. Fireplace, including masonry and labor
10. Plumbing, both rough and finished
11. Wiring, both rough and finished
12. Windows
13. Roofing, including sheetmetal and vents
14. Insulation
15. Drywall or plaster
16. Siding
17. Gutters, downspouts, sheet metal, and rain drains
18. Concrete flatwork and gravel
19. Heating
20. Garage and exterior doors
21. Painting and decorating
22. Trim and finish interior doors, including material and labor
23. Underlayment
24. Carpeting, including the amount of carpet and padding, and the cost of labor
25. Vinyl floor covering amount and cost of labor
26. Formica
27. Fixtures and hardware
28. Cabinets
29. Appliances
30. Intercom/stereo system
31. Vacuum system
32. Burglar alarm
33. Weatherstripping and venting
34. Final grading, cleanup, and landscaping
35. Supervision, overhead, and profit
36. Subtotal of land costs
37. Financing costs

▼ CONSTRUCTION INSPECTIONS

When the architect or designer is responsible for the supervision of the construction project, then it is necessary to work closely with the building contractor to get the proper inspections at the necessary times. There are two types of inspections that occur most frequently. The regularly scheduled code inspections that are required during specific phases of construction help ensure that the construction methods and materials meet local and national code requirements. The general intent of these inspections is to protect the safety of the occupants and the public. Another type of inspection is often conducted by the lender during certain phases of construction. The purpose of these inspections is to ensure that the materials and methods described in the plans and specifications are being used. The lender has a valuable interest here. If the materials and methods are inferior or not of the standard expected, then the value of the structure may not be what the lender had considered when a preliminary appraisal was made. Another reason for these inspections, and probably the reason that the builder likes best, is the disbursement inspection. These inspections may be requested at various times, such as monthly, or they may be related to a specific disbursement schedule of four times during construction, for example. The intended result of these inspections is the release of funds for payment of work completed.

When the architect or designer supervises the total construction, then he or she must work closely with the contractor to ensure that the project is completed in a timely manner. When a building project remains idle, the overhead costs, such as construction interest, begin to add up quickly. A contractor that bids a job high but builds quickly may be able to save money in the final analysis. Some overhead costs go on daily even when work has stopped or slowed. The supervisor should also have a good knowledge of scheduling so that inspections can be obtained at the proper time. If an inspection is requested when not ready, the building official may charge a fee for excess time spent. Always try to develop a good rapport with building officials so each encounter goes as smoothly as possible.

FORM No. 748
STEVENS-NESS LAW PUBLISHING CO., PORTLAND, OR. 97204
1976

COMPLETION NOTICE

Notice hereby is given that the building or structure on the following described premises, to-wit (insert legal description including street address, if known):

has been completed.

All persons claiming a lien upon the same under Oregon's Construction Lien Law hereby are notified to file a claim of lien as required by ORS 87.035.

Dated _____ , 19 ___ .

....
Owner or Mortgagee

By ....

P. O. Address ....

....

STATE OF _____   )
) ss.
County of ....   )

I, ... , being first duly sworn, depose and say:

That on my behalf or as agent for ....

I did on the .... day of.. ,
19........ , duly post a notice of which the above is a true copy, in a conspicuous place upon the land or upon the improvement situated thereon described in said notice, to-wit: by posting, nailing, tacking, pasting, fastening or .... (indicate which) such notice at the front entrance on the building or improvement constructed, altered or repaired on the above described land. (If no building, state in what manner posted.)

....

....

....

....

Subscribed and sworn to before me this........

day of.. , 19........ .

....
Notary Public for _____ .

My commission expires:

(SEAL)

Record with recording officer within 5 days after posting
—ORS 87.045 (3).

STATE OF _____   )
) ss.
County of _____   )

I certify that the within instrument was filed in my office on the .... day of.. , 19........ , at .... o'clock ....M., and recorded in book .... on page .... or as file/reel number .... of the Construction Lien Book of said County.

Witness my hand and seal of County affixed.

....
Recording Officer.

By ....
                                      Deputy.

Figure 43–2 Completion notice. *Courtesy Stevens-Ness Law Publishing Co.*

CHAPTER

43 Construction Supervision Procedures Test

DIRECTIONS

Answer the questions with short, complete statements or drawings as needed on an 8½ × 11 sheet of notebook paper as follows:

1. Letter your name and the date at the top of the sheet.
2. Letter the question number and provide the answer. You do not need to write out the question.

QUESTIONS

Question 43–1 Define and list the elements of the following construction-related documents:
a. loan applications
b. contracts
c. building permits
d. completion notice
e. bids
f. change orders

Question 43–2 Outline and briefly discuss the required building construction inspections.

DIRECTIONS

1. All forms must be neatly hand-lettered or typed, unless otherwise specified by your instructor.

PROBLEMS

Problem 43–1 Copy and then complete the building permit application on page 673 based on this information.

○ Project location address: 3456 Barrington Drive, your city and state
○ Nearest cross street: Washington Street
○ Subdivision name: Barrington Heights
○ Township: 2S
○ Range: 1E
○ Section: 36
○ Tax lot: 2400
○ Lot size: 15000 sq ft.
○ Building area: 2000 sq ft.
○ Basement area: None
○ Garage area: 576 sq ft.
○ Stories: 1
○ Bedrooms: 3
○ Water source: Public
○ Sewage disposal: Public
○ Estimated cost of labor and materials: $88,500
○ Plans and specifications made by: You
○ Owner's name: Your teacher; address and phone may be fictitious
○ Builder's name: You, your address and phone number
○ You sign as applicant

○ Homebuilder's registration no.: Your social security number or another fictitious number
○ Date: Today's date

Problem 43–2 Copy and then complete the building contract in Figure 43–1 based on this information:

○ Today's date.
○ Name yourself as the contractor and your teacher as the owner. (Give complete first name, middle initial, and last name where names are required.)
○ ARTICLE I:
Construction of approximate 2000-square-foot house to be located at 3456 Barrington Drive, your city and state. Also known as Lot 8, Block 2, Barrington Heights, your county and state.
○ Drawing and specifications prepared by you.
○ All said work to be done under the direction of you.
○ ARTICLE II:
Commence work within 10 days and substantially complete on or before:
Give a date four months from the date of the contract.
○ ARTICLE III:
$88,500
Payable at the following times:
1. One-third upon completion of the foundation.
2. One-third upon completion of drywall.
3. One-third 30 days after posting the completion notice.
○ ARTICLE X:
Insurance $250,000; $500,000; and $50,000.
○ ARTICLE XVIII:
Give the owner's name, address, and phone number. (A fictitious address and phone number may be used if preferred.) Give your name and address as the contractor.
○ ARTICLE XX:
This contract amount is valid for a period of 60 days. If, for reasons out of the contractor's control, construction has not begun by the end of this 60-day period, contractor has the right to rebid and revise the contract.

Problem 43–3 Copy and then complete the change order form on page 674 based on this information:

○ Today's date
○ Project Name: Your teacher's name
○ Location: 3456 Barrington Drive, your city and state. Also known as Lot 8, Block 2, Barrington Heights, your county and state
○ Description of change: Add a 24″ × 48″ insulated tempered flat glass skylight in the vaulted ceiling centered in the entry foyer
○ Additional cost: $850
○ Adjusted total project cost: $89,350
○ Signatures

Problem 43–4 Copy and then complete the completion notice (top half of form above double line only; the rest of the information is for official Notary Public and recording officer) in Figure 43–2 based on this information:

○ Today's date
○ Owner or mortgagee: Your name
○ P.O. address: Your address

BUILDING PERMIT APPLICATION

Amount Due _____

Project Location (Address) _____

Nearest Cross Street _____

Subdivision Name _____ Lot _____ Block _____

Township _____ Range _____ Section _____ Tax Lot _____

Lot Size _____ (Sq. Ft.) Building Area _____ (Sq. Ft.) Basement Area _____ (Sq. Ft.) Garage Area _____ (Sq. Ft.)

Stories _____ Bedrooms _____ Water Source _____ Sewage Disposal _____

Estimated Cost of Labor and Material _____

Plans and Specifications made by _____ accompany this application.

Owner's Name _____ Builder's Name _____

Address _____ Address _____

City _____ State _____ City _____ State _____

Phone _____ Zip _____ Phone _____ Zip _____

I certify that I am registered under the provisions of ORS Chapter 701 and my registration is in full force and effect. I also agree to build according to the above description, accompanying plans and specifications, the State of Oregon Building Code, and to the conditions set forth below.

_____ _____
APPLICANT   HOMEBUILDER'S REGISTRATION NO.   DATE

I agree to build according to the above description, accompanying plans and specifications, the State of Oregon Building Code, and to the conditions set forth below.

_____   _____
APPLICANT   DATE

TO BE FILLED IN BY APPLICANT

Problem 43-1

CHANGE ORDER

Date: _____

Project Name: _____

_____

Location: _____

_____

_____

Description of Change:

Additional Cost: _____

Reduction in cost: _____

Adjusted total project cost: _____

_____
OWNER

_____
BUILDER

_____
LENDER

Problem 43–3

SECTION XIII

COMMERCIAL DRAFTING

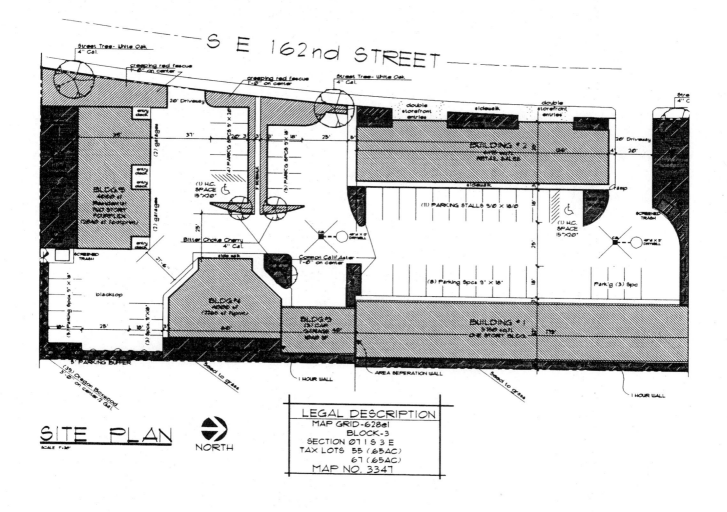

SITE PLAN
SCALE 1"=20'

NORTH

LEGAL DESCRIPTION
MAP GRID-628e1
BLOCK-3
SECTION 01 1 S 3 E
TAX LOTS 55 (.65AC)
67 (.65AC)
MAP NO. 3347

CHAPTER 44
Commercial Construction Projects

▼ OFFICE PRACTICE

Although the design process is similar for commercial projects, the actual office practice for them is different. As a new employee, you will be given jobs that are typically not the office favorites. These might include making corrections or lettering drawings that were drawn by others on the staff. As your linework and lettering speed and quality improve, so will the variety of drafting projects that you work on. No matter what your drawing level, you will be required to research your drawing project. The two major tools that you will be using will be Sweets catalogs and the building code covering your area. Chapter 45 will introduce you to building codes as they apply to commercial projects.

Sweets catalogs are a collection of vendors' brochures that are used in nearly every architectural and engineering office. In residential construction, a drafter may be required to conduct a limited amount of research. In commercial projects, because of the various materials that are available, the drafter usually spends several hours a week researching product information.

Figure 44–1 shows an example from the Simpson Strong-Tie catalog of Sweets. This column cap (CC) is a common connector used to fasten beams to wood post. A common method of specifying this connection in the calculations might include only the use of a CC cap. The drafter would be required to determine the size of the post and the size of the beam to be supported, and then select the proper cap.

▼ TYPES OF DRAWINGS

Most of the information to which you have been introduced while learning about residential architecture can be applied to commercial projects as well. There will be differences, of course, but the basic principles and procedures of drafting will be the same. Many of the differences will be discussed in this chapter.

Calculations

One of the prime differences of a commercial project from residential drafting is your need to work with engineers' calcuations. No matter if you work for a designer, architect, or engineer, you will need to use a set of calculations as you draw the plans for a commercial project. In residential drafting, engineer's calcs are used to determine sizes of retaining walls and seismic and wind loads. In commercial drafting, a set of calculations is provided for the entire structure, detailing everything from beam sizes to the size and number of nails to use in a wall.

Although called "engineer's calcs," these calculations may be prepared by an engineer or an architect. By state law, commercial calculations are required to be signed by a licensed architect or engineer. Because of the complexity of many concrete and steel structures, architects typically design and coordinate the drafting of the project but hire a consulting engineer to design the structural members. The architect typically designs the structure and decides where structural columns and beams will be located, and an engineer determines the stress and required member to resist this stress.

Figure 44–2 shows a page from the engineer's calcs. Most calcs are divided into three areas. The calc for a specific area usually begins with a statement of the problem. In Figure 44–2 the engineer is determining the size of the roof sheathing to be used for a warehouse. The second area of the calcs is the mathematical formula used to determine the stress and needed reaction to that stress. This area of the calculations is of little meaning to the inexperienced drafter. But since the part of the calcs that you will need is the solution area, you will note that the engineer placed the solution to the math problem in a box for easy reference. This is the information that the drafter needs to place on the plans.

Most calculations start at the highest level of the structure and work down to the foundation level. This allows loads to be accumulated as formulas are worked out so that loads from a past solution can be used on a lower level. Figure 44–3 shows the calculations that were used to determine the loads on a column.

Figure 44–4 shows the drawing that the drafter produced from the calcs. Notice that there are items on the drawing that were not on the specifications. This is where the experience of the construction process is used. Based on past experience and a knowledge of the construction process, the drafter is able to draw such items as the shim without having it specified. Depending on the engineer and the experience level of the drafting team, the calcs may or may not contain sketches. If the engineer is specifying a common construction technique to a skilled drafting team, usually a sketch would not be required. The information to draw this detail was

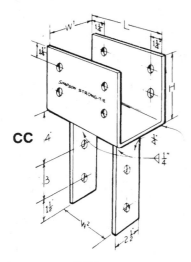

CC COLUMN CAP

FACTORY VALUES! Precision factory gang-punched holes speed installation and insure full bolt values.

SPECIFICATIONS
1. Special corrosion protection Linear Polymer Formula (Simpson Gray).
2. Straps are fillet-welded both sides to bottom of cap.
3. Straps are centered upon the cap unless otherwise specified
4. For complete CC values consult Approval No. 1211.
5. For CCOB beam column cap values, utilize Table or consult Approval No. 1211, applying values no greater than the lesser element employed.
6. **MATERIAL:** ¼" hot-rolled steel.

For special, custom or rough lumber sizes, provide dimensions.

*Note: Any W² dimension may be specified in combination with any column cap size given. For example, specify as "CC65" for a 5" column and 6" (nominal) beam width requirement. **COLUMN CAP ONLY** may be specified for field-welding to pipe or other column condition by specifying as "CCO—". **SPECIAL COLUMN CAPS** with W¹, "L", "H", and hole schedules different from above may be special ordered. CCOB—Any two CCO's may be specified for back-to-back welding to create the CCOB cross beam connector. For end conditions specify ECC column caps and provide dimensions in accordance with Table.

*Column straps may be rotated 90° on special orders where W¹ is greater than W².

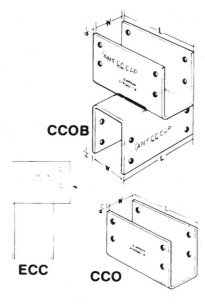

Model No.	W¹	W²*	L*	H	Holes for Cap Bolt	Holes for Strap Bolt	Bolt Values	Seat Load Vertical**
CC44	3⅝"	3⅝"	7"	4"	(2) ⅝ MB	(2) ⅝ MB	3024	9430
CC3¼-4	3¼"	3⅝"	11"	6½"	(4) ⅝ MB	(2) ⅝ MB	6050	15470
CC3¼-6	3¼"	5½"	11"	6½"	(4) ⅝ MB	(2) ⅝ MB	6050	15470
CC5¼-6	5¼"	5½"	13"	8"	(4) ¾ MB	(2) ¾ MB	9310	29980
CC5¼-8	5¼"	7½"	13"	8"	(4) ¾ MB	(2) ¾ MB	9310	29980
CC64	5½"	3⅝"	11"	6½"	(4) ⅝ MB	(2) ⅝ MB	6050	23290
CC46	3⅝"	5½"	11"	6½"	(4) ⅝ MB	(2) ⅝ MB	6050	14820
CC66	5½"	5½"	11"	6½"	(4) ⅝ MB	(2) ⅝ MB	6050	23290
CC68	5½"	7½"	11"	6½"	(4) ⅝ MB	(2) ⅝ MB	6050	23290
CC76	6⅞"	5½"	13"	8"	(4) ¾ MB	(2) ¾ MB	9625	40220
CC7-7	6⅞"	6⅞"	13"	8"	(4) ¾ MB	(2) ¾ MB	9625	40220
CC7-8	6⅞"	7½"	13"	8"	(4) ¾ MB	(2) ¾ MB	9625	40220
CC86	7½"	5½"	13"	8"	(4) ¾ MB	(2) ¾ MB	9625	40220
CC88	7½"	7½"	13"	8"	(4) ¾ MB	(2) ¾ MB	9400	37540
CC96	8⅞"	5½"	13"	8"	(4) ¾ MB	(2) ¾ MB	9400	43790
CC98	8⅞"	7½"	13"	8"	(4) ¾ MB	(2) ¾ MB	9400	37540
CC106	9½"	5½"	13"	8"	(4) ¾ MB	(2) ¾ MB	9400	47550

**Subject to:
As limited by nominal beam sizes @ 385 psi or normal Glulam sizes @ 450 psi of seat area.
End bearing value of post, L/R of post, or other values to be deducted.
*ECC Models are approximately 4" shorter than the "L" dimension given in Table, that provides for half the number of cap bolts with consequent decrease in bolt and seat load values.

ACCEPTED—see Research Recommendation No. 1211 of the International Conference of Building Officials (Uniform Building Code).

Figure 44–1 Vendors' catalogs are needed to determine many components shown on details. This page, from the Simpson Strong-Tie Company, can be used to determine the cap dimensions at post-to-beam details.

compiled from the foundation, grading, roof, and floor plans, as well as the sections and other areas of the calcs.

Floor Plans

Commercial projects are similar to residential projects in their components. Each uses similar types of drawings to display basic information. The floor plan of a commercial project is used to show the locations of materials in the same way that a residential floor plan does. The difference in the floor plan comes in the type of material being specified. Figure 44–5 shows an example of a floor plan for college classrooms.

Because most commercial projects are so complex, specific areas of the structure are often clarified by grid reference markers. Figure 44–5 shows grid references 6 through 12 and A through K. The print reader can use these references to relate each drawing and detail back to the floor plan. Commercial floor plans are usually larger than residential plans to clarify complicated areas. Figure 44–6 shows an enlarged view of the bath areas shown in Figure 44–5. To keep the floor plans uncluttered and to aid clarity, schedules, notes, and symbols are used extensively throughout commercial drawings. Figure 44–7 shows an example of the finish schedule for the structure in Figure 44–5.

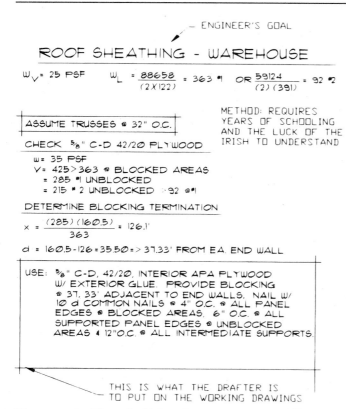

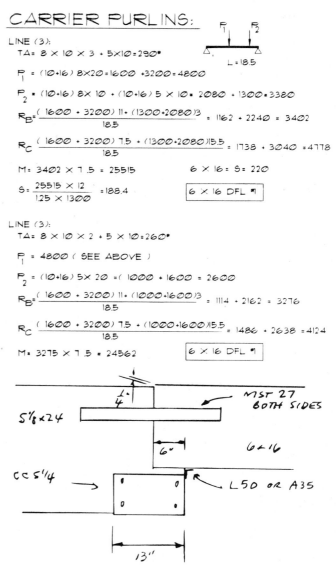

Figure 44-2 The calculations for a structure typically contain a problem to be solved, the mathematical solution, and the specification to be placed on the drawing by a drafter. *Courtesy Kenneth D. Smith, AIA, StructureForm Masters, Inc.*

Figure 44-3 An example of the calculations to determine the loading criteria at a beam connection over a column. The drafter draws the detail, using proper line and lettering quality. *Courtesy Kenneth D. Smith, AIA, StructureForm Masters, Inc.*

Electrical Plans

On a set of residential plans, an electrical plan is drawn to provide information for the locations of outlets, switches, light fixtures, and other related information for the electrician. A similar electrical plan is provided for a set of commercial drawings. Information for the electrical plans is usually specified by the owner or by major tenants. This information is given to an electrical engineer, who will specify how the circuitry is to be installed. The engineer typically marks the needed information on a print of the floor plan. This print is then given to a drafter who transfers the information to an electrical plan. Because of the complexity of the electrical needs for commercial projects, a separate plan is often drawn to show the power outlet needs, as seen in Figure 44-8 along with a plan for the lighting needs, as seen in Figure 44-9. A plan is typically provided for roof-mounted or other exterior equipment. Schematic drawings may also be provided by the electrical engineer and must be incorporated into the working drawings. See Figure 44-10.

Reflected Ceiling Plan

Acoustical ceilings are typically used on commercial projects and supported by wires in what is called a *t-bar system*. A reflected ceiling plan shows how the ceiling tiles are to be placed. The plan also typically shows the location of light panels or ceiling-mounted heat registers. Figure 44-11 shows an example of a reflected ceiling plan. The drafter must work with information provided by both the electrical and the mechanical engineers to draw this plan properly.

Mechanical Plans

The mechanical plans for commercial projects are used in the same manner as in residential projects. These plans show the location of heating and cooling equipment as well as the duct runs. As with other plans, the mechanical contractor marks all equipment sizes and

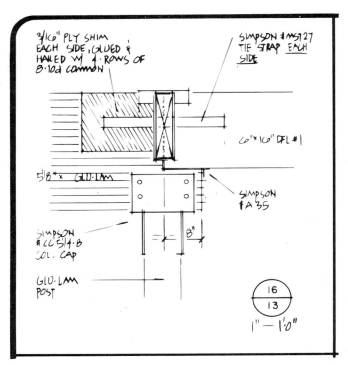

Figure 44–4 A detail drawn to convey the information in the engineer's calculations shown in Figure 44–3.

duct runs on a print, and a drafter redraws the information on the mechanical plans. Figure 44–12 shows an example of a mechanical plan. Figure 44–13 and 44–14 show examples of details that supplement the plan view.

Plumbing Plan

As with other plans that are generated from the floor plan, the information for the plumbing plan is supplied by the engineer and redrawn by a drafter. Depending on the complexity of the project, the plumbing plan may be divided into sewer and fresh water plans. Figure 44–15 shows an example of a plumbing plan that combines both systems on one plan. Notice that this plan shows far more detail and dimensions than the plans used for residential projects. Because of the complexity of commercial projects, more details are usually placed on the plumbing plan. Figure 44–16 shows examples of the details that the drafter was required to place on the plan view.

Foundation Plans

Foundations in commercial work are typically concrete slabs because of their durability and low labor cost. As seen in Figure 44–17, the foundation for a commercial project is similar to that of a residential project. The size of the footings will vary, but the same type of stresses affect a commercial project that affect a residence. They differ only in their magnitude. In commercial projects it is common to have both a slab and a foundation plan. The slab, or slab-on-grade plan as it is sometimes called, is a

plan view of the construction of the floor system and specifies the size and location of concrete pours. A slab-on-grade plan can be seen in Figure 44–18. The foundation plan is used to show below-grade concrete work. An example of a foundation plan can be seen in Figure 44–19.

Elevations

Elevations for commercial projects are very similar to the elevations required for a residential project. Figure 44–20 shows an example of an elevation for a restaurant. Notice that the elevation shows the same types of information as a residential elevation but uses much more detail. Because the size of many commercial structures is so large, commercial projects are sometimes drawn at a scale of $\frac{1}{8}'' = 1'-0''$ or smaller and have very little detail added. Material is clarified through the use of details.

Sections

The drawing of sections for commercial projects is similar to residential projects, but the sections often have a different use in commercial projects. In residential projects, sections are used to show major types of construction. This use is also made of sections on commercial plans, but the sections are used primarily as a reference map of the structure. Sections are usually drawn at a scale of $\frac{1}{4}'' = 1'-0''$ or smaller, and details of specific intersections are referenced to the sections. Figure 44–21 shows an example of this type of section. Another common method of drawing sections is to draw several partial sections, as seen in Figure 44–22.

Details

On a residential plan, you may draw details of a few special connections or of stock items such as fireplaces or footings. In commercial projects, drawing details is one of the primary jobs of the drafter. Because there are so many variables in construction techniques and materials, details are used to explain a specific area of construction. Figure 44–23 shows some of the details that were required to supplement the section.

Roof Plans

Roof plans and roof drainage plans are typically used more in commercial projects than in residential work. Because the roof system is usually flat and contains concentrated loads from mechanical equipment, a detailed placement of beams and interior supports is usually required. Two of the most common types of roof systems used on commercial projects are the truss and the panelized systems. The use of trusses was discussed in Chapter 25. A panelized roof consists of large beams supporting smaller beams, which in turn support 4 × 8 roofing panels. The system will be further discussed in Chapter 46. The roof plan can be seen in Figure 44–24 and the roof framing plan can be seen in Figure 44–25.

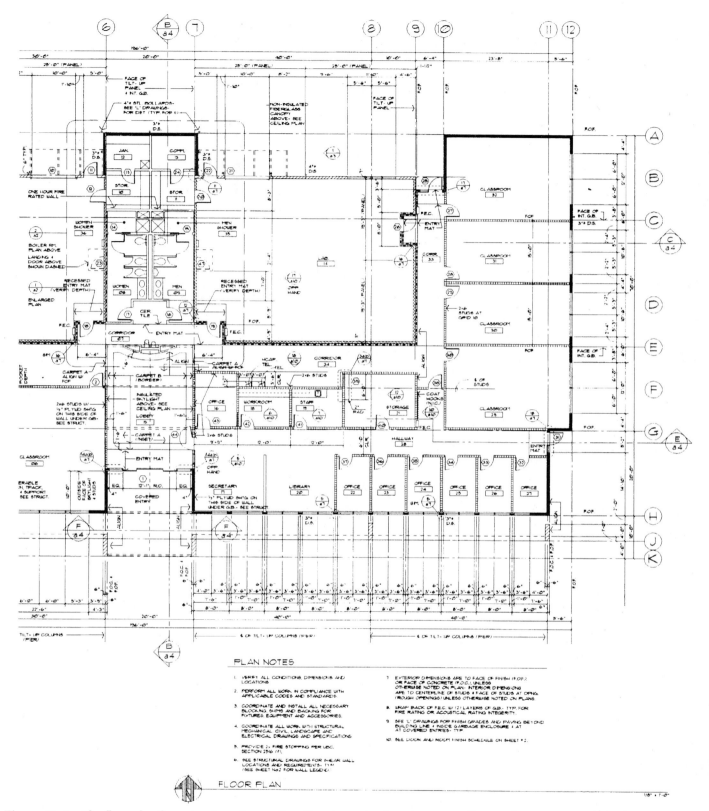

Figure 44–5 The floor plan for a structure shows all wall and opening locations, as well as a reference map to help the print reader understand the relationship of other drawings to the floor plan. *Courtesy Architects Barrentine, Bates, & Lee, AIA.*

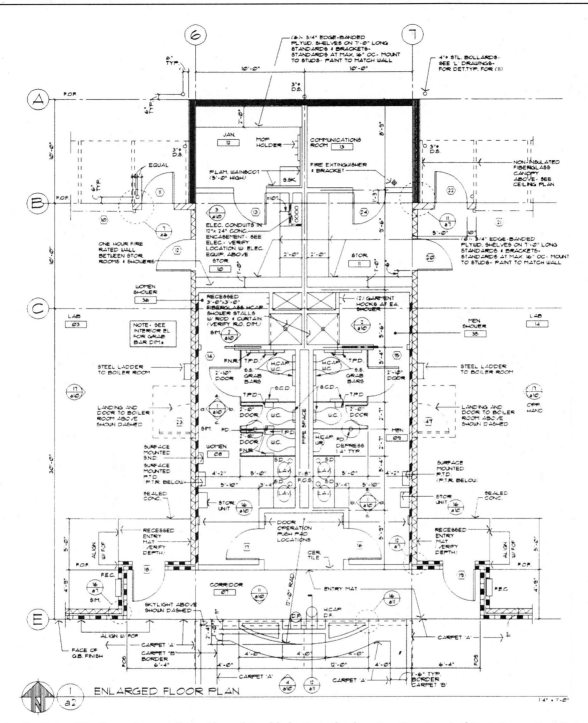

Figure 44–6 Portions of the floor plan are often enlarged to add clarity to the drawing. *Courtesy Architects Barrentine, Bates & Lee, AIA.*

Interior Elevations

The kind of interior elevations drawn for a commercial project will depend on the type of project to be drawn. In an office or warehouse building, cabinets are minimal and usually not drawn. For the office setting, an interior decorator may be used to coordinate and design interior spaces. On projects such as a restaurant, the drafter probably will work closely with the supplier of the kitchen equipment to design an area that will suit both the users and the equipment. This requires coordinating the floor plan with the interior elevations. Figure 44–26 shows the elevations for the bathroom in Figure 44–6.

Plot Plans

The *plot* or *site plan* for a commercial project resembles the plot plan for a residential project. One major

ROOM FINISH SCHEDULE

NO.	ROOM NAME	FLOOR	BASE	WALLS N	E	S	W	CEILING	HEIGHT	REMARKS
01	LOUNGE	4	1	1	1	1	1	1/6	VARIES	
02	CORRIDOR	1/2	1	1	1	1	1	3	7'-6"/9'-0"	
03	LAB	1/6	1	3	3/3	3	4		VARIES	BASE AT G.B. WALLS ONLY
04	VOCATIONAL CLASSROOM	2	1	1	1	1	2		9'-0"	
05	CLASSROOM	2	1	1	1	1	2		9'-0"	
06	CLASSROOM	2	1	1	1	1	2		9'-0"	
07	CORRIDOR	1/6	1	1	1	*1	3		7'-6"	* ENTRY KIOSK
08	WOMEN	5	2	2	2	2	2		8'-0"	
09	MEN	5	2	2	2	2	2		8'-0"	
10	STORAGE	6	1	1	1	1/4	1		8'-8"	
11	STORAGE	6	1	1	1/4	1	1		8'-8"	
12	JANITOR	6	1	2	2/6	2/6	2		VARIES	
13	COMMUNICATIONS ROOM	6	1	1	1	1	1		VARIES	
14	LAB	1/6	1	1/3	3	1/3	3	4	VARIES	BASE AT G.B. WALLS ONLY
15	LOBBY	2/5	1	*1	1	1	6		VARIES	* ENTRY KIOSK
16	OFFICE	3	1	1	1	1	2		9'-0"	
17	SECRETARY	2	1	1	1	1	2		9'-0"	
18	WORKROOM	4	1	1	1	1	2		9'-0"	
19	STAFF	4	1	1	1	1	2		9'-0"	
20	LIBRARY	2	1	1	1	1	2		9'-0"	
21	STORAGE	4	1	1	1	1	2		9'-0"	
22	OFFICE	3	1	1	1	1	2		9'-0"	
23	OFFICE	3	1	1	1	1	2		9'-0"	
24	OFFICE	3	1	1	1	1	2		9'-0"	
25	OFFICE	3	1	1	1	1	2		9'-0"	
26	OFFICE	3	1	1	1	1	2		9'-0"	
27	OFFICE	3	1	1	1	1	2		9'-0"	
28	HALLWAY	1/2	1	1	1	1	1		9'-0"	
29	CLASSROOM	2	1	1	1	1	2		9'-0"	
30	CLASSROOM	2	1	1	1	1	2		9'-0"	
31	CLASSROOM	2	1	1	1	1	2		9'-0"	
32	CLASSROOM	2	1	1	1	1	2		9'-0"	
33	CORRIDOR	1/2	1	1	1	1	3		9'-0"	
34	CORRIDOR	2	1	1	1	1	3		9'-0"	
35	MEN SHOWER	5	2	2	2	2	2	1	8'-0"	
36	WOMEN SHOWER	5	2	2	2	2	2	1	8'-0"	
37	BOILER ROOM	6	1	4	4	2	5		VARIES	
38	ELECTRICAL ROOM	6	1	4	2	2	5		VARIES	

ROOM FINISH KEY

FLOORS
1. ENTRY MAT
2. CARPET 'A'
3. CARPET 'B'
4. SHEET VINYL
5. CERAMIC TILE (COLOR No.1)
6. SEALED CONC.

BASE
1. 4" COVED RUBBER BASE
2. 5" COVED CER. TILE (SEE INTERIOR ELEVATIONS) THINSET OVER WR G.B. (COLOR No.2)

NOTES
* SEE INT. ELEVATIONS
— NO WORK REQUIRED

WALLS
1. GYPSUM BOARD PAINT
2. WATER RESISTANT GYPSUM BOARD PAINT
3. EXPOSED CONCRETE - ELASTOMERIC PAINT
4. EXPOSED CONCRETE W/ SEALER
5. 5'-0" HIGH PLAM WAINSCOT OVER WR G.B. W/ METAL EDGES

CEILINGS
1. GYPSUM BOARD ON 2x FRAMING AT 16" OC/1 HOUR FIRE RATED + WATER RESISTANT G.B. AT ROOMS 08, 09, 35 & 36)
2. 2x4 SUSP. ACOUSTIC TILE (NON-RATED)
3. GYPSUM BOARD ON EACH SIDE OF 2x CEILING JOISTS AT 24" OC (ONE HOUR FIRE RATED AT CORRIDORS)
4. EXPOSED ROOF STRUCTURE- PAINT
5. (2) LAYERS OF 5/8" GYPSUM BOARD OVER MET. HAT CHANNELS AT 16" OC.- (ONE HOUR FIRE RATED)
6. INSULATED FIBERGLASS SKYLIGHT

WALL LEGEND

EXTERIOR WALLS

1. EXTERIOR INSULATION & FINISH SYSTEM (EIFS) OVER 1/2" PLYWD. WALL SHTG. OVER 2x6 STUDS AT 16" OC. W/ R-19 FOILFACE FIBERGLASS INSUL. & 5/8" GYP. BOARD INTERIOR FINISH (CONTINUE DOUBLE LAYER OF FINISH FROM INTERIOR WALLS AT OFFSETS IN EXTERIOR WALLS)

2. EXTERIOR INSULATION & FINISH SYSTEM (EIFS) OVER 5-1/2" TILT-UP CONC. WALLS W/ PAINTED EXPOSED CONC. ON GARBAGE ENCLOSURE SIDE ONLY - (8'-10" HIGH WALL)

3. EXTERIOR INSULATION & FINISH SYSTEM (EIFS) OVER 5-1/2" TILT-UP CONC. WALLS W/ SEALED OR PAINTED INTERIOR FINISH

4. 12" TILT-UP CONC. COLUMNS- PAINT ALL EXPOSED SURFACES

INTERIOR WALLS

5. 2x4 STUDS AT 16" OC. W/ 3" ACOUSTIC INSUL. IN STUD CAVITY - G.B. ON ONE SIDE (LEAVE AIR SPACE BETWEEN STUDS AND/OR TILT-UP CONC. WALL) EXTEND FROM FLOOR LINE UP TO ROOF FRAMING (ONE HOUR FIRE RATED CONSTRUCTION)

6. 5-1/2" TILT-UP CONC. WALL W/ EXTERIOR INSUL. & FINISH SYSTEM (AT EXTERIOR) ABOVE ALL LOW ROOF AREAS-SEAL OR PAINT ALL EXPOSED CONC. SURFACES

7. 2x4 STUDS AT 16" OC. W/ 3" ACOUSTIC INSUL.- (2) LAYERS OF G.B. ON BOTH SIDES EXTEND FROM FLOOR LINE UP TO ROOF FRAMING (END SECOND LAYER OF G.B. 6" ABOVE CEILING LINE)- 2x6 STUDS AS NOTED ON FLOOR PLAN (ONE HOUR FIRE RATED CONSTRUCTION AT LOBBY (15) & CORRIDORS)

8. 2x4 STUDS AT 16" OC. W/ 3" ACOUSTIC INSUL.- G.B. ON BOTH SIDES OF WALL EXTEND FROM FLOOR LINE UP TO ROOF FRAMING, UNLESS OTHERWISE NOTED (ONE HOUR FIRE RATED AS NOTED ON PLANS)

9. 2x4 STUDS AT 16" OC. W/ 3" ACOUSTIC INSUL.- (1/2" RESILIENT CHANNELS AT 16" OC. W/ G.B. ON ONE SIDE & (2) LAYERS OF G.B. ON OPPOSITE SIDE- EXTEND FROM FLOOR LINE UP TO 6" ABOVE CEILING LINE

10. METAL HAT CHANNELS AT 16" OC. FURRING OVER CONC. WALL OR CONC. CONDUIT ENCASEMENT W/ G.B. FINISH- EXTEND FROM FLOOR LINE UP TO FLOOR FRAMING

11. 2x4 STUDS AT 16" OC. W/ 3" ACOUSTIC INSUL. & (2) LAYERS OF G.B. ON ONE SIDE (LEAVE AIR SPACE BETWEEN STUDS) EXTEND FROM FLOOR LINE TO ROOF FRAMING (ONE HOUR FIRE RATED CONSTRUCTION)

COMPNAME

difference is the need to draw parking spaces, driveways, curbs, and walkways. In commercial projects, the plot plan is often drawn as part of the preliminary design study. Parking spaces are determined by the square footage of the building and the type of usage the building will receive. Another major consideration is locating all existing utilities. Water lines and meters, sewer laterals and manholes, and underground electric and communication services adjacent to the property must be shown. Figure 44–27 shows an example of a plot plan for an office building.

Grading Plan

A *grading plan* shows the existing soil contours and any changes that must be made to the site to accommodate the structure and related improvements. Finish-grade contours representing 1 and 5' (300/1525 mm) intervals are typically shown, along with retaining walls and underground drainage facilities. Contours also reflect surface depression used to divert water away from the structure to drainage grates. A drafter working for a civil engineer will then translate field notes into a grading plan, as seen in Figure 43–28.

Landscape Plan

The size of the project determines how the landscape plan is drawn. On a small project, a drafter may be given the job of drawing the landscape plan. When this is the case, the drafter must determine suitable plants for the area and the job site and then specify the plants by their proper Latin name on the plan. On typical commercial projects, a copy of the plot plan is given to a landscape architect, who will specify the size, type, and location of plantings. A method of maintaining the plantings is also typically shown on the plan, as seen in Figure 44–29.

Figure 44–7 Schedules, notes, and special symbols are typically used to keep the floor plan from becoming cluttered. *Courtesy Architects Barrentine, Bates, & Lee, AIA.*

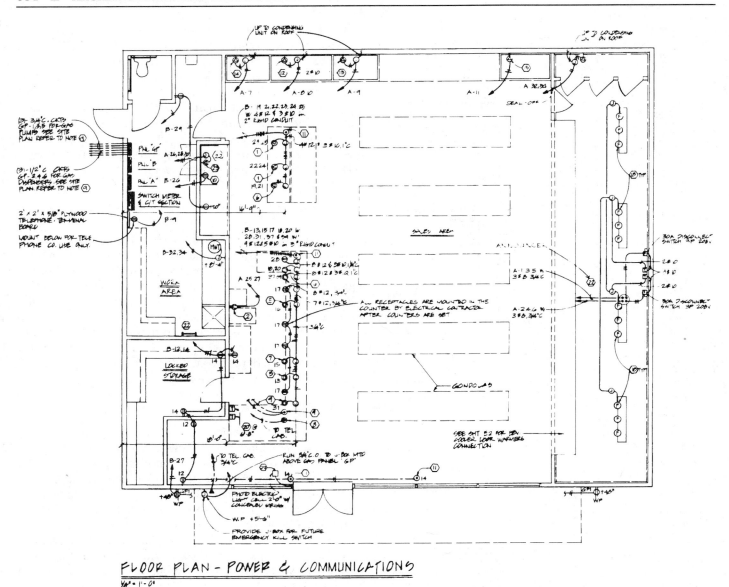

FLOOR PLAN - POWER & COMMUNICATIONS
¼" = 1'-0"

Figure 44–8 A plan is drawn to show power supplies. *Courtesy The Southland Corporation.*

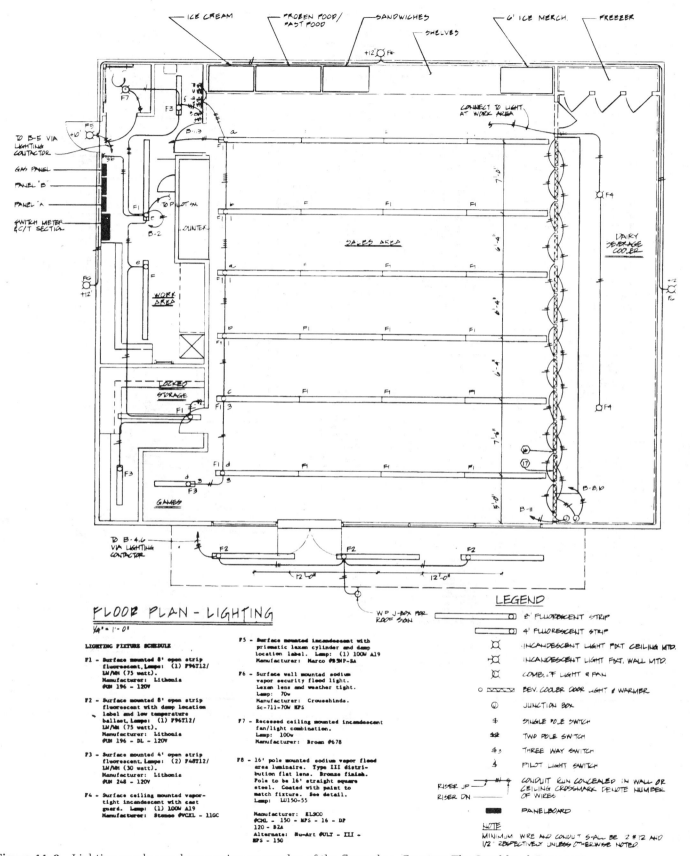

FLOOR PLAN - LIGHTING

1/4" = 1'-0"

LIGHTING FIXTURE SCHEDULE

F1 - Surface mounted 8' open strip fluorescent, Lamps: (1) F96T12/LW/WM (75 watt). Manufacturer: Lithonia #UN 196 - 120V

F2 - Surface mounted 8' open strip fluorescent with damp location label and low temperature ballast, Lamps: (1) F96T12/LW/WM (75 watt). Manufacturer: Lithonia #UN 196 - DL - 120V

F3 - Surface mounted 4' open strip fluorescent, Lamps: (2) F48T12/LW/WM (30 watt). Manufacturer: Lithonia #UN 248 - 120V

F4 - Surface ceiling mounted vapor-tight incandescent with cast guard. Lamp: (1) 100W A19 Manufacturer: Stonco #VCKL - 11GC

F5 - Surface mounted incandescent with prismatic lexan cylinder and damp location label. Lamp: (1) 100W A19 Manufacturer: Marco #B3NP-5A

F6 - Surface wall mounted sodium vapor security flood light. Lexan lens and weather tight. Lamp: 70w Manufacturer: Crousehinds. Sc-711-70W MPS

F7 - Recessed ceiling mounted incandescent fan/light combination. Lamp: 100w Manufacturer: Broan #678

F8 - 16' pole mounted sodium vapor flood area luminaire. Type III distribution flat lens. Bronze finish. Pole to be 16' straight square steel. Coated with paint to match fixture. See detail. Lamp: LU150-55

Manufacturer: ELSCO #CML - 150 - MPS - 16 - DF 120 - B2A Alternate: Hu-Art #ULT - III - MPS - 150

LEGEND

▭▭▭	8' FLUORESCENT STRIP
▭▭	4' FLUORESCENT STRIP
✕	INCANDESCENT LIGHT FIXT. CEILING MTD.
⊗	INCANDESCENT LIGHT FIXT. WALL MTD.
✕	COMB. OF LIGHT & FAN
○ ▭▭▭	BEV. COOLER DOOR LIGHT & WARMER
⊙	JUNCTION BOX
$	SINGLE POLE SWITCH
$$	TWO POLE SWITCH
$₃	THREE WAY SWITCH
$	PILOT LIGHT SWITCH

RISER UP
RISER DN

CONDUIT RUN CONCEALED IN WALL OR CEILING CROSSMARK DENOTE NUMBER OF WIRES

◼◼◼ PANELBOARD

NOTE
MINIMUM WIRE AND CONDUIT SHALL BE 2 #12 AND 1/2" RESPECTIVELY UNLESS OTHERWISE NOTED

Figure 44-9   Lighting needs are shown using an overlay of the floor plan. *Courtesy The Southland Corporation.*

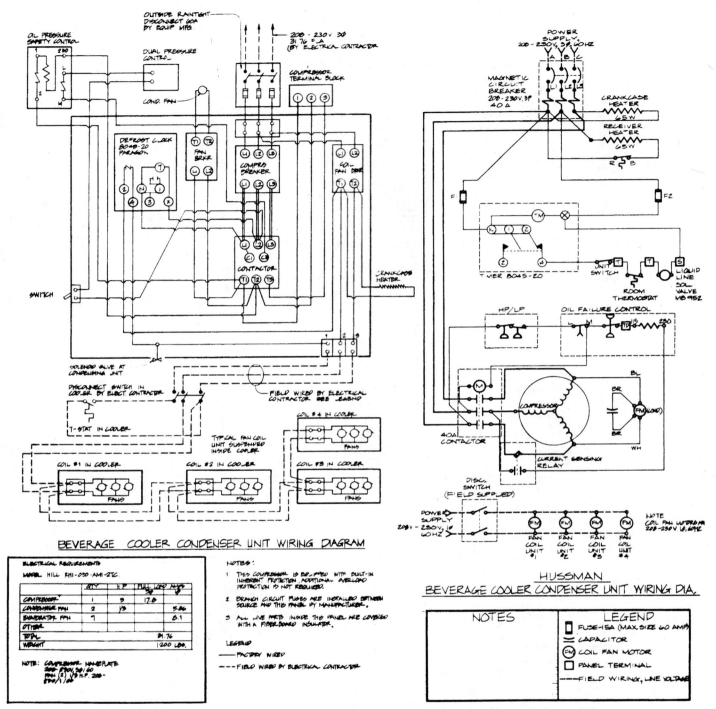

Figure 44–10 A drafter is often required to draw schematic diagrams to help explain electrical needs. *Courtesy The Southland Corporation.*

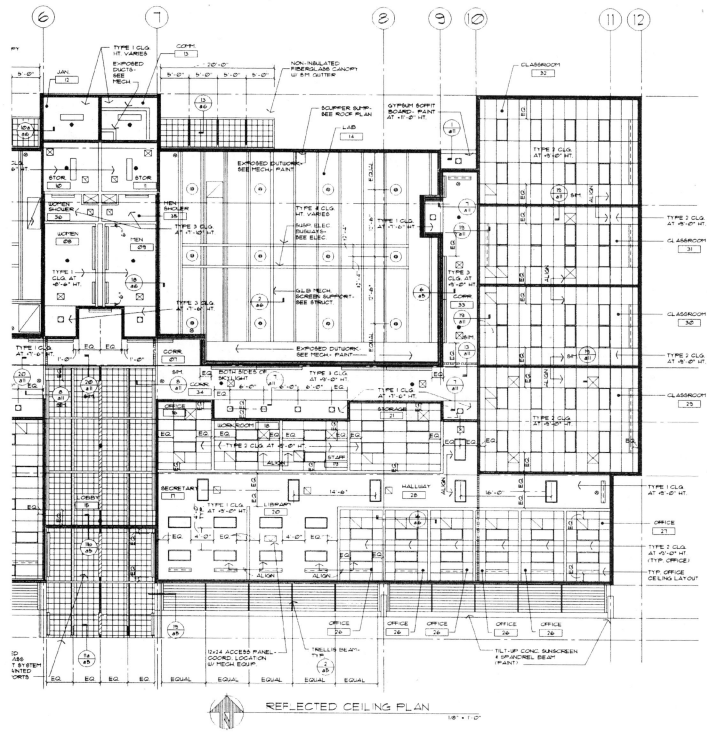

REFLECTED CEILING PLAN

1/8" = 1'-0"

Figure 44–11   A reflected ceiling plan is used to show the layout for the suspended ceiling system. *Courtesy Architects Barrentine, Bates & Lee, AIA.*

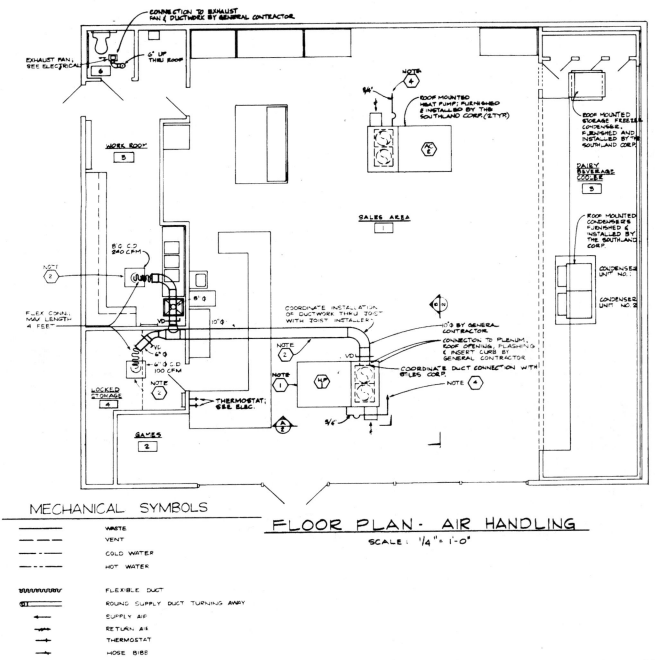

Figure 44-12 A mechanical plan is drawn to show the heating and air-conditioning requirements. *Courtesy The Southland Corporation.*

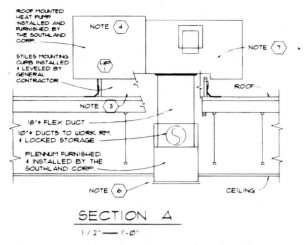

ROOF MOUNTED HEAT PUMP INSTALLED AND FURNISHED BY THE SOUTHLAND CORP.

STILES MOUNTING CURB INSTALLED & LEVELED BY GENERAL CONTRACTOR

NOTE ④

NOTE ⑦

ROOF

NOTE ③

18"⌀ FLEX DUCT
10"⌀ DUCTS TO WORK RM. & LOCKED STORAGE

PLENUM FURNISHED & INSTALLED BY THE SOUTHLAND CORP.

NOTE ⑥

CEILING

SECTION A
1/2" ——— 1'-0"

Figure 44–13   Details are required to show how HVAC equipment is to be installed. *Courtesy The Southland Corporation.*

NOTES:

① "GENERAL ELECTRIC (TRANE)" HEAT PUMP FURNISHED AND INSTALLED BY THE SOUTHLAND CORP. (SEE SCHEDULE)

② DUCTWORK TO WORK RM. & LOCKED STORAGE FURNISHED AND INSTALLED BY THE GENERAL CONTRACTOR. DIFFUSERS FURNISHED AND INSTALLED BY THE SOUTHLAND CORP.

③ "STILES" INSERT ROOF CURB FURNISHED AND INSTALLED BY THE GENERAL CONTRACTOR. SHIM AS REQUIRED TO LEVEL. (STILES CORP - P.O. BOX 2116, 1675 CRANE WAY, SPARKS, NEVADA 89431.

④ CONDENSATE DRAIN LINE FURNISHED AND INSTALLED BY GENERAL CONTRACTOR.

⑤ "STILES" FLEXIBLE DUCT FURNISHED AND INSTALLED BY THE SOUTHLAND CORP.

⑥ "STILES" SUPPLY AND RETURN AIR PLENUM AND DIFFUSER FURNISHED AND INSTALLED BY THE SOUTHLAND CORP.

⑦ "STILES" AIR MIXING PLENUM FURNISHED AND INSTALLED BY THE SOUTHLAND CORP.

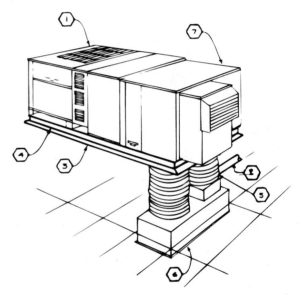

Figure 44–14   HVAC details and notes. *Courtesy The Southland Corporation.*

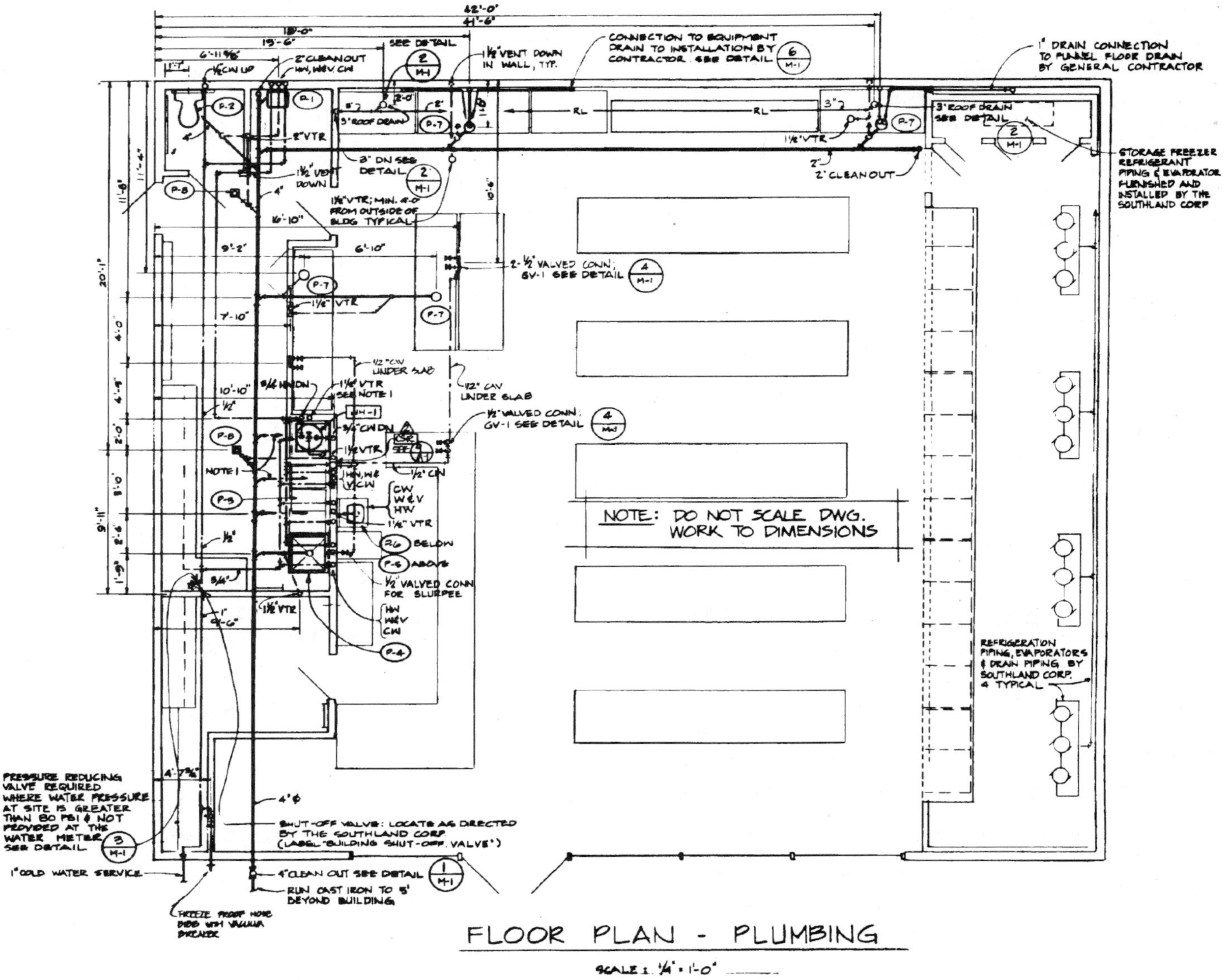

Figure 44–15 Required plumbing information is usually shown on an overlay of the floor plan. *Courtesy The Southland Corporation.*

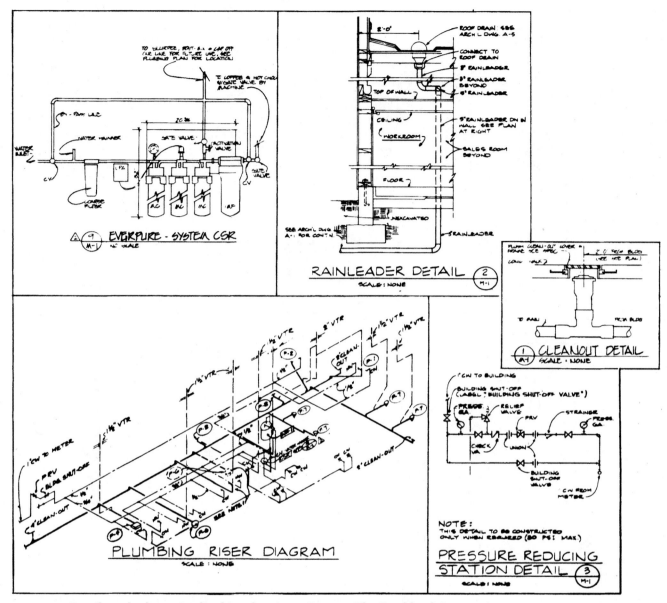

Figure 44–16   Details and schematic plumbing drawings. *Courtesy The Southland Corporation.*

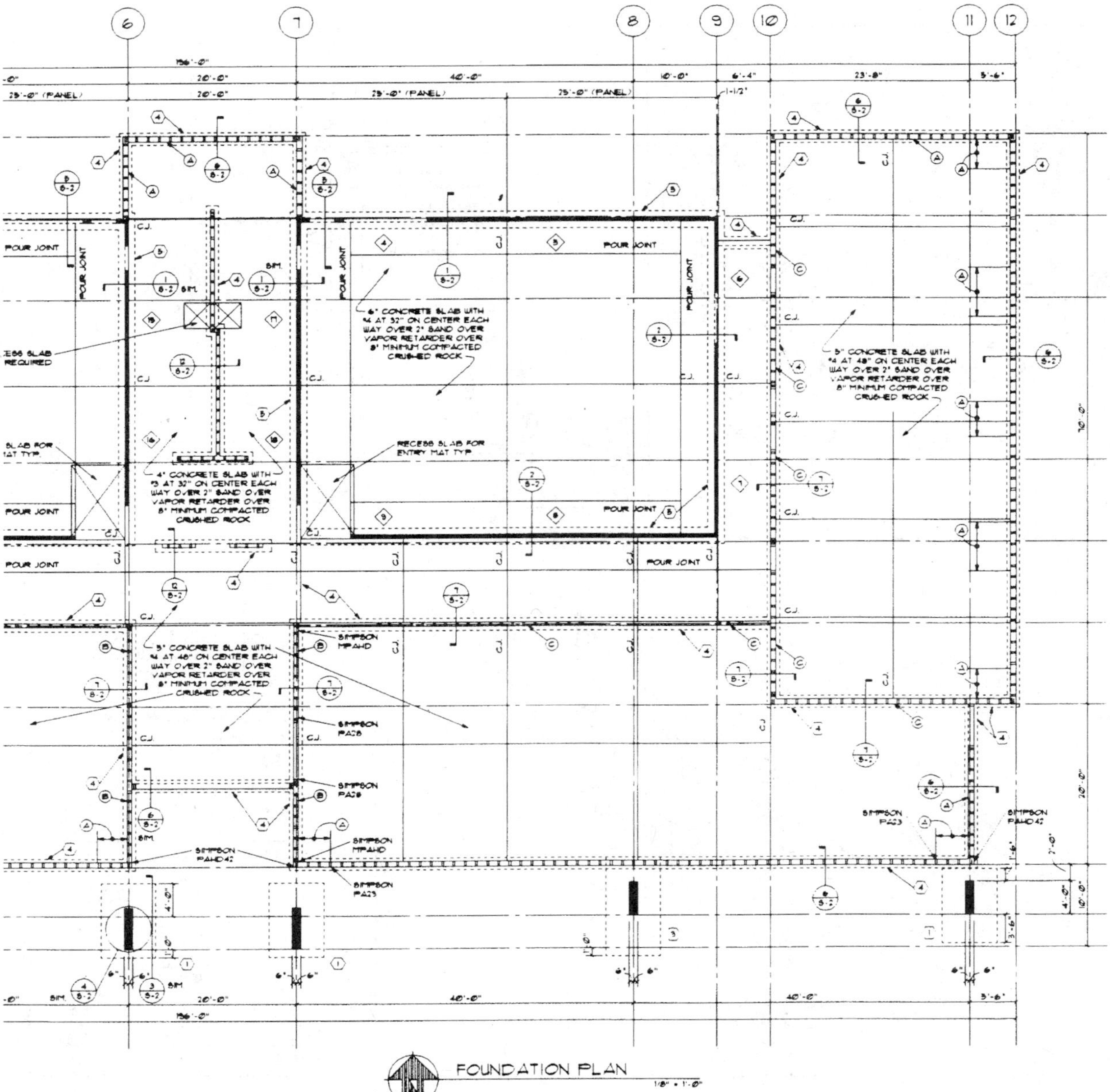

FOUNDATION PLAN
1/8" = 1'-0"

SHEAR WALL SCHEDULE		
MK	SHEATHING	NAILING
Ⓐ	1/2" PLYWOOD ONE SIDE	8d AT 6" ON CENTER EDGES, 12" ON CENTER IN FIELD
Ⓑ	1/2" PLYWOOD ONE SIDE	8d AT 4" ON CENTER EDGES, 12" ON CENTER IN FIELD
Ⓒ	5/8" GYP. WALLBOARD ONE SIDE	6d COOLER OR WALLBOARD NAILS AT 7" ON CENTER TO ALL SUPPORTS (UNBLOCKED)

NOTES: 1. USE COMMON NAILS U.O.N.
2. PROVIDE BLOCKING AT ALL UNSUPPORTED PLYWOOD EDGES.
3. WALLS NOTED ON PLAN ARE TYPE NOTED FULL LENGTH OF WALL (OR LENGTH SHOWN BY DIM. LINES)

FOOTING SCHEDULE		
MK	SIZE	REINFORCING
①	6'-6" x 9'-0" x 20" THICK	LONGIT.: (6) #4 TOP (7) #5 BTM. TRANS.: (8) #4 TOP (8) #5 BTM.
②	6'-6" x 10'-0" x 20" THICK	LONGIT.: (6) #4 TOP (7) #5 BTM. TRANS.: (9) #4 TOP (9) #5 BTM.
③	6'-6" x 12'-0" x 20" THICK	LONGIT.: (6) #4 TOP (7) #5 BTM. TRANS.: (11) #4 TOP (8) #5 BTM.
④	18" WIDE x 8" THICK x CONTINUOUS	(2) #4 CONTINUOUS BOTTOM
⑤	24" WIDE x 16" THICK x CONTINUOUS	(2) #5 CONTINUOUS TOP AND BOTTOM
⑥	2'-0" x 2'-0" x 18" THICKENED SLAB	UNREINFORCED

Figure 44–17 A concrete slab foundation is typically used for commercial structures. *Courtesy Architects Barrentine, Bates & Lee, AIA.*

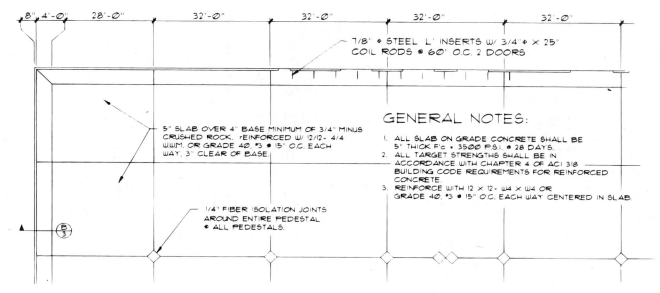

Figure 44–18 A slab-on-grade plan shows the size and location of all concrete pours. *Courtesy Structureform Masters, Inc.*

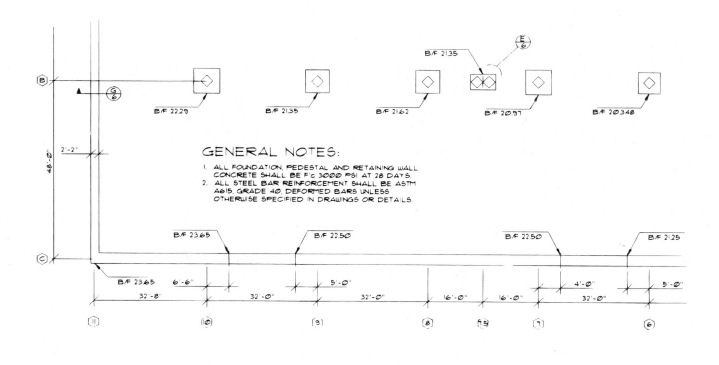

FOUNDATION PLAN
SCALE: 3/32"=1'-0"

Figure 44–19 A foundation plan is used to show below-grade concrete work. *Courtesy StructureForm Masters, Inc.*

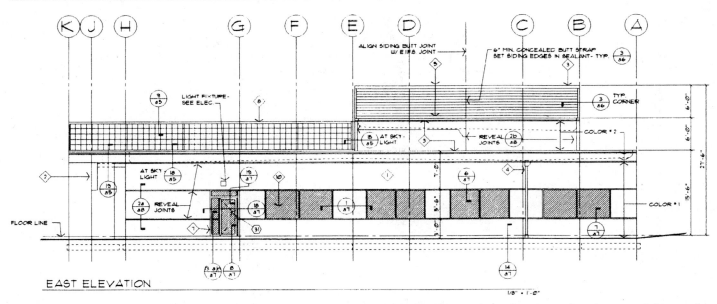

EAST ELEVATION

1/8" · 1'-0"

Figure 44–20 Elevations for commercial projects are similar to the elevations for residential projects. Exterior materials must be shown, as seen in this drawing. *Courtesy Architects Barrentine, Bates, & Lee, AIA.*

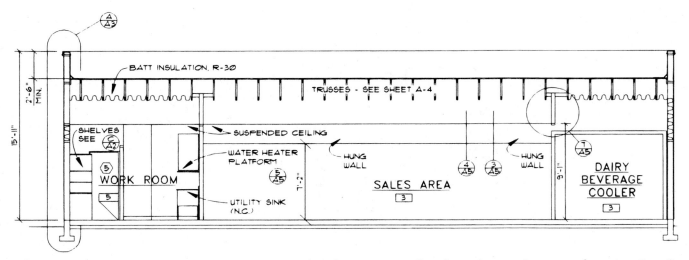

Figure 44–21 Sections for commercial projects are often drawn at a small scale to show major types of construction. Specific information is usually shown in details. *Courtesy The Southland Corporation.*

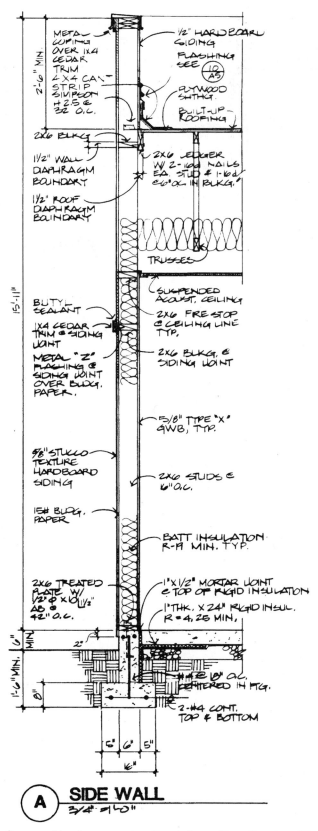

Figure 44–22 Partial sections are often used to show construction information. *Courtesy The Southland Corporation.*

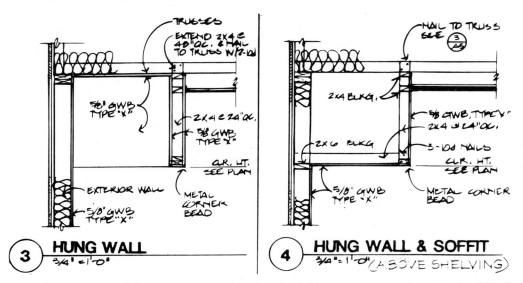

Figure 44–23 Drawing details is one of the most common jobs performed by drafters. *Courtesy The Southland Corporation.*

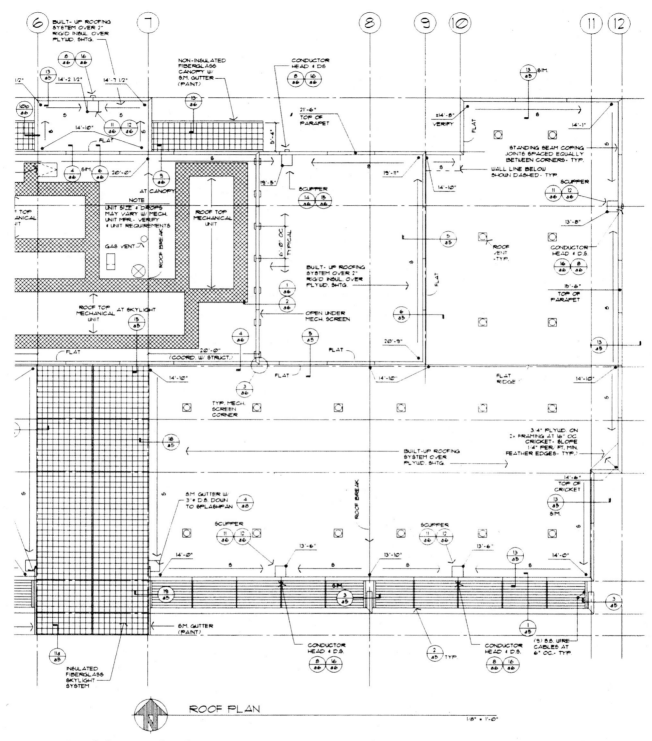

ROOF PLAN

1/8" = 1'-0"

Figure 44–24 A roof plan is used to show construction components at the roof level. *Courtesy Architects Barrentine, Bates & Lee, AIA.*

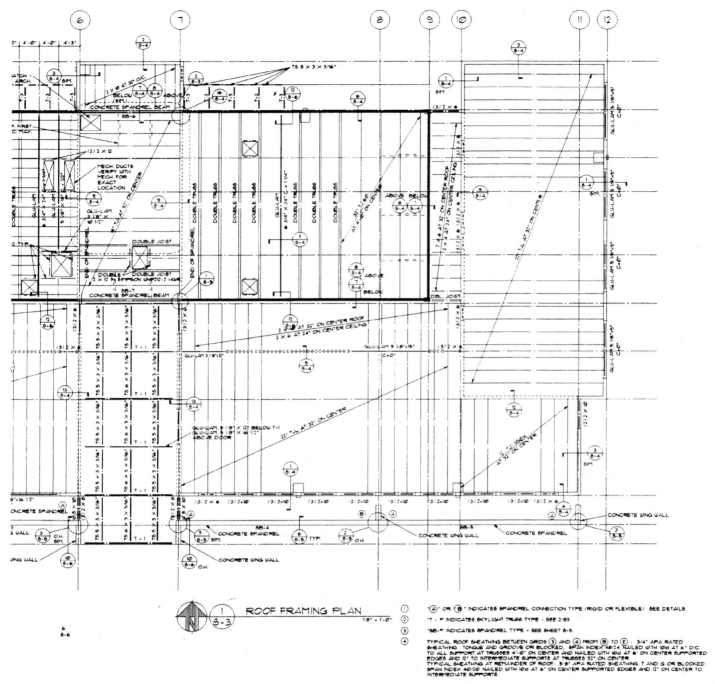

Figure 44–25 A roof framing plan is used to show structural components at the roof level. *Courtesy Architects Barrentine, Bates & Lee, AIA.*

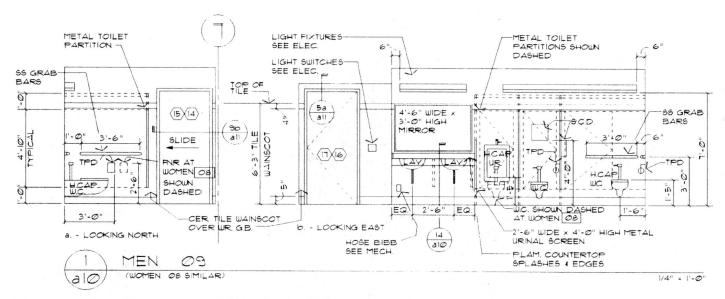

Figure 44–26   In addition to structural details, the drafter may be required to draw interior elevations and details. *Courtesy Architects, Barrentine, Bates & Lee, AIA.*

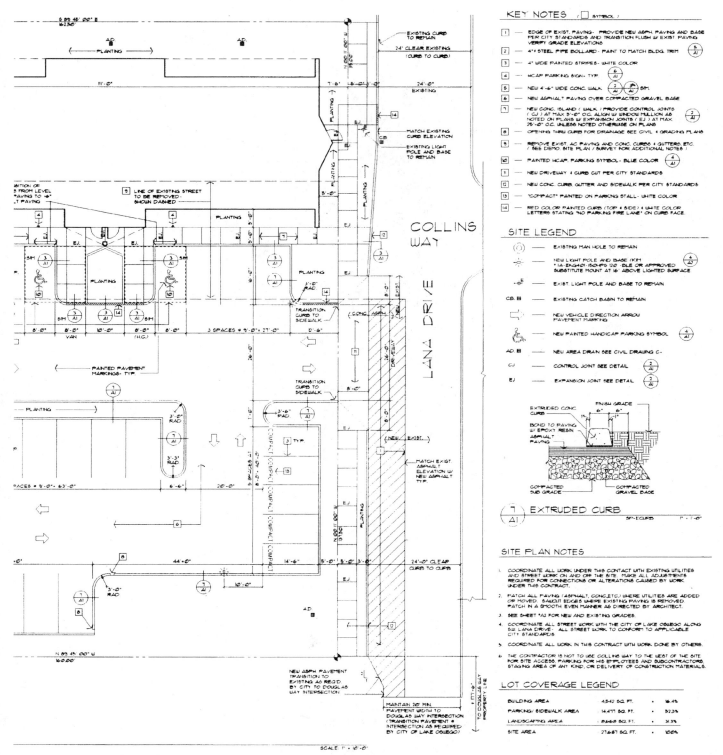

Figure 44–27 A plot, or site, plan for a commercial project is similar to a residential plan. Parking information is a major addition to a commercial plan. *Courtesy Architects Barrentine, Bates & Lee, AIA.*

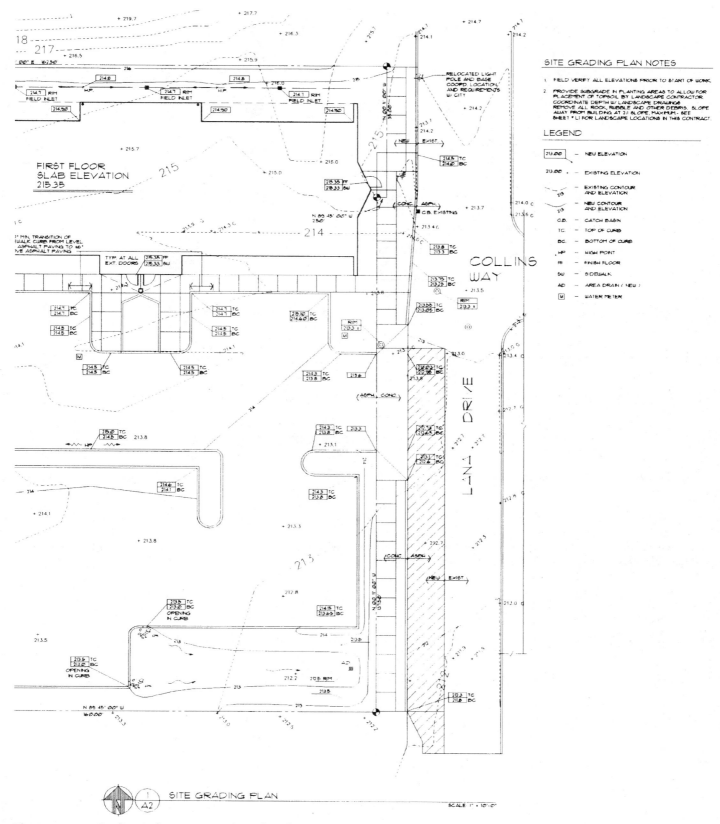

Figure 44–28   Grading information is often placed on an overlay of the plot plan to show cut and fill requirements of a job site. *Courtesy Architects Barrentine, Bates & Lee, AIA.*

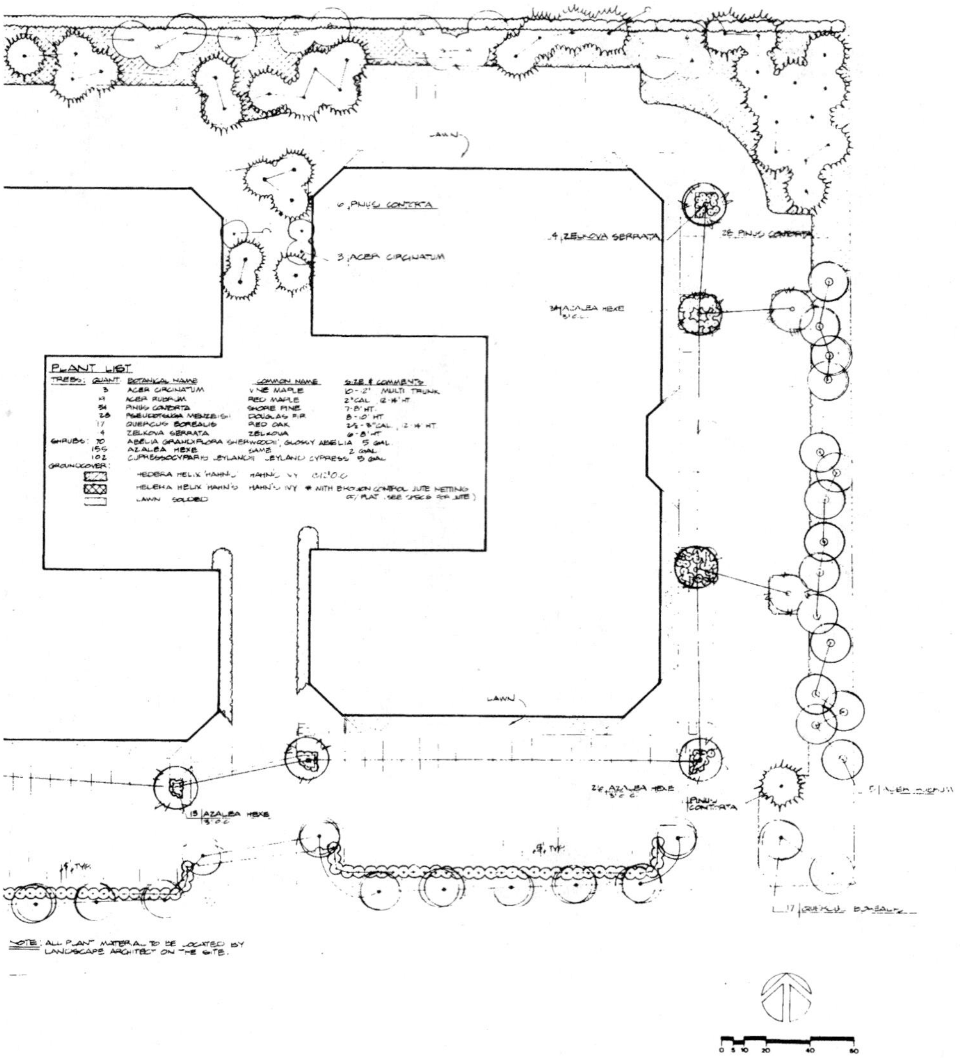

Figure 44–29   The planting or landscape plan is drawn as an overlay to the plot plan to show how the site will be planted and maintained. *Courtesy Guthrie, Slusarenko, and Leeb, Architects.*

◤ CADD Applications

USING CADD FOR COMMERCIAL AND MULTIFAMILY PROJECTS

In most commercial architectural projects, the architect works with consultants in the fields of mechanical, electrical, and structural engineering; plumbing systems; and interior design and space planning. These professionals use background drawings supplied by the architect to create their individual designs and plans. Background drawings are a subset of the information contained in the architect's floor plans.

Background drawings sent to individual consultants may contain different information. For example, the mechanical, electrical, and plumbing engineers need background drawings with sinks, toilets, showers, and utilities shown; these items may be omitted, to avoid confusion, on the structural background drawing. In multistory architectural projects, the background drawings may be used to coordinate floor-to-floor stacking relationships.

Multifamily housing projects provide a perfect application for CADD. Most of these projects have three or more unit plans, which may be copied, mirrored, and rotated to create a complete multifamily plan. After the unit plan is drawn, it can be edited using mirror, copy, or rotated to create the building floor plan.

When using CADD for architectural projects, it is important to relate one drawing to the next for accurate overlay purposes. The datum point method relates all plan drawings in the set through a common point. The datum point may be any convenient point on the drawing. Many drafters select the lower left corner of the drawing, represented by X and Y coordinates 0,0, as the datum point. Then the lower right corner of each additional drawing is coordinated the same way. The relationship of the floor plans is established in relationship to the datum point, so each plan is aligned, one over the other.

CHAPTER

44   Commercial Construction Projects Test

DIRECTIONS

Answer the questions with short, complete statements or drawings.

1. Letter your name and the date at the top of the sheet.
2. Letter the question number and provide the answer. You do not need to write out the question.

QUESTIONS

Question 44–1   List three types of offices in which you might be able to work as a drafter.

Question 44–2   List two major research tools that you will be using as a commercial drafter.

Question 44–3   What is a reflected ceiling plan?

Question 44–4   What are engineer's calculations?

Question 44–5   What two types of plumbing plans may be drawn for a commercial project?

Question 44–6   Describe two methods of drawing sections in commercial drawings.

Question 44–7   List and describe two types of plans that are usually based on the plot plan.

Question 44–8   What are the three parts of a set of calcs? How do they affect a drafter?

Question 44–9   Describe three plans that can be used to describe site-related material.

Question 44–10   Explain the difference between two types of plans that are used to describe work done at or below the finished grade.

CHAPTER 45

Building Codes and Commercial Design

One of the biggest differences you will notice between residential and commercial drafting is the increased dependency you will have on the building codes. You were introduced to residential building codes in Chapter 7 and have seen their effect on construction throughout the entire book. In residential drafting you may have been able to use a shortened version of the code such as *Dwelling Construction Under the Uniform Building Code.* If so, commercial construction drafting will be your first introduction to the full building code. As you draw commercial projects, the building codes will become much more influential. To be an effective drafter, you must be able to use properly the code that governs your area. Although there are many different codes in use throughout the country, most will be similar to the Uniform Building Code (UBC), Basic National Building Code (BOCA), or the Standard Building Code (SBC).

If you haven't done so already, check with the building department in your area and determine the code that will cover your drafting. Typically you will spend approximately $75.00 to purchase a code book. This may seem like a large expenditure to you now, but it will be one of your most useful drafting tools. Your code book should be considered a drafting tool just as pencils and vellum are. If possible, purchase the looseleaf version of the code rather than the hardcover edition. The hardcover edition may look nicer sitting on your shelf, but this will be one book that will not just sit. The looseleaf edition will allow you to update your codes with yearly amendments, as well as state and local additions.

Purchase a code book as soon as possible in your drafting career, and don't hesitate to mark or place tabs on certain pages to identify key passages. You won't need to memorize the book, but you will need to refer to certain areas again and again. Page tabs will help you find needed tables or formulas quickly.

▼ DETERMINE THE CATEGORIES

To effectively use a building code during the initial design stage, you must determine classifications used to define a structure: the occupancy group, location on the property, type of construction, floor area, height, and occupant load.

Occupancy Groups

The *occupancy group* specifies by whom or how the structure will be used. To protect the public adequately, buildings used for different purposes are designed to meet the hazards of that usage. Think of the occupancy group with which you are most familiar, the R occupancy. The R grouping not only covers the single-family residence, but also includes duplexes, apartments, lodging areas, and hotels. This one group covers single-family residences and multistory apartments with hundreds of occupants. Obviously, the safety requirements for the multileveled apartment with hundreds of occupants should be different from a single-family residence.

The code that you are using will affect how the occupancy is listed. The letter of the occupancy listing generally is the first letter of the word that it represents. Figure 45–1 shows a detailed listing of occupancy categories from the UBC.

Structures often have more than one occupancy group within the structure. Separation must be provided at the wall or floor dividing these different occupancies. Figure 45–2 shows a listing of required separations based on UBC requirements. If you are working on a structure that has a 10,000 sq ft office area over an enclosed parking garage, the two areas must be separated. By using the table in Figure 45–1 you can see that the office area is a B occupancy and the garage is an S4 occupancy. By using the table in Figure 45–2 you can determine that a wall with a one-hour rating must be provided between these two areas. The hourly rating is given to construction materials specifying the length of time that the item must resist structural damage that would be caused by a fire. Various ratings are assigned by the codes for materials ranging from nonrated to a four-hour rating. Whole chapters are assigned by each code to cover specific construction requirements for various materials. Chapter 7 of the UBC serves as an introduction to the fire resistance of materials.

Table 3-A—DESCRIPTION OF OCCUPANCIES BY GROUP AND DIVISION

GROUP AND DIVISION	SECTION	DESCRIPTION OF OCCUPANCY
A–1	303.1.1	A building or portion of a building having an assembly room with an occupant load of 1,000 or more and a legitimate stage.
A–2		A building or portion of a building having an assembly room with an occupant load of less than 1,00 and a legitimate stage.
A–2.1		A building or portion of a building having an assembly with an occupant load of 300 or more without a legitimate stage, including such buildings used for educational purposes and not classed as a Group E or Group B Occupancy.
A–3		Any building or portion of a building having an assembly room with an occupant load of less than 300 without a legitimate stage, including such buildings used for educational purposes and not classes as a Group E or Group B Occupancy.
A–4		Stadiums, reviewing stands and amusement park structures not included within other Group A Occupancies.
B	304.1	A building or structure, or a portion thereof, for office, professional or service-type transactions, including storage of records and accounts, and eating and drinking establishments with an occupant load of less than 50.
E–1	305.1	Any building used for educational purposes through the 12th grade by 50 or more persons for more than 12 hours per week or four hours in any one day.
E–2		Any building used for educational purposes through the 12th grade by less than 50 persons for more than 12 hours per week or four hours in any one day.
E–3		Any building or portion thereof used for day-care purposes for more than six persons.
F–1	306.1	Moderate-hazard factory and industrial occupancies include factory and industrial uses not classified as Group F, Division 2 Occupancies.
F–2		Low-hazard factory and industrial occupancies include facilities producing noncombustible or nonexplosive materials which during finishing, packing or processing do not involve a significant fire hazard.
H–1	307.1	Occupancies with a quantity of material in the building in excess of those listed in Table 3–D which present a high explosion hazard as listed in Section 307.1.1.
H–2		Occupancies with a quantity of material in the building in excess of those listed in Table 3–D which present a moderate explosion hazard or a hazard from accelerated burning as listed in Section 307.1.1.
H–3		Occupancies with a quantity of material in the building in excess of those listed in Table 3–D which present a high fire or physical hazard as listed in Section 307.1.1.
H–4		Repair garages not classified as Group S, Division 3 Occupancies.
H–5		Aircraft repair hangars not classified as Group S, Division 5 Occupancies and heliports.
H–6	307.1 and 307.11	Semiconductor fabrication facilities and comparable research and development areas when the facilities in which hazardous production materials are used, and the aggregate quantity of material is in excess of those listed in Table 3–D or 3–E.
H–7	307.1	Occupancies having quantities of materials in excess of those listed in Table 3–E that are health hazards as listed in Section 307.1.1.
I–1.1	308.1	Nurseries for the full-time care of children under the age of six (each accommodating more than five children), hospitals, sanitariums, nursing homes with nonambulatory patients and similar buildings (each accommodating more than five patients).
I–1.2		Health-care centers for ambulatory patients receiving outpatient medical care which may render the patient incapable of unassisted self-preservation (each tenant space accommodating more than five such patients).
I–2		Nursing homes for ambulatory patients, homes for children six years of age or over (each accommodating more than five persons).
I–3		Mental hospitals, mental sanitariums, jails, prisons, reformatories and buildings where personal liberties of inmates are similarly restrained.
M	309.1	A building or structure, or a portion thereof, for the display and sale of merchandise, and involving stocks of goods, wares or merchandise, incidental to such purposes and accessible to the public.
R–1	310.1	Hotels and apartment houses, congregate residences (each accommodating more than 10 persons).
R–3		Dwellings, lodging houses, congregate residences (each accommodating 10 or fewer persons).
S-1	311.1	Moderate hazard storage occupancies include buildings or portions of buildings used for storage of combustible maerials not classified as Group S, Division 2 or Group H Occupancies.
S-2		Low-hazard storage occupancies include buildings or portions of buildings used for storage of noncombustible materials.
S-3		Repair garages and parking garages not classified as Group S, Division 4 Occupancies.
S-4		Open parking garages.
S–5		Aircraft hangars and helistops.
U–1	312.1	Private garages, carports, sheds and agricultural buildings.
U–2		Fences over 6 feet (182.9 mm) high, tanks and towers.

Figure 45–1 Common occupancy listings. *Reproduced from the 1994 edition of the* Uniform Building Code, *Vol. 1 and 2, © 1994 with permission of the publisher, the International Conference of Building Officials.*

Table 3–B Required Separation in Buildings of Mixed Occupancy[1] (Hours)

	A-1	A-2	A-2.1	A-3	A-4	B	E	F-1	F-2	H-2	H-3	H-4,5	H-6,7[2]	I	M	R-1	R-3	S-1	S-2	S-3	S-5	U-1[3]
A-1		N	N	N	N	3	N	3	3	4	4	4	4	3	3	1	1	3	3	4[5]	3	1
A-2			N	N	N	1	N	1	1	4	4	4	4	3	1	1	1	1	1	3[5]	1	1
A-2.1				N	N	1	N	1	1	4	4	4	4	3	1	1	1	1	1	3[5]	1	1
A-3					N	N	N	N	N	4	4	4	3	2	N	1	1	N	1	3[5]	1	1
A-4						1	N	1	1	4	4	4	4	3	1	1	1	1	1	3[5]	1	1
B							1	N[6]	N	2	1	1	1	2	N	1	1	N	1	1	1	1
E								1	1	4	4	4	3	1	1	1	1	1	1	3[5]	1	1
F-1									1	2	1	1	1	3	N[6]	1	1	N	1	1	1	1
F-2										2	1	2	1	2	1	1	1	N	1	1	1	1
H-1	NOT PERMITTED IN MIXED OCCUPANCIES. SEE CHAPTER 10.																					
H-2										1	1	2	4	2	4	4	2	2	2	2	1	
H-3											1	1	4	1	3	3	1	1	1	1	1	
H-4.5												1	4	1	3	3	1	1	1	1	1	
H-6,7[2]													4	1	4	4	1	1	1	1	3	
I														2	1	1	2	4	4[5]	3	1	
M															1	1	1[4]	1[4]	1	1	1	
R-1																N	3	1	3[5]	1	1	
R-3																	1	1	1	1	1	
S-1																		1	1	1	1	
S-2																			1	1	N	
S-3																				1	1	
S-4	OPEN PARKING GARAGES ARE EXCLUDED EXCEPT AS PROVIDED IN SECTION 311.2.																					
S-5																						N

N—No requirements for fire resistance.

[1]For detailed requirements and exceptions, see Section 302.4.

[2]For special provisions on highly toxic materials, see the Fire Code.

[3]For agricultural buildings, see also Appendix Chapter 3.

[4]See Section 310.2.1 for exception.

[5]Group S, Division 3 Occupancies used exclusively for parking or storage of pleasure-type motor vehicles and provided no repair or refueling is done may have the occupancy separation reduced one hour.

[6]For Group F, Division 1 woodworking establishments with more than 2,500 square feet, the occupancy separation shall be one hour.

Figure 45–2 If a structure contains more than one occupancy grouping, the areas will need to be separated by protective construction methods to reduce the spread of fire. The fire rating of the separating construction can be determined by comparing the occupancy of each type of construction. *Reproduced from the 1994 edition of the* Uniform Building Code, *Vol. 1 and 2, © 1994 with permission of the publisher, the International Conference of Building Officials.*

Once the occupancy grouping is determined, turn to the section of the building code that introduces specifics on that occupancy grouping. Studens often become frustrated with the amount of page turning that is required to find particular information in the codes. Often you will look in one section of the code, only to be referred to another section of the code and then on to another. It may be frustrating, but the design of a structure is often complicated and requires patience.

Building Location and Size

When the occupancy of the structure was determined in Figure 45–1, information on the location of the structure to the property lines was also given. The type of occupancy will affect the location of a building on the property. The location to the property lines will also affect the size and amount of openings that are allowed in a wall. Notice that the reader is referred to specific chapters within the code for information about openings based on size, location, and type of construction in Figure 45–1.

Type of Construction

Once the occupancy of the structure has been determined, the type of construction used to protect the occupants can be determined. The type of construction will determine the kind of building materials that can or cannot be used in the construction of the building. The kind of material used in construction will determine the ability of the structure to resist fire. Five general types of construction are typically specified by building codes and are represented by the numbers 1 through 5.

Construction in building types 1 and 2 requires the structural elements such as walls, floors, roofs, and exits to be constructed or protected by approved noncombustible materials such as steel, iron, concrete, or masonry. Construction in types 3, 4, and 5 can be steel, iron, concrete, masonry, or wood. In addition to specifying the structural framework, the type of construction will also dictate the material used for interior partitions, exit specifications, and a wide variety of other requirements for the building. Figure 45–3 shows the construction requirements for various areas of a structure based on the type of construction. Specific areas of the code must now be researched to determine how these requirements can be met.

Building Area

Once the type of construction has been determined, the size of the structure can be determined. Figure 45–4 shows a listing of allowable floor area based on the type of construction to be used. These are basic sq ft sizes, which may be altered depending on the different construction techniques that can be used. If you are working on the plans for an office building with 20,000 sq ft, this table can be used to determine the type of construction required. First determine the occupancy. By using Figure 45–1 you can find that the structure is a B occupancy. Enter Figure 45–4 on the right side using type V-n. For a Group B occupancy using type V-n construction, the structure would be limited to 8,000 sq ft. Type II F.R. or Type I F.R. construction methods would need to be used unless the structure is divided into smaller segments by the use of fire rated walls. You will now need to turn to the areas of the code that cover these types of construction to determine types of materials that could be used.

Determine the Height

The occupancy and type of construction will determine the maximum height of the structure. Figure 45–4 can be used to determine the allowable height of a structure. The theoretical 20,000 sq ft office building could be 12 stories (or 160') high if constructed of type 2 construction materials, or of unlimited height if built with type 1 materials. Remember that this height is based on building requirements governing fire and public safety. Zoning regulations for a specific area may further limit the height of the structure.

Table 6-A—TYPES OF CONSTRUCTION—FiRE-RESISTIVE REQUIREMENTS (In Hours)
For details, see occupancy section in Chapter 3, type of construction sections in this chapter and sections referenced in this table.

| BUILDING ELEMENT | TYPE I | TYPE II | | | TYPE III | | TYPE IV | TYPE V | |
| | | Noncombustible | | | Combustible | | | | |
	Fire-resistive	Fire-resistive	1-Hr.	N	1-Hr.	N	H.T.	1-Hr.	N
1. Bearing walls—exterior	4 Sec. 602.3.1	4 Sec. 603.3.1	1	N	4 Sec. 604.3.1	4 Sec. 604.3.1	4 Sec. 605.3.1	1	N
2. Bearing walls—interior	3	2	1	N	1	N	1	1	N
3. Nonbearing walls—exterior	4 Sec. 602.3.1	4 Sec. 603.3.1	1 Sec. 603.3.1	N	4 Sec. 604.3.1	4 Sec. 604.3.1	4 Sec. 605.3.1	1	N
4. Structural frame[1]	3	2	1	N	1	N	1 or H.T.	1	N
5. Partitions—permanent	1[2]	1[2]	1[2]	N	1	N	1 or H.T.	1	N
6. Shaft enclosures[3]	2	2	1	1	1	1	1	1	1
7. Floors and floor-ceilings	2	2	1	N	1	N	H.T.	1	N
8. Roofs and roof-ceilings	2 Sec. 602.5	1 Sec. 603.5	1 Sec. 603.5	N	1	N	H.T.	1	N
9. Exterior doors and windows	Sec. 602.3.2	Sec. 603.3.2	Sec. 603.3.2	Sec. 603.3.2	Sec. 604.3.2	Sec. 604.3.2	Sec. 605.3.2	Sec. 606.3	Sec. 606.3
10. Stairway construction	Sec. 602.4	Sec. 603.4	Sec. 603.4	Sec. 603.4	Sec. 604.4	Sec. 604.4	Sec. 605.4	Sec. 606.4	Sec. 606.4

N—No general requirements for fire resistance. H.T.—Heavy timber.

[1]Structural frame elements in an exterior wall that is located where openings are not permitted or where protection of openings is required, shall be protected against external fire exposure as required for exterior bearing walls or the structural frame, whichever is greater.

[2]Fire-retardant-treated wood (see Section 207) may be used in the assembly, provided fire-resistance requirements are maintained. See Sections 602 and 603.

[3]For special provisions, see Sections 304.6, 306.6 and 711.

Figure 45–3 Construction requirements for various parts of a structure based on the UBC. *Reproduced from the 1994 edition of the Uniform Building Code, Vol. 1 and 2, © 1994 with permission of the publisher, the International Conference of Building Officials.*

Table 5-B—BASIC ALLOWABLE BUILDING HEIGHTS AND BASIC ALLOWABLE FLOOR AREA FOR BUILDINGS ONE STORY IN HEIGHT[1]

TYPE OF CONSTRUCTION		I F.R.	II F.R.	II One-hour	II N	III One-hour	III N	IV H.T.	V One-hour	V N
Use Group	Height	UL	160 (48 768 mm)	65 (19 812 mm)	55 (16 764 mm)	65 (19 812 mm)	55 (16 764 mm)	65 (19 812 mm)	50 (15 240 mm)	40 (12 192 mm)
						× 0.0929 for m²				
A-1	H	UL	4	Not Permitted						
	A	UL	29,900							
A-2, 2.1[2]	H	UL	4	2	NP	2	NP	2	2	NP
	A	UL	29,900	13,500	NP	13,500	NP	13,500	10,500	NP
A-3, 4[2]	H	UL	12	2	1	2	1	2	2	1
	A	UL	29,900	13,500	9,100	13,500	9,100	13,500	10,500	6,100
B,F-1, M, S-1, S-3, S-5	H	UL	12	4	2	4	2	4	3	2
	A	UL	39,900	18,000	12,000	18,000	12,000	18,000	14,000	8,000
E-1, 2, 3[4]	H	UL	4	2	1	2	1	2	2	1
	A	UL	45,200	20,200	13,500	20,200	13,500	20,200	15,700	9,100
F-2, S-2[3]	H	UL	12	4	2	4	2	4	3	2
	A	UL	59,900	27,000	18,000	27,000	18,000	27,000	21,000	12,000
H-1[5]	H	1	1	1	1	Not Permitted				
	A	15,000	12,400	5,600	3,700					
H-2[5]	H	UL	2	1	1	1	1	1	1	1
	A	15,000	12,400	5,600	3,700	5,600	3,700	5,600	4,400	2,500
H-3, 4, 5[5]	H	UL	5	2	1	2	1	2	2	1
	A	UL	24,800	11,200	7,500	11,200	7,500	11,200	8,800	5,100
H-6, 7	H	3	3	3	2	3	2	3	3	1
	A	UL	39,900	18,000	12,000	18,000	12,000	18,000	14,000	8,000
I-1.1[6], 1.2	H	UL	3	1	NP	1	NP	1	1	NP
	A	UL	15,100	6,800	NP	6,800	NP	6,800	5,200	NP
I-2	H	UL	3	2	NP	2	NP	2	2	NP
	A	UL	15,100	6,800	NP	6,800	NP	6,800	5,200	NP
I-3	H	UL	2	Not Permitted[7]						
	A	UL	15,100							
R-1	H	UL	12	4	2[9]	4	2[9]	5	3	2[9]
	A	UL	29,900	13,500	9,100[9]	13,500	9,100[9]	13,500	10,500	6,000[9]
R-3	H	UL	3	3	3	3	3	3	3	3
	A	Unlimited								
S-4	See Table 3-H									
U[8]	H	See Chapter 3								
	A									

Figure 45–4 The basic allowable height and floor area of a structure based on the UBC. Once the occupancy group is determined, the basic allowable floor area can be determined based on the fire resistance of the construction materials to be used. Type I material has the highest resistance to fire. Type V-N construction is most prone to damage from fire. The allowable height in floors is above each listing of floor area. *Reproduced from the 1994 edition of the* Uniform Building Code, *Vol. 1 and 2, © 1994 with permission of the publisher, the International Conference of Building Officials.*

Determine the Occupant Load

You have already determined how the structure will be used. Now you must determine how many people may use the structure. Figure 45–5 is an example of the table used to compute the occupant load. In each occupancy, the intended size of the structure is divided by the occupant load factor to determine the occupant load. In a 20,000 sq ft office structure, the occupant load is 200. This is found by dividing the size of 20,000 sq ft by the occupant factor of 100. The occupant load will be needed to determine the number of exits, the size and locations of doors, and many other construction requirements.

▼ USING THE CODES

The need to determine the five classifications of a building may seem senseless to you as a beginning drafter. Although these are procedures that the designer, architect, or engineer will perform in the initial design stage, the drafter must be aware of these classifications as basic

10-A—MINIMUM EGRESS REQUIREMENTS[1]

USE[2]	MINIMUM OF TWO EXITS OTHER THAN ELEVATORS ARE REQUIRED WHERE NUMBER OF OCCUPANTS IS AT LEAST	OCCUPANT LOAD FACTOR[3] (square feet) × 0.0929 for m²
1. Aircraft hangars (no repair)	10	500
2. Auction rooms	30	7
3. Assembly areas, concentrated use (without fixed seats) Auditoriums Churches and chapels Dance floors Lobby accessory to assembly occupancy Lodge rooms Reviewing stands Stadiums	50	7
Waiting area	50	3
4. Assembly areas, less-concentrated use Conference rooms Dining rooms Drinking establishments Exhibit rooms Gymnasiums Lounges Stages	50	15
5. Bowling alley (assume no occupant load for bowling lanes)	50	4
6. Children's homes and homes for the aged	6	80
7. Classrooms	50	20
8. Congregate residences	10	200
9. Courtrooms	50	40
10. Dormitories	10	50
11. Dwellings	10	300
12. Exercising rooms	50	50
13. Garage, parking	30	200
14. Hospitals and sanitariums— Health-care center Nursing homes Sleeping rooms Treatment rooms	10 6 10	80 80 80
15. Hotels and apartments	10	200
16. Kitchen—commercial	30	200
17. Library reading room	50	50
18. Locker rooms	30	50
19. Malls (see Chapter 4)	—	—
20. Manufacturing areas	30	200
21. Mechanical equipment room	30	300
22. Nurseries for children (day care)	7	35
23. Offices	30	100
24. School shops and vocational rooms	50	50
25. Skating rinks	50	50 on the skating area; 15 on the deck
26. Storage and stock rooms	30	300
27. Stores—retail sales rooms Basements and ground floor Upper floors	50 50	30 60
28. Swimming pools	50	50 for the pool area; 15 on the deck
29. Warehouses	30	500
30. All others	50	100

[1]Access to, and egress from, buildings for persons with disabilities shall be provided as specified in Chapter 11.

[2]For additional provisions on number of exits from Groups H and I Occupancies and from rooms containing fuel-fired equipment or cellulose nitrate, see Sections 1018, 1019 and 1020, respectively.

[3]This table shall not be used to determine working space requirements per person.

[4]Occupant load based on five persons for each alley, including 15 feet (4572 mm) of runway.

Figure 45–5 The occupant load of a structure is determined by dividing the intended size of the structure by the occupant load factor. *Reproduced from the 1994 edition of the* Uniform Building Code, *Vol. 1 and 2, © 1994 with permission of the publisher, the International Conference of Building Officials.*

drafting functions are performed on the project. Many of the problems that the drafter will need the code to solve will require knowledge of five basic code limitations. Practice using the tables presented in this text as well as similar tables of the code that govern your area.

Once you feel comfortable determining the basic categories of a structure, study the chapter of your building code that presents basic building requirements for a structure. Chapters 7–9 introduce fire protection, and chapters 10–12 introduce occupant needs.

Once you feel at ease reading through the chapters dealing with general construction, work on the chapters of the code that deal with uses of different types of material. Start with the chapter that deals with the most common building material in your area. For most of you, this will be section 23, which governs the use of wood. Although you will be familiar with the use of wood in residential structures, the building codes will open many new areas that were not required in residential construction.

CHAPTER

 Building Codes and Commercial Design Test

DIRECTIONS

Answer the questions with short, complete statements.

1. Letter your name and the date at the top of the sheet.
2. Letter the question number and provide the answer. You do not need to write out the question.

QUESTIONS

Question 45–1 What is the occupancy rating of a room used to train 50 or more junior college drafters?

Question 45–2 What fire rating would be required for a wall separating a single-family residence from a garage?

Question 45–3 What is the height limitation on an office building of type 2-N construction?

Question 45–4 What is the maximum allowable floor area of a public parking facility using type 4 construction methods?

Question 45–5 What fire resistance is required for the exterior walls of a structure used for full-time care of children that is within 5′ of a property line?

Question 45–6 What fire rating is required between a wall separating an A-1 occupancy from a B occupancy?

Question 45–7 What is the maximum height of a residence of type 2 through 5 construction?

Question 45–8 You are drafting a 12,000 sq ft retail sales store. Determine the occupancy, the least restrictive type of construction to be used, maximum height and square footage, and the occupant load for each possible level.

CHAPTER 46

Common Commercial Construction Materials

As you work on commercial projects, you will be exposed to new types of structures and codes different from the ones you were using with residential drawing. You will also be using different types of materials than are typically used in residential construction. Most of the materials used in commercial construction can be used in residential construction, but they are not because of their cost and the associated cost of labor with each material. Wood is the exception. Common materials used in commercial construction are wood, concrete block, poured concrete, and steel.

▼ WOOD

Wood is used in many types of commercial buildings in a manner similar to its use in residential construction. The western platform system, which was covered in Chapter 24, is a common framing method for multifamily and office buildings as well as others. Heavy wood timbers are also used for some commercial construction. As a drafter, you will need to be familiar with both types of construction.

Platform Construction

Platform construction methods in commercial projects are similar to residential methods. Wood is rarely used at the foundation level but is a common material for walls and intermediate and upper-level floor systems. Trusses or truss joists are also common floor joist materials for commercial projects. Although each is used in residential construction, their primary usage is in commercial construction.

Walls. Wood is used to frame walls on many projects. The biggest difference in wood wall construction is not in the framing method but in the covering materials. Depending on the type of occupancy and the type of construction required, wood framed walls may require a special finish to achieve a required fire protection. Figure 46–1 shows several different methods of finishing a wood wall to achieve various fire ratings.

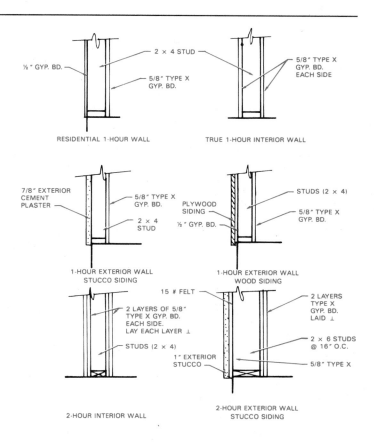

Figure 46–1 Walls often require special treatment to achieve the needed fire rating for certain types of construction. (See Figure 45–2.) Consult your local building code for complete descriptions of coverings.

Roofs. Wood is also used to frame the roof system of many commercial projects. Joists, trusses, and panelized systems are the most typically used framing systems. Both joist and truss systems have been discussed in Sections VII and VIII. These systems usually allow the joists or trusses to be placed at 24 or 32″ (600 or 800 mm) o.c. for commercial uses. An example of a truss system can be seen in Figure 46–2. The panelized roof system typically uses beams placed approximately 20 to 30′ (6100 to 9100 mm) apart. Smaller beams called purlins are then placed

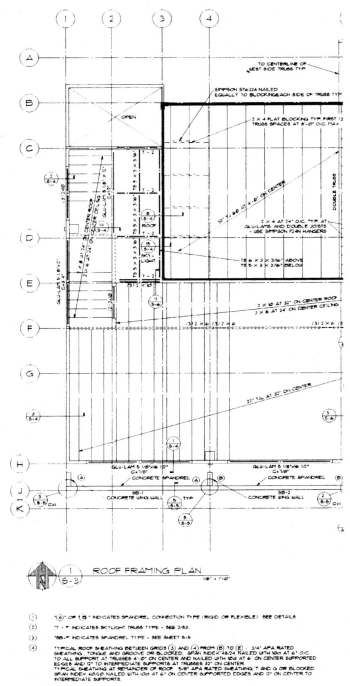

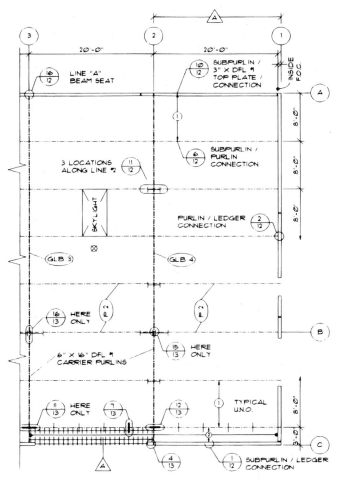

Figure 46–3 A panelized roof framing plan. *Courtesy Ken Smith and Associates, AIA.*

Figure 46–2 A roof plan using trusses to span between supports. *Courtesy Van Domelen/Looijenga/McGarrigle/Knauf Consulting Engineers.*

Heavy Timber Construction

In addition to standard uses of wood in the western platform system, large wood members are sometimes used for the structural framework of a building. This method of construction consists of wood members 5″ (130 mm) wide and is typically used for both appearance and structural reasons. Heavy timbers have excellent structural and fire retardant qualities. In a fire, heavy wood members will char on their exterior surfaces but will maintain their structural integrity long after an equal-sized steel beam will have failed. A roof framed with heavy timbers will resemble the plan in Figure 46–3.

Laminated Beams

Because of the difficulty of producing large beams in long lengths from solid wood, large beams are typically constructed from smaller members laminated together to form the larger beam. Laminated beams are a common material for buildings such as gymnasiums and churches that require large amounts of open spaces. Three of the

between the main beams, typically using an 8′ (2400 mm) spacing. Joists that are 2 or 3″ (50 or 75 mm) wide are then placed between these purlins at 24″ (600 mm) o.c. The roof is then covered with plywood sheathing. Figure 46–3 shows an example of a roof framing plan for a panelized roof system. The availability of materials, labor practices, and use of the building determines which type of roof will be used.

most common types of laminated beams are the single span, Tudor arch, and the three-hinged arch beams. Examples of each can be seen in Figure 46–4.

The single span beam is often used in standard platform framing methods. These beams are typically referred to as glu-lams. Because of their increased structural qualities, a laminated beam can be used to replace a much larger sawed beam. Laminated beams often have a curve, or camber, built into the beam. The camber is designed into the beam to help resist the loads to be carried.

The Tudor and three-hinged arch members are a post-and-beam system combined into one member. These beams are specified on plans in a method similar to other beams. The drafter's major responsibility when working with either heavy timber construction or laminated beams is in the drawing of connection details. Beam to beam, beam to column, and column to support are among the most common details drawn by drafters. When the size of the beam does not match premanufactured connectors, the drafter will be required to draw the fabrication details for a connector. Figure 46–5 shows an example of such a detail.

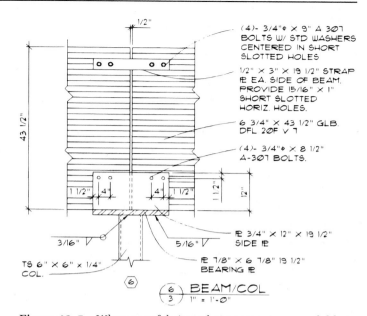

Figure 46–5 When a prefabricated connector is unavailable in the proper size, a drafter will need to work from an engineer's sketches to draw the required connectors. *Courtesy Ken Smith and Associates, AIA.*

▼ CONCRETE BLOCK

Concrete block is used in some areas for above-ground construction, but it is primarily used as a foundation material in residential construction. In commercial construction, concrete blocks are used to form the wall system for many types of buildings. Concrete blocks provide a durable and relatively inexpensive material to install and maintain. Blocks are typically manufactured in 8 × 8 × 16 (200 × 200 × 400) modules. The sizes listed are width, height, then length. Other common sizes are 4 × 8 × 16, 6 × 8 × 16, and 12 × 8 × 16 (100 × 200 × 400, 150 × 200 × 400, and 300 × 200 × 400). The actual size of the block is smaller than the nominal size so that mortar joints can be included. Although the person responsible for the design determines the size of the structure, it is important that the drafter be aware of the modular principles of concrete block construction. Lengths of walls, locations of openings in a wall, and the heights of walls and openings must be based on the modular size of the block being used. Failure to maintain the

modular layout can result in a tremendous increase in the cost of labor to cut and lay the blocks.

The drafter's major responsibility when working with concrete blocks is to detail steel reinforcing patterns. Concrete blocks are often reinforced with a wire mesh at every other course of blocks. Where the risk of seismic danger is great, concrete blocks are required to have reinforcing steel placed within the wall. These bars are placed in a grid pattern throughout the wall to help tie the blocks together. The steel is placed in a block that has a channel running through it. This cell is then filled with grout to form a header or bond beam within the wall. Figure 46–6 shows an example of a detail that a drafter would be required to draw to specify the bond beam.

When the concrete blocks are required to support a load from a beam, a pilaster is often placed in the wall to

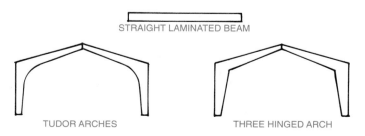

Figure 46–4 Common laminated beam shapes.

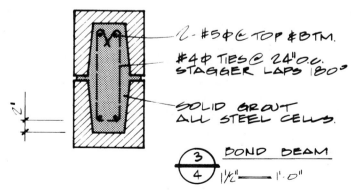

Figure 46–6 When working with concrete block, the drafter is typically required to detail reinforcing.

help transfer beam loads down into the footing. Pilasters are also used to provide vertical support to the wall when the wall is required to span long distances. An example of a detail that the drafter might be required to draw can be seen in Figure 46–7. Wood can be attached to concrete in several ways. Two of the most common methods are by the use of a seat, as seen in Figure 46–8, or a metal connector, as seen in Figure 46–9.

▼ POURED CONCRETE

Concrete is a common building material composed of sand and gravel bonded together with cement and water. One of the most common types of cement is portland cement. Portland cement contains pulverized particles of limestone, cement rock, oyster shells, silica sand, shale, iron ore, and gypsum. The gypsum controls the time required for the cement to set.

Concrete can be poured in place at the job site, formed off site and delivered ready to be erected into place, or formed at the job site and lifted into place. Your area of the country, the office that you work in, and the type of structure to be built dictate which of these concrete construction methods you will be using.

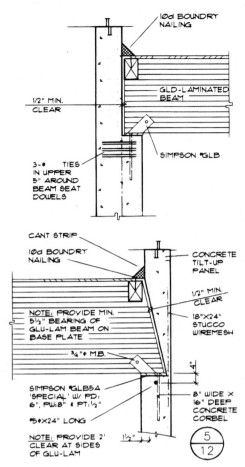

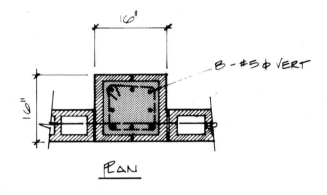

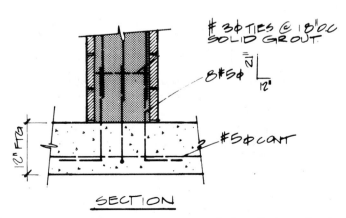

Figure 46–7 Pilasters are a common feature that require steel reinforcing.

Figure 46–8 Wood beams are often allowed to rest on a ledge. *Courtesy Structureform Masters, Inc.*

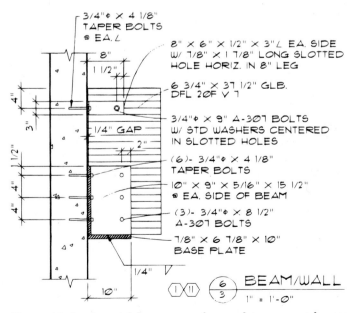

Figure 46–9 A metal hanger can be used to support beams when they intersect a wall. *Courtesy Ken Smith and Associates, AIA.*

Types of Drawings

As you work on concrete drawings, you will find that two types will typically be needed. Structural or engineering drawings are one type that is similar to other drawings that you have been exposed to. These drawings show general information that is required for sales, marketing, engineering, or erection purposes. Shop drawings, the other type, are needed to specify the fabrications process of the part. Depending on the complexity of the component and the structure, these two types of drawings may be combined. Figure 46–10 shows an example of an engineering drawing. An example of shop drawings can be seen in Figure 46–11. As a drafter in an architectural or engineering firm, you typically will be drawing engineering drawings. The fabrication drawings are drawn by engineers and drafters at the fabricating company.

Cast in Place

You have been exposed to the methods of cast-in-place concrete for residential foundations and retaining walls. Commercial applications are similar, but the size of the casting and amount of reinforcing vary greatly. In addition to foundation and floor systems, concrete is often used for walls, columns, and floors above ground. Walls and columns are usually constructed by setting steel reinforcing in place and then surrounding it by wooden forms to contain the concrete. Once the concrete has been poured and allowed to set, the forms can be removed. As a drafter, you will be required to draw details, showing not only sizes of the part to be constructed but also steel placement within the wall or column. This will typically consist of drawing the vertical steel and the horizontal ties. Ties are wrapped around the vertical steel to keep

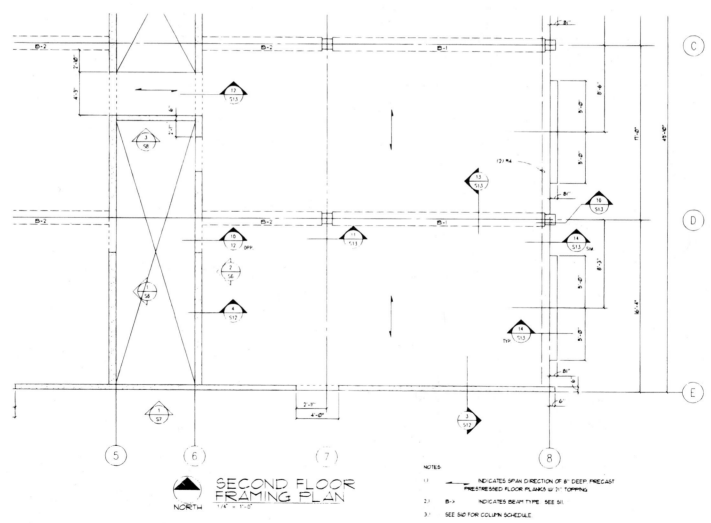

Figure 46–10 A framing plan for a precast concrete structure shows the locations for all concrete beams and panels and serves as a reference map for details. *Courtesy KPFF Consulting Engineers.*

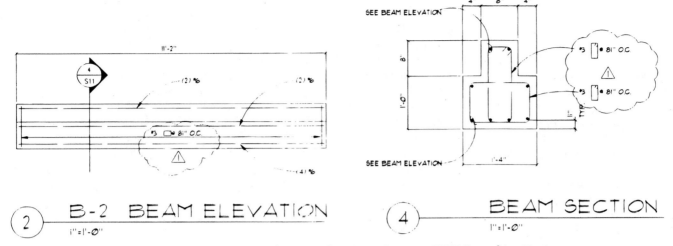

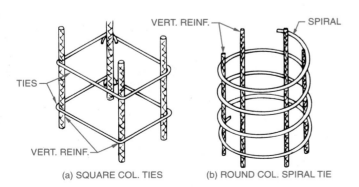

(a) SQUARE COL. TIES (b) ROUND COL. SPIRAL TIE

Figure 46–12 Common methods of reinforcing poured concrete columns.

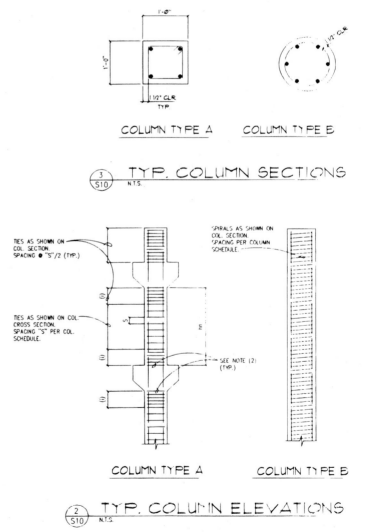

Figure 46–13 Common methods of reinforcing poured concrete columns. The horizontal and spiral ties are attached to the vertical bars with wire ties to resist movement during pouring. *Courtesy KPFF Consulting Engineers.*

Figure 46–11 A precast concrete beam elevation and section drawings. *Courtesy KPFF Consulting Engineers.*

the column from separating when placed under a load. Figure 46–12 shows two examples of column reinforcing methods. Figure 46–13 shows the drawings required to detail the construction of a rectangular concrete column. Depending on the complexity of the object to be formed, the drafter may be required to draw the detail for the column and for the forming system.

Concrete is also used on commercial projects to form an above-grade floor. The floor slab can be supported by a steel deck or be entirely self-supporting. The steel deck system is typically used on structures constructed with a steel frame. Two of the most common poured-in-place concrete floor systems are the ribbed and waffle floor methods. Each can be seen in Figure 46–14.

The ribbed system is used in many office buildings. The ribs serve as floor joists to support the slab but are actually part of the slab. Spacing of the ribs depends on the span and the reinforcing material. The waffle system is used to provide added support for the floor slab and is typically used in the floor system of parking garages.

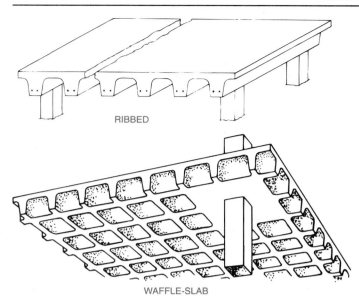

Figure 46–14 Two common concrete floor systems are the ribbed and waffle systems.

Precast Concrete

Precast concrete construction consists of forming walls or other components off site and transporting the part to the job site. Figure 46–15 shows an example of a precast beam being lifted into place. In addition to detailing how precast members will be constructed, drawings must also include methods of transporting and lifting the part into place. Precast parts typically have an exposed metal flange so that the part can be connected to other parts. Figure 46–16 shows common details used for wall connections.

In addition to being precast, many concrete products are also prestressed. Concrete is prestressed by placing steel cables held in tension between the concrete forms while the concrete is poured around them. Once the concrete has hardened, the tension on the cables is released.

Figure 46–15 Precast concrete beams and panels are often formed offsite, delivered to the job site, and then set into place by a crane.

LOCATE CONNECTION AT MID-HEIGHT AND 1'-0" BELOW TOP OF FRAMING (2 CONNECTIONS PER PANEL JOINT) WHERE PANEL JOINT OCCURS AT BEAM LINE, LOCATE ONE PANEL CONNECTION AT MID-HEIGHT ONLY.

Ø'-3" X 3/8" X 1'-0" FLAT BAR WITH 2- 1/2" DIA. X 1'-6" NELSON DEFORMED BARS BENT AROUND EXTRA VERTICAL #4 BAR BY 4'-0" LONG.

EACH SIDE. DO NOT OVERHEAT.

Ø'-8" LONG FILLER ROD BY JOINT WIDTH + 1/8"

2 1/2" X 2 1/2" X 3/8" X 1'-0" ANGLE WITH 2- 1/2" DIA. BY 1'-6" NELSON DEFORMED BARS BENT OVER EXTRA #4 BY 4'-0" VERTICAL

3/4" TYPICAL   Ø'-1" CLEAR CRITICAL

EXTERIOR FACE U. O. N.

ADDITIONAL NOTES AT DETAIL (8/S-2)

(9/S-2) PANEL CORNER CONNECTION   DETL?   1 1/2" = 1'-0"

Figure 46–16 A detail showing the connection of two precast wall panels. *Courtesy Van Domelen/Looijenga/McGarrigle/ Knauf Consulting Engineers.*

As the cables attempt to regain their original shape, compression pressure is created within the concrete. The compression in the concrete helps prevent cracking and deflection and often allows the size of the member to be reduced. Figure 46–17 shows common shapes that are typically used in prestressed construction.

Tilt-up

Tilt-up construction is a method using preformed wall panels, which are lifted into place. Pnels may be formed at the job or off site. Forms for a wall are constructed in a horizontal position, and the required steel is placed in the form. Concrete is then poured around the steel and allowed to harden. Once the panel has reached its design strength, it can be lifted into place. When using this type of construction, the drafter will usually be drawing a plan view to specify the panel locations, as seen in Figure 46–18. Figure 46–19 shows an example of a typical steel placement drawing.

▼ STEEL CONSTRUCTION

Steel construction can be divided into the three categories: steel studs, prefabricated steel structures, and steel-framed structures.

Steel Studs

Prefabricated steel studs are used in many types of commercial structures to help meet the requirements of

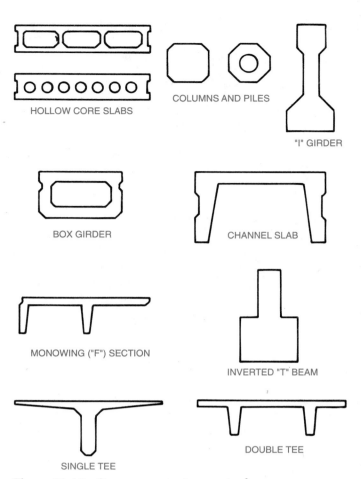

HOLLOW CORE SLABS

COLUMNS AND PILES

"I" GIRDER

BOX GIRDER

CHANNEL SLAB

MONOWING ("F") SECTION

INVERTED "T" BEAM

SINGLE TEE

DOUBLE TEE

Figure 46–17   Common precast concrete shapes.

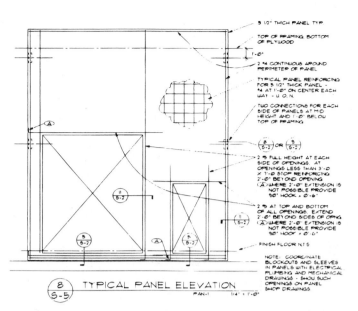

TYPICAL PANEL ELEVATION

Figure 46–19   Steel reinforcing in panels is shown in a panel elevation and referenced to a plan view. *Courtesy Van Domeleh/Looijenga/McGarrigle/Knauf Consulting Engineers.*

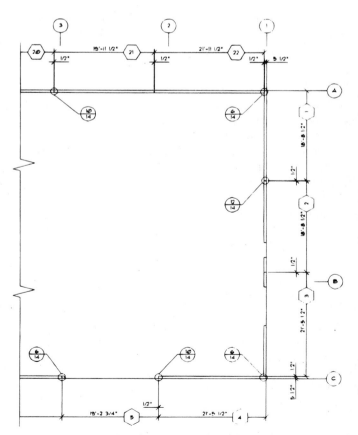

Figure 46–18   A plan view is used to show the location of precast panels. *Courtesy Ken Smith and Associates, AIA.*

Figure 46–20   Steel studs are often used where noncombustible construction is required. *Courtesy Mike Jefferis.*

types 1, 2, and 3 construction methods. Steel studs offer lightweight, noncombustible, corrosion-resistant framing for interior walls, and load-bearing exterior walls up to four stories high. Steel members are available for use as studs or joists. Members are designed for rapid assembly and are predrilled for electrical and plumbing conduits. The standard 24″ (610 mm) spacing reduces the number of studs required by about one-third when compared with common studs spacing of 16″ (406 mm) o.c. Widths of studs can range from 3⅝ to 10″ (93 to 254 mm) but can be manufactured in any width. The material used to produce studs ranges from 12 to 20 gage steel, depending on the loads to be supported. Figure 46–20 shows steel studs as they are typically used. Steel studs are mounted in a channel track at the top and bottom of the wall. This channel is similar to the top and bottom plates of a standard stud wall. Horizontal bridging is placed through the

predrilled holes in the studs and then welded to the stud serving a function similar to solid blocking in a stud wall. Figure 46–21 shows components of steel stud framing.

Prefabricated Steel Structures

Prefabricated or metal buildings have become a common type of structure for commercial structures in many parts of the country. Because these structures are premanufactured, a drafter probably will not be drawing these in an architectural or engineering office. Although a drafter usually completes the plans for a metal building, the drafter is usually employed by the building manufacturer.

Steel-Framed Buildings

Steel-framed buildings require engineering and shop drawings similar to those used for concrete structures. As

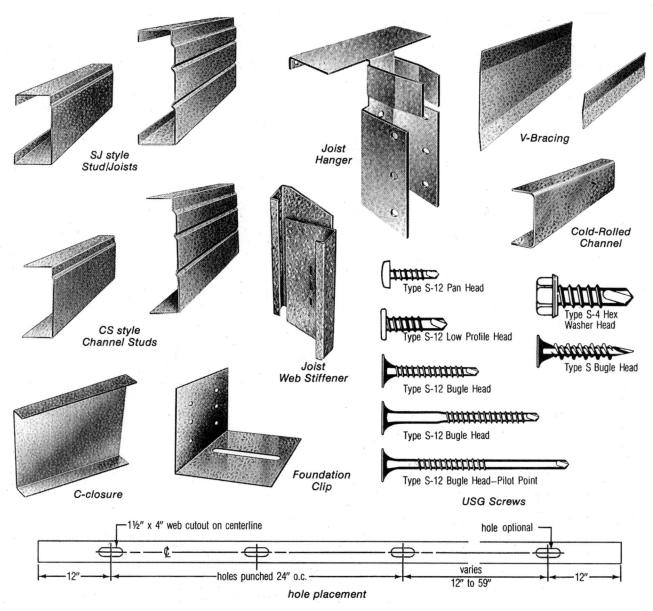

Figure 46–21 Common components of steel stud construction. *Courtesy United States Gypsum Company.*

a drafter in an engineering or architectural firm, you will most likely be drawing engineering drawings similar to the one in Figure 46–22. An engineer or drafter working for the steel fabricator will typically develop the shop drawings similar to the one in Figure 46–23.

As a drafter working on steel-framed structures, you will need to become familiar with the *Manual of Steel Construction,* published by the American Institute of Steel Construction (AISC). In addition to your building code, this manual will be one of your prime references since it will be helpful in determining dimensions and properties of common steel shapes.

Common Steel Products

Structural steel is typically identified as a plate, a bar or by its shape. Plates are flat pieces of steel of various thickness used at the intersection of different members. Figure 46–24 shows an example of a steel connector that uses top, side, and bottom plates. Plates are typically

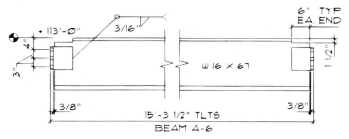

Figure 46–23 Structural steel fabrication drawings are made for each steel component to specify exact sizes, drilling and cutting locations. *Courtesy Van Domelen/Looijenga/McGarrigle/Knauf Consulting Engineers.*

specified on a drawing by giving the thickness, width, and length, in that order. The symbol ℙ is often used to specify plate material.

Bars are the smallest of structural steel products. Bars are round, square, rectangular, or hexagonal when seen in cross section. Bars are often used as supports or braces for other steel parts.

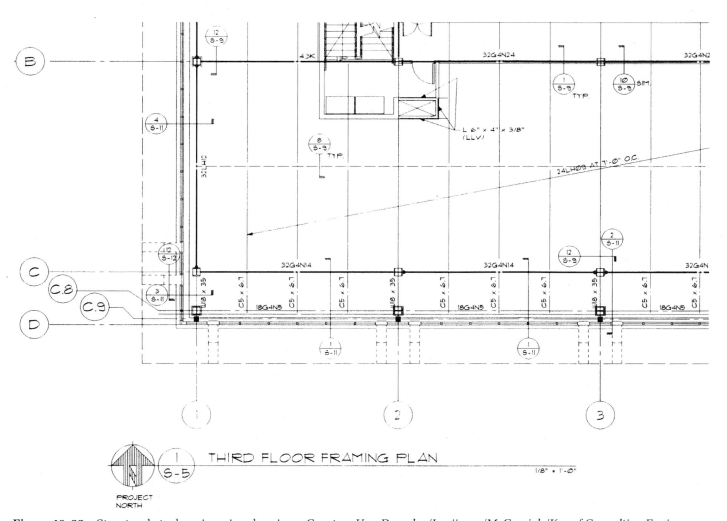

Figure 46–22 Structural steel engineering drawings. *Courtesy Van Domelen/Looijenga/McGarrigle/Knauf Consulting Engineers.*

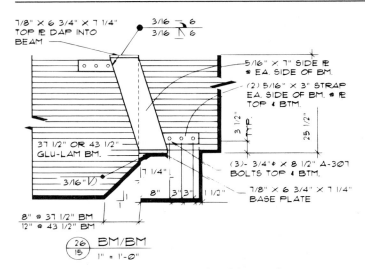

Figure 46–24 Steel plates are used to fabricate beam connectors. *Courtesy Ken Smith and Associates, AIA.*

Structural steel is typically produced in the shapes that are seen in Figure 46–25. M, S, and W are the names that are given to steel shapes that have a cross-sectional area in the shape of the letter I. The three differ in the width of their flanges. The flange is the horizontal leg of the I shape and the vertical leg is the web. In addition to varied flange widths, the S shape flanges also vary in depth.

Angles are structural steel components that have an L shape. The legs of the angle may be either equal or unequal in length but are usually equal in thickness. Channels have a squared C cross-sectional area and are represented by the letter C when they are specified in note form. Structural tees are cut from W, S, and M steel shapes by cutting the webs. Common designations are WT, ST, and MT.

Structural tubing is manufactured in square, rectangular, and round cross-sectional configurations. These members are used as columns to support loads from other members. Tubes are specified by the size of the outer wall followed by the thickness of the wall.

▼ COMMON CONNECTION METHODS

Bolts

Bolts are used for many connections in lumber and steel construction. The diameter and length of the bolt should be specified on plans. Washers or plates are also specified so that the bolt head and nut will not be pulled through the hole made for the bolt. The strength of the bolt also needs to be specified. Bolts are classified as to their strength by the American Society for Testing and Materials (ASTM).

Welds

Welds are classified according to the type of joint on which they are used. Four common welds used in construction are the fillet, back, plug or slot, and groove weld. Each type of weld can be seen in Figure 46–26. A

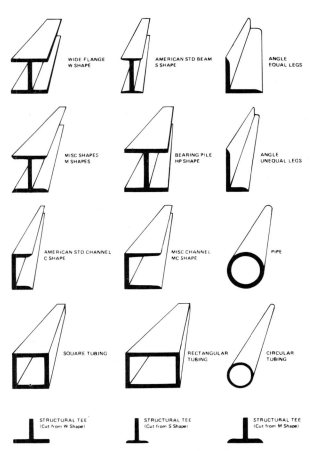

Figure 46–25 Standard structural steel shapes.

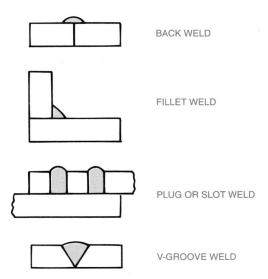

Figure 46–26 Common structural weld symbols.

BASIC WELD SYMBOLS FOR DRAFTING									
BACK	FILLET	PLUG OR SLOT	GROOVE						
			J	V	U	SQUARE	BEVEL	FLARE V	FLARE BEVEL
⌓	◺	☐	Ͱ	V	Y	‖	⌵	⋎	⼁⌵

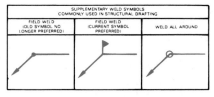

Figure 46–27 Basic welding symbols for structural drafting.

symbol is used to designate the type of weld. Figure 46–27 shows common welding symbols used in construction. Welds are represented in details as shown in Figure 46–28. The horizontal line is a reference line connected to an angled leader line. Each line should be drawn using tools and not drawn freehand.

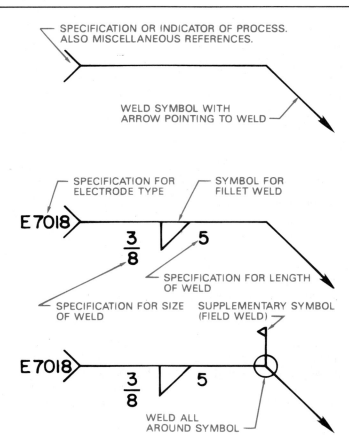

EXPLANATION OF TYPICAL WELD SYMBOLS

Figure 46–28 Common elements of a weld symbol for structural drafting.

CADD Applications

USING CADD TO DRAW STRUCTURAL DETAILS

CADD has increased productivity in drawing structural details. One advantage is the use of standard details. Once a detail has been drawn, it may be automatically inserted into any drawing at any time. The detail may be inserted and used as is, or it may be inserted and modified to provide specific information that changes from one drawing to the next. Every time a CADD drafter draws a detail, it is saved in a drawing file for later use. With this process it is not necessary to draw the same drawing more than once. CADD structural packages are available that provide a variety of standard structural details. Most of these packages are based on parametric design, which means that the standard structural detail may be altered by supplying the computer with data. For example, a set of structural stairs may be drawn automatically by providing the computer with total rise, number of risers, tread dimension, reinforcing size and spacing, and railing information.

Even when creating nonstandard original drawings, CADD is easier than manual drafting. Many CADD software packages have custom structural tablet menu overlays that provide standard structural shapes that may be inserted on the drawing at any time. This reduces the time for a drafter to individually draw these symbols with a pencil and manual template.

CHAPTER

46 Common Commercial Construction Materials Test

DIRECTIONS

Answer the questions with short, complete statements.

1. Letter your name and the date at the top of the sheet.
2. Letter the question number and provide the answer. You do not need to write out the question.

QUESTIONS

Question 46–1 Under what conditions can wood be used to frame a wall that requires a two-hour fire rating?

Question 46–2 Why are the materials used to frame commercial buildings not typically used on residential projects?

Question 46–3 Describe how a panelized roof is framed.

Question 46–4 How does the F_b value for a DF/DF glu-lam beam compare with a DFL #1 beam?

Question 46–5 Why is heavy timber a better construction material than steel in areas of high fire risk?

Question 46–6 Describe two types of drawings typically required for steel and concrete projects.

Question 46–7 How is concrete prestressed?

Question 46–8 How are precast concrete panels usually connected to each other?

Question 46–9 What is the recommended height limit for load-bearing steel studs?

Question 46–10 Sketch the proper symbol for a 3/16″ fillet weld, opposite side, field weld.

CHAPTER 47

Structural Drafting

This chapter introduces basic concepts and drawing methods expected of an entry-level drafter in an architect's or engineer's office. The project example at the end of this chapter will be working drawings for a concrete tilt-up warehouse.

There are two main drawing objectives: to coordinate various plans and to draw coordinated details that correspond to different areas of the building.

▼ PLAN THE DRAWING

Before you begin drawing, search all plan views for similarities. Start at the upper level and work to the lowest. This should show you the loads that must be supported as your work progresses. As you work from top to bottom, you should see beams in the roof supported by columns on the floor plan, pedestals on the slab plan, and footings at the foundation. Had you begun at the foundation level, it might not have been clear why a given pier was in a particular location.

As you work on the various plans, you should also be studying details that relate to the plan. Details will give you a better understanding, whiich should result in faster drawing speed.

A you draw, you may not be able to get your questions answered as quickly as they arise. This is typical of work situations; it is often difficult to find an engineer to answer your questions. Using a nonreproducible blue pencil, keep a list of your questions right on your drawing.

▼ PROCEDURE—FROM CALCULATIONS TO WORKING DRAWING

As you work on projects for an architect or engineer, you will be given a set of calculations, sketches, and similar drawings that have been done by the office. The calculations are the mathematical solutions to particular problems. Calculations are printed in color following the related problem. As a drafter, you will be given a set of specifications and be expected to determine the goal of the engineer. You will then translate the results into working drawings. Typically, this will mean skipping over the math work. For the project you will be working on, the math has been omitted and only the material to be placed on the drawings has been shown.

Sketches an engineer provides are usually tools to help in the solution of the calculations. A drafter is expected to use the sketch as a guide to help determine size and material to be used.

NOTE: *The sketches contained in this chapter are to be used as a guide only.* As you progress through the project, you will find that some parts of a drawing do not match things that have already been drawn. It is your responsibility to coordinate the material that you draw with other drawings for this project. If you are unsure of what to draw, consult your instructor.

In addition to the calculations and sketches for this specific job, a drafter is expected to study similar jobs and details for common elements that can be used in the new drawing. This includes many common notes as well as common connection methods. A drafter would also be expected to consult vendor catalogs for specific details of prefabricated material.

Drawing Layout

There is no one totally correct layout for this project. Usually a project will be drawn so details and plans that relate to a given construction crew will be grouped together. Therefore, the roofer will never see the foundation plan. Often this technique of grouping details and plans requires material to be specified on plans more than once. As you lay out this project, two common methods might be used.

1. Group all plan views together, then all roof details, all concrete details, etc., in the same order that the plan views are presented.
2. Group a plan view with all related details; for instance, show the roof plan and then the roof details; the slab plan and then the slab details.

The layout format will not really be important until you have the plan views complete and you start to lay out details. By that time, you should have a much better understanding of the project.

One frustration associated with this project and this type of drawing is that it cannot be completed without interruptions. You will be required to draw the basic drawing, stop, then solve a related problem at another level. You will be working on several drawings at the same time. The good news is that all drawings belong to the same project. It is common in an office to switch from one job to another several times a day depending on an engineer's or a client's appointments. On this project, you will be expected to shift from one drawing to another, keep them all coordinated, and still keep your sanity.

Material

Drafters in structural firms often work with several different drawing materials; ink on vellum or Mylar, graphite on vellum, polyester on vellum, any combination of these, or CADD. The plan views, elevations, and panel elevations are well suited to CADD drafting because there are so many layering possibilities. Verify the drawing medium with your instructor whenever you start a drawing.

The material is not as important as the end result. Excellent line quality reflecting consistent width and density are a must for a structural detailer. Many of the details in this text are similar to what you will be drawing. The details feature several different line weights as should your drawings.

▼ DETAIL COORDINATION

As you draw each detail, provide room for a title, scale, and detail marker under the detail. The title will typically tell the function of the drawing, such as BEAM/WALL CONNECTION, and should be neatly printed using ¼″ (6 mm) high lettering. The scale should be placed under the title in ⅛″ (3 mm) high lettering. A line should separate the two and should have a detail reference circle on the right end of the line. The reference circle should be about ¾″ (19 mm) in diameter with the line passing through the center of the circle. There will eventually be a detail number on the top of the line. On the bottom of the line, place the page number where the detail was drawn. Do not fill in the circles until all the drawing is complete.

Throughout the plans, you will be referring to a specific detail location for more information. Typically, circles ⅜″–½″ (9–13 mm) in diameter are used so the reference circle does not totally dominate the plan.

▼ LETTERING

Lettering is an important skill to be mastered, even in this age of computers. You will probably get your initial job interview based solely on your lettering. Other than this minor detail, excellent lettering is essential for easy-to-use drawings. The best structural lettering has simple shapes, is easy to read, and is quick to produce. Lettering must be uniform in size, angle, and spacing if you are to advance in the field. Practice lettering that has uniform angles of vertical strokes, uniform spacing, and proper placement of notes. Place notes so they require the shortest leader line without getting in the way of the drawing. Notes are often placed in the drawing or detail. Be careful to keep the vertical strokes of letters away from lines in the drawing.

▼ ORDER OF PRECEDÉNCE

This project may be unlike others you have done. Not only does it have sketches and engineer's calculations (calcs); it also has some very large errors. Most are so obvious (at the beginning) that you have no trouble finding them. They get tougher as you gain more knowledge. The errors are here to help you learn to think as a drafter. Because engineers are not perfect, you sometimes need to sort through conflicting information to solve a problem. Usually though, the engineer is correct, and you will find that you have misinterpreted the information. If you think you have found an error, do not make changes in the drawings until you have discussed them with the engineer. (In this case, discuss them with your instructor.) To help you sort through conflicting information, use the following order of precedence.

1. Written changes by the engineer (your instructor) as change orders.
2. Verbal changes given by the engineer (your instructor).
3. Engineer's calcs (in color).
4. Sketches by engineer.
5. Lecture notes and sketches.
6. Your own decision (should it come to this, please check with your instructor).

▼ OCCUPANCY

The space between grid 1—7.5 will be used for manufacturing wood cabinets including storage of stains, varnish, and glue. The space between grid 7.5—11 will be used for sales and office space. Before you start drawing, review chapter 45 and determine occupancy of each area, the least restrictive types of construction to be used, maximum height, square footage, and occupant load of each portion.

CHAPTER
47 Structural Drafting Test

DIRECTIONS

Students do not usually have adequate time to draw the entire set of plans required to construct this building. Your instructor may wish to add or delete drawings or have you work individually or in teams. Verify with your instructor what problems will be drawn.

Problem 47–1   Roof Drainage Plan.

   Goal   The purpose of this drawing is to show the elevations of the roof supports. By changing the elevation of supports, you can control the flow of water and direct it to downspouts.

   Problem   The building on which you will work will be nearly as large as a football field. Therefore, you have a large area that collects water. If the water is not quickly drained from the roof, it must be treated as a live load, and the beam sizes must be increased. This plan will show the slope of structural members and the roof, and it will show overflow drain locations. Use the sketch as a guide.

   Method   At a scale of 1/16″ = 1′–0″ draw:

1. The outline of the entire structure with 6″ wide walls. Determine size of building from other sketches.
2. The center of the glu-lams at grid B.
3. Four roof drains, overflow drains, and scuppers equally spaced on grids A and C; assume drains to be 8″ in diameter.
4. Dimension as per sketch.
5. Letter all elevations at drain.
6. Specify drains, overflows, and scuppers.

7. General notes: (1) Roof and overflow drains to be general-purpose type with nonferrous domes and 4″ diameter outlets. (2) Overflow drains to be set with inlet 2″ above drain inlet and shall be independently connected to drain lines. (3) Scuppers to be 4″ high × 7″ wide with 4″ rectangular corrugated downspouts. Provide a 6 × 9 conductor head at the top of down spouts.

Problem 47–2   Roof Framing Plan.

   Goal   The purpose of this drawing is to show roof framing used to support the finished roofing. This structure will have a series of laminated beams, which extend through the center of the structure at grid B from grid 1 through 11. It will also have a wood ledger bolted to each wall at grid A and C. Trusses will span between the ledger and the glu-lams. Blocking will be shown at each end of the roof between the trusses. Above the trusses will be sheets of plywood. Four skylights will be equally placed in each bay. This plan will also show how the separation wall intersects the beams.

   Problem   The building on which you are working is approximately 300′ × 100′ × 27′ high. The walls will act like a sail. To keep the structure rigid, the roof framing will be connected to the walls to form a rigid connection. In addition, a diaphragm will be built into each end of the structure. As the walls bend from wind or seismic pressure, the roof will shift in the direction of the pressure. As the pressure is decreased, the walls will return to their original shape. The rigid connections ensure that when the walls spring back to their original position, the roof will also return to its original position. If the connections at this level fail, the roof will fall from its supports. The roof framing plan is used to show how and where these items are located. Both

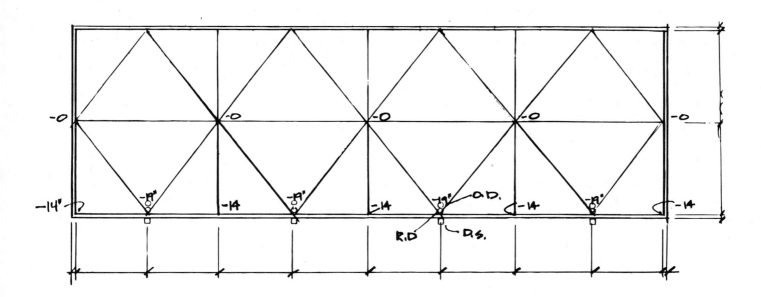

ROOF DRAINAGE PLAN
1/16″ = 1′-0″

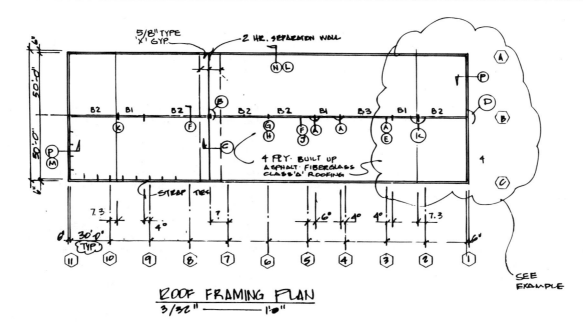

5/8" TYPE 'X' GYP.

2 HR. SEPARATION WALL

4 PLY. BUILT UP ASPHALT FIBERGLASS CLASS 'A' ROOFING

STRAP TIES

SEE EXAMPLE

ROOF FRAMING PLAN

3/32" ———— 1'-0"

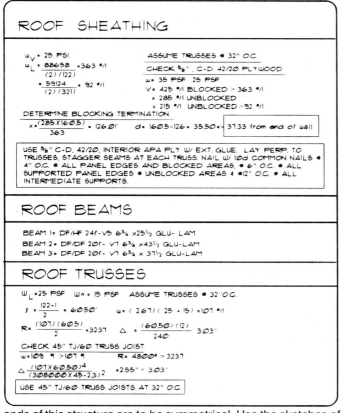

ROOF SHEATHING

W_V = 25 PSI
W_L = $\frac{88658}{(2)(122)}$ = 363 #/1
= $\frac{59124}{(2)(321)}$ = 92 #/1

ASSUME TRUSSES @ 32" O.C.
CHECK 3/8", C-D 42/20 PLYWOOD
W = 35 PSF 25 PSF
V = 425 #/1 BLOCKED > 363 #/1
= 285 #/1 UNBLOCKED
= 215 #/1 UNBLOCKED > 92 #/1

DETERMINE BLOCKING TERMINATION
X = $\frac{(285 \times 160.5)}{363}$ = 126.01' d = 160.5-126 = 35.50 ⟶ 37.33 from end of wall

USE 3/8" C-D, 42/20, INTERIOR APA PLY W/ EXT. GLUE. LAY PERP. TO TRUSSES, STAGGER SEAMS AT EACH TRUSS. NAIL W/ 10d COMMON NAILS @ 4" O.C. @ ALL PANEL EDGES AND BLOCKED AREAS, @ 6" O.C. @ ALL SUPPORTED PANEL EDGES @ UNBLOCKED AREAS & @12" O.C. @ ALL INTERMEDIATE SUPPORTS.

ROOF BEAMS

BEAM 1: DF/HF 24f-V5 6¾ x25½ GLU-LAM
BEAM 2: DF/DF 20f- V7 6¾ x43½ GLU-LAM
BEAM 3: DF/DF 20f- V7 6¾ x 37½ GLU-LAM

ROOF TRUSSES

W_L = 25 PSF W_A = 15 PSF ASSUME TRUSSES @ 32"O.C.
$l = \frac{122-1}{2}$ = 60.50' w = (2.67)(25 + 15) = 107 #/1
R = $\frac{(107)(60.5)}{2}$ = 3237 Δ = $\frac{(60.50)(12)}{240}$ = 3.03"
CHECK 45" TJ/60 TRUSS JOIST
w = 109 # > 107 # R = 4800# > 3237
Δ = $\frac{(107 \times 60.50)^4}{(308000 \times 45-23)^2}$ = 2.55" < 3.03"

USE 45" TJ/60 TRUSS JOISTS AT 32" O.C.

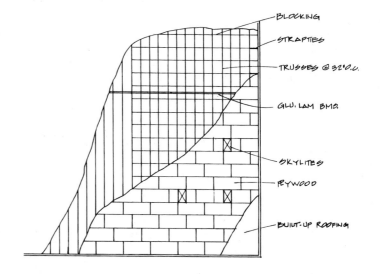

BLOCKING
STRAPTIES
TRUSSES @ 32'-0.C.
GLU-LAM BMS.
SKYLITES
PLYWOOD
BUILT-UP ROOFING

ends of this structure are to be symmetrical. Use the sketches of the roof as a guide.

Method Use a scale of ³⁄₃₂″ = 1′–0″. If you're unsure of what you are drawing, look through the roof framing and truss details.
1. Lay out and draw the walls.
2. Lay out and draw the locations of beams.
3. Dimension and label all grids.
4. Lay out truss, blocking, skylights, strap ties, and 4 × 8 plywood as per sketch and roof details.
5. Draw plywood over blocking and trusses and strap ties and skylights.
6. Draw 2-hour wall and roof area.

7. Letter general notes: (1) Beam sizes. Specify beam size and type by beam, or provide a beam schedule. See engineer's calcs for size of beams 1–3. (2) Bay referencing. Use ¼″ hex and print letter or number centered in hex. (3) Do not place detail markers on your plan at this time. These markers should be added to each plan as the details are drawn.
8. Local notes (verify notes with calculations).
   1. Bristolite 3069-A.S.-DD-CC/WTH-C.P. double domed curb-mounted skylights with 4 per bay (80 total).
   2. 45″ TJ/60 Truss joist at 32″ o.c.
   3. 3 × 4 dfl std. and better solid blocking at 48″ o.c. 36′ out from each end wall.

4. Simpson MST 27 strap ties at 8′–0″ o.c. for entire perimeter. See Detail ***. (Place detail marker in note for future reference).

5. ⅝″ C-D 42/20 . . . . (See engineer's calcs for complete specifications and provide required information on drawing.)

Problem 47–3   Slab on Grade

Goal   This drawing will provide directions for pouring the concrete floor. Major items shown on this plan will include the floor slab and control joints, pedestal footings to support steel columns, loading docks, and doors.

Problem   The concrete slab is so large that the concrete crew will pour it in stages. This will account for the 30 × 25 grids. These control joints will also serve to minimize cracking. The 4′ wide strip at the perimeter is to allow for movement of the concrete walls. The small squares at grid B are to allow for movement of the slab from the steel columns, which support the roof.

Method   Your goals on this drawing are to determine door locations in the walls, slab control joints, and L connectors which will support the slab where it is over fill. Use a scale of ³⁄₃₂ = 1′–0″. Use the sketch as a guide for your drawing.

1. Lay out grids A and C and 1–11.
2. Lay out the exterior walls. Keep the drawing in the upper right corner of the page to allow for possible detail placement around the plan.
3. Lay out door openings using sketches of the panel elevations for locations.
4. Locate the loading docks. See slab details for the wall thickness. Locate the dock so that there is an equal amount of space at each side of the door and the dock walls. This is not the same as centering the dock on the center of the doors.
5. Lay out control joints at grid B, grids 2–10, 4′ perimeter, and pedestals. See wall details for the pedestal size. Draw at an appropriate scale.
6. Draw walls and control joints. Use varied line weight to distinguish between the two.
7. Draw the structural connectors using bold lines.
8. Dimension as per sketch.
9. Label the grids, elevations, and notes.
10. General notes: (1) All slab on grade concrete shall be 5″ thick F′c = 3500 p.s.i. @ 28 days. (2) All target strengths shall be in accordance with Chapter 4 of ACI 318 Building Code Requirements for reinforced concrete. (3) Reinforce with 12 × 12-w4 × w4 or grd 40, #3 @ 15″ o.c. each way centered in slab.
11. Local notes:

On Sketch	Note Should Read
Doweled joints. . .	Doweled joints w/#5 × 15″ smooth dowels @ 12″ o.c.
¼″ joint. . .	¼″ fiber isolation joints around entire pedestal.
5″ slab. . .	5″ slab over 4″ base minimum of ¾″ minus crushed rock. Reinforce w/ 12/12- 4/4 w.w.m. or grade 40, # 3 @ 15″ o.c. each way, 3″ clear of base.
L connectors. . .	¾″ dia. 'L' structural connection inserts w/ ¾″ diameter × 25′ coil rods @ 5′-7′ o.c. Provide 2″ min. rod penetration into inserts 2¾″ from top of slab.

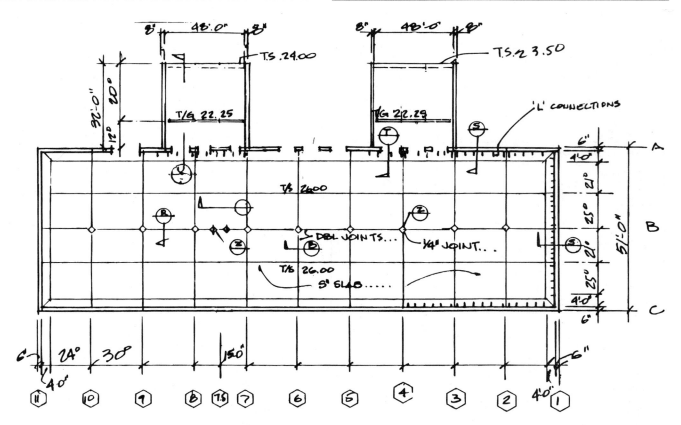

SLAB ON GRADE PLAN
3/32″ ———— 1′0″

ADD PERSONAL DOORS SEE PANEL ELEVATIONS FOR LOCATIONS.

SEE PANEL ELEV. TO LOCATE ALL OPENINGS. DON'T DIMEN. OPENTNGS ON this PLAN

Problem 47–4 Foundation Plan.

Goal The purpose of this drawing is to show concrete supports for the walls and columns. Supports will consist of a continuous footing at the perimeter of the structure and individual piers placed under the pedestals that support the columns.

Problem This drawing will show the size and location of the footings, as well as any change in their elevation. All elevations should be given using the proper symbol.

Method Use a scale of ³⁄₃₂″ = 1′–0″ for this drawing. Place the drawing in the upper right-hand corner of the page. Use the sketch of the foundation as a guide.

1. Lay out grids A and C and 1–11.
2. Lightly lay out the exterior face of the concrete walls and loading docks. This is for your reference only since only the retaining walls will be darkened on this plan.

3. Determine the size of the exterior footing, and lay it out in the proper location. Refer to the foundation details for size information.
4. Draw the location of all elevation changes in the footing.
5. Determine the size of the pedestal footings. Since all footings are close in size, given the scale of the drawing, lay out all footings at the *average* size.
6. Draw all footings.
7. Dimension drawing as per sketch.
8. Label all grids, elevations, and notes.
9. General Notes. (1) All foundation, pedestal and retaining wall concrete shall be F′c 3000 PSI at 28 days. (2) All steel bar reinforcement shall be ASTM A615, Grade 40, deformed bars unless otherwise specified in drawings or details.
10. Local notes. None required.

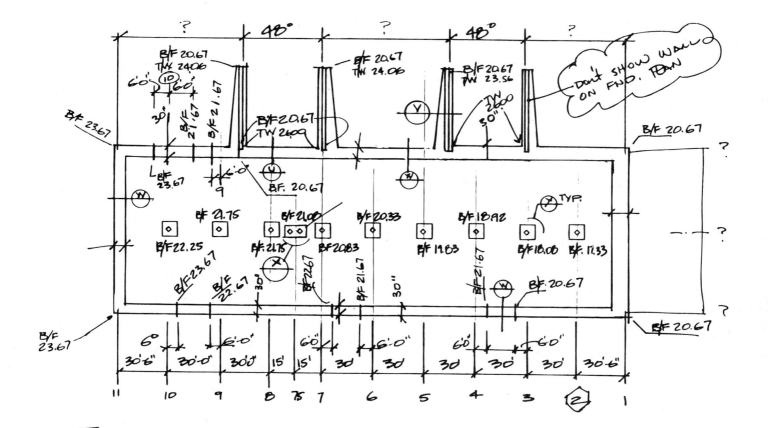

FOUNDATION PLAN
3/32"= 1'-0"

Problem 47–5   Roof Beam Details.

Goal   The purpose of these details is to show how the glu-lam beams are attached in end-to-end connections, to columns, and to the wall at grids 1 and 11. Use the sketches and calcs as a guide for your drawings.

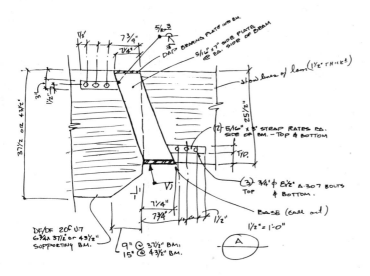

❍ DETAIL A: BEAM TO BEAM. Use of a saddle to hang 25.5″ beam from large beams.
❍ DETAIL B: BEAM TO WALL. Beams supported on columns with 2-hour wall between. Gyp. bd. to remain unbroken.
❍ DETAIL C: WALL TO ROOF. Similar to "B" but shows roof between beam and concrete walls.
❍ DETAIL D: BEAM TO WALL. Top and side views to show plates for beam support.
❍ DETAIL E & F: BEAM TO COLUMN. Both show bearing-plate connections to steel column.
❍ DETAIL G: BEAM TO COLUMN. Intersection of two beams over a steel column. All other beam intersections are in midspan and are not over a column.

Method   These drawings should be done in two stages. Lay out and final drawing with lettering and dimensions. Use the calcs, sketches, and the vendor's material from vendor's brochures or Sweet's catalogs.

1. Determine the location of details. Details should be drawn with either the roof drainage and framing plans or on a separate sheet. In addition, determine what order the details will be presented. Place the details in an order that reflects their relationship to the building, not the order that the engineer thought of them.

2. Lightly block out each detail, allowing approximately 15 minutes per detail. Given this brief amount of time, draw only major features, such as beam sizes, metal brackets, and bolt centerlines.

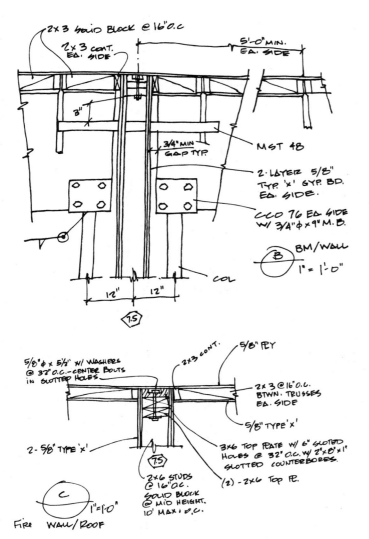

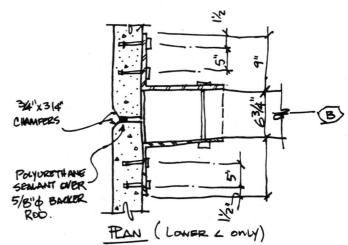

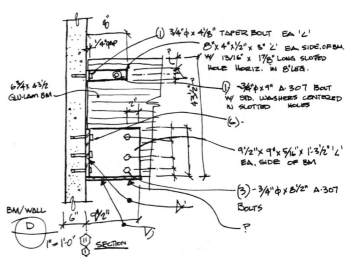

3. Draw beams with finish-line quality. Use several line types or colors. Draw steel plates and brackets as boldest lines, beam outlines slightly thinner, followed by bolts or other connectors, followed by glu-lam lines. Lam-lines should be lighter than other lines—almost a cross between a construction and an object line. Represent steel plates in side view with pairs of parallel thin lines at 45°.

4. Dimension all information from the sketch or calcs. Remember, some information can be placed within the drawing. Keep extension lines approximately 1–1½" long. Interrupt lam lines as required for easy reading.

5. Place weld information on details as required. Be sure you understand what you are connecting before you order a weld. Also, be mindful of symbols for "this side, or opposite side."

6. Notes. Place notes in or near the drawing as required. Similar information in two details can be covered in one note carefully placed between the two details.

7. Place detail markers, title, and scale below each detail. Do not place any information in the circle yet.

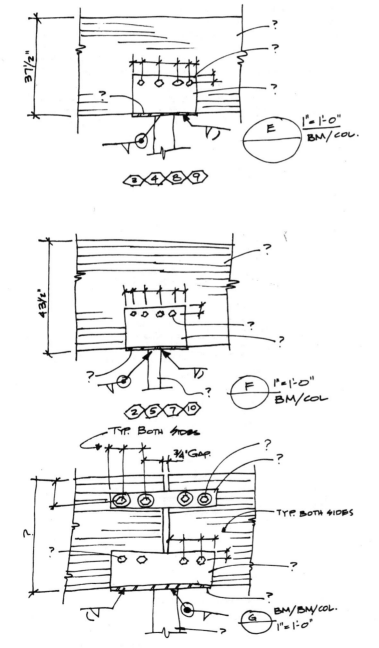

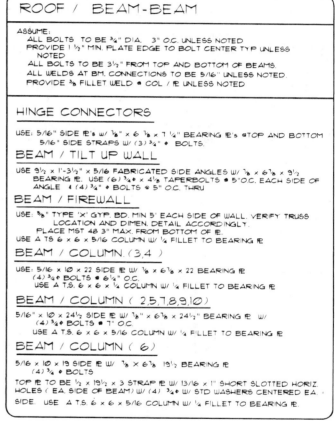

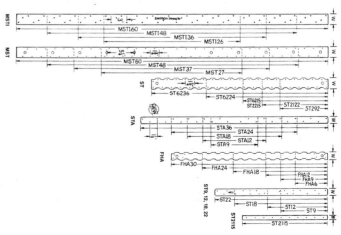

Problem 47–6 Truss Details.

Goal The purpose of these details is to show how the trusses are connected in end-to-end connections, to beams, to ledgers, and to the wall at grids 1 and 11 at A and C.

○ DETAIL H: Basic truss-to-beam connection. Information in this detail should be reflected in all roof details.

○ DETAIL I: Remember that I, O, and Q are never used in the final detail callouts.

○ DETAIL J: Truss-to-beam at a column. Braces are added from the truss to the glu-lam so that the beam will not roll off the column as a result of wind pressure against the walls at grid A and C.

○ DETAIL K: Truss-to-truss connection at the inner edge of the roof diaphragm. This bolting will, in effect, make two trusses function as one. Make use of vendor specs for sizes.

○ DETAIL L: Ledger splice at grid A and C.

○ DETAIL M: Ledger splice at grid 1 and 11.

○ DETAIL N: Truss-to-ledger connection at A and C, perpendicular.

○ DETAIL P: Truss-to-ledger connection at 1 and 11, parallel.

Method These drawings should be done in two stages. Lay out and final drawing with lettering and dimensions. Use the calcs and the vendor's material to supply needed information.

1. Determine the location of details. Details should be drawn with the beam details or on a separate sheet. In addition, determine in what order the details will be presented. Place the details in an order that reflects their relationship to the building, not the order that the engineer thought of them.

2. Lay out each detail, allowing approximately 15 minutes per detail. Given this brief amount of time, draw only major features, such as beam or truss sizes, metal brackets, and bolt center lines. As you lay out the trusses, assume the web is 1″ diameter aluminum placed at a 45° angle starting 6″ from the end of the truss. The bottom chord of the truss *never* touches the beam. Also, carefully study the truss diagram provided by the vendor. The top chord, which supports the entire load on the truss, never touches the beam. The truss is supported by a metal connector.

3. Draw beams with finish-line quality. Use several line types or colors. Draw steel plates and brackets as boldest lines, beam outlines slightly thinner, followed by bolts or other connectors, followed by glu-lam lines. Represent steel plates in side view with pairs of two parallel thin lines at 45°.

4. Dimension all information from the sketch or calcs. Remember, some information can be placed within the drawing. Interrupt lam lines as required for easy reading.

5. Place weld information on details as required. Be sure you understand what you are connecting before you order a weld. Also, be mindful of symbols for "this side, or opposite side."

6. Notes. Place notes in or near the drawing as required. Similar information in two details can be covered in one note carefully placed between the two details.

7. Place detail markers and title and scale below each detail. Do not yet place any information in the circle, however.

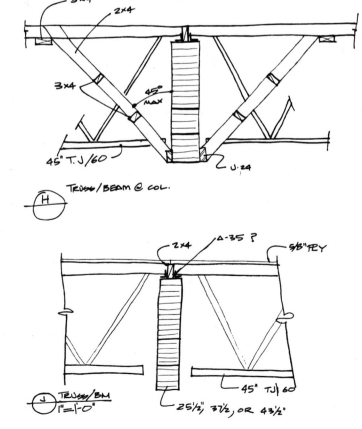

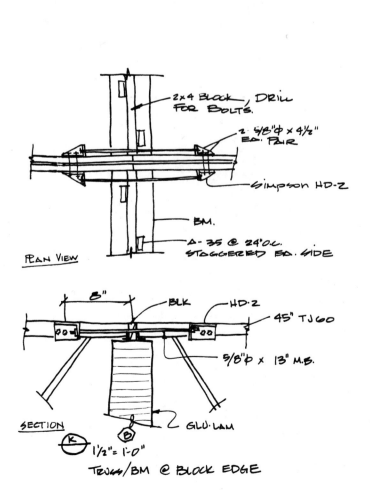

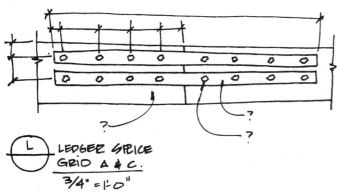

L LEDGER SPLICE
GRID A & C.
3/4" = 1'-0"

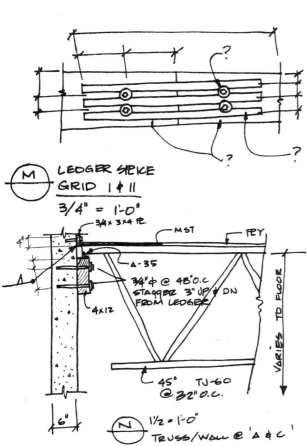

M LEDGER SPLICE
GRID I & II
3/4" = 1'-0"

3/4 x 3x4 PL.
MST   PLY
4"
A-35
3/4"φ @ 48"O.C.
STAGGER 3" UP & DN
FROM LEDGER
4x12
45" TJ-60
@ 32"O.C.
VARIES TO FLOOR
6"
N 1½ = 1'-0"
TRUSS/WALL @ 'A & C'

TRUSS / BEAM DETAILS

ROOF LEDGERS: USE 4x12 D.F.L. #2
LEDGER SPLICES: USE (2) ½ x 3 STRAP PL EXTENDING 4 BOLTS ON
EACH SIDE OF SPLICE. USE ¾"φ x 8" BOLTS @
20" O.C. W/ 2 ROWS @ 3½ O.C.

LEDGER SPLICE/ NON BEARING WALLS (I & II):
USE (3) SIMPSON MST48 STRAP TIES CENTERED OVER SPLICE @
3½"O.C.- 1¾ DOWN FROM LEDGER TOP. PROVIDE (4) ¾"φ x 8" BOLTS
THRU ¼ x 3"φ WASHERS CENTERED BETWEEN PL's.

TRUSSES : USE 45" DEEP TJ/60 TRUSSJOIST @ 32" O.C.
ASSUME: 2.3" DEEP TOP AND BOTTOM CHORD.
1" φ ALUM. WEBS @ 45° ± 6" IN FROM END OF TOP CHORD.
BTM. CHORD TO BE 2" MIN. CLEAR OF BM.

TRUSS/BM CONNECTION:
PROVIDE 2x4 SOLID BLOCKING BTWN. TRUSSES OVER BEAM.
ATTACH TO GLU-LAM W/ SIMPSON CO. A-35 @ 32" O.C. @ EA. BLOCK
ALTERNATE SIDES & PROVIDE (2) EA. BLK. NAIL W/ n8 ALL HOLES.

TRUSS-BM/COLUMN PROVIDE 2x4 BRACE ALONG BEAM FOR
(2) TRUSS SPACES EA. SIDE OF COLUMN. USE 3x4 SOLID BLOCK BTWN.
BRACES. NAIL BLOCK TO BRACE W/ (5) 16d NAILS. PROVIDE 3x4x10"
NAILER @ TRUSSES. NAIL W/ (2)16d's EA. TRUSS. USE SIMPSON U-24 HGR
2" UP FOM BTM. OF BEAM TO BRACE.   SET BRACE AT 45° MAX. FROM
BEAM.

TRUSS-BM @ BLOCKING EDGE: USE (2) SIMPSON HD-2
HOLDOUNS EITHER SIDE OF SPLICE W/ (2) ⅝"φ x 4½" BOLTS THRU
TRUSS, AND ⅝"φ x 15" BOLTS ACROSS SPLICE.

TRUSS/WALL (A & C): USE 4x12 DFL #2 TREATED LEDGER W/
¾"φ x8" BOLTS @48" O.C. THRU ¼"x3φ WASHERS. SOLID BLOCK BTWN.
TRUSSES @ LEDGER W/ 2x4 BLOCKS W/ (2) SIMPSON A-35 EA. BLK.
PROVIDE ¾" x 3"x4 PL ABOVE PLYWOOD SHEATH. BOLT TO WALL W/
(1) ¾" X 4⅛ BOLTS AT EA. TRUSS. WELD SIMPSON MST 27 STRAP
TO PL W/ ⅛" FILLET WELD AND NAIL TO TRUSSES W/ n8 EA. HOLE.

TRUSS/WALL (I, II):
USE 4x12 DFL TREATED LEDGER W/ ¾"φ x 8" BOLTS @ 6'-0" O.C. THRU
⅛"φ x3" WASHERS STAGGERED 3" UP/DN. USE 3x4PL (SEE ABOVE) @ EA.
3rd. BLOCK W/ MST27 STRAP. PROVIDE 3x4 SOLID BLOCK @ 48" O.C.
FOR 37.3' OUT FROM WALL (SEE PLYWOOD DIAPHRAM SPECS.)
CONNECT BLOCK TO TRUSS W/ SIMPSON Z-44 HGR.

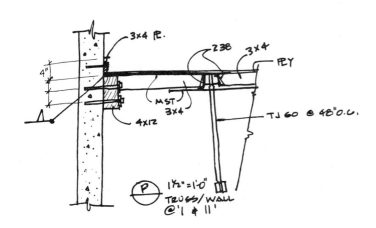

3x4 PL.
4"
238 3x4
PLY
MST
3x4
4x12
TJ 60 @ 48"O.C.
P 1½" = 1'-0"
TRUSS/WALL
@'I & II'

Problem 47–7 Slab Details.

Goal These drawings will provide information required to complete the floor slab, showing intersections in the concrete slab and door joints, and elevations.

Method Use a scale as indicated on each detail. Place details on the same sheet as the slab, on grade plan, or on a separate sheet with all other concrete details.

1. Determine in what order the details will be presented. Place the details in an order that reflects their relationship to the building, not the order that the engineer thought of them.
2. Lay out each detail, allowing approximately 15 minutes per detail. Given this brief amount of time, draw only major features such as slab outlines, and mesh and steel center lines.
3. Draw all lines. Be careful that all rebar is drawn with the same type of line quality. Draw mesh as a thin line with the "X" spaced at 4″ o.c.
4. Dimension as per sketches.
5. Place detail markers on the slab plan and section.
6. Label as required.

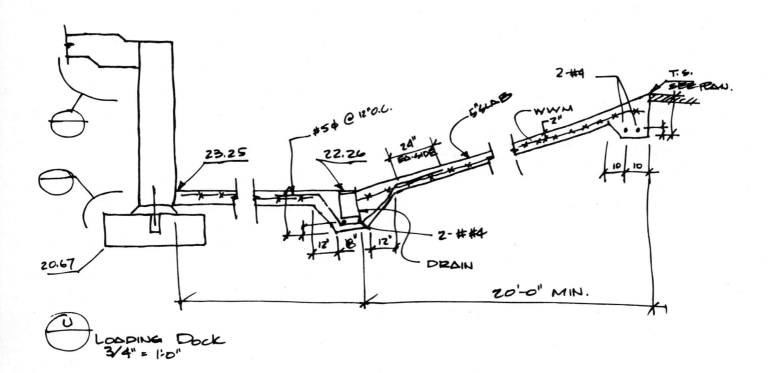

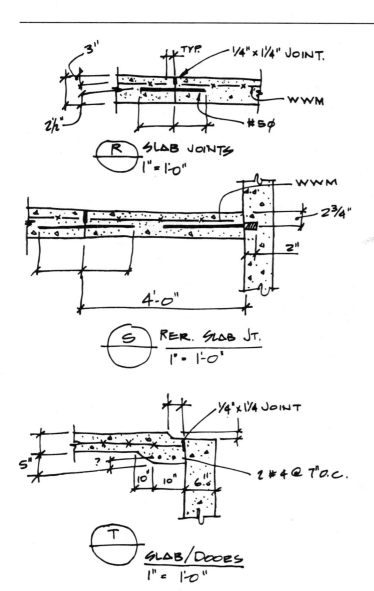

3"

TYP.

1/4" x 1 1/4" JOINT.

WWM

2 1/2"

5 ⌀

R SLAB JOINTS
1" = 1'-0"

WWM

2 3/4"

2"

4'-0"

S RER. SLAB JT.
1" = 1'-0"

1/4" x 1 1/4" JOINT

5"

?

10"   10"   6"

2 # 4 @ 7" O.C.

T SLAB/DOORS
1" = 1'-0"

SLAB/ DOCK

SLAB:
ALL SLAB ON GRADE CONC. TO BE F'c = 3500 PSI @ 28 DAYS. ALL TARGET STRENGTHS SHALL BE IN ACCORDANCE W/ CHAPTER 4 OF ACI 318 BUILDING CODE REQUIREMENTS FOR REINFORCED CONCRETE.   SLABS TO BE 5" THICK UNLESS NOTED W/ WWF 12x12 -W4xW4 OR GRD 40 # 3 @ 15" O.C. EA. WAY CENTERED IN SLAB.

JOINTS:
PROVIDE 1/4" FIBER INSOLATION JOINTS W/ NEOPRENE JOINT SEALANT OVER 3/8"⌀ BACKER BEAD AT ALL CONTROL JOINTS AND AROUND ALL FOUNDATION PEDESTALS.
PROVIDE # 5 x 15" LONG SMOOTH DOWELLS @ 12"O.C. AT ALL CONTROL JOINTS UNLESS NOTED (SEE PER. JTS.) COVER W/ PAPER SLEEVES ON EA SIDE OF JOINT.

KEEP WWM 2" MIN. CLEAR OF JOINT. KEEP # 5's 2 1/2" CLR OF BOTTOM OF SLAB.

PERIMETER JOINTS.
SAME AS ABOVE EXCEPT: ALL SMOOTH DOWELLS TO BE # 5x 30" LONG.

PROVIDE 3/4"⌀ 'L' STRUCTURAL CONNECTION INSERTS W/ 3/4"⌀ x 25"LONG COIL RODS @ 5'-7" O.C. PROVIDE 2" MIN ROD PENETRATION INTO INSERTS.

DOORS:
PROVIDE 3" WIDE x 3/4 x 3/4 CHAMFER IN SLAB AT ALL DOORS.

THICKEN SLAB AT ALL DOORS TO 10" AND PROVIDE (2) # 4 2" CLR OF BTM. OF SLAB.

LOADING DOCK:
PROVIDE 12' WIDE BASE PLATFORM W/ 1/4"/12" SLOPE TO DRAIN. DRAIN TO BE 8" x 12" DEEP POLYESTER CONC. DRAIN W/ C.I. GRATE. THICKEN SLAB @ GRATE TO 18" DP. W/ (2) # 4 @ 15" O.C. 3" CLR OF BTM OF FTG.

PROVIDE 20' MIN SLOPE IN DOCK. THICKEN SLAB TO 10" @ ENTRY W/ 2 # 4 @ 6" O.C. 3" CLR OF BTM OF FTG.

Problem 47–8 Foundation Details.

Goal These details will provide information on the concrete at the foundation level, showing either the exterior foundation or the pedestals.

Method Use a scale as indicated on each detail. Place details on the same sheet as the foundation plan or on a separate sheet with all other conrete details. Use the sketches and the calcs as a guide for your drawing.

1. Determine in what order the details will be presented. Place the details in an order that reflects their relationship to the building, not the order that the engineer thought of them.
2. Determine the missing dimensions from the calculations prior to layout work.
3. Lay out each detail, allowing approximately 15 minutes per detail. Given this brief amount of time, draw only major features such as concrete outlines and mesh and steel center lines. Be sure to allow room for schedules by each detail where they are required. Be careful that all rebar is drawn with the same type of line quality as used in the slab details.
4. Dimension as per sketches.
5. Place detail markers on the foundation plan and section.
6. Label as required.

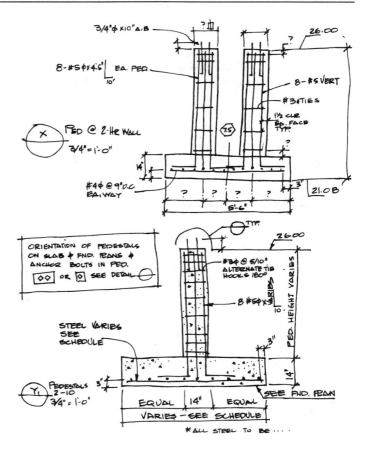

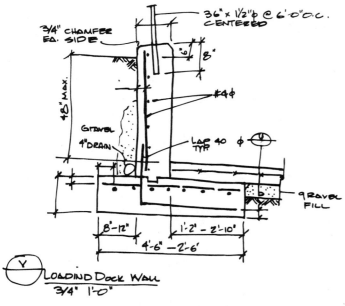

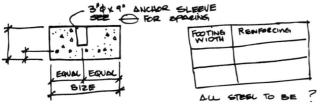

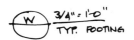

FOOTING	PEDESTAL HEIGHT	SIZE	REINFORCING STEEL				BTM OF FTG.
B-2	7.50	7'-0" φ	#	φ	@	O.C.	17.33
B-3							
B-4							
B-5							
B-6							
B-7							
B-7.5							
B-8							
B-9							
B-10							
ALL STEEL TO BE GRADE ..... UNLESS NOTED							

FOUNDATION

ASSUME: ALL FOUNDATION, PEDESTAL, AND RETAINING WALL CONC. TO BE F'c =3000 PSI @ 28 DAYS.

ALL STEEL BAR REINFORCEMENT SHALL BE ASTM A615, GRADE 40, DEFORMED BARS UNLESS OTHERWISE SPECIFIED IN DRAWINGS OR DETAILS.

USE 9 1/2 × 9 1/2" × 7/8" STEEL BASE ℗ W/ 4-7/8"⌀ HOLES. USE 1/4" FILLET ● ℗/COL. SUPPORT ℗ ON 1" NON-SHRINK NON-METALLIC GROUT.

PEDESTALS USE 14"⌀ W/ (8)- GRADE 40 #5 VERT. RE-BAR 1½" CLR OF PEDESTAL FACE (HEIGHT VARIES, SEE FOUNDATION PLAN AND COMPLETE SCHEDULE). USE #3 HORIZ. TIES W/ TOP AND BOTTOM (3) TIES ● 5" O.C. & INTERMEDIATES ● 10" O.C. HORIZ. TIES TO BE WITHIN 1½ OF TOP AND BOTTOM. PROVODE (4) ¾"⌀ × 10" A.B. IN 6½" GRID W/ 2½" BOLT PROJECTION.

PEDESTALS @ 7.5 SIMILAR TO TYPICAL PEDESTALS (ABOVE) UNLESS NOTED. PROVIDE #3 HORIZ TIES (2) ● 5"O.C. WITHIN 1½" OF TOP OF PEDESTAL, AND BALANCE ● 10"O.C. W/ LAST TIE WITHIN 1½" OF BTM OF PED. USE 5'-6" × 3'-6" × 18" DEEP FOOTING.

PEDESTAL FOUNDATIONS (@B-6) USE A 6'⌀ ×14" DP. W/GRD. 40, #5 ●10"O.C. EAWAY 3" CLEAR OF BASE

PEDESTAL FOUND. (@ 2,5,7,10) USE 7'⌀ × 14" DP. W/ GRD. 40, #6 ● 10"O.C. EAWAY 3" CLR OF BASE.

PEDESTAL FOUND. (@ 3,4,8,9,) USE 65⌀ × 14" DP. W/ GRADE 40, #6 ● 12"O.C. EACH WAY 3" CLEAR OF BASE.

FOUND.- LOAD BEAR.: 36" × 14" DP. W/ (3) GRD. 40,5● 9"O.C. 3" CLR OF FTG. BTM.

FOUND.- NON-BEAR(1& 11) USE 26" × 14" DP. FOOTING W/ (2) GRD. 40, #5 ● 12"O.C. 3" CLR OF FTG. BTM.

RETAINING WALL: USE 8" THICK WALL W/GRD. 40, #4 ● 10" O.C. EA. WAY, 2" CLR OF SOIL SIDE OF WALL. USE 14" THICK FOOTING × 45' WIDE W/ GRADE 40, #4 ● 10" O.C. EA. WAY, 2" CLR OF TOP OF FOOTING. PROVIDE #5 ● 10" O.C. EA. WAY 3" CLR OF BTM OF FTG. W/ #5 ● 10 O.C. 'L' SHAPED BAR W/ 24" MIN. PROJECTION INTO FOOTING. USE #4⌀ CONT. ● "L". FOOTING TO TAPER FROM 8"-12" PROJECTION FROM WALL ON SOIL SIDE AND BTWN. 1'-2"- 2'-10" ON DOCK SIDE. PROVIDE 4" ⌀ DRAIN IN 12" × 24" MIN. ¾" MINUS GRAVEL BED.

Problem 47–9 Typical cross section.

Goal The purpose of this drawing is to show the vertical relationship of structural members. It functions as a reference map and is not intended to provide a detailed explanation of the connections. Most of the details that will be drawn will be referenced on this drawing.

Method Using a scale of ³⁄₁₆″ = 1′–0″, draw the section. Do not draw the entire section.

1. Place a section marker in the same location on the roof drainage, roof framing, slab, and foundation plans before starting the drawing.
2. Establish base lines to represent the top of the slab floor and the wall at C and the beams at B.
3. Determine the elevations of the footings at the wall and pedestal from the foundation plan. Your detail must match your reference marker, not your neighbor's or the engineer's sketch. The location of the reference marker will affect your wall size.
4. Establish vertical heights at B and C, based on panel elevation.
5. Lay out the ledger, trusses, and beams.
6. Draw all items. Be careful to distinguish between concrete and wood members.
7. Dimension as per the sketch.
8. Label grids and local notes.

On Problem	Note Should Read
⅝″ ply. . .	See note on roof calcs.
T.J.I. . . .	45″ TJ/60 joist at 24″ o.c.
6″ walls. . .	6″ tilt up conc. walls/
Roofing. . .	4 ply built-up asphaltic fiberglass class 'A' roofing.
Beam. . .	6¾″ × 25.5, 37.5 or 43.5 glu-lam beam

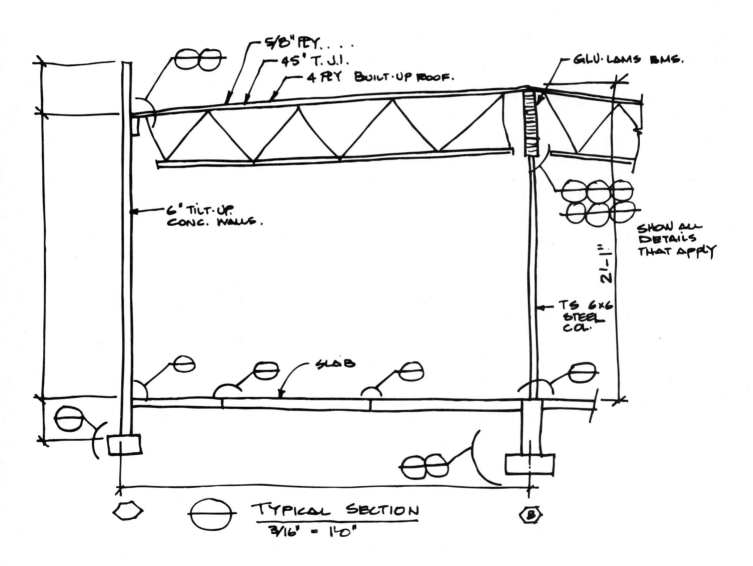

Problem 47–10 Exterior Elevations.

Goal These drawings will provide an external view of the building to be constructed.

Method Typically, an elevation of each side of a structure would be drawn. Verify with your instructor which elevations will be drawn with your project. Use a scale of ³⁄₃₂″ = 1′–0″ and follow the procedures outlined in chapters 21 and 22. For best use of a page, place the elevations in the upper portion of a sheet to allow room for other drawings. If you are using CADD, compare the elevations and the panel elevations so both drawings can be done in layers.

Layout

1. Draw a reference line to represent the slab. This line will never be on the finished drawing but will help locate the bottom of all doors.
2. Establish each end of the structure and each grid.
3. Establish the top of the walls using information on the typical corss-sections and details.
4. Establish finish grade elevations.
5. Lay out all door locations using the locations shown on the panel elevation sketches.
6. Locate loading docks based on the foundation plan.
7. Determine the locations of three strips, and sign as per sketch.
8. Draw the grade and loading dock using bold lines.
9. Draw the outline of the structure.
10. Draw grid lines as two, thin parallel lines.
11. Draw doors.
12. Draw stripes and sign.
13. Draw concrete symbols on some of the walls.
14. Label as per sketch.

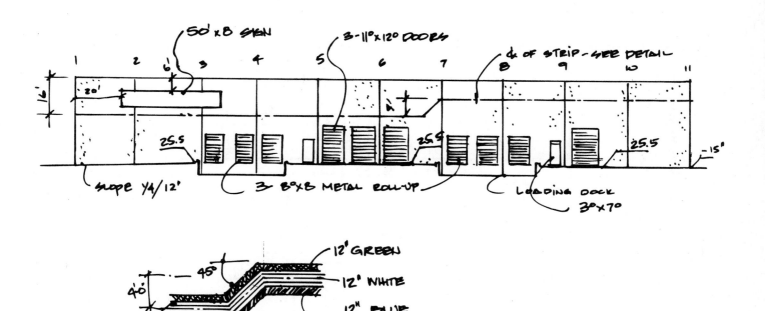

Problem 47–11 Panel Elevations.

Goal These drawings will provide an internal view of the walls to be constructed, showing wall shape and openings.

Method Typically, an elevation of each panel of a structure would be drawn. Verify with your instructor if more than the north face will be drawn for your project. Use a scale of ³⁄₃₂″ = 1′–0″. Due to the length of the structure, the sketch of the north panel elevations is shown on two separate drawings. Your drawings should reflect the true layout of the structure and include all information specified in the calcs. Place this drawing on the same sheet as the elevations. Use a print of the elevations to speed the layout, or set this up as a layer to be used on the exterior elevations if using CADD.

Layout

1. Draw a reference line to represent the slab. This line will never be on the finished drawing but will help locate the bottom of each panel.

2. Establish each end of the structure and each grid.

3. Establish the top of the walls using information on the sections and elevations.

4. Determine the foundation elevations and lay out the top of the footing.

5. Lay out all door locations.

6. Draw the outline of the panels.

7. Draw the grid lines.

8. Draw door openings.

9. Provide dimensions to describe all wall shapes and all door openings. Keep size location separate from location dimension.

10. Print all notes.

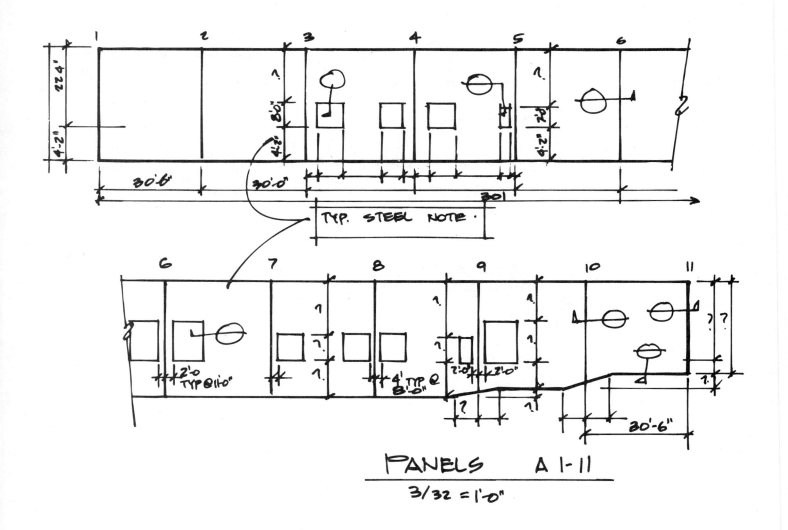

PANELS   A 1-11

3/32 = 1'-0"

Problem 47–12 Panel Details.

Goal The purpose of these drawings is to show steel placement around any openings in the tilt-up walls. Steel could be shown on the panel elevations, but more clarity is gained by providing details.

Method A detail will be provided for each size opening, to show typical steel placement. Be aware that all steel specified by the doors is in addition to any steel specified in the general note for the walls. Typical wall steel is usually only specified in a note and not shown. This allows special steel patterns to be shown

and easily specified. Use a scale of ¼″ = 1′–0″ minimum. Place these details on the same sheet as the panel elevations. Reference these details to the section, and slab on grade plan.

1. Draw the door opening allowing approximately 7′–0″ on all four sides.
2. Lay out steel placements based on dimensions specified on the panel elevation sketches and calcs.
3. Draw the steel much bolder than the opening outline.
4. Dimension each opening as per the sketch.
5. Label all steel.

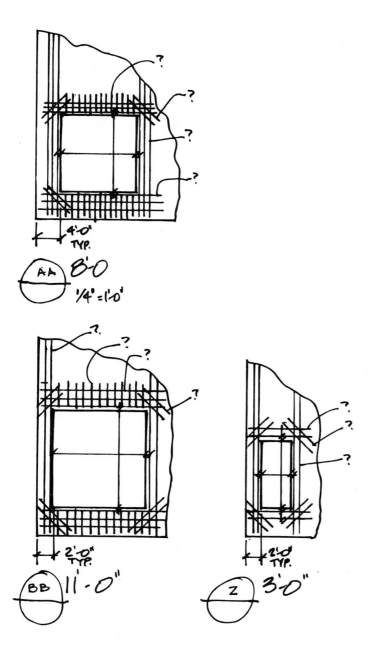

WALL PANELS

PANELS: ALL WALLS TO BE 6" THICK, 4000 PSI CONCRETE W/ GRADE 60-#5 @ 10"O.C. VERTICALLY & GRADE 40-#4 @ 12" O.C. HORIZONTALLY CENTERED IN WALL. EXTEND ALL WALL STEEL TO WITHIN 2" OF TOP AND BOTTOM.

ALL PANELS TO HAVE EXPOSED AGGREGATE FINISH UNLESS NOTED

PROVIDE (2) GRADE 40, #5 x 4'-0" 45° DIAG. STEEL TYPICAL TOP AND BOTTOM OF ALL DOORS. TIE FIRST DIAG. TO INTERSECTION OF HORIZ. AND VERT. STEEL. W/ SECOND BAR AT 8" O.C.

PANEL/PANEL: USE (2) ¾"φ x 2 ½" COIL INSERTS THRU 1"φ HOLES IN 11"h x 10"w x ⅜" ℙ. INSERTS TO BE 1½" MIN. FROM ℙ EDGES, 8'O.C. AND 3" MIN. FROM EDGE OF CONCRETE PANEL.

PROVIDE 3x 13 x ⅜" ℙ INSET FLUSH INTO CONC. 3" MIN. FROM PANEL EDGE W/ (2) ¾"φ x3" HEADED CONCRETE ANCHORS @ 10" O.C. 1½" MIN. FROM ℙ ENDS. WELD TO ℙ W/ ⅛" FILLET ALL AROUND @ FIELD. PROVIDE 1½" MIN. LAP OF 11x10 ℙ OVER 3x13 ℙ & WELD W/ ¼" FILLET WELD.

ALL CONNECTOR ℙ TO BE @ 5'-0" O.C. MAX. 12" MIN. FROM TOP AND BTM OF PANELS.

PANEL CORNERS: SIMILAR TO PANEL/PANEL JOINT UNLESS NOTED. USE ⅜"x 5"x3" x11" ℙ.

ALL CORNER ℙ TO BE @ 6'-0" O.C. MAX. 12" MIN. FROM TOP AND BTM. OF PANEL EDGES.

PANEL/ FND.: ⅞" φ x14" STRUCTURAL CONNECTOR BY RICHMOND OR EQUAL, 8" MIN. INTO FND. WITH 5" MIN WALL PENETRATION. SET IN NON-SHRINK GROUT IN 3x9 ANCHOR SLEEVE IN FOUNDATION. USE: 6 PER PANEL @ GRID 11

9	"	"	@	"	1
8	"	"	@	"	C
8	"	"	@	"	1-4
7	"	"	@	"	4-7
6	"	"	@	"	7-11

11'-0" DOOR: IN ADDITION TO WALL STEEL PROVIDE:
(2) GRADE 60, #6 @ 8"O.C. 1" CLEAR OF EACH FACE VERTICALLY FOR FULL HEIGHT OF WALL.
(3) GRADE 60, #4 @ 12" O.C. HORIZONTALLY ABOVE AND BELOW DOORS, EXTEND 24" MIN EA. SIDE OF DOOR.

ABOVE AND BELOW DOOR PROVIDE GRD. 40, #4 @ 12"O.C VERT. EXTEND VERT. 36" ABOVE AND BELOW DOOR.

···· ALL STEEL PATTERNS TO START WITHIN 1½" OF WALL OPENINGS.

8'-0" DOOR: IN ADDITION TO WALL STEEL PROVIDE:
(3) GRADE 40, #5φ @ 8" O.C. @ 1" CLR OF EACH FACE VERTICALLY FOR FULL HEIGHT OF WALL.
(3) GRADE 60, #4 @ 12" O.C. HORIZONTALLY ABOVE AND BELOW DOORS, EXTEND 24" MIN EA. SIDE OF DOOR.

ABOVE AND BELOW DOOR PROVIDE GRD. 40, #4 @ 12"O.C. VERT. EXTEND VERT. 36" ABOVE AND BELOW DOOR.

···· ALL STEEL PATTERNS TO START WITHIN 1½" OF WALL OPENINGS.

3'-0" DOOR: IN ADDITION TO WALL STEEL PROVIDE:
(2) GRADE 60, #6 @ 6" O.C. 1" CLR. OF EA. FACE VERTICALLY FOR FULL HEIGHT OF WALL.

(2) GRD 40, #5 x 7'-0" TOP AND BTM. @ 8" O.C.

Problem 47–13   Wall Details.

Goal   The purpose of the wall details is to aid in the erection and connection of the concrete wall panels and to show connections between the steel columns and pedestals.

Method   Use the scale as indicated on each sketch. Place the details with the panel elevations, section, or with other concrete details.

1. Determine in what order the details will be presented. Place the details in an order that reflects their relationship to the building.
2. Lightly block out each detail, allowing approximately 15 minutes per detail. Given this brief amount of time, draw only major items such as concrete, steel connector, and column outlines.
3. Draw details with finish-line quality. Use several line types or colors to distinguish clearly concrete, steel plates, and bolts.
4. Dimension all information based on sketches and calcs. Do not share dimension locations of bolts.
5. Place all weld symbols as required.
6. Place notes as required. Notes can be shared between details.
7. Place detail markers, title, and scale below each detail. Do not yet place information in the circle.

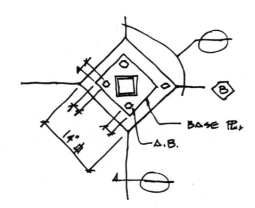

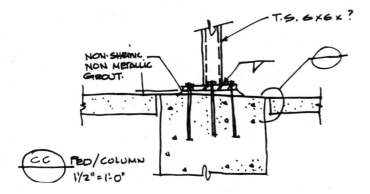

Problem 47–14 Project Coordination.

Goal Up to this point, you have drawn plan views, elevations, sections, and details. These drawings must now be placed in a logical presentation order and be referenced to each other.

Method

1. Place all drawings in order. Suggested order: roof plan, roof details, slab plan, slab details, etc. through to the foundation level, sections, and elevations. Or you can place all plan views from drainage to foundation in descending order, sections, elevations, panels, and details.

2. Once the page order has been determined, place a page number in the title block. Structural drawings will always reflect the total number of pages in the title block as well as the actual page being viewed. In the title block, the numbers 2 of 7 would be placed on the second of seven pages.

3. Assign a page number to all details. All details drawn on page 1 would have a 1 placed in the lower half of the detail reference circle. All details drawn on page 2 would have a 2 placed in the lower half of the reference circle. This pattern should be repeated until all details have a page number.

4. Assign a reference letter to all details. Two methods may be used. Step A is the same for each method.

A. Starting on page 1, label each detail A, B, C, etc., until all details on page 1 have been assigned a letter. Start in one corner and work across or up and down the page. Do not skip all over the page as you assign reference letters.

B. On page 2, assign each detail a letter starting with the letter A, and work through the alphabet. Use the same pattern that was used on page 1. Repeat this process for all pages with details.

or

B. On page 2, instead of starting over at A, use the letter that would follow the last letter used on the preceding detail. Never use the letters I, O, or Q. After detail Z, use AA, BB, etc.

C. After all details have been referenced, make sure that each detail referenced on the plan views and sections, etc., has the correct reference symbol. Some details will be referenced on more than one plan. Most will be on the section.

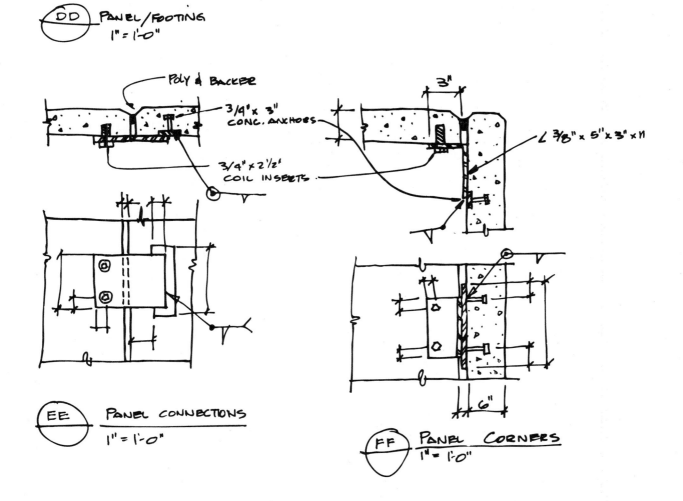

APPENDIX A

METRIC IN CONSTRUCTION

The following tables are courtesy *Metric in Construction Newsletter*, published by the Construction Metrication Council of National Institute of Building Sciences, Washington, D.C., fax: 202–289–1092.

▼ UNIT CONVERSION TABLES

SI SYMBOLS AND PREFIXES

Quantity	Unit	Symbol
BASE UNITS		
Length	Meter	m
Mass	Kilogram	kg
Time	Second	s
Electric current	Ampere	A
Thermodynamic temperature	Kelvin	K
Amount of substance	Mole	mol
Luminous intensity	Candela	cd
SI SUPPLEMENTARY UNITS		
Plane angle	Radian	rad
Solid angle	Steradian	sr

SI PREFIXES

Multiplication Factor	Prefix	Symbol
$1\ 000\ 000\ 000\ 000\ 000\ 000 = 10^{18}$	exa	E
$1\ 000\ 000\ 000\ 000\ 000 = 10^{15}$	peta	P
$1\ 000\ 000\ 000\ 000 = 10^{12}$	tera	T
$1\ 000\ 000\ 000 = 10^{9}$	giga	G
$1\ 000\ 000 = 10^{6}$	mega	M
$1\ 000 = 10^{3}$	kilo	k
$100 = 10^{2}$	hecto	h
$10 = 10^{1}$	deka	da
$0.1 = 10^{-1}$	deci	d
$0.01 = 10^{-2}$	centi	c
$0.001 = 10^{-3}$	milli	m
$0.000\ 001 = 10^{-6}$	micro	μ
$0.000\ 000\ 001 = 10^{-9}$	nano	n
$0.000\ 000\ 000\ 000 = 10^{-12}$	pico	p
$0.000\ 000\ 000\ 000\ 000 = 10^{-15}$	femto	f
$0.000\ 000\ 000\ 000\ 000\ 000 = 10^{-18}$	atto	a

CONVERSION FACTORS

To convert	to	multiply by
LENGTH		
1 mile (U.S. statute)	km	1.609 344
1 yd	m	0.9144
1 ft	m	0.3048
	mm	304.8
1 in	mm	25.4
AREA		
1 mile^2 (U.S. statute)	km^2	2.589 998
1 acre (U.S. survey)	ha	0.404 6873
	m^2	4,046,873
1 yd^2	m^2	0.836 1274
1 ft^2	m^2	0.092 903 04
1 in^2	mm^2	645.16
VOLUME, MODULUS OF SECTION		
1 acre ft	m^3	1,233.489
1 yd^3	m^3	0.764 5549
100 board ft	m^3	0.235 9737
1 ft^3	m^3	0.028 316 85
	L (dm^3)	28.3168
1 in^3	mm^3	16 387.06
	mL (cm^3)	16.3871
1 barrel (42 U.S. gallons)	m^3	0.158 9873
(FLUID) CAPACITY		
1 gal (U.S. liquid)*	L**	3.785 412
1 qt (U.S. liquid)	mL	946.3529
1 pt (U.S. liquid)	mL	473.1765
1 fl oz (U.S.)	mL	29.5735
1 gal (U.S. liquid)	m^3	0.003 785 412

*1 gallon (UK) approx. 1.2 gal (U.S.)
**1 liter approx. 0.001 cubic meters

To convert	to	multiply by
SECOND MOMENT OF AREA		
1 in^4	mm^4	4,162,314
	m^4	$4{,}162{,}314 \times 10^{-7}$
PLANE ANGLE		
1° (degree)	rad	0.017 453 29
	mrad	17.453 29
1′ (minute)	urad	290.8882
1″ (second)	urad	4.848 137

CONVERSION FACTORS—(Continued)

To convert	to	multiply by
VOLUME RATE OF FLOW		
1 ft³/s	m³/s	0.028 316 85
1 ft³/min	L/s	0.471 9474
1 gal/min	L/s	0.063 0902
1 gal/min	m³/min	0.0038
1 gal/h	mL/s	1.051 50
1 million gal/d	L/s	43.8126
1 acre ft/s	m³/s	1233.49
TEMPERATURE INTERVAL		
1° F	°C or K	0.555 556
		⁵⁄₉°C = ⁵⁄₉K
MASS		
1 ton (short)	metric ton	0.907 185
	kg	907.1847
1 lb	kg	0.453 5924
1 oz	g	28.349 52
1 long ton (2,240 lb)	kg	1,016.047
MASS PER UNIT AREA		
1 lb/ft²	kg/m²	4.882 428
1 oz/yd²	g/m²	33.905 75
1 oz/ft²	g/m²	305.1517
DENSITY (MASS PER UNIT VOLUME)		
1 lb/ft³	kg/m³	16.01846
1 lb/yd³	kg/m³	0.593 2764
1 ton/yd³	t/m³	1.186 553
FORCE		
1 tonf (ton-force)	kN	8.896 44
1 kip (1,000 lbf)	kN	4.448 22
1 lbf (pound-force)	N	4.448 22
MOMENT OF FORCE, TORQUE		
1 lbf · ft	N · m	1.355 818
1 lbf · in	N · m	0.112 9848
1 tonf · ft	kN · m	2.711 64
1 kip · ft	kN · m	1.355 82
FORCE PER UNIT LENGTH		
1 lbf/ft	N/m	14.5939
1 lbf/in	N/m	175.1268
1 tonf/ft	kN/m	29.1878
PRESSURE, STRESS, MODULUS OF ELASTICITY (FORCE PER UNIT AREA) (1 Pa = 1 N/m²)		
1 tonf/in²	MPa	13.7895
1 tonf/ft²	kPa	95.7605
1 kip/in²	MPa	6.894 757
1 lbf/in²	kPa	6.894 757
1 lbf/ft²	Pa	47.8803
Atmosphere	kPa	101.3250
1 inch mercury	kPa	3.376 85
1 foot (water column at 32°F.)	kPa	2.988 98

CONVERSION FACTORS—(Continued)

To convert	to	multiply by
WORK, ENERGY, HEAT (1J = 1N · m = 1W · s)		
1 kWh (550ft · lbf/s)	MJ	3.6
1 Btu (Int. Table)	kJ	1.055 056
	J	1,055.056
1 ft · lbf	J	1.355 818
COEFFICIENT OF HEAT TRANSFER		
1 Btu/(ft² · h · °F)	W/(m² · K)	5.678 263
THERMAL CONDUCTIVITY		
1 Btu/(ft · h · °F)	W/(m · K)	1.730 735
ILLUMINANCE		
1 lm/ft² (footcandle)	lx (lux)	10.763 91
LUMINANCE		
1 cd/ft²	cd/m²	10.7639
1 foot lambert	cd/m²	3.426 259
1 lambert	kcd/m²	3.183 099

▼ CONVERSION AND ROUNDING

The conversion of inch-pound units to metric is an important part of the metrication process. But conversion can seem deceptively simple because most measurements have implied, not expressed, tolerances and many products (like 2-by-4s) are designated in rounded, easy-to-remember "nominal" sizes, not actual ones. For instance, if anchor bolts are to be imbedded in masonry to a depth of 8 inches, what should this depth be in millimeters? A strict conversion (using 1 inch = 25.4 mm) results in an exact dimension of 203.2 mm. But this implies an accuracy of 0.1 mm (1/254 inch) and a tolerance of ± 0.05 mm (1/508 inch), far beyond any reasonable measure for field use. Similarly, 203 mm is overly precise, implying an accuracy of 1 mm (about 1/25 inch) and a tolerance of ± 0.5 mm (about 1/50 inch). As a practical matter, ± 3 mm (1/8 inch) is well within the tolerance for setting anchor bolts. Applying ± 3 mm to 203.2 mm, the converted dimension should be in the range of 200 mm to 206 mm. Metric measuring devices emphasize 10 mm increments and masons work on a 200 mm module, so the selection of 200 mm would be a convenient dimension for masons to use in the field. Thus, a reasonable metric conversion for 8 inches, *in this case*, is 200 mm.

This example may sound complicated but in fact we mentally round to easy-to-use numbers all the time and think nothing of it. What the example does illustrate is the need for experience, common sense, and consideration of how measures are used. Much has been written about conversion but the basic points to remember are these:

○ Understand the allowable tolerances for the measurements you are converting.

○ Always convert with the end application or use in mind. Remember that dimensional tolerances at the work site are rarely less than a few millimeters so it is easiest for field personnel to measure in 10 mm or 5 mm increments.

○ The most common conversion error is under-rounding, which implies a level of precision that is not inherent in the inch-pound number. If your linear conversions are accurate to 0.1 mm or even 1 mm, you are probably doing them incorrectly; any dimension over a few inches usually can be rounded to the nearest 5 mm (1/5 inch) and anything over a few feet to the nearest 10 mm (2/5 inch)—or more.

For example, the NFPA Life Safety Code rounds dimensions under 30 inches to the nearest millimeter, from 30 inches to 10 feet to the nearest 10 mm, from 10 to 50 feet to the nearest 100 mm, and over 100 feet to the nearest meter. Accessibility standards uniformly round dimensions under 1/2 inch to the nearest 0.5 mm, from 1/2 inch to 2 feet to the nearest millimeter, from 2 to 40 feet to the nearest 5 mm, and over 40 feet to the nearest meter. ASTM standards vary in their rounding practices depending on the subject involved, but many use the rounded conversion factor 1 inch = 25 mm.

ROUNDING TABLE, 1/32 INCH TO 4 INCHES

Underline denotes exact conversion
Shaded figures are too exact for most uses

Inches	Nearest 0.1 mm (1/254")	Nearest 1 mm (1/25")	Nearest 5 mm (1/5")
1/32"	0.8	1	
1/16"	1.6	2	
3/32"	2.4	2	
1/8"	3.2	3	
3/16"	4.8	5	
1/4"	6.4	6	
5/16"	7.9	8	
3/8"	9.5	10	
7/16"	11.1	11	
1/2"	_12.7_	13	
9/16"	14.3	14	
5/8"	15.9	16	
3/4"	19.0	19	
7/8"	22.2	22	
1"	_25.4_	25	25
1-1/4"	_31.8_	32	30
1-1/2"	_38.1_	38	40
1-3/4"	_44.4_	44	45
2"	_50.8_	51	50
2-1/4"	_57.2_	57	55
2-1/2"	_63.5_	64	65
2-3/4"	_69.8_	70	70
3"	_76.2_	76	75
3-1/4"	_82.6_	83	85
3-1/2"	_88.9_	89	90
3-3/4"	_95.2_	95	95
4"	_101.6_	102	100

ROUNDING TABLE, 4 INCHES TO 100 FEET

Underline denotes exact conversion
Shaded figures are too exact for most uses

Inches and Feet	Nearest 0.1 mm (1/254")	Nearest 1 mm (1/25")	Nearest 5 mm (1/5")	Nearest 10 mm (2/5")	Nearest 50 mm (2")	Nearest 100 mm (4")	1"= 25 mm exactly
4"	101.6	102	100	100			100
5"	127	127	125	130			125
6"	152.4	152	150	150			150
7"	177.8	178	180	180			175
8"	203.2	203	205	200			200
9"	228.6	229	230	230			225
10"	254	254	255	250			250
11"	279.4	279	280	280			275
1'-0"	304.8	305	305	300	300		300
2'-0"	609.6	610	610	610	600		600
3'-0"	914.4	914	915	910	900		900
4'-0"	1219.2	1219	1220	1220	1200		1200
5'-0"		1524	1525	1520	1500		1500
6'-0"		1829	1830	1830	1850		1800
7'-0"		2134	2135	2130	2150		2100
8'-0"		2438	2440	2440	2450		2400
9'-0"		2743	2745	2740	2750		2700
10'-0"		3048	3050	3050	3050	3000	3000
15'-0"		4572	4570	4570	4550	4600	4500
20'-0"		6096	6095	6100	6100	6100	6000
25'-0'		7620	7620	7620	7600	7600	7500
30'-0"		9144	9145	9140	9150	9100	9000
40'-0"		12 192	12 190	12 190	12 200	12 200	12 000
50'-0"		15 240	15 240	15 240	15 250	15 200	15 000
75'-0"		22 860	22 860	22 860	22 850	22 900	22 500
100'-0"		30 480	30 480	30 480	30 500	30 500	30 000

Layer Titles

Partial CAD File List: C8713-1.DWG, C8713-2.DWG, S8713-1.DWG, A8713-2.DWG, A8713-3.DWG, A8713-4.DWG, A8713-9.DWG, M8713-1.DWG

Partial Drawing List: Civil Site Plan, Site Details, Foundation Plan, Architectural Floor Plan, Reflected Ceiling Plan, Finish Plan, Roof/Framing Plan, Elevations, Sections, Schedules and Details, HVAC Plan, Power Plan, Lighting Plan, Plumbing Plan

(Legend: ● = solid mark, ○ = open mark)

Layer Names	Comments	Civil Site Plan	Site Details	Foundation Plan	Arch. Floor Plan	Reflected Ceiling Plan	Finish Plan	Roof/Framing Plan	Elevations	Sections	Schedules & Details	HVAC Plan	Power Plan	Lighting Plan	Plumbing Plan
A-WALL	Walls				●	○	○					○	○	○	○
A-WALL-EXST	Existing walls				●										
A-WALL-DEMO	Demolished walls				●										
A-WALL-HEAD	Door headers					●							○	○	
A-WALL-PATT	Wall poche				●	○	○					○	○	○	○
A-DOOR	Doors				●	○	○					○	○	○	○
A-DOOR-EXST	Existing doors				●										
A-DOOR-DEMO	Demolished doors				●										
A-DOOR-IDEN	Door numbers, hardware group				●										
A-GLAZ	Windows				●	○	○					○	○	○	○
A-GLAZ-EXST	Existing windows				●										
A-GLAZ-DEMO	Demolished windows				●										
A-FLOR-CASE	Casework				●		○							○	○
A-FLOR-IDEN	Room numbers, names, tags				●	○	○					○	○	○	○
A-FLOR-OVHD	Overhead platform				●										
A-FLOR-PATT	Floor pavers						●								
A-EQPM	General equipment				●								○	○	
A-CLNG	General ceiling linework					●							○	○	
A-CLNG-GRID	Ceiling 2'x2' tiles					●									○
A-CLNG-OPEN	Ceiling /roof penetrations					●							○	○	
A-CLNG-PATT	Drywall ceiling poche					●								○	
A-CLNG-TEES	Ceiling main tees					●								○	
A-ROOF	Roof plan line work							●							
A-ROOF-GUTT	Gutters							●							
A-ROOF-PATT	Shingles hatch and stipple poche							●							
A-ELEV	Building elevations line work								●						
A-ELEV-OTLN	Building outlines								●	○					
A-ELEV-NOTE	Building elevations reference notes								●						
A-ELEV-SYMB	Building elevations drafting symbols								●						
A-ELEV-DIMS	Building elevations dimensions								●						
A-ELEV-PATT	Building elevations texture								●						
A-ELEV-TTLB	Building elevations title block info								●						
A-SECT	Wall sections line work									●					
A-SECT-NOTE	Wall sections reference notes									●					
A-SECT-SYMB	Wall sections drafting symbols									●					
A-SECT-DIMS	Wall sections dimensions									●					
A-SECT-PATT	Wall sections hatching									●					
A-SECT-TTLB	Wall sections title block info									●					
A-DETL	Construction details line work										●				
A-DETL-NOTE	Construction details reference notes										●				
A-DETL-SYMB	Construction details drafting symbols										●				
A-DETL-DIMS	Construction details dimensions										●				
A-DETL-PATT	Insulation										●				
A-DETL-TTLB	Construction details title block info										●				
A-SHBD	Architectural border and title block	●	○		○	○	○	○	○	○	○				
A-PLCL-NOTE	Reflected ceiling plan reference notes					●									
A-PLCL-SYMB	Reflected ceiling plan drafting symbols					●									
A-PLCL-DIMS	Reflected ceiling plan dimensions					●									
A-PLCL-TTLB	Reflected ceiling plan title block info					●									
A-MFN-NOTE	Floor plan reference notes				●										
A-MFN-SYMB	Floor plan drafting symbols				●										
A-MFN-DIMS	Floor plan dimensions				●										
A-MFN-TTLB	Floor plan title block info				●										
A-PLFN-NOTE	Finish plan reference notes						●								
A-PLFN-SYMB	Finish plan drafting symbols						●								
A-PLFN-DIMS	Finish plan dimensions						●								
A-PLFN-TTLB	Finish plan title block info						●								
A-PROF-NOTE	Roof plan reference notes							●							
A-PROF-SYMB	Roof plan drafting symbols							●							
A-PROF-DIMS	Roof plan dimensions							●							
A-PROF-TTLB	Roof plan title block info							●							
A-SCHD-TEXT	Schedules text and line work										●				
S-FNDN	General foundation linework			●											
S-SLAB	General slab linework			●											
S-FRAM	Framing plan beams, joists							●							
S-SHBD	Structural sheet border and title block			●											
S-PFND-NOTE	Foundation plan reference notes			●											
S-PFND-SYMB	Foundation plan drafting symbols			●											
S-PFND-DIMS	Foundation plan dimensions			●											
S-PFND-TTLB	Foundation plan title block info			●											
M-EXHS-DUCT	Exhaust system ductwork											●			
M-HVAC-CDFF	HVAC ceiling diffusers											●			
M-HVAC-DUCT	HVAC ductwork											●			
M-HVAC-EQPM	HVAC equipment											●			
M-HOTW-PIPE	Hot water piping											●			
M-CWAT-PIPE	Chilled water piping											●			
M-SHBD	Mechanical sheet border and title block											●	○	○	○
M-PHVA-NOTE	HVAC plan reference notes											●			
M-PHVA-SYMB	HVAC plan drafting symbols											●			
M-PHVA-DIMS	HVAC plan dimensions											●			
M-PHVA-TTLB	HVAC plan title block info											●			
P-DOMW-EQPM	Domestic hot & cold water equipment														●
P-DOMW-PIPE	Domestic hot & cold water piping														●
P-SANR-PIPE	Sanitary piping														●
P-SANR-FIXT	Plumbing fixtures				●	○						○			●
P-SANR-FLDR	Floor drains														●
P-PPLM-NOTE	Plumbing plan reference notes														●
P-PPLM-SYMB	Plumbing plan drafting symbols														●
P-PPLM-DIMS	Plumbing plan dimensions														●
P-PPLM-TTLB	Plumbing plan title block info														●
E-LITE	Light fixtures					●								○	
E-LITE-EXST	Existing light fixtures					●								○	
E-LITE-SWCH	Lighting switches													●	
E-LITE-CIRC	Lighting circuits													●	
E-POWR-WALL	Power wall outlets and receptacles												●		
E-POWR-PANL	Power panels												●		
E-POWR-CIRC	Power circuits												●		
E-POWR-NUMB	Power circuit numbers												●		
E-PPOW-NOTE	Power plan reference notes												●		
E-PPOW-SYMB	Power plan drafting notes												●		
E-PPOW-DIMS	Power plan dimensions												●		
E-PPOW-TTLB	Power plan title block info												●		
E-PLIT-NOTE	Lighting plan reference notes													●	
E-PLIT-SYMB	Lighting plan drafting symbols													●	
E-PLIT-DIMS	Lighting plan dimensions													●	
E-PLIT-TTLB	Lighting plan title block info													●	
C-BLDG	New building footprint and poche	●													
C-BLDG-DEMO	Existing building to be removed	●													
C-BLDG-EXST	Existing building footprints	●													
C-BLDG-PATT	New building area poche	●													
C-DETL	Site details line work		●												
C-DETL-NOTE	Site details reference notes		●												
C-DETL-SYMB	Site details drafting symbols		●												
C-DETL-DIMS	Site details dimensions		●												
C-DETL-TTLB	Site details title block info		●												
C-PKNG	New parking lot stripes and precast curbs	●													
C-PKNG-DEMO	Existing parking lots to be removed	●													
C-PKNG-PATT	New parking lot poche	●													
C-PROP	Property and setback lines	●		○	○			○					○		○
C-ROAD	New roads	●		○	○			○					○		○
C-ROAD-EXST	Existing roads	●													
C-ROAD-PATT	New roads poche	●													
C-PSIT-NOTE	Site plan reference notes	●													
C-PSIT-SYMB	Site plan drafting symbols	●													
C-PSIT-DIMS	Site plan dimensions	●													
C-PSIT-TTLB	Site plan title block info	●													
L-SITE-DEMO	Existing fencing to be removed	●													
L-WALK	New walks and steps	●		○	○			○						○	○
L-WALK-EXST	Existing walks and steps	●		○	○			○						○	○
L-WALK-PATT	New walks and steps poche	●			○										

Courtesy: American Institute of Architects.

APPENDIX C

CABO Tables

**ALLOWABLE SPAN FOR GIRDERS AND REQUIRED SIZE OF COLUMNS AND FOOTINGS
TO SUPPORT ROOFS, INTERIOR BEARING PARTITIONS AND FLOORS**

SIZE OF GIRDER REQUIRED		SPACING OF GIRDER[2] "S"	TYPE OF LOADING[3]			SIZE OF COLUMNS REQUIRED[4]		SIZE OF PLAIN CONCRETE FOOTING REQUIRED[4]
Wood[1]			A	B	C	Steel	Wood	
4" × 12"	6" × 10"	10' 15' 20'	5'-6" 4'-0" —	— — —	— — —	3" Steel pipe[5]	4" × 4"	2' × 2' × 8"
(7)	6" × 12"	10' 15' 20'	8'-6" 6'-0" 4'-6"	5'-0" 4'-0" —	— — —			
(7)	(7)	10' 15' 20'	12'-0" 10'-0" 8'-0"	9'-0" 8'-0" 7'-0"	8'-0" 7'-0" 6'-0"		6" × 6"	4' × 4' × 16"[6]
(7)	(7)	10' 15' 20'	16'-0" 13'-6" 12'-0"	12'-6" 10'-6" 9'-6"	11'-0" 10'-0" 8'-0"		8" × 8"	4'-3" × 4'-3" × 17"[6]
(7)	(7)	10' 15' 20'	20'-0" 17'-0" 15'-0"	16'-0" 13'-6" 12'-0"	13'-6" 11'-6" 10'-0"			

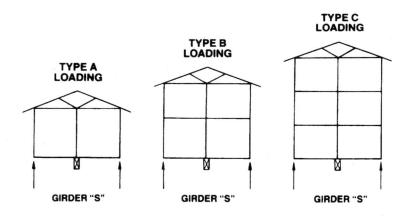

TYPE A LOADING — GIRDER "S"

TYPE B LOADING — GIRDER "S"

TYPE C LOADING — GIRDER "S"

[1]Spans for wood girders are based on No. 2 grade lumber. No. 3 grade may be used with appropriate design.

[2]The spacing "S" is the tributary load in the girder. It is found by adding the unsupported spans of the floor joists on each side which are supported by the girder and dividing by 2.

[3]Figures under "type of loading" columns are the allowable girder spans.

[4]Required size of columns is based on girder support from two sides. Size of footing is based on allowable soil pressure of 2,000 pounds per square foot.

[5]Standard weight.

[6]Footing thickness is based on the use of plain concrete with a specified compressive strength of not less than 2,500 pounds per square inch at 28 days. If approved, the footing thickness may be reduced based on an engineering design utilizing higher-strength concrete and/or reinforcement.

[7]Girder will require an approved design.

FASTENER SCHEDULE FOR STRUCTURAL MEMBERS

DESCRIPTION OF BUILDING MATERIALS	NUMBER & TYPE OF FASTENER 1 2 3	SPACING OF FASTENERS
Joist to sill or girder, toe nail	3-8d	—
1" × 6" subfloor or less to each joist, face nail	2-8d / 2 staples, 1¾"	—
Wider than 1" × 6" subfloor to each joist, face nail	3-8d / 4 staples, 1¾"	—
2" subfloor to joist or girder, blind and face nail	2-16d	—
Sole plate to joist or blocking, face nail	16d	16" o.c.
Top or sole plate to stud, end nail	2-16d	—
Stud to sole plate, toe nail	3-8d or 2-16d	—
Double studs, face nail	16d	24" o.c.
Double top plates, face nail	16d	24" o.c.
Top plates, taps and intersections, face nail	2-16d	—
Continued header, two pieces	16d	16" o.c. along each edge
Ceiling joists to plate, toe nail	3-8d	—
Continuous header to stud, toe nail	4-8d	—
Ceiling joist, taps over partitions, face nail	3-16d	—
Ceiling joist to parallel rafters, face nail	3-16d	—
Rafter to plate, toe nail	2-16d	—
1" brace to each stud and plate, face nail	2-8d / 2 staples, 1¾"	—
1" × 6" sheathing to each bearing, face nail	2-8d / 2 staples, 1¾"	—
1" × 8" sheathing to each bearing, face nail	2-8d / 3 staples, 1¾"	—
Wider than 1" × 8" sheathing to each bearing, face nail	3-8d / 4 staples, 1¾"	—
Built-up corner studs	16d	24" o.c.
Built-up girder and beams	16d	32" o.c. at top and bottom and staggered two 20d at ends and at each splice
2" planks	2-16d	At each bearing
Roof rafters to ridge, valley or hip rafters: toe nail / face nail	4-16d / 3-16d	—
Rafter ties to rafters, face nail	3-8d	—

DESCRIPTION OF BUILDING MATERIALS	DESCRIPTION OF FASTENER 3 5	SPACING OF FASTENERS Edges	SPACING OF FASTENERS Intermediate Supports4
Plywood and particleboard, roof and wall sheathing to frame			
5/16"-1/2"	6d / Staple 16 ga.	6"	12"
19/32"-3/4"	8d smooth or 6d deformed	6"	12"
7/8"-1"	8d	6"	12"
11/8"-11/4"	10d smooth or 8d deformed	6"	12"
Other wall sheathing7			
1/2" fiberboard sheathing6	1½" galvanized roofing nail / 6d common nail / Staple 16 ga., 1⅛" long	3"	6"
25/32" fiberboard sheathing6	1¾" galvanized roofing nail / 8d common nail / Staple 16 ga., 1½" long	3"	6"
1/2" gypsum sheathing	1½" galvanized roofing nail / 6d common nail / Staple 16 ga., 1½" long	4"	8"
Particleboard roof and wall sheathing 5/16"-1/2"	6d common nail	6"	12"
5/8"-3/4"	8d common nail / Staple 16 ga., 1½" long	6"	12"
Plywood and particleboard, combination subfloor-underlayment to framing			
3/4" and less	6d deformed	6"	12"
7/8"-1"	8d deformed	6"	12"
11/8"-11/4"	10d smooth or 8d deformed	6"	6"

[1] All nails are smooth-common, box or deformed shanks except where otherwise stated.

[2] Nail is a general description and may be T-head, modified round head or round head.

[3] Staples are No. 16 gauge wire and have a minimum 7/16-inch O.D. crown width.

[4] Nails shall be spaced at not more than 6 inches o.c. at all supports where spans are 48 inches or greater.

[5] The number of fasteners required for connections not included in this table shall be based on values set forth in Table No. R-402.3a(I).

[6] Four-foot by 8-foot or 4-foot by 9-foot panels shall be applied vertically.

[7] Gypsum sheathing shall conform to ASTM C79 listed in Section S-26.402. Fiberboard sheathing shall conform to AHA 194.1, ASTM D2277, and ASTM C208 listed in Section S-26.402. Other sheathing materials shall be approved by the building official.

ALTERNATE ATTACHMENTS

NOMINAL MATERIAL THICKNESS	DESCRIPTION[1][2] OF FASTENER & LENGTH	SPACING[3] OF FASTENERS	
		Edges	Intermediate Supports
Plywood or Particleboard Subfloor, Roof and Wall Sheathing to Framing			
5/16"	.097 - .099 Nail 1 1/2" Staple 15 ga. 1 3/8"	6"	12"
3/8"	Staple 15 ga. 1 3/8"	6"	12"
	.097 - .099 Nail 1 1/2"	4"	10"
15/32" and 1/2"	Staple 15 ga. 1 1/2"	6"	12"
	.097 - .099 Nail 1 5/8"	3"	6"
19/32" and 5/8"	.113 Nail 1 7/8" Staple 15 and 16 ga. 1 5/8"	6"	12"
	.097 - .099 Nail 1 3/4"	3"	6"
23/32" and 3/4"	Staple 14 ga. 1 3/4"	6"	12"
	Staple 15 ga. 1 3/4"	5"	10"
	.097 - .099 Nail 1 7/8"	3"	6"
1"	Staple 14 ga. 2"	5"	10"
	.113 Nail 2 1/4" Staple 15 ga. 2"	4"	8"
	.097 - .099 Nail 2 1/8"	3"	6"
Floor Underlayment; Plywood—Hardboard—Particleboard			
1/4" and 5/16"	.097 - .099 Nail 1 1/2" Staple 15 and 16 ga. 1 1/4"	6"	12"
	.080 Nail 1 1/4"	5"	10"
	Staple 18 ga. 3/16 crown 7/8"	3"	6"
3/8"	.097 - .099 Nail 1 1/2" Staple 15 and 16 ga. 1 3/8"	6"	12"
	.080 Nail 1 3/8"	5"	10"
1/2"	.113 Nail 1 7/8" Staple 15 and 16 ga. 1 1/2"	6"	12"
	.097 - .099 Nail 1 3/4"	5"	10"

[1]Nail is a general description and may be T-head, modified round head, or round head.

[2]Staples shall have a minimum crown width of 7/16-inch o.d. except as noted.

[3]Nails or staples shall be spaced at not more than 6 inches o.c. at all supports where spans are 48 inches or greater. Nails or staples shall be spaced at not more than 10 inches o.c. at intermediate supports for floors.

Joist Connectors

SIMPSON Strong-Tie CONNECTORS

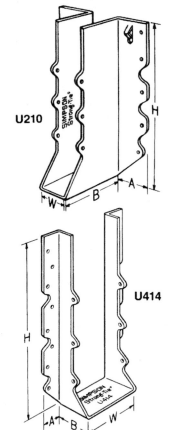

U210

U414

Courtesy Simpson Strong-Tie.

U STANDARD JOIST HANGERS

MATERIAL: 16 gauge steel
FINISH: Galvanized
OPTIONS AVAILABLE: Sloped and/or skewed models, see SLOPED AND/OR SKEWED U AND HU HANGERS on page 29.
CODE ACCEPTANCE: ICBO accepted; see Evaluation Report No. 1258.

MODEL NO.	JOIST SIZE	DIMENSIONS				FASTENERS		AVG. ULT.	ALLOWABLE LOADS [1]			
									NORMAL		MAXIMUM	
		W	H	B	A	HEADER	JOIST		10d	16d	10d	16d
U24	2x4	1 9/16	3 1/8	1 1/2	7/8	4 – 16d	2 – 10dx1 1/2	2575	435	535	540	670
U26	2x6,8	1 9/16	4 3/4	2	1 1/4	6 – 16d	4 – 10dx1 1/2	3680	650	805	815	1005
U210	2x10,12,14	1 9/16	7 13/16	2	1 1/4	10 – 16d	6 – 10dx1 1/2	6200	1085	1270	1355	1420
U214	2x14,16	1 9/16	10	2	1 1/4	12 – 16d	8 – 10dx1 1/2	7200	1300	1525	1630	1750
U24R	ROUGH 2x4	2	3 5/8	2	1 1/4	4 – 16d	2 – 10dx1 1/2	2575	435	755	540	755
U26R	ROUGH 2x6,8	2	5 5/8	2	1 1/4	8 – 16d	4 – 10dx1 1/2	3680	650	1160	815	1160
U210R	ROUGH 2x10,12,14	2	9 1/8	2	1 1/4	14 – 16d	6 – 10dx1 1/2	6200	1085	1270	1355	1420
U34	3x4	2 9/16	3 3/8	2	1 1/4	4 – 16d	2 – 10d	2600	470	535	590	670
U36	3x6,8	2 9/16	5 3/8	2	1 1/4	8 – 16d	4 – 10d	5000	945	1070	1180	1345
U310	3x10,12	2 9/16	8 7/8	2	1 1/4	14 – 16d	6 – 10d	9800	1650	1875	2065	2350
U314	3x14,16	2 9/16	10 1/2	2	1 1/4	16 – 16d	6 – 10d	11000	1890	2145	2360	2690
U24-2	(2) 2x4	3 1/8	3	2	1 1/4	4 – 16d	2 – 10d	2600	470	535	590	670
U26-2	(2) 2x6,8	3 1/8	5	2	1 1/4	8 – 16d	4 – 10d	5000	945	1070	1180	1345
U210-2	(2) 2x10,12,14	3 1/8	8 1/2	2	1 1/4	14 – 16d	6 – 10d	9800	1650	1875	2065	2350
U44	4x4	3 9/16	2 7/8	2	1 1/4	4 – 16d	2 – 10d	2600	470	535	590	670
U46	4x6,8	3 9/16	4 7/8	2	1 1/4	8 – 16d	4 – 10d	5000	945	1070	1180	1345
U410	4x10,12	3 9/16	8 3/8	2	1 1/4	14 – 16d	6 – 10d	9800	1650	1875	2065	2350
U414	4x14,16	3 9/16	10	2	1 1/4	16 – 16d	6 – 10d	11000	1890	2145	2360	2690
U44R	ROUGH 4x4	4	2 5/8	2	1 1/4	4 – 16d	2 – 16d	2600	470	535	590	670
U46R	ROUGH 4x6,8	4	4 5/8	2	1 1/4	8 – 16d	4 – 16d	5000	945	1070	1180	1345
U410R	ROUGH 4x10,12,14	4	8 1/8	2	1 1/4	14 – 16d	6 – 16d	9800	1650	1875	2065	2350
U26-3	(3) 2x6,8	4 5/8	4 1/4	2	7/8	8 – 16d	2 – 10d	5000	945	1070	1180	1345
U210-3	(3) 2x10,12,14	4 5/8	7 3/4	2	7/8	14 – 16d	6 – 10d	9800	1650	1875	2065	2350
U66	6x6,8	5 1/2	5	2	1 1/4	8 – 16d	4 – 10d	5000	945	1070	1180	1345
U610	6x10	5 1/2	8 1/2	2	1 1/4	14 – 16d	6 – 10d	9800	1650	1875	2065	2350
U66R	ROUGH 6x6,8	6	5	2	1 1/4	8 – 16d	4 – 16d	5000	945	1070	1180	1345
U610R	ROUGH 6x10,12,14	6	8 1/2	2	1 1/4	14 – 16d	6 – 16d	9800	1650	1875	2065	2350

1. The allowable loads under the "10d" column heading are the loads allowed when 10d common nails are used instead of the 16d header nails.

HU/HHU HEAVY DUTY JOIST HANGERS

6.6/Sim

MODEL NO.	JOIST SIZE	DIMENSIONS				FASTENERS[1]		AVG. ULT.	ALLOWABLE LOADS		
		W	H	B	A	HEADER	JOIST		UPLIFT	NORM.	MAX.
HU26	2x4,6	1⁹⁄₁₆	3¹⁄₁₆	2	1¹⁄₈	4 – 16d	2 – 10dx1¹⁄₂	2600	210	535	670
HU28	2x8	1⁹⁄₁₆	5¹⁄₄	2	1¹⁄₈	6 – 16d	4 – 10dx1¹⁄₂	3700	420	805	1010
HU210	2x10	1⁹⁄₁₆	7¹⁄₈	2	1¹⁄₈	8 – 16d	4 – 10dx1¹⁄₂	4900	420	1070	1345
HU212	2x12	1⁹⁄₁₆	9	2	1¹⁄₈	10 – 16d	6 – 10dx1¹⁄₂	6200	630	1340	1680
HU214	2x14	1⁹⁄₁₆	10¹⁄₈	2¹⁄₂	1¹⁄₈	12 – 16d	6 – 10dx1¹⁄₂	8500	630	1610	2015
HU34	3x4	2⁹⁄₁₆	3³⁄₈	2	1¹⁄₈	4 – 16d	2 – 10d	2600	210	535	670
HU36	3x6	2⁹⁄₁₆	5³⁄₈	2	1¹⁄₈	8 – 16d	4 – 10d	9474	420	1070	1345
HU38	3x8	2⁹⁄₁₆	7¹⁄₈	2	1¹⁄₈	10 – 16d	4 – 10d	11383	420	1340	1680
HU310	3x10	2⁹⁄₁₆	8⁷⁄₈	2	1¹⁄₈	14 – 16d	6 – 10d	14566	630	1875	2350
HU312	3x12	2⁹⁄₁₆	10⁵⁄₈	2¹⁄₂	1¹⁄₄	16 – 16d	6 – 10d	14316	630	2145	2690
HU314	3x14	2⁹⁄₁₆	12³⁄₈	2¹⁄₂	1¹⁄₄	18 – 16d	8 – 10d	14900	840	2410	3010
HU316	3x16	2⁹⁄₁₆	14¹⁄₈	2¹⁄₂	1¹⁄₄	20 – 16d	8 – 10d	14900	840	2680	3360
HU24 – 2	(2) 2x4	3¹⁄₈	3¹⁄₁₆	2	1¹⁄₈	4 – 16d	2 – 10d	2600	210	535	670
HU26 – 2	(2) 2x6	3¹⁄₈	5¹⁄₁₆	2	1¹⁄₈	8 – 16d	4 – 10d	9474	420	1070	1345
HHU26 – 2	(2) 2x6	3¹⁄₈	5¹⁄₁₆	2¹⁄₂	1¹⁄₄	8 – N20AN	4 – N20AN	9474	695	1390	1740
HU28 – 2	(2) 2x8	3¹⁄₈	6¹³⁄₁₆	2	1¹⁄₈	10 – 16d	4 – 10d	11383	420	1340	1680
HHU28 – 2	(2) 2x8	3¹⁄₈	6¹³⁄₁₆	2¹⁄₂	1¹⁄₄	10 – N20AN	4 – N20AN	11383	695	1740	2170
HU210 – 2	(2) 2x10	3¹⁄₈	8⁹⁄₁₆	2	1¹⁄₈	14 – 16d	6 – 10d	14566	630	1875	2350
HHU210 – 2	(2) 2x10	3¹⁄₈	8⁹⁄₁₆	2¹⁄₂	1¹⁄₄	14 – N20AN	6 – N20AN	14566	1045	2435	3040
HU212 – 2	(2) 2x12	3¹⁄₈	10⁵⁄₁₆	2¹⁄₂	1¹⁄₄	16 – 16d	6 – 10d	14316	630	2145	2690
HHU212 – 2	(2) 2x12	3¹⁄₈	10⁵⁄₁₆	2¹⁄₂	1¹⁄₄	16 – N20AN	6 – N20AN	14316	1045	2780	3475
HU214 – 2	(2) 2x14	3¹⁄₈	12¹⁄₁₆	2¹⁄₂	1¹⁄₄	18 – 16d	8 – 10d	14900	840	2410	3010
HHU214 – 2	(2) 2x14	3¹⁄₈	12¹⁄₁₆	2¹⁄₂	1¹⁄₄	18 – N20AN	8 – N20AN	14900	1390	3130	3910
HHU216 – 2	3 ¹⁄₈ LAM.	3 ¹⁄₈	13⁷⁄₈	2¹⁄₂	1¹⁄₄	20 – N20A	8 – N20A	14900	1390	3475	4345
HU44	4x4	3⁹⁄₁₆	2⁷⁄₈	2	1¹⁄₈	4 – 16d	2 – 10d	2600	210	535	670
HU46	4x6	3⁹⁄₁₆	4⁷⁄₈	2	1¹⁄₈	8 – 16d	4 – 10d	9474	420	1070	1345
HHU46	4x6	3⁹⁄₁₆	4⁷⁄₈	2¹⁄₂	1¹⁄₄	8 – N20AN	4 – N20AN	9474	695	1390	1740
HU48	4x8	3⁹⁄₁₆	6⁵⁄₈	2	1¹⁄₈	10 – 16d	4 – 10d	11383	420	1340	1680
HHU48	4x8	3⁹⁄₁₆	6⁵⁄₈	2¹⁄₂	1¹⁄₄	10 – N20AN	4 – N20AN	11383	695	1740	2170
HU410	4x10	3⁹⁄₁₆	8³⁄₈	2	1¹⁄₈	14 – 16d	6 – 10d	14566	630	1875	2350
HHU410	4x10	3⁹⁄₁₆	8³⁄₈	2¹⁄₂	1¹⁄₄	14 – N20AN	6 – N20AN	14566	1045	2435	3040
HU412	4x12	3⁹⁄₁₆	10¹⁄₈	2¹⁄₂	1¹⁄₄	16 – 16d	6 – 10d	14316	630	2145	2690
HHU412	4x12	3⁹⁄₁₆	10¹⁄₈	2¹⁄₂	1¹⁄₄	16 – N20AN	6 – N20AN	14316	1045	2780	3475
HU414	4x14	3⁹⁄₁₆	11⁷⁄₈	2¹⁄₂	1¹⁄₄	18 – 16d	8 – 10d	14900	840	2410	3010
HHU414	4x14	3⁹⁄₁₆	11⁷⁄₈	2¹⁄₂	1¹⁄₄	18 – N20AN	8 – N20AN	14900	1390	3130	3910
HU416	4x16	3⁹⁄₁₆	13⁵⁄₈	2¹⁄₂	1¹⁄₄	20 – 16d	8 – 10d	14900	840	2680	3360
HHU416	4x16	3⁹⁄₁₆	13⁵⁄₈	2¹⁄₂	1¹⁄₄	20 – N20AN	8 – N20AN	14900	1390	3475	4345
HHU5.125/12	5¹⁄₈ LAM.	5¹⁄₄	10¹⁄₈	2¹⁄₂	1¹⁄₄	16 – N20A	6 – N20A	14316	1045	2780	3475
HHU5.125/16	5¹⁄₈ LAM.	5¹⁄₄	13⁵⁄₈	2¹⁄₂	1¹⁄₄	20 – N20A	8 – N20A	14900	1390	3475	4345
HU66	6x6	5¹⁄₂	5	2	1¹⁄₄	8 – 16d	4 – 16d	9474	420	1070	1345
HHU66	6x6	5¹⁄₂	5	2	1¹⁄₄	8 – N20AN	4 – N20AN	9474	695	1390	1740
HU68	6x8	5¹⁄₂	6⁵⁄₈	2	1¹⁄₄	10 – 16d	4 – 16d	11383	420	1340	1680
HHU68	6x8	5¹⁄₂	6⁵⁄₈	2	1¹⁄₄	10 – N20AN	4 – N20AN	11383	695	1740	2170
HU610	6x10	5¹⁄₂	8³⁄₈	2	1¹⁄₄	14 – 16d	6 – 16d	14566	630	1875	2350
HHU610	6x10	5¹⁄₂	8³⁄₈	2	1¹⁄₄	14 – N20AN	6 – N20AN	14566	1045	2435	3040
HU612	6x12	5¹⁄₂	10¹⁄₈	2¹⁄₂	1¹⁄₄	16 – 16d	6 – 16d	14316	630	2145	2690
HHU612	6x12	5¹⁄₂	10¹⁄₈	2¹⁄₂	1¹⁄₄	16 – N20AN	6 – N20AN	14316	1045	2780	3475
HU614	6x14	5¹⁄₂	11⁷⁄₈	2¹⁄₂	1¹⁄₄	18 – 16d	8 – 16d	14900	840	2410	3010
HHU614	6x14	5¹⁄₂	11⁷⁄₈	2¹⁄₂	1¹⁄₄	18 – N20AN	8 – N20AN	14900	1390	3130	3910
HU616	6x16	5¹⁄₂	13⁵⁄₈	2¹⁄₂	1¹⁄₄	20 – 16d	8 – 16d	14900	840	2680	3360
HHU616	6x16	5¹⁄₂	13⁵⁄₈	2¹⁄₂	1¹⁄₄	20 – N20AN	8 – N20AN	14900	1390	3475	4345

Courtesy Simpson Strong-Tie.

SIMPSON Strong-Tie CONNECTORS

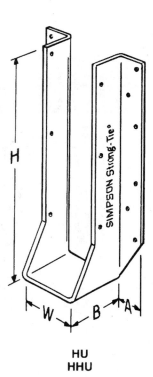

HU HHU

Projection seat on most models for maximum bearing and section economy.

1. N20AN and N20A fasteners are included with HHU models.

APA Plywood Specifications

▼ **METRIC CONVERSIONS**

Metric equivalents of nominal thicknesses and common sizes of wood structural panels are tabulated below (1 inch = 25.4 millimeters):

PANEL NOMINAL THICKNESS

in.	mm
1/4	6.4
5/16	7.9
11/32	8.7
3/8	9.5
7/16	11.1
15/32	11.9
1/2	12.7
19/32	15.1
5/8	15.9
23/32	18.3
3/4	19.1
7/8	22.2
1	25.4
1-3/32	27.8
1-1/8	28.6

PANEL NOMINAL DIMENSIONS
(Width × Length)

ft.	mm	m (approx.)
4×8	1219×2438	1.22×2.44
4×9	1219×2743	1.22×2.74
4×10	1219×3048	1.22×3.05

Courtesy American Plywood Association.

(1) Indicates structural use:
   B - Simple span bending member.
   C - Compression member.
   T - Tension member.
   CB - Continuous or cantilevered span bending member.
(2) Mill number.
(3) Identification of ANSI Standard A190.1, Structural Glued Laminated Timber.
(4) Code recognition of *American Wood Systems* as a quality assurance agency for glued structural members.
(5) Applicable laminating specification.
(6) Applicable combination number.
(7) Species of lumber used.
(8) Designates appearance grade, INDUSTRIAL, ARCHITECTURAL, PREMIUM.

Table 2
GUIDE TO APA PERFORMANCE RATED PANELS (A) (B)
FOR APPLICATION RECOMMENDATIONS, SEE FOLLOWING PAGES.

APA Rated Sheathing Typical Trademark	APA RATED SHEATHING 24/16 7/16 INCH SIZED FOR SPACING EXPOSURE 1 000 PRP-108 HUD-UM-40C	Specially designed for subflooring and wall and roof sheathing. Also good for a broad range of other construction and industrial applications. Can be manufactured as plywood, as a composite, or as OSB. EXPOSURE DURABILITY CLASSIFICATIONS: Exterior, Exposure 1, Exposure 2. COMMON THICKNESSES: 5/16, 3/8, 7/16, 15/32, 1/2, 19/32, 5/8, 23/32, 3/4.
APA Structural I Rated Sheathing(c) Typical Trademark	APA RATED SHEATHING 32/16 15/32 INCH SIZED FOR SPACING EXPOSURE 1 000 PS 1-83 C-D PRP-108 / APA RATED SHEATHING 32/16 15/32 INCH SIZED FOR SPACING EXPOSURE 1 000 STRUCTURAL 1 RATED DIAPHRAGMS-SHEAR WALLS PANELIZED ROOFS PRP-108	Unsanded grade for use where shear and cross-panel strength properties are of maximum importance, such as panelized roofs and diaphragms. Can be manufactured as plywood, as a composite, or as OSB. EXPOSURE DURABILITY CLASSIFICATIONS: Exterior, Exposure 1. COMMON THICKNESSES: 5/16, 3/8, 7/16, 15/32, 1/2, 19/32, 5/8, 23/32, 3/4.
APA Rated Sturd-I-Floor Typical Trademark	APA RATED STURD-I-FLOOR 20 OC 19/32 INCH SIZED FOR SPACING T&G NET WIDTH 47-1/2 EXPOSURE 1 000 PRP-108 HUD-UM-40C	Specially designed as combination subfloor-underlayment. Provides smooth surface for application of carpet and pad and possesses high concentrated and impact load resistance. Can be manufactured as plywood, as a composite, or as OSB. Available square edge or tongue-and-groove. EXPOSURE DURABILITY CLASSIFICATIONS: Exterior, Exposure 1, Exposure 2. COMMON THICKNESSES: 19/32, 5/8, 23/32, 3/4, 1, 1-1/8.
APA Rated Siding Typical Trademark	APA RATED SIDING 24 OC 19/32 INCH SIZED FOR SPACING EXTERIOR 000 PRP-108 HUD-UM-40C / APA RATED SIDING 303-18-S/W 16 OC 11/32 INCH GROUP 1 SIZED FOR SPACING EXTERIOR 000 PS 1-83 FHA-UM-64 PRP-108	For exterior siding, fencing, etc. Can be manufactured as plywood, as a composite or as an overlaid OSB. Both panel and lap siding available. Special surface treatment such as V-groove, channel groove, deep groove (such as APA Texture 1-11), brushed, rough sawn and overlaid (MDO) with smooth- or texture-embossed face. Span Rating (stud spacing for siding qualified for APA Sturd-I-Wall applications) and face grade classification (for veneer-faced siding) indicated in trademark. EXPOSURE DURABILITY CLASSIFICATION: Exterior. COMMON THICKNESSES: 11/32, 3/8, 7/16, 15/32, 1/2, 19/32, 5/8.

(a) Specific grades, thicknesses and exposure durability classifications may be in limited supply in some areas. Check with your supplier before specifying.

(b) Specify Performance Rated Panels by thickness and Span Rating. Span Ratings are based on panel strength and stiffness. Since these properties are a function of panel composition and configuration as well as thickness, the same Span Rating may appear on panels of different thickness. Conversely, panels of the same thickness may be marked with different Span Ratings.

(c) All plies in Structural I plywood panels are special improved grades and panels marked PS 1 are limited to Group 1 species. Other panels marked Structural I Rated qualify through special performance testing.

Structural II plywood panels are also provided for, but rarely manufactured. Application recommendations for Structural II plywood are identical to those for APA RATED SHEATHING plywood.

Table 3
GUIDE TO APA SANDED & TOUCH-SANDED PLYWOOD PANELS (A) (B) (C)
FOR APPLICATION RECOMMENDATIONS, SEE FOLLOWING PAGES.

APA A-A

Typical Trademark

| A-A • G-1 • EXPOSURE 1-APA • 000 • PS1-83 |

Use where appearance of both sides is important for interior applications such as built-ins, cabinets, furniture, partitions; and exterior applications such as fences, signs, boats, shipping containers, tanks, ducts, etc. Smooth surfaces suitable for painting. EXPOSURE DURABILITY CLASSIFICATIONS: Interior, Exposure 1, Exterior. COMMON THICKNESSES: 1/4, 11/32, 3/8, 15/32, 1/2, 19/32, 5/8, 23/32, 3/4.

APA A-B

Typical Trademark

| A-B • G-1 • EXPOSURE 1-APA • 000 • PS1-83 |

For use where appearance of one side is less important but where two solid surfaces are necessary. EXPOSURE DURABILITY CLASSIFICATIONS: Interior, Exposure 1, Exterior. COMMON THICKNESSES: 1/4, 11/32, 3/8, 15/32, 1/2, 19/32, 5/8, 23/32, 3/4.

APA A-C

Typical Trademark

APA
A-C GROUP 1
EXTERIOR
000
PS 1-83

For use where appearance of only one side is important in exterior or interior applications, such as soffits, fences, farm buildings, etc.[f] EXPOSURE DURABILITY CLASSIFICATION: Exterior. COMMON THICKNESSES: 1/4, 11/32, 3/8, 15/32, 1/2, 19/32, 5/8, 23/32, 3/4.

APA A-D

Typical Trademark

APA
A-D GROUP 1
EXPOSURE 1
000
PS 1-83

For use where appearance of only one side is important in interior applications, such as paneling, built-ins, shelving, partitions, flow racks, etc.[f] EXPOSURE DURABILITY CLASSIFICATIONS: Interior, Exposure 1. COMMON THICKNESSES: 1/4, 11/32, 3/8, 15/32, 1/2, 19/32, 5/8, 23/32, 3/4.

APA B-B

Typical Trademark

| B-B • G-2 • EXPOSURE 1-APA • 000 • PS1-83 |

Utility panels with two solid sides. EXPOSURE DURABILITY CLASSIFICATIONS: Interior, Exposure 1, Exterior. COMMON THICKNESSES: 1/4, 11/32, 3/8, 15/32, 1/2, 19/32, 5/8, 23/32, 3/4.

APA B-C

Typical Trademark

APA
B-C GROUP 1
EXTERIOR
000
PS 1-83

Utility panel for farm service and work buildings, boxcar and truck linings, containers, tanks, agricultural equipment, as a base for exterior coatings and other exterior uses or applications subject to high or continuous moisture.[f] EXPOSURE DURABILITY CLASSIFICATION: Exterior. COMMON THICKNESSES: 1/4, 11/32, 3/8, 15/32, 1/2, 19/32, 5/8, 23/32, 3/4.

Table 3
Continued

APA B-D Typical Trademark	_APA_ B-D GROUP 2 EXPOSURE 1 000 PS 1-83	Utility panel for backing, sides of built-ins, industry shelving, slip sheets, separator boards, bins and other interior or protected applications.[f] EXPOSURE DURABILITY CLASSIFICATIONS: Interior, Exposure 1. COMMON THICKNESSES: 1/4, 11/32, 3/8, 15/32, 1/2, 19/32, 5/8, 23/32, 3/4.
APA Underlayment Typical Trademark	_APA_ UNDERLAYMENT GROUP 1 EXPOSURE 1 000 PS 1-83	For application over structural subfloor. Provides smooth surface for application of carpet and pad and possesses high concentrated and impact load resistance. For areas to be covered with resilient flooring, specify panels with "sanded face."[e] EXPOSURE DURABILITY CLASSIFICATIONS: Interior, Exposure 1. COMMON THICKNESSES[d]: 1/4, 11/32, 3/8, 15/32, 1/2, 19/32, 5/8, 23/32, 3/4.
APA C-C Plugged(g) Typical Trademark	_APA_ C-C PLUGGED GROUP 2 EXTERIOR 000 PS 1-83	For use as an underlayment over structural subfloor, refrigerated or controlled atmosphere storage rooms, pallet fruit bins, tanks, boxcar and truck floors and linings, open soffits, and other similar applications where continuous or severe moisture may be present. Provides smooth surface for application of carpet and pad and possesses high concentrated and impact load resistance. For areas to be covered with resilient flooring, specify panels with "sanded face."[e] EXPOSURE DURABILITY CLASSIFICATION: Exterior. COMMON THICKNESSES[d]: 11/32, 3/8, 15/32, 1/2, 19/32, 5/8, 23/32, 3/4.
APA C-D Plugged Typical Trademark	_APA_ C-D PLUGGED GROUP 2 EXPOSURE 1 000 PS 1-83	For open soffits, built-ins, cable reels, separator boards and other interior or protected applications. Not a substitute for Underlayment or APA Rated Sturd-I-Floor as it lacks their puncture resistance. EXPOSURE DURABILITY CLASSIFICATIONS: Interior, Exposure 1. COMMON THICKNESSES: 3/8, 15/32, 1/2, 19/32, 5/8, 23/32, 3/4.

(a) Specific plywood grades, thicknesses and exposure durability classifications may be in limited supply in some areas. Check with your supplier before specifying.

(b) Sanded Exterior plywood panels, C-C Plugged, C-D Plugged and Underlayment grades can also be manufactured in Structural I (all plies limited to Group 1 species).

(c) Some manufacturers also produce plywood panels with premium N-grade veneer on one or both faces. Available only by special order. Check with the manufacturer.

(d) Panels 1/2 inch and thicker are Span Rated and do not contain species group number in trademark.

(e) Also available in Underlayment A-C or Underlayment B-C grades, marked either "touch sanded" or "sanded face."

(f) For nonstructural floor underlayment, or other applications requiring improved inner ply construction, specify panels marked either "plugged inner plies" (may also be designated plugged crossbands under face or plugged crossbands or core); or "meets underlayment requirements."

(g) Also may be designated APA Underlayment C-C Plugged.

Table 4
GUIDE TO APA SPECIALTY PLYWOOD PANELS (A)
FOR APPLICATION RECOMMENATIONS, SEE FOLLOWING PAGES.

APA Decorative Typical Trademark	APA DECORATIVE GROUP 2 INTERIOR 000 PS 1-83	Rough-sawn, brushed, grooved, or striated faces. For paneling, interior accent walls, built-ins, counter facing, exhibit displays. Can also be made by some manufacturers in Exterior for exterior siding, gable ends, fences and other exterior applications. Use recommendations for Exterior panels vary with the particular product. Check with the manufacturer. EXPOSURE DURABILITY CLASSIFICATIONS: Interior, Exposure 1, Exterior. COMMON THICKNESSES: 5/16, 3/8, 1/2, 5/8.
APA High Density Overlay (HDO)(b) Typical Trademark HDO · A-A · G-1 · EXT-APA · 000 · PS 1-83		Has a hard semi-opaque resin-fiber overlay on both faces. Abrasion resistant. For concrete forms, cabinets, countertops, signs, tanks. Also available with skid-resistant screen-grid surface. EXPOSURE DURABILITY CLASSIFICATION: Exterior. COMMON THICKNESSES: 3/8, 1/2, 5/8, 3/4.
APA Medium Density Overlay (MDO)(b) Typical Trademark	APA M. D. OVERLAY GROUP 1 EXTERIOR 000 PS 1-83	Smooth, opaque, resin-fiber overlay on one or both faces. Ideal base for paint, both indoors and outdoors. For exterior siding, paneling, shelving, exhibit displays, cabinets, signs. EXPOSURE DURABILITY CLASSIFICATION: Exterior. COMMON THICKNESSES: 11/32, 3/8, 15/32, 1/2, 19/32, 5/8, 23/32, 3/4.
APA Marine Typical Trademark MARINE · A-A · EXT-APA · PS 1-83		Ideal for boat hulls. Made only with Douglas-fir or western larch. Subject to special limitations on core gaps and face repairs. Also available with HDO or MDO faces. EXPOSURE DURABILITY CLASSIFICATION: Exterior. COMMON THICKNESSES: 1/4, 3/8, 1/2, 5/8, 3/4.
APA B-B Plyform Class I(b) Typical Trademark	APA PLYFORM B-B CLASS I EXTERIOR 000 PS 1-83	Concrete form grades with high reuse factor. Sanded both faces and mill-oiled unless otherwise specified. Special restrictions on species. Also available in HDO for very smooth concrete finish, and with special overlays. EXPOSURE DURABILITY CLASSIFICATION: Exterior. COMMON THICKNESSES: 19/32, 5/8, 23/32, 3/4.
APA Plyron Typical Trademark PLYRON · EXPOSURE 1-APA · 000		Hardboard face on both sides. Faces tempered, untempered, smooth or screened. For countertops, shelving, cabinet doors, flooring. EXPOSURE DURABILITY CLASSIFICATIONS: Interior, Exposure 1, Exterior. COMMON THICKNESSES: 1/2, 5/8, 3/4.

(a) Specific plywood grades, thicknesses and exposure durability classifications may be in limited supply in some areas. Check with your supplier before specifying.

(b) Can also be manufactured in Structural I (all plies limited to Group 1 species).

Wood Beam Safe Load Table

This appendix shows a partial listing of span tables from *Wood Structural Design Data*, 1986 edition, courtesy of National Forest Products Association. Once the W, F_b, E, and F_v are known, spans can be determined for a simple beam of spans that are normally found in a residence. Refer to *Wood Structural Design Data* for a complete listing of all lumber sizes.

SIZE OF BEAM		900	1,000	F_b 1,100	1,200	1,300	SIZE OF BEAM		900	1,000	F_b 1,100	1,200	1,300
				6'—0''							7'—0''		
4 x 4	W	714	793	873	952	1,032	4 x 6	W	1,512	1,680	1,848	2,016	2,184
	w	119	132	145	158	172		w	216	240	264	288	312
	F_v	43	48	53	58	63		F_v	58	65	72	78	85
	E	1,388	1,542	1,697	1,851	2,005		E	1,030	1,145	1,259	1,374	1,489
4 x 6	W	1,764	1,960	2,156	2,352	2,548	4 x 8	W	2,628	2,920	3,212	3,504	3,796
	w	294	326	359	392	424		w	375	417	458	500	542
	F_v	68	76	84	91	99		F_v	77	86	94	103	112
	E	883	981	1,079	1,175	1,276		E	782	868	955	1,042	1,129
4 x 8	W	3,066	3,406	3,747	4,088	4,428	4 x 10	W	4,278	4,753	5,228	5,704	6,179
	w	511	567	624	616	738		w	611	679	746	814	882
	F_v	90	100	110	120	130		F_v	99	110	121	132	143
	E	670	744	819	893	968		E	612	681	749	817	885
6 x 6	W	2,772	3,081	3,389	3,389	4,005	6 x 6	W	2,376	2,640	2,904	3,169	3,433
	w	462	513	564	616	667		w	339	377	414	452	490
	F_v	68	76	84	91	99		F_v	58	65	72	78	85
	E	883	981	1,079	1,178	1,276		E	1,030	1,145	1,259	1,374	1,489
6 x 8	W	5,156	5,729	6,302	6,875	7,447	6 x 8	W	4,419	4,910	5,401	5,892	6,383
	w	859	954	1,050	1,145	1,241		w	631	701	771	841	911
	F_v	93	104	114	125	135		F_v	80	89	98	107	116
	E	648	719	792	864	935		E	755	839	924	1,007	1,091

WOOD BEAM SAFE LOAD TABLE

WOOD BEAM SAFE LOAD TABLE

SIZE OF BEAM		900	1,000	F_b 1,100	1,200	1,300	SIZE OF BEAM		900	1,000	F_b 1,100	1,200	1,300
				8'–0''							9'–0''		
4 x 6	W	1,323	1,470	1,617	1,764	1,911	4 x 8	W	2,044	2,271	2,498	2,725	2,952
	w	165	183	202	220	238		w	227	252	277	302	328
	F_v	51	57	63	68	74		F_v	60	67	73	80	87
	E	1,178	1,309	1,439	1,570	1,701		E	1,005	1,117	1,228	1,340	1,452
4 x 8	W	2,299	2,555	2,810	3,066	3,321	4 x 10	W	3,327	3,697	4,066	4,436	4,806
	w	287	319	351	383	415		w	369	410	451	492	534
	F_v	67	75	83	90	98		F_v	77	85	94	102	111
	E	893	993	1,092	1,191	1,291		E	788	875	963	1,050	1,138
4 x 10	W	3,743	4,159	4,575	4,991	5,407	4 x 12	W	3,921	5,468	6,015	6,562	7,109
	w	467	519	571	623	675		w	546	607	668	729	789
	F_v	86	96	105	115	125		F_v	93	104	114	125	135
	E	700	778	856	934	1,011		E	648	720	792	863	936
4 x 12	W	5,537	6,152	6,767	7,382	7,998	4 x 14	W	7,087	7,875	8,662	9,450	10,237
	w	692	769	845	922	999		w	787	875	962	1,050	1,137
	F_v	105	117	128	140	152		F_v	112	125	137	150	162
	E	576	640	703	768	832		E	540	600	660	720	780
4 x 14	W	7,973	8,859	9,745	10,361	11,517	6 x 8	W	3,437	3,819	4,201	4,583	4,965
	w	996	1,107	1,218	1,328	1,439		w	381	424	466	509	551
	F_v	126	140	154	168	182		F_v	62	69	76	83	90
	E	480	533	586	640	693		E	972	1,079	1,187	1,295	1,403
6 x 6	W	2,079	2,310	2,541	2,772	3,003	6 x 10	W	5,515	6,128	6,740	7,353	7,966
	w	259	288	317	346	375		w	612	680	748	817	885
	F_v	51	57	63	68	74		F_v	79	87	96	105	114
	E	1,178	1,309	1,439	1,570	1,701		E	767	852	937	1,023	1,108
6 x 8	W	3,867	4,296	4,726	5,156	5,585	6 x 12	W	8,081	8,979	9,877	10,775	11,673
	w	483	537	590	644	698		w	897	997	1,097	1,197	1,297
	F_v	70	78	86	93	101		F_v	95	106	117	127	138
	E	864	960	1,055	1,152	1,247		E	633	704	774	845	915
6 x 10	W	6,204	6,894	7,583	8,272	8,962	6 x 14	W	11,137	12,375	13,612	14,850	16,087
	w	775	861	947	1,034	1,120		w	1,237	1,375	1,512	1,650	1,787
	F_v	89	98	108	118	128		F_v	112	125	137	150	162
	E	682	757	833	909	985		E	540	600	659	720	779
6 x 12	W	9,092	10,102	11,112	12,122	13,133							
	w	1,136	1,262	1,389	1,515	1,641							
	F_v	107	119	131	143	155							
	E	563	626	688	751	813							
6 x 14	W	12,529	13,921	15,314	16,706	18,098							
	w	1,566	1,740	1,914	2,088	2,262							
	F_v	126	140	154	168	182							
	E	480	533	586	640	693							

WOOD BEAM SAFE LOAD TABLE

10'-0"

SIZE OF BEAM		F_b 900	1,000	1,100	1,200	1,300
4 x 10	W	2,994	3,327	3,660	3,992	4,325
	w	299	332	366	399	432
	F_v	69	77	84	92	100
	E	875	972	1,070	1,167	1,264
4 x 12	W	4,429	4,921	5,414	5,906	6,398
	w	442	492	541	590	639
	F_v	84	93	103	112	121
	E	720	800	880	960	1,040
4 x 14	W	6,378	7,087	7,796	8,505	9,213
	w	637	708	779	850	921
	F_v	101	112	123	135	146
	E	600	666	733	800	866
6 x 8	W	3,093	3,437	3,781	4,125	4,468
	w	309	343	378	412	446
	F_v	56	62	68	75	81
	E	1,079	1,199	1,319	1,439	1,559
6 x 10	W	4,963	5,515	6,066	6,618	7,169
	w	496	551	606	661	716
	F_v	71	79	87	95	102
	E	852	947	1,042	1,136	1,231
6 x 12	W	7,273	8,081	8,890	9,698	10,506
	w	727	808	889	969	1,050
	W	86	95	105	115	124
	E	704	782	860	939	1,017
6 x 14	W	10,023	11,137	12,251	13,365	14,478
	w	1,002	1,113	1,225	1,336	1,447
	F_v	101	112	123	135	146
	E	600	666	733	800	866

11'-0"

SIZE OF BEAM		F_b 900	1,000	1,100	1,200	1,300
4 x 8	W	1,672	1,858	2,044	2,229	2,415
	w	152	168	185	202	219
	F_v	49	54	60	65	71
	E	1,228	1,365	1,502	1,638	1,775
4 x 10	W	2,722	3,024	3,327	3,629	3,932
	w	247	274	302	329	357
	F_v	63	70	77	84	91
	E	963	1,070	1,177	1,284	1,391
4 x 12	W	4,026	4,474	4,921	5,369	5,816
	w	366	406	447	488	528
	F_v	76	85	93	102	110
	E	792	879	968	1,055	1,143
4 x 14	W	5,798	6,443	7,087	7,731	8,376
	w	527	585	644	702	761
	F_v	92	102	112	122	132
	E	660	733	806	880	953
6 x 8	W	2,812	3,125	3,437	3,750	4,062
	w	255	284	312	340	369
	F_v	51	56	62	68	73
	E	1,187	1,319	1,451	1,585	1,715
6 x 10	W	4,512	5,013	5,515	6,016	6,518
	w	410	455	501	546	592
	F_v	64	71	79	86	93
	E	937	1,042	1,146	1,250	1,350
6 x 12	W	6,612	7,347	8,081	8,816	9,551
	w	601	667	734	801	868
	F_v	78	87	95	104	113
	E	774	860	946	1,033	1,119
6 x 14	W	9,112	10,125	11,137	12,150	13,162
	w	828	920	1,012	1,104	1,196
	F_v	92	102	112	122	132
	E	659	733	806	880	953

WOOD BEAM SAFE LOAD TABLE

SIZE OF BEAM		900	1,000	F_b 1,100	1,200	1,300	SIZE OF BEAM		900	1,000	F_b 1,100	1,200	1,300
				12'–0"							13'–0"		
4 x 8	W	1,533	1,703	1,873	2,044	2,214	4 x 8	W	1,415	1,572	1,729	1,886	2,044
	w	127	141	156	170	184		w	108	120	133	145	157
	F_v	45	50	55	60	65		F_v	41	46	51	55	60
	E	1,340	1,489	1,638	1,787	1,936		E	1,452	1,613	1,775	1,936	2,097
4 x 10	W	2,495	3,772	3,050	3,327	3,604	4 x 10	W	2,303	2,559	2,815	3,071	3,327
	w	207	231	254	277	300		w	177	196	216	236	255
	F_v	57	64	70	77	83		F_v	53	59	65	71	77
	E	1,050	1,167	1,284	1,401	1,517		E	1,138	1,264	1,391	1,517	1,644
4 x 12	W	3,691	4,101	4,511	4,921	5,332	4 x 12	W	3,407	3,786	4,164	4,543	4,921
	w	307	341	375	410	444		w	262	291	320	349	378
	F_v	70	78	85	93	101		F_v	64	72	79	86	93
	E	863	960	1,055	1,151	1,247		E	936	1,039	1,143	1,247	1,351
4 x 14	W	5,315	5,906	6,496	7,087	7,678	4 x 14	W	4,906	5,451	5,997	6,542	7,087
	w	442	492	541	590	639		w	377	419	461	503	545
	F_v	84	93	103	112	121		F_v	77	86	95	103	112
	E	720	800	880	960	1,040		E	780	866	953	1,039	1,126
6 x 8	W	2,578	2,864	3,151	3,437	3,723	6 x 8	W	2,379	2,644	2,908	3,173	3,437
	w	214	238	262	286	310		w	183	203	223	244	264
	F_v	46	52	57	62	67		F_v	43	48	52	57	62
	E	1,295	1,439	1,583	1,727	1,871		E	1,403	1,559	1,715	1,871	2,027
6 x 10	W	4,136	4,596	5,055	5,515	5,974	6 x 10	W	3,818	4,242	4,666	5,091	5,515
	w	344	383	421	459	497		w	293	326	358	391	424
	F_v	59	65	72	79	85		F_v	54	60	66	73	79
	E	1,023	1,136	1,250	1,364	1,477		E	1,108	1,231	1,354	1,477	1,601
6 x 12	W	6,061	6,734	7,408	8,081	8,755	6 x 12	W	5,595	6,216	6,838	7,460	8,081
	w	505	561	617	673	729		w	430	478	526	573	621
	F_v	71	79	87	95	103		F_v	66	73	81	88	95
	E	845	939	1,033	1,126	1,220		E	915	1,017	1,119	1,220	1,322
6 x 14	W	8,353	9,281	10,209	11,137	12,065	6 x 14	W	7,710	8,567	9,424	10,280	11,137
	w	696	773	850	928	1,005		w	593	659	724	790	856
	F_v	84	93	103	112	121		F_v	77	86	95	103	112
	E	720	800	880	960	1,040		E	779	866	953	1,039	1,126

WOOD BEAM SAFE LOAD TABLE

SIZE OF BEAM		900	1,000	F_b 1,100	1,200	1,300	SIZE OF BEAM		900	1,000	F_b 1,100	1,200	1,300
				14'-0"							15'-0"		
4 x 8	W	1,314	1,460	1,606	1,752	1,898	4 x 8	W	1,226	1,362	1,499	1,635	1,771
	w	93	104	114	125	135		w	81	90	99	109	118
	F_v	38	43	47	51	56		F_v	36	40	44	48	52
	E	1,564	1,737	1,911	2,085	2,259		E	1,675	1,862	2,048	2,234	2,420
4 x 10	W	2,139	2,376	2,614	2,852	3,089	4 x 10	W	1,996	2,218	2,440	2,661	2,883
	w	152	169	186	203	220		w	133	147	162	177	192
	F_v	49	55	60	66	71		F_v	46	51	56	61	66
	E	1,225	1,362	1,498	1,634	1,770		E	1,313	1,459	1,605	1,751	1,897
4 x 12	W	3,164	3,515	3,867	4,218	4,570	4 x 12	W	2,953	3,281	3,609	3,937	4,265
	w	226	251	276	301	326		w	196	218	240	262	284
	F_v	60	66	73	80	87		F_v	56	62	68	75	81
	E	1,008	1,119	1,231	1,343	1,455		E	1,079	1,199	1,319	1,439	1,559
4 x 14	W	4,556	5,062	5,568	6,075	6,581	4 x 14	W	4,252	4,725	5,197	5,670	6,142
	w	325	361	397	433	470		w	283	315	346	378	409
	F_v	72	80	88	96	104		F_v	67	75	82	90	97
	E	840	933	1,026	1,119	1,213		E	900	1,000	1,099	1,199	1,299
6 x 8	W	2,209	2,455	2,700	2,946	3,191	6 x 8	W	2,062	2,291	2,520	2,750	2,979
	w	157	175	192	210	227		w	137	152	168	183	198
	F_v	40	44	49	53	58		F_v	37	41	45	50	54
	E	1,511	1,679	1,847	2,015	2,183		E	1,619	1,799	1,979	2,159	2,339
6 x 10	W	3,545	3,939	4,333	4,727	5,121	6 x 10	W	3,309	3,676	4,044	4,412	4,779
	w	253	281	309	337	365		w	220	245	269	294	318
	F_v	50	56	62	67	73		F_v	47	52	58	63	68
	E	1,193	1,326	1,458	1,591	1,724		E	1,278	1,421	1,563	1,705	1,847
6 x 12	W	5,195	5,772	6,350	6,927	7,504	6 x 12	W	4,849	5,387	5,926	6,465	7,004
	w	371	412	453	494	536		w	323	359	395	431	466
	F_v	61	68	75	82	88		F_v	57	63	70	76	83
	E	986	1,095	1,205	1,314	1,424		E	1,056	1,173	1,291	1,408	1,526
6 x 14	W	7,159	7,955	8,750	9,546	10,341	6 x 14	W	6,682	7,425	8,167	8,910	9,652
	w	511	568	625	681	738		w	445	495	544	594	643
	F_v	72	80	88	96	104		F_v	67	75	82	90	97
	E	839	933	1,026	1,119	1,213		E	900	1,000	1,099	1,199	1,299

WOOD BEAM SAFE LOAD TABLE

SIZE OF BEAM		900	1,000	F_b 1,100	1,200	1,300	SIZE OF BEAM		900	1,000	F_b 1,100	1,200	1,300
				16'–0''									
4 x 8	W	1,149	1,277	1,405	1,533	1,660							
	w	71	79	87	95	103							
	F_v	33	37	41	45	49							
	E	1,787	1,986	2,184	2,383	2,582							
4 x 10	W	1,871	2,079	2,287	2,495	2,703							
	w	116	129	142	155	168							
	F_v	43	48	52	57	62							
	E	1,401	1,556	1,712	1,868	2,023							
4 x 12	W	2,768	3,076	3,383	3,691	3,999							
	w	173	192	211	230	249							
	F_v	52	58	64	70	76							
	E	1,151	1,279	1,407	1,535	1,663							
4 x 14	W	3,986	4,429	4,872	5,315	5,758							
	w	249	276	304	332	359							
	F_v	63	70	77	84	91							
	E	960	1,066	1,173	1,279	1,386							
6 x 8	W	1,933	2,148	2,363	2,578	2,792							
	w	120	134	147	161	174							
	F_v	35	39	42	46	50							
	E	1,727	1,919	2,111	2,303	2,495							
6 x 10	W	3,102	3,447	3,791	4,196	4,481							
	w	193	215	236	258	280							
	F_v	44	49	54	59	64							
	E	1,364	1,515	1,667	1,818	1,970							
6 x 12	W	4,546	5,051	5,556	6,061	6,566							
	w	284	315	347	378	410							
	F_v	53	59	65	71	77							
	E	1,126	1,252	1,377	1,502	1,627							
6 x 14	W	6,264	6,960	7,657	8,353	9,049							
	w	391	435	478	522	565							
	F_v	63	70	77	84	91							
	E	960	1,066	1,173	1,279	1,386							

Abbreviations

Drafters and designers print many abbreviations to conserve space. Using standard abbreviations ensures that drawings are interpreted accurately.

Here are three guidelines for proper use:

1. Abbreviations are in capital letters.
2. A period is used only when the abbreviation may be confused with a word.
3. Several words use the same abbreviation; use is defined by the location.

access panel	**ap**
acoustic	**ac.**
acoustic plaster	**ac. pl.**
actual	**act.**
addition	**add.**
adhesive	**adh.**
adjustable	**adj.**
aggregate	**aggr.**
air conditioning	**A.C.**
alternate	**alt.**
alternating current	**ac**
aluminum	**alum.**
American Institute of Architects	**A.I.A.**
American Institute of Building Designers	**A.I.B.D.**
American Institute of Steel Construction	**A.I.S.C.**
American Institute of Timber Construction	**A.I.T.C.**
American National Standards Institute	**A.N.S.I.**
American Plywood Association	**A.P.A.**
American Society of Civil Engineers	**A.S.C.E.**
American Society of Heating, Refrigerating, and Air Conditioning Engineers	**A.S.H.R.A.E.**
American Society of Landscape Architects	**A.S.L.A.**
American Society for Testing and Materials	**A.S.T.M.**
amount	**amt.**
ampere	**amp.**
anchor bolt	**A.B.**
angle	$\angle$
approximate	**approx.**
approved	**appd.**
architectural	**arch.**
area	**a**
asbestos	**asb.**
asphalt	**asph.**
asphaltic concrete	**asph. conc.**
at	**@**
automatic	**auto.**
avenue	**ave.**
average	**avg.**
balcony	**balc.**
base	**b**

basement	**basm.**
Basic National Building Code	**BOCA**
batten	**batt.**
bathroom	**b**
bathtub	**bt.**
beam	**bm.**
bearing	**br.**
bedroom	**br**
bench mark	**B.M.**
bending moment	**bm**
better	**btr.**
between	**btwn.**
beveled	**bev.**
bidet	**bdt.**
block	**blk.**
blocking	**blkg.**
blower	**blo.**
board	**bd.**
board feet	**bd ft**
both sides	**B.S.**
both ways	**B.W.**
bottom	**btm.**
bottom of footing	**B.F.**
boulevard	**blvd.**
brass	**br.**
brick	**brk.**
British thermal unit	**Btu**
bronze	**brz.**
broom closet	**bc**
building	**bldg.**
building line	**bL**
built-in	**blt-in**
buzzer	**buz.**
by	✕
cabinet	**cab.**
cast concrete	**c conc.**
cast iron	**C.I.**
catalog	**cat.**
catch basin	**C.B.**
caulking	**calk.**
ceiling	**clg.**
ceiling diffuser	**C.D.**
ceiling joist	**C.J. or ceil. jst.**
cement	**cem.**
center	**ctr.**
center to center	**C-C**
centerline	**₵**
centimeter	**cm.**
ceramic	**cer**
chamfer	**cham.**
channel	**C**
check	**chk.**
cinder block	**cin. blk.**
circle	**cir.**
circuit	**cir.**
circuit breaker	**cir bkr**
class	**cl.**
cleanout	**C.O.**
clear	**clr**
coated	**ctd.**
cold water	**C.W.**
column	**col.**
combination	**comb.**
common	**com.**
composition	**comp**
computer-aided drafting	**CAD**
concrete	**conc.**
concrete masonry unit	**C.M.U.**
conduit	**cnd.**
construction	**const.**
Construction Standards Institute	**CSI**
continuous	**cont.**
contractor	**contr.**
control joint	**C.J.**
copper	**cop.**
corridor	**corr.**
corrugate	**corr.**
countersink	**csk.**
courses	**c.**
cubic	**cu.**
cubic feet	**cu ft**
cubic feet per minute	**cfm**
cubic inch	**cu. in.**
cubic yard	**cu yd**
damper	**dpr.**
dampproofing	**dp.**
dead load	**DL**
decibel	**db.**
decking	**dk.**
deflection	**d.**
degree	**° or deg.**
design	**dsgn**
detail	**det.**
diagonal	**diag.**
diameter	**φ or dia.**
diffuser	**dif.**
dimension	**dimen.**
dining room	**dr.**
dishwasher	**D/W**
disposal	**disp.**
ditto	**″ or do**
division	**div.**
door	**dr.**
double	**dbl.**
double hung	**D.H.**
Douglas fir	**DF**
down	**dn.**
downspout	**D.S.**
drain	**D**
drawing	**dwg.**
drinking fountain	**D.F.**

dryer	**d.**	front	**fnt.**
drywall	**D.W.**	full size	**fs**
		furnace	**furn.**
each	**ea.**	furred ceiling	**fc**
each face	**E.F.**	future	**fut.**
each way	**E.W.**		
east	**e.**	gallon	**gal**
elbow	**el.**	galvanized	**galv.**
electrical	**elect.**	galvanized iron	**g.i.**
elevation	**elev.**	gage	**ga.**
enamel	**enam.**	garage	**gar.**
engineer	**engr.**	gas	**g.**
entrance	**ent.**	girder	**gird.**
Environmental Protection		glass	**gl.**
Agency	**EPA**	glue laminated	**glu-lam**
Equal	**eq**	grade	**gr.**
Equipment	**equip.**	grade beam	**gr. bm.**
estimate	**est.**	grating	**grtg.**
excavate	**exc.**	gravel	**gvl.**
exhaust	**exh.**	grille	**gr.**
existing	**exist.**	ground	**gnd.**
expansion joint	**exp. jt.**	ground fault circuit interrupter	**gfci**
exposed	**expo.**	grout	**gt.**
extension	**extn.**	gypsum	**gyp.**
exterior	**ext.**	gypsum board	**gyp. bd.**
fabricate	**fab.**	hardboard	**hdb.**
face brick	**F.B.**	hardware	**hdw.**
face of studs	**f.o.s.**	hardwood	**hdwd.**
Fahrenheit	**F**	head	**hd.**
Federal Housing Administration	**F.H.A.**	header	**hdr.**
feet / foot	**' or ft**	heater	**htr.**
feet per minute	**fpm**	heating	**htg.**
finished	**fin.**	heating/ventilating/air	
finished floor	**fin. fl.**	conditioning	**HVAC**
finished grade	**fin. gr.**	height	**ht.**
finished opening	**F.O.**	hemlock	**hem**
firebrick	**fbrk.**	hemlock-fir	**hem-fir**
fire hydrant	**F.H.**	hollow core	**h.c.**
fireproof	**f.p.**	horizontal	**horiz.**
fixture	**fix.**	horsepower	**H.P.**
flammable	**flam.**	hose bibb	**H.B.**
flange	**flg.**	hot water	**H.W.**
flashing	**fl.**	hot water heater	**H.W.H.**
flexible	**flex.**	hundred	**c**
floor	**flr.**		
floor drain	**F.D.**	illuminate	**illum.**
floor joist	**fl. jst.**	incandescent	**incan.**
floor sink	**F.S.**	inch	**" or in.**
fluorescent	**fluor.**	inch pounds	**in. lb.**
folding	**fldg.**	incinerator	**incin.**
foot	**(') ft.**	inflammable	**infl.**
foot candle	**fc.**	inside diameter	**I.D.**
footing	**ftg.**	inside face	**I.F.**
foot pounds	**ft. lb.**	inspection	**insp.**
forced air unit	**F.A.U.**	install	**inst.**
foundation	**fnd.**	insulate	**ins.**

insulation	insul.	molding	mldg.
interior	int.	motor	mot.
International Conference of Building Officials	ICBO	mullion	mull.
iron	i	National Association of Home Builders	N.A.H.B.
		National Bureau of Standards	N.B.S.
jamb	jmb.	natural	nat.
joint	jt.	natural grade	nat. gr.
joist	jst.	noise reduction coefficient	N.R.C.
junction	jct.	nominal	nom.
junction box	J-box	not applicable	n.a.
		not in contract	N.I.C.
kiln dried	K.D.	not to scale	N.T.S.
kilowatt	kW	number	# or no.
kilowatt hour	kWh		
Kip (1,000 lb.)	K	obscure	obs.
kitchen	kit.	on center	O.C.
knockout	K.O.	opening	opg.
		opposite	opp.
laboratory	lab.	ounce	oz
laminated	lam.	outside diameter	O.D.
landing	ldg.	outside face	O.F.
laundry	lau.	overhead	ovhd.
lavatory	lav.		
length	lgth.	painted	ptd.
level	lev.	pair	pr.
light	lt.	panel	pnl.
linear feet	lin. ft	parallel	// or par.
linen closet	L. cL.	part	pt.
linoleum	lino.	partition	part.
live load	LL	pavement	pvmt.
living room	liv.	penny	d
long	lg.	per	/
louver	lv.	perforate	perf.
lumber	lum.	perimeter	per.
		permanent	perm.
machine bolt	M.B.	perpendicular	⊥ or perp.
manhole	M.H.	pi (3.1416)	π
manufacturer	manuf.	plaster	pls.
marble	mrb.	plasterboard	pls. bd.
masonry	mas.	plastic	plas.
material	mat.	plate	℗ or pl.
maximum	max.	platform	plat.
mechanical	mech.	plumbing	plmb.
medicine cabinet	M.C.	plywood	ply.
medium	med.	polished	pol.
membrane	memb.	polyethelyne	poly.
metal	mtl.	polyvinyl chloride	PVC
meter	m	position	pos.
mile	mi	pound	# or lb
minimum	min.	pounds per square foot	psf
minute	(') min.	pounds per square inch	psi
mirror	mirr.	precast	prcst.
miscellaneous	misc.	prefabricated	prefab.
mixture	mix.	preferred	pfd.
model	mod.	preliminary	prelim.
modular	mod.		

pressure treated	**p.t.**	socket	**soc.**
property	**prop.**	soil pipe	**S.P.**
pull chain	**P.C.**	solid block	**sol. blk.**
pushbutton	**P.B.**	solid core	**S.C.**
		Southern Building Code	**S.B.C.**
quality	**qty.**	Southern pine	**SP**
quantity	**qty.**	Southern Pine Inspection	
		Bureau	**S.P.I.B.**
radiator	**rad.**	specifications	**specs.**
radius	**r or rad.**	spruce-pine-fir	**SPF**
random length & width	**R L & W**	square	**□ or sq**
range	**r.**	square feet	**# or sq ft**
receptacle	**recp.**	square inch	**# or sq in**
recessed	**rec.**	stainless steel	**sst.**
redwood	**rdwd.**	stand pipe	**st. p.**
reference	**ref.**	standard	**std.**
refrigerator	**refr.**	steel	**stl.**
register	**reg.**	stirrup	**stir.**
reinforcing	**reinf.**	stock	**stk.**
reinforcing bar	**rebar**	storage	**sto.**
reproduce	**repro.**	storm drain	**S.D.**
required	**reqd.**	street	**st.**
resistance moment	**rm**	structural	**str.**
return	**ret.**	structural clay tile	**S.C.T.**
revision	**rev.**	substitute	**sub.**
ridge	**rdg.**	supply	**sup.**
riser	**ris.**	surface	**sur.**
roof	**rf.**	surface four sides	**S4S**
roof drain	**R.D.**	surface two sides	**S2S**
roofing	**rfg.**	suspended ceiling	**susp. clg.**
room	**rm.**	switch	**sw.**
rough	**rgh.**	symbol	**sym.**
rough opening	**R.O.**	symmetrical	**sym.**
round	**φ or rd**	synthetic	**syn.**
		system	**sys.**
safety	**saf.**	tangent	**tan.**
sanitary	**san.**	tar and gravel	**t & g**
scale	**sc.**	tee	**t**
schedule	**sch.**	telephone	**tel.**
screen	**scrn.**	television	**tv.**
screw	**scr.**	temperature	**temp.**
second	**sec.**	terra-cotta	**T.C.**
section	**sect.**	terrazzo	**tz.**
select	**sel.**	thermostat	**thrm.**
select structural	**sel. st.**	thickness	**thk.**
self-closing	**s.c.**	thousand	**m**
service	**serv.**	thousand board feet	**MBF**
sewer	**sew.**	threshold	**thr.**
sheathing	**shtg.**	through	**thru**
sheet	**sht.**	toilet	**tol.**
shower	**sh.**	tongue and groove	**T & G**
side	**s.**	top of wall	**T.W.**
siding	**sdg.**	total	**tot.**
sill cock	**S.C.**	tread	**tr.**
similar	**sim.**	tubing	**tub.**
single hung	**S.H.**	typical	**typ.**

Underwriters' Laboratories, Inc.	UL	wainscot	wsct.
unfinished	unfin.	wall vent	W.V.
Uniform Building Code	UBC	washing machine	wm
United States Department of		waste stack	W.S.
Housing and Urban		water closet	W.C.
Development	H.U.D.	water heater	W.H.
urinal	ur.	waterproof	W.P.
utility	util.	watt	W
		weather stripping	ws
V-joint	V-jt.	weatherproof	wp.
valve	v	weep hole	wh
vanity	van.	weight	wt.
vapor barrier	v.b.	welded wire fabric	W.W.F.
vapor proof	vap. prf.	west	w
ventilation	vent.	white pine	WP
vent pipe	vp	wide flange	W or W̶
vent stack	V.S.	width	w
vent through roof	V.T.R.	window	wdw.
vertical	vert.	with	w
vertical grain	vert. gr.	without	w/o
vinyl	vin.	wood	wd
vinyl asbestos tile	V.A.T.	wrought iron	W.I.
vinyl base	V.B.		
vinyl tile	V.T.	yard	yd
vitreous	vit.	yellow pine	YP
vitreous clay tile	V.C.T.		
volt	v	zinc	zn.
volume	vol.		

Glossary

ACOUSTICS The science of sound and sound control.

ADOBE A heavy clay soil used in many southwestern states to make sun-dried bricks.

AGGREGATE Stone, gravel, cinder, or slag used as one of the components of concrete.

AIR-DRIED LUMBER Lumber that has been stored in yards or sheds for a period of time after cutting. Building codes typically assume a 19 percent moisture content when determining joist and beams of air-dried lumber.

AIR DUCT A pipe, typically made of sheet metal, that carries air from a source such as a furnace or air conditioner to a room within a structure.

AIR TRAP A "U"-shaped piped placed in wastewater lines to prevent backflow of sewer gas.

ALCOVE A small room adjoining a larger room, often separated by an archway.

AMPERE (AMPS) A measure of electrical current.

ANCHOR A metal tie or strap used to tie building members to each other.

ANCHOR BOLT A threaded bolt used to fasten wooden structural members to masonry.

ANGLE IRON A structural piece of steel shaped to form a 90 degree angle.

APPRAISAL The estimated value of a piece of property.

APRON The inside trim board placed below a window sill. The term is also used to apply to a curb around a driveway or parking area.

AREAWAY A subsurface enclosure to admit light and air to a basement. Sometimes called a window well.

ASBESTOS A mineral that does not burn or conduct heat; it is usually used for roofing material.

ASHLAR MASONRY Squared masonry units laid with a horizontal bed joint.

ASH PIT An area in the bottom of the firebox of a fireplace to collect ash.

ASPHALT An insoluble material used for making floor tile and for waterproofing walls and roofs.

ASPHALTIC CONCRETE A mixture of asphalt and aggregate which is used for driveways.

ASPHALT SHINGLE Roof shingles made of asphalt-saturated felt and covered with mineral granules.

ASSESSED VALUE The value assigned by governmental agencies to determine the taxes to be assessed on structures and land.

ATTIC The area formed between the ceiling joists and rafters.

ATRIUM An inside courtyard of a structure which may be either open at the top or covered with a roof.

AWNING WINDOW A window that is hinged along the top edge.

BACKFILL Earth, gravel, or sand placed in the trench around the footing and stem wall after the foundation has cured.

BAFFLE A shield, usually made of scrap material, to keep insulation from plugging eave vents. Also used to describe wind- or sound-deadening devices.

BALCONY A deck or patio that is above ground level.

BALLOON FRAMING A building construction method that has vertical wall members that extend uninterrupted from the foundation to the roof.

BAND JOIST A joist set at the edge of the structure that runs parallel to the other joist. Also called a rim joist.

BANISTER A handrail beside a stairway.

BASEBOARD The finish trim where the wall and floor intersect, or an electric wall heater that extends along the floor.

BASE COURSE The lowest course in brick or concrete masonary unit construction.

BASE LINE A reference line.

BASEMENT A level of a structure that is built either entirely below grade level (full basement) or partially below grade (daylite basement).

BATT A blanket insulation usually made of fiberglass to be used between framing members.

BATTEN A board used to hide the seams when other boards are joined together.

BATTER BOARD A horizontal board used to form footings.

BAY A division of space within a building, usually divided by beams or columns.

BEAM A horizontal structural member that is used to support roof or wall loads. Often called a header.

BEAMED CEILING A ceiling that has support beams that are exposed to view.

BEARING PLATE A support member, often a steel plate used to spread weight over a larger area.

BEARING WALL A wall that supports vertical loads in addition to its own weight.

BENCH MARK A reference point used by surveyors to establish grades and construction heights.

BENDING One of three major forces acting on a beam. It is the tendency of a beam to bend or sag between its supports.

BENDING MOMENT A measure of the forces that cause a beam to break by bending. Represented by (M).

BEVELED SIDING Siding that has a tapered thickness.

BIBB An outdoor faucet which is threaded so that a hose may be attached.

BILL OF MATERIAL A part of a set of plans that lists all of the material needed to construct a structure.

BIRD BLOCK A block placed between rafters to maintain a uniform spacing and to keep animals out of the attic.

BIRD'S MOUTH A notch cut into a rafter to provide a bearing surface where the rafter intersects the top plate.

BLIND NAILING Driving nails in such a way that the heads are concealed from view.

BLOCKING Framing members, typically wood, placed between joist, rafters or studs to provide rigidity. Also called bridging.

BOARD AND BATTEN A type of siding using vertical boards with small wood strips (battens) used to cover the joints of the boards.

BOARD FOOT The amount of wood contained in a piece of lumber 1 in. thick by 12 in. wide by 12 in. long.

BOND The morter joint between two masonry units, or a pattern in which masonry units are arranged.

BOND BEAM A reinforced concrete beam used to strengthen masonry walls.

BOTTOM CHORD The lower, usually horizontal, member of a truss.

BOX BEAM A hollow built-up structural unit.

BRIDGING Cross blocking between horizontal members used to add stiffness. Also called blocking.

BREEZEWAY A covered walkway with open sides between two different parts of a structure.

BROKER A representative of the seller in property sales.

Btu British thermal unit. A unit used to measure heat.

BUILDING CODE Legal requirements designed to protect the public by providing guidelines for structural, electrical, plumbing, and mechanical areas of a structure.

BUILDING LINE An imaginary line determined by zoning departments to specify on which area of a lot a structure may be built (also known as a setback).

BUILDING PAPER A waterproofed paper used to prevent the passage of air and water into a structure.

BUILDING PERMIT A permit to build a structure issued by a governmental agency after the plans for the structure have been examined and the structure is found to comply with all building code requirements.

BUILT-UP BEAM A beam built of smaller members that are bolted or nailed together.

BUILT-UP ROOF A roof composed of three or more layers of felt, asphalt, pitch, or coal tar.

BULLNOSE Rounded edges of cabinet trim.

BUTT JOINT The junction where two members meet in a square-cut joint; end to end, or edge to edge.

BUTTRESS A projection from a wall often located below roof beams to provide support to the roof loads and to keep long walls in the vertical position.

CABINET WORK The interior finish woodwork of a structure, especially cabinetry.

CANTILEVER Projected construction that is fastened at only one end.

CANT STRIP A small built-up area between two intersecting roof shapes to divert water.

CARPORT A covered automobile parking structure that is typically not fully enclosed.

CARRIAGE The horizontal part of a stair stringer that supports the tread.

CASEMENT WINDOW A hinged window that swings outward.

CASING The metal, plastic, or wood trim around a door or a window.

CATCH BASIN An underground reservoir for water drained from a roof before it flows to a storm drain.

CATHEDRAL WINDOW A window with an upper edge which is parallel to the roof pitch.

CAULKING A soft, waterproof material used to seal seams and cracks in construction.

CAVITY WALL A masonry wall formed with an air space between each exterior face.

CEILING JOIST The horizontal member of the roof which is used to resist the outward spread of the rafters and to provide a surface on which to mount the finished ceiling.

CEMENT A powder of alumina, silica, lime, iron oxide, and magnesia pulverized and used as an ingredient in mortar and concrete.

CENTRAL HEATING A heating system in which heat is distributed throughout a structure from a single source.

CESSPOOL An underground catch basin for the collection and dispersal of sewage.

CHAMFER A beveled edge formed by removing the sharp corner of a piece of material.

CHANNEL A standard form of structural steel with three sides at right angles to each other forming the letter C.

CHASE A recessed area of column formed between structural members for electrical, mechanical, or plumbing materials.

CHECK Lengthwise cracks in a board caused by natural drying.

CHECK VALVE A valve in a pipe that permits flow in only one direction.

CHIMNEY An upright structure connected to a fireplace or furnace that passes smoke and gases to outside air.

CHORD The upper and lower members of a truss which are supported by the web.

CINDER BLOCK A block made of cinder and cement used in construction.

CIRCUIT BREAKER A safety device which opens and closes an electrical circuit.

CLAPBOARD A tapered board used for siding that overlaps the board below it.

CLEARANCE A clear space between building materials to allow for air flow or access.

CLERESTORY A window or group of windows which are placed above the normal window height, often between two roof levels.

COLLAR TIES A horizontal tie between rafters near the ridge to help resist the tendency of the rafters to separate.

COLONIAL A style of architecture and furniture adapted from the American colonial period.

COLUMN A vertical structural support, usually round and made of steel.

COMMON WALL The partition that divides two different dwelling units.

COMPRESSION A force that crushes or compacts.

COMPUTER AIDED DRAFTING Using a computer as a drafting aid.

COMPUTER INPUT Information placed into a computer.

COMPUTER MAINFRAME The main computer that controls many smaller desk terminals.

COMPUTER OUTPUT Information which is displayed or printed by a computer.

COMPUTER SOFTWARE Programs used to control a computer.

CONCENTRATED LOAD A load centralized in a small area. Usually the weight supported by a post results in a concentrated load.

CONCRETE A building material made from cement, sand, gravel, and water.

CONCRETE BLOCKS Blocks of concrete that are precast. The standard size is 8 × 8 × 16.

CONDENSATION The formation of water on a surface when warm air comes in contact with a cold surface.

CONDUCTOR Any material that permits the flow of electricity. A drain pipe that diverts water from the roof (a downspout).

CONDUIT A bendable pipe or tubing used to encase electrical wiring.

CONSTRUCTION LOAN A mortgage loan to provide cash to the builder for material and labor that is typically repaid when the structure is completed.

CONTINUOUS BEAM A single beam that is supported by more than two supports.

CONTOURS A line that represents land formations.

CONTRACTOR The manager of a construction project, or one specific phase of it.

CONTROL JOINT An expansion joint in a masonry wall formed by raking mortar from the vertical joint.

CONVENIENCE OUTLET An electrical receptacle through which current is drawn for the electrical system of an appliance.

COPING A masonry cap placed on top of a block wall to protect it from water penetration.

CORBEL A ledge formed in a wall by building out successive courses of masonry.

CORNICE The part of the roof that extends out from the wall. Sometimes referred to as the eave.

COUNTERFLASH A metal flashing used under normal flashing to provide a waterproof seam.

COURSE A continuous row of building material such as shingles, stone or brick.

COURT An unroofed space surrounded by walls.

CRAWL SPACE The area between the floor joists and the ground.

CRICKET A diverter built to direct water away from an area of a roof where it would otherwise collect such as behind a chimney.

CRIPPLE A wall stud that is cut at less than full length.

CROSS BRACING Boards fastened diagonally between structural members such as floor joists to provide rigidity.

CROSSHATCH Thin lines drawn about 1/16″ apart, typically at 45 degrees, to show masonry in plan view or wood that has been sectioned.

CUL-DE-SAC A dead end street with no outlet which provides a circular turn-around.

CULVERT An underground passageway for water, usually part of a drainage system.

CUPOLA A small structure built above the main roof level to provide light or ventilation.

CURE The process of concrete drying to its maximum design strength, usually taking 28 days.

CURTAIN WALL An exterior wall which provides no structural support.

DAMPER A movable plate that controls the amount of draft for a woodstove, fireplace, or furnace.

DATUM A reference point for starting a survey.

DEADENING Material used to control the transmission of sound.

DEAD LOAD The weight of building materials or other unmoveable objects in a structure.

DECKING A wood material used to form the floor or roof, typically used in 1 and 2 in. thicknesses.

DENSITY The number of people allowed to live in a specific area of land or to work in a specific area of a structure.

DEPRECIATION Loss of monetary value.

DESIGNER A person who designs buildings, but is not licensed as is an architect.

DIGITIZER An input tool for computer aided drafting used to draw on a flat bed plotter. The device translates images to numbers for transmission to the computer.

DISK Computer storage units that store information or programs used to run the computer.

DIVERTER A metal strip used to divert water.

DORMER A structure which projects from a sloping roof to form another roofed area. This new area is typically used to provide a surface to install a window.

DOUBLE HUNG A type of window in which the upper and lower halves slide past each other to provide an opening at the top and bottom of the window.

DOWNSPOUT A pipe which carries rain water from the gutters of the roof to the ground.

DRAIN A collector for a pipe that carries water.

DRESSED LUMBER Lumber that has been surfaced by a planing machine to give the wood a smooth finish.

DRY ROT A type of wood decay caused by fungi that leaves the wood a soft powder.

DRYWALL An interior wall covering installed in large sheets made from gypsum board.

DRY WELL A shallow well used to disperse water from the gutter system.

DUCTS Pipes, typically made of sheet metal, used to conduct hot or cold air of the HVAC system.

DUPLEX OUTLET A standard electrical convenience outlet with two receptacles.

DUTCH DOOR A type of door that is divided horizontally in the center so that each half of the door may be opened separately.

DUTCH HIP A type of roof shape that combines features of a gable and a hip roof.

EASEMENT An area of land that cannot be built upon because it provides access to a structure or to utilities such as power or sewer lines.

EAVE The lower part of the roof that projects from the wall. See cornice.

EGRESS A term used in building codes to describe access.

ELASTIC LIMIT The extent a material can be bent and still return to its original shape.

ELBOW An L-shaped plumbing pipe.

ELEVATION The height of a specific point in relation to another point. The exterior views of a structure.

ELL An extension of the structure at a right angle to the main structure.

EMINENT DOMAIN The right of a government to condemn private property so that it may be obtained for public use.

ENAMEL A paint that produces a hard, glossy, smooth finish.

EQUITY The value of real estate in excess of the balance owed on the mortgage.

ERGONOMICS The study of human space and movement needs as they relate to a given work area, such as a kitchen.

EXCAVATION The removal of soil for construction purposes.

EXPANSION JOINT A joint installed in concrete construction to reduce cracking and to provide workable areas.

FABRICATION Work done on a structure away from the job site.

FACADE The exterior covering of a structure.

FACE BRICK Brick that is used on the visible surface to cover other masonry products.

FACE GRAIN The pattern in the visible veneer of plywood.

FASCIA A horizontal board nailed to the end of rafters or trusses to conceal their ends.

FEDERAL HOUSING ADMINISTRATION (FHA) A governmental agency that insures home loans made by private lending institutions.

FELT A tar-impregnated paper used for water protection under roofing and siding materials. Sometimes used under concrete slabs for moisture resistance.

FIBER BENDING STRESS The measurement of structural members used to determine their stiffness.

FIBERBOARD Fibrous wood products that have been pressed into a sheet. Typically used for the interior construction of cabinets and for a covering for the subfloor.

FILL Material used to raise an area for construction. Typically gravel or sand is used to provide a raised, level building area.

FILLED INSULATION Insulation material that is blown or poured into place in attics and walls.

FILLET WELD A weld between two surfaces that butt at 90° to each other with the weld filling the inside corner.

FINISHED LUMBER Wood that has been milled with a smooth finish suitable for use as trim and other finish work.

FINISHED SIZE Sometimes called the dressed size, the finished size represents the actual size of lumber after all milling operations and is typically about ½ in. smaller than the nominal size, which is the size of lumber before planing.

NOMINAL SIZE (IN.)	FINISHED SIZE (IN.)
1	3/4
2	1 1/2
4	3 1/2
6	5 1/2
8	7 1/4
10	9 1/4
12	11 1/4
14	13 1/4

FIREBRICK A refractory brick capable of withstanding high temperatures and used for lining fireplaces and furnaces.

FIREBOX The combustion chamber of the fireplace where the fire occurs.

FIREBRICK A brick made of a refractory material that can withstand great amounts of heat and is used to line the visible face of the firebox.

FIRE CUT An angular cut on the end of a joist or rafter that is supported by masonry. The cut allows the wood member to fall away from the wall without damaging a masonry wall when the wood is damaged by fire.

FIRE DOOR A door used between different types of construction which has been rated as being able to withstand fire for a certain amount of time.

FIREPROOFING Any material that is used to cover structural materials to increase their fire rating.

FIRE RATED A rating given to building materials to specify the amount of time the material can resist damage caused by fire.

FIRE-STOP Blocking placed between studs or other structural members to resist the spread of fire.

FIRE WALL A wall constructed of materials resulting in a specified time that the wall can resist fire before structural damage will occur.

FLAGSTONE Flat stones used typically for floor and wall coverings.

FLASHING Metal used to prevent water leaking through surface intersections.

FLAT ROOF A roof with a minimal roof pitch, usually about ⅛″ per 12″.

FLOOR PLUG A 110 convenience outlet located in the floor.

FLUE A passage inside of the chimney to conduct smoke and gases away from a firebox to outside air.

FLUE LINER A terra-cotta pipe used to provide a smooth flue surface so that unburned materials will not cling to the flue.

FOOTING The lowest member of a foundation system used to spread the loads of a structure across supporting soil.

FOOTING FORM The wooden mold used to give concrete its shape as it cures.

FOUNDATION The system used to support a building's loads and made up of stem walls, footings, and piers. The term is used in many areas to refer to the footing.

FRAME The structural skeleton of a building.

FROST LINE The depth to which soil will freeze.

FURRING Wood strips attached to structural members that are used to provide a level surface for finishing materials when different-sized structural members are used.

GABLE A type of roof with two sloping surfaces that intersect at the ridge of the structure.

GABLE END WALL The triangular wall that is formed at each end of a gable roof between the top plate of the wall and the rafters.

GALVANIZED Steel products that have had zinc applied to the exterior surface to provide protection from rusting.

GAMBREL A type of roof formed with two planes on each side. The lower pitch is steeper than the upper portion of the roof.

GIRDER A horizontal support member at the foundation level.

GLUED-LAMINATED TIMBER (GLU-LAM) A structural member made up of layers of lumber that are glued together.

GRADE The designation of the quality of a manufactured piece of wood.

GRADING The moving of soil to effect the elevation of land at a construction site.

GRAVEL STOP A metal strip used to retain gravel at the edge of built-up roofs.

GREEN LUMBER Lumber that has not been kiln-dried and still contains moisture.

GROUND FAULT CIRCUIT INTERRUPTER (GFCI OR GFI) A 110 convenience outlet with a built-in circuit breaker. GFCI outlets are to be used within 5'–0" of any water source.

GROUT A mixture of cement, sand, and water used to fill joints in masonry and tile construction.

GUARDRAIL A horizontal protective railing used around stairwells, balconies, and changes of floor elevation greater than 30 in.

GUSSET A plate added to the side of intersecting structural members to help form a secure connection and to reduce stress.

GUTTER A metal or plastic drainage system for collecting and disposing of water from roofs.

GYPSUM BOARD An interior finishing material made of gypsum and fiberglass and covered with paper which is installed in large sheets.

HALF-TIMBER A frame construction method where spaces between wood members are filled with masonry.

HANGER A metal support bracket used to attach two structural members.

HARDBOARD Sheet material formed of compressed wood fibers used as an underlayment for flooring.

HEAD The upper portion of a door or window frame.

HEADER A horizontal structural member used to support other structural members over openings such as doors and windows.

HEADER COURSE A horizontal masonry course with the end of each masonry unit exposed.

HEADROOM The vertical clearance in a room over a stairway.

HEARTH The fire-resistant floor within and extending a minimum of 18 in. in front of the firebox.

HEARTWOOD The inner core of a tree trunk.

HIP The exterior edge formed by two sloping roof surfaces.

HIP ROOF A roof shape with four sloping sides.

HORIZONTAL SHEAR One of three major forces acting on a beam, it is the tendency of the fibers of a beam to slide past each other in a horizontal direction.

HOSE BIBB A water outlet that is threaded to receive a hose.

HUMIDIFIER A mechanical device that controls the amount of moisture inside of a structure.

I BEAM The generic term for a wide flange or American standard steel beam with a cross section in the shape of the letter I.

INDIRECT LIGHTING Mechanical lighting that is reflected off a surface.

INSULATION Material used to restrict the flow of heat, cold, or sound from one surface to another.

ISOMETRIC A drawing method which enables three surfaces of an object to be seen in one view, with the base of each surface drawn at 30° to the horizontal plane.

JACK RAFTER A rafter which is cut shorter than the other rafters to allow for an opening in the roof.

JACK STUD A wall member which is cut shorter than other studs to allow for an opening such as a window. Also called a cripple stud.

JALOUSIE A type of window made of thin horizontal panels that can be rotated between the open and closed position.

JAMB The vertical members of a door or window frame.

JOIST A horizontal structural member used in repetitive patterns to support floor and ceiling loads.

KILN A heating unit for the removal of moisture from wood.

KILN DRIED A method of drying lumber in a kiln or oven. Kiln dried lumber has a reduced moisture content when compared to lumber that has been air dried.

KING STUD A full-length stud placed at the end of a header.

KIP Used in some engineering formulas to represent 1,000 pounds.

KNEE WALL A wall of less than full height.

KNOT A branch or limb of a tree that is cut through in the process of manufacturing lumber.

LALLY COLUMN A vertical steel column that is used to support floor or foundation loads.

LAMINATED Several layers of material that have been glued together under pressure.

LANDING A platform between two flights of stairs.

LATERAL Sideways action in a structure caused by wind or seismic forces.

LATH Wood or sheet metal strips which are attached to the structural frame to support plaster.

LATTICE A grille made by criss-crossing strips of material.

LAVATORY A bathroom sink, or the room which is equipped with a washbasin.

LEDGER A horizontal member which is attached to the side of wall members to provide support for rafters or joists.

LIEN A monetary claim on property.

LINTEL A horizontal steel member used to provide support for masonry over an opening.

LIVE LOAD The loads from all movable objects within a structure including loads from furniture and people. External loads from snow and wind are also considered live loads.

LOAD-BEARING WALL A support wall which holds floor or roof loads in addition to its own weight.

LOOKOUT A beam used to support eave loads.

LOUVER An opening with horizontal slats to allow for ventilation.

MANSARD A four-sided, steep-sloped roof.

MANTEL A decorative shelf above the opening of a fireplace.

MARKET VALUE The amount that property can be sold for.

MESH A metal reinforcing material placed in concrete slabs and masonry walls to help resist cracking.

METAL TIES A manufactured piece of metal for joining two structural members together.

METAL WALL TIES Corrugated metal strips used to bond brick veneer to its support wall.

MILLWORK Finished woodwork that has been manufactured in a milling plant. Examples are window and door frames, mantels, moldings, and stairway components.

MINERAL WOOL An insulating material made of fibrous foam.

MODULE A standardized unit of measurement.

MODULAS OF ELASTICITY (E) The degree of stiffness of beam.

MOISTURE BARRIER Typically a plastic material used to restrict moisture vapor from penetrating into a structure.

MOLDING Decorative strips, usually made of wood, used to conceal the seam in other finishing materials.

MOMENT The tendency of a force to rotate around a certain point.

MONOLITHIC Concrete construction created in one pouring.

MONUMENT A boundary marker set to mark property corners.

MORTAR A combination of cement, sand, and water used to bond masonry units together.

MORTGAGE A contract pledging the sale of property upon full payment of purchase price.

MORTGAGEE The lender of the money to the mortgagor for the purchase of property.

MORTGAGOR The buyer of property who is paying off the mortgage to the mortgagee.

MUD ROOM A room or utility entrance where soiled clothing can be removed before entering the main portion of the residence.

MUDSILL The horizontal wood member that rests on concrete to support other wood members.

MULLION A horizontal or vertical divider between sections of a window.

MUNTIN A horizontal or vertical divider within a section of a window.

NAILER A wood member bolted to concrete or steel members to provide a nailing surface for attaching other wood members.

NEWEL The end post of a stair railing.

NOMINAL SIZE An approximate size achieved by rounding the actual material size to the nearest larger whole number.

NONBEARING WALL A wall which supports no loads other than its own weight. Some building codes consider walls which support only ceiling loads as nonbearing.

NONFERROUS METAL Metal, such as copper or brass, that contains no iron.

NOSING The rounded front edge of a tread which extends past the riser.

OBSCURE GLASS Glass that is not transparent.

ON CENTER A measurement taken from the center of one member to the center of another member.

ORIENTATION The locating of a structure on property based on the location of the sun, prevailing winds, view, and noise.

OUTLET An electrical receptacle which allows for current to be drawn from the system.

OUTRIGGER A support for roof sheathing and the fascia which extends past the wall line perpendicular to the rafters.

OVERHANG The horizontal measurement of the distance the roof projects from a wall.

OVERLAY DRAFTING The practice of drawing a structure in several layers which must be combined to produce the finished drawing.

PAD An isolated concrete pier.

PARAPET A portion of wall that extends above the edge of the roof.

PARGING A thin coat of plaster used to smooth a masonry surface.

PARQUET FLOORING Wood flooring laid to form patterns.

PARTITION An interior wall.

PARTY WALL A wall dividing two adjoining spaces such as apartments or offices.

PENNY The length of a nail represented by the letter "d". "16d" is read as "sixteen penny".

PERSPECTIVE A drawing method which provides the illusion of depth by the use of vanishing points.

PHOTODRAFTING The use of photography to produce a base drawing on which additional drawings can be added.

PIER A concrete or masonry foundation support.

PILASTER A reinforcing column built into or against a masonry wall.

PILING A vertical foundation support driven into the ground to provide support on stable soil or rock.

PITCH A description of roof angle comparing the vertical rise to the horizontal run.

PLANK Lumber which is 1½ to 3½ in. in thickness.

PLASTER A mix of sand, cement, and water, used to cover walls and ceilings.

PLAT A map of an area of land which shows the boundaries of individual lots.

PLATE A horizontal member at the top (top plate) or bottom (sole plate or sill) of walls used to connect the vertical wall members.

PLENUM An air space for transporting air from the HVAC system.

PLOT A parcel of land.

PLOTTER An output device used in computer aided drafting to draw lines and symbols.

PLUMB True vertical.

PLYWOOD Wood composed of three or more layers, with the grain of each layer placed at 90° to each other and bonded with glue.

POCHÉ A shading method using graphite applied with a soft tissue in a rubbing motion.

PORCH A covered entrance to a structure.

PORTICO A roof supported by columns instead of walls.

PORTLAND CEMENT A hydraulic cement made of silica, lime, and aluminum that has become the most common cement used in the construction industry because of its strength.

POST A vertical wood structural member usually 4 × 4 or larger.

PRECAST A concrete component which has been cast in a location other than the one in which it will be used.

PREFABRICATED Buildings or components that are built away from the job site and transported ready to be used.

PRESTRESSED A concrete component that is placed in compression as it is cast to help resist deflection.

PRINCIPAL The original amount of money loaned before interest is applied.

PROGRAM A set of instructions which controls the functions of a computer.

PURLIN A horizontal roof member which is laid perpendicular to rafters to help limit deflection.

PURLIN BRACE A support member which extends from the purlin down to a load-bearing wall or header.

QUAD A courtyard surrounded by the walls of buildings.

QUARRY TILE An unglazed, machine-made tile.

QUARTER ROUND Wood molding that has the profile of one-quarter of a circle.

RABBET A rectangular groove cut on the edge of a board.

RAFTER The inclined structural member of a roof system designed to support roof loads.

RAFTER/CEILING JOIST An inclined structural member which supports both the ceiling and the roof materials.

RAKE JOINT A recessed mortar joint.

REACTION The upward forces acting at the supports of a beam.

REBAR Reinforcing steel used to strengthen concrete.

REFERENCE BUBBLE A symbol used to designate the origin of details and sections.

REGISTER An opening in a duct for the supply of heated or cooled air.

REINFORCED CONCRETE Concrete that has steel rebar placed in it to resist tension.

RHEOSTAT An electrical control device used to regulate the current reaching a light fixture. A dimmer switch.

RELATIVE HUMIDITY The amount of water vapor in the atmosphere compared to the maximum possible amount at the same temperature.

RENDERING An artistic process applied to drawings to add realism.

RESTRAINING WALL A masonry wall supported only at the bottom by a footing that is designed to resist soil loads.

RETAINING WALL A masonry wall supported at the top and bottom, designed to resist soil loads.

R-FACTOR A unit of thermal resistance applied to the insulating value of a specific building material.

RIBBON A structural wood member framed into studs to support joists or rafters.

RIDGE The uppermost area of two intersecting roof planes.

RIDGE BOARD A horizontal member that rafters are aligned against to resist their downward force.

RIDGE BRACE A support member used to transfer the weight from the ridge board to a bearing wall or beam. The brace is typically spaced at 48 in. O.C., and may not exceed a 45° angle from vertical.

RIM JOIST A joist at the perimeter of a structure that runs parallel to the other floor joist.

RISE The amount of vertical distance between one tread and another.

RISER The vertical member of stairs between the treads.

ROLL ROOFING Roofing material of fiber or asphalt that is shipped in rolls.

ROOF DRAIN A receptacle for removal of roof water.

ROUGH FLOOR The subfloor, usually hardboard, which serves as a base for the finished floor.

ROUGH HARDWARE Hardware used in construction, such as nails, bolts, and metal connectors, that will not be seen when the project is complete.

ROUGH IN To prepare a room for plumbing or electrical additions by running wires or piping for a future fixture.

ROUGH LUMBER Lumber that has not been surfaced but has been trimmed on all four sides.

ROUGH OPENING The unfinished opening between framing members allowed for doors, windows, or other assemblies.

ROWLOCK A pattern for laying masonry units so that the end of the unit is exposed.

RUN The horizontal distance of a set of steps or the measurement describing the depth of one step.

SADDLE A small gable-shaped roof used to divert water from behind a chimney.

SASH An individual frame around a window.

SCAB A short member that overlaps the butt joint of two other members used to fasten those members.

SCALE A measuring instrument used to draw materials at reduced size.

SCHEDULE A written list of similar components such as windows or doors.

SCRATCH COAT The first coat of stucco which is scratched to provide a good bonding surface for the second coat.

SEASONING The process of removing moisture from green lumber by either air- (natural) or kiln-drying.

SECTION A type of drawing showing an object as if it had been cut through to show interior construction.

SEISMIC Earthquake-related forces.

SEPTIC TANK A tank in which sewage is decomposed by bacteria and dispersed by drain tiles.

SERVICE CONNECTION The wires that run to a structure from a power pole or transformer.

SETBACK The minimum distance required between the structure and the property line.

SHAKE A hand-split wooden roof shingle.

SHEAR The stress that occurs when two forces from opposite directions are acting on the same member. Shearing stress tends to cut a member just as scissors cut paper.

SHEAR PANEL A plywood panel applied to walls to resist wind and seismic forces by keeping the studs in a vertical position.

SHEATHING A covering material placed over walls, floors and roofs which serves as a backing for finishing materials.

SHIM A piece of material used to fill a space between two surfaces.

SHIPLAP A siding pattern of overlapping rabbeted edges.

SILL A horizontal wood member placed at the bottom of walls and openings in walls.

SKYLIGHT An opening in the roof to allow light and ventilation that is usually covered with glass or plastic.

SLAB A concrete floor system typically poured at ground level.

SLEEPERS Strips of wood placed over a concrete slab in order to attach other wood members.

SMOKE CHAMBER The portion of the chimney located directly over the firebox which acts as a funnel between the firebox and the chimney.

SMOKE SHELF A shelf located at the bottom of the smoke chamber to prevent down-drafts from the chimney from entering the firebox.

SOFFIT A lowered ceiling, typically found in kitchens, halls, and bathrooms to allow for recessed lighting or HVAC ducts.

SOIL STACK The main vertical waste-water pipe.

SOLDIER A masonry unit laid on end with its narrow surface exposed.

SOLE PLATE The plate placed at the bottom of a wall.

SPACKLE The covering of sheetrock joints with joint compound.

SPAN The horizontal distance between two supporting members.

SPECIFICATIONS Written descriptions or requirements to specify how a structure is to be constructed.

SPLICE Two similar members that are joined together in a straight line usually by nailing or bolting.

SPLIT-LEVEL A house that has two levels, one about a half a level above or below the other.

SQUARE An area of roofing covering 100 square feet.

STACK A vertical plumbing pipe.

STAIR WELL The opening in the floor where a stair will be framed.

STILE A vertical member of a cabinet, door, or decorative panel.

STIRRUP A U-shaped metal bracket used to support wood beams.

STOCK Common sizes of building materials as they are sold.

STOP A wooden strip used to hold windows in place.

STRESS A live or dead load acting on a structural member. Stress results as the fibers of a beam resist an external force.

STRESSED-SKIN PANEL A hollow, built-up member typically used as a beam.

STRETCHER A course of masonry laid horizontally with the end of the unit exposed.

STRINGER The inclined support member of a stair that supports the risers and treads.

STUCCO A type of plaster made from Portland cement, sand, water, and a coloring agent that is applied to exterior walls.

STUD The vertical framing member of a wall which is usually 2 × 4 or 2 × 6 in size.

SUBFLOOR The flooring surface which is laid on the floor joist and serves as a base layer for the finished floor.

SUMP A recessed area in a basement floor to collect water so that it can be removed by a pump.

SURFACED LUMBER Lumber that has been smoothed on at least one side.

SWALE A recessed area formed in the ground to help divert ground water away from a structure.

TAMP To compact soil or concrete.

TENSILE STRENGTH The resistance of a material or beam to the tendency to stretch.

TENSION Forces that cause a material to stretch or pull apart.

TERMITE SHIELD A strip of sheet metal used at the intersection of concrete and wood surfaces near ground level to prevent termites from entering the wood.

TERRA-COTTA Hard-baked clay typically used as a liner for chimneys.

THERMAL CONDUCTOR A material suitable for transmitting heat.

THERMAL RESISTANCE Represented by the letter R, resistance measures the ability of a material to resist the flow of heat.

THERMOSTAT A mechanical device for controlling the output of HVAC units.

THRESHOLD The beveled member directly under a door.

THROAT The narrow opening to the chimney that is just above the firebox. The throat of a chimney is where the damper is placed.

TILT UP A method of construction in which concrete walls are cast in a horizontal position and then lifted into place.

TIMBER Lumber with a cross-sectional size of 4 × 6 inches or larger.

TOENAIL Nails driven into a member at an angle.

TONGUE AND GROOVE A joint where the edge of one member fits into a groove in the next member.

TRANSOM A window located over a door.

TRAP A U-shaped pipe below plumbing fixtures which holds water to prevent odor and sewer gas from entering the fixture.

TREAD The horizontal member of a stair on which the foot is placed.

TRIMMER Joist or rafters that are used to frame an opening in a floor, ceiling, or roof.

TRUSS A prefabricated or job-built construction member formed of triangular shapes used to support roof or floor loads over long spans.

TRUSS A framework made in triangular-shaped segments used for spanning distances greater than is possible using standard components and methods.

ULTIMATE STRENGTH The unit stress within a member just before it breaks.

UNIT STRESS The maximum permissible stress a structural member can resist without failing. Represented by (f).

VALLEY The internal corner formed between two intersecting roof surfaces.

VAPOR BARRIER Material that is used to block the flow of water vapor into a structure. Typically 6 mil (.006 in.) black plastic.

VAULT An inclined ceiling area.

VENEER A thin outer covering or nonload bearing masonry face material.

VENTILATION The process of supplying and removing air from a structure.

VENT PIPE Pipes that provide air into the waste lines to allow drainage by connecting each plumbing fixture to the vent stack.

VENT STACK A vertical pipe of a plumbing system used to equalize pressure within the system and to vent sewer gases.

VERTICAL SHEAR One of three major stresses acting on a beam, it is the tendency of a beam to drop between its supports.

VESTIBULE A small entrance or lobby.

WAINSCOT A paneling applied to the lower portion of a wall.

WALLBOARD Large flat sheets of gypsum, typically ½ or ⅝ in. thick, used to finish interior walls.

WARP Variation from true shape.

WATERPROOF Material or a type of construction that prevents the absorption of water.

WEATHER STRIP A fabric or plastic material placed along the edges of doors, windows, and skylites to reduce air infiltration.

WEEP HOLE An opening normally in the bottom course of a masonry to allow for drainage.

WORK TRIANGLE The triangular area created in the kitchen by drawing a line from the sink, to the refrigerator, and to the cooking area.

WYTHE A single unit thickness of a masonry wall.

Index